I0023240

Quantum Chemistry

Third Edition

Quantum Chemistry

Third Edition

John P. Lowe
Department of Chemistry
The Pennsylvania State University
University Park, Pennsylvania

Kirk A. Peterson
Department of Chemistry
Washington State University
Pullman, Washington

ELSEVIER
ACADEMIC
PRESS

Amsterdam • Boston • Heidelberg • London • New York • Oxford
Paris • San Diego • San Francisco • Singapore • Sydney • Tokyo

Acquisitions Editor: *Jeremy Hayhurst*
Project Manager: *A. B. McGee*
Editorial Assistant: *Desiree Marr*
Marketing Manager: *Linda Beattie*
Cover Designer: *Julio Esperas*
Composition: *Integra Software Services*
Cover Printer: *Phoenix Color*
Interior Printer: *Maple-Vail Book Manufacturing Group*

Elsevier Academic Press
30 Corporate Drive, Suite 400, Burlington, MA 01803, USA
525 B Street, Suite 1900, San Diego, CA 92101-4495, USA
84 Theobald's Road, London WC1X 8RR, UK

This book is printed on acid-free paper. ♾

Copyright © 2006, Elsevier Inc. All rights reserved.

No part of this publication may be reproduced or transmitted in any form or by any means, electronic or mechanical, including photocopy, recording, or any information storage and retrieval system, without permission in writing from the publisher.

Permissions may be sought directly from Elsevier's Science & Technology Rights Department in Oxford, UK: telephone: (+44) 1865 843830, fax: (+44) 1865 853333, e-mail: permissions@elsevier.co.uk. You may also complete your request on-line via the Elsevier homepage (http://www.elsevier.com), by selecting "Customer Support" and then "Obtaining Permissions."

Library of Congress Cataloging-in-Publication Data
Lowe, John P.
Quantum chemistry. -- 3rd ed. / John P. Lowe, Kirk A. Peterson.
p. cm.
Includes bibliographical references and index.
ISBN 0-12-457551-X
1. Quantum chemistry. I. Peterson, Kirk A. II. Title.
QD462.L69 2005
541'.28--dc22
2005019099

British Library Cataloguing in Publication Data
A catalogue record for this book is available from the British Library

ISBN-13: 978-0-12-457551-6
ISBN-10: 0-12-457551-X

For all information on all Elsevier Academic Press publications
visit our Web site at www.books.elsevier.com

Printed in the United States of America
05 06 07 08 09 10 9 8 7 6 5 4 3 2 1

Working together to grow
libraries in developing countries

www.elsevier.com | www.bookaid.org | www.sabre.org

ELSEVIER BOOK AID International Sabre Foundation

To

Nancy

-J. L.

THE MOLECULAR CHALLENGE

Sir Ethylene, to scientists fair prey,
(Who dig and delve and peek and push and pry,
And prove their findings with equations sly)
Smoothed out his ruffled orbitals, to say:
"I stand in symmetry. Mine is a way
Of mystery and magic. Ancient, I
Am also deemed immortal. Should I die,
Pi would be in the sky, and Judgement Day
Would be upon us. For all things must fail,
That hold our universe together, when
Bonds such as bind me fail, and fall asunder.
Hence, stand I firm against the endless hail
Of scientific blows. I yield not." Men
And their computers stand and stare and wonder.

W.G. LOWE

Contents

Preface to the Third Edition

We have attempted to improve and update this text while retaining the features that make it unique, namely, an emphasis on physical understanding, and the ability to estimate, evaluate, and predict results without blind reliance on computers, while still maintaining rigorous connection to the mathematical basis for quantum chemistry. We have inserted into most chapters examples that allow important points to be emphasized, clarified, or extended. This has enabled us to keep intact most of the conceptual development familiar to past users. In addition, many of the chapters now include multiple choice questions that students are invited to solve in their heads. This is not because we think that instructors will be using such questions. Rather it is because we find that such questions permit us to highlight some of the definitions or conclusions that students often find most confusing far more quickly and effectively than we can by using traditional problems. Of course, we have also sought to update material on computational methods, since these are changing rapidly as the field of quantum chemistry matures.

This book is written for courses taught at the first-year graduate/senior undergraduate levels, which accounts for its implicit assumption that many readers will be relatively unfamiliar with much of the mathematics and physics underlying the subject. Our experience over the years has supported this assumption; many chemistry majors are exposed to the requisite mathematics and physics, yet arrive at our courses with poor understanding or recall of those subjects. That makes this course an opportunity for such students to experience the satisfaction of finally seeing how mathematics, physics, and chemistry are intertwined in quantum chemistry. It is for this reason that treatments of the simple and extended Hückel methods continue to appear, even though these are no longer the methods of choice for serious computations. These topics nevertheless form the basis for the way most non-theoretical chemists understand chemical processes, just as we tend to think about gas behavior as "ideal, with corrections."

Preface to the Second Edition

The success of the first edition has warranted a second. The changes I have made reflect my perception that the book has mostly been used as a teaching text in introductory courses. Accordingly, I have removed some of the material in appendixes on mathematical details of solving matrix equations on a computer. Also I have removed computer listings for programs, since these are now commonly available through commercial channels. I have added a new chapter on MO theory of periodic systems—a subject of rapidly growing importance in theoretical chemistry and materials science and one for which chemists still have difficulty finding appropriate textbook treatments. I have augmented discussion in various chapters to give improved coverage of time-dependent phenomena and atomic term symbols and have provided better connection to scattering as well as to spectroscopy of molecular rotation and vibration. The discussion on degenerate-level perturbation theory is clearer, reflecting my own improved understanding since writing the first edition. There is also a new section on operator methods for treating angular momentum. Some teachers are strong adherents of this approach, while others prefer an approach that avoids the formalism of operator techniques. To permit both teaching methods, I have placed this material in an appendix. Because this edition is more overtly a text than a monograph, I have not attempted to replace older literature references with newer ones, except in cases where there was pedagogical benefit.

A strength of this book has been its emphasis on physical argument and analogy (as opposed to pure mathematical development). I continue to be a strong proponent of the view that true understanding comes with being able to "see" a situation so clearly that one can solve problems in one's head. There are significantly more end-of-chapter problems, a number of them of the "by inspection" type. There are also more questions inviting students to explain their answers. I believe that thinking about such questions, and then reading explanations from the answer section, significantly enhances learning.

It is the fashion today to focus on state-of-the-art methods for just about everything. The impact of this on education has, I feel, been disastrous. Simpler examples are often needed to develop the insight that enables understanding the complexities of the latest techniques, but too often these are abandoned in the rush to get to the "cutting edge." For this reason I continue to include a substantial treatment of simple Hückel theory. It permits students to recognize the connections between MOs and their energies and bonding properties, and it allows me to present examples and problems that have maximum transparency in later chapters on perturbation theory, group theory, qualitative MO theory, and periodic systems. I find simple Hückel theory to be educationally indispensable.

Much of the new material in this edition results from new insights I have developed in connection with research projects with graduate students. The work of all four of my students since the appearance of the first edition is represented, and I am delighted to thank Sherif Kafafi, John LaFemina, Maribel Soto, and Deb Camper for all I have learned from them. Special thanks are due to Professor Terry Carlton, of Oberlin College, who made many suggestions and corrections that have been adopted in the new edition.

Doubtless, there are new errors. I would be grateful to learn of them so that future printings of this edition can be made error-free. Students or teachers with comments, questions, or corrections are more than welcome to contact me, either by mail at the Department of Chemistry, 152 Davey Lab, The Pennsylvania State University, University Park, PA 16802, or by e-mail directed to JL3 at PSUVM.PSU.EDU.

Preface to the First Edition

My aim in this book is to present a reasonably rigorous treatment of molecular orbital theory, embracing subjects that are of practical interest to organic and inorganic as well as physical chemists. My approach here has been to rely on physical intuition as much as possible, first solving a number of specific problems in order to develop sufficient insight and familiarity to make the formal treatment of Chapter 6 more palatable. My own experience suggests that most chemists find this route the most natural.

I have assumed that the reader has at some time learned calculus and elementary physics, but I have not assumed that this material is fresh in his or her mind. Other mathematics is developed as it is needed. The book could be used as a text for undergraduate or graduate students in a half or full year course. The level of rigor of the book is somewhat adjustable. For example, Chapters 3 and 4, on the harmonic oscillator and hydrogen atom, can be truncated if one wishes to know the nature of the solutions, but not the mathematical details of how they are produced.

I have made use of appendixes for certain of the more complicated derivations or proofs. This is done in order to avoid having the development of major ideas in the text interrupted or obscured. Certain of the appendixes will interest only the more theoretically inclined student. Also, because I anticipate that some readers may wish to skip certain chapters or parts of chapters, I have occasionally repeated information so that a given chapter will be less dependent on its predecessors. This may seem inelegant at times, but most students will more readily forgive repetition of something they already know than an overly terse presentation.

I have avoided early usage of bra-ket notation. I believe that simultaneous introduction of new concepts and unfamiliar notation is poor pedagogy. Bra-ket notation is used only after the ideas have had a change to jell.

Problem solving is extremely important in acquiring an understanding of quantum chemistry. I have included a fair number of problems with hints for a few of them in Appendix 14 and answers for almost all of them in Appendix 15.[1]

It is inevitable that one be selective in choosing topics for a book such as this. This book emphasizes ground state MO theory of molecules more than do most introductory texts, with rather less emphasis on spectroscopy than is usual. Angular momentum is treated at a fairly elementary level at various appropriate places in the text, but it is never given a full-blown formal development using operator commutation relations. Time-dependent phenomena are not included. Thus, scattering theory is absent,

[1] In this Second Edition, these Appendices are numbered Appendix 12 and 13.

although selection rules and the transition dipole are discussed in the chapter on time-independent perturbation theory. Valence-bond theory is completely absent. If I have succeeded in my effort to provide a clear and meaningful treatment of topics relevant to modern molecular orbital theory, it should not be difficult for an instructor to provide for excursions into related topics not covered in the text.

Over the years, many colleagues have been kind enough to read sections of the evolving manuscript and provide corrections and advice. I especially thank L. P. Gold and O. H. Crawford, who cheerfully bore the brunt of this task.

Finally, I would like to thank my father, Wesley G. Lowe, for allowing me to include his sonnet, "The Molecular Challenge."

Chapter 1

Classical Waves and the Time-Independent Schrödinger Wave Equation

1-1 Introduction

The application of quantum-mechanical principles to chemical problems has revolutionized the field of chemistry. Today our understanding of chemical bonding, spectral phenomena, molecular reactivities, and various other fundamental chemical problems rests heavily on our knowledge of the detailed behavior of electrons in atoms and molecules. In this book we shall describe in detail some of the basic principles, methods, and results of quantum chemistry that lead to our understanding of electron behavior.

In the first few chapters we shall discuss some simple, but important, particle systems. This will allow us to introduce many basic concepts and definitions in a fairly physical way. Thus, some background will be prepared for the more formal general development of Chapter 6. In this first chapter, we review briefly some of the concepts of classical physics as well as some early indications that classical physics is not sufficient to explain all phenomena. (Those readers who are already familiar with the physics of classical waves and with early atomic physics may prefer to jump ahead to Section 1-7.)

1-2 Waves

1-2.A Traveling Waves

A very simple example of a traveling wave is provided by cracking a whip. A pulse of energy is imparted to the whipcord by a single oscillation of the handle. This results in a wave which travels down the cord, transferring the energy to the popper at the end of the whip. In Fig. 1-1, an idealization of the process is sketched. The shape of the disturbance in the whip is called the *wave profile* and is usually symbolized $\psi(x)$. The wave profile for the traveling wave in Fig. 1-1 shows where the energy is located at a given instant. It also contains the information needed to tell how much energy is being transmitted, because the height and shape of the wave reflect the vigor with which the handle was oscillated.

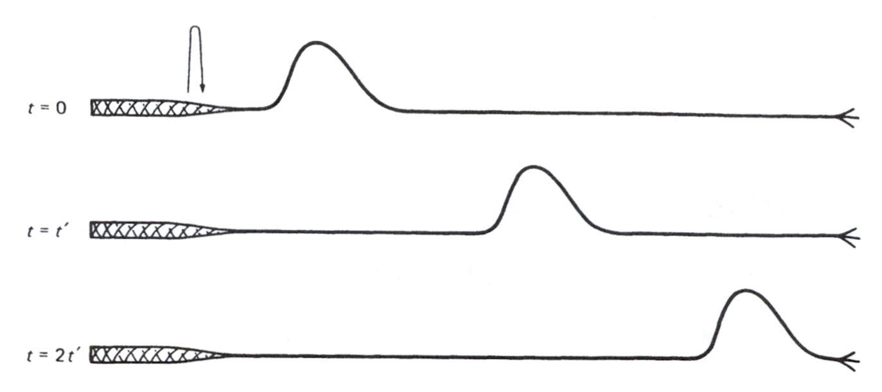

Figure 1-1 ▸ Cracking the whip. As time passes, the disturbance moves from left to right along the extended whip cord. Each segment of the cord oscillates up and down as the disturbance passes by, ultimately returning to its equilibrium position.

The feature common to all traveling waves in classical physics is that energy is transmitted through a medium. The medium itself undergoes no permanent displacement; it merely undergoes local oscillations as the disturbance passes through.

One of the most important kinds of wave in physics is the *harmonic* wave, for which the wave profile is a sinusoidal function. A harmonic wave, at a particular instant in time, is sketched in Fig. 1-2. The maximum displacement of the wave from the rest position is the *amplitude* of the wave, and the *wavelength* λ is the distance required to enclose one complete oscillation. Such a wave would result from a harmonic[1] oscillation at one end of a taut string. Analogous waves would be produced on the surface of a quiet pool by a vibrating bob, or in air by a vibrating tuning fork.

At the instant depicted in Fig. 1-2, the profile is described by the function

$$\psi(x) = A\sin(2\pi x/\lambda) \tag{1-1}$$

($\psi = 0$ when $x = 0$, and the argument of the sine function goes from 0 to 2π, encompassing one complete oscillation as x goes from 0 to λ.) Let us suppose that the situation in Fig. 1-2 pertains at the time $t = 0$, and let the velocity of the disturbance through the medium be c. Then, after time t, the distance traveled is ct, the profile is shifted to the right by ct and is now given by

$$\Psi(x,t) = A\sin[(2\pi/\lambda)(x - ct)] \tag{1-2}$$

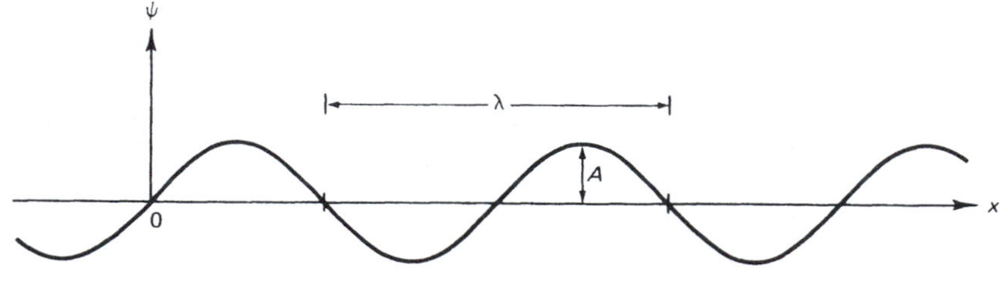

Figure 1-2 ▸ A harmonic wave at a particular instant in time. A is the amplitude and λ is the wavelength.

[1]A harmonic oscillation is one whose equation of motion has a sine or cosine dependence on time.

A capital Ψ is used to distinguish the time-dependent function (1-2) from the time-independent function (1-1).

The *frequency* ν of a wave is the number of individual repeating wave units passing a point per unit time. For our harmonic wave, this is the distance traveled in unit time c divided by the length of a wave unit λ. Hence,

$$\nu = c/\lambda \tag{1-3}$$

Note that the wave described by the formula

$$\Psi'(x,t) = A \sin[(2\pi/\lambda)(x - ct) + \epsilon] \tag{1-4}$$

is similar to Ψ of Eq. (1-2) except for being displaced. If we compare the two waves at the same instant in time, we find Ψ' to be shifted to the left of Ψ by $\epsilon\lambda/2\pi$. If $\epsilon = \pi, 3\pi, \ldots$, then Ψ' is shifted by $\lambda/2, 3\lambda/2, \ldots$ and the two functions are said to be exactly out of phase. If $\epsilon = 2\pi, 4\pi, \ldots$, the shift is by $\lambda, 2\lambda, \ldots$, and the two waves are exactly in phase. ϵ is the *phase factor* for Ψ' relative to Ψ. Alternatively, we can compare the two waves at the same point in x, in which case the phase factor causes the two waves to be displaced from each other in time.

1-2.B Standing Waves

In problems of physical interest, the medium is usually subject to constraints. For example, a string will have ends, and these may be clamped, as in a violin, so that they cannot oscillate when the disturbance reaches them. Under such circumstances, the energy pulse is unable to progress further. It cannot be absorbed by the clamping mechanism if it is perfectly rigid, and it has no choice but to travel back along the string in the opposite direction. The reflected wave is now moving into the face of the primary wave, and the motion of the string is in response to the demands placed on it by the two simultaneous waves:

$$\Psi(x,t) = \Psi_{\text{primary}}(x,t) + \Psi_{\text{reflected}}(x,t) \tag{1-5}$$

When the primary and reflected waves have the same amplitude and speed, we can write

$$\begin{aligned}\Psi(x,t) &= A \sin[(2\pi/\lambda)(x - ct)] + A \sin[(2\pi/\lambda)(x + ct)] \\ &= 2A \sin(2\pi x/\lambda) \cos(2\pi ct/\lambda)\end{aligned} \tag{1-6}$$

This formula describes a *standing wave*—a wave that does not appear to travel through the medium, but appears to vibrate "in place." The first part of the function depends only on the x variable. Wherever the sine function vanishes, Ψ will vanish, regardless of the value of t. This means that there are places where the medium does not ever vibrate. Such places are called *nodes*. Between the nodes, $\sin(2\pi x/\lambda)$ is finite. As time passes, the cosine function oscillates between plus and minus unity. This means that Ψ oscillates between plus and minus the value of $\sin(2\pi x/\lambda)$. We say that the x-dependent part of the function gives the maximum displacement of the standing wave, and the t-dependent part governs the motion of the medium back and forth between these extremes of maximum displacement. A standing wave with a central node is shown in Fig. 1-3.

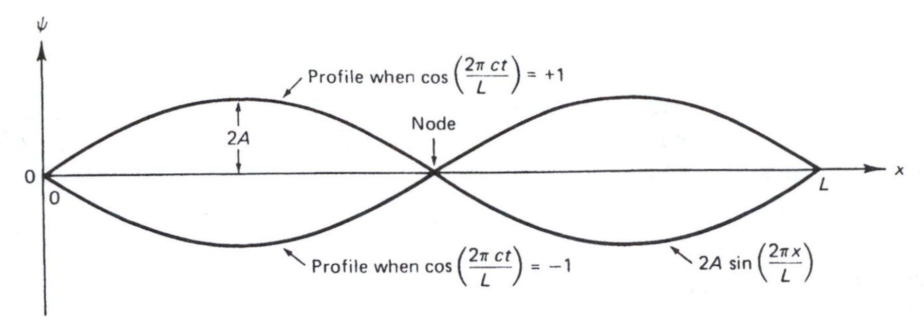

Figure 1-3 ▶ A standing wave in a string clamped at $x = 0$ and $x = L$. The wavelength λ is equal to L.

Equation (1-6) is often written as

$$\Psi(x, t) = \psi(x)\cos(\omega t) \tag{1-7}$$

where

$$\omega = 2\pi c/\lambda \tag{1-8}$$

The profile $\psi(x)$ is often called the *amplitude function* and ω is the *frequency factor.*

Let us consider how the energy is stored in the vibrating string depicted in Fig. 1-3. The string segments at the central node and at the clamped endpoints of the string do not move. Hence, their kinetic energies are zero at all times. Furthermore, since they are never displaced from their equilibrium positions, their potential energies are likewise always zero. Therefore, the total energy stored at these segments is always zero as long as the string continues to vibrate in the mode shown. The maximum kinetic and potential energies are associated with those segments located at the wave peaks and valleys (called the *antinodes*) because these segments have the greatest average velocity and displacement from the equilibrium position. A more detailed mathematical treatment would show that the total energy of any string segment is proportional to $\psi(x)^2$ (Problem 1-7).

1-3 The Classical Wave Equation

It is one thing to draw a picture of a wave and describe its properties, and quite another to predict what sort of wave will result from disturbing a particular system. To make such predictions, we must consider the physical laws that the medium must obey. One condition is that the medium must obey Newton's laws of motion. For example, any segment of string of mass m subjected to a force F must undergo an acceleration of F/m in accord with Newton's second law. In this regard, wave motion is perfectly consistent with ordinary particle motion. Another condition, however, peculiar to waves, is that each segment of the medium is "attached" to the neighboring segments so that, as it is displaced, it drags along *its* neighbor, which in turn drags along its neighbor,

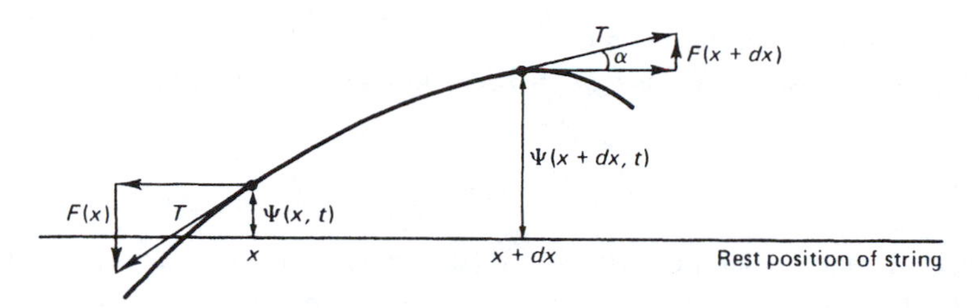

Figure 1-4 ▶ A segment of string under tension T. The forces at each end of the segment are decomposed into forces perpendicular and parallel to x.

etc. This provides the mechanism whereby the disturbance is propagated along the medium.[2]

Let us consider a string under a tensile force T. When the string is displaced from its equilibrium position, this tension is responsible for exerting a restoring force. For example, observe the string segment associated with the region x to $x+dx$ in Fig. 1-4. Note that the tension exerted at either end of this segment can be decomposed into components parallel and perpendicular to the x axis. The parallel component tends to stretch the string (which, however, we assume to be unstretchable), the perpendicular component acts to accelerate the segment toward or away from the rest position. At the right end of the segment, the perpendicular component F divided by the horizontal component gives the slope of T. However, for small deviations of the string from equilibrium (that is, for small angle α) the horizontal component is nearly equal in length to the vector T. This means that it is a good approximation to write

$$\text{slope of vector } T = F/T \quad \text{at } x+dx \tag{1-9}$$

But the slope is also given by the derivative of Ψ, and so we can write

$$F_{x+dx} = T\,(\partial\Psi/\partial x)_{x+dx} \tag{1-10}$$

At the other end of the segment the tensile force acts in the opposite direction, and we have

$$F_x = -T(\partial\Psi/\partial x)_x \tag{1-11}$$

The net perpendicular force on our string segment is the resultant of these two:

$$F = T\left[(\partial\Psi/\partial x)_{x+dx} - (\partial\Psi/\partial x)_x\right] \tag{1-12}$$

The difference in slope at two infinitesimally separated points, divided by dx, is by definition the second derivative of a function. Therefore,

$$F = T\,\partial^2\Psi/\partial x^2\,dx \tag{1-13}$$

[2]Fluids are of relatively low viscosity, so the tendency of one segment to drag along its neighbor is weak. For this reason fluids are poor transmitters of *transverse* waves (waves in which the medium oscillates in a direction perpendicular to the direction of propagation). In *compression* waves, one segment displaces the next by pushing it. Here the requirement is that the medium possess elasticity for compression. Solids and fluids often meet this requirement well enough to transmit compression waves. The ability of rigid solids to transmit both wave types while fluids transmit only one type is the basis for using earthquake-induced waves to determine how deep the solid part of the earth's mantle extends.

Equation (1-13) gives the force on our string segment. If the string has mass m per unit length, then the segment has mass $m\,dx$, and Newton's equation $F = ma$ may be written

$$T\,\partial^2\Psi/\partial x^2 = m\,\partial^2\Psi/\partial t^2 \tag{1-14}$$

where we recall that acceleration is the second derivative of position with respect to time.

Equation (1-14) is the wave equation for motion in a string of uniform density under tension T. It should be evident that its derivation involves nothing fundamental beyond Newton's second law and the fact that the two ends of the segment are linked to each other and to a common tensile force. Generalizing this equation to waves in three-dimensional media gives

$$\left(\frac{\partial^2}{\partial x^2} + \frac{\partial^2}{\partial y^2} + \frac{\partial^2}{\partial z^2}\right)\Psi(x, y, z, t) = \beta\frac{\partial^2\Psi(x, y, z, t)}{\partial t^2} \tag{1-15}$$

where β is a composite of physical quantities (analogous to m/T) for the particular system.

Returning to our string example, we have in Eq. (1-14) a *time-dependent* differential equation. Suppose we wish to limit our consideration to standing waves that can be separated into a space-dependent amplitude function and a harmonic time-dependent function. Then

$$\Psi(x,t) = \psi(x)\cos(\omega t) \tag{1-16}$$

and the differential equation becomes

$$\cos(\omega t)\frac{d^2\psi(x)}{dx^2} = \frac{m}{T}\psi(x)\frac{d^2\cos(\omega t)}{dt^2} = -\frac{m}{T}\psi(x)\omega^2\cos(\omega t) \tag{1-17}$$

or, dividing by $\cos(\omega t)$,

$$d^2\psi(x)/dx^2 = -(\omega^2 m/T)\psi(x) \tag{1-18}$$

This is the classical *time-independent* wave equation for a string.

We can see by inspection what kind of function $\psi(x)$ must be to satisfy Eq. (1-18). ψ is a function that, when twice differentiated, is reproduced with a coefficient of $-\omega^2 m/T$. One solution is

$$\psi = A\sin\left(\omega\sqrt{m/T}x\right) \tag{1-19}$$

This illustrates that Eq. (1-18) has sinusoidally varying solutions such as those discussed in Section 1-2. Comparing Eq. (1-19) with (1-1) indicates that $2\pi/\lambda = \omega\sqrt{m/T}$. Substituting this relation into Eq. (1-18) gives

$$d^2\psi(x)/dx^2 = -(2\pi/\lambda)^2\psi(x) \tag{1-20}$$

which is a more useful form for our purposes.

For three-dimensional systems, the classical time-independent wave equation for an isotropic and uniform medium is

$$(\partial^2/\partial x^2 + \partial^2/\partial y^2 + \partial^2/\partial z^2)\psi(x, y, z) = -(2\pi/\lambda)^2\psi(x, y, z) \tag{1-21}$$

where λ depends on the elasticity of the medium. The combination of partial derivatives on the left-hand side of Eq. (1-21) is called the *Laplacian*, and is often given the shorthand symbol ∇^2 (del squared). This would give for Eq. (1-21)

$$\nabla^2\psi(x, y, z) = -(2\pi/\lambda)^2\psi(x, y, z) \tag{1-22}$$

1-4 Standing Waves in a Clamped String

We now demonstrate how Eq. (1-20) can be used to predict the nature of standing waves in a string. Suppose that the string is clamped at $x = 0$ and L. This means that the string cannot oscillate at these points. Mathematically this means that

$$\psi(0) = \psi(L) = 0 \tag{1-23}$$

Conditions such as these are called *boundary conditions*. Our question is, "What functions ψ satisfy Eq. (1-20) and also Eq. (1-23)?" We begin by trying to find the most general equation that can satisfy Eq. (1-20). We have already seen that $A\sin(2\pi x/\lambda)$ is a solution, but it is easy to show that $A\cos(2\pi x/\lambda)$ is also a solution. More general than either of these is the linear combination[3]

$$\psi(x) = A\sin(2\pi x/\lambda) + B\cos(2\pi x/\lambda) \tag{1-24}$$

By varying A and B, we can get different functions ψ.

There are two remarks to be made at this point. First, some readers will have noticed that other functions exist that satisfy Eq. (1-20). These are $A\exp(2\pi ix/\lambda)$ and $A\exp(-2\pi ix/\lambda)$, where $i = \sqrt{-1}$. The reason we have not included these in the general function (1-24) is that these two exponential functions are mathematically equivalent to the trigonometric functions. The relationship is

$$\exp(\pm ikx) = \cos(kx) \pm i\sin(kx). \tag{1-25}$$

This means that any trigonometric function may be expressed in terms of such exponentials and vice versa. Hence, the set of trigonometric functions and the set of exponentials is redundant, and no additional flexibility would result by including exponentials in Eq. (1-24) (see Problem 1-1). The two sets of functions are *linearly dependent*.[4]

The second remark is that for a given A and B the function described by Eq. (1-24) is a single sinusoidal wave with wavelength λ. By altering the ratio of A to B, we cause the wave to shift to the left or right with respect to the origin. If $A = 1$ and $B = 0$, the wave has a node at $x = 0$. If $A = 0$ and $B = 1$, the wave has an antinode at $x = 0$.

We now proceed by letting the boundary conditions determine the constants A and B. The condition at $x = 0$ gives

$$\psi(0) = A\sin(0) + B\cos(0) = 0 \tag{1-26}$$

---

[3]Given functions $f_1, f_2, f_3, \ldots$. A *linear combination* of these functions is $c_1 f_1 + c_2 f_2 + c_3 f_3 + \cdots$, where $c_1, c_2, c_3, \ldots$ are numbers (which need not be real).

[4]If one member of a set of functions $(f_1, f_2, f_3, \ldots)$ can be expressed as a linear combination of the remaining functions (i.e., if $f_1 = c_2 f_2 + c_3 f_3 + \cdots$), the set of functions is said to be linearly dependent. Otherwise, they are linearly independent.

However, since $\sin(0)=0$ and $\cos(0)=1$, this gives

$$B=0 \tag{1-27}$$

Therefore, our first boundary condition forces B to be zero and leaves us with

$$\psi(x)=A\sin(2\pi x/\lambda) \tag{1-28}$$

Our second boundary condition, at $x=L$, gives

$$\psi(L)=A\sin(2\pi L/\lambda)=0 \tag{1-29}$$

One solution is provided by setting A equal to zero. This gives $\psi=0$, which corresponds to no wave at all in the string. This is possible, but not very interesting. The other possibility is for $2\pi L/\lambda$ to be equal to $0, \pm\pi, \pm 2\pi, \ldots, \pm n\pi, \ldots$ since the sine function vanishes then. This gives the relation

$$2\pi L/\lambda=n\pi, \quad n=0,\pm 1,\pm 2,\ldots \tag{1-30}$$

or

$$\lambda=2L/n, \quad n=0,\pm 1,\pm 2,\ldots \tag{1-31}$$

Substituting this expression for λ into Eq. (1-28) gives

$$\psi(x)=A\sin(n\pi x/L), \quad n=0,\pm 1,\pm 2,\ldots \tag{1-32}$$

Some of these solutions are sketched in Fig. 1-5. The solution for $n=0$ is again the uninteresting $\psi=0$ case. Furthermore, since $\sin(-x)$ equals $-\sin(x)$, it is clear that the set of functions produced by positive integers n is not physically different from the set produced by negative n, so we may arbitrarily restrict our attention to solutions with positive n. (The two sets are linearly dependent.) The constant A is still undetermined. It affects the amplitude of the wave. To determine A would require knowing how much energy is stored in the wave, that is, how hard the string was plucked.

It is evident that there are an infinite number of acceptable solutions, each one corresponding to a different number of half-waves fitting between 0 and L. But an even larger infinity of waves has been excluded by the boundary conditions—namely, all waves having wavelengths not divisible into $2L$ an integral number of times. The result

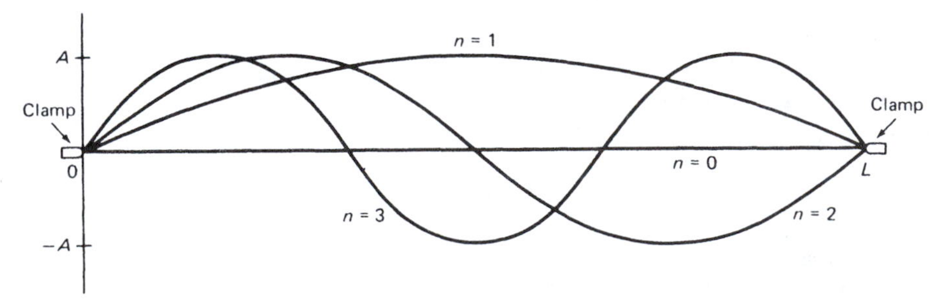

Figure 1-5 ▶ Solutions for the time-independent wave equation in one dimension with boundary conditions $\psi(0)=\psi(L)=0$.

of applying boundary conditions has been to restrict the allowed wavelengths to certain discrete values. As we shall see, this behavior is closely related to the quantization of energies in quantum mechanics.

The example worked out above is an extremely simple one. Nevertheless, it demonstrates how a differential equation and boundary conditions are used to define the allowed states for a system. One could have arrived at solutions for this case by simple physical argument, but this is usually not possible in more complicated cases. The differential equation provides a systematic approach for finding solutions when physical intuition is not enough.

1-5 Light as an Electromagnetic Wave

Suppose a charged particle is caused to oscillate harmonically on the z axis. If there is another charged particle some distance away and initially at rest in the xy plane, this second particle will commence oscillating harmonically too. Thus, energy is being transferred from the first particle to the second, which indicates that there is an oscillating electric field emanating from the first particle. We can plot the magnitude of this electric field at a given instant as it would be felt by a series of imaginary test charges stationed along a line emanating from the source and perpendicular to the axis of vibration (Fig. 1-6).

If there are some magnetic compasses in the neighborhood of the oscillating charge, these will be found to swing back and forth in response to the disturbance. This means that an oscillating *magnetic* field is produced by the charge too. Varying the placement of the compasses will show that this field oscillates in a plane perpendicular to the axis of vibration of the charged particle. The combined electric and magnetic fields traveling along one ray in the xy plane appear in Fig. 1-7.

The changes in electric and magnetic fields propagate outward with a characteristic velocity c, and are describable as a traveling wave, called an electromagnetic wave. Its frequency ν is the same as the oscillation frequency of the vibrating charge. Its wavelength is $\lambda = c/\nu$. Visible light, infrared radiation, radio waves, microwaves, ultraviolet radiation, X rays, and γ rays are all forms of electromagnetic radiation, their only difference being their frequencies ν. We shall continue the discussion in the context of light, understanding that it applies to all forms of electromagnetic radiation.

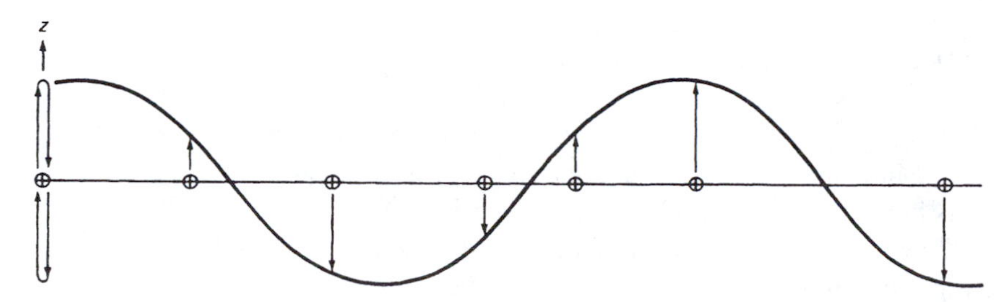

Figure 1-6 ▶ A harmonic electric-field wave emanating from a vibrating electric charge. The wave magnitude is proportional to the force felt by the test charges. The charges are only imaginary; if they actually existed, they would possess mass and under acceleration would absorb energy from the wave, causing it to attenuate.

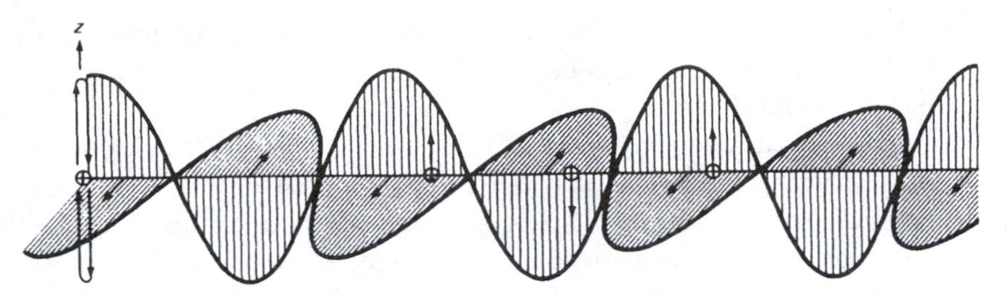

Figure 1-7 ▶ A harmonic electromagnetic field produced by an oscillating electric charge. The arrows without attached charges show the direction in which the north pole of a magnet would be attracted. The magnetic field is oriented perpendicular to the electric field.

If a beam of light is produced so that the orientation of the electric field wave is always in the same plane, the light is said to be plane (or linearly) polarized. The plane-polarized light shown in Fig. 1-7 is said to be z polarized. If the plane of orientation of the electric field wave rotates clockwise or counterclockwise about the axis of travel (i.e., if the electric field wave "corkscrews" through space), the light is said to be right or left circularly polarized. If the light is a composite of waves having random field orientations so that there is no resultant orientation, the light is unpolarized.

Experiments with light in the nineteenth century and earlier were consistent with the view that light is a wave phenomenon. One of the more obvious experimental verifications of this is provided by the interference pattern produced when light from a point source is allowed to pass through a pair of slits and then to fall on a screen. The resulting interference patterns are understandable only in terms of the constructive and destructive interference of waves. The differential equations of Maxwell, which provided the connection between electromagnetic radiation and the basic laws of physics, also indicated that light is a wave.

But there remained several problems that prevented physicists from closing the book on this subject. One was the inability of classical physical theory to explain the intensity and wavelength characteristics of light emitted by a glowing "blackbody." This problem was studied by Planck, who was forced to conclude that the vibrating charged particles producing the light can exist only in certain discrete (separated) energy states. We shall not discuss this problem. Another problem had to do with the interpretation of a phenomenon discovered in the late 1800s, called the *photoelectric effect*.

1-6 The Photoelectric Effect

This phenomenon occurs when the exposure of some material to light causes it to eject electrons. Many metals do this quite readily. A simple apparatus that could be used to study this behavior is drawn schematically in Fig. 1-8. Incident light strikes the metal dish in the evacuated chamber. If electrons are ejected, some of them will strike the collecting wire, giving rise to a deflection of the galvanometer. In this apparatus, one can vary the potential difference between the metal dish and the collecting wire, and also the intensity and frequency of the incident light.

Suppose that the potential difference is set at zero and a current is detected when light of a certain intensity and frequency strikes the dish. This means that electrons

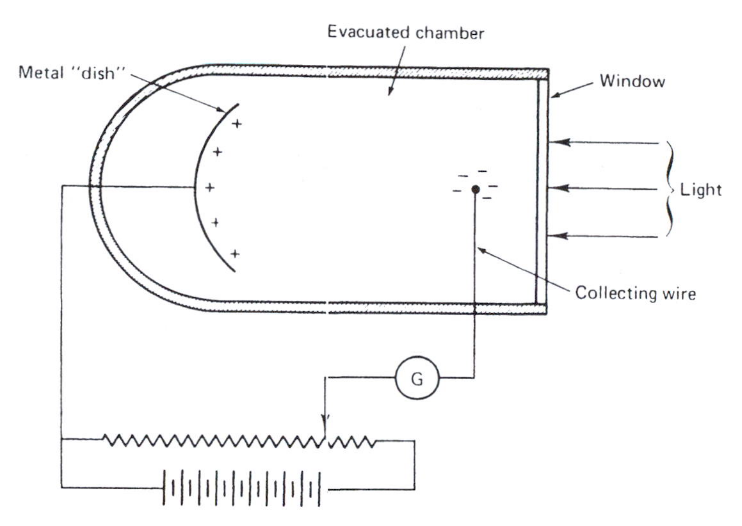

Figure 1-8 ▶ A phototube.

are being emitted from the dish with finite kinetic energy, enabling them to travel to the wire. If a retarding potential is now applied, electrons that are emitted with only a small kinetic energy will have insufficient energy to overcome the retarding potential and will not travel to the wire. Hence, the current being detected will decrease. The retarding potential can be increased gradually until finally even the most energetic photoelectrons cannot make it to the collecting wire. This enables one to calculate the maximum kinetic energy for photoelectrons produced by the incident light on the metal in question.

The observations from experiments of this sort can be summarized as follows:

1. Below a certain cutoff frequency of incident light, no photoelectrons are ejected, no matter how intense the light.
2. Above the cutoff frequency, the number of photoelectrons is directly proportional to the intensity of the light.
3. As the frequency of the incident light is increased, the maximum kinetic energy of the photoelectrons increases.
4. In cases where the radiation intensity is extremely low (but frequency is above the cutoff value) photoelectrons are emitted from the metal without any time lag.

Some of these results are summarized graphically in Fig. 1-9. Apparently, the kinetic energy of the photoelectron is given by

$$\text{kinetic energy} = h(\nu - \nu_0) \tag{1-33}$$

where h is a constant. The cutoff frequency ν_0 depends on the metal being studied (and also its temperature), but the slope h is the same for all substances.

We can also write the kinetic energy as

$$\text{kinetic energy} = \text{energy of light} - \text{energy needed to escape surface} \tag{1-34}$$

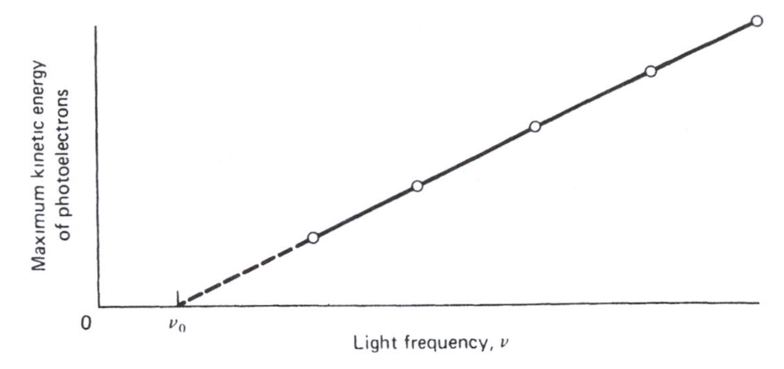

Figure 1-9 ▶ Maximum kinetic energy of photoelectrons as a function of incident light frequency, where ν_0 is the minimum frequency for which photoelectrons are ejected from the metal in the absence of any retarding or accelerating potential.

The last quantity in Eq. (1-34) is often referred to as the *work function* W of the metal. Equating Eq. (1-33) with (1-34) gives

$$\text{energy of light} - W = h\nu - h\nu_0 \tag{1-35}$$

The material-dependent term W is identified with the material-dependent term $h\nu_0$, yielding

$$\text{energy of light} \equiv E = h\nu \tag{1-36}$$

where the value of h has been determined to be 6.626176×10^{-34} J sec. (See Appendix 10 for units and conversion factors.)

Physicists found it difficult to reconcile these observations with the classical electromagnetic field theory of light. For example, if light of a certain frequency and intensity causes emission of electrons having a certain maximum kinetic energy, one would expect increased light *intensity* (corresponding classically to a greater electromagnetic field amplitude and hence greater energy density) to produce photoelectrons of higher kinetic energy. However, it only produces more photoelectrons and does not affect their energies. Again, if light is a wave, the energy is distributed over the entire wavefront and this means that a low light intensity would impart energy at a very low rate to an area of surface occupied by one atom. One can calculate that it would take years for an individual atom to collect sufficient energy to eject an electron under such conditions. No such induction period is observed.

An explanation for these results was suggested in 1905 by Einstein, who proposed that the incident light be viewed as being comprised of discrete units of energy. Each such unit, or *photon*, would have an associated energy of $h\nu$, where ν is the frequency of the oscillating emitter. Increasing the intensity of the light would correspond to increasing the number of photons, whereas increasing the frequency of the light would increase the energy of the photons. If we envision each emitted photoelectron as resulting from a photon striking the surface of the metal, it is quite easy to see that Einstein's proposal accords with observation. But it creates a new problem: If we are to visualize light as a stream of photons, how can we explain the wave properties of light, such as the double-slit diffraction pattern? What is the physical meaning of the electromagnetic wave?

Essentially, the problem is that, in the classical view, the square of the electromagnetic wave at any point in space is a measure of the energy density at that point. Now the square of the electromagnetic wave is a continuous and smoothly varying function, and if energy is continuous and infinitely divisible, there is no problem with this theory. But if the energy cannot be divided into amounts smaller than a photon—if it has a particulate rather than a continuous nature—then the classical interpretation cannot apply, for it is not possible to produce a smoothly varying energy distribution from energy *particles* any more than it is possible to produce, at the microscopic level, a smooth density distribution in gas made from atoms of matter. Einstein suggested that the square of the electromagnetic wave at some point (that is, the sum of the squares of the electric and magnetic field magnitudes) be taken as the *probability density* for finding a photon in the volume element around that point. The greater the square of the wave in some region, the greater is the probability for finding the photon in that region. Thus, the classical notion of energy having a definite and smoothly varying distribution is replaced by the idea of a smoothly varying probability density for finding an atomistic packet of energy.

Let us explore this probabilistic interpretation within the context of the two-slit interference experiment. We know that the pattern of light and darkness observed on the screen agrees with the classical picture of interference of waves. Suppose we carry out the experiment in the usual way, except we use a light source (of frequency ν) so weak that only $h\nu$ units of energy per second pass through the apparatus and strike the screen. According to the classical picture, this tiny amount of energy should strike the screen in a delocalized manner, producing an extremely faint image of the entire diffraction pattern. Over a period of many seconds, this pattern could be accumulated (on a photographic plate, say) and would become more intense. According to Einstein's view, our experiment corresponds to transmission of one photon per second and each photon strikes the screen at a localized point. Each photon strikes a new spot (not to imply the same spot cannot be struck more than once) and, over a long period of time, they build up the observed diffraction pattern. If we wish to state in advance where the next photon will appear, we are unable to do so. The best we can do is to say that the next photon is more likely to strike in one area than in another, the relative probabilities being quantitatively described by the square of the electromagnetic wave.

The interpretation of electromagnetic waves as probability waves often leaves one with some feelings of unreality. If the wave only tells us relative probabilities for finding a photon at one point or another, one is entitled to ask whether the wave has "physical reality," or if it is merely a mathematical device which allows us to analyze photon distribution, the photons being the "physical reality." We will defer discussion of this question until a later section on electron diffraction.

EXAMPLE 1-1 A retarding potential of 2.38 volts just suffices to stop photoelectrons emitted from potassium by light of frequency $1.13 \times 10^{15}\ s^{-1}$. What is the work function, W, of potassium?

SOLUTION ▶ $E_{light} = h\nu = W + KE_{electron}$, $W = h\nu - KE_{electron} = (4.136 \times 10^{-15}\text{eV s})(1.13 \times 10^{15}\ s^{-1}) - 2.38\,\text{eV} = 4.67\,\text{eV} - 2.38\,\text{eV} = 2.29\,\text{eV}$ [Note convenience of using h in units of eV s for this problem. See Appendix 10 for data.] ◀

EXAMPLE 1-2 Spectroscopists often express ΔE for a transition between states in wavenumbers , e.g., m^{-1}, or cm^{-1}, rather than in energy units like J or eV. (Usually cm^{-1} is favored, so we will proceed with that choice.)
a) What is the physical meaning of the term wavenumber?
b) What is the connection between wavenumber and energy?
c) What wavenumber applies to an energy of 1.000 J? of 1.000 eV?

SOLUTION ► a) Wavenumber is the number of waves that fit into a unit of distance (usually of one centimeter). It is sometimes symbolized $\tilde{\nu}$. $\tilde{\nu} = 1/\lambda$, where λ is the wavelength in centimeters.
b) Wavenumber characterizes the light that has photons of the designated energy. $E = h\nu = hc/\lambda = hc\tilde{\nu}$. (where c is given in cm/s).
c) $E = 1.000$ J $= hc\tilde{\nu}$; $\tilde{\nu} = 1.000$ J$/hc = 1.000$ J $/[(6.626 \times 10^{-34}$ J s$)(2.998 \times 10^{10}$ cm/s$)] = 5.034 \times 10^{22}$ cm^{-1}. Clearly, this is light of an extremely short wavelength since more than 10^{22} wavelengths fit into 1 cm. For 1.000 eV, the above equation is repeated using h in eV s. This gives $\tilde{\nu} = 8065\, cm^{-1}$. ◄

1-7 The Wave Nature of Matter

Evidently light has wave and particle aspects, and we can describe it in terms of photons, which are associated with waves of frequency $\nu = E/h$. Now photons are rather peculiar particles in that they have zero rest mass. In fact, they can exist only when traveling at the speed of light. The more normal particles in our experience have nonzero rest masses and can exist at any velocity up to the speed-of-light limit. Are there also waves associated with such normal particles?

Imagine a particle having a finite rest mass that somehow can be made lighter and lighter, approaching zero in a continuous way. It seems reasonable that the existence of a wave associated with the motion of the particle should become more and more apparent, rather than the wave coming into existence abruptly when $m = 0$. De Broglie proposed that all material particles are associated with waves, which he called "matter waves," but that the existence of these waves is likely to be observable only in the behaviors of extremely light particles.

De Broglie's relation can be reached as follows. Einstein's relation for photons is

$$E = h\nu \tag{1-37}$$

But a photon carrying energy E has a relativistic mass given by

$$E = mc^2 \tag{1-38}$$

Equating these two equations gives

$$E = mc^2 = h\nu = hc/\lambda \tag{1-39}$$

or

$$mc = h/\lambda \tag{1-40}$$

A normal particle, with nonzero rest mass, travels at a velocity v. If we regard Eq. (1-40) as merely the high-velocity limit of a more general expression, we arrive at an equation relating particle momentum p and associated wavelength λ:

$$mv = p = h/\lambda \tag{1-41}$$

or

$$\lambda = h/p \tag{1-42}$$

Here, m refers to the rest mass of the particle plus the relativistic correction, but the latter is usually negligible in comparison to the former.

This relation, proposed by de Broglie in 1922, was demonstrated to be correct shortly thereafter when Davisson and Germer showed that a beam of electrons impinging on a nickel target produced the scattering patterns one expects from interfering waves. These "electron waves" were observed to have wavelengths related to electron momentum in just the manner proposed by de Broglie.

Equation (1-42) relates the de Broglie wavelength λ of a matter wave to the momentum p of the particle. A higher momentum corresponds to a shorter wavelength. Since

$$\text{kinetic energy } T = mv^2 = (1/2m)(m^2v^2) = p^2/2m \tag{1-43}$$

it follows that

$$p = \sqrt{2mT} \tag{1-44}$$

Furthermore, Since $E = T + V$, where E is the total energy and V is the potential energy, we can rewrite the de Broglie wavelength as

$$\lambda = \frac{h}{\sqrt{2m(E-V)}} \tag{1-45}$$

Equation (1-45) is useful for understanding the way in which λ will change for a particle moving with constant total energy in a varying potential. For example, if the particle enters a region where its potential energy increases (e.g., an electron approaches a negatively charged plate), $E - V$ decreases and λ increases (i.e., the particle slows down, so its momentum decreases and its associated wavelength increases). We shall see examples of this behavior in future chapters.

Observe that if $E \geq V, \lambda$ as given by Eq. (1-45) is real. However, if $E < V, \lambda$ becomes imaginary. Classically, we never encounter such a situation, but we will find it is necessary to consider this possibility in quantum mechanics.

EXAMPLE 1-3 A He^{2+} ion is accelerated from rest through a voltage drop of 1.000 kilovolts. What is its final deBroglie wavelength? Would the wavelike properties be very apparent?

SOLUTION ▶ Since a charge of two electronic units has passed through a voltage drop of 1.000×10^3 volts, the final kinetic energy of the ion is 2.000×10^3 eV. To calculate λ, we first convert from eV to joules: $KE \equiv p^2/2m = (2.000 \times 10^3 \text{ eV})(1.60219 \times 10^{-19} \text{ J/eV}) = 3.204 \times 10^{-16}$ J. $m_{He} = (4.003 \text{ g/mol})(10^{-3} \text{ kg/g})(1 \text{ mol}/6.022 \times 10^{23} \text{atoms}) = 6.65 \times 10^{-27}$ kg; $p = \sqrt{2m_{He} \cdot KE} = [2(6.65 \times 10^{-27} \text{ kg})(3.204 \times 10^{-16} \text{ J})]^{1/2} = 2.1 \times 10^{-21}$ kg m/s. $\lambda = h/p = (6.626 \times 10^{-34} \text{ Js})/(2.1 \times 10^{-21} \text{ kg m/s}) = 3.2 \times 10^{-13}$ m $= 0.32$ pm. This wavelength is on the order of 1% of the radius of a hydrogen atom–too short to produce observable interference results when interacting with atom-size scatterers. For most purposes, we can treat this ion as simply a high-speed particle. ◀

1-8 A Diffraction Experiment with Electrons

In order to gain a better understanding of the meaning of matter waves, we now consider a set of simple experiments. Suppose that we have a source of a beam of monoenergetic electrons and a pair of slits, as indicated schematically in Fig. 1-10. Any electron arriving at the phosphorescent screen produces a flash of light, just as in a television set. For the moment we ignore the light source near the slits (assume that it is turned off) and inquire as to the nature of the image on the phosphorescent screen when the electron beam is directed at the slits. The observation, consistent with the observations of Davisson and Germer already mentioned, is that there are alternating bands of light and dark, indicating that the electron beam is being diffracted by the slits. Furthermore, the distance separating the bands is consistent with the de Broglie wavelength corresponding to the energy of the electrons. The variation in light intensity observed on the screen is depicted in Fig. 1-11a.

Evidently, the electrons in this experiment are displaying wave behavior. Does this mean that the electrons are spread out like waves when they are detected at the screen? We test this by reducing our beam intensity to let only one electron per second through the apparatus and observe that each electron gives a localized pinpoint of light, the entire diffraction pattern building up gradually by the accumulation of many points. Thus, the square of de Broglie's matter wave has the same kind of statistical significance that Einstein proposed for electromagnetic waves and photons, and the electrons really are localized particles, at least when they are detected at the screen.

However, if they are really particles, it is hard to see how they can be diffracted. Consider what happens when slit b is closed. Then all the electrons striking the screen must have come through slit a. We observe the result to be a single area of light on the screen (Fig. 1-11b). Closing slit a and opening b gives a similar (but displaced)

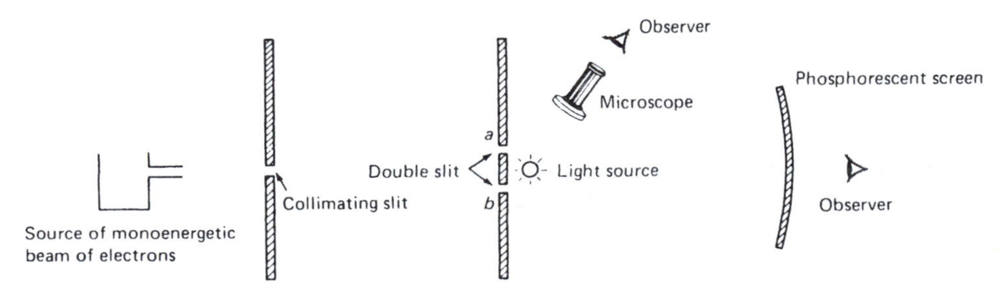

Figure 1-10 ► The electron source produces a beam of electrons, some of which pass through slits a and/or b to be detected as flashes of light on the phosphorescent screen.

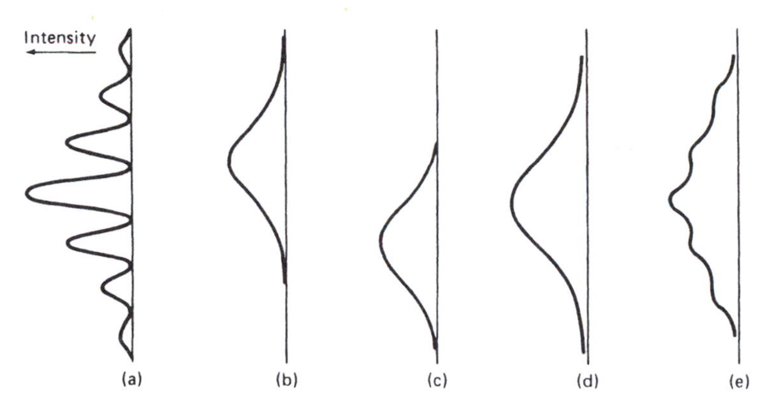

Figure 1-11 ▶ Light intensity at phosphorescent screen under various conditions: (a) a and b open, light off; (b) a open, b closed, light off; (c) a closed, b open, light off; (d) a and b open, light on, λ short; (e) a and b open, light on, λ longer.

light area, as shown in Fig. 1-11c. These patterns are just what we would expect for particles. Now, with both slits open, we expect half the particles to pass through slit a and half through slit b, the resulting pattern being the *sum* of the results just described. Instead we obtain the diffraction pattern (Fig. 1-11a). How can this happen? It seems that, somehow, an electron passing through the apparatus can sense whether one or both slits are open, even though as a particle it can explore only one slit or the other. One might suppose that we are seeing the result of simultaneous traversal of the two slits by two electrons, the path of each electron being affected by the presence of an electron in the other slit. This would explain how an electron passing through slit a would "know" whether slit b was open or closed. But the fact that the pattern builds up even when electrons pass through at the rate of one per second indicates that this argument will not do. Could an electron be coming through both slits at once?

To test this question, we need to have detailed information about the positions of the electrons as they pass through the slits. We can get such data by turning on the light source and aiming a microscope at the slits. Then photons will bounce off each electron as it passes the slits and will be observed through the microscope. The observer thus can tell through which slit each electron has passed, and also record its final position on the phosphorescent screen. In this experiment, it is necessary to use light having a wavelength short in comparison to the interslit distance; otherwise the microscope cannot resolve a flash well enough to tell which slit it is nearest. When this experiment is performed, we indeed detect each electron as coming through one slit or the other, and not both, but we also find that the diffraction pattern on the screen has been lost and that we have the broad, featureless distribution shown in Fig. 1-11d, which is basically the sum of the single-slit experiments. What has happened is that the photons from our light source, in bouncing off the electrons as they emerge from the slits, have affected the momenta of the electrons and changed their paths from what they were in the absence of light. We can try to counteract this by using photons with lower momentum; but this means using photons of lower E, hence longer λ. As a result, the images of the electrons in the microscope get broader, and it becomes more and more ambiguous as to which slit a given electron has passed through or that it really passed through only one slit. As we become more and more uncertain about the path of each electron as it moves past the slits, the accumulating diffraction pattern becomes more and more pronounced (Fig. 1-11e). (Since this is a "thought experiment," we can ignore the inconvenient fact that our "light" source must produce X rays or γ rays in order to have a wavelength short in comparison to the appropriate interslit distance.)

This conceptual experiment illustrates a basic feature of microscopic systems—we cannot measure properties of the system without affecting the future development of the system in a nontrivial way. The system with the light turned off is significantly different from the system with the light turned on (with short λ), and so the electrons arrive at the screen with different distributions. No matter how cleverly one devises the experiment, there is some minimum necessary disturbance involved in any measurement. In this example with the light off, the problem is that we know the momentum of each electron quite accurately (since the beam is monoenergetic and collimated), but we do not know anything about the way the electrons traverse the slits. With the light on, we obtain information about electron position just beyond the slits but we change the momentum of each electron in an unknown way. The measurement of particle position leads to loss of knowledge about particle momentum. This is an example of the *uncertainty principle* of Heisenberg, who stated that the product of the simultaneous uncertainties in "conjugate variables," a and b, can never be smaller than the value of Planck's constant h divided by 4π:

$$\Delta a \cdot \Delta b \geq h/4\pi \tag{1-46}$$

Here, Δa is a measure of the uncertainty in the variable a, etc. (The easiest way to recognize conjugate variables is to note that their dimensions must multiply to joule seconds. Linear momentum and linear position satisfies this requirement. Two other important pairs of conjugate variables are energy–time and angular momentum–angular position.) In this example with the light off, our uncertainty in momentum is small and our uncertainty in position is unacceptably large, since we cannot say which slit each electron traverses. With the light on, we reduce our uncertainty in position to an acceptable size, but subsequent to the position of each electron being observed, we have much greater uncertainty in momentum.

Thus, we see that the appearance of an electron (or a photon) as a particle or a wave depends on our experiment. Because *any* observation on so small a particle involves a significant perturbation of its state, it is proper to think of the electron plus apparatus as a single system. The question, "Is the electron a particle or a wave?" becomes meaningful only when the apparatus is defined on which we plan a measurement. In some experiments, the apparatus and electrons interact in a way suggestive of the electron being a wave, in others, a particle. The question, "What is the electron when were not looking?," cannot be answered experimentally, since an experiment is a "look" at the electron. In recent years experiments of this sort have been carried out using single atoms.[5]

EXAMPLE 1-4 The lifetime of an excited state of a molecule is 2×10^{-9} s. What is the uncertainty in its energy in J? In cm^{-1}? How would this manifest itself experimentally?

[5]See F. Flam [1].

SOLUTION ▶ The Heisenberg uncertainty principle gives, for minimum uncertainty $\Delta E \cdot \Delta t = h/4\pi$. $\Delta E = (6.626 \times 10^{-34}\text{ J s})/[(4\pi)(2 \times 10^{-9}\text{ s})] = 2.6 \times 10^{-26}\text{ J}$ $(2.6 \times 10^{-26}\text{J})\ (5.03 \times 10^{22}\text{ cm}^{-1}\text{J}^{-1}) = 0.001\text{ cm}^{-1}$ (See Appendix 10 for data.) Larger uncertainty in E shows up as greater line-width in emission spectra. ◀

1-9 Schrödinger's Time-Independent Wave Equation

Earlier we saw that we needed a wave equation in order to solve for the standing waves pertaining to a particular classical system and its set of boundary conditions. The same need exists for a wave equation to solve for matter waves. Schrödinger obtained such an equation by taking the classical time-independent wave equation and substituting de Broglie's relation for λ. Thus, if

$$\nabla^2\psi = -(2\pi/\lambda)^2\psi \tag{1-47}$$

and

$$\lambda = \frac{h}{\sqrt{2m(E-V)}} \tag{1-48}$$

then

$$\left[-(h^2/8\pi^2 m)\nabla^2 + V(x,y,z)\right]\psi(x,y,z) = E\psi(x,y,z) \tag{1-49}$$

Equation (1-49) is Schrödinger's time-independent wave equation for a single particle of mass m moving in the three-dimensional potential field V.

In classical mechanics we have separate equations for wave motion and particle motion, whereas in quantum mechanics, in which the distinction between particles and waves is not clear-cut, we have a single equation—the Schrödinger equation. We have seen that the link between the Schrödinger equation and the classical *wave* equation is the de Broglie relation. Let us now compare Schrödinger's equation with the classical equation for *particle* motion.

Classically, for a particle moving in three dimensions, the total energy is the sum of kinetic and potential energies:

$$(1/2m)(p_x^2 + p_y^2 + p_z^2) + V = E \tag{1-50}$$

where p_x is the momentum in the x coordinate, etc. We have just seen that the analogous Schrödinger equation is [writing out Eq. (1-49)]

$$\left[\frac{-h^2}{8\pi^2 m}\left(\frac{\partial^2}{\partial x^2} + \frac{\partial^2}{\partial y^2} + \frac{\partial^2}{\partial z^2}\right) + V(x,y,z)\right]\psi(x,y,z) = E\psi(x,y,z) \tag{1-51}$$

It is easily seen that Eq. (1-50) is linked to the quantity in brackets of Eq. (1-51) by a relation associating classical momentum with a partial differential operator:

$$p_x \longleftrightarrow (h/2\pi i)(\partial/\partial x) \tag{1-52}$$

and similarly for p_y and p_z. The relations (1-52) will be seen later to be an important postulate in a formal development of quantum mechanics.

The left-hand side of Eq. (1-50) is called the *hamiltonian* for the system. For this reason the operator in square brackets on the LHS of Eq. (1-51) is called the *hamiltonian operator*[6] H. For a given system, we shall see that the construction of H is not difficult. The difficulty comes in solving Schrödinger's equation, often written as

$$H\psi = E\psi \tag{1-53}$$

The classical and quantum-mechanical wave equations that we have discussed are members of a special class of equations called *eigenvalue equations*. Such equations have the format

$$\text{Op } f = cf \tag{1-54}$$

where Op is an operator, f is a function, and c is a constant. Thus, eigenvalue equations have the property that operating on a function regenerates the same function times a constant. The function f that satisfies Eq. (1-54) is called an *eigenfunction* of the operator. The constant c is called the *eigenvalue* associated with the eigenfunction f. Often, an operator will have a large number of eigenfunctions and eigenvalues of interest associated with it, and so an index is necessary to keep them sorted, viz.

$$\text{Op } f_i = c_i f_i \tag{1-55}$$

We have already seen an example of this sort of equation, Eq. (1-19) being an eigenfunction for Eq. (1-18), with eigenvalue $-\omega^2 m/T$.

The solutions ψ for Schrödinger's equation (1-53), are referred to as eigenfunctions, wavefunctions, or state functions.

EXAMPLE 1-5 a) Show that $\sin(3.63x)$ is not an eigenfunction of the operator d/dx.
b) Show that $\exp(-3.63ix)$ is an eigenfunction of the operator d/dx. What is its eigenvalue?
c) Show that $\frac{1}{\pi}\sin(3.63x)$ is an eigenfunction of the operator $((-h^2/8\pi^2 m)d^2/dx^2)$. What is its eigenvalue?

SOLUTION ▶ a) $\frac{d}{dx}\sin(3.63x) = 3.63\cos(3.63x) \neq$ constant times $\sin(3.63x)$.
b) $\frac{d}{dx}\exp(-3.63ix) = -3.63i\exp(-3.63ix) =$ constant times $\exp(-3.63ix)$. Eigenvalue $= -3.63i$.
c)

$$\begin{aligned}((-h^2/8\pi^2 m)d^2/dx^2)\tfrac{1}{\pi}\sin(3.63x) &= (-h^2/8\pi^2 m)(1/\pi)(3.63)\tfrac{d}{dx}\cos(3.63x)\\ &= [(3.63)^2 h^2/8\pi^2 m]\cdot(1/\pi)\sin(3.63x)\\ &= \text{constant times } (1/\pi)\sin(3.63x).\\ \text{Eigenvalue} &= (3.63)^2 h^2/8\pi^2 m.\end{aligned}$$

◀

[6]An *operator* is a symbol telling us to carry out a certain mathematical operation. Thus, d/dx is a differential *operator* telling us to differentiate anything following it with respect to x. The function $1/x$ may be viewed as a multiplicative operator. Any function on which it operates gets multiplied by $1/x$.

1-10 Conditions on ψ

We have already indicated that the square of the electromagnetic wave is interpreted as the probability density function for finding photons at various places in space. We now attribute an analogous meaning to ψ^2 for *matter* waves. Thus, in a one-dimensional problem (for example, a particle constrained to move on a line), the probability that the particle will be found in the interval dx around the point x_1 is taken to be $\psi^2(x_1)\,dx$. If ψ is a complex function, then the *absolute square*, $|\psi|^2 \equiv \psi^*\psi$ is used instead of ψ^2.[7] This makes it mathematically impossible for the average mass distribution to be negative in any region.

If an eigenfunction ψ has been found for Eq. (1-53), it is easy to see that $c\psi$ will also be an eigenfunction, for any constant c. This is due to the fact that a multiplicative constant commutes[8] with the operator H, that is,

$$H(c\psi) = cH\psi = cE\psi = E(c\psi) \tag{1-56}$$

The equality of the first and last terms is a statement of the fact that $c\psi$ is an eigenfunction of H. The question of which constant to use for the wavefunction is resolved by appeal to the probability interpretation of $|\psi|^2$. For a particle moving on the x axis, the probability that the particle is between $x = -\infty$ and $x = +\infty$ is unity, that is, a certainty. This probability is also equal to the sum of the probabilities for finding the particle in each and every infinitesimal interval along x, so this sum (an integral) must equal unity:

$$c^*c \int_{-\infty}^{+\infty} \psi^*(x)\psi(x)\,dx = 1 \tag{1-57}$$

If the selection of the constant multiplier c is made so that Eq. (1-57) is satisfied, the wavefunction $\psi' = c\psi$ is said to be *normalized*. For a three-dimensional function, $c\psi(x, y, z)$, the normalization requirement is

$$c^*c \int_{-\infty}^{+\infty}\int_{-\infty}^{+\infty}\int_{-\infty}^{+\infty} \psi^*(x, y, z)\psi(x, y, z)\,dx\,dy\,dz \equiv |c|^2 \int_{\text{all space}} |\psi|^2\,dv = 1 \tag{1-58}$$

As a result of our physical interpretation of $|\psi|^2$ plus the fact that ψ must be an eigenfunction of the hamiltonian operator H, we can reach some general conclusions about what sort of mathematical properties ψ can or cannot have.

First, we require that ψ be a *single-valued* function because we want $|\psi|^2$ to give an unambiguous probability for finding a particle in a given region (see Fig. 1-12). Also, we reject functions that are infinite in any region of space because such an infinity will always be infinitely greater than any finite region, and $|\psi|^2$ will be useless as a measure of comparative probabilities.[9] In order for $H\psi$ to be defined everywhere, it is necessary that the second derivative of ψ be defined everywhere. This requires that the first derivative of ψ be *piecewise continuous* and that ψ itself be *continuous* as in Fig.1d. (We shall see an example of this shortly.)

[7] If $f = u + iv$, then f^*, the complex conjugate of f, is given by $u - iv$, where u and v are real functions.

[8] a and b are said to *commute* if $ab = ba$.

[9] There are cases, particularly in relativistic treatments, where ψ is infinite at single points of zero measure, so that $|\psi|^2\,dx$ remains finite. Normally we do not encounter such situations in quantum chemistry.

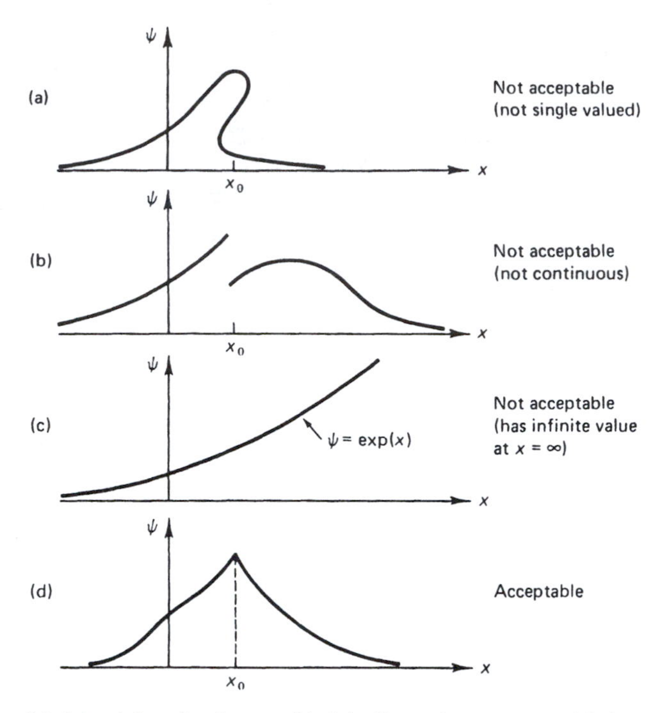

Figure 1-12 ▶ (a) ψ is triple valued at x_0. (b) ψ is discontinuous at x_0. (c) ψ grows without limit as x approaches $+\infty$ (i.e., ψ "blows up," or "explodes"). (d) ψ is continuous and has a "cusp" at x_0. Hence, first derivative of ψ is discontinuous at x_0 and is only piecewise continuous. This does not prevent ψ from being acceptable.

Functions that are single-valued, continuous, nowhere infinite, and have piecewise continuous first derivatives will be referred to as *acceptable* functions. The meanings of these terms are illustrated by some sample functions in Fig. 1-12.

In most cases, there is one more general restriction we place on ψ, namely, that it be a normalizable function. This means that the integral of $|\psi|^2$ over all space must not be equal to zero or infinity. A function satisfying this condition is said to be *square-integrable*.

1-11 Some Insight into the Schrödinger Equation

There is a fairly simple way to view the physical meaning of the Schrödinger equation (1-49). The equation essentially states that E in $H\psi = E\psi$ depends on two things, V and the second derivatives of ψ. Since V is the potential energy, the second derivatives of ψ must be related to the kinetic energy. Now the second derivative of ψ with respect to a given direction is a measure of the rate of change of slope (i.e., the curvature, or "wiggliness") of ψ in that direction. Hence, we see that a more wiggly wavefunction leads, through the Schrödinger equation, to a higher kinetic energy. This is in accord with the spirit of de Broglie's relation, since a shorter wavelength function is a more wiggly function. But the Schrödinger equation is more generally applicable because we can take second derivatives of any acceptable function, whereas wavelength is defined

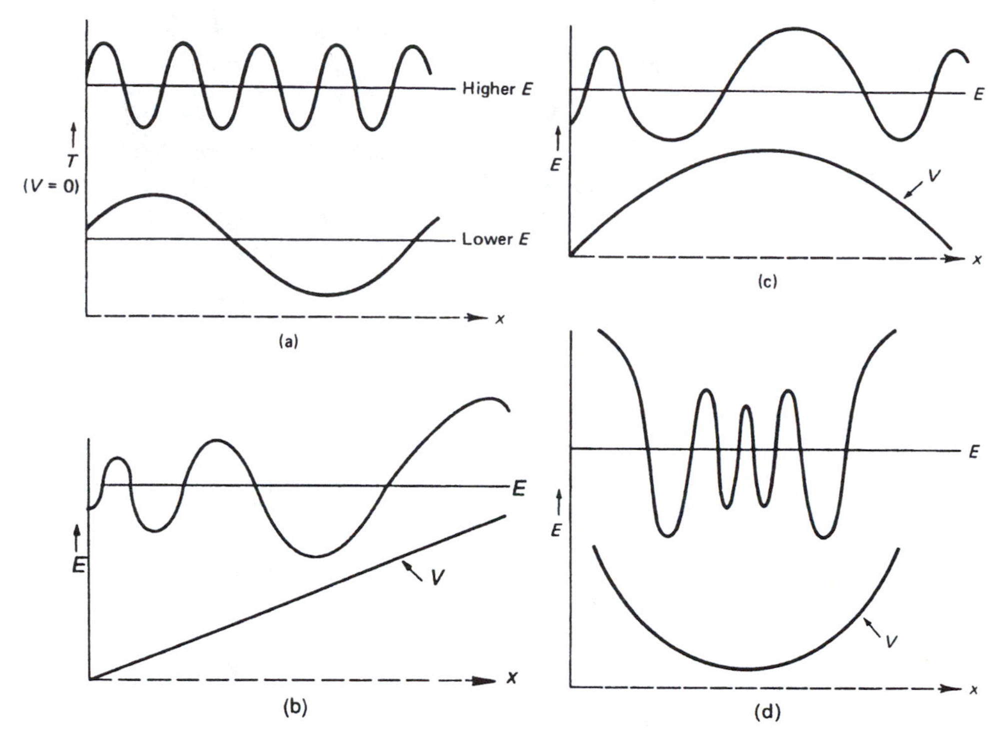

Figure 1-13 ► (a) Since $V = 0$, $E = T$. For higher T, ψ is more wiggly, which means that λ is shorter. (Since ψ is periodic for a free particle, λ is defined.) (b) As V increases from left to right, ψ becomes less wiggly. (c)–(d) ψ is most wiggly where V is lowest and T is greatest.

only for periodic functions. Since E is a constant, the solutions of the Schrödinger equation must be more wiggly in regions where V is low and less wiggly where V is high. Examples for some one-dimensional cases are shown in Fig. 1-13.

In the next chapter we use some fairly simple examples to illustrate the ideas that we have already introduced and to bring out some additional points.

1-12 Summary

In closing this chapter, we collect and summarize the major points to be used in future discussions.

1. Associated with any particle is a wavefunction having wavelength related to particle momentum by $\lambda = h/p = h/\sqrt{2m(E-V)}$.

2. The wavefunction has the following physical meaning; its absolute square is proportional to the probability density for finding the particle. If the wavefunction is normalized, its square is *equal* to the probability density.

3. The wavefunctions ψ for time-independent states are eigenfunctions of Schrödinger's equation, which can be constructed from the classical wave equation by requiring $\lambda = h/\sqrt{2m(E-V)}$, or from the classical particle equation by replacing p_k with $(h/2\pi i)\partial/\partial k$, $k = x, y, z$.

4. For ψ to be acceptable, it must be single-valued, continuous, nowhere infinite, with a piecewise continuous first derivative. For most situations, we also require ψ to be square-integrable.

5. The wavefunction for a particle in a varying potential oscillates most rapidly where V is low, giving a high T in this region. The low V plus high T equals E. In another region, where V is high, the wavefunction oscillates more slowly, giving a low T, which, with the high V, equals the same E as in the first region.

1-12.A Problems[10]

1-1. Express $A\cos(kx) + B\sin(kx) + C\exp(ikx) + D\exp(-ikx)$ purely in terms of $\cos(kx)$ and $\sin(kx)$.

1-2. Repeat the standing-wave-in-a-string problem worked out in Section 1-4, but clamp the string at $x = +L/2$ and $-L/2$ instead of at 0 and L.

1-3. Find the condition that must be satisfied by α and β in order that $\psi(x) = A\sin(\alpha x) + B\cos(\beta x)$ satisfy Eq. (1-20).

1-4. The apparatus sketched in Fig. 1-8 is used with a dish plated with zinc and also with a dish plated with cesium. The wavelengths of the incident light and the corresponding retarding potentials needed to just prevent the photoelectrons from reaching the collecting wire are given in Table P1-4. Plot incident light frequency versus retarding potential for these two metals. Evaluate their work functions (in eV) and the proportionality constant h (in eV s).

TABLE P1-4 ▶

	Retarding potential (V)	
λ(Å)	Cs	Zn
6000	0.167	—
3000	2.235	0.435
2000	4.302	2.502
1500	6.369	1.567
1200	8.436	6.636

1-5. Calculate the de Broglie wavelength in nanometers for each of the following:

a) An electron that has been accelerated from rest through a potential change of 500 V.
b) A bullet weighing 5 gm and traveling at 400 m s^{-1}.

1-6. Arguing from Eq. (1-7), what is the time needed for a standing wave to go through one complete cycle?

[10]Hints for a few problems may be found in Appendix 12 and answers for almost all of them appear in Appendix 13.

1-7. The equation for a standing wave in a string has the form

$$\Psi(x,t) = \psi(x)\cos(\omega t)$$

a) Calculate the time-averaged potential energy (PE) for this motion. [*Hint*: Use $\text{PE} = -\int F\,d\Psi$; $F = ma$; $a = \partial^2\Psi/\partial t^2$.]
b) Calculate the time-averaged kinetic energy (KE) for this motion. [*Hint*: Use $\text{KE} = 1/2mv^2$ and $v = \partial\Psi/\partial t$.]
c) Show that this harmonically vibrating string stores its energy *on the average* half as kinetic and half as potential energy, and that $E(x)_{\text{av}}\alpha\psi^2(x)$.

1-8. Indicate which of the following functions are "acceptable." If one is not, give a reason.

a) $\psi = x$
b) $\psi = x^2$
c) $\psi = \sin x$
d) $\psi = \exp(-x)$
e) $\psi = \exp(-x^2)$

1-9. An acceptable function is never infinite. Does this mean that an acceptable function must be square integrable? If you think these are not the same, try to find an example of a function (other than zero) that is never infinite but is not square integrable.

1-10. Explain why the fact that $\sin(x) = -\sin(-x)$ means that we can restrict Eq. (1-32) to nonnegative n without loss of physical content.

1-11. Which of the following are eigenfunctions for d/dx?

a) x^2
b) $\exp(-3.4x^2)$
c) 37
d) $\exp(x)$
e) $\sin(ax)$
f) $\cos(4x) + i\sin(4x)$

1-12. Calculate the minimum de Broglie wavelength for a photoelectron that is produced when light of wavelength 140.0 nm strikes zinc metal. (Workfunction of zinc = 3.63 eV.)

Multiple Choice Questions

(Intended to be answered without use of pencil and paper.)

1. A particle satisfying the time-independent Schrödinger equation must have

a) an eigenfunction that is normalized.
b) a potential energy that is independent of location.
c) a de Broglie wavelength that is independent of location.

d) a total energy that is independent of location.
e) None of the above is a true statement.

2. When one operates with d^2/dx^2 on the function $6\sin(4x)$, one finds that

a) the function is an eigenfunction with eigenvalue -96.
b) the function is an eigenfunction with eigenvalue 16.
c) the function is an eigenfunction with eigenvalue -16.
d) the function is not an eigenfunction.
e) None of the above is a true statement.

3. Which one of the following concepts did Einstein propose in order to explain the photoelectric effect?

a) A particle of rest mass m and velocity v has an associated wavelength λ given by $\lambda = h/mv$.
b) Doubling the intensity of light doubles the energy of each photon.
c) Increasing the wavelength of light increases the energy of each photon.
d) The photoelectron is a particle.
e) None of the above is a concept proposed by Einstein to explain the photoelectric effect.

4. Light of frequency ν strikes a metal and causes photoelectrons to be emitted having maximum kinetic energy of $0.90\,h\nu$. From this we can say that

a) light of frequency $\nu/2$ will not produce any photoelectrons.
b) light of frequency 2ν will produce photoelectrons having maximum kinetic energy of $1.80\,h\nu$.
c) doubling the intensity of light of frequency ν will produce photoelectrons having maximum kinetic energy of $1.80\,h\nu$.
d) the work function of the metal is $0.90\,h\nu$.
e) None of the above statements is correct.

5. The reason for normalizing a wavefunction ψ is

a) to guarantee that ψ is square-integrable.
b) to make $\psi^*\psi$ equal to the probability distribution function for the particle.
c) to make ψ an eigenfunction for the Hamiltonian operator.
d) to make ψ satisfy the boundary conditions for the problem.
e) to make ψ display the proper symmetry characteristics.

Reference

[1] F. Flam, *Making Waves with Interfering Atoms. Physics Today*, 921–922 (1991).

Chapter 2

Quantum Mechanics of Some Simple Systems

2-1 The Particle in a One-Dimensional "Box"

Imagine that a particle of mass m is free to move along the x axis between $x = 0$ and $x = L$, with no change in potential (set $V = 0$ for $0 < x < L$). At $x = 0$ and L and at all points beyond these limits the particle encounters an infinitely repulsive barrier ($V = \infty$ for $x \leq 0$, $x \geq L$). The situation is illustrated in Fig. 2-1. Because of the shape of this potential, this problem is often referred to as a *particle in a square well* or a *particle in a box* problem. It is well to bear in mind, however, that the situation is really like that of a particle confined to movement along a finite length of wire.

When the potential is discontinuous, as it is here, it is convenient to write a wave equation for each region. For the two regions beyond the ends of the box

$$\frac{-h^2}{8\pi^2 m}\frac{d^2\psi}{dx^2} + \infty\psi = E\psi, \quad x \leq 0, x \geq L \tag{2-1}$$

Within the box, ψ must satisfy the equation

$$\frac{-h^2}{8\pi^2 m}\frac{d^2\psi}{dx^2} = E\psi, \quad 0 < x < L \tag{2-2}$$

It should be realized that E must take on the same values for both of these equations; the eigenvalue E pertains to the entire range of the particle and is not influenced by divisions we make for mathematical convenience.

Let us examine Eq. (2-1) first. Suppose that, at some point within the infinite barrier, say $x = L + dx$, ψ is finite. Then the second term on the left-hand side of Eq. (2-1) will be infinite. If the first term on the left-hand side is finite or zero, it follows immediately that E is infinite at the point $L + dx$ (and hence everywhere in the system). Is it possible that a solution exists such that E is finite? One possibility is that $\psi = 0$ at all points where $V = \infty$. The other possibility is that the first term on the left-hand side of Eq. (2-1) can be made to cancel the infinite second term. This might happen if the second derivative of the wavefunction is infinite at all points where $V = \infty$ and $\psi \neq 0$. For the second derivative to be infinite, the first derivative must be discontinuous, and so ψ itself must be nonsmooth (i.e., it must have a sharp corner; see Fig. 2-2). Thus, we see that it may be possible to obtain a finite value for both E and ψ at $x = L + dx$, provided that ψ is nonsmooth there. But what about the next point, $x = L + 2dx$, and all the other points outside the box? If we try to use the same device, we end up with the requirement that ψ be nonsmooth at every point where $V = \infty$. A function that is

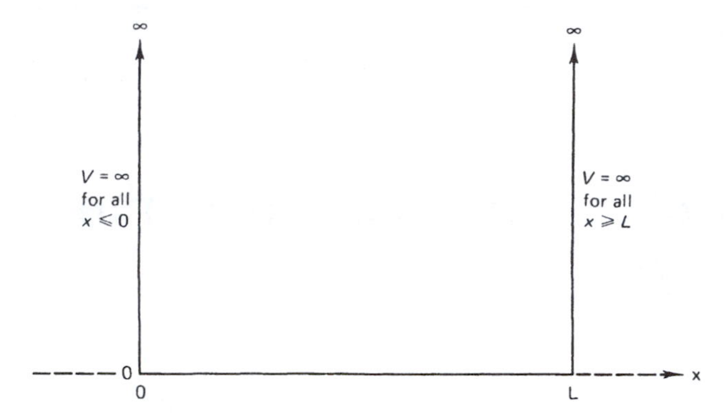

Figure 2-1 ▶ The potential felt by a particle as a function of its x coordinate.

continuous but which has a point-wise discontinuous first derivative is a contradiction in terms (i.e., a continuous f cannot be 100% corners. To have recognizable corners, we must have some (continuous) edges. We say that the first derivative of ψ must be *piecewise continuous*.) Hence, if $V = \infty$ *at a single point*, we might find a solution ψ which is finite at that point, with finite energy. If $V = \infty$ over a finite range of connected points, however, either E for the system is infinite, and ψ is finite over this region or E is not infinite (but is indeterminate) and ψ is zero over this region.

We are not concerned with particles of infinite energy, and so we will say that the solution to Eq. (2-1) is $\psi = 0$.[1]

Turning now to Eq. (2-2) we ask what solutions ψ exist in the box having associated eigenvalues E that are finite and positive. Any function that, when twice differentiated, yields a negative constant times the selfsame function is a possible candidate for ψ. Such functions are $\sin(kx)$, $\cos(kx)$, and $\exp(\pm ikx)$. But these functions are not all independent since, as we noted in Chapter 1,

$$\exp(\pm ikx) = \cos(kx) \pm i \sin(kx) \tag{2-3}$$

We thus are free to express ψ in terms of $\exp(\pm ikx)$ or else in terms of $\sin(kx)$ and $\cos(kx)$. We choose the latter because of their greater familiarity, although the final answer must be independent of this choice.

The most general form for the solution is

$$\psi(x) = A \sin(kx) + B \cos(kx) \tag{2-4}$$

where A, B, and k remain to be determined. As we have already shown, ψ is zero at $x \leq 0, x \geq L$ and so we have as boundary conditions

$$\psi(0) = 0 \tag{2-5}$$

$$\psi(L) = 0 \tag{2-6}$$

Mathematically, this is precisely the same problem we have already solved in Chapter 1 for the standing waves in a clamped string. The solutions are

$$\begin{aligned} \psi(x) &= A \sin(n\pi x/L), \quad n = 1, 2, \ldots, \quad 0 < x < L \\ \psi(x) &= 0, \quad x \leq 0, x \geq L \end{aligned} \tag{2-7}$$

[1] Thus, the particle never gets into these regions. It is meaningless to talk of the energy of the particle in such regions, and our earlier statement that E must be identical in Eqs. (2-1) and (2-2) must be modified; E is constant in all regions where ψ is finite.

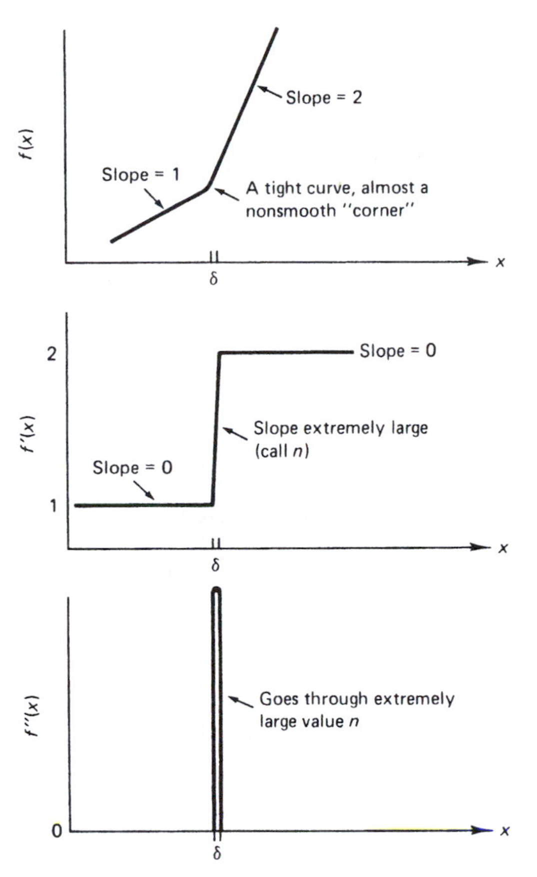

Figure 2-2 ▶ As the function $f(x)$ approaches being nonsmooth, δ approaches zero (the width of one point) and n approaches infinity.

One difference between Eq. (2-7) and the string solutions is that we have rejected the $n = 0$ solution in Eq. (2-7). For the string, this solution was for no vibration at all—a physically realizable circumstance. For the particle-in-a-box problem, this solution is rejected because it is not square-integrable. (It gives $\psi = 0$, which means *no* particle on the x axis, contradicting our starting premise. One could also reject this solution for the classical case since it means no energy in the string, which might contradict a starting premise depending on how the problem is worded.)

Let us check to be sure these functions satisfy Schrödinger's equation:

$$\begin{aligned} H\psi(x) &= \frac{-h^2}{8\pi^2 m}\frac{d^2[A\sin(n\pi x/L)]}{dx^2} \\ &= \frac{-h^2}{8\pi^2 m}\left[-A\frac{n^2\pi^2}{L^2}\sin\left(\frac{n\pi x}{L}\right)\right] \\ &= \frac{n^2h^2}{8mL^2}\left[A\sin\left(\frac{n\pi x}{L}\right)\right] = E\psi(x) \end{aligned} \tag{2-8}$$

This shows that the functions (2-7) are indeed eigenfunctions of H. We note in passing that these functions are acceptable in the sense of Chapter 1.

The only remaining parameter is the constant A. We set this to make the probability of finding the particle in the well equal to unity:

$$\int_0^L \psi^2(x)dx = A^2 \int_0^L \sin^2(n\pi x/L)dx = 1 \tag{2-9}$$

This leads to (Problem 2-2)

$$A = \sqrt{2/L} \tag{2-10}$$

which completes the solving of Schrödinger's time-independent equation for the problem. Our results are the normalized eigenfunctions

$$\psi_n(x) = \sqrt{(2/L)}\sin(n\pi x/L), \quad n = 1, 2, 3, \ldots \tag{2-11}$$

and the corresponding eigenvalues, from Eq. (2-8),

$$E_n = n^2h^2/8mL^2, \quad n = 1, 2, 3, \ldots \tag{2-12}$$

Each different value of n corresponds to a different stationary state of this system.

2-2 Detailed Examination of Particle-in-a-Box Solutions

Having solved the Schrödinger equation for the particle in the infinitely deep square-well potential, we now examine the results in more detail.

2-2.A Energies

The most obvious feature of the energies is that, as we move through the allowed states ($n = 1, 2, 3, \ldots$), E skips from one discrete, well-separated value to another (1, 4, 9 in units of $h^2/8mL^2$). Thus, the particle can have only certain discrete energies—the energy is *quantized*. This situation is normally indicated by sketching the allowed *energy levels* as horizontal lines superimposed on the potential energy sketch, as in Fig. 2-3a. The fact that each energy level is a horizontal line emphasizes the fact that E is a constant and is the same regardless of the x coordinate of the particle. For this reason, E is called a *constant of motion*. The dependence of E on n^2 is displayed in the increased spacing between levels with increasing n in Fig. 2-3a. The number n is called a *quantum number*.

We note also that E is proportional to L^{-2}. This means that the more tightly a particle is confined, the greater is the spacing between the allowed energy levels. Alternatively, as the box is made wider, the separation between energies decreases and, in the limit of an infinitely wide box, disappears entirely. Thus, we associate quantized energies with spatial confinement.

For some systems, the degree of confinement of a particle depends on its total energy. For example, a pendulum swings over a longer trajectory if it has higher energy. The potential energy for a pendulum is given by $V = \frac{1}{2}kx^2$ and is given in Fig. 2-3b. If one solves the Schrödinger equation for this system (see Chapter 3), one finds that the energies are proportional to n rather than n^2. We can rationalize this by thinking of

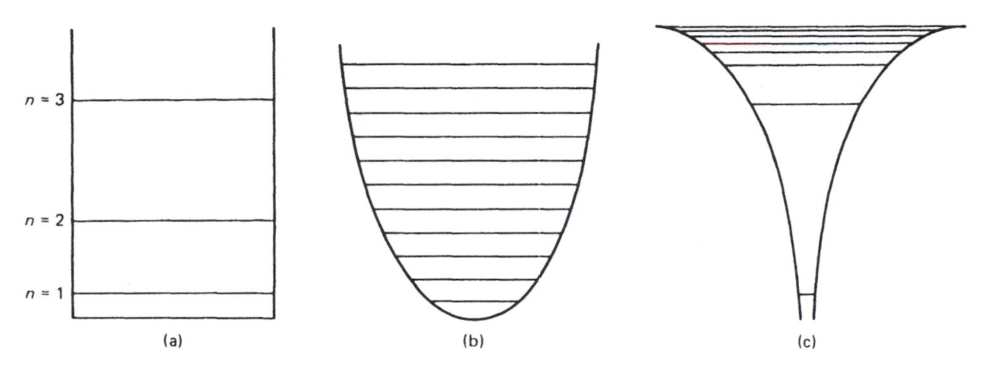

Figure 2-3 ▶ Allowed energies for a particle in various one-dimensional potentials. (a) "box" with infinite walls. (b) quadratic potential, $V = \frac{1}{2}kx^2$. (c) $V = -1/|x|$. Tendency for higher levels in (b) and (c) not to diverge as in (a) is due to larger "effective box size" for higher energies in (b) and (c).

the particle as occupying successively bigger boxes as we go to higher energies. This counteracts the n^2 increase in energy levels found for constant box width. For the potential $V = -1/|x|$ (which is the one-dimensional analog of a hydrogen atom) E varies as $1/n^2$ (Fig. 2-3c), and this is also consistent with the effective increase in L with increasing E.

The energy is proportional to $1/m$. This means that the separation between allowed energy levels decreases as m increases. Ultimately, for a macroscopic object, m is so large that the levels are too closely spaced to be distinguished from the continuum of levels expected in classical mechanics. This is an example of the *correspondence principle*, which, in its most general form, states that the predictions of quantum mechanics must pass smoothly into those of classical mechanics whenever we progress in a continuous way from the microscopic to the macroscopic realm.

Notice that the lowest possible energy for this system occurs for $n = 1$ and is $E = h^2/8mL^2$. This remarkable result means that a constrained particle (i.e., L not infinite) can never have an energy of zero. Evidently, the particle continues to move about in the region 0 to L, even at a temperature of absolute zero. For this reason, $h^2/8mL^2$ is called the *zero-point energy* for this system. In general, a finite zero-point energy occurs in any system having a restriction for motion in any coordinate. (Note that *finite* here means *not equal to zero*.)

It is possible to show that, for $L \neq \infty$, our particle in a box would have to violate the Heisenberg uncertainty principle to achieve an energy of zero. For, suppose the energy is precisely zero. Then the momentum must be precisely zero too. (In this system, all energy of the particle is kinetic since $V = 0$ in the box.) If the momentum p_x is *precisely* zero, however, our *uncertainty* in the value of the momentum Δp_x is also zero. If Δp_x is zero, the uncertainty principle [Eq. (2-46)] requires that the uncertainty in position Δx be infinite. But we know that the particle is between $x = 0$ and $x = L$. Hence, our uncertainty is on the order of L, not infinity, and the uncertainty principle is not satisfied. However, when $L = \infty$ (the particle is unconstrained), it is possible for the uncertainty principle to be satisfied simultaneously with having $E = 0$, and this is in satisfying accord with the fact that $E = h^2/8mL^2$ goes to zero as L approaches infinity.

Finally, we note that each separate value of n leads to a different energy. Thus, no two states have the same energy, and the states are said to be *nondegenerate* with respect to energy.

EXAMPLE 2-1 Consider an electron in a one-dimensional box of length 258 pm. a) What is the zero-point energy (ZPE) for this system? For a mole of such systems? b) What electronic speed classically corresponds to this ZPE? Compare to the speed of light.

SOLUTION ▶ a)

$$ZPE = E_{lowest} = E_{n=1}$$
$$= 1^2h^2/8mL^2 = \frac{(1)^2(6.626 \times 10^{-34}\,\text{J s})^2}{8(9.11 \times 10^{-31}\,\text{kg})(258 \times 10^{-12}\,\text{m})^2}$$
$$= 9.05 \times 10^{-19}\,\text{J};$$

Per mole, this equals

$$(9.05 \times 10^{-19}\,\text{J})(6.022 \times 10^{23}\,\text{mol}^{-1})(1\,\text{kJ}/10^3\,\text{J}) = 54.5\,\text{kJ mol}^{-1}$$

b) E is all kinetic energy since $V = 0$ in the box, so $E = mv^2/2$. Hence,

$$v = \left[\frac{2E}{m}\right]^{1/2} = \left[\frac{2(9.05 \times 10^{-19}\,\text{J})}{9.11 \times 10^{-31}\,\text{kg}}\right]^{1/2} = 1.41 \times 10^6\,\text{m s}^{-1}$$

Compared to the speed of light, this is $\frac{1.41\times10^6\,\text{m s}^{-1}}{2.998\times10^8\,\text{m s}^{-1}} = 0.0047$, or about 0.5% of the speed of light. ◀

2-2.B Wavefunctions

We turn now to the eigenfunctions (2-11) for this problem. These are typically displayed by superimposing them on the energy levels as shown in Fig. 2-4 for the three lowest-energy wavefunctions. (It should be recognized that the energy units of the vertical axis do *not* pertain to the amplitudes of the wavefunctions.)

It is apparent from Fig. 2-4 that the allowed wavefunctions for this system could have been produced merely by placing an integral number of half sine waves in the

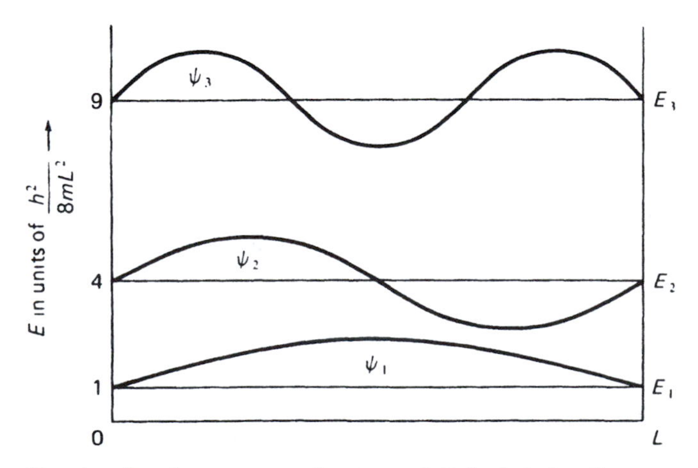

Figure 2-4 ▶ The eigenfunctions corresponding to $n = 1, 2, 3$, plotted on the corresponding energy levels. The energy units of the ordinate do not refer to the wavefunctions ψ. Each wavefunction has a zero value wherever it intersects its own energy level, and a maximum value of $\sqrt{2/L}$.

range 0–L. The resulting wavelengths would then yield the energy of each state through application of de Broglie's relation (1-42). Thus, by inspection of Fig. 2-4, the allowed wavelengths are

$$\lambda = 2L/n, \quad n = 1, 2, 3, \ldots \tag{2-13}$$

Therefore

$$p = h/\lambda = nh/2L \tag{2-14}$$

and

$$E = p^2/2m = n^2h^2/8mL^2 \tag{2-15}$$

in agreement with Eq. (2-12). As pointed out in Section 1-11, the wavefunctions having higher kinetic energy oscillate more rapidly. (Here $V = 0$, and E is all kinetic energy.)

Let us now consider the physical meaning of the eigenfunctions ψ. According to our earlier discussion, ψ^2 summarizes the results of many determinations of the position of the particle. Suppose that we had a particle-in-a-box system that we had somehow prepared in such a way that we knew it to be in the state with $n = 1$. We can imagine some sort of experiment, such as flashing a powerful γ-ray flashbulb and taking an instantaneous photograph, which tells us where the particle was at the instant of the flash. Now, suppose we wish to determine the position of the particle again. We want this second determination to be for the $n = 1$ state also, but we cannot use our original system for this because we have "spoiled" it by our first measurement process. Hitting the particle with one or more γ-ray photons has knocked it into some other state, and we do not even know which one. Therefore, we must either reprepare our system, or else use a separate system whose preparation is identical to that of the first system. In general we shall assume that we have an inexhaustible supply of identically prepared systems. Therefore, we take a second photograph (on our second system) using the same photographic plate. Then we photograph a third system, a fourth, etc., until we have amassed a large number of images of the particle on the film. The distribution of these images is given by ψ_1^2. (Since ψ is always a real function for this system, we do not need to bother with $\psi^*\psi$.) Other states, like ψ_2, ψ_3, will lead to different distributions of images. The results for the several states are depicted in Fig. 2-5. It is obvious that the probability for finding the particle near the center of the box is predicted to be much larger for the $n = 1$ than the $n = 2$ state.

The probability for finding the particle at the midpoint of the "wire" in the $n = 2$ state approaches zero in the limit of our measurement becoming precise enough to observe a single point. This troubles many students at first encounter because they worry about how the particle can get from one side of the box to the other in the $n = 2$ state. In fact, this question can be raised for any state whose wavefunction has any nodes. However, our discussion in the preceding paragraph shows that this question, like the question, "Is an electron a particle or a wave when we are not looking?" has no meaningful answer because no experiment can be conceived that would answer it. To test whether or not the particle does travel from one side of the box to the other, we would have to prepare the system in the $n = 2$ state and measure the position of the particle enough times so that we either (a) always find it on the same side (requires many measurements for confidence), or (b) find it on different sides (requires at least two measurements).

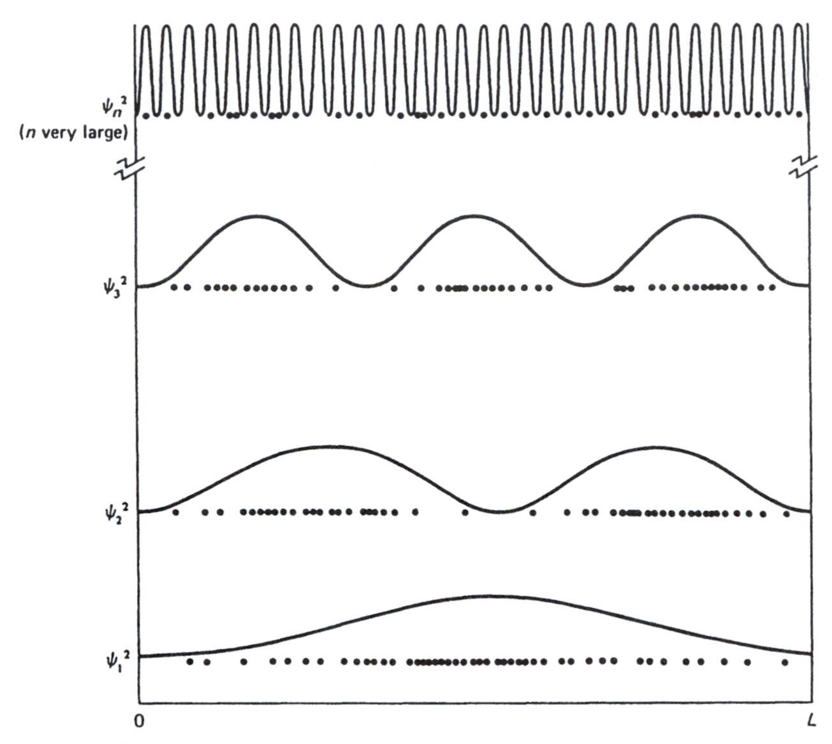

Figure 2-5 ▶ ψ^2 and observed particle distribution for the three lowest-energy and one high-energy state of the particle in a one-dimensional box.

But for our question to be answered, the system must be in the $n = 2$ state throughout this entire experiment, and we have seen that the process of measuring particle position prevents this. (If we find the particle first on the left and later on the right, we cannot be sure it did not travel across the midpoint while the system was perturbed by the first measurement.) Thus, the sketches in Fig. 2-5 are most safely regarded as a summary of the results of measurements on an *ensemble* of systems.

Classically, since the particle has constant energy, hence constant speed, we would expect the particle to spend equal time in each line segment dx between 0 and L, but Fig. 2-5 shows that the quantum system with $n = 1$ predicts that the particle spends more time in segments near the center. It is characteristic of lower-energy states of quantum-mechanical systems to display "anti-classical" distributions. With higher quantum numbers, the distribution becomes difficult to distinguish from the distribution predicted by classical physics (see Fig. 2-5). This is another example of the tendency of quantum-mechanical predictions to approach classical predictions when one goes toward the macroscopic realm (here large n and therefore large E).

EXAMPLE 2-2 For a particle in the $n = 2$ state in a one-dimensional box of length L,
a) *estimate* the probability, ρ, for finding the particle between $x = 0$ and $x = 0.20L$.
b) *calculate* the probability that you estimated in part a.
c) what probability for finding the particle between $x = 0$ and $x = 0.20L$ is predicted by classical physics?

SOLUTION ▶ a) The sketch of ψ_2^2 in Fig. 2-5 makes it clear that the probability for finding the particle in the range $0 \leq x \leq L/4$ is 0.25 (equal to the area under the curve). The range $0 \leq x \leq 0.20L$ is 20% shorter, and the missing 20% of the range is associated with a relatively large probability—almost double the average in the range. That means we are missing nearly 40% of the probablity, so slightly more than 60% remains. 60% of 0.25 is 0.15, so the probability, ρ, in the range $0 - 0.20L$ is slightly larger than 0.15.

b)

$$\begin{aligned}
\rho &= \int_0^{0.2L} \psi_2^2 dx = \frac{2}{L}\int_0^{0.2L} \sin^2\left(\frac{2\pi x}{L}\right) dx \\
&= \left(\frac{2}{L}\right)\left(\frac{L}{2\pi}\right)\int_0^{0.2L} \sin^2\left(\frac{2\pi x}{L}\right) d\left(\frac{2\pi x}{L}\right) \\
&= \frac{1}{\pi}\int_0^{0.4\pi} \sin^2 y dy \\
&= \text{(from Appendix 1)} \frac{1}{\pi}\left\{\frac{y}{2} - \frac{1}{4}\sin 2y|_0^{0.4L}\right\} \\
&= \frac{1}{\pi}\left\{0.2\pi - \frac{1}{4}\sin 0.8\pi\right\} \\
&= 0.20 - \frac{1}{4\pi}\sin 0.8\pi = 0.153
\end{aligned}$$

c) The classical particle travels with constant speed, hence has a constant probability function. Therefore, the probability for finding the particle in *any* 20% of the box is 0.20. ◀

2-2.C Symmetry of Wavefunctions

Inspection of Fig. 2-5 shows that the particle has equal probabilities for being observed in the left half and right half of the box, regardless of state. This seems reasonable because there is no physical factor discriminating between these halves. We shall now show that the hamiltonian operator is invariant for a reflection through the box center, and that a necessary consequence of this is that ψ has certain symmetry properties.

First, we show that H is invariant. Reflection through the box center is accomplished by replacing x by $-x + L$. We can define a reflection operator R such that $Rf(x) = f(-x + L)$; i.e., R reflects any function through a plane normal to x at $x = L/2$ (see Fig. 2-6).

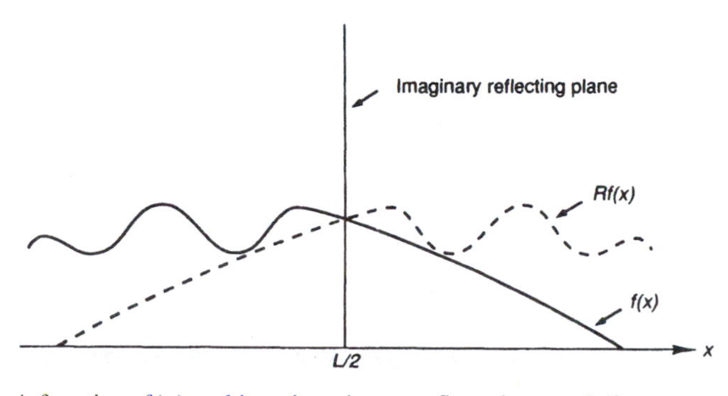

Figure 2-6 ▶ A function $f(x)$ and its mirror image reflected at $x = L/2$.

The kinetic part of the hamiltonian, T, is unchanged by R:

$$RT = R\left[-\frac{h^2}{8\pi^2 m}\frac{d^2}{dx^2}\right] = \frac{-h^2}{8\pi^2 m}\frac{d^2}{d(-x+L)^2}$$
$$= \frac{-h^2}{8\pi^2 m}\frac{d^2}{dx^2} = T \qquad (2\text{-}16)$$

where we have used the fact that L is constant and $d/d(-x) = -d/dx$. That the potential part of H is unchanged by reflection through $L/2$ is easily seen; the identical infinite barriers merely interchange position. Therefore, $RT = T$ and $RV = V$, and $RH = R(T+V) = RT + RV = T + V = H$.

Now let us see what this means for eigenfunctions of H. Assume we have a normalized eigenfunction ψ

$$H\psi = E\psi \qquad (2\text{-}17)$$

The two sides of Eq. (2-17) will still be equal if we reflect our coordinate system throughout the equation. (If two functions are identical in one coordinate system, say a right handed system, then they are identical in any coordinate system.) Therefore,[2]

$$(RH)(R\psi) = (RE)(R\psi) \qquad (2\text{-}18)$$

But E is simply a constant, and so it is immune to R. Furthermore, we have just seen that $RH = H$. Therefore,

$$H(R\psi) = E(R\psi) \qquad (2\text{-}19)$$

which shows that the function $R\psi$ is an eigenfunction of H with the same eigenvalue as ψ.

We have already mentioned that the eigenfunctions of this system are nondegenerate with respect to energy. This is equivalent to saying that no two linearly independent eigenfunctions having the same eigenvalue exist for this system. But we have just shown that ψ and $R\psi$ are both eigenfunctions having the same eigenvalue E. Therefore, we are forced to conclude that ψ and $R\psi$ are linearly dependent, that is,

$$R\psi = c\psi \qquad (2\text{-}20)$$

where c is a constant. A moment's thought shows that $R\psi$ must still be normalized (since reflecting a function does not change its area or the area under its square), and it also must still be real (since reflecting a real function does not introduce imaginary character). Therefore,

$$\int_0^L (R\psi)^2 dx = 1 = \int_0^L (c\psi)^2\, dx = c^2 \int_0^L \psi^2 dx = c^2 \qquad (2\text{-}21)$$

where we have made use of the fact that ψ is normalized. If $c^2 = 1$, then $c = \pm 1$ and

$$R\psi = \pm\psi \qquad (2\text{-}22)$$

[2]The parentheses in Eq. (2-18) are meant to show the restricted extent of operation of R. This is a departure from the usual mathematical notational convention, but it is hoped that this temporary departure results in greater clarity for the student.

When $R\psi = +\psi$, as is the case for ψ_1 or ψ_3 (Fig. 2-4), ψ is said to be *symmetric*, or *even*, for reflection. If $R\psi = -\psi$, as for ψ_2, ψ is said to be *antisymmetric*, or *odd*. (A function that is neither symmetric nor antisymmetric is said to be *unsymmetric*, or *asymmetric*. Be careful to avoid confusing "asymmetric" with "antisymmetric.")

We have proved a very important property of wavefunctions. *In general, if ψ is the wavefunction for a nondegenerate state, it must be symmetric or antisymmetric under any transformation that leaves H unchanged.*

2-2.D Orthogonality of Wavefunctions

It is possible to show that integration over the product of two different particle-in-a box eigenfunctions, ψ_n and ψ_m, must give zero as the result:

$$\int_0^L \psi_n \psi_m dx = 0, \quad n \neq m \tag{2-23}$$

When functions have this property—that their product gives zero when integrated over the entire range of coordinates—they are said to be *orthogonal*. (Since the "box" eigenfunctions vanish for $x \leq 0$ or $x \geq L$, integration from 0 to L suffices.)

We can use symmetry arguments to demonstrate orthogonality among certain pairs of "box" eigenfunctions, for example, ψ_1 and ψ_2. Figure 2-7 shows that, since ψ_1 is symmetric and ψ_2 is antisymmetric for reflection, the product of these functions is antisymmetric. (In fact, it is not difficult to show in general that the product of two symmetric or of two antisymmetric functions is symmetric, and that an antisymmetric function times a symmetric function gives an antisymmetric product. See Problem 2-7.) Integration over an antisymmetric function *must give zero* as the result since an antisymmetric function has to have equal amounts of positive and negative area. Therefore, ψ_1 and ψ_2 are orthogonal "by symmetry" as, indeed, are all the symmetric–antisymmetric pairs of wavefunctions. Since all ψ's having odd quantum number n are symmetric, and all ψ's having even n are antisymmetric, we have used symmetry to prove ψ_n and

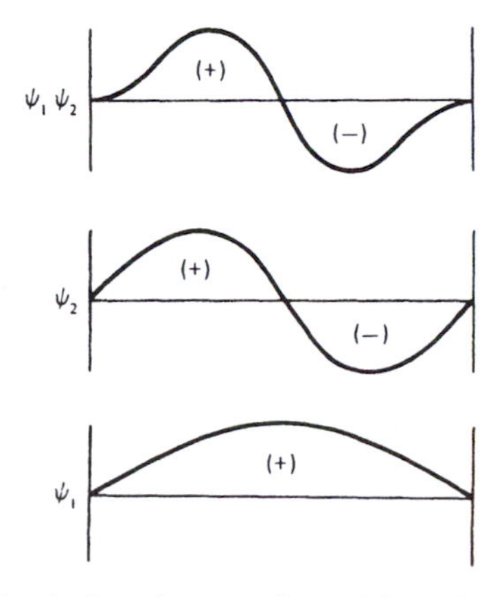

Figure 2-7 ▶ ψ_1 is symmetric, ψ_2 is antisymmetric, and $\psi_1\psi_2$ is antisymmetric. The total signed area bounded by the odd functions is zero since complete cancellation of positive and negative components occurs.

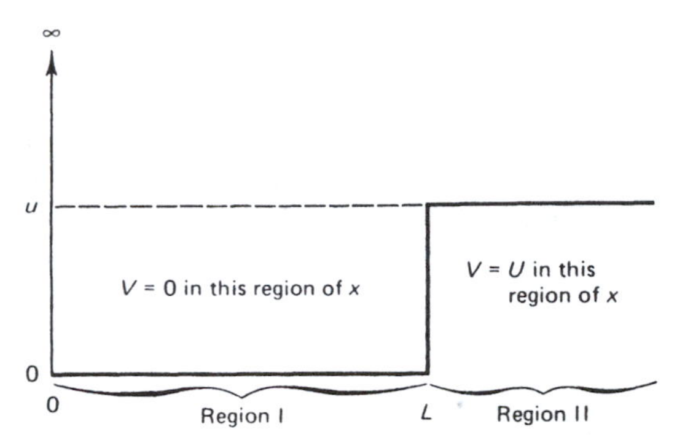

Figure 2-8 ▶ The potential for a one-dimensional "box" with one infinite barrier at $x \leq 0$ and a barrier of $V = U$ at $x \geq L$.

ψ_m orthogonal for n even and m odd. To show orthogonality for n and m both even or both odd requires doing the integral out explicitly (Problem 2-9).

The eigenfunctions (2-11) are orthogonal to each other and individually normalized, and we refer to them as *orthonormal* functions. Mathematically, this is summarized as

$$\int_0^L \psi_n \psi_m dx = \left\{ \begin{matrix} 0, \ n \neq m \\ 1, \ n = m \end{matrix} \right\} \equiv \delta_{n,m} \tag{2-24}$$

The quantity $\delta_{n,m}$ is called the *Kronecker delta function*, and it is merely a convenient shorthand for the information in the braces.

EXAMPLE 2-3 ψ_1 and ψ_3 are both symmetric functions. Therefore their product, $\psi_1\psi_3$, is symmetric. How, then, can the integral of this product vanish?

SOLUTION ▶ A sketch of the product of these functions shows it to be symmetric, with negative values in the central region and wings of positive values on each side. Since ψ_1 and ψ_3 are known to be orthogonal, the negative region must exactly cancel the sum of the two positive regions (though we wouldn't know this for sure from a sketch). This shows that, whereas the integral over an antismmetric integrand *must* equal zero, an integral over a symmetric (or unsymmetric) integrand may or may not equal zero. ◀

2-3 The Particle in a One-Dimensional "Box" with One Finite Wall

Let us now modify the system just discussed by lowering the potential on one side of the "box" to some finite value U. The resulting potential is shown in Fig. 2-8. We can think of a bead on a wire encountering infinite repulsion at $x = 0$ and finite repulsion for $x \geq L$. As before, it is convenient to break up the problem into separate regions of x. For the region $x \leq 0$ where V is infinite, ψ must be zero for the same reasons as before (Section 2-1).

When the particle is in region I of Fig. 2-8, $V = 0$ and all is identical to our earlier box. Therefore, in this region we will have harmonic waves of the general form

$$\psi_{\mathrm{I}} = A_{\mathrm{I}} \sin(2\pi x/\lambda_{\mathrm{I}}) + B_{\mathrm{I}} \cos(2\pi x/\lambda_{\mathrm{I}}) \tag{2-25}$$

where we have used the form (1-24) in which the wavelength appears explicitly. As before, the boundary condition that ψ vanish at $x = 0$ forces B_{I} to vanish, leaving

$$\psi_{\mathrm{I}} = A_{\mathrm{I}} \sin(2\pi x/\lambda_{\mathrm{I}}) \tag{2-26}$$

For the moment we have no other boundary condition on ψ_{I} because we do not know that ψ equals zero at the finite barrier. We do know, however, that the wavelength λ_{I}, whatever it turns out to be, will be related to the energy through

$$\lambda_{\mathrm{I}} = h/\sqrt{2m(E - V_{\mathrm{I}})} \tag{2-27}$$

and, since $V_{\mathrm{I}} = 0$ (in region I),

$$\lambda_{\mathrm{I}} = h/\sqrt{2mE} \tag{2-28}$$

which is a real number for positive E.

We now turn to region II. Since V is constant here also, ψ will again be a harmonic wave. As before, we have our choice of two general forms:

$$\psi_{\mathrm{II}} = A_{\mathrm{II}} \sin(2\pi x/\lambda_{\mathrm{II}}) + B_{\mathrm{II}} \cos(2\pi x/\lambda_{\mathrm{II}}) \tag{2-29}$$

or [see Eq. (2-3)]

$$\psi_{\mathrm{II}} = C_{\mathrm{II}} \exp(+2\pi i x/\lambda_{\mathrm{II}}) + D_{\mathrm{II}} \exp(-2\pi i x/\lambda_{\mathrm{II}}) \tag{2-30}$$

There are two possibilities for the energy of the particle: $E \leq U$ and $E > U$. The first of these corresponds to the classical situation where the particle has insufficient energy to escape from the box and get into region II. Let us see what quantum mechanics says about this case in region II.

For this case, λ_{II} is imaginary since

$$\lambda_{\mathrm{II}} = h/\sqrt{2m(E - U)} \tag{2-31}$$

and $E - U$ is negative. Because λ_{II} is imaginary, it is more convenient to use the general form (2-30) because then the i in the exponential argument can combine with the i of λ_{II} to produce a real argument. Let us assume that λ_{II} is equal to i times a *positive* number. (This will not affect our results.)

Let us now examine the properties of the two exponential functions in Eq. (2-30). The first exponential has an argument that is *real* (because the i's cancel) and positive (because of our above assumption). As x increases, this exponential increases rapidly, approaching infinity. Since acceptable functions do not blow up like this, we set C_{II} equal to zero to prevent it. The second exponential has a negative, real argument, so it decays exponentially toward zero as x approaches infinity. This is acceptable behavior, and we are left with

$$\psi_{\mathrm{II}} = D_{\mathrm{II}} \exp(-2\pi i x/\lambda_{\mathrm{II}}) \tag{2-32}$$

We now have formulas describing fragments of the wavefunction for the two regions. All that remains is to join these together at $x = L$ in such a way that the resulting wavefunction is continuous at $x = L$ and has a continuous first derivative there. (Recall from Section 2-1 that this second requirement results from the fact that the potential is finite at $x = L$. Hence, ψ must be smooth at $x = L$.)

The continuity requirement gives

$$A_{\mathrm{I}} \sin(2\pi L/\lambda_{\mathrm{I}}) = D_{\mathrm{II}} \exp(-2\pi i L/\lambda_{\mathrm{II}}) \tag{2-33}$$

Taking the derivatives of ψ_{I} and ψ_{II} and setting these equal at $x = L$ (to force smoothness) gives

$$(2\pi/\lambda_{\mathrm{I}}) A_{\mathrm{I}} \cos(2\pi L/\lambda_{\mathrm{I}}) = (-2\pi i/\lambda_{\mathrm{II}}) D_{\mathrm{II}} \exp(-2\pi i L/\lambda_{\mathrm{II}}) \tag{2-34}$$

The exponential term is common to both Eqs. (2-33) and (2-34), providing the basis for another equality:

$$A_{\mathrm{I}} \sin(2\pi L/\lambda_{\mathrm{I}}) = (-A_{\mathrm{I}}\lambda_{\mathrm{II}}/i\lambda_{\mathrm{I}}) \cos(2\pi L/\lambda_{\mathrm{I}}) \tag{2-35}$$

or

$$\tan(2\pi L/\lambda_{\mathrm{I}}) = i\lambda_{\mathrm{II}}/\lambda_{\mathrm{I}} \tag{2-36}$$

Substituting for λ_{I} and λ_{II} as indicated by Eqs. (2-28) and (2-31) gives

$$\tan(2\pi L\sqrt{2mE}/h) = -\sqrt{E}/\sqrt{U - E} \tag{2-37}$$

The only unknown in Eq. (2-37) is the total energy E. For given values of L, m, and U, only certain values of $E < U$ will satisfy Eq. (2-37). Thus, the particle can have only certain energies when it is trapped in the "box." These allowed energies can be found by graphing the left-hand side and right-hand side of Eq. (2-37) as functions of E. The values of E where the plots intersect satisfy Eq. (2-37). Figure 2-9 illustrates the graphical solution of Eq. (2-37) for a particular set of values for L, m, and U.

Once a value of E is selected, λ_{I} and λ_{II} are known [from Eqs. (2-28) and (2-31)] and it remains only to find A_{I} and D_{II}. The ratio $A_{\mathrm{I}}/D_{\mathrm{II}}$ may be found from Eq. (2-33). The numerical values of A_{I} and D_{II} will then be obtainable if we require that the wavefunction be normalized. A set of such solutions is shown in Fig. 2-10.

Before solving for the case where $E > U$, let us discuss in detail the results just obtained.

In the first place, the energies are quantized, much as they were in the infinitely deep square well. There is some difference, however. In the infinitely deep well or box, the energy levels increased with the square of the quantum number n. Here they increase less rapidly (the dashed lines in Fig. 2-10 show the allowed energy levels which result when $U = \infty$) because the barrier becomes effectively less restrictive for particles with higher energies (see the following). For the lowest solution, for example, slightly less than one-half a sine wave is needed in one box width of distance. Thus, the wavelength here is slightly longer than in an infinitely deep well of equal width, and so, by de Broglie's relation, the energy is slightly lower. Notice that the effect of lowering the height of one wall is least for the levels lying deepest in the well.

The solutions sketched in Fig. 2-10 indicate that there is a finite probability for finding the particle in the region $x > L$ even though it must have a negative kinetic energy there. Thus, quantum mechanics allows the particle to penetrate into regions where classical mechanics claims it cannot go. Notice that the penetration becomes more appreciable as the energy of the particle approaches that of the barrier. This results

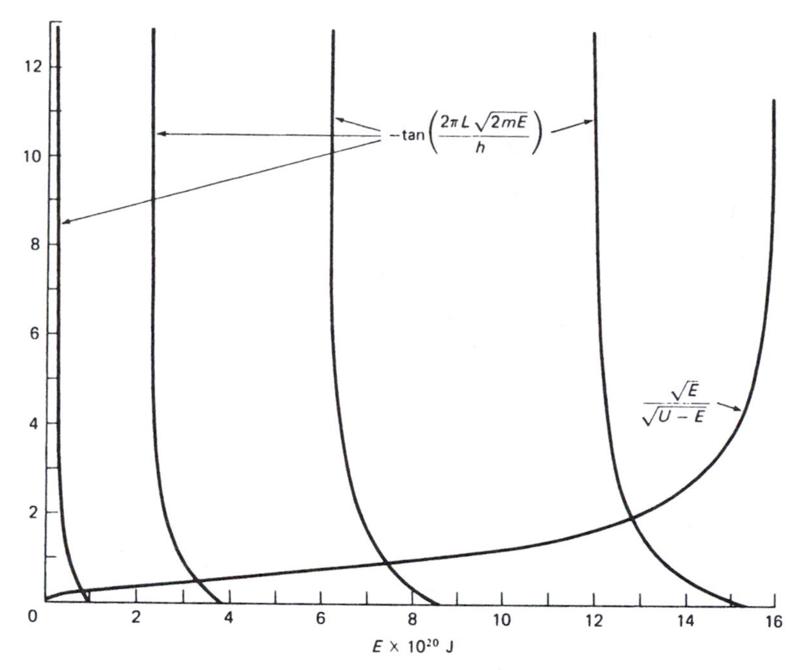

Figure 2-9 ▶ Graphical solution of the equation $-\tan(2\pi L\sqrt{2mE}/h) = \sqrt{E}/\sqrt{U-E}$. Here $L = 2.50$ nm, $m = 9.11 \times 10^{-31}$ kg, $U = 1$ eV $= 16.02 \times 10^{-20}$ J. Intersections occur at $E = 0.828 \times 10^{-20}$ J, 3.30×10^{-20} J, 7.36×10^{-20} J and 12.8×10^{-20} J.

from the fact that $E - U$ determines the rate at which the exponential in ψ_{II} decays [see Eqs. (2-31) and (2-32)]. In the limit that $U \to \infty$, the wavefunction vanishes at the barrier, in agreement with the results of the infinite square well of Section 2-1.

If the barrier in Fig. 2-8 has finite thickness (V becomes zero again at, say, $x = 2L$), then there is a finite probability that a particle in the well will penetrate through the barrier and appear on the other side. This phenomenon is called *quantum-mechanical tunneling*, and this is the way, for example, an α particle escapes from a nucleus even though it classically lacks sufficient energy to overcome the attractive nuclear forces. We emphasize that the tunneling referred to in this example is really not a stationary state phenomenon. We have an *initial* condition (particle in the well) and ask what the half-life is for the escape of the particle—a time-dependent problem.

We saw earlier that the energy quantization for the particle in the infinitely deep well could be thought of as resulting from fitting integral numbers of half sine waves into a fixed width. Most sine waves just will not fit perfectly, and so most energies are not allowed. In this problem the waves are allowed to leak past one of the well walls, but we can still see why only certain energies are allowed. Suppose that we pick some arbitrary energy E for the particle. We know that ψ must be zero at the left wall of the well where $V = \infty$. Starting there, we can draw a sine wave of wavelength determined by E across the well to the right wall, as shown in Fig. 2-11. When the wave hits the right wall, it must join on smoothly to a decaying exponential, which also depends on E. Most of the time, it will be impossible to effect a smooth junction, and that particular value of E will be disallowed.

Let us now consider the case where $E > U$. In region I, the considerations are the same as before. Then, ψ_{I} is a sine wave that can be drawn from the left wall and has a

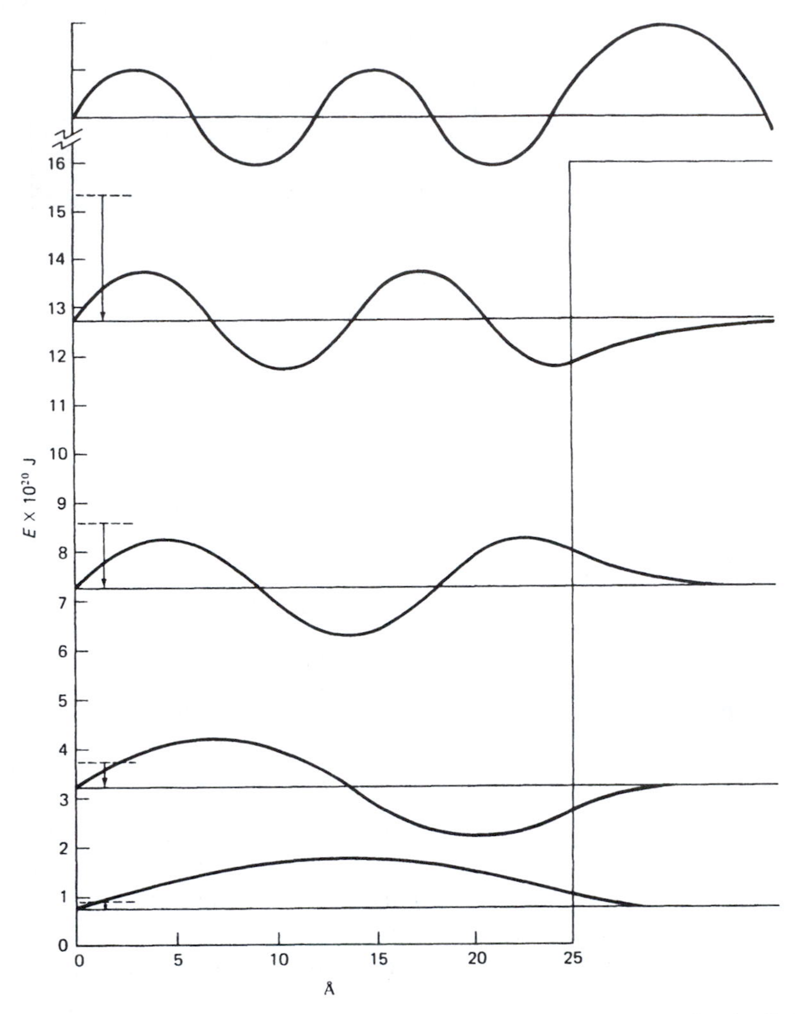

Figure 2-10 ▶ Solutions for particle in well with one finite wall (see Fig. 2-9 for details). Dashed lines correspond to energy levels which would exist if $U = \infty$.

wavelength determined by $E(=T)$ from de Broglie's relation. This sine wave arrives at $x = L$ with a certain magnitude and a certain derivative (assuming that the multiplier A_I has been fixed at some arbitrary value). In region II, we also have a solution of the usual form

$$\psi_{\mathrm{II}}(x) = A_{\mathrm{II}} \sin(2\pi x/\lambda_{\mathrm{II}}) + B_{\mathrm{II}} \cos(2\pi x/\lambda_{\mathrm{II}}) \tag{2-38}$$

where λ_{II} is real and determined by $E - U$, which is now positive. The question is, can we always adjust λ_{II} (by changing A_{II} and B_{II}) so that it has the same value and slope at $x = L$ that ψ_I has? A little thought shows that such adjustment is indeed always possible. The two adjustments allowed in ψ_{II} correspond to a change of *phase* for ψ_{II} (a shift in the horizontal direction) and a change in *amplitude* for ψ_{II}. The only thing about ψ_{II} we cannot change is the wavelength, since this is determined by $E - U$. This is just a physical description of the mathematical circumstance in which we have two adjustable parameters and two requirements to fit—a soluble problem. The essential difference between this case and that of the trapped particle is that here we have fewer

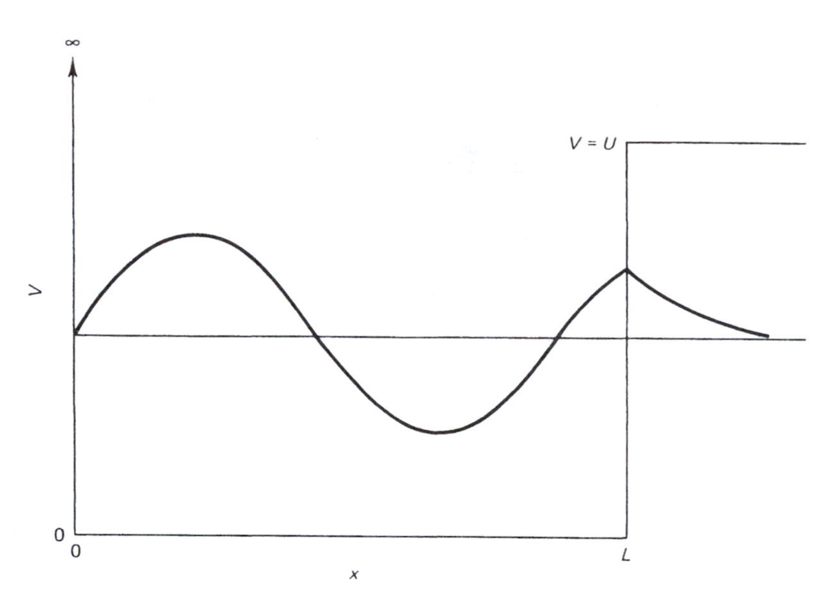

Figure 2-11 ▶ An example of partial wavefunctions for an arbitrary energy E. These functions cannot be joined smoothly at $x = L$ and so this value of E is not allowed.

boundary conditions. Before, our square-integrability requirement was used to remove a positive exponential term. That requirement is, in effect, a boundary condition—ψ must vanish at $x = \infty$—and it led to energy quantization. Then we used the normality requirement to achieve unique values for A_{I} and D_{II}. In this case we cannot get a square-integrable solution. ψ_{II} goes on oscillating as $x \to \infty$, and so we have no boundary condition there. As a result, E is not quantized and ψ is not normalizable, so that only ratios of A_{I}, A_{II}, and B_{II} are obtainable.

The energy scheme for the particle in the potential well with one finite wall, then, is discrete when $E < U$, and continuous when $E > U$.

Notice the way in which the wavelengths vary in Fig. 2-10. We have already seen that the time-independent Schrödinger equation states that the total energy for a particle in a stationary state is the same at all particle positions (i.e., a *constant of motion*). The kinetic and potential energies must vary together, then, in such a way that their sum is constant. This is reflected by the fact that the wavelength of an unbound solution is shorter in region I than it is in region II. In region I, $V = 0$, so that all energy of the particle is kinetic ($T = E$). In region II, $V > 0$, so that the kinetic energy ($T = E - V$) is less than it was in region I. Therefore, the de Broglie wavelength, which is related to *kinetic* energy, must be greater in region II.

EXAMPLE 2-4 For the system described in the caption for Fig. 2-9, calculate the percentage drop of the lowest-energy state that results from barrier penetration.

SOLUTION ▶ For this, we need to solve the problem for the simple particle-in-a-box system (for which $U = \infty$).

$$E_1 = \frac{n^2h^2}{8mL^2} = \frac{(1)^2(6.626 \times 10^{-34}\,\mathrm{J\,s})^2}{8(9.11 \times 10^{-31}\,\mathrm{kg})(2.500 \times 10^{-10}\,\mathrm{m})^2} = 9.64 \times 10^{-21}\,\mathrm{J}$$

compared to 8.28×10^{-21} J. Barrier penetration lowers E_1 by 14%. ◀

2-4 The Particle in an Infinite "Box" with a Finite Central Barrier

Another example of barrier penetration in a stationary state of a system is provided by inserting a barrier of finite height and thickness at the midpoint of the infinite square well of Section 2-1 (see Fig. 2-12).

The boundary conditions for this problem are easily obtained by obvious extensions of the considerations already discussed. Rather than solve this case directly, we shall make use of our insights from previous systems to deduce the main characteristics of the solutions. Let us begin by considering the case where the barrier is infinitely high.

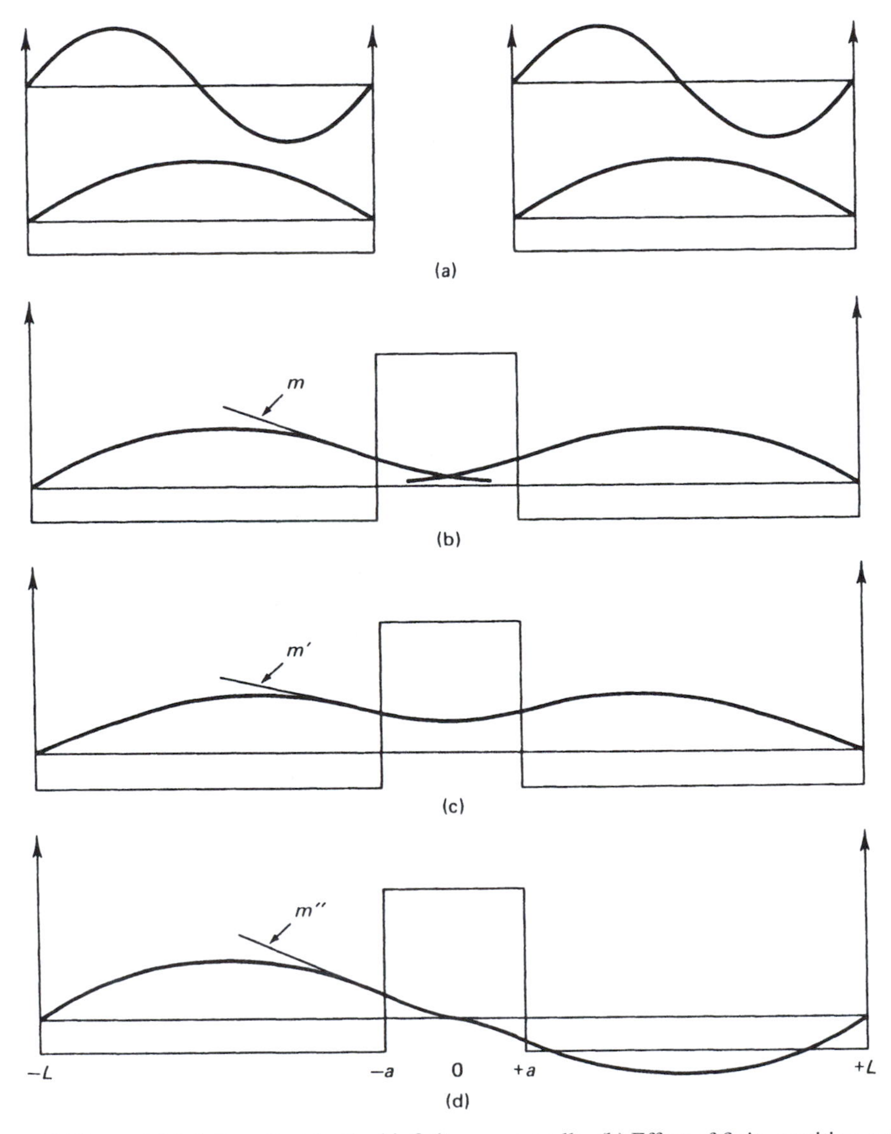

Figure 2-12 ▶ (a) Solutions for identical infinite square wells. (b) Effect of finite partition on half waves. (c) Symmetric combination of half waves. (d) Antisymmetric combination of half waves.

Then the problem becomes merely that of two isolated infinite square wells, each well having solutions as described in Sections 2-1 and 2-2.

Now, as the height of the barrier is lowered from infinity, what happens? The levels lying deepest in the two sections should be least affected by the change. They must still vanish at the outer walls but now they can penetrate slightly into the finite barrier. Thus, the lowest state in, say, the left-hand section of the well will begin to look as given in Fig. 2-12b. The solution on the right side will do likewise, of course. As this happens, their energies will decrease slightly since their wavelengths increase. However, since the two wells are no longer separated by an infinite barrier, they are no longer independent. We can no longer talk about separate solutions for the two halves. Each solution for the Schrödinger equation is now a solution for the whole system from $x=-L$ to $+L$. Furthermore, symmetry arguments state that, since the hamiltonian for this problem is symmetric for reflection through $x=0$, the solutions, if nondegenerate, must be either symmetric or antisymmetric through $x=0$.

This requirement must be reconciled with the barrier-penetration behavior indicated by Fig. 2-12b, which is also occurring. One way to accomplish this is by summing the two half waves as shown in Fig. 2-12c, giving a symmetric wavefunction. Alternatively, subtraction gives the antisymmetric form shown in Fig. 2-12d. Both of these solutions will be lower in energy than their infinite-well counterparts, because the wavelengths in Fig. 2-12c and d continue to be longer than in 2-12a. Will their energies be equal to each other? Not quite. By close inspection, we can figure out which solution will have the lower energy.

In Figs. 2-12b to 2-12d, the slopes of the half wave, the symmetric, and the antisymmetric combinations at the finite barrier are labeled respectively m, m', and m''. What can we say about their relative values? The slope m' should be less negative than m because the decaying exponential producing m has an increasing exponential added to it when producing m'. Slope m'' should be more negative than m since the decaying exponential has an increasing exponential subtracted from it in case d, causing it to decay faster. This means that the sine curve on the left-hand side of Fig. 2-12c cannot be identical with that on the left side of Fig. 2-12d since they must arrive at the barrier with different slopes. (The same is true for the right-hand sides, of course.) How can we make the sine wave arrive with a less negative slope m'?—by increasing the wavelength slightly so that not quite so much of the sine wave fits into the left well (see Fig. 2-13a). Increasing the wavelength slightly means, by de Broglie's relation, that the energy of the particle is decreased. Similarly, the sine curve in Fig. 2-12d must be shortened so that it will arrive at the barrier with slope m'', which corresponds to an energy increase. Of course, now that the energy has changed outside the barrier, it must change inside the barrier too. This would require going back and modifying the exponentials inside the barrier. But the first step is sufficient to indicate the qualitative results: The symmetric solution has lower energy. In Fig. 2-13a is a detailed sketch of the final solution for the two lowest states.

There is a simpler way to decide that the symmetric solution has lower energy. As the barrier height becomes lower and lower, the two solutions become more and more separated in energy, but they always remain symmetric or antisymmetric with respect to reflection since the hamiltonian always has reflection symmetry. In the limit when the barrier completely disappears we have a simple square well again (but larger), the lowest solution of which is symmetric. (See Fig. 2-13b.) This lowest symmetric solution must

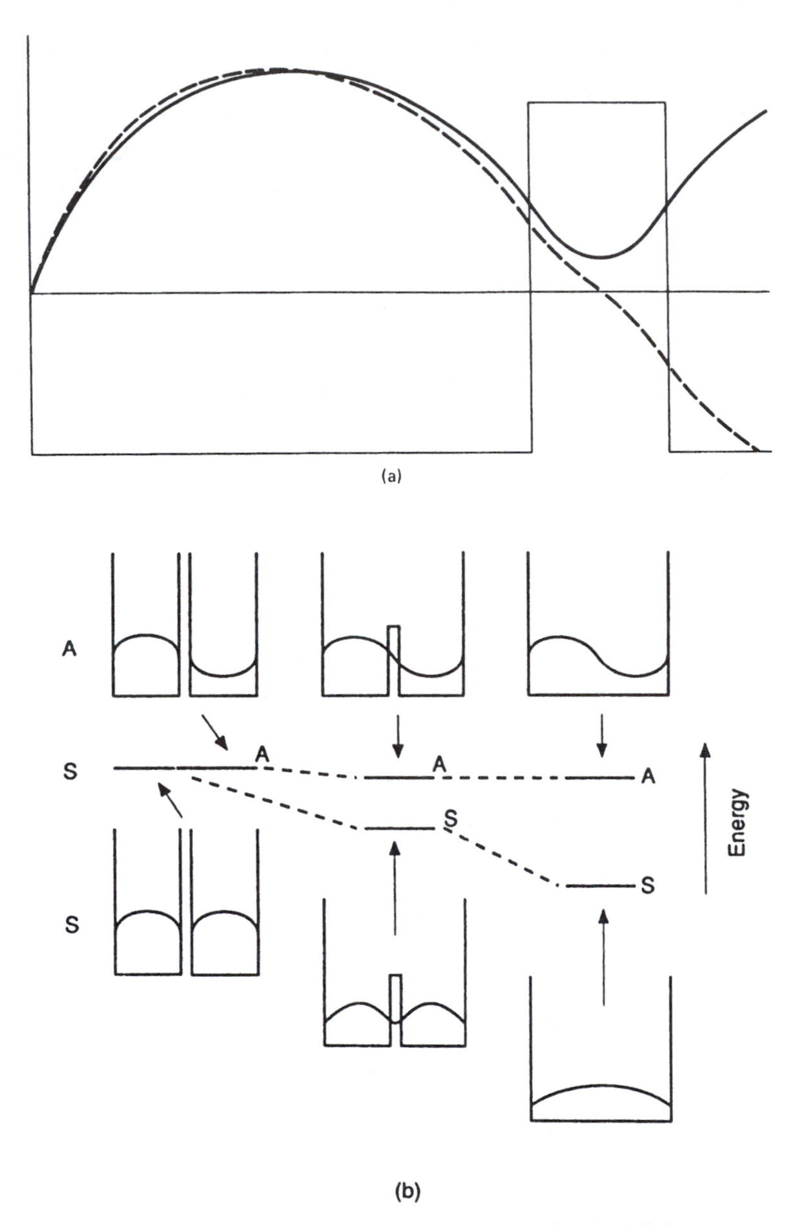

Figure 2-13 ▶ (a) Detailed sketch of the two lowest solutions for the infinite square-well divided by a finite barrier at the midpoint. The waves are sketched from a common energy value for ease of comparison. Actually, the symmetric wave has a lower energy. (b) A correlation diagram relating energies when the barrier is infinite (left side) with those when the barrier vanishes. Letters A and S refer to antisymmetric and symmetric solutions, respectively.

"come from" the symmetric combination of smaller-well wavefunctions sketched at the left of Fig. 2-13b; similarly, the second lowest, antisymmetric solution of the large well correlates with the antisymmetric small-well combination (also at left in Fig. 2-13b). A figure of the kind shown in Fig. 2-13b is called a *correlation diagram*. It shows how the energy eigenvalues change throughout a continuous, symmetry-conserving process.

We shall see that the correlation of wavefunction symmetries in such a manner as this is a powerful technique in understanding and predicting chemical behavior.

The splitting of energy levels resulting from barrier penetration is an extremely pervasive phenomenon in quantum chemistry. It occurs regardless of whether the barrier separates identical or nonidentical potential regions, i.e., regardless of whether the final system is symmetric or unsymmetric. When two atoms (N and N, or C and O) interact to form a molecule, the original atomic wavefunctions combine to form molecular wavefunctions in much the same way as was just described. One of these molecular wavefunctions may have an energy markedly lower than those in the corresponding atoms. Electrons having such a wavefunction will stabilize the molecule relative to the separated atoms.

Another case in which energy level splitting occurs is in the *vibrational* spectrum of ammonia. Ammonia is most stable in a pyramidal configuration, but is capable of inverting through a higher-energy planar configuration into an equivalent "mirror image" pyramid. Thus, vibrations tending to flatten out the ammonia molecule occur in a potential similar to the double well, except that in ammonia the potential is not discontinuous. The lowest vibrational energy levels are not sufficiently high to allow classical inversion of ammonia. However, these vibrational levels are split by interaction through barrier penetration just as quantum mechanics predicts. The energy required to excite ammonia from the lowest of these sublevels to its associated sublevel can be accurately measured through microwave spectroscopy. Knowledge of the level splittings in turn allows a precise determination of the height of the barrier to inversion in ammonia (see Fig. 2-14).

It is easy to anticipate the appearance of the solutions for the square well with central barrier for energies greater than the partition height. They will be sinusoidal waves, symmetric or antisymmetric in the well, and vanishing at the walls. Their wavelengths will be somewhat longer in the region of the partition than elsewhere because some of the kinetic energy of the particle is transformed to potential energy there. A sketch of the final results is given in Fig. 2-15.

EXAMPLE 2-5 Fig. 2-15 shows energy levels for states when the barrier has finite height. When the barrier is made infinitely high, the levels at E_1 and E_2 merge into one level. Where does the energy of that one level lie—below E_1, between E_1 and E_2, or above E_2—and why?

SOLUTION ▶ It lies above E_2. When the barrier is finite, there is always some penetration, so λ is always at least a little larger than is the case for the infinite barrier. If λ is larger, E is lower. ◀

2-5 The Free Particle in One Dimension

Suppose a particle of mass m moves in one dimension in a potential that is everywhere zero. The Schrödinger equation becomes

$$\frac{-h^2}{8\pi^2 m}\frac{d^2\psi}{dx^2} = E\psi \qquad (2\text{-}39)$$

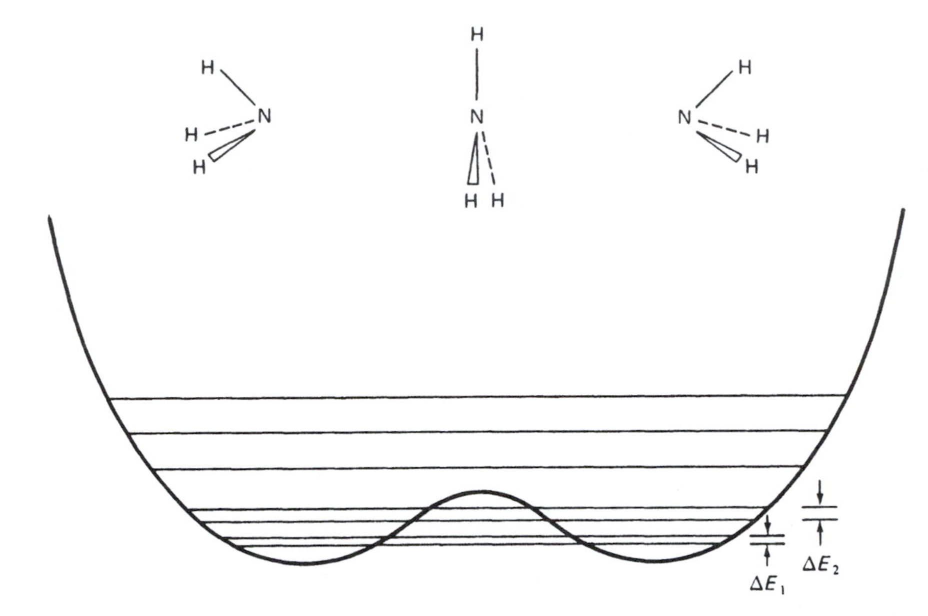

Figure 2-14 ▸ Sketch of potential for inversion vibrational mode in ammonia. The lowest levels are split by tunneling. The low energy transition ΔE_1 is visible in the microwave region whereas the second transition ΔE_2 is visible in the infrared. $\Delta E_1 = 0.16 \times 10^{-22}$ J; $\Delta E_2 = 7.15 \times 10^{-22}$ J.

which has as solutions

$$\psi = A \exp(\pm 2\pi i \sqrt{2mE}x/h) \tag{2-40}$$

or alternatively, trigonometric solutions

$$\psi = A' \sin(2\pi \sqrt{2mE}x/h), \quad \psi = A' \cos(2\pi \sqrt{2mE}x/h) \tag{2-41}$$

As is most easily seen from the exponential forms (2-40), if E is negative, ψ will blow up at either $+\infty$ or $-\infty$, and so we reject negative energies. Since there are no boundary conditions, it follows that E can take on any positive value; the energies of the free particle are not quantized. This result would be expected from our earlier results on constrained particles. There we saw that quantization resulted from spatial constraints, and here we have none.

The constants A and A' of Eqs. (2-40) and (2-41) cannot be evaluated in the usual way, since the solutions do not vanish at $x = \pm\infty$. Sometimes it is convenient to evaluate them to correspond to some experimental situation. For instance, suppose that one was working with a monoenergetic beam of electrons having an intensity of one electron every 10^{-6} m. Then we could normalize ψ of Eq. (2-40) so that

$$\int_0^{10^{-6}\,\mathrm{m}} |\psi|^2 \, dx = 1$$

There is a surprising difference in the particle distributions predicted from expressions (2-40) and (2-41). The absolute square $\psi^*\psi$ of the exponentials is a constant (A^*A), whereas the squares of the trigonometric functions are fluctuating functions

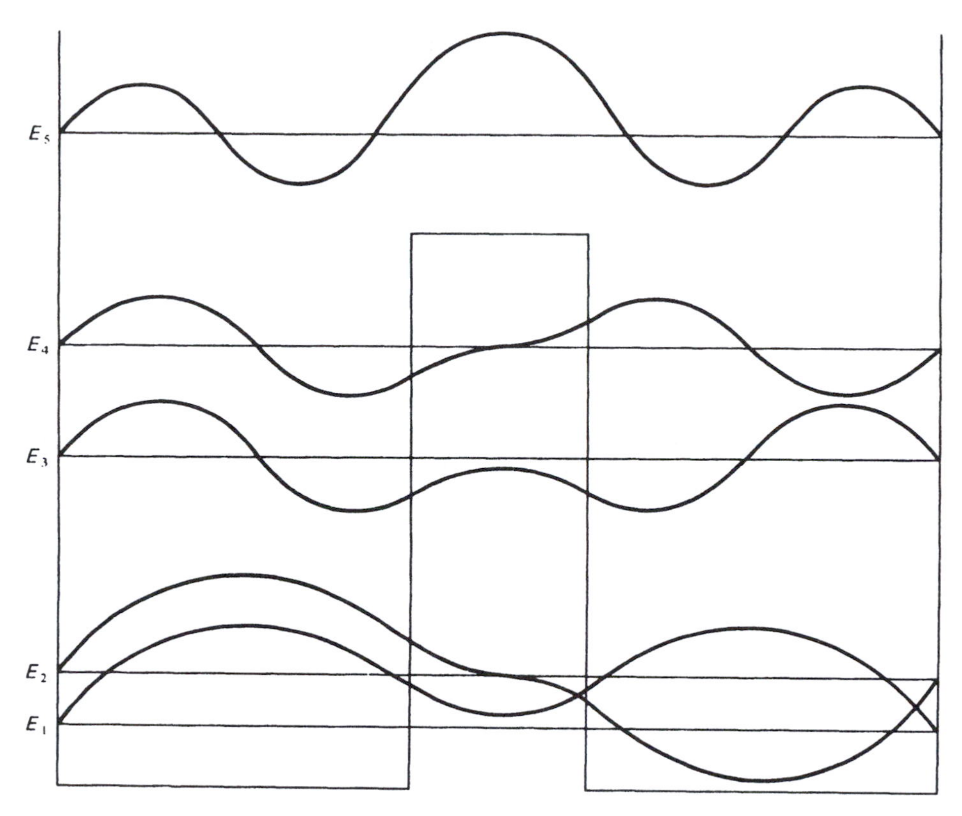

Figure 2-15 ▶ Wavefunctions for the infinite square well with finite partition.

of x. It seems sensible for a particle moving without restriction to have a constant probability distribution, and it seems absurd for it to have a varying probability distribution. What causes this peculiar behavior? It results from there being two independent solutions for each value of E (except $E=0$). This degeneracy with respect to energy means that from a degenerate pair, ψ and ψ', one can produce any number of new eigenfunctions, $\psi'' = a\psi + b\psi'$ (Problem 2-11). In such a situation, the symmetry proof of Section 2-2 does not hold. However, there will always be an independent pair of degenerate wavefunctions that *will* satisfy certain symmetry requirements. Thus, in the problem at hand, we have one pair of solutions, the exponentials, which do have the proper symmetry since their absolute squares are constant. From this pair we can produce any number of linear combinations [one set being given by Eq. (2-41)], but these need not display the symmetry properties anymore.

The exponential solutions have another special attribute: A particle whose state is described by one of the exponentials has a definite linear momentum, whereas, when described by a trigonometric function, it does not. In Section 1-9, it was shown that the connection between classical and wave mechanics could be made if one related the classical momentum, p_x, with a quantum mechanical operator $(h/2\pi i)d/dx$. Now, for a particle to have a definite (sharp) value p for its momentum really means that, if we measure the momentum at some instant, there is no possibility of getting any value other than p. This means that the particle in the state described by ψ *always* has momentum p, no matter where it is in x; i.e., its momentum is a constant of motion, just as its energy is. This corresponds to saying that there is an eigenvalue equation for momentum, just as for energy. Thus

$$\frac{h}{2\pi i}\frac{d\psi}{dx} = p\psi \tag{2-42}$$

The statement made earlier, that the exponential solutions correspond to the particle having sharp momentum, means that the exponentials (2-40) must be solutions to Eq. (2-42). This is easily verified:

$$\frac{h}{2\pi i}\frac{d}{dx}\left[A\exp\left(\frac{\pm 2\pi i\sqrt{2mE}x}{h}\right)\right] = \pm\sqrt{2mE}\left[A\exp\left(\frac{\pm 2\pi i\sqrt{2mE}x}{h}\right)\right]$$

Thus, the positive and negative exponential solutions correspond to momentum values of $+\sqrt{2mE}$ and $-\sqrt{2mE}$, respectively, and are interpreted as referring to particle motion toward $+\infty$ and $-\infty$ respectively. Since energy is related to the square of the momentum, these two solutions have identical energies. (The solution for $E = 0$ corresponds to no momentum at all, and the directional degeneracy is removed.) A mixture of these states contains contributions from two different momenta but only one energy, so linear combinations of the exponentials fail to maintain a sharp value for momentum but do maintain a sharp value for energy.

EXAMPLE 2-6 An electron is accelerated along the x axis towards $x = \infty$ from rest through a potential drop of 1.000 kV.
a) What is its final momentum?
b) What is its final de Broglie wavelength?
c) What is its final wavefunction?

SOLUTION ▶ a) $p_x = \sqrt{2mE} = [2(9.105 \times 10^{-31}\ \text{kg})(1.602 \times 10^{-16}\,\text{J})]^{1/2} = 1.708 \times 10^{-23}\ \text{kg m s}^{-1}$
b) $\lambda = \frac{h}{p} = \frac{6.626\times 10^{-34}\,\text{J s}}{1.708\times 10^{-23}\,\text{kg m s}^{-1}} = 3.879 \times 10^{-11}\ \text{m}$
c) $\psi = A\exp(+2\pi i\sqrt{2mE}x/h)$(choose + because moving towards $x = +\infty$) $= A\exp(2\pi i p/h)x = A\exp(2\pi i x/\lambda) = A\exp[2\pi i(2.578 \times 10^{10}\ \text{m}^{-1})x]$. ◀

2-6 The Particle in a Ring of Constant Potential

Suppose that a particle of mass m is free to move around a ring of radius r and zero potential, but that it requires infinite energy to get off the ring. This system has only one variable coordinate—the angle ϕ. In classical mechanics, the useful quantities and relationships for describing such circular motion are those given in Table 2-1.

Comparing formulas for linear momentum and angular momentum reveals that the variables mass and linear velocity are analogous to moment of inertia and angular velocity in circular motion, where the coordinate ϕ replaces x. The Schrödinger equation for circular motion, then, is

$$\frac{-h^2}{8\pi^2 I}\frac{d^2\psi(\phi)}{d\phi^2} = E\psi(\phi) \tag{2-43}$$

TABLE 2-1 ▶

Quantity	Formula	Units
Moment of inertia	$I = mr^2$	$\mathrm{g\,cm^2}$ or $\mathrm{kg\,m^2}$
Angular velocity	$\omega = \Delta\phi/\Delta t = v/r$	$\mathrm{s^{-1}}$
Angular momentum (linear momentum times orbit radius)	$mvr = I\omega$	$\mathrm{g\,cm^2/s}$ or erg s or J s

which has, as solutions

$$A\exp(\pm ik\phi) \tag{2-44}$$

or alternatively

$$A'\sin(k\phi) \tag{2-45}$$

and

$$A'\cos(k\phi) \tag{2-46}$$

where [substituting Eq. (2-45) or (2-46) into (2-43) and operating]

$$k = 2\pi\sqrt{2IE}/h \tag{2-47}$$

Let us solve the problem first with the trigonometric functions. Starting at some arbitrary point on the ring and moving around the circumference with a sinusoidal function, we shall eventually reencounter the initial point. In order that our wavefunction be single valued, it is necessary that ψ repeat itself every time ϕ changes by 2π radians. Thus, for ϕ given by Eq. (2-45),

$$\sin(k\phi) = \sin[k(\phi + 2\pi)] \tag{2-48}$$

Similarly, for ψ given by Eq. (2-49)

$$\cos(k\phi) = \cos(k\phi + 2k\pi) \tag{2-49}$$

Either of these relations is satisfied only if k is an integer. The case in which $k = 0$ is not allowed for the sine function since it then vanishes everywhere and is unsuitable. However, $k = 0$ is allowed for the cosine form. The normalized solutions are, then,

$$\begin{aligned}\psi &= (1/\sqrt{\pi})\sin(k\phi), \quad k = 1, 2, 3, \ldots \\ \psi &= (1/\sqrt{\pi})\cos(k\phi), \quad k = 1, 2, 3, \ldots \\ \psi &= (1/\sqrt{2\pi}) \quad \text{(from the } k = 0 \text{ case for the cosine)}\end{aligned} \tag{2-50}$$

Now let us examine the exponential form of ψ (Eq. 2-44). The requirement that ψ repeat itself for $\phi \to \phi + 2\pi$ gives

$$A\exp(\pm ik\phi) = A\exp[\pm ik(\phi + 2\pi)] = A\exp(\pm ik\phi)\exp(\pm 2\pi ik)$$

or

$$\exp(\pm 2\pi ik) = 1$$

Taking the positive case and utilizing Eq. (2-3), we obtain

$$\cos(2\pi k) + i\sin(2\pi k) = 1 \quad \text{or} \quad \cos(2\pi k) = 1 \quad \text{and} \quad \sin(2\pi k) = 0$$

Again, k must be an integer. (The same result arises by requiring that $d\psi/d\phi$ repeat for $\phi \to \phi + 2\pi$.) Thus, an alternative set of normalized solutions is

$$\psi = \left(1/\sqrt{2\pi}\right)\exp(ik\phi) \quad k = 0, \pm 1, \pm 2, \pm 3, \ldots \tag{2-51}$$

The energies for the particle in the ring are easily obtained from Eq. (2-47):

$$E = k^2h^2/8\pi^2 I, \quad k = 0, \pm 1, \pm 2, \pm 3, \ldots \tag{2-52}$$

The energies increase with the square of k, just as in the case of the infinite square well potential. Here we have a single state with $E = 0$, and doubly degenerate states above, whereas, in the square well, we had no solution at $E = 0$, and all solutions were nondegenerate. The solution at $E = 0$ means that there is no finite zero point energy to be associated with free rotation, and this is in accord with uncertainty principle arguments since there is no constraint in the coordinate ϕ.

The similarity between the particle in a ring and the free particle problems is striking. Aside from the fact that in the ring the energies are quantized and the solutions are normalizable, there are few differences. The exponential solutions (2-51) are eigenfunctions for the *angular momentum operator* $(h/2\pi i)d/d\phi$. The two angular momenta for a pair of degenerate solutions correspond to particle motion clockwise or counterclockwise in the ring. (The nondegenerate solution for $E = 0$ has no angular momentum, hence no ability to achieve degeneracy through directional behavior.) The particle density predicted by the exponentials is uniform in the ring, while that for the trigonometric solutions is not. Since the trigonometric functions tend to localize the particle into part of the ring, thereby causing $\Delta\phi \neq \infty$, it is consistent that they are impure momentum states (Δ ang. mom. $\neq 0$). (Infinite uncertainty in the coordinate ϕ means that all values of ϕ in the range 0–2π are equally likely.)

EXAMPLE 2-7 Demonstrate that any two degenerate exponential eigenfunctions for a particle in a ring are orthogonal.

SOLUTION ► Such a pair of degenerate wavefunctions can be written as $\psi_+ = \frac{1}{\sqrt{2\pi}}\exp(ik\phi)$, $\psi_- = \frac{1}{\sqrt{2\pi}}\exp(-ik\phi)$. These are orthogonal if $\int_0^{2\pi}\psi_+^*\psi_-\,d\phi = 0$ where we must use the complex conjugate of either one of the wavefunctions since ψ is complex. But $\psi_+^* = \psi_-$, so

$$\begin{aligned}
\int_0^{2\pi}\psi_-\psi_-\,d\phi &= \frac{1}{2\pi}\int_0^{2\pi}\exp(-2i\phi)\,d\phi = \frac{1}{2\pi}\int_0^{2\pi}[\cos(2\phi) - i\sin(2\phi)]d\phi \\
&= \frac{1}{2\pi}\left[\sin(2\phi)|_0^{2\pi} - i(-\cos(2\phi)|_0^{2\pi}\right] \\
&= \frac{1}{2\pi}[\sin(4\pi) - \sin(0) + i\cos(4\pi) - i\cos(0)] \\
&= \frac{1}{2\pi}[0 - 0 + i - i] = 0.
\end{aligned}$$

◄

2-7 The Particle in a Three-Dimensional Box: Separation of Variables

Let us now consider the three-dimensional analog of the square well of Section 2-1. This would be a three-dimensional box with zero potential inside and infinite potential outside. As before, the particle has no probability for penetrating beyond the box. Therefore, the Schrödinger equation is just

$$\frac{-h^2}{8\pi^2 m}\left(\frac{\partial^2}{\partial x^2}+\frac{\partial^2}{\partial y^2}+\frac{\partial^2}{\partial z^2}\right)\psi = E\psi \tag{2-53}$$

and ψ vanishes at the box edges.

The hamiltonian operator on the left side of Eq. (2-53) can be written as a sum of operators, one in each variable (e.g., $H_x = (-h^2/8\pi^2 m)\partial^2/\partial x^2$). Let us assume for the moment that ψ can be written as a product of three functions, each one being a function of a different variable, x, y, or z (i.e., $\psi = X(x)Y(y)Z(z)$). If we can show that such a ψ satisfies Eq. (2-53), we will have a much simpler problem to solve. Using this assumption, Eq. (2-53) becomes

$$(H_x + H_y + H_z)XYZ = EXYZ \tag{2-54}$$

This can be expanded and then divided through by XYZ to obtain

$$\frac{H_x XYZ}{XYZ}+\frac{H_y XYZ}{XYZ}+\frac{H_z XYZ}{XYZ} = E \quad \text{(a constant)} \tag{2-55}$$

Now, since H_x, for example, operates only on functions of x, but not y or z, we can carry out some limited cancellation. Those functions that are *not* operated on in a numerator can be canceled against the denominator. Those that *are* operated on cannot be canceled since these are differential operators [e.g., in $(1/x)dx^2/dx$ it is not permissible to cancel $1/x$ against x^2 before differentiating: $(1/x)dx^2/dx \neq dx/dx$]. Such cancellation gives

$$\frac{H_x X}{X}+\frac{H_y Y}{Y}+\frac{H_z Z}{Z} = E \quad \text{(a constant)} \tag{2-56}$$

Now, suppose the particle is moving in the box parallel to the x axis so that the variables y and z are not changing. Then, of course, the functions Y and Z are also not changing, so $H_y Y/Y$ and $H_z Z/Z$ are both constant. Only $H_x X/X$ can vary—but *does* it vary? Not according to Eq. (2-56), which reduces under these conditions to

$$\frac{H_x X}{X} + \text{constant} + \text{constant} = E \quad \text{(a constant)} \tag{2-57}$$

Therefore, even though the particle is moving in the x direction, $H_x X/X$ must also be a constant, which we shall call E_x. Similar reasoning leads to analogous constants E_y and E_z. Furthermore, the behavior of $H_x X/X$ must really be independent of whether the particle is moving parallel to the y and z axes. Even if y and z do change, they do not appear in the quantity $H_x X/X$ anyway. Thus we may write, without restriction,

$$\frac{H_x X}{X} = E_x, \qquad \frac{H_y Y}{Y} = E_y, \qquad \frac{H_z Z}{Z} = E_z \tag{2-58}$$

and, from Eq. (2-56),

$$E_x + E_y + E_z = E \tag{2-59}$$

Our original equation in three variables has been separated into three equations, one in each variable. The first of these equations may be rewritten

$$H_x X = E_x X \tag{2-60}$$

which is just the Schrödinger equation for the particle in the one-dimensional square well, which we have already solved. For a rectangular box with $L_x \neq L_y \neq L_z$ we have the general solution

$$\psi = XYZ = \sqrt{2/L_x}\sin(n_x \pi x/L_x)\sqrt{2/L_y}\sin\left(n_y \pi y/L_y\right)\sqrt{2/L_z}\sin(n_z \pi z/L_z) \tag{2-61}$$

and

$$E = E_x + E_y + E_z = (h^2/8m)\left(n_x^2/L_x^2 + n_y^2/L_y^2 + n_z^2/L_z^2\right) \tag{2-62}$$

For a cubical box, $L_x = L_y = L_z = L$, and the energy expression simplifies to

$$E = (h^2/8mL^2)\left(n_x^2 + n_y^2 + n_z^2\right) \tag{2-63}$$

The lowest energy occurs when $n_x = n_y = n_z = 1$, and so

$$E_1(1) = 3h^2/8mL^2 \tag{2-64}$$

Thus, the cubical box has three times the zero point energy of the corresponding one-dimensional well, one-third coming from each independent coordinate for motion (i.e., "degree of freedom"). The one in parentheses indicates that this level is nondegenerate. The next level is produced when one of the quantum numbers n has a value of two while the others have values of one. There are three independent ways of doing this; therefore, the second level is triply degenerate, and $E_2(3) = 6h^2/8mL^2$. Proceeding, $E_3(3) = 9h^2/8mL^2$, $E_4(3) = 11h^2/8mL^2$, $E_5(1) = 12h^2/8mL^2$, $E_6(6) = 14h^2/8mL^2$, etc. Apparently, the energy level scheme and degeneracies of these levels do not proceed in the regular manner which is found in the one-dimensional cases we have studied.

EXAMPLE 2-8 Verify that E_6 is six-fold degenerate.

SOLUTION ▶ $E_6 = 14h^2/8mL^2$, so $n_x^2 + n_y^2 + n_z^2 = 14$. There is only one combination of integers that satisfies this relation, namely 1, 2, and 3. So we simply need to deduce how many unique ways we can assign these integers to n_x, n_y, and n_z. There are three ways to assign 1. For each of these three choices, there remain but two ways to assign 2, and then there is only one way to assign 3. So the number of unique ways is $3 \times 2 \times 1 = 6$. (Or one can simply write down all of the possibilities and observe that there are six of them.) ◀

We shall now briefly consider what probability distributions for the particle are predicted by these solutions. The lowest-energy solution has its largest value at the box center where all three sine functions are simultaneously largest. The particle

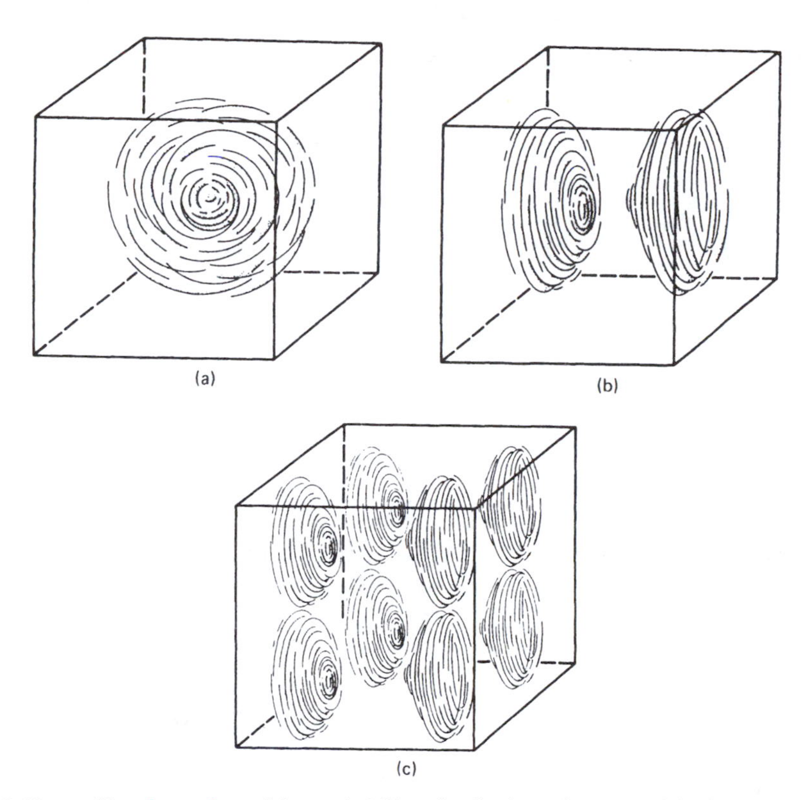

Figure 2-16 ▶ Sketches of particle probability distributions for a particle in a cubical box. (a) $n_x = n_y = n_z = 1$. (b) $n_x = 2$, $n_y = n_z = 1$. (c) $n_x = n_y = n_z = 2$.

distribution is sketched in Fig. 2-16a. The second level may be produced by setting $n_x = 2$, and $n_y = n_z = 1$. Then there will be a nodal plane running through the box perpendicular to the x axis, producing the split distribution shown in Fig. 2-16b. Since there are three ways this node can be oriented to produce distinct but energetically equal distributions, this energy level is triply degenerate. The particle distribution for the state where $n_x = n_y = n_z = 2$ is sketched in Fig. 2-16c. It is apparent that, in the high energy limit, the particle distribution becomes spread out uniformly throughout the box in accord with the classical prediction.

The separation of variables technique which we have used to convert our three-dimensional problem into three independent one-dimensional problems will recur in other quantum-chemical applications. Reviewing the procedure makes it apparent that this technique will work whenever the hamiltonian operator can be cleanly broken into parts dependent on completely different coordinates. This is always possible for the kinetic energy operator in cartesian coordinates. However, the potential energy operator often prevents separation of variables in physical systems of interest.

It is useful to state the general results of separation of variables. Suppose we have a hamiltonian operator, with associated eigenfunctions and eigenvalues:

$$H\psi_i = E_i\psi_i \tag{2-65}$$

Suppose this hamiltonian can be separated, for example,

$$H(\alpha, \beta) = H_\alpha(\alpha) + H_\beta(\beta) \tag{2-66}$$

where α and β stand for two different coordinates or groups of coordinates. Then it follows that

$$\psi_{j,k} = f_j(\alpha)\, g_k(\beta) \tag{2-67}$$

where

$$H_\alpha f_j = a_j f_j \tag{2-68}$$

and

$$H_\beta g_k = b_k g_k \tag{2-69}$$

Furthermore,

$$E_{j,k} = a_j + b_k \tag{2-70}$$

In other words, if a hamiltonian is separable, then the eigenfunctions will be *products* of eigenfunctions of the subhamiltonians, and the eigenvalues will be *sums* of the subeigenvalues.

2-8 The Scattering of Particles in One Dimension

Consider the potential shown in Fig. 2-17a. We imagine that a beam of particles, each having energy E, originates from the left and travels toward $x = \infty$, experiencing a constant potential everywhere except at the potential step at $x = 0$. We are interested in what becomes of these particles—what fraction makes it all the way to the "end" (some kind of particle trap to the right of the step) and what fraction is reflected back toward $x = -\infty$. Problems of this type are related to scattering experiments where electrons, for example, travel through potential jumps produced by electronic devices or through a dilute gas where potential changes occur in the neighborhood of atoms.

This kind of problem differs from most of those discussed earlier because the particle is not trapped (classically), so all nonnegative energy values are possible. We already know what the form of the eigenfunctions is for the constant potentials to the left and right of the step for any choice of E. On the left they are linear combinations of $\exp(\pm i\sqrt{2mE}x/\hbar)$, where $\hbar \equiv h/2\pi$, and to the right they are linear combinations of $\exp(\pm i\sqrt{2m(E-U)}x/\hbar)$. The only thing we do not yet know is which linear combinations to take. That is, we need to find A, B, C, and D in

$$\psi_{\text{left}} = A\exp(ikx/\hbar) + B\exp(-ikx/\hbar), \quad x < 0, \quad k = \sqrt{2mE} \tag{2-71}$$

$$\psi_{\text{right}} = C\exp(ik'x/\hbar) + D\exp(-ik'x/\hbar), \quad x > 0, \quad k' = \sqrt{2m(E-U)} \tag{2-72}$$

We have seen earlier that the exponentials having positive arguments correspond to particles traveling from left to right, etc. We signify this with arrows in Fig. 2-17a.

The nature of ψ_{right} is qualitatively different depending on whether E is larger or smaller than the step height U. If $E < U$, the exponential arguments become real. One of the exponentials decays and the other explodes as x increases, just as we saw in Section 2-3. We discard the exploding exponential. The decaying exponential on the right must now be made to join smoothly onto ψ_{left} at the step. That is, ψ_{left} and

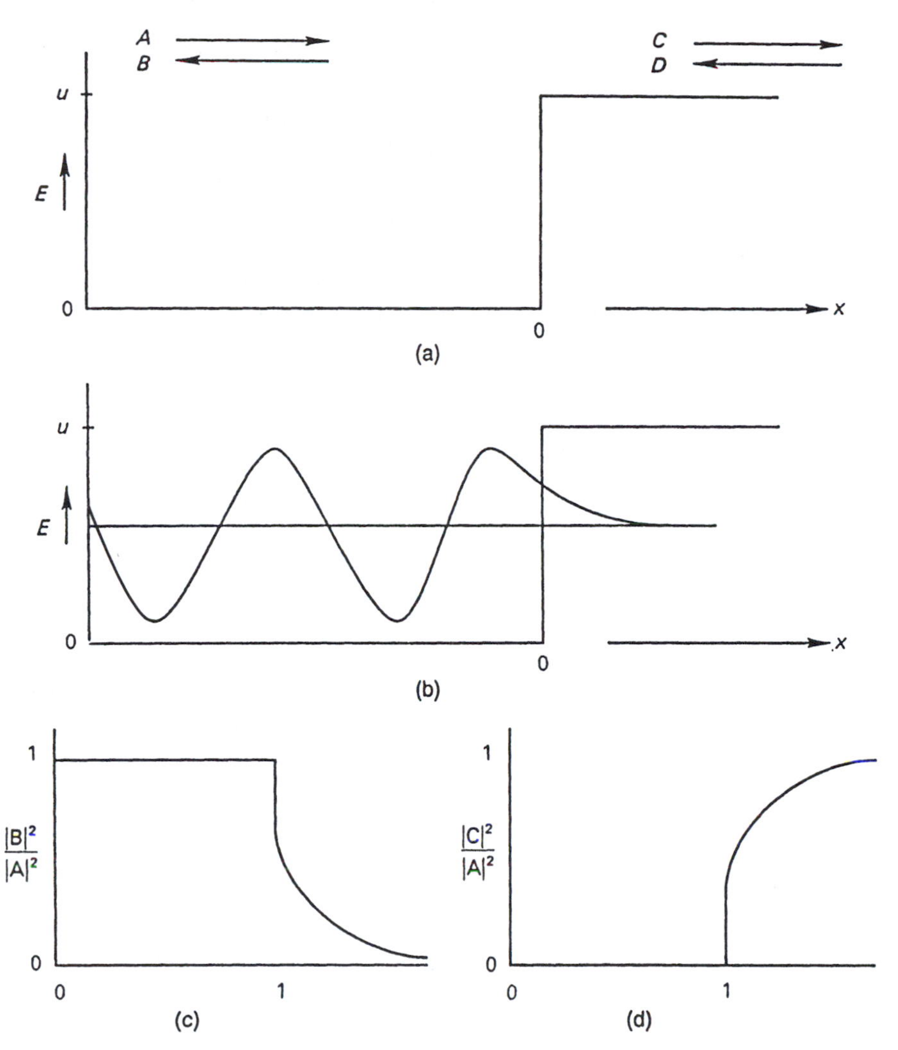

Figure 2-17 ▶ (a) A potential step of height U at $x = 0$. (b) A wavefunction having $E = U/2$. (c) Fraction of particles reflected as a function of E. (d) Fraction of particles transmitted as a function of E. (Note: The vertical line at left of parts (a) and (b) is not a barrier. It is merely an energy ordinate.)

ψ_{right} must have the same value and slope at $x = 0$ (Fig. 2-17b). This means that (fch2:eqn2-72 must have *real* value and slope at the step, which forces it to be a trigonometric wave. Because there is no additional boundary condition farther left, this trigonometric wave can always be shifted in phase and amplitude to join smoothly onto the decaying exponential at the right. (Compare Fig. 2-17b to Fig. 2-11.) The final values of A and B are simply those that give the appropriate phase and amplitude.

The ratio A^*A/B^*B is the relative fluxes of particles traveling toward the right or left in the region to the left of the step. If $|A| = |B|$, the fluxes are equal, corresponding to total reflection of the beam from the step potential.

It is not difficult to show (Problem 2-24) that $|A| = |B|$ whenever ψ_{left} has real value and slope at any point, i.e., for any trigonometric wave, and so the potential of Fig. 2-17a gives total reflection if $E < U$. (The fact that some particle density exists at $x > 0$ due to barrier penetration does not affect this conclusion. The evaluation of extent of reflection assumes that a time-independent (steady-state) situation has been achieved, so the penetration population remains constant and none of the new particles entering from the left are "lost" due to barrier penetration.)

The situation changes when we consider cases for $E > U$. For any such E, we now have two acceptable exponential functions on both the right and the left. We proceed by realizing that the function with coefficient D in ψ_{right} should be rejected since it corresponds to particles traveling from right to left, i.e., to particles that have been reflected from the trap. But we assume the trap to be 100% effective, so once again we have only one term in ψ_{right} As before, we set about forcing the values and slopes of Eqs. (2-71) and (2-72) (with $D = 0$) to be equal at $x = 0$. This time, however, there is no requirement that these values be real. We arrive at the relations (Problem 2-25)

$$\frac{B}{A} = \frac{k - k'}{k + k'} \text{ and } \frac{C}{A} = \frac{2k}{k + k'} \tag{2-73}$$

The extent of reflection is equal to

$$\frac{|B|^2}{|A|^2} = \frac{(k - k')^2}{(k + k')^2} \tag{2-74}$$

This can be seen to range from zero, when $k' = k$, to one, when $k' = 0$. $k' = k$ when $E = E - U$, i.e., when U is negligible compared to E. So zero reflection (total transmission) is approached in the high-energy limit. Only when $E = U$ does $k' = 0$, so total reflection occurs only when E equals the barrier height (or, as we saw previously, is lower). A plot of the fraction of particles reflected versus E/U appears in Fig. 2-17c. The transmission, equal to 1.0–reflection, is plotted in Fig. 2-17d.

The approach represented by this scattering problem is to identify the two terms that can contribute to the wavefunction in each region; then to recognize that one of the terms in one region is lost, either because it explodes or because it corresponds to reflection from the particle trap; then to force a smooth junction at the position of the discontinuity in the potential; and finally to draw conclusions about reflection and transmission from the values of the absolute squares of the coefficients. Notice that, for $E > U$, we could just as well have postulated the beam to be coming from the right, with the trap at the left. This would lead us to set $A = 0$ in Eq. (2-71). Even in cases like this, where the particles are passing over the edge of a potential cliff, there is backscatter (Problem 2-26).

An additional feature appears when we consider potentials that change at two points in x, as in Fig. 2-18a. The solution now involves three regions and two places ($x = \pm a$) where ψ and $d\psi/dx$ must be made equal. As before, we decide where the particle source and trap are and set one coefficient equal to zero (say G). Detailed solution of this problem is tedious.[3] For our purposes, it is the nature of the result that is important. The extent of transmission as a function of E/U is plotted in Fig. 2-18b. There are two obvious ways this differs from the step-potential transmission function of Fig. 2-17d: First, some of the particles are transmitted even when their energy is less than U. This is the result of barrier penetration leading to finite particle density at the right-hand side of the barrier transmission due to tunneling. (The extent of tunneling transmission depends on the thickness of the barrier.) Second, there are oscillations

[3]See Merzbacher [2, Chapter 6].

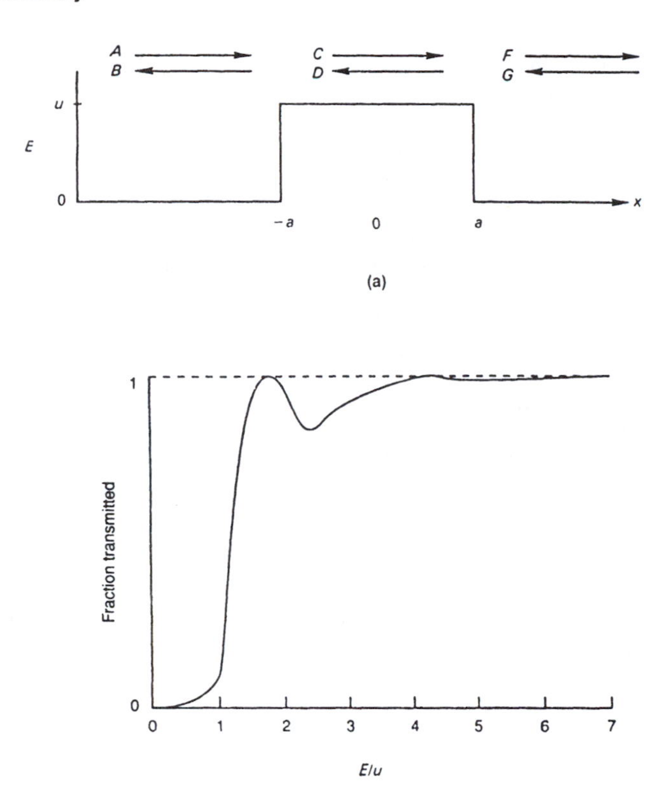

Figure 2-18 ▶ (a) A rectangular-hill potential of height U and width $2a$. (b) Fraction of beam transmitted as a function of E/U, where $U = h^2/2\pi^2 ma^2$.

in the transmission function for $E > U$, with 100% transmission occurring at intervals in E instead of only in the infinite limit. These come about because of interference between waves reflecting off the front and back edges of the barrier. This is most easily understood by recognizing that 100% transmission corresponds to no reflection, so then $B = 0$. This occurs when the wave reflecting back from $x = -a$ is of opposite phase to that reflecting back from $x = +a$, and this happens whenever there is an integral number of de Broglie half-wavelengths between $x = -a$ and a. At energies where this happens, the beam behaves as though the potential barrier is not there.

The variation of reflection from thin films (e.g., soap bubbles) of light of different wavelengths results in the perception of colors and is a familiar example of scattering interference. Less familiar is the variation in reflection of a particle beam, outlined above. However, once we recognize the wave nature of matter, we must expect particles to manifest the same sort of wave properties we associate with light.

2-9 Summary

In this chapter we have discussed the following points:

1. A particle constrained in the classical sense (i.e., lacking the energy to overcome barriers preventing its motion over the entire coordinate range) will have quantized energy levels and a finite zero-point energy. In the mathematical analysis, this arises from requirements on ψ at boundaries.

2. ψ can be nonsmooth, or cusped, where V is infinite at a point. If V is infinite over a finite range, ψ must be zero there.
3. Nondegenerate eigenfunctions of H must be symmetric or antisymmetric for any operation that leaves H unchanged.
4. $|\psi|^2$ may be regarded as a statistical measure—a summary of many measurements of position on independent, but identically prepared, systems.
5. Quantum-mechanical predictions approach classical predictions in the limits of large E, or large mass, or very high quantum number values.
6. Integrals with antisymmetric integrands must vanish.
7. $|\psi|^2$ does not vanish in regions where $V > E$ if V is finite. This is called "barrier penetration."
8. One-dimensional motion of a free particle has a continuum of energy levels. Except for $E = 0$, the states are doubly degenerate. Therefore, any mixture of such a pair of states is still an eigenfunction of H. But only two eigenfunctions (for a given $E \neq 0$) are also eigenfunctions for the momentum operator. These are the exponential functions. Since they correspond to different momenta, mixing them produces functions that are not eigenfunctions for the momentum operator.
9. Motion of a particle on a ring has quantum-mechanical solutions very similar to those for free-particle motion in one dimension. In both cases, there is no zero-point energy. Both are doubly degenerate for $E > 0$ because two directional possibilities are present. Both have a set of exponential solutions that are eigenfunctions for momentum. The main difference is that the particle-in-a-ring energies are quantized, due to head-to-tail "joining conditions" on ψ.
10. Increasing the dimensionality of a particle's range of motion increases the number of quantum numbers needed to define the wavefunctions. In cases where the hamiltonian operator can be written as a sum of operators for different coordinates (i.e., is "separable"), the problem greatly simplifies; the wavefunctions become products, and the energies become sums.
11. Scattering problems are treated by selecting an energy of interest from the continuum of possibilities, removing functions that describe nonphysical processes such as backscatter from the trap, and matching wave values and slopes at region boundaries. Resulting wavefunctions show wave interference effects similar to those observed for light.

2-9.A Problems

2-1. Ascertain that the expression (2-12) for energy has the proper dimensions.

2-2. Solve Eq. (2-9) for A.

2-3. There is a simple way to show that A in Eq. (2-9) must equal $\sqrt{2/L}$. It involves sketching ψ^2, recognizing that $\sin^2 x + \cos^2 x = 1$, and asking what A must equal in order to make the area under ψ^2 equal 1. Show this for $n = 1$, and argue why it must give the same result for all n.

2-4. Evaluate the probability for finding a particle in the middle third of a one-dimensional box in states with $n = 1, 2, 3, 10^4$. Compare your answers with the sketches in Fig. 2-5 to see if they are reasonable.

2-5. a) *Estimate* the probability for finding a particle in the $n = 1$ state in the line element Δx centered at the midpoint of a one-dimensional box if $\Delta x = 0.01L$. How does this compare to the classical probability?

b) Repeat the problem, but with Δx centered one third of the way from a box edge.

2-6. a) Use common sense to evaluate the following integral for the particle in a one-dimensional box, assuming that ψ is normalized.

$$\int_0^{L/5} \psi_5^2 dx$$

b) How does this value compare to that for the integral over the same range, but using ψ_1 instead of ψ_5? (Larger, smaller, or equal?) Use a sketch to defend your answer.

2-7. Let S and A be respectively symmetric and antisymmetric functions for the operator R. Evaluate the following, where R operates on every function to its right: (a) RS (b) RA (c) RSS (d) RAA (e) RAS (f) RAASASSA (g) RAASASAA. Can you think of a simple general rule for telling when a product of symmetric and antisymmetric functions will be antisymmetric?

2-8. Using the concept of odd and even functions, ascertain *by inspection of sketches* whether the following need be identically zero:

a) $\int_0^{\pi} \sin\theta \cos\theta \, d\theta$
b) $\int_{-\pi}^{\pi} \sin\theta \cos\theta \, d\theta$
c) $\int_{-1}^{1} x \cos x \, dx$
d) $\int_{-a}^{a} \cos y \sin^2 y \, dy$
e) $\int_0^{\pi} \sin^3\theta \cos^2\theta \, d\theta$
f) $\int_0^{\pi} \sin^2\theta \cos^3\theta \, d\theta$
g) $\int_{-1}^{1} \int_{-1}^{1} x^2 y \, dx \, dy$
h) $\int_{-\pi}^{\pi} x \sin x \cos x \, dx$
i) $\int_0^{\pi} \sin x \frac{d}{dx} \sin^2 x \, dx$
j) $\int_{-\pi}^{\pi} \sin^2 x \frac{d}{dx} \sin x \, dx$

2-9. Verify Eq. (2-23) for the general case $n \neq m$ by explicit integration.

2-10. For the potential of Fig. 2-8, when $E < U$ the energies are discrete, and when $E > U$, they are continuous. Is there a solution with $E = U$? What special requirements are there, if any, for such a solution to exist?

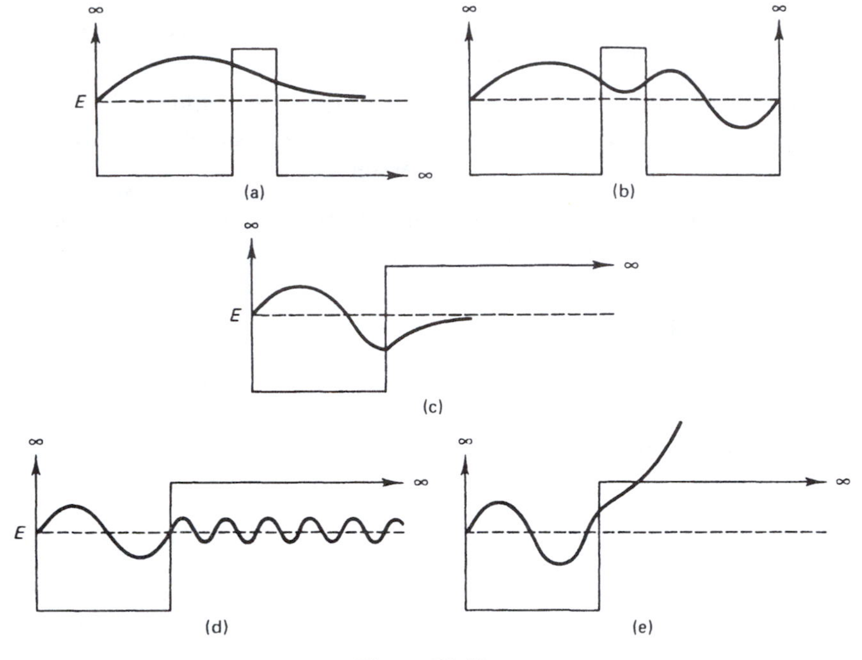

Figure P2-12 ▶

2-11. Prove the following statement: any linear combination of degenerate eigenfunctions of H is also an eigenfunction of H.

2-12. In a few words, indicate what is wrong with the wavefunctions sketched in the potentials shown in Fig. P2-12. If the solution appears to be acceptable, indicate this fact.

2-13. A double-well potential ranges from $x = 0$ to $x = 2L$ and has a thin (width = $0.01L$) rectangular barrier of finite height centered at L.

a) Sketch the wavefunction that goes with the fourth energy level in this system, assuming that its energy is less than the height of the barrier.
b) *Estimate* the energy of this level for a particle of mass m.

2-14. Use the simple approach presented in Problem 2-3 to demonstrate that $A = 1/\sqrt{\pi}$ for the trigonometric particle-in-a-ring eigenfunctions and $1/\sqrt{2\pi}$ for the exponential eigenfunctions.

2-15. Explain why $(2\pi)^{-1/2}\exp(i\sqrt{2}\phi)$ is unacceptable as a wavefunction for the particle in a ring.

2-16. For a particle in a ring, an eigenfunction is $\psi = (1/\sqrt{\pi})\cos(3\phi)$.

a) Write down H.
b) Evaluate $H\psi$ and identify the energy.
c) Is this a state for which angular momentum is a constant of motion? Demonstrate that your answer is correct.

2-17. Consider two related systems—a particle in a ring of constant potential and another just like it except for a very thin, infinitely high barrier inserted at $\phi = 0$. When this barrier is inserted,

a) are any *energies* added or lost?
b) do any *degeneracies* change?
c) are exponential and sine–cosine forms both still acceptable for eigenfunctions?
d) is angular momentum still a constant of motion?

2-18. Consider a particle of mass m in a two-dimensional box having side lengths L_x and L_y with $L_x = 2L_y$ and $V = 0$ in the box, ∞ outside.

a) Write an expression for the allowed energy levels of this system.
b) What is the zero-point energy?
c) Calculate the energies and degeneracies for the lowest eight energy levels.
d) Sketch the wavefunction for the fourth level.
e) Suppose $V = 10$ J in the box. What effect has this on (i) the eigenvalues? (ii) the eigenfunctions?

2-19. Consider the particle in a three-dimensional rectangular box with $L_x = L_y = L_z/2$. What would be the energy when $n_x = 1, n_y = 2, n_z = 2$? For $n_x = 1, n_y = 1, n_z = 4$? Can you guess the meaning of the term "accidental degeneracy?"

2-20. Consider a particle of mass m in a cubical box with $V = 0$ at $0 < x, y, z < L$.

a) Is $\left(1/\sqrt{2}\right)\left(\psi_{5,1,1} - \psi_{3,3,3}\right)$ an eigenfunction for this system? Explain your reasoning.
b) Estimate the probability for finding the particle in a volume element $\Delta V = 0.001V$ at the box center when the system is in its lowest-energy state. What is the classical value?

2-21. Kuhn [1] has suggested that the mobile π electrons in polymethine dyes can be modeled after the one-dimensional box. Consider the symmetric carbocyanine dyes (I) where the positive charge "resonates" between the two nitrogen atoms. The zigzag polymethine path along which the π electrons are relatively free to move extends along the conjugated system between the two nitrogens. Kuhn assumed a box length L equal to this path length plus one extra bond length on each end (so that the nitrogens would not be at the very edge of the box where they would be prevented from having any π-electron charge). This gives $L = (2n + 10)l$ where l is 1.39 Å, the bond length of an intermediate (i.e., between single and double) C–C bond. The number of π electrons in the polymethine region is $2n + 10$. Assume that each energy level in the box is capable of holding no more than two electrons and that the electronic transition responsible for the dye color corresponds to the promotion of an electron from the highest filled to the lowest empty level, the levels having initially been filled starting with the lowest, as shown in Fig. P2-13. Calculate ΔE and λ for the cases $n = 0, 1, 2, 3$ and compare with the observed values of maximum absorption of about 5750, 7150, 8180, and 9250 Å, respectively.

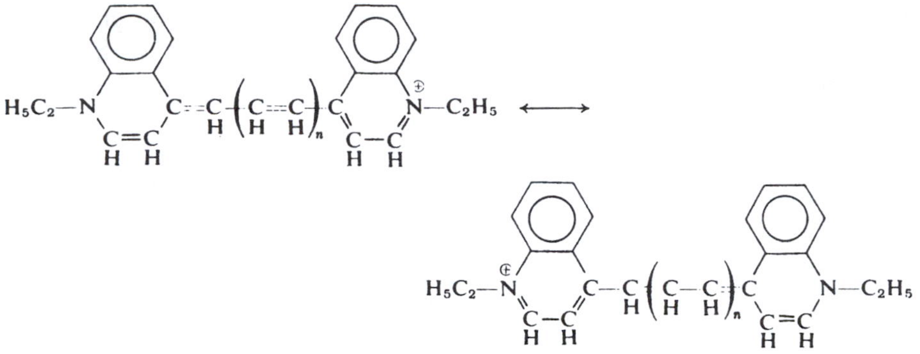

(I)

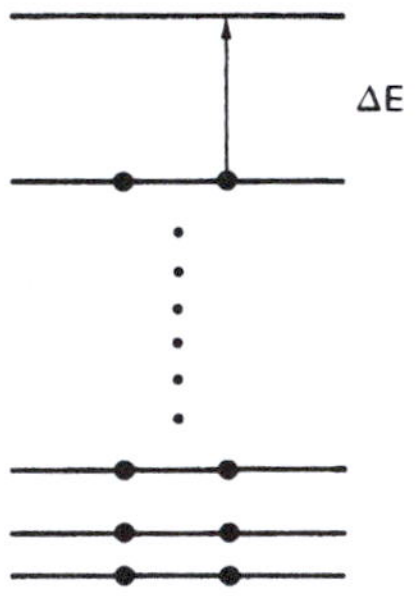

Figure P2-13 ▶

2-22. Show whether momentum in the x direction is a constant of motion for a free particle of mass m in states described by the following functions. In cases where it is a constant of motion, give its value. In all cases, evaluate the kinetic energy of the particle.

a) $\psi = \sin 3x$
b) $\psi = \exp(3ix)$
c) $\psi = \cos 3x$
d) $\psi = \exp(-3ix)$

2-23. Show whether angular momentum perpendicular to the plane of rotation is a constant of motion for a particle of mass m moving in a ring of constant potential in states described by the following functions. In cases where it is a constant of motion, give its value. In all cases, evaluate the kinetic energy of the particle.

a) $\psi = (1/\sqrt{\pi}) \cos 3\phi$
b) $\psi = (1/\sqrt{2\pi}) \exp(-3i\phi)$
c) $\psi = (1/\sqrt{\pi}) \sin 3\phi$
d) $\psi = (1/\sqrt{2\pi}) \exp(3i\phi)$

2-24. Demonstrate that the requirement that $\psi = A\exp(ikt) + B\exp(-ikt)$ have *real* value and slope at a point in x suffices to make $|A| = |B|$.

2-25. Derive relations (2-73) and (2-74) by matching wave values and slopes at the step.

2-26. Solve the problem for the step potential shown in Fig. 2-17a, but with the beam traveling from a source at right toward a trap at left.

2-27. The reflection coefficient is defined as $|B|^2/|A|^2$ for a beam originating from the left of Fig. 2-17a. The transmission coefficient is defined as $(k'/k)\,|C|^2/|A|^2$.

a) Why is the factor k'/k needed?
b) Show that the sum of these coefficients equals unity, consistent with a steady-state situation.

2-28. For scattering from a potential such as that in Fig. 2-18a, 100% transmission occurs at various finite particle energies. Find the lowest two values of E/U for which this occurs for particles of mass m, barrier width d, and barrier height $U = 2h^2/\pi^2 m d^2$.

2-29. Calculate "frequencies" in cm^{-1} needed to accomplish the transitions ΔE_1 and ΔE_2 in Fig. 2-14.

Multiple Choice Questions

(Try to answer these by inspection.)

1. The integral $\int_{-a}^{a} \cos(x)\sin(x)\,dx$

a) equals zero for any value of a, and $\cos(x)$ is antisymmetric in the range of the integral.
b) is unequal to zero except for certain values of a, and $\cos(x)$ is symmetric in the range of the integral.
c) equals zero for any value of a and $\cos(x)$ is symmetric in the range of the integral.
d) is unequal to zero except for certain values of a, and $\sin(x)$ is antisymmetric in the range of the integral.
e) equals zero for any value of a, and $\sin(x)$ is symmetric in the range of the integral.

2. $\int_0^{2\pi} x\sin(x)\cos(x)\,dx$

Which one of the following statements is true about the above integral and the three functions in its integrand?

a) All three functions are antisymmetric in the range and the integral equals zero.
b) Two functions are antisymmetric and one is symmmetric in the range, and the integral is unequal to zero.
c) Two functions are symmetric and one is antisymmetric in the range, and the integral is equal to zero.
d) One function is symmetric, one is antisymmetric, and one is unsymmetric in the range, and the integral is unequal to zero.
e) None of the above is a true statement.

3. In solving the particle in a one-dimensional box problem with infinite repulsive walls at $x = 0$ and L, we started with the function $A\sin(kx) + B\cos(kx)$. Which one of the following is a true statement?

a) The value of k is found by requiring that the solution be normalized.
b) Adding $C\exp(ikx)$ to the above function would prevent it from being an eigenfunction of the hamiltonian operator.
c) It is necessary that this function equal L when $x = 0$.
d) The boundary condition at $x = L$ is used to show that $B = 0$.
e) None of the above is a true statement.

4. It is found that a particle in a one-dimensional box of length L can be excited from the $n = 1$ to the $n = 2$ state by light of frequency ν. If the box length is doubled, the frequency needed to produce the $n = 1$ to $n = 2$ transition becomes

a) $\nu/4$
b) $\nu/2$
c) 2ν
d) 4ν
e) None of the above is correct.

5. For a particle in a one-dimensional box with infinite walls at $x = 0$ and L, and in the $n = 3$ state, the probability for finding the particle in the range $0 \leq x \leq L/4$ is

a) greater than $1/3$.
b) exactly $1/6$.
c) exactly $1/3$.
d) less than $1/6$.
e) None of the above is correct.

6. A student calculates the probability for finding a particle in the left-most 10% of a one-dimensional box for the $n = 1$ state. Which one of the following answers could be correct?

a) 1.742
b) 0.024
c) 0.243
d) 0.100
e) None of the above is reasonable.

7. Which one of the following statements is true about the particle in a one-dimensional box with infinite walls? (All integrals range over the full length of the box.)

a) $\int \psi_3^2\,dx = 0$ because these wavefunctions are orthogonal.
b) $\int \psi_1\psi_3\,dx = 0$ because both of these wavefunctions are symmetric.
c) $\int \psi_1\psi_2\,dx = 0$ because these wavefunctions are normalized.
d) $\int \psi_1\psi_3\,dx = 0$ because these wavefunctions are orthogonal.
e) None of the above is a true statement.

8. Which one of the following is the correct formula for the lowest-energy eigenfunction for a particle in a one-dimensional box having infinite barriers at $x = -L/2$ and $L/2$?

a) $\sqrt{\frac{2}{L}}\sin(\pi x/L)$
b) $\sqrt{\frac{2}{L}}\cos(\pi x/L)$
c) $\sqrt{\frac{2}{L}}\exp(i\pi x/L)$
d) $\sqrt{\frac{2}{L}}\exp(-i\pi x/L)$
e) None of the above is correct.

9. A particle is free to move in the x dimension without constraint (i.e., under the influence of a constant potential, which we assume to be zero). For this system, the wavefunction $\psi(x) = \exp(3.4x)$ is not acceptable because

a) it is not an eigenfunction of the hamiltonian operator.
b) it is multi-valued.
c) it is discontinuous.
d) it approaches infinity as x approaches infinity.
e) it goes to zero as x approaches minus infinity.

10. Consider two identical one-dimensional square wells connected by a *finite* barrier. Which one of the following statements about the quantum-mechanical time-independent solutions for this system is true when two equivalent "half-solutions" in the two wells are joined together to produce two overall solutions?

a) The combination that is symmetric for reflection through the central barrier always has the lower energy of the two.
b) The combination that places a node in the barrier always has the lower energy of the two.
c) The sum of the two half-solutions always has the lower energy of the two.
d) The difference of the two half-solutions always has the lower energy of the two.
e) None of the above is a true statement.

11. For a single particle-in-a-ring system having energy $9h^2/8\pi^2 I$ we can say that the angular momentum, when measured, will equal

a) $3\hbar$
b) $\sqrt{12}\hbar$
c) either $3\hbar$ or $-3\hbar$
d) zero
e) None of the above is a true statement.

12. A particle in a ring has wavefunctions that

a) result from placing an integral number of half-waves in the circumference of the ring.
b) must be eigenfunctions for $(h/2\pi i)d/d\phi$.
c) are all doubly degenerate, due to two rotational directions.

d) correspond to energies that increase with the square of the quantum number.
e) None of the above is a true statement.

13. The hamiltonian operator for a system is

$H = -(h^2/8\pi^2 m)\nabla^2 + x^2 + y^2 + z^2$

For this system we should expect

a) two quantum numbers at most.
b) eigenfunctions that are sums of functions, each depending on only one of the variables.
c) eigenvalues that are products of eigenvalues of separated equations.
d) eigenvalues that are sums of eigenvalues of separated equations.
e) None of the above is a correct statement.

14. For a particle in a one-dimensional box with one infinite barrier and one finite barrier of height U,

a) $\psi = 0$ at both barriers if E is less than U.
b) barrier penetration is smallest for the lowest-energy state.
c) no more than one quantized state can exist, and its energy is less than U.
d) only states having E greater than U can exist.
e) None of the above statements is correct.

15. For a particle of mass m in a cubical box having edge length L,

a) the zero point energy is $3h^2/8mL^2$.
b) the probability density has its maximum value at the box center for ψ_{211}.
c) the degeneracy of the ground state is 3.
d) ψ_{111} has three nodal planes.
e) None of the above is a true statement.

References

[1] H. Kuhn, *J. Chem. Phys.* **17**, 1198 (1949).

[2] B. Merzbacher, *Quantum Mechanics*, 3rd ed. Wiley, New York, 1998.

Chapter 3

The One-Dimensional Harmonic Oscillator

3-1 Introduction

In Chapter 2 we examined several systems with discontinuous potential energies. In this chapter we consider the simple harmonic oscillator—a system with a continuously varying potential. There are several reasons for studying this problem in detail. First, the quantum-mechanical harmonic oscillator plays an essential role in our understanding of molecular vibrations, their spectra, and their influence on thermodynamic properties. Second, the qualitative results of the problem exemplify the concepts we have presented in Chapters 1 and 2. Finally, the problem provides a good demonstration of mathematical techniques that are important in quantum chemistry. Since many chemists are not overly familiar with some of these mathematical concepts, we shall deal with them in detail in the context of this problem.

3-2 Some Characteristics of the Classical One-Dimensional Harmonic Oscillator

A pendulum consisting of a large mass hanging by an almost weightless wire, and swinging through a very small angle, is a close approximation to a classical harmonic oscillator. It is an oscillator since its motion is back and forth over the same path. It is harmonic to the extent that the restoring force on the mass is proportional to the horizontal component of the displacement of the mass from its rest position. This force law, known as *Hooke's law*, is the common first approximation made in the analysis of a system vibrating about an equilibrium position. If we let the x axis be the coordinate of displacement of the mass, with $x=0$ as the rest position, then we may write the restoring force as

$$F=-kx, \tag{3-1}$$

where k is the *force constant*. The minus sign assures that the force on the displaced mass is always directed toward the rest position.

We can use this force expression to determine an *equation of motion* for the mass, that is, an equation relating its location in space, x, to its location in time, t:

$$F=-kx(t)=ma=m\frac{d^2x(t)}{dt^2} \tag{3-2}$$

or

$$\frac{d^2x(t)}{dt^2} = (-k/m)x(t) \tag{3-3}$$

According to this equation, $x(t)$ is a function that, when differentiated twice, is regenerated with the multiplier $-k/m$. A general solution is

$$x(t) = a \sin\left(\sqrt{\frac{k}{m}}\,t\right) + b \cos\left(\sqrt{\frac{k}{m}}\,t\right) \tag{3-4}$$

If we require that $x(t)$ be at its maximum value L at $t = 0$ (as though the pendulum is held at its position of maximum displacement and then released at $t = 0$), then it follows that $b = L$. Since the pendulum is also motionless at $t = 0$, $(dx(t)/dt)_{t=0} = 0$, and so $a = 0$. Hence,

$$x\,(t) = L \cos\left(\sqrt{\frac{k}{m}}\,t\right) \tag{3-5}$$

Thus, the equation of motion [Eq. (3-3)] leads to a function, $x(t)$, that describes the trajectory of the oscillator. This function has the trigonometric time dependence characteristic of *harmonic* motion. From this expression, we see that $x(t)$ repeats itself whenever the argument of the cosine increases by 2π. This will require a certain time interval t'. Thus, the pendulum makes one complete back and forth motion in a time t' given by

$$\sqrt{\frac{k}{m}}\,t' = 2\pi \quad t' = 2\pi\sqrt{\frac{m}{k}} \tag{3-6}$$

so the *frequency* of the oscillation ν is

$$\nu = 1/t' = \frac{1}{2\pi}\sqrt{\frac{k}{m}} \tag{3-7}$$

Suppose that one were to take a multiflash photograph of a swinging pendulum from above. The result would look as shown in Fig. 3-1a, the number of images being much greater near the termini of the swing (called the "turning points") than at the middle because the pendulum is moving fastest as it crosses the middle. This, in turn, results from the fact that all the potential energy of the mass has been converted to kinetic energy at the middle point. We thus arrive at a classical prediction for the time-averaged distribution of the projection of the harmonic oscillator in the displacement coordinate: This *distribution function* is greatest in regions where the potential energy is highest (Fig. 3-1b) (Problem 3-1).

Let us calculate and compare the time-averaged potential and kinetic energies for the classical harmonic oscillator. When the particle is at some instantaneous displacement x', its potential energy is

$$\begin{aligned} V\,(x') &= \text{(applied force times distance to return to } x = 0\text{)} \\ &= \int_0^{x'} kx\,dx = \frac{1}{2}kx'^2 \end{aligned} \tag{3-8}$$

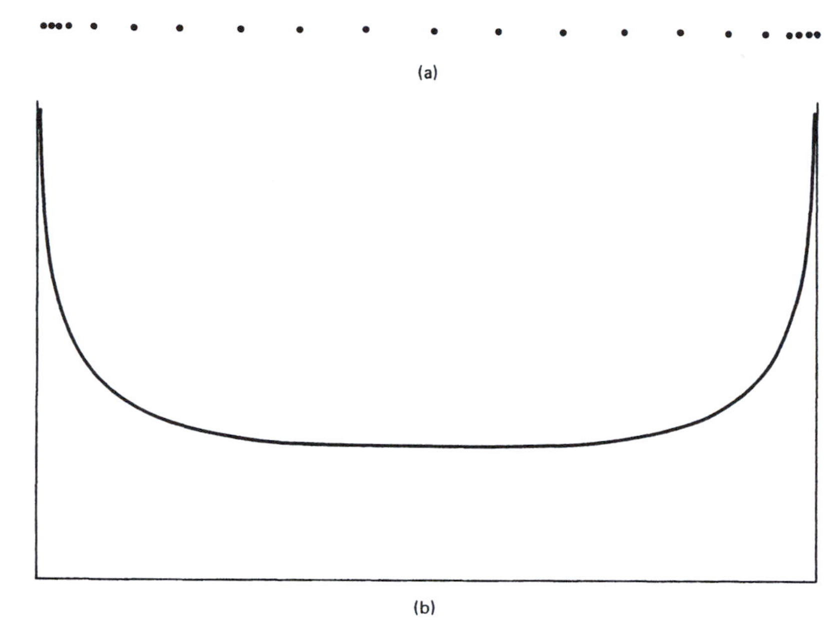

Figure 3-1 ▶ (a) Results of a uniform multiflash photograph of a swinging pendulum as photographed from above. (b) Distribution function corresponding to the continuous limit of discrete distribution shown in (a).

The cumulative value of the potential energy over one complete oscillation, V_c, is given by the integral

$$V_c\left(t'-0\right)=\int_0^{t'} V\left(t\right)\,dt=\frac{1}{2}k\int_0^{t'} x\left(t\right)^2\,dt \tag{3-9}$$

Substituting for $x(t)$ as indicated in Eq. (3-5)

$$\begin{aligned} V_c\left(t'-0\right) &= \frac{1}{2}kL^2\int_0^{t'}\cos^2\left(\sqrt{\frac{k}{m}}\,t\right)dt \\ &= \frac{1}{2}kL^2\sqrt{m/k}\int_0^{t'}\cos^2\left(\sqrt{\frac{k}{m}}\,t\right)d\left(\sqrt{\frac{k}{m}}\,t\right) \end{aligned} \tag{3-10}$$

When $t=t'$, $\sqrt{k/m}t=2\pi$, and we may rewrite Eq. (3-10) as

$$V_c\left(t'-0\right)=\frac{1}{2}kL^2\sqrt{m/k}\int_0^{2\pi}\cos^2 y\,dy=\left(\pi/2\right)kL^2\sqrt{m/k} \tag{3-11}$$

If we now divide by t' to get the average potential energy per unit time, we find

$$\bar{V}=\frac{V_c\left(t'-0\right)}{t'}=\frac{(\pi/2)kL^2\sqrt{m/k}}{2\pi\sqrt{m/k}}=\frac{kL^2}{4} \tag{3-12}$$

Thus, we have the average potential energy. If we knew the total energy, which is a constant of motion, we could get the average kinetic energy by taking the difference.

It is easy to evaluate the total energy by taking advantage of its constancy over time. Since we can choose any point in time to evaluate it, we select the moment of release ($t=0$ and $x=L$). The mass is motionless so that the total energy is identical to the potential energy:

$$E=\frac{1}{2}kL^2 \tag{3-13}$$

Comparing Eqs. (3-12) and (3-13) we see that, $\bar{T}=kL^2/4$. On the average, then, the classical harmonic oscillator stores half of its total energy as potential energy and half as kinetic energy.

EXAMPLE 3-1 A 10.00 g mass on a Hooke's law spring with force constant $k=0.0246\ \mathrm{N\ m^{-1}}$ is pulled from rest at $x=0$ to $x=0.400\ \mathrm{m}$ and released.
a) What is its total energy?
b) What is its frequency of oscillation?
c) How do these answers change if the mass weighs 40.00 g?

SOLUTION ▶ a) $E=T+V$. At $x=0.400\ \mathrm{m}$, $T=0$, $E=V=kx^2/2=0.50(0.0246\ \mathrm{N\ m^{-1}})(0.400\ \mathrm{m})^2=0.0020\ \mathrm{J}$
b) $\nu=\frac{1}{2\pi}\left(\frac{k}{m}\right)^{1/2}=\frac{1}{2\pi}\left(\frac{0.0246\ \mathrm{N\ m^{-1}}}{0.0100\ \mathrm{kg}}\right)^{1/2}=0.25\ \mathrm{s^{-1}}$
c) E is unchanged, and ν is halved to $0.125\ \mathrm{s^{-1}}$. The energy depends on the force constant and the displacement of the oscillator, but not on the mass of the oscillator. The frequency depends not only on the force constant, but also on the mass because the frequency is affected by the *inertia* of the oscillator. A greater mass has a lower frequency of oscillation for the same force constant. ◀

3-3 The Quantum-Mechanical Harmonic Oscillator

We have already seen [Eq. (3-8)] that the potential energy of a harmonic oscillator is given by

$$V(x)=\frac{1}{2}kx^2 \tag{3-14}$$

so we can immediately write down the one-dimensional Schrödinger equation for the harmonic oscillator:

$$\left[(-h^2/8\pi^2m)(d^2/dx^2)+\frac{1}{2}kx^2\right]\psi(x)=E\psi(x) \tag{3-15}$$

The detailed solution of this differential equation is taken up in the next section. Here we show that we can understand a great deal about the nature of the solutions to this equation by analogy with the systems studied in Chapter 2.

In Fig. 3-2a are shown the potential, some eigenvalues, and some eigenfunctions for the harmonic oscillator. The potential function is a parabola [Eq. (3-14)] centered at $x=0$ and having a value of zero at its lowest point. For comparison, similar information

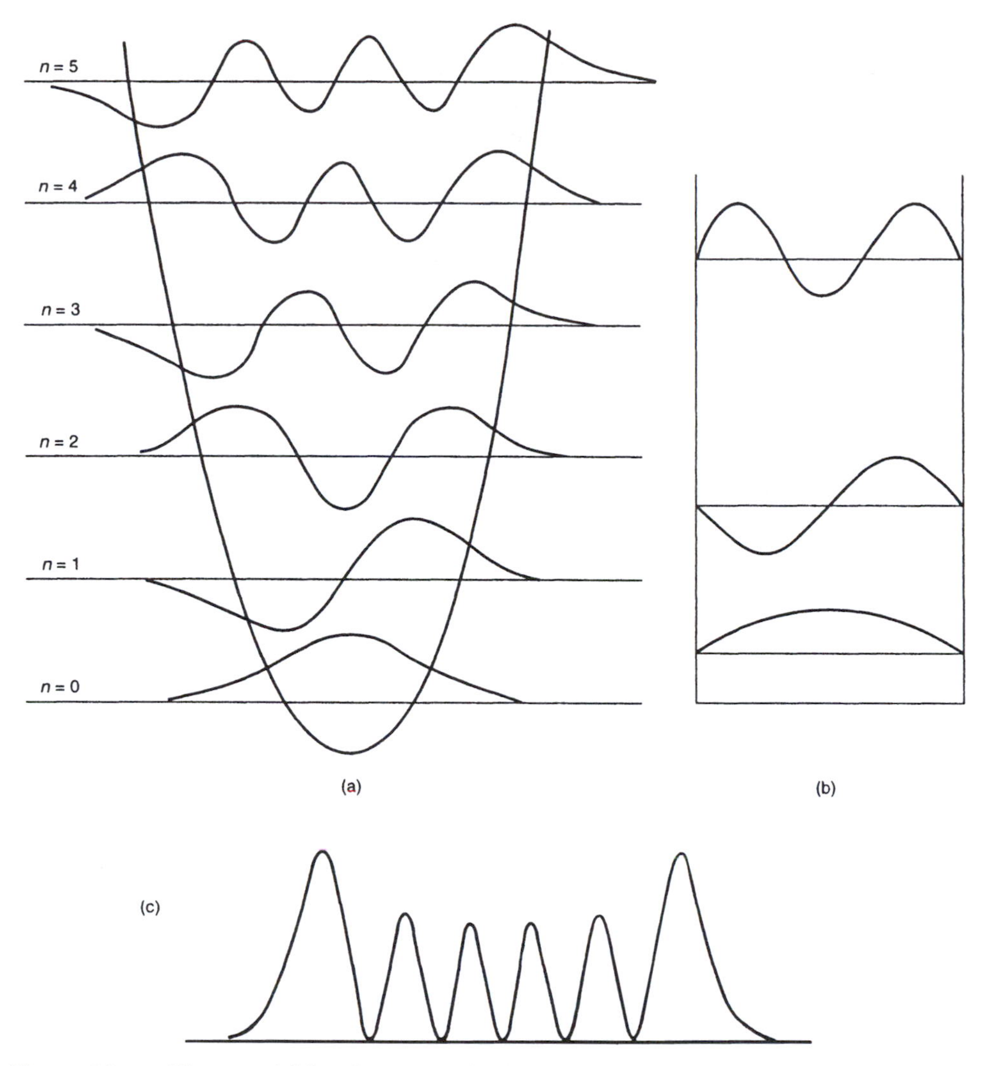

Figure 3-2 ▶ The potential function, energy levels, and wavefunctions for (a) the harmonic oscillator, (b) the particle in the infinitely deep square well. (c) ψ^2 for the harmonic oscillator in the state $n = 5$.

is graphed in Fig. 3-2b for the particle in a box with infinitely high walls. Some important features of the harmonic oscillator are:

1. *The energy-level spacing for the harmonic oscillator is constant.* As mentioned in Chapter 2, we expect the energy levels for the harmonic oscillator to diverge less rapidly than those for the square well because the higher energy states in the harmonic oscillator have effectively larger "boxes" than do the lower states (that is, the more energetic the oscillator, the more widely separated are its classical turning points). That the spacing for the oscillator grows less rapidly than that for the box is therefore reasonable. That it is actually *constant* is something we will show mathematically in the next section.

2. *The wavefunctions for the harmonic oscillator are either symmetric or antisymmetric under reflection through* $x = 0$. This is necessary because the hamiltonian is invariant

to reflection through $x=0$ *and* because the eigenfunctions are nondegenerate. Both of these conditions apply also to the box problem. If we imagine moving from the box potential to the parabolic potential by a process of continuous deformation, the symmetry is not altered and there is no reason to expect nondegenerate box levels to come together and become degenerate. Therefore, the oscillator wavefunction symmetries are not surprising. As a consequence, we have at once that the symmetric wavefunctions are automatically orthogonal to antisymmetric wavefunctions, as discussed earlier. (Actually, all the wavefunctions are orthogonal to each other. This is proved in Section 3-4.)

3. *The harmonic oscillator has finite zero-point energy.* (The evidence for this in Fig. 3-2a is the observation that the line for the lowest ($n=0$) energy level lies above the lowest point of the parabola, where $V=0$.) This is expected since the change from square well to parabolic well does not remove the restrictions on particle position; it merely changes them.

4. *The particle has a finite probability of being found beyond the classical turning points; it penetrates the barrier.* This is to be expected on the basis of earlier considerations since the barrier is not infinite at the classical turning point. (The potential becomes infinite only at $x=\pm\infty$.)

5. *In the lowest-energy state the probability distribution favors the particle being in the low-potential central region of the well, while at higher energies the distribution approaches more nearly the classical result of favoring the higher potential regions (Fig. 3-2c).*

3-4 Solution of the Harmonic Oscillator Schrödinger Equation

3-4.A Simplifying the Schrödinger Equation

Equation (3-15) is simplified by substituting in the following relations:

$$\alpha \equiv 8\pi^2 mE/h^2 \tag{3-16}$$

$$\beta^2 \equiv 4\pi^2 mk/h^2 \tag{3-17}$$

The quantities α and β have units of m^{-2}. We will assume that β is the positive root of β^2. The quantity α is necessarily positive. The Schrödinger equation now can be written

$$\frac{d^2\psi(x)}{dx^2} + \left(\alpha - \beta^2 x^2\right)\psi(x) = 0 \tag{3-18}$$

3-4.B Establishing the Correct Asymptotic Behavior

At very large values of $|x|$, the quantity α (which is a constant since E is a constant) becomes negligible compared to $\beta^2 x^2$. That is, the Schrödinger equation (3-18) approaches more and more closely the asymptotic form

$$d^2\psi(x)/dx^2 = \beta^2 x^2 \psi(x), \quad |x| \to \infty \tag{3-19}$$

What we need, then, are solutions $\psi(x)$ that approach the solutions of Eq. (3-19) at large values of $|x|$. The solutions for Eq. (3-19) can be figured out from the general rule for differentiating exponentials:

$$(d/dx)\exp(u(x)) = \exp(u(x))du(x)/dx \tag{3-20}$$

Then

$$(d^2/dx^2)\exp(u(x)) = [(du(x)/dx)^2 + d^2u(x)/dx^2]\exp(u(x)) \tag{3-21}$$

We want the term in square brackets to become equal to β^2x^2 in the limit of large $|x|$. We can arrange for this to happen by setting

$$u(x) = \pm\beta x^2/2 \tag{3-22}$$

for then

$$(d^2/dx^2)\exp(u(x)) = (\beta^2x^2 \pm \beta)\exp(\pm\beta x^2/2) \tag{3-23}$$

At large values of $|x|$, β is negligible compared to β^2x^2, and so $\exp(\pm\beta x^2/2)$ are *asymptotic solutions* for Eq. (3-19). As $|x|$ increases, the positive exponential increases rapidly whereas the negative exponential dies away. We have seen that, for the wavefunction to be physically meaningful, we must reject the solution that blows up at large $|x|$. On the basis of these considerations, we can say that, if ψ contains $\exp(-\beta x^2/2)$, it will have the correct asymptotic behavior *if no other term is present that dominates at large* $|x|$. Therefore,

$$\psi(x) = q(x)\exp(-\beta x^2/2) \tag{3-24}$$

and it remains to find the function $q(x)$.

3-4.C The Differential Equation for *q(x)*

Substituting Eq. (3-24) into the Schrödinger equation (3-18) gives

$$\exp(-\beta x^2/2)\left[-\beta q(x) - 2\beta x\frac{dq(x)}{dx} + \frac{d^2q(x)}{dx^2} + \alpha q(x)\right] = 0 \tag{3-25}$$

This equation is satisfied only if the term in brackets is zero:

$$\frac{d^2q(x)}{dx^2} - 2\beta x\frac{dq(x)}{dx} + (\alpha - \beta)q(x) = 0 \tag{3-26}$$

Thus, we now have a differential equation for $q(x)$.

At this point it is convenient to transform variables to put the equation into a simpler form. Let

$$y = \sqrt{\beta}x \tag{3-27}$$

Then

$$d/dy = d/d\left(\sqrt{\beta}x\right) = \left(1/\sqrt{\beta}\right)d/dx \tag{3-28}$$

so that

$$d/dx = \sqrt{\beta}d/dy \tag{3-29}$$

Similarly

$$\frac{d^2}{dx^2} = \frac{\beta d^2}{dy^2} \tag{3-30}$$

and

$$x = \frac{y}{\sqrt{\beta}} \tag{3-31}$$

Substituting Eqs. (3-29)–(3-31) into (3-26), and defining $f(y)$ as

$$f(y) \equiv f\left(\sqrt{\beta}x\right) = q(x) \tag{3-32}$$

we obtain (after dividing by β)

$$\frac{d^2 f(y)}{dy^2} - 2y\frac{df(y)}{dy} + [(\alpha/\beta) - 1]\, f(y) = 0 \tag{3-33}$$

3-4.D Representing *f* as a Power Series

Now $f(y)$ is some function of y that must be single valued, continuous, and smooth (i.e., have a continuous first derivative), if ψ is to be properly behaved. Can we think of any functions that satisfy these properties? Of course, we can think of a limitless number of them. For example, 1, y, y^2, y^3, y^4, etc., are all single valued, continuous, and have continuous derivatives, and so is any linear combination of such functions (e.g., $4y^3 - y + 3$). Other examples are $\sin(y)$ and $\exp(y)$. These functions can be expressed as infinite sums of powers of y, however, and so they are included, in principle, in the first example. Thus

$$\sin(y) = y - y^3/3! + y^5/5! - y^7/7! + \cdots \tag{3-34}$$

and

$$\exp(y) = \sum_{n=0}^{\infty} y^n/n! \tag{3-35}$$

that is, $\sin(y)$ and the set of all positive powers of y are *linearly dependent*. Because the powers of y can be combined linearly to reproduce certain other functions, the powers of y are called a *complete set* of functions. However, we must exercise some care with the concept of completeness. The positive powers of y cannot be used to reproduce a discontinuous function, or a function with discontinuous derivatives. Hence, there are certain restrictions on the nature of functions $g(y)$, that satisfy the relation

$$g(y) = \sum_{n=0}^{\infty} c_n y^n \tag{3-36}$$

These restrictions define a class of functions, and the *powers of y are a complete set only within this class.* The positive powers of y, then, form a complete set, but if we remove one of the members of the set, say 1 (the zero power of y), then the set is no longer complete. This means that the remaining members of the set cannot compensate for the role played by the missing member. In other words, the missing member cannot be expressed as a linear combination of the remaining members. In this example

$$1 \neq \sum_{n=1}^{\infty} c_n y^n \tag{3-37}$$

This is easily demonstrated to be true since the left-hand side of Eq. (3-37) is unity whereas the right-hand side must be zero when $y = 0$ for any choice of coefficients. Thus, the powers of y are *linearly independent* functions (no one of them can be expressed as a linear combination of all the others).

The function $f(y)$ involved in ψ should be a member of the class of functions for which the powers of y form a complete set. Therefore, we may write

$$f(y) = \sum_{n=0}^{\infty} c_n y^n \tag{3-38}$$

and seek an expression for the unknown multipliers c_n.

3-4.E Establishing a Recursion Relation for *f*

Notice that if

$$f(y) = c_0 + c_1 y + c_2 y^2 + c_3 y^3 + c_4 y^4 + \cdots \tag{3-39}$$

then

$$df(y)/dy = c_1 + 2c_2 y + 3c_3 y^2 + 4c_4 y^3 + \cdots \tag{3-40}$$

and

$$d^2 f(y)/dy^2 = 2c_2 + 2 \cdot 3c_3 y + 3 \cdot 4c_4 y^2 + \cdots \tag{3-41}$$

Thus, substituting Eq. (3-38) into (3-33) gives

$$\begin{aligned} &1 \cdot 2c_2 + 2 \cdot 3c_3 y + 3 \cdot 4c_4 y^2 + \cdots - 2c_1 y - 2 \cdot 2c_2 y^2 - 2 \cdot 3c_3 y^3 - \cdots \\ &+ [(\alpha/\beta) - 1]c_0 + [(\alpha/\beta) - 1]c_1 y + [(\alpha/\beta) - 1]c_2 y^2 + \cdots \\ &= 0 \end{aligned} \tag{3-42}$$

Now we will take advantage of the fact that the various powers of y form a linearly independent set. Equation (3-42) states that the expression on the LHS equals zero for all values of y. There are two ways this might happen. One of these is that minus the constant part of the expression is always exactly equal to the y-dependent part, no matter what the value of y. This would require a relationship like Eq. (3-37) (except with an equality), which we have seen is not possible for independent functions. The remaining possibility is that the various independent parts of Eq. (3-42) are individually equal to zero—the constant is zero, the coefficient for y is zero, etc. This gives us a whole set of equations. Setting the constant term equal to zero gives

$$2c_2 + [(\alpha/\beta) - 1]c_0 = 0 \cdots \quad m = 0 \tag{3-43}$$

Setting the coefficient for the first power of y to zero gives

$$2 \cdot 3c_3 - 2c_1 + [(\alpha/\beta) - 1]c_1 = 0 \cdots \quad m = 1 \tag{3-44}$$

The y^2 term gives

$$3 \cdot 4c_4 - 2 \cdot 2c_2 + [(\alpha/\beta) - 1]c_2 = 0 \cdots \quad m = 2 \tag{3-45}$$

By inspecting this series, we can arrive at the general result

$$(m+1)(m+2)c_{m+2} + [(\alpha/\beta) - 1 - 2m]c_m = 0 \tag{3-46}$$

or

$$c_{m+2} = \frac{-[(\alpha/\beta) - 2m - 1]}{(m+1)(m+2)} c_m \tag{3-47}$$

Equation (3-47) is called a *recursion relation*. If we knew c_0, we could produce c_2, c_4, c_6, etc., by continued application of Eq. (3-47). Similarly, knowledge of c_1 would lead to c_3, c_5, etc. Thus, it appears that the coefficients for even powers of y and those for odd powers of y form separate sets. Choosing c_0 determines one set, and choosing c_1 determines the other, and the choices for c_0 and c_1 seem independent. This separation into two sets is reasonable when we recall that our final solutions must be symmetric or antisymmetric in x, hence also in y. The asymptotic part of ψ, $\exp(-\beta x^2/2)$, is symmetric about $x = 0$, and so we expect the remainder of ψ, $f(y)$, to be either symmetric (even powers of $y = \sqrt{\beta}x$) or antisymmetric (odd powers). Thus, we can anticipate that some of our solutions will have $c_0 \neq 0$, $c_2 \neq 0$, $c_4 \neq 0$, ... and $c_1 = c_3 = c_5 = \cdots = 0$. This will produce symmetric solutions. The remaining solutions will have $c_0 = c_2 = c_4 = \cdots = 0$ and $c_1 \neq 0$, $c_3 \neq 0, \ldots$, and be antisymmetric.

EXAMPLE 3-2 Evaluate c_3 and c_5 if $c_1 = 1$, for arbitrary α and β. What ratio α/β will make $c_3 = 0$? What ratio α/β will make $c_5 = 0$ but $c_3 \neq 0$?

SOLUTION ▸ $c_1 = 1$, For c_3, we use $m = 1$, $c_3 = \frac{-[\alpha/\beta - 2 - 1]}{(2)(3)} \times 1 = \frac{3 - \alpha/\beta}{6}$.
For c_5 we use $m = 3$, $c_5 = \frac{-[\alpha/\beta - 6 - 1]}{(4)(5)} \times \frac{[3 - \alpha/\beta]}{6} = \frac{(7 - \alpha/\beta)(3 - \alpha/\beta)}{120}$.
$\alpha/\beta = 3$ makes $c_3 = 0$ (and also c_5). $\alpha/\beta = 7$ makes $c_5 = 0$, but not c_3. ◂

3-4.F Preventing *f(y)* from Dominating the Asymptotic Behavior

We now examine the asymptotic behavior of $f(y)$. Recall that, at very large values of $|y|$, $f(y)$ must become insignificant compared to $\exp(-\beta x^2/2) \equiv \exp(-y^2/2)$. We will show that $f(y)$ fails to have this behavior if its power series expression is infinitely long. That is, we will show that $f(y)$ behaves asymptotically like $\exp(y^2)$, which dominates $\exp(-y^2/2)$.

We know that the series expression for $\exp(y^2)$ is

$$\exp(y^2) = 1 + y^2 + \frac{y^4}{2!} + \frac{y^6}{3!} + \cdots + \frac{y^n}{(n/2)!} + \frac{y^{n+2}}{(n/2+1)!} + \cdots \tag{3-48}$$

The series for $f(y)$ has terms

$$\cdots + c_n y^n + c_{n+2} y^{n+2} + c_{n+4} y^{n+4} + \cdots \tag{3-49}$$

The ratio between coefficients for two adjacent terms high up in the series (large n) for $\exp(y^2)$ is, from Eq. (3-48)

$$\frac{\text{coeff for y}^{n+2}}{\text{coeff for y}^n} = \frac{(n/2)!}{[(n/2)+1]!} = \frac{1}{(n/2)+1} \xrightarrow{\text{large } n} \frac{2}{n} \tag{3-50}$$

and for $f(y)$ it is, from Eqs. (3-49) and (3-47),

$$\frac{\text{coeff for y}^{n+2}}{\text{coeff for y}^n} = \frac{c_{n+2}}{c_n} = \frac{-(\alpha/\beta)+1+2n}{n^2+3n+2} \xrightarrow{\text{large } n} \frac{2}{n} \tag{3-51}$$

This means that, at large values of y, when the higher-order terms in the series dominate, $f(y)$ behaves like $\exp(y^2)$. Then

$$\lim_{y\to\infty} \psi(y) = \lim_{y\to\infty} f(y)\exp(-y^2/2) = \exp(y^2/2) \longrightarrow \infty \tag{3-52}$$

The asymptotic behavior of ψ is ruined. We can overcome this problem by requiring the series for $f(y)$ to terminate at some finite power. In other words, $f(y)$ must be a polynomial. This condition is automatically fulfilled if any one of the coefficients in a given series (odd or even) is zero since Eq. (3-47) guarantees that all the higher coefficients in that series will then vanish. Therefore, we require that some coefficient vanish:

$$c_{n+2} = 0 \tag{3-53}$$

Assuming that this is the lowest zero coefficient (i.e., $c_n \neq 0$), Eq. (3-47) gives

$$(\alpha/\beta) - 2n - 1 = 0 \tag{3-54}$$

or

$$\alpha = \beta(2n+1) \tag{3-55}$$

3-4.G The Nature of the Energy Spectrum

Now we have a recipe for producing acceptable solutions for the Schrödinger equation for the harmonic oscillator. If we desire a symmetric solution, we set $c_1 = 0$ and $c_0 = 1$. If we want the polynomial to terminate at y^n, we require that α and β be related as in Eq. (3-55). In this way we can generate an unlimited number of symmetric solutions, one for each even value of n that can be chosen for the terminal value. Similarly, an unlimited set of antisymmetric solutions results from setting $c_0 = 0$ and $c_1 = 1$ and allowing the highest contributing value of n to take on various odd integer values.

(Solved Problem 3-2 provides an example by showing that c_5 becomes the first odd-power-coefficient equal to zero if $\alpha/\beta = 7$.)

Since we now know that an acceptable solution satisfies Eq. (3-55), we can examine the energy spectrum. Substituting into Eq. (3-55) the expressions for α and β [Eqs. (3-16) and (3-17)], we obtain

$$8\pi^2 mE/h^2 = (2\pi\sqrt{mk}/h)(2n+1) \tag{3-56}$$

or

$$E = h\left(n+\frac{1}{2}\right)\left(\frac{1}{2\pi}\right)\sqrt{k/m} = \left(n+\frac{1}{2}\right)h\nu \tag{3-57}$$

where the classical definition of ν [Eq. (3-7)] has been used.

This result shows that, whenever n increases by unity, the energy increases by $h\nu$, so the energy levels are evenly spaced as shown in Fig. 3-2. At absolute zero, the system will lose its energy to its surroundings insofar as possible. However, since $n = 0$ in the lowest permissible state for the system, there will remain a zero-point energy of $\frac{1}{2}h\nu$.

3-4.H Nature of the Wavefunctions

The lowest energy solution corresponds to $n = 0$. This means that c_0 is the highest nonzero coefficient in the power series expansion for $f(y)$. Hence, we must set $c_2 = 0$. (The odd-powered series coefficients are all zero for this case.) Thus, for $n = 0$, we can write the unnormalized wavefunction as [from Eq. (3-24)]

$$\psi_0 = c_0 \exp(-y^2/2) = c_0 \exp(-\beta x^2/2) \tag{3-58}$$

This is just a constant times a Gauss error function or simple "gaussian-type" function. This wavefunction is sketched in Fig. 3-2 and is obviously symmetric.

The next solution has $n = 1$, $c_1 \neq 0$ but $c_3 = c_5 = \cdots = 0$. (The even-powered series coefficients are all zero for this case.) The unnormalized wavefunction for $n = 1$ is then

$$\psi_1(y) = c_1 y \exp(-y^2/2) \tag{3-59}$$

The exponential is symmetric and y is antisymmetric, and so their product, $\psi_1(y)$, is antisymmetric (Fig. 3-2).

To get ψ_2 we need to use the recursion relation (3-47). We know that odd-index coefficients are all zero and that only c_0 and c_2 of the even-index coefficients are nonzero. Assuming $c_0 = 1$, we have (using (Eq. 3-47) with $m = 0$)

$$c_2 = \frac{-[(\alpha/\beta) - 2\cdot 0 - 1]}{(1)(2)}\cdot 1 = -\frac{(\alpha/\beta) - 1}{2} \tag{3-60}$$

But the ratio α/β is determined by the requirement that $c_4 = 0$. Referring to Eq. (3-55), this gives (for $n = 2$) $\alpha/\beta = 5$, and so

$$c_2 = -4/2 = -2 \tag{3-61}$$

and the unnormalized wavefunction is

$$\psi_2(y) = (1 - 2y^2)\exp(-y^2/2) \tag{3-62}$$

Polynomials like $(1-2y^2)$, which are solutions to the differential equation (3-33), are known as Hermite (her·*meet*) polynomials, $H_n(y)$. In addition to the recursion relation (3-47), which we have derived, other definitions are known. One of these involves successive differentiation:

$$H_n(y)=(-1)^n \exp(y^2)\frac{d^n \exp(-y^2)}{dy^n} \tag{3-63}$$

Thus, if we want $H_2(y)$, we just set $n=2$ in Eq. (3-63) and evaluate that expression to get

$$H_2(y)=4y^2-2 \tag{3-64}$$

which differs from our earlier result by a factor of -2. Yet another means of producing Hermite polynomials is by using the generating function

$$G(y,u)=\exp[y^2-(u-y)^2]\equiv\sum_{n=0}^{\infty}(H_n(y)/n!)u^n \tag{3-65}$$

We use this expression as follows:

1. Express the exponential in terms of its power series, writing down a few of the leading terms. There will exist, then, various powers of u and y and factorial coefficients.
2. Collect together all the terms containing u^2.
3. The coefficient for this term will be equal to $H_2(y)/2$.

This is a fairly clumsy procedure for producing polynomials, but Eq. (3-65) is useful in determining general mathematical properties of these polynomials. For instance, Eq. (3-65) will be used in showing that the harmonic oscillator wavefunctions are orthogonal.

3-4.I Orthogonality and Normalization

We will now show that the harmonic oscillator wavefunctions are orthogonal, i.e., that

$$\int_{-\infty}^{+\infty}\psi_n(y)\psi_m(y)\,dy=\int_{-\infty}^{+\infty}H_n(y)H_m(y)\exp(-y^2)\,dy=0\cdots \quad n\neq m \tag{3-66}$$

Consider the integral involving two generating functions and the exponential of y^2:

$$\int_{-\infty}^{+\infty}G(y,u)G(y,v)\exp(-y^2)dy$$
$$=\sum_n\sum_m u^n v^m \underbrace{\int_{-\infty}^{+\infty}\frac{H_n(y)H_m(y)}{n!m!}\exp(-y^2)\,dy}_{c_{nm}} \tag{3-67}$$

where we label the integral c_{nm} for convenience. The left-hand side of Eq. (3-67) may also be written as [using Eq. (3-65)]

$$\int_{-\infty}^{+\infty}\exp[-(y-u-v)^2]\exp(2uv)dy=\exp(2uv)\int_{-\infty}^{+\infty}\exp[-(y-u-v)^2]dy \tag{3-68}$$

However, we can add constants to the differential element without affecting the integral value, and u and v are constants when only y varies. Therefore, (3-68) becomes (see Appendix 1 for a table of useful integrals)

$$\exp(2uv)\int_{-\infty}^{+\infty}\exp[-(y-u-v)^2]d(y-u-v)=\exp(2uv)\sqrt{\pi}$$
$$=\sqrt{\pi}\{1+2uv+4u^2v^2/2!+8u^3v^3/3!+\cdots+2^nu^nv^n/n!+\cdots\} \quad (3\text{-}69)$$

This expression is equal to the right-hand side of Eq. (3-67). Comparing Eq. (3-67) with (3-69), we see that $c_{11}=2\sqrt{\pi}$ since the term u^1v^1 is multiplied by 2π in Eq. (3-69) and by c_{11} in Eq. (3-67). Similarly $c_{22}=4\sqrt{\pi}/2!$ But $c_{12}=0$. Hence, we arrive at the result

$$c_{nm}=\int_{-\infty}^{+\infty}\frac{H_n(y)\,H_m(y)}{n!m!}\exp\left(-y^2\right)dy=\sqrt{\pi}\left(2^n/n!\right)\delta_{n,m} \quad (3\text{-}70)$$

($\delta_{n,m}$ is the "Kronecker" delta. It is a discontinuous function having a value of unity when $n=m$ but zero when $n\neq m$.) So

$$\int_{-\infty}^{+\infty}\psi_n(y)\,\psi_m(y)\,dy=\sqrt{\pi}m!2^n\delta_{n,m} \quad (3\text{-}71)$$

This proves the wavefunctions to be orthogonal and also provides us with a normalizing factor. Normality refers to integration in x, rather than in $y=\sqrt{\beta}x$, so we must change the differential element in Eq. (3-71):

$$\int_{-\infty}^{+\infty}\psi_n^2(y)\,dy=\sqrt{\beta}\int_{-\infty}^{+\infty}\psi_n^2(y)\,dx=\sqrt{\pi}n!2^n \quad (3\text{-}72)$$

Requiring that $\int_{-\infty}^{+\infty}\psi_n^2(y)\,dx=1$ leads to the expression for the normalized wavefunctions:

$$\psi_n(y)=\left(\sqrt{\frac{\beta}{\pi}}\frac{1}{2^nn!}\right)^{1/2}H_n(y)\exp\left(-y^2/2\right),\qquad n=0,1,2,\ldots \quad (3\text{-}73)$$

The first members of the set of Hermite polynomials are

$$H_0(y)=1,\quad H_1(y)=2y,\quad H_2(y)=4y^2-2,\quad H_3(y)=8y^3-12y$$
$$H_4(y)=16y^4-48y^2+12,\quad H_5(y)=32y^5-160y^3+120y \quad (3\text{-}74)$$

3-4.J Summary of Solution of Harmonic-Oscillator Schrödinger Equation

The detailed solution is so long that the reader may have lost the broad outline.

The basic steps were:

1. Determine the asymptotic behavior of the Schrödinger equation and of ψ. This produces a gaussian factor $\exp(-y^2/2)$ times a function of y, $f(y)$.
2. Obtain a differential equation for the rest of the wavefunction, $f(y)$.
3. Represent $f(y)$ as a power series in y, and find a recursion relation for the coefficients in the series. The symmetries of the wavefunctions are linked to the symmetries of the series.
4. Force the power series to be finite (i.e., polynomials) so as not to spoil the asymptotic behavior of the wavefunctions. This leads to a relation between α and β that produces uniformly spaced, quantized energy levels.
5. Recognize the polynomials as being Hermite polynomials, and utilize some of the known properties of these functions to establish orthogonality and normalization constants for the wavefunctions.

EXAMPLE 3-3 Which of the following expressions are, by inspection, unacceptable eigenfunctions for the Schrödinger equation for the one-dimensional harmonic oscillator?

a) $(64y^6 - 480y^4 + 720y^2 - 120)\exp(y^2/2)$

b) $(64y^6 - 480y^5 + 720y^3 - 120)\exp(-y^2/2)$

c) $(64y^6 - 480y^4 + 720y^2 - 120)\exp(-y^2/2)$

SOLUTION ▶ a) is unacceptable because $\exp(y^2/2)$ blows up at large $|y|$.

b) is unacceptable because the polynomial contains terms of both even and odd powers.

c) is acceptable. ◀

3-5 Quantum-Mechanical Average Value of the Potential Energy

We showed in Section 3-2 that the classical harmonic oscillator stores, on the average, half of its energy as kinetic energy, and half as potential. We now make the analogous comparison in the quantum-mechanical system for the ground ($n = 0$) state.

The wavefunction is

$$\psi_0(x) = (\beta/\pi)^{1/4}\exp(-\beta x^2/2) \tag{3-75}$$

and the probability distribution of the particle along the x coordinate is given by $\psi_0^2(x)$. The total energy is constant and equal to

$$E_0 = \frac{1}{2}h\nu = (h/4\pi)\sqrt{k/m} \tag{3-76}$$

and the potential energy as a function of x is

$$V(x) = \frac{1}{2}kx^2 \tag{3-77}$$

The probability for finding the oscillating particle in the line element dx around some point x_1 is $\psi_0^2(x_1)\,dx$, since ψ_0 of Eq. (3-75) is normalized. Hence, the average value for the potential energy is simply the sum of all the potential energies due to all the elements dx, each weighted by the probability for finding the particle there:

$$\bar{V} = \int_{-\infty}^{\infty} [\text{prob. to be in } dx][V \text{ at } dx]\,dx = \int_{-\infty}^{+\infty} \psi_0^2(x)V(x)\,dx \tag{3-78}$$

This is

$$\bar{V} = (\beta/\pi)^{1/2} \cdot \frac{1}{2}k \int_{-\infty}^{+\infty} \exp\left(-\beta x^2\right) x^2\,dx = \sqrt{\beta/\pi} \cdot \frac{1}{2}k \cdot \frac{1}{2}\sqrt{\pi/\beta^3} \tag{3-79}$$

where we have referred to Appendix 1 to evaluate the integral. Using the definition of β^2 (Eq. 3-17) we have

$$\bar{V} = k/4\beta = (k/4) \cdot h/(2\pi\sqrt{mk}) = (h/8\pi)\sqrt{k/m} \tag{3-80}$$

which is just one half of the total energy [Eq. (3-76)]. This means that the average value of the kinetic energy must also equal half of the total energy, since $\bar{V} + \bar{T} = E$. We thus arrive at the important result that the ratio of average potential and kinetic energies is the same in the classical harmonic oscillator and the ground state of the quantum-mechanical system. This result is also true for the higher states. For other kinds of potential, the storage need not be half and half, but whatever it is, it will be the same for the classical and quantum-mechanical treatments of the system. We discuss this point in more detail later when we examine the virial theorem (Chapter 11 and Appendix 8).

3-6 Vibrations of Diatomic Molecules

Two atoms bonded together vibrate back and forth along the internuclear axis. The standard first approximation is to treat the system as two nuclear masses, m_1 and m_2 oscillating harmonically with respect to the center of mass. The force constant k is determined by the "tightness" of the bond, with stronger bonds usually having larger k.

The two-mass problem can be transformed to motion of one *reduced* mass, μ, vibrating harmonically with respect to the center of mass. μ is equal to $m_1m_2/(m_1+m_2)$. The force constant for the vibration of the reduced mass remains identical to the force constant for the two masses, and the distance of the reduced mass from the center of mass remains identical to the distance between m_1 and m_2. Thus we have a very convenient simplification: We can use the harmonic oscillator solutions for a single oscillating mass μ as solutions for the two-mass problem. All of the wavefunctions and energy formulas are just what we have already seen except that m is replaced by μ. The practical consequence of this is that we can use the spectroscopically measured energy spacings between molecular vibrational levels to obtain the value of k for a molecule.

EXAMPLE 3-4 There is a strong absorption in the infrared spectrum of $H^{35}Cl$ at $2992\,cm^{-1}$, which corresponds to an energy of 5.941×10^{-20} J. This light energy, E, is absorbed in order to excite HCl from the $n=0$ to the $n=1$ vibrational state. What is the value of k, the force constant, in HCl?

SOLUTION ▶ The vibrational spacing $h\nu$ must be equal to 5.941×10^{-20} J. We know that, since μ replaces m, $\nu = (1/2\pi)\sqrt{k/\mu}$, which means that $k = 4\pi^2 E^2 \mu / h^2$. The formula for μ is $m_H m_{Cl}/(m_H + m_{Cl}) = 1.614 \times 10^{-27}$ kg. It follows that k is equal to $4\pi^2(5.941 \times 10^{-20}\,J)^2$ $(1.614 \times 10^{-27}\,kg)/(6.626 \times 10^{-34}\,J\,s)^2 = 512\,N\,m^{-1}$. ◀

3-7 Summary

In this chapter we have discussed the following points:

1. The energies for the quantum-mechanical harmonic oscillator are given by the formula $E_n = (n + 1/2)h\nu$, $n = 0, 1, 2, \ldots$, where $\nu = (1/2\pi)\sqrt{k/m}$. This gives nondegenerate energy levels separated by equal intervals ($h\nu$) and a zero-point energy of $h\nu/2$.

2. The wavefunctions for this system are symmetric or antisymmetric for reflection through $x = 0$. This symmetry alternates as n increases and is related to the presence of even or odd powers of y in the Hermite polynomial in ψ.

3. Each wavefunction is orthogonal to all of the others, even in cases where the symmetries are the same.

4. The harmonic oscillator wavefunctions differ from particle-in-a-box wavefunctions in two important ways: They penetrate past the classical turning points (i.e., past the values of x where $E = V$), and they have larger distances available to them as a result of the opening out of the parabolic potential function at higher energies. This gives them more room in which to accomplish their increasing number of wiggles as n increases, and so the energy does not rise as quickly as it otherwise would.

5. The manner in which the total energy is partitioned into average potential and kinetic parts is the same for classical and quantum-mechanical harmonic oscillators, namely, half and half.

6. Vibrations in molecules are usually approximately harmonic. Mass is replaced by reduced mass in the energy formula. Measuring the energy needed to excite a molecular vibration allows one to calculate the harmonic force constant for that particular vibrational mode.

3-7.A Problems

3-1. From the equation of motion (3-5) show that the classical distribution function is proportional to $(1 - x^2/L^2)^{-1/2}$.

3-2. A classical harmonic oscillator with mass of 1.00 kg and operating with a force constant of $2.00 \text{ kg s}^{-2} = 2.00 \text{ J m}^{-2}$ is released from rest at $t = 0$ and $x = 0.100$ m.

a) What is the function $x(t)$ describing the trajectory of the oscillator?
b) Where is the oscillating mass when $t = 3$ seconds?
c) What is the total energy of the oscillator?
d) What is the potential energy when $t = 3$ seconds?
e) What is the *time-averaged* potential energy?
f) What is the time-averaged kinetic energy?
g) How fast is the oscillator moving when $t = 3$ seconds?
h) Where are the turning points for the oscillator?
i) What is the frequency of the oscillator?

3-3. Find the expression for the classical turning points for a one-dimensional harmonic oscillator in terms of n, m, h, and k.

3-4. a) Equation (3-73) for $\psi_n(y)$ is a rather formidable expression. It can be broken down into three portions, each with a certain purpose. Identify the three parts and state the role that each plays in meeting the mathematical requirements on ψ.
b) Produce expressions for the normalized harmonic oscillator with $n = 0, 1, 2$.

3-5. Operate explicitly on ψ_0 with H and show that ψ_0 is an eigenfunction having eigenvalue $h\nu/2$.

3-6. a) At what values of y does ψ_2 have a node?
b) At what values of y does ψ_2^2 have its maximum value?

3-7. Give a simple reason why $(2 + y - 3y^2)\exp(-y^2/2)$ cannot be a satisfactory wavefunction for the harmonic oscillator. What about $2y\exp(+y^2/2)$? You should be able to answer by inspection, without calculation and without reference to tabulations.

3-8. Consider the function $(32x^5 - 160x^3 + 120x)\exp(-x^2/2)$.

a) How does this function behave at large values of $\pm x$? Explain your answer.
b) What can you say about the symmetry of this function?
c) What are the *value* and *slope* of this function at $x = 0$?

3-9. Let $f(x) = 3\cos x + 4$. $f(x)$ is expressed as a power series in x: $f(x) = \sum c_n x^n$, with $n = 0, 1, 2, \ldots$.

a) What is the value of c_0?
b) What is the value of c_1?

3-10. Only one of the following is $H_5(y)$, a Hermite polynomial. Which ones are not, and why?

a) $16y^5 + 130y$
b) $24y^5 - 110y^3 + 90y - 18$
c) $32y^5 - 160y^3 + 120y$

3-11. *Sketch* the function $(1-2x^2)\exp(-x^2)$ versus x.

3-12. Given that $\int_0^\infty x\exp(-x^2)\,dx = 1/2$, and $\int_0^\infty x^2\exp(-x^2)\,dx = \sqrt{\pi}/4$, evaluate

a) $\int_{-\infty}^{\infty} x\exp(-x^2)\,dx$
b) $\int_{-\infty}^{\infty} x^2\exp(-x^2)\,dx$

3-13. Evaluate $\int_{-\infty}^{\infty}(x+4x^3)\exp(-5x^2)\,dx$.

3-14. For the $n=1$ state of the harmonic oscillator:

a) Calculate the values of the classical turning points.
b) Calculate the values of the positions of maximum probability density.
c) What is the probability for finding the oscillator between $x=0$ and $x=\infty$?
d) Estimate the probability for finding the oscillator in a line increment Δx equal to 1% of the distance between classical turning points and centered on one of the positions of maximum probability density.

3-15. Calculate the probability for finding the ground state harmonic oscillator *beyond* its classical turning points.

3-16. Use the differential expression (3-63) for Hermite polynomials to produce $H_2(y)$.

3-17. Use the generating function (3-65) to produce $H_2(y)$.

3-18. Write down the Schrödinger equation for a three-dimensional (isotropic) harmonic oscillator. Separate variables. What will be the zero-point energy for this system? What will be the degeneracy of the energy level having a value of $(9/2)h\nu$? $(5/2)h\nu$? Sketch (roughly) each of the solutions for the latter case and note their similarity to case (b) in Fig. 2-16.

3-19. Suppose $V(x) = (1/2)kx^2$ for $x \geq 0$, and ∞ for $x < 0$. What can you say about the eigenfunctions and eigenvalues for this system?

3-20. a) Evaluate $H_3(x)$ at $x=2$.
b) Evaluate $H_2(\sin\theta)$ at $\theta = 30°$.

3-21. What is the average potential energy for a harmonic oscillator when $n=5$? What is the average kinetic energy?

3-22. Each degree of translational or rotational freedom can contribute up to $\frac{1}{2}R$ to the molar heat capacity of an ideal diatomic gas, whereas the vibrational degree of freedom can contribute up to R. Explain.

3-23. Calculate the force constants for vibration in $H^{19}F$, $H^{35}Cl$, $H^{81}Br$, and $H^{127}I$, given that the infrared absorptions for the $n=0$ to $n=1$ transitions are seen, respectively, at 4138, 2991, 2649, and 2308 cm^{-1}. What do these force constants imply about the relative bond strengths in these molecules?

Multiple Choice Questions

(Try to answer these without referring to the text.)

1. Which one of the following statements conflicts with the quantum mechanical results for a one-dimensional harmonic oscillator?

a) The smaller the mass of the oscillating particle, the greater will be its zero-point energy, for a fixed force constant.
b) The frequency is the same as that of a classical oscillator with the same mass and force constant.
c) Increasing the force constant increases the spacing between adjacent energy levels.
d) The spacing between adjacent energy levels is unaffected as the vibrational quantum number increases.
e) The vibrational potential energy is a constant of motion.

2. A quantum-mechanical harmonic oscillator

a) spends most of its time near its classical turning points in its lowest-energy state.
b) has $\psi = 0$ at its classical turning points.
c) has doubly degenerate energy levels.
d) has energy levels that increase with the square of the quantum number.
e) None of the above is a correct statement.

3. Light of wavelength 4.33×10^{-6} m excites a quantum-mechanical harmonic oscillator from its ground to its first excited state. Which one of the following wavelengths would accomplish this same transition if i) the force constant only was doubled? ii) the mass only was doubled?

a) 4.33×10^{-6} m
b) 2.16×10^{-6} m
c) 3.06×10^{-6} m
d) 6.12×10^{-6} m
e) 8.66×10^{-6} m

4. For a classical harmonic oscillator, the probability for finding the oscillator in the middle 2% of the oscillation range is

a) greater than 0.02.
b) equal to 0.02.
c) less than 0.02.
d) unknown since it depends on the force constant.
e) unknown since it depends on the amplitude.

Chapter 4

The Hydrogenlike Ion, Angular Momentum, and the Rigid Rotor

4-1 The Schrödinger Equation and the Nature of Its Solutions

4-1.A The Schrödinger Equation

Consider the two-particle system composed of an electron (charge $-e$) and a nucleus having atomic number Z and charge Ze. (See Appendix 12 for values of physical constants, such as e.) Let x_1, y_1, z_1 be the coordinates of the nucleus and x_2, y_2, z_2 be those for the electron. The distance between the particles is, then, $[(x_1 - x_2)^2 + (y_1 - y_2)^2 + (z_1 - z_2)^2]^{1/2}$. The potential energy is given by the product of the charges divided by the distance between them. If e is expressed in coulombs, C, the potential energy in joules is

$$V = \frac{-Ze^2}{4\pi\,\varepsilon_0\left[(x_1 - x_2)^2 + (y_1 - y_2)^2 + (z_1 - z_2)^2\right]^{1/2}} \tag{4-1}$$

where ε_0 is the vacuum permittivity (8.8542×10^{-12} $\mathrm{J^{-1}\ C^2\ m^{-1}}$). The time-independent Schrödinger equation for this system is

$$\begin{aligned}&\left[\frac{-h^2}{8\pi^2 M}\left(\frac{\partial^2}{\partial x_1^2} + \frac{\partial^2}{\partial y_1^2} + \frac{\partial^2}{\partial z_1^2}\right) - \frac{h^2}{8\pi^2 m_\mathrm{e}}\left(\frac{\partial^2}{\partial x_2^2} + \frac{\partial^2}{\partial y_2^2} + \frac{\partial^2}{\partial z_2^2}\right)\right.\\ &\left. - \frac{Ze^2}{4\pi\varepsilon_0\left[(x_1 - x_2)^2 + (y_1 - y_2)^2 + (z_1 - z_2)^2\right]^{1/2}}\right]\psi(x_1, y_1, z_1, x_2, y_2, z_2)\\ &= E\psi(x_1, y_1, z_1, x_2, y_2, z_2)\end{aligned} \tag{4-2}$$

where M and m_e are the masses of the nucleus and electron, respectively. The hamiltonian operator in brackets in Eq. (4-2) has three terms, corresponding to a kinetic energy operator for the nucleus, a kinetic energy operator for the electron, and a potential term for the pair of particles.

Equation (4-2) has eigenfunctions ψ that are dependent on the positions of both the electron and the nucleus. It is possible to convert to center-of-mass coordinates and then to separate Eq. (4-2) into two equations, one for the motion of the center of mass and *one for a particle of reduced mass moving around a fixed center to which it is attracted in exactly the same way the electron is attracted to the nucleus*. Because this conversion is rather tedious, we will not perform it in this book,[1] but merely discuss the results. The first of the two resulting equations treats the center of mass as a free particle moving through field-free space; its eigenvalues are simply *translational* energies of the ion. For us, the interesting equation is the second one, which is

$$\left\{\frac{-h^2}{8\pi^2\mu}\left(\frac{\partial^2}{\partial x^2}+\frac{\partial^2}{\partial y^2}+\frac{\partial^2}{\partial z^2}\right)-\frac{Ze^2}{4\pi\varepsilon_0\left(x^2+y^2+z^2\right)^{1/2}}\right\}\psi(x,y,z)$$
$$=E\psi(x,y,z) \tag{4-3}$$

The quantity μ is the *reduced mass* for the particle in our center-of-mass system, and is given by

$$\mu=m_\mathrm{e}M/(m_\mathrm{e}+M) \tag{4-4}$$

The coordinates x, y, and z are the coordinates of the reduced-mass particle *with respect to the center of mass of the system.*

Even without going through the detailed procedure of converting to center-of-mass coordinates, we can show that Eq. (4-3) makes sense. In the idealized case in which M is infinitely greater than m_e, μ equals m_e, and Eq. (4-3) becomes just the Schrödinger equation for the motion of an electron about a fixed nucleus at the coordinate origin. For real atoms or ions this would not be a bad approximation because, even in the case of the lightest nucleus (i.e., the hydrogen atom), M is nearly 2000 times m_e, and so μ is very close to m_e, and the center of mass is very near the nucleus. Therefore, the result of using center-of-mass coordinates to separate the Schrödinger equation is almost identical to making the assumption at the outset that the nucleus is fixed, and simply writing down the one-particle Schrödinger equation:

$$\left\{\frac{-h^2}{8\pi^2m_\mathrm{e}}\left(\frac{\partial^2}{\partial x^2}+\frac{\partial^2}{\partial y^2}+\frac{\partial^2}{\partial z^2}\right)-\frac{Ze^2}{4\pi\varepsilon_0\left(x^2+y^2+z^2\right)^{1/2}}\right\}\psi(x,y,z)=E\psi(x,y,z) \tag{4-5}$$

The use of m_e instead of μ [i.e., Eq. (4-5) instead of (4-3)] has no effect on the qualitative nature of the solutions. However, it does produce small errors in eigenvalues—errors that are significant in the very precise measurements and calculations of atomic spectroscopy (Problem 4-1). In what follows we shall use μ, but for purposes of discussion we will pretend that the nucleus and center of mass coincide.

Equation (4-3) can be transformed into spherical polar coordinates. (Some important relationships between spherical polar and Cartesian coordinates are given in Fig. 4-1). The result is

$$[(-h^2/8\pi^2\mu)\nabla^2-(Ze^2/4\pi\varepsilon_0 r)]\psi(r,\theta,\phi)=E\psi(r,\theta,\phi) \tag{4-6}$$

Here ∇^2 is understood to be in spherical polar coordinates. In these coordinates it looks quite complicated:[2]

$$\nabla^2=\frac{1}{r^2}\frac{\partial}{\partial r}\left(r^2\frac{\partial}{\partial r}\right)+\frac{1}{r^2\sin\theta}\frac{\partial}{\partial\theta}\left(\sin\theta\frac{\partial}{\partial\theta}\right)+\frac{1}{r^2\sin^2\theta}\frac{\partial^2}{\partial\phi^2} \tag{4-7}$$

[1] See, for example, Eyring *et al.* [1, Chapter VI] or Levine [2, pp. 127–130].

[2] See, e.g., Eyring et al. [1, Appendix III].

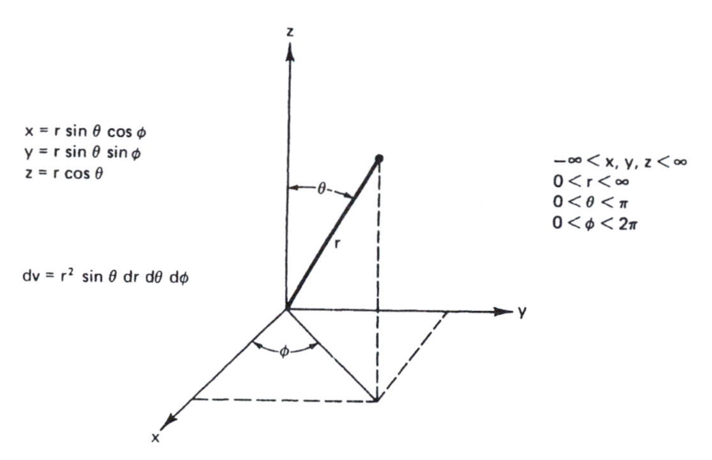

Figure 4-1 ▶ The spherical polar coordinate system. The angle ϕ is called the azimuthal angle. Notice that the differential volume element is *not* equal to $dr\,d\theta\,d\phi$ and that the ranges of values for r, θ, ϕ are *not* $-\infty$ to $+\infty$.

However, this coordinate system is the natural one for this system and leads to the easiest solution despite this rather formidable looking operator.

Notice that the potential term, $-Ze^2/4\pi\varepsilon_0 r$, has no θ or ϕ dependence. The potential is *spherically symmetric*. However, θ and ϕ dependence does enter the hamiltonian through ∇^2, so the eigenfunctions ψ may be expected to show angular dependence.

Next we describe the solutions of the Schrödinger equation (4-6), relegating to later sections the mathematical details of how the solutions are obtained.

EXAMPLE 4-1 Using the spherical polar coordinate system of Fig. 4-1, calculate the volume occupied by the skin of a spherical shell, where the inside radius of the skin is 100.0 mm and the thickness of the skin is 1.000 mm.

SOLUTION ▶ One way to solve this problem is to calculate the volume inside the entire sphere, including the skin, and then to subtract the volume of the sphere occupying the space inside the skin. The formula for the volume of a sphere of radius r can be calculated from dv by integrating r from 0 to r, θ from 0 to π, and ϕ from 0 to 2π:

$$V = \int_0^r r^2 dr \int_0^\pi \sin\theta d\theta \int_0^{2\pi} d\phi = \frac{r^3}{3}\Big|_0^r \cdot -\cos\theta|_0^\pi \cdot \phi|_0^{2\pi}$$
$$= \frac{r^3}{3}(-(-1-1)) \cdot 2\pi = \frac{4}{3}\pi r^3$$

(You presumably already knew this formula, but it is useful to review how it comes out of this integration.) Proceeding,

$$V_{skin} = V(r = 101 \text{ mm}) - V(r = 100 \text{ mm})$$
$$= \frac{4}{3}\pi[(101 \text{ mm})^3 - (100 \text{ mm})^3] = 1.269 \times 10^5 \text{mm}^3$$

Another way (slightly less exact) to approximate this volume is to calculate the area of the spherical shell ($4\pi r^2$) and multiply by its thickness:

$$V \sim 4\pi r^2 \Delta r = 4\pi(100 \text{ mm})^2 \cdot 1.00 \text{ mm}$$
$$= 1.257 \times 10^5 \text{ mm}^3$$

For a skin whose thickness is small compared to its radius, we see that this approximation is very good. ◀

4-1.B The Nature of the Eigenvalues

The potential energy, $-Ze^2/4\pi\varepsilon_0 r$, becomes negatively infinite when $r = 0$ and approaches zero as r becomes very large. This potential is sketched in Fig. 4-2 for the case in which $Z = 1$. We expect the energy levels to diverge less rapidly here than was the case for the harmonic oscillator since the "effective box size" increases more rapidly with increasing energy in this case than in the harmonic oscillator case. (See Fig. 2-3 for the one-dimensional analogs.) Since the harmonic oscillator levels are separated by a *constant* ($h\nu$, for one- *or* three-dimensional cases), the hydrogenlike ion levels should *converge* at higher energies. Figure 4-2 shows that this is indeed the case. Furthermore, by analogy with the case of the particle in a box with one finite wall, we expect the allowed energies to form a discrete set for the classically trapped electron ($E < 0$) and a continuum for the unbound cases ($E > 0$). Thus, the spectrum of eigenvalues sketched in Fig. 4-2 is in qualitative accord with understandings developed earlier.

The lowest allowed energy for the system is far above the low-energy limit ($-\infty$) of the potential well. This corresponds to the finite zero point energy which we have seen

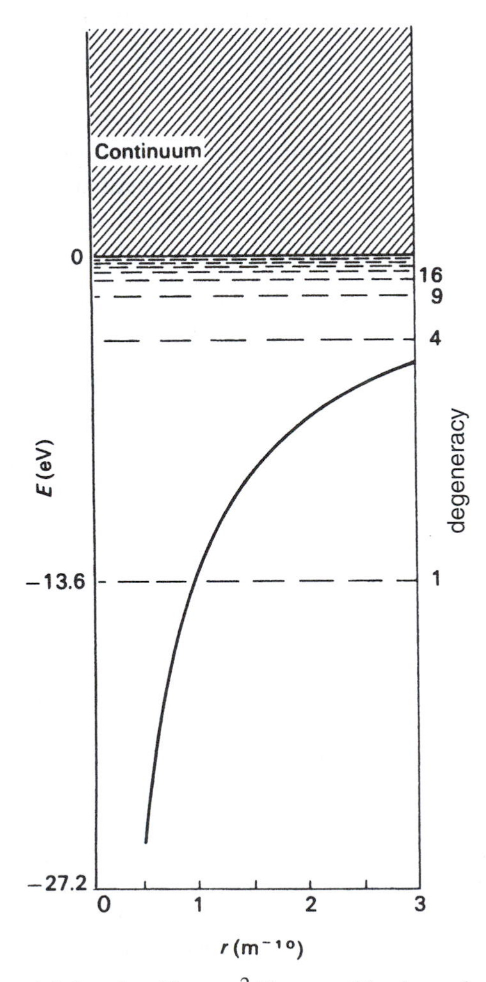

Figure 4-2 ▸ The potential function $V = -e^2/4\pi\varepsilon_0 r$ with eigenvalues superimposed (dashed lines). Degeneracies for the first few levels are noted on the right.

in other systems where particle motion is constrained. Here it means that, at absolute zero, the electron does not come to rest on the nucleus (which would give $T=0$, $V=-\infty$, $E=-\infty$), but rather continues to move about with a finite total energy.

All of the energy levels of Fig. 4-2 are degenerate except for the lowest one. The order of the degeneracy is listed next to each of the lowest few levels in Fig. 4-2. This degeneracy is not surprising since we are dealing here with a three- dimensional system, and we have earlier seen that, in such cases (e.g., the cubic box), the physical equivalence of different directions in space can produce degeneracies (called "spatial degeneracies"). We shall see later that some of the degeneracies in this system do indeed result from directional equivalence (here, spherical symmetry), whereas others do not.

The discrete, negative eigenvalues are given by the formula

$$E_n = \frac{-\mu Z^2 e^4}{8\varepsilon_0^2 h^2 n^2} = (-13.6058\,\text{eV})\frac{Z^2}{n^2}, \quad n = 1, 2, 3, \ldots \tag{4-8}$$

EXAMPLE 4-2 Calculate the ionization energy (IE) of C^{5+} in its ground state, in electron volts.

SOLUTION ▶ The ionization energy equals the negative of the ground state energy. $Z=6$ and $n=1$, so $IE = (13.6058\,\text{eV})\frac{36}{1} = 489.808\,\text{eV}$. ◀

4-1.C The Lowest-Energy Wavefunction

We will now discuss the lowest-energy eigenfunction of Eq. (4-6) in some detail, since an understanding of atomic wavefunctions is crucial in quantum chemistry. The derivation of formulas for this and other wavefunctions will be discussed in later sections, but it is not necessary to labor through the mathematical details of the exact solution of Eq. (4-6) to be able to understand most of the essential features of the eigenfunctions.

The formula for the normalized, lowest-energy solution of Eq. (4-6) is

$$\psi(r) = (1/\sqrt{\pi})(Z/a_0)^{3/2}\exp(-Zr/a_0) \tag{4-9}$$

where $a_0 = 5.2917706 \times 10^{-11}$ m (called the *Bohr radius*) and Ze is the nuclear charge. A sketch of ψ versus r for $Z=1$ is superimposed on the potential function in Fig. 4-3a. It is apparent that the electron penetrates the potential barrier (Problem 4-3).

The square of the wavefunction (4-9) tells us how the electron is distributed about the nucleus. In Fig. 4-3b is plotted $\psi^2(r)$ as a function of r. We refer to ψ^2 as the *electron probability density function*. In this case, the probability density is greatest at the nucleus ($r=0$) and decays to zero as r becomes infinite.

It is important for the chemist to be able to visualize the electron distributions, or charge "clouds," in atoms and molecules, and various methods of depicting electron distributions have been devised. In Fig. 4-3 a few of these are presented for the lowest-energy wavefunction. The dot picture (Fig. 4-3c) represents what one would expect if one took a multiflash photograph of a magnified, slowed-down hydrogenlike ion (assuming no disturbance of the ion by the photographing process). Each dot represents an instantaneous electron position, and the density of these dots is greatest at the nucleus.

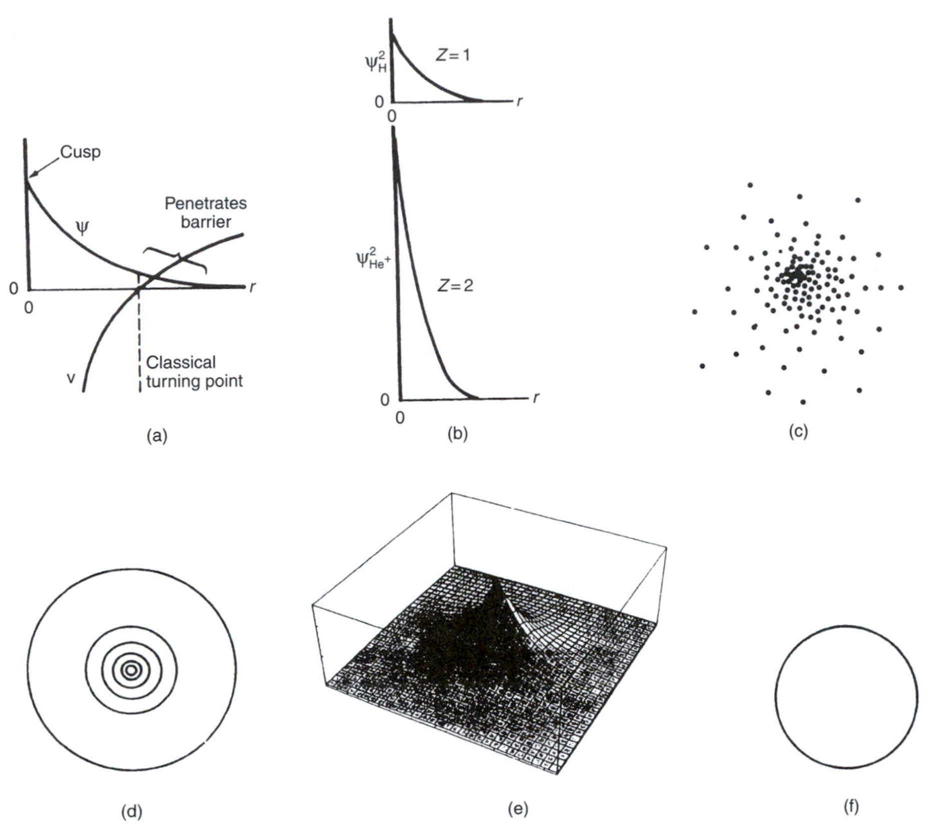

Figure 4-3 ▶ (a) H-atom wavefunction superposed on $-e^2/r$ potential curve. (b) Wavefunction squared for H and He^+. (c) Dot picture of electron distribution. (d) Contour diagram of electron distribution. (e) Computer-generated graphic version of (d). (f) Single-contour representation of electron distribution.

An alternative way of picturing the charge is to draw a contour diagram, each contour indicating that the density has increased or decreased by a certain amount (Fig. 4-3d). A more striking version of the contour plot for ψ_{1s}^2 is shown in Fig. 4-3e. Perhaps the simplest representation is a sketch of the single contour that encloses a certain amount (say 90%) of the electronic charge (Fig. 4-3f). (We have been describing the electron as a point charge moving rapidly about the nucleus. However, for most purposes it is just as convenient to picture the electron as being smeared out into a charge cloud like those sketched in Fig. 4-3. Thus, the statements *the electron spends 90% of its time inside this surface*, and *90% of the electronic charge is contained within this surface*, are equivalent.)

The multiflash "photograph" sketched in Fig. 4-3c shows the electron probability density to be greatest at the nucleus. Suppose that we were just to take a single flash photograph. Then the electron would appear as a single dot. At what distance from the nucleus would this dot be most likely to occur? Surprisingly, the answer is not zero. Despite the fact that the *probability density* is a maximum at $r = 0$, the *probability* for finding the electron in a volume element at the nucleus approaches zero. This is because the probability density, $\psi^2(r)$, is the measure of the probability *per unit volume* for the electron being at various distances from the nucleus. When we compare a tiny volume

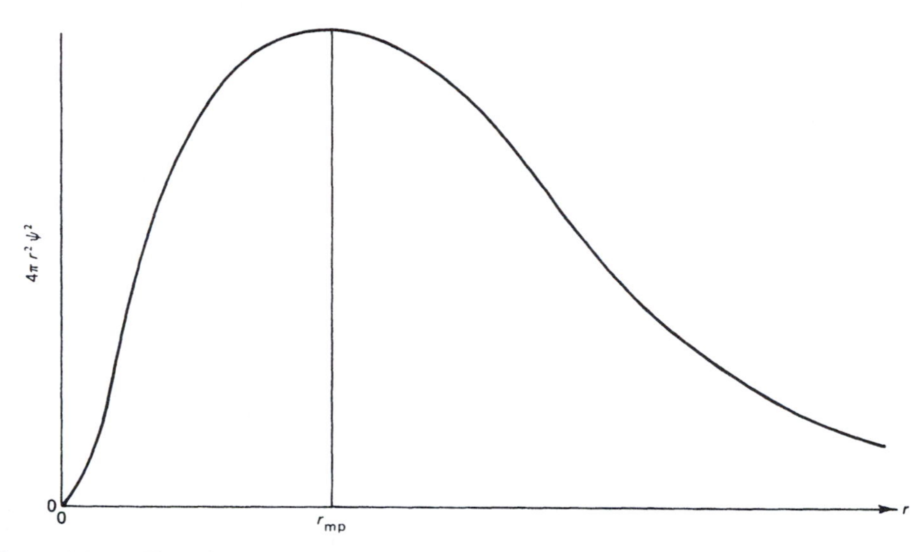

Figure 4-4 ▶ The volume-weighted probability density for the lowest-energy eigenfunction of the hydrogenlike ion. The most probable value of r occurs at r_{mp}.

element near the nucleus with an identical one farther out, we see from Fig. 4-3 that there is indeed more likelihood for the electron being in the volume element nearer the nucleus. *But there are more volume elements associated with the larger distance*. (The number of identical volume elements varies as the area of the surface of the sphere, $4\pi r^2$.) Hence, the probability for the electron being in a radial element dr at a given *distance* r from the nucleus is given by the number of volume elements at r times the probability density per unit volume element. The reason for the near-zero probability for finding the electron in a volume element at the nucleus is that the number of volume elements associated with $r = 0$ is vanishingly small compared to the number associated with larger r values. Figure 4-4 is a graph of $4\pi r^2 \psi^2$, the volume-weighted probability density. It is apparent that the most probable value of r, r_{mp}, occurs at a nonzero distance from the nucleus. [The reader is familiar with analogous distinctions. Rhode Island has a higher *population density* than Texas, but the population of Texas (density times area) is greater. Again, matter in the universe has a much higher *mass density* in stars and planets than in intergalactic gas or dust, but the total *mass* of the latter far exceeds that of the former due to the much greater volume of "empty" space.]

EXAMPLE 4-3 Estimate, for the hydrogen atom in the 1s state, the amount of electronic charge located in a spherical shell that is 1.00 pm thick and which has a radius of 60.0 pm.

SOLUTION ▶ Recall from Example 4-1 that, for relatively thin shells like this, we can estimate the volume of the shell by taking its area times its thickness. Also, we expect the charge density, ψ^2, to change very little over the 1.00 pm thickness of the shell, so we can take its value at 60.0 pm as constant. Then charge density $= \psi^2(r = 60.0\,\text{pm}) = (\pi a_0^3)^{-1} \exp(-2(60.0\,\text{pm})/a_0)$. Recalling that $a_0 = 52.9\,\text{pm}$, this gives $0.0329 a_0^{-3}$. This gives density per cubic bohr. To proceed, we can either convert this to density per cubic picometer by dividing by $(52.9\,\text{pm/bohr})^3$ or by converting r to bohr, by dividing by 52.9 pm/bohr. We choose the latter. Then $r = 60.0\,\mathrm{pm}/52.9\,\mathrm{pm/bohr} = 1.13a_0$, and $\Delta r = 1.00\,\mathrm{pm}/52.9\,\mathrm{pm/bohr} = 0.0189a_0$, so volume of shell $= 4\pi r^2 \Delta r = 4\pi(1.13a_0)^2(0.0189a_0) = 0.3028a_0^3$. Volume times density $= (0.3028a_0^3)(0.0329a_0^{-3}) = 0.010$. So 1% of the electronic charge resides in this shell. ◄

We can calculate the value of r_{mp} by finding which r value gives the maximum value of $4\pi r^2 \psi^2$. Recall that we can do this by finding the value of r that causes the first derivative of $4\pi r^2 \psi^2$ to vanish, that is, we require

$$(d/dr)[4\pi r^2 \pi^{-1}(Z/a_0)^3 \exp(-2Zr/a_0)] = 0 \tag{4-10}$$

This gives

$$\text{constants} \cdot [2r - (2Zr^2/a_0)] \exp(-2Zr/a_0) = 0 \tag{4-11}$$

The term in brackets vanishes when $r = a_0/Z$, so this is the value of r_{mp}. For $Z = 1$, $r_{mp} = a_0$; a_0 is the most probable distance of the electron from the nucleus in the hydrogen atom. For the He^+ ion ($Z = 2$), the most probable distance is only half as great, consistent with a more contracted charge cloud.

Of more interest, often, is the *average value* of the distance of the electron from the nucleus. If we could sample the instantaneous distance of the electron from the nucleus a large number of times and calculate the average value, what sort of result would we obtain? The probability for finding the electron at any given distance r is given by the volume-weighted probability density of Fig. 4-4. Inspection of that figure suggests that the average value of the position of the electron $\bar{r}$ is greater than r_{mp}, the most probable value. But exactly how much bigger is $\bar{r}$ than r_{mp}? How should we compute the average value? The reader is familiar with the way an average test score is calculated from a collection of scores. For example, suppose the series of scores to be averaged is 0, 2, 6, 6, 7, 7, 7, 10, and that 0 and 10 are the minimum and maximum possible scores. The average is given by

$$\begin{aligned} \text{average} &= \frac{\text{sum of scores}}{\text{number of scores}} \\ &= \frac{0+2+6+6+7+7+7+10}{8} = \frac{45}{8} \end{aligned} \tag{4-12}$$

Another way to write this is

$$\begin{aligned} \text{average} &= \frac{\text{frequency of score} \times \text{score}}{\text{sum of frequencies}} \\ &= \frac{1\cdot 0+0\cdot 1+1\cdot 2+0\cdot 3+0\cdot 4+0\cdot 5+2\cdot 6+3\cdot 7+0\cdot 8+0\cdot 9+1\cdot 10}{1+0+1+0+0+0+2+3+0+0+1} \end{aligned}$$

or

$$average = \frac{\sum_{i=0}^{10} (\text{frequency of } i) \cdot i}{\sum_{i=0}^{10} (\text{frequency of } i)} \tag{4-13}$$

The same idea is used to compute a quantum-mechanical average. For the average value of r we take each possible value of r times its frequency (given by $\psi^2\, dv$) and sum over all these values.[3] For a continuous variable like r, we must resort to integration to accomplish this. We divide by the "sum" of frequencies by dividing by $\int \psi^2\,dv$. Thus

$$\bar{r} = \frac{\int^{\text{all space}} r\psi^2 dv}{\int^{\text{all space}} \psi^2 dv} = \frac{\int_0^{2\pi} d\phi \int_0^{\pi} \sin\theta\, d\theta \int_0^{\infty} r\psi^2 r^2 dr}{\int^{\text{all space}} \psi^2 dv} \tag{4-14}$$

The denominator is unity since ψ is normalized. The integrals over θ and ϕ involve parts of the volume element dv, and not ψ^2, because this wavefunction (4-9) does not depend on θ or ϕ. Continuing,

$$\bar{r} = \phi|_0^{2\pi} \cdot -\cos\theta|_0^{\pi} \cdot \pi^{-1} (Z/a_0)^3 \int_0^{\infty} r^3 \exp(-2Zr/a_0)\, dr \tag{4-15}$$

Utilizing the information in Appendix 1 for the integral over r, this becomes

$$\bar{r} = \frac{2\pi[-(-1)+1](1/\pi)(Z/a_0)^3 3!}{(2Z/a_0)^4} \tag{4-16}$$

$$= 4\pi \cdot (1/\pi)(Z/a_0)^3 \cdot 6a_0^4/16Z^4 = \frac{3a_0}{2Z} \tag{4-17}$$

(It is useful to remember that integration over the ϕ and θ parts of dv gives 4π as the result *if no other angle-dependent functions occur in the integral.*) Comparing (4-17) with our expression for r_{mp} indicates that $\bar{r}$ is 1.5 times greater than r_{mp}.

Notice that the lowest-energy eigenfunction is finite at $r=0$ even though V is infinite there. This is allowed by our arguments in Chapter 2 because the infinity in V occurs at only one point, so it can be cancelled by a discontinuity in the derivative of ψ. This is possible only if ψ has a corner or cusp at $r=0$ (see Fig. 4-3a and *e*).

4-1.D Quantum Numbers and Nomenclature

There are three quantum numbers, n, l, and m (all integers), characterizing each solution of the Schrödinger equation (4-6). Of these, only n enters the energy formula (4-8), so all solutions having the same values of n but different values of l and m will be energetically degenerate. As is shown in following sections, these quantum numbers are related in their allowed values. The l quantum number must be nonnegative and smaller than n. The m number may be positive, negative or zero, but its absolute value cannot exceed l. Thus,

$$l = 0, 1, 2, \ldots, n-1 \tag{4-18}$$

$$|m| \le l \tag{4-19}$$

For the lowest-energy wavefunction we have already described, $n=1$, so $l=m=0$. No other choices are possible, so this level is nondegenerate. The convention (from atomic spectroscopy) is to refer to an $l=0$ solution as an "s function," or "s orbital." (For $l=0, 1, 2, 3, 4, 5$, the spectroscopic designation goes s, p, d, f, g, h.) Because n equals unity, the wavefunction is labeled 1s.

[3] The $4\pi r^2$ part of $4\pi r^2\psi^2$ in Fig. 4-4 is included in dv, as will be seen shortly.

When $n = 2$, there are four sets of l and m quantum numbers satisfying rules (4-18) and (4-19). They are listed below with their spectroscopic labels:

$$\begin{array}{llll} l=0, m=0 & 2\mathrm{s} & l=1, m=0 & 2\mathrm{p}_0 \\ l=1, m=-1 & 2\mathrm{p}_{-1} & l=1, m=+1 & 2\mathrm{p}_{+1} \end{array} \tag{4-20}$$

Extending these rules to the $n = 3$ energy level produces nine functions designated 3s, $3p_{-1}$, $3p_0$, $3p_{+1}$, $3d_{-2}$, $3d_{-1}$, $3d_0$, $3d_{+1}$, $3d_{+2}$. In general, the degeneracy of the energy level characterized by n is n^2.

EXAMPLE 4-4 Explain how it comes about from the quantum number rules that the degeneracy equals n^2.

SOLUTION ► The rules indicate that, for each value of n, there are n values of l (e.g., $n = 1, l = 0; n = 2, l = 0$ and $1, \ldots$), and each value of l is associated with $2l + 1$ m_l values (e.g., $l = 2$, $m_l = -2, -1, 0, 1, 2$, which is five values). Note that $2l + 1$ must be an odd number of member states, and the odd number within each set keeps increasing as l increases. Thus, for $n = 4, l = 0, 1, 2, 3$, leading to degenerate sets of states, respectively having 1, 3, 5, and 7 members. **But a sequence of n odd numbers, starting from 1, is always equal to n^2**. QED ◄

4-1.E Nature of the Higher-Energy Solutions

The second energy level is associated with the 2s, $2p_{-1}$, $2p_0$ and $2p_{+1}$ orbitals. The wavefunction for the 2s state is

$$\psi_{2s} = \frac{1}{4\sqrt{2\pi}}\left(\frac{Z}{a_0}\right)^{3/2}\left(2 - \frac{Zr}{a_0}\right)\exp\left(\frac{-Zr}{2a_0}\right) \tag{4-21}$$

Since this is a function of r only, it is a spherically symmetric function. (In fact, all s orbitals are spherically symmetric.) The 2s orbital is more "spread out" than the 1s orbital because the exponential in ψ_{2s} decays more slowly and because the exponential is multiplied by Zr/a_0 (the 2 becomes negligible compared to Zr/a_0 at large r). As a result, the charge cloud associated with the 2s orbital is more diffuse. (For this reason, when we approximate a polyelectronic atom like beryllium by putting electrons in 1s and 2s orbitals the 2s electrons are referred to as "outer" and the 1s electrons are called "inner.")

At small values of r, $(2 - Zr/a_0)$ is positive, and at large distances it is negative, so ψ_{2s} has a spherical *radial node* (a zero in the r coordinate). In Fig. 4-5 the first three s orbitals are plotted. We see that the nth s orbital has $(n - 1)$ spherical nodal surfaces dividing regions where the wavefunctions have different sign. The appearance of more and more nodes in the radial coordinate as the energy increases is certainly familiar from previous examples. Notice how the wavefunctions oscillate most rapidly and nodes are most closely spaced in the regions near the nucleus where the electron classically would have its greatest kinetic energy.

The 2s orbital is orthogonal to the 1s orbital, and also to all higher s orbitals. This would not be possible if there were no radial nodes. The product $\psi_{1s}\psi_{2s}$ will vanish upon integration only if it either vanishes everywhere or else has positive and negative regions that cancel on integration. Since ψ_{1s} and ψ_{2s} are almost everywhere finite, the former condition does not occur. Since ψ_{1s} has the same sign everywhere, their product can have positive and negative regions only if ψ_{2s} has positive and negative regions, and hence, a node.

Let us now consider the 2p functions. They are[4]

$$\psi_{2p_0} = \frac{1}{4\sqrt{2\pi}}\left(\frac{Z}{a_0}\right)^{3/2}\frac{Zr}{a_0}\exp\left(\frac{-Zr}{2a_0}\right)\cos\theta \tag{4-22}$$

$$\psi_{2p_{\pm1}} = \mp\frac{1}{8\sqrt{\pi}}\left(\frac{Z}{a_0}\right)^{3/2}\frac{Zr}{a_0}\exp\left(\frac{-Zr}{2a_0}\right)\sin\theta\exp(\pm i\phi) \tag{4-23}$$

All of these functions have the same radial exponential decay as the 2s orbital, so we can say that the 2s and 2p orbitals are about equal in size. However, since the 2p orbitals contain the factor Zr/a_0 where the 2s contains $(2 - Zr/a_0)$, the 2p orbitals vanish at the nucleus and not at any intermediate r value; they have no radial nodes. The 2p orbitals are endowed with directional properties by their angular dependences. The $2p_0$ orbital is particularly easy to understand because the factors $r\cos\theta$ behave exactly like the z Cartesian coordinate. Hence, we can rewrite $2p_0$ (also called $2p_z$) in mixed coordinates as

$$\psi_{2p_z} = \frac{1}{4\sqrt{2\pi}}\left(\frac{Z}{a_0}\right)^{5/2} z\exp\left(\frac{-Zr}{2a_0}\right) \tag{4-24}$$

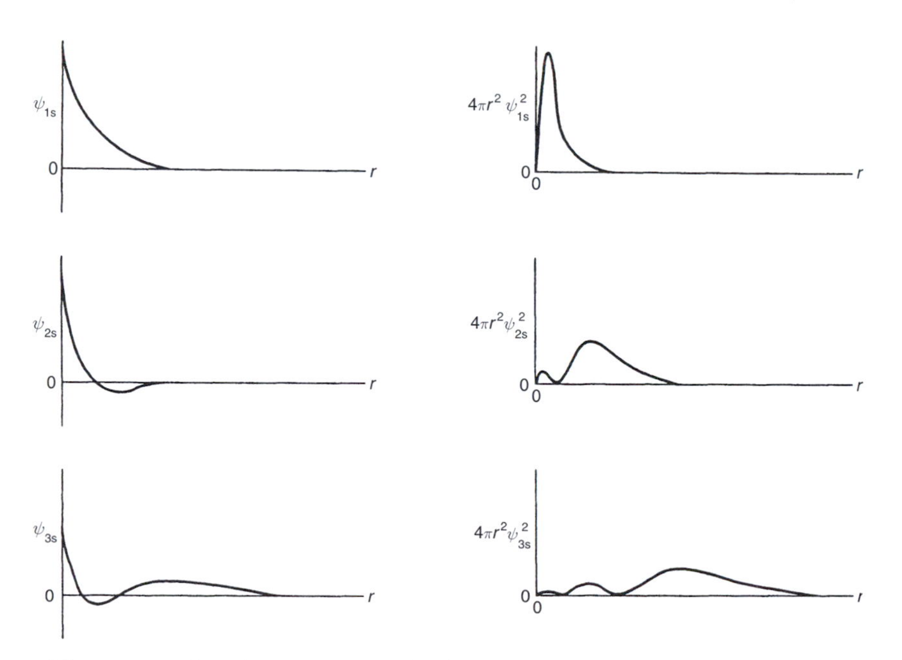

Figure 4-5 ▶ s wavefunctions versus r and volume-weighted electron densities versus r for the hydrogenlike ion.

[4]The $\mp$ factor in $\psi_{2p_{\pm1}}$ results from a phase factor that is omitted from many textbooks. It has no effect on our discussion here, but is consistent with an implicit choice of sign for certain integrals appearing in Appendix 4. See, e.g., Zare [4, Chapter 1].

(Be careful not to confuse the *atomic number Z* with the *coordinate z*.) The exponential term in Eq. (4-24) is spherically symmetric, resembling a diffuse 1s orbital. The function z vanishes in the xy plane and becomes increasingly positive or negative as we move away from the plane in either direction. As a result, ψ_{2p_z} looks as sketched in Fig. 4-6. It has nearly spherical lobes, one with positive phase and one with negative phase. When ψ_{2p_z} is squared, the contour lines remain unchanged in relative position, but the function becomes everywhere positive in sign.

The $2p_{\pm 1}$ orbitals are more difficult to visualize since they are complex functions. The charge distributions associated with these orbitals must be real, however. These are given by $\psi^*\psi$, where ψ^* is the complex conjugate of ψ. (Recall that, for complex wavefunctions, $\psi^*\psi$ must be used for probability distributions, rather than ψ^2.) One obtains the complex conjugate of a function by merely reversing the signs of all the i's in the function. It is evident from Eq. (4-23) that $\psi^*_{2p_{+1}} = -\psi_{2p_{-1}}$ and $\psi^*_{2p_{-1}} = -\psi_{2p_{+1}}$. Hence $\psi^*_{2p_{+1}}\psi_{2p_{+1}} = \psi^*_{2p_{-1}}\psi_{2p_{-1}} = -\psi_{2p_{+1}}\psi_{2p_{-1}}$: both the $2p_{+1}$ and the $2p_{-1}$ orbitals give the same charge distribution. This distribution is the same as that for the $2p_z$ orbital except that the angle dependence is $\frac{1}{2}\sin^2\theta$ instead of $\cos^2\theta$. However, since $\sin^2\theta + \cos^2\theta = 1$, it follows that the sum of $2p_{+1}$ and $2p_{-1}$ charge clouds must be such

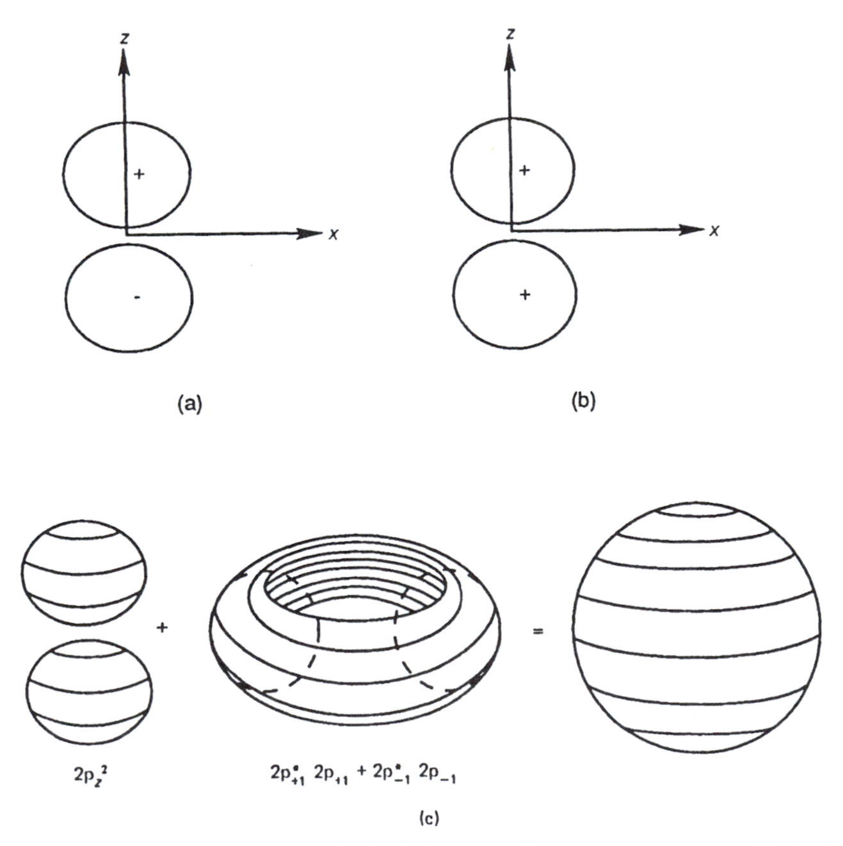

Figure 4-6 ▶ (a) Drawing of the $2p_z$ orbital. (b) Drawing of the square of the $2p_z$ orbital. (c) Drawing of $\psi^2_{2p_z} + \psi^*_{2p_{-1}}\psi_{2p_{-1}} + \psi^*_{2p_{+1}}\psi_{2p_{+1}}$ = spherically symmetric distribution. The curved lines in (c) are a visualization aid and are not mathematically significant.

that, when added to the charge cloud for $2p_0$ a spherical charge cloud results (since the angular dependence is removed). We already know that the $2p_z$ distribution is dumbbell shaped, so it follows that $2p_{+1}$ and $2p_{-1}$ produce doughnut-shaped distributions. (A sphere minus a dumbbell equals a doughnut. See Fig. 4-6.) The shape can also be inferred from the $\sin^2\theta$-dependence, which is a maximum in the xy plane.

When solving the particle-in-a-ring problem, we saw that we could arrive at either a set of real trigonometric solutions or a set of complex exponential solutions. Since the ϕ dependence of the $2p_{+1}$ orbitals is identical to that for $m = \pm 1$ solutions of the particle in the ring, the same situation holds here. Because $\psi_{2p_{+1}}$ and $\psi_{2p_{-1}}$ are energetically degenerate eigenfunctions, any linear combination of them is also an eigenfunction of the hamiltonian (Problem 2-11). Therefore, let us find linear combinations that are entirely real. The complex part of $\psi_{2p_{\pm 1}}$, $\exp(\pm i\phi)$, satisfies the relation

$$\exp(\pm i\phi) = \cos\phi \pm i \sin\phi \tag{4-25}$$

so that

$$\exp(+i\phi) + \exp(-i\phi) = 2\cos\phi \tag{4-26}$$

and

$$i^{-1}[\exp(+i\phi) - \exp(-i\phi)] = 2\sin\phi \tag{4-27}$$

Thus, we have two linear combinations of $\exp(\pm i\phi)$ that are real. It follows immediately that

$$\psi_{2p_x} = \frac{1}{\sqrt{2}}\left[\psi_{2p_{-1}} - \psi_{2p_{+1}}\right] = \frac{1}{4\sqrt{2\pi}}\left(\frac{Z}{a_0}\right)^{3/2}\frac{Zr}{a_0}\exp\left(\frac{-Zr}{2a_0}\right)\sin\theta\cos\phi \tag{4-28}$$

$$\psi_{2p_y} = \frac{i}{\sqrt{2}}\left[\psi_{2p_{-1}} + \psi_{2p_{+1}}\right] = \frac{1}{4\sqrt{2\pi}}\left(\frac{Z}{a_0}\right)^{3/2}\frac{Zr}{a_0}\exp\left(\frac{-Zr}{2a_0}\right)\sin\theta\sin\phi \tag{4-29}$$

where the factor $2^{-1/2}$ is used to maintain normality. Since $r\sin\theta\cos\phi$ and $r\sin\theta\sin\phi$ are equivalent to the Cartesian coordinates x and y, respectively, Eqs. (4-28) and (4-29) are commonly referred to as the $2p_x$ and $2p_y$ orbitals. They are exactly like the $2p_z$ orbital except that they are oriented along the x and y axes (merely replace the z in Eq. (4-24) with x or y).

The 2s, $2p_x$, $2p_y$, and $2p_z$ orbitals are all orthogonal to one another. This is easily shown from symmetry considerations. Each 2p orbital is antisymmetric for reflection in its nodal plane, whereas 2s is symmetric for all reflections. Hence, the product $\psi_{2s}\psi_{2p}$ is always antisymmetric with respect to some reflection so its integral vanishes. The 2p functions are mutually orthogonal because, if one 2p orbital is antisymmetric for some reflection, the others are always symmetric for that reflection. Hence, the product is antisymmetric for that reflection. Another way to show that the 2p orbitals are mutually orthogonal is to note that they behave like x, y, and z vectors and that these vectors are orthogonal (i.e., perpendicular; orthogonality in functions is equivalent to perpendicularity in vectors). Sometimes the orthogonality of functions is most clearly seen if we sketch out the product and note whether the positive and negative regions cancel by symmetry (Fig. 4-7).

The $n = 3$ level has nine solutions associated with it. The 3s orbital, plotted in Fig. 4-5, has one more node than the 2s orbital and is more diffuse. The 3p orbitals have the same angular terms as did the 2p orbitals so they can be written as real functions having the same directional properties as x, y and z vectors. The 3p orbitals differ from the 2p orbitals in that they possess a radial node and are more diffuse (see Fig. 4-8). The remaining five levels, 3d levels, may also be written in either complex or real form. The real orbitals are given by the formulas

$$\left.\begin{array}{r} 3d_{z^2} = \\ 3d_{x^2-y^2} = \\ 3d_{xy} = \\ 3d_{xz} = \\ 3d_{yz} = \end{array}\right\} \frac{2}{\sqrt{2592\pi}} \left(\frac{Z}{a_0}\right)^{3/2} \left(\frac{2Zr}{3a_0}\right)^2 \exp\left(\frac{-Zr}{3a_0}\right) \left\{\begin{array}{l} \left(1/\sqrt{3}\right)\left(3\cos^2\theta - 1\right) \\ \sin^2\theta\cos 2\phi \\ \sin^2\theta\sin 2\phi \\ \sin 2\theta\cos\phi \\ \sin 2\theta\sin\phi \end{array}\right. \tag{4-30}$$

These angular factors, times r^2, have directional properties identical to the Cartesian subscripts on the left, except that $3d_{z^2}$ is a shorthand for $3d_{3z^2-r^2}$. These orbitals are sketched in Fig. 4-8. It is obvious from these figures that $3d_{x^2-y^2}$ has the same symmetry and orientation as the sum of the two vectors x^2 and $-y^2$, and that the other 3d orbitals have a similar connection with the notation (except for $3d_{z^2}$). The 3d functions are

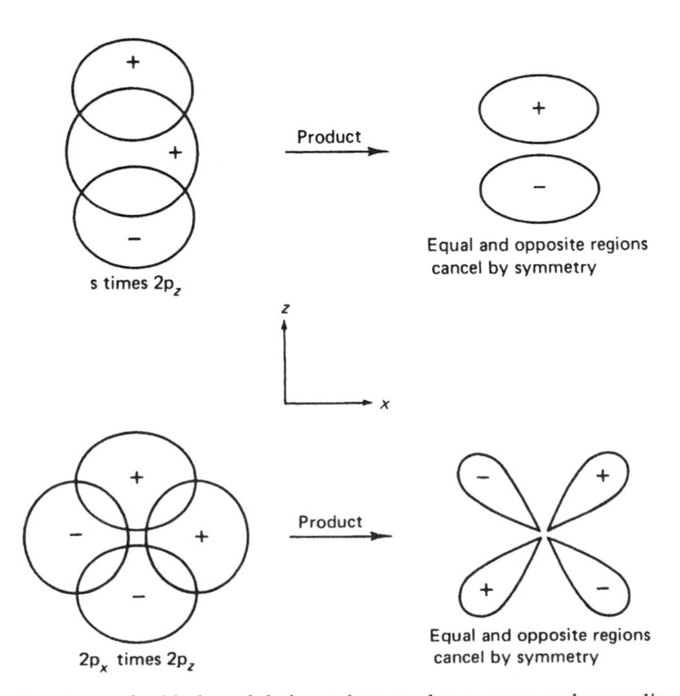

Figure 4-7 ▸ Drawings of orbitals and their products to demonstrate orthogonality.

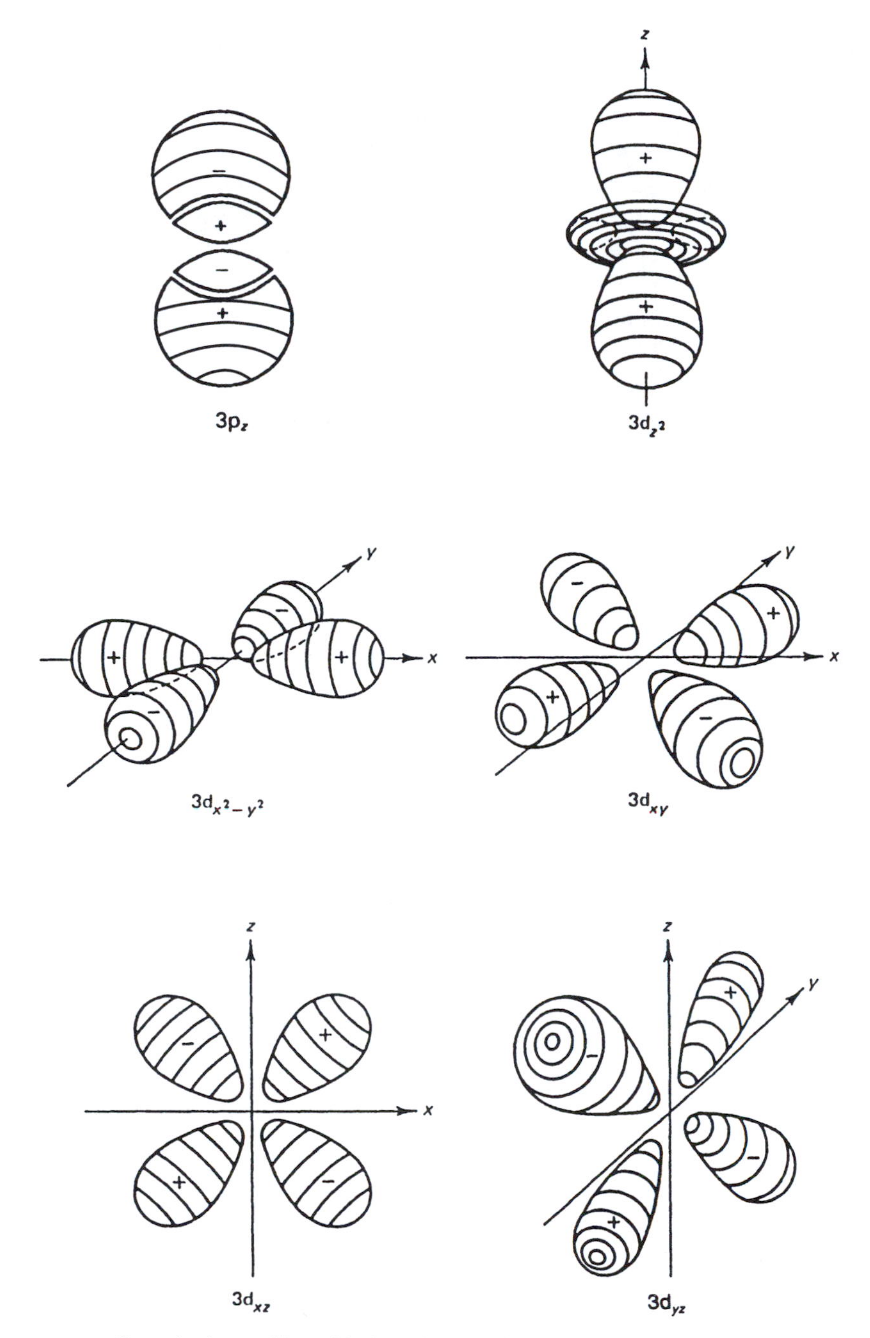

Figure 4-8 ▶ Some hydrogenlike orbitals at the $n = 3$ level.

about the same size as the 3p and 3s functions, but have no radial nodes at intermediate r values.

A general pattern emerges when we examine the nodal properties of the orbitals at various energies. At the lowest energy we have no nodes and the level is nondegenerate. At the next level, we find that each function possesses a single node. There is one way to put in a radial node and so we get one 2s orbital. Or we can put in a planar node. But we have three choices for orthogonal orientations of this plane leading to three independent p orbitals. At the $n = 3$ level we find orbitals containing two nodes. The possibilities are: two radial nodes (3s), a radial node and a planar node ($3p_x$, $3p_y$, $3p_z$), two planar nodes ($3d_{xy}$, $3d_{xz}$, $3d_{yz}$, $3d_{x^2-y^2}$, $3d_{z^2-x^2}$, $3d_{z^2-y^2}$). (But, since $z^2 - y^2 - (z^2 - x^2) = x^2 - y^2$, the last three 3d orbitals are not linearly independent. Hence the last two are combined to form $3d_{z^2}$: $z^2 - x^2 + z^2 - y^2 = 3z^2 - (x^2 + y^2 + z^2) = 3z^2 - r^2$. This function can be seen to correspond to a positive dumbbell encircled by a small negative doughnut, or "belly band." The two nodes in this orbital arising from nodes in θ are conical surfaces rather than planes.) It is apparent that the degeneracies between various 2p orbitals, or 3d orbitals, are spatial degeneracies, due only to the physical equivalence of various directions in space. The degeneracy between 2s and 2p, or 3s, 3p, and 3d is not due to spatial symmetry. The fact that an angular node is energetically equivalent to a radial node is a peculiarity of the particular potential $(-Ze^2/4\pi\varepsilon_0 r)$ for this problem. This degeneracy is removed for noncoulombic central-field potentials.

The eigenfunctions corresponding to states in the energy continuum, like the bound states, can be separated into radial and angular parts. The radial parts of the spherically symmetric eigenfunctions at two nonnegative energies are given in Fig. 4-9. Note that the rate of oscillation of these functions is greatest at the nucleus, where the local kinetic energy is largest, in accord with the ideas presented in Chapters 1 and 2. Unbound-state

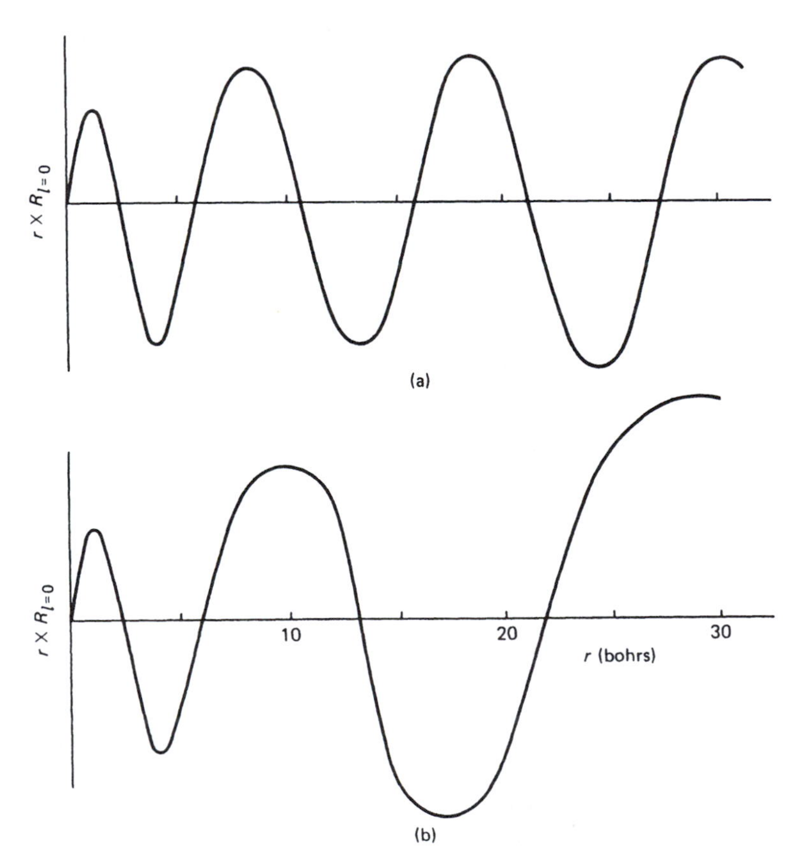

Figure 4-9 ▶ Radial part of unbound H-atom states (times r) versus r at two energies: (a) $E = 13.6\text{eV}$; (b) $E = 0$.

wavefunctions are not used in most quantum chemical applications, so we will not discuss them further in this book.

EXAMPLE 4-5 A nodal plane is one through which a wavefunction is antisymmetric for reflection. Consider the $3d_{xy}$ and $3d_{yz}$ orbitals of Fig. 4-8. Through which planes do these two orbitals have different reflection symmetries?

SOLUTION ▶ $3d_{xy}$ is antisymmetric for reflection only through the x, z and y, z planes. $3d_{yz}$ is antisymmetric for reflection only through the x, z and x, y planes. Hence, these orbitals differ in their symmetries for reflection through the y, z and x, y planes. ◀

4-2 Separation of Variables

We shall indicate in some detail the way in which the Schrödinger equation (4-6) is solved. Recall the strategy of separating variables which we used in Section 2-7:

1. Express ψ as a product of functions, each depending on only one variable.
2. Substitute this product into the Schrödinger equation and try to manipulate it so that the equation becomes a sum of terms, each depending on a single variable. These terms must sum to a constant.
3. Since terms for different variables are independent of each other, the terms for each variable must equal a constant. This enables one to set up an equation in each variable. If this can be done, the initial assumption (1) is justified.

In this case we begin by assuming that

$$\psi(r,\theta,\phi) = R(r)\Theta(\theta)\Phi(\phi) \tag{4-31}$$

Substituting into Eq. (4-6) gives

$$\frac{-h^2}{8\pi^2\mu r^2}\left[\Theta\Phi\frac{d}{dr}\left(r^2\frac{dR}{dr}\right) + R\Phi\frac{1}{\sin\theta}\frac{d}{d\theta}\left(\sin\theta\frac{d\Theta}{d\theta}\right) + R\Theta\frac{1}{\sin^2\theta}\frac{d^2\Phi}{d\phi^2}\right] - \frac{Ze^2}{4\pi\varepsilon_0 r}R\Theta\Phi = ER\Theta\Phi \tag{4-32}$$

Since each derivative operator now acts on a function of a single coordinate, we use total, rather than partial, derivative notation.

Let us first see if we can isolate the ϕ dependence. Multiplying Eq. (4-32) by $(-8\pi\mu r^2\sin^2\theta/h^2R\Theta\Phi)$ and rearranging gives

$$\frac{\sin^2\theta}{R}\frac{d}{dr}\left(r^2\frac{dR}{dr}\right) + \frac{8\pi^2\mu r^2\sin^2\theta}{h^2}\left(E + \frac{Ze^2}{4\pi\varepsilon_0 r}\right) + \frac{\sin\theta}{\Theta}\frac{d}{d\theta}\left(\sin\theta\frac{d\Theta}{d\theta}\right) + \frac{1}{\Phi}\frac{d^2\Phi}{d\phi^2} = 0 \tag{4-33}$$

The r and θ dependence is still mixed in the first two terms, but we now have a rather simple term in the coordinate ϕ. Now we can argue, as in Section 2-7, that, as ϕ alone changes, the first three terms in Eq. (4-33) do not change. That is, if only ϕ changes, Eq. (4-33) may be written

$$\text{constant} + \text{constant} + \text{constant} + (1/\Phi)(d^2\Phi/d\phi^2) = 0 \tag{4-34}$$

so that

$$(1/\Phi)(d^2\Phi/d\phi^2) = -\text{m}^2 \text{ (a constant)} \tag{4-35}$$

We call the constant $-m^2$ for future mathematical convenience. We can rearrange Eq. (4-35) into the more familiar form for an eigenvalue equation:

$$d^2\Phi/d\phi^2 = -m^2\Phi \tag{4-36}$$

We arrived at Eq. (4-36) by assuming that only ϕ changes while r and θ are constant. However, it should be obvious that the behavior of the term in ϕ is uninfluenced by changes in r and θ since it has no dependence on these coordinates. Thus, by establishing that this term is constant under certain circumstances, we have actually shown that it must be constant under all circumstances, and we have produced an eigenvalue equation for Φ.

We can now proceed with further separation of variables. Since we know that the last term in Eq. (4-33) is a constant, we can write

$$\frac{1}{R}\frac{d}{dr}\left(r^2\frac{dR}{dr}\right) + \frac{8\pi^2\mu r^2}{h^2}\left(E + \frac{Ze^2}{4\pi\varepsilon_0 r}\right) + \frac{1}{\Theta\sin\theta}\frac{d}{d\theta}\left(\sin\theta\frac{d\Theta}{d\theta}\right) - \frac{m^2}{\sin^2\theta} = 0 \tag{4-37}$$

Note that we have separated the θ and r dependences by dividing through by $\sin^2\theta$. We now have two terms wholly dependent on r and two wholly dependent on θ, their sum being zero. Hence, as before, the sum of the two r-dependent terms must equal a constant, β, and the sum of the θ-dependent terms must equal $-\beta$. Thus

$$\frac{d}{dr}\left(r^2\frac{dR}{dr}\right) + \frac{8\pi^2\mu r^2}{h^2}\left(E + \frac{Ze^2}{4\pi\varepsilon_0 r}\right)R = \beta R \tag{4-38}$$

$$\frac{1}{\sin\theta}\frac{d}{d\theta}\left(\sin\theta\frac{d\Theta}{d\theta}\right) - \frac{m^2\Theta}{\sin^2\theta} = -\beta\Theta \tag{4-39}$$

where we have multiplied through by R in the first equation and by Θ in the second.

The assumption that $\psi = R\Theta\Phi$ has led to separate equations for R, Θ, and Φ. This indicates that the assumption of separability was valid. However, there is some linkage between R and Θ *via* β, and between Θ and Φ *via* m.

4-3 Solution of the R, Θ, and Φ Equations

4-3.A The Φ Equation

The solution of Eq. (4-36) is similar to that of the particle in a ring problem of Section 2-6. The normalized solutions are

$$\Phi = (1/\sqrt{2\pi})\exp(im\phi), \quad m = 0, \pm1, \pm2, \ldots \tag{4-40}$$

As shown in Section 2-6, the constant m must be an integer if Φ is to be a single-valued function.

4-3.B The Θ Equation

There is great similarity between the mathematical techniques used in solving the R and Θ equations and those used to solve the one-dimensional harmonic oscillator problem of Chapter 3. Hence, we will only summarize the steps involved in these solutions and make a few remarks about the results. More detailed treatments are presented in many texts.[5]

The Θ equation can be solved as follows:

1. Change the variable to obtain a more convenient form for the differential equation.
2. Express the solution as a power series and obtain a recursion relation.
3. Observe that the series diverges for certain values of the variables, producing nonsquare-integrable wavefunctions. Correct this by requiring that the series terminate. This requires that the truncated series be either symmetric or antisymmetric in the variable and also that β of Eq. (4-38) and (4-39) be equal to $l(l+1)$ with l an integer.
4. Recognize these truncated series as being the associated Legendre functions.
5. Return to the original variable to obtain an expression for Θ in terms of the starting coordinate.

Reference to the end of Section 3-4 will illustrate the similarity between this and the harmonic oscillator case.

The final result for $m \geqslant 0$ is

$$\Theta_{l,m}(\theta) = (-1)^m \left[\frac{(2l+1)}{2}\frac{(l-|m|)!}{(l+|m|)!}\right]^{1/2} P_l^{|m|}(\cos\theta) \tag{4-41}$$

For $m < 0$ the phase factor $(-1)^m$ should be omitted.[6] The term in square brackets is a normalizing function, and $P_l^{|m|}(\cos\theta)$ represents some member of the series of associated Legendre functions. When $m = 0$, these become the ordinary Legendre polynomials. The first few ordinary Legendre polynomials are

$$\begin{aligned} &P_0(x) = 1, \quad P_1(x) = x, \quad P_2(x) = \frac{1}{2}(3x^2 - 1) \\ &P_3(x) = \frac{1}{2}(5x^3 - 3x) \end{aligned} \tag{4-42}$$

The first few associated Legendre functions are

$$\begin{aligned} &P_1^1(x) = (1-x^2)^{1/2}, \quad P_2^1(x) = 3(1-x^2)^{1/2}x, \\ &P_2^2(x) = 3(1-x^2), \quad P_3^1(x) = \tfrac{3}{2}(1-x^2)^{1/2}(5x^2-1), \\ &P_3^2(x) = 15(1-x^2)x, \quad P_3^3(x) = 15(1-x^2)^{3/2} \end{aligned} \tag{4-43}$$

It is also true that

$$P_l^{|m|}(x) = 0 \quad \text{if } |m| > l \tag{4-44}$$

[5] See, e.g., Pauling and Wilson [3, Chapter 5].

[6] This is the same phase factor that we saw earlier for $\psi_{2p_{\pm 1}}$.

Thus, $\Theta(\theta)$, and hence $\psi(r, \theta, \phi)$, vanishes unless $|m| \leq l$, giving us one of our quantum number rules [Eq. (4-19)].

The associated Legendre functions satisfy an orthogonality relation:

$$\int_{-1}^{+1} P_l^{|m|}(x) P_{l'}^{|m|}(x)\, dx = \frac{2}{(2l+1)} \frac{(l+|m|)!}{(l-|m|)!} \delta_{ll'} \qquad (4\text{-}45)$$

For a further discussion of these functions, the reader should consult a more advanced text on quantum mechanics.

4-3.C The *R* Equation

The R equation can be solved as follows:

1. Assume that E is negative (this restricts us to bound states), and note that $\beta = l(l+1)$ from the previous solving of the Θ equation.

2. Change variables for mathematical convenience.

3. Find the asymptotic solution pertaining to the large r limit, where the R equation becomes simplified.

4. Express the wavefunction as a product of the asymptotic solution and an unknown function. Express this unknown function as a power series and (after dealing with some singularities) obtain a recursion relation.

5. Note that the power series overpowers the asymptotic part of the solution unless the series is truncated. This leads to the requirement that n be an integer and hence that E be quantized. It also requires that $n > l$.

6. Recognize the truncated series to be associated Laguerre polynomials times ρ^l, where ρ is defined below.

The resulting solution is, if $\mu = m_e$,

$$R_{nl} = -\left[\left(\frac{2Z}{na_0}\right)^3 \frac{(n-l-1)!}{2n\,[(n+l)!]^3}\right]^{1/2} \exp(-\rho/2)\, \rho^l L_{n+l}^{2l+1}(\rho) \qquad (4\text{-}46)$$

where $\rho = 2Zr/na_0$ and $a_0 = \epsilon_0 h^2/\pi m_e e^2 = 5.2917706 \times 10^{-11}$ m. The term in brackets is a normalizing function. The exponential term is the asymptotic solution and it guarantees that $R(r)$ will approach zero as r approaches infinity. The third term, ρ^l, is produced in the course of removing singularities (i.e., places where parts of a differential equation become infinite). The last term, $L(\rho)$, symbolizes the various members of the set of associated Laguerre polynomials. Like the Legendre functions, these are mathematically well characterized. A few of the low-index associated Laguerre polynomials are

$$L_1^1(\rho) = 1, \qquad L_2^1(\rho) = 2\rho - 4,$$
$$L_3^1(\rho) = -3\rho^2 + 18\rho - 18, \quad L_3^3(\rho) = -6 \tag{4-47}$$

4-4 Atomic Units

It is convenient to define a system of units that is more natural for working with atoms and molecules. The commonly accepted system of atomic units for some important quantities is summarized in Table 4-1. [Note: the symbol $\hbar$ ("h-cross or h-bar") is often used in place of $h/2\pi$.] Additional data on values of physical quantities, units, and conversion factors can be found in Appendix 10.

In terms of these units, Schrödinger's equation and its resulting eigenfunctions and eigenvalues for the hydrogenlike ion become much simpler to write down. Thus, the

TABLE 4-1 ► Atomic Units

Quantity	Atomic unit in cgs or other units	Values of some atomic properties in atomic units (a.u.)
Mass	$m_e = 9.109534 \times 10^{-28}$ g	Mass of electron = 1 a.u.
Length	$a_0 = 4\pi\varepsilon_0\hbar^2/m_e e^2$ $= 0.52917706 \times 10^{-10}$ m (= 1 bohr)	Most probable distance of 1s electron from nucleus of H atom = 1 a.u
Time	$\tau_0 = a_0\hbar/e^2$ $= 2.4189 \times 10^{-17}$ s	Time for 1s electron in H atom to travel one bohr = 1 a.u.
Charge	$e = 4.803242 \times 10^{-10}$ esu $= 1.6021892 \times 10^{-19}$ coulomb	Charge of electron = −1 a.u.
Energy	$e^2/4\pi\varepsilon_0 a_0 = 4.359814 \times 10^{-18}$ J (= 27.21161 eV ≡ 1 hartree)	Total energy of 1s electron in H atom = −1/2 a.u.
Angular momentum	$\hbar = h/2\pi$ $= 1.0545887 \times 10^{-34}$ J s	Angular momentum for particle in ring = 0, 1, 2, ... a.u.
Electric field strength	$e/a_0^2 = 5.1423 \times 10^9$ V/cm	Electric field strength at distance of 1 bohr from proton = 1 a.u.

TABLE 4-2 ▶ Eigenfunctions for the Hydrogenlike Ion in Atomic Units

Spectroscopic symbol	Formula
1s	$(1/\sqrt{\pi})Z^{3/2}\exp(-Zr)$
2s	$(1/4\sqrt{2\pi})Z^{3/2}(2-Zr)\exp(-Zr/2)$
2p_x	$(1/4\sqrt{2\pi})Z^{5/2}r\exp(-Zr/2)\sin\theta\cos\phi$
2p_y	$(1/4\sqrt{2\pi})Z^{5/2}r\exp(-Zr/2)\sin\theta\sin\phi$
2p_z	$(1/4\sqrt{2\pi})Z^{5/2}r\exp(-Zr/2)\cos\theta$
3s	$(1/81\sqrt{3\pi})Z^{3/2}(27-18Zr+2Z^2r^2)\exp(-Zr/3)$
3p_x	$(\sqrt{2}/81\sqrt{\pi})Z^{5/2}r(6-Zr)\exp(-Zr/3)\sin\theta\cos\phi$
3p_y	$(\sqrt{2}/81\sqrt{\pi})Z^{5/2}r(6-Zr)\exp(-Zr/3)\sin\theta\sin\phi$
3p_z	$(\sqrt{2}/81\sqrt{\pi})Z^{5/2}r(6-Zr)\exp(-Zr/3)\cos\theta$
$3\text{d}_{z^2}\,(\equiv 3\text{d}_{3z^2-r^2})$	$(1/81\sqrt{6\pi})Z^{7/2}r^2\exp(-Zr/3)(3\cos^2\theta-1)$
$3\text{d}_{x^2-y^2}$	$(1/81\sqrt{2\pi})Z^{7/2}r^2\exp(-Zr/3)\sin^2\theta\cos 2\phi$
3d_{xy}	$(1/81\sqrt{2\pi})Z^{7/2}r^2\exp(-Zr/3)\sin^2\theta\sin 2\phi$
3d_{xz}	$(1/81\sqrt{2\pi})Z^{7/2}r^2\exp(-Zr/3)\sin 2\theta\cos\phi$
3d_{yz}	$(1/81\sqrt{2\pi})Z^{7/2}r^2\exp(-Zr/3)\sin 2\theta\sin\phi$

Schrödinger equation in atomic units is (assuming infinite nuclear mass, so that $\mu = m_\text{e}$)

$$\left(-\frac{1}{2}\nabla^2 - \frac{Z}{r}\right)\psi = E\psi \tag{4-48}$$

The energies are

$$E_n = -\frac{Z^2}{2n^2} \tag{4-49}$$

The lowest-energy solution is

$$\psi_{1\text{s}} = \sqrt{Z^3/\pi}\exp(-Zr) \tag{4-50}$$

The formulas for the hydrogenlike ion solutions (in atomic units) of most interest in quantum chemistry are listed in Table 4-2. The tabulated functions are all in real, rather than complex, form. Problems involving atomic orbitals are generally far easier to solve in atomic units.

4-5 Angular Momentum and Spherical Harmonics

We have now discussed three problems in which a particle is free to move over the entire range of one or more coordinates with no change in potential. The first case was the free particle in one dimension. Here we found the eigenfunctions to be simple trigonometric or exponential functions of x. The trigonometric form is identical to the harmonic amplitude function of a standing wave in an infinitely long string. We might refer to such functions as "linear harmonics." The second case was the particle-in-a-ring problem, which again has solutions that may be expressed either as sine-cosine or exponential functions of the angle ϕ. By analogy with linear motion, we could refer to these as "circular harmonics." Finally, we have described the hydrogenlike ion, where the particle can move over the full ranges of θ and ϕ (i.e., over the surface of a sphere) with no change in potential. The solutions we have just described—the products $\Theta_{l,m}(\theta)\Phi_m(\phi)$—are called *spherical harmonics* and are commonly symbolized $Y_{l,m}(\theta,\phi)$. Thus for $m \geqslant 0$

$$Y_{l,m}(\theta,\phi) = (-1)^m \left[\frac{(2l+1)}{4\pi}\frac{(l-|m|)!}{(l+|m|)!}\right]^{1/2} P_l^{|m|}(\cos\theta)\exp(im\phi) \tag{4-51}$$

and for $m < 0$ the factor $(-1)^m$ is omitted. Because so many physical systems have spherical symmetry, spherical harmonics are very important in classical and quantum mechanics.

Closely linked with spherical harmonics is *angular momentum*. Angular momentum is an important physical property because it is conserved in an isolated dynamical system; it is a *constant of motion* for the system. Angular momentum is described by magnitude and direction, so it is a vector quantity.[7] The classical system, in the absence of external forces, is constrained to move in such a way as to preserve both the direction and the magnitude of this vector. For a mass of m kg moving in a circular orbit of radius r m with an angular velocity of ω radians per second, the angular momentum has magnitude $mr^2\omega$ kg m^2/s (or, alternatively, joule seconds). The direction of the vector is given by the right-hand rule: the index finger of the right hand points along the particle trajectory and the extended thumb points along the angular momentum vector (see Fig. 4-10). (Alternatively, in a right-handed coordinate system, motion of a mass in the xy plane from $+x$ toward $+y$ produces angular momentum in the $+z$ direction.) In vector notation, $\mathbf{L} = \mathbf{r} \times \mathbf{p}$, where $\mathbf{L}$ is angular momentum, $\mathbf{r}$ is the position vector, and $\mathbf{p}$ is the linear momentum.

Some of the more interesting properties of angular momentum relate to the situation where circular motion occurs *in the presence of an external field*. A familiar example is a gyroscope mounted on a pivot and experiencing the gravitational field of the earth. The gyroscope flywheel is usually started with the gyroscope in an almost vertical position. After release, the gyroscope precesses about the axis of field direction. As time passes, the tilt of the gyroscope away from the field direction (which we take to be the z direction) increases (see Fig. 4-11). If there were no friction in the bearings, the angle of tilt would not change, and the gyroscope would precess about z indefinitely, maintaining whatever angle of tilt it found itself with initially. Notice that, in such

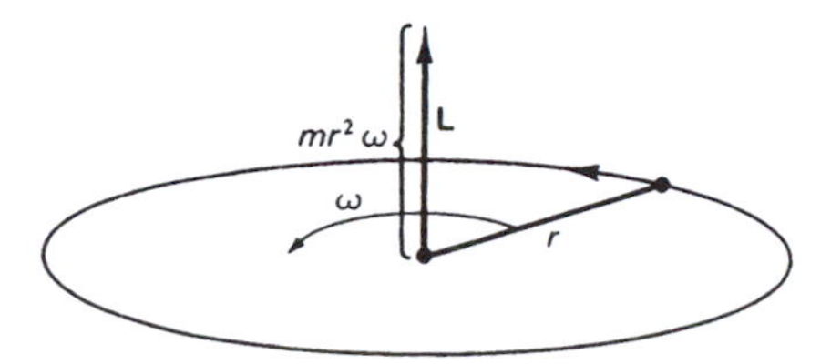

Figure 4-10 ▶ The angular momentum vector **L** for a particle of mass m moving with angular velocity ω about a circular orbit of radius r in the direction indicated.

[7] Strictly speaking, angular momentum is a pseudovector—it is dual to a second order antisymmetric tensor. However, for the remainder of this book, we can and shall ignore this distinction.

a case, the angular momentum due to the flywheel, $\mathbf{L}_f$, is conserved *in magnitude only*. Its direction is constantly changing. Thus, $\mathbf{L}_f$ is *not* a constant of motion in the presence of a z-directed field. Neither are the components $\mathbf{L}_{f_x}$ and $\mathbf{L}_{f_y}$, which change in magnitude as the gyroscope precesses. However, $\mathbf{L}_{f_z}$ *is* a constant of motion if the angle of tilt does not change. If we add on to $\mathbf{L}_f$ the angular momentum $\mathbf{L}_g$ due to the precession of the gyroscope as a whole (including the center of mass of the flywheel but ignoring its rotation), we find that the total angular momentum for the gyroscope (including flywheel motion), $\mathbf{L}$ and its components L_x, L_y, and L_z behave similarly to $\mathbf{L}_f$ and its components (see Fig. 4-12). We may summarize these observations from classical physics as follows: a rotating rigid body conserves $\mathbf{L}$ (hence, L_x, L_y, L_z) in the absence of external forces. In the presence of a z-directed, time-independent external force, L_z and $|\mathbf{L}|$, the magnitude of $\mathbf{L}$ (but not its direction) are conserved. Furthermore, in a system comprising several moving parts, the total angular momentum is the sum of the individual angular momenta, and the z component is the sum of the individual z components:

$$\mathbf{L} = \sum_i \mathbf{L}_i \tag{4-52}$$

$$L_z = \sum_i L_{zi} \tag{4-53}$$

Many characteristics of the classical situation are maintained in quantum mechanics. In particular, it can be shown that a hydrogenlike ion eigenfunction can always be associated with "sharp" values for L_z, but not for L_x or L_y, and that the *magnitude* of $\mathbf{L}$ is sharp, but not its direction. We have indicated several times in this book that a sharp value (i.e., a constant of motion) exists when a state function is an eigenfunction for an operator associated with the property. For example, all of our hydrogenlike ion wavefunctions are eigenfunctions for the hamiltonian operator, so all are associated with sharp energies. We introduced the operator for the z-component of angular momentum, in Section 2-6, as $(h/2\pi i)d/d\phi$. Generalizing to situations with several variables requires switching to partial derivative notation. Then our operator,

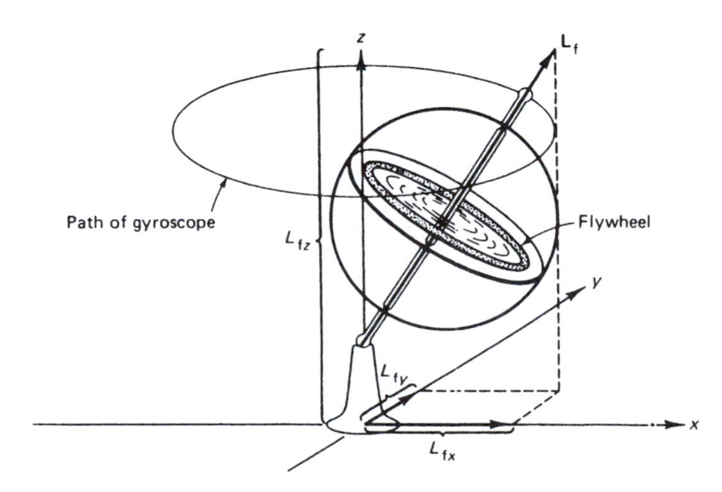

Figure 4-11 ▶ A gyroscope with the angular momentum of the *flywheel*, $\mathbf{L}_f$, together with x, y, and z components of $\mathbf{L}_f$, at a given instant.

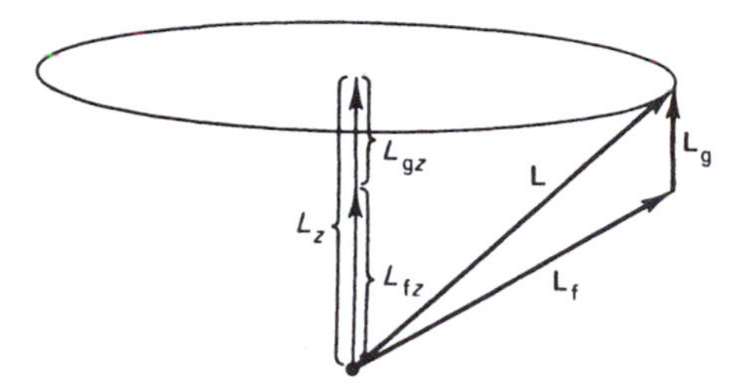

Figure 4-12 ▶ The total angular momentum of the gyroscope **L** is shown as the sum of $\mathbf{L}_f$, the angular momentum of the flywheel, and $\mathbf{L}_g$, the angular momentum of the gyroscope. **L** precesses, so only L_z and the magnitude of **L** are constants of motion.

symbolized $\hat{L}_z$, is $(\hbar/i)\partial/\partial\phi$. In atomic units, $\hat{L}_z$ is $(1/i)\partial/\partial\phi$. This operator was introduced in Section 2-6, where it was given the symbol p_ϕ. (A general discussion on operators will be given in Chapter 6. The carat symbol is frequently used to denote an operator.) Our statement that hydrogenlike eigenfunctions have sharp L_z means that we expect $\hat{L}_z\psi_{n,l,m}(r,\theta,\phi) = \text{constant}\cdot\psi_{n,l,m}(r,\theta,\phi)$. Since all these eigenfunctions have $\exp(im\phi)$ as their only ϕ-dependent term, it follows immediately that

$$\hat{L}_z\psi_{n,l,m} = m\hbar\psi_{n,l,m} \tag{4-54}$$

or, equivalently,

$$\hat{L}_z Y_{l,m}(\theta,\phi) = m\hbar Y_{l,m}(\theta,\phi) \tag{4-55}$$

Hence, the quantum number m is equal to the z component of angular momentum in units of $\hbar$ for the state in question. This means that the angular momentum associated with an s state ($l = 0$, so $m = 0$) has a zero z component, while a p state ($l = 1$, so $m = -1, 0, +1$) can have a z component of $-\hbar$, 0, or $\hbar$.

The other quantity that we have stated is conserved in these systems is the magnitude of **L**. In quantum mechanics, it is convenient to deal with the square of this magnitude $\mathbf{L}^2$. The quantum-mechanical operator associated with this quantity is

$$\begin{aligned}\hat{L}^2 &= -\hbar^2\left[(\partial^2/\partial\theta^2) + \cot\theta(\partial/\partial\theta) + (1/\sin^2\theta)(\partial^2/\partial\phi^2)\right]\\ &= -\hbar^2\left[(1/\sin\theta)(\partial/\partial\theta)\sin\theta(\partial/\partial\theta) + (1/\sin^2\theta)(\partial^2/\partial\phi^2)\right]\end{aligned} \tag{4-56}$$

The result of operating on $Y_{l,m}(\theta,\phi)$ with this operator is

$$\hat{L}^2 Y_{l,m}(\theta,\phi) = l(l+1)\hbar^2 Y_{l,m}(\theta,\phi) \tag{4-57}$$

This means that the *square* of the magnitude of the total angular momentum equals $l(l+1)\hbar^2$. Hence, for an s state it is zero, for a p state it is $2\hbar^2$, for a d state it is $6\hbar^2$, etc.

One can construct vector diagrams to parallel these relationships. A few of these are sketched in Fig. 4-13.

Operators for $\hat{L}_x$ and $\hat{L}_y$ can also be constructed. They are

$$\hat{L}_x = i\hbar[\sin\phi(\partial/\partial\theta) + \cot\theta\cos\phi(\partial/\partial\phi)] \tag{4-58}$$

$$\hat{L}_y = -i\hbar[\cos\phi(\partial/\partial\theta) - \cot\theta\sin\phi(\partial/\partial\phi)] \tag{4-59}$$

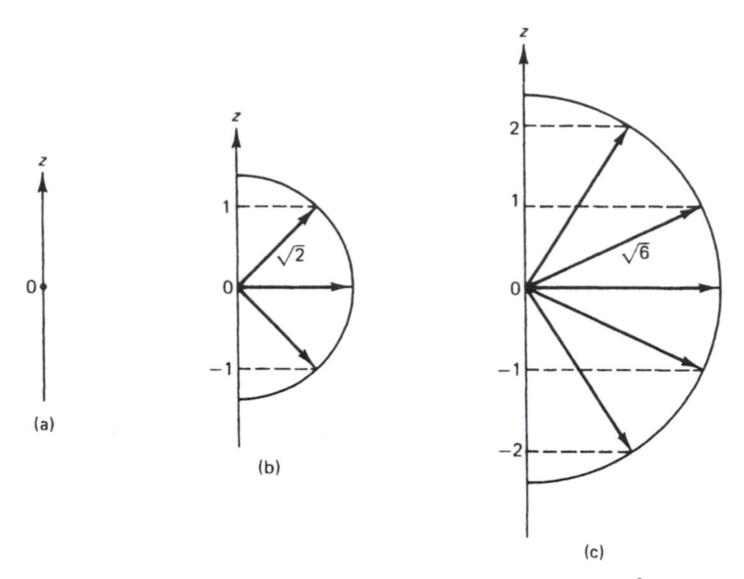

Figure 4-13 ▸ Vector relationships that satisfy the following rules: $\mathbf{L}^2 = l(l+1)$, $L_z = m$, $m = -l, -l+1, \ldots, 0, \ldots, l-1, l$. The quantum rules correspond to a classical analog where the gyroscope can have only certain discrete angles of tilt. (a) s; $l = 0$; $l(l+1) = 0$; $m = 0$. (b) p; $l = 1$; $l(l+1) = 2$; $m = -1, 0, +1$. (c) d; $l = 2$; $l(l+1) = 6$; $m = -2, -1, 0, +1, +2$. (all in atomic units)

The hydrogenlike eigenfunctions are not necessarily eigenfunctions for either of these operators (Problem 4-23).

It is interesting to consider the physical meaning of these results. If a quantity has a sharp value, it means that we will always get that value no matter when we measure that property for systems in the state being considered. Thus, the z component of angular momentum for hydrogen atoms in the $2p_{+1}$ state will always be measured to be $+1\hbar$, or one atomic unit. For the x or y component, however, repeated measurements (on an ensemble of $2p_{+1}$ atoms) will yield a spread of values. We can measure (or compute) an *average* value of L_x or L_y but not a sharp value. In terms of our mental model (a gyroscope) this seems sensible enough except for one thing. Our hydrogenlike eigenfunctions are solutions for a central field potential with *no* external field. Under such conditions, $\mathbf{L}$, L_x, L_y, and L_z are classically all constants of motion. Why, then, are they not all sharp quantum mechanically? The answer is that quantum-mechanical state functions never contain more information than is, in principle, extractable by measurement. To measure a component of angular momentum in a system always means, in practice, subjecting the system to some sort of external force. Furthermore, this system must obey the limitations decreed by the uncertainty principle. The hydrogenlike ion wavefunctions cannot simultaneously be eigenfunctions for $\hat{L}_x$, $\hat{L}_y$, and $\hat{L}_z$ because that would give simultaneous sharp values (i.e., no uncertainty) for the conjugate variables angular momentum vector length and angular momentum vector orientation. This would violate the uncertainty principle, which is in turn a reflection of limitations on our ability to measure one variable without affecting another (see Section 1-8).

It is possible, working only with the quantum-mechanical operators, to generate the eigenvalues of $\hat{L}_z$ and $\hat{L}^2$. This approach is a deviation from our main line of development and is contained in Appendix 4. (It is recommended that Chapter 6 be completed before reading Appendix 4.) We give only the results. They are

$$\hat{L}_z f_{l,m} = m\hbar f_{l,m}, \quad m = -l, -l+1, \ldots, l-1, l \tag{4-60}$$

$$\hat{L}^2 f_{l,m} = l(l+1)\hbar^2 f_{l,m} \tag{4-61}$$

These look like the results already given in Eqs. (4-55) and (4-57). There is a difference, however. Here there is no indication that m is an integer, whereas in Eq. (4-55) m must be an integer, as indicated by the presence of zero in its value list. There are two ways in which we can have a sequence of the form $-l, -l+1, \ldots, l-1, l$. One way is to have an integer series, for example, $-2, -1, 0, +1, +2$, which must contain zero. The other way is to have a half-integer series, for example, $-\frac{3}{2}, -\frac{1}{2}, +\frac{1}{2}, +\frac{3}{2}$, which skips zero. If we work only with the properties of the operators, we find that either possibility is allowed. But if we assume that the as-yet-unspecified eigenfunctions $f_{l,m}$ are separable into θ- and ϕ-dependent parts, we find ourselves restricted to the integer series. For *orbital* angular momentum (due to motion of the electron in the atomic orbital), the z component must be (in atomic units) an *integer*, for we have seen that the state functions ψ contain the spherical harmonics $Y_{l,m}$, which are indeed separable. *Spin* angular momentum for an electron (to be discussed in more detail in the next chapter), has half-integer z components of angular momentum, and the eigenfunctions corresponding to spin cannot be expressed with spherical harmonics.

4-6 Angular Momentum and Magnetic Moment

If a charged particle is accelerated, a magnetic field is produced. Since circular motion of constant velocity requires acceleration (classically) it follows that *a charged particle having angular momentum will also have a magnetic moment.* The magnetic moment is proportional to the angular momentum, colinear with it, and oriented in the same direction if the charge is positive. For an electron, the magnetic moment is given by

$$\mu = -\beta_e \mathbf{L} \tag{4-62}$$

where β_e, the *Bohr magneton*, has a value of $9.274078 \times 10^{-24}\ \mathrm{J\,T^{-1}}$ (equal to $\frac{1}{2}$a.u.), where T is magnetic field strength in Tesla. (β_e is defined to contain the $\hbar$ that belongs to $\mathbf{L}$, so it is only the $\sqrt{l(l+1)}$ part of $\mathbf{L}$ that is used in the calculation.)

EXAMPLE 4-6 What is the magnitude of the orbital magnetic moment for an electron in a 3d state of a hydrogen atom? In a 4d state of He^+?

SOLUTION ▶ For any d state, $l = 2$, so, in a.u., $|\mu| = \beta_e \mathbf{L} = \beta_e\sqrt{l(l+1)} = \sqrt{6}\beta_e = 2.27 \times 10^{-23}\mathrm{JT^{-1}}$. (Since we want magnitude, we can ignore the minus sign.) The value does not depend on the quantum number n nor on atomic number Z, so it is the same for He^+. ◀

Applying a magnetic field of strength B defines a z-direction about which the magnetic moment vector precesses. The z-component, μ_z, of the precessing vector interacts with the applied field B. The interaction energy is

$$E = -\mu_z B = \beta_e L_z B = \beta_e m B \tag{4-63}$$

This means that some of the degeneracies among the energy levels of the hydrogenlike ion will be removed by imposing an external magnetic field. For instance, the $2p_{+1}$ and $2p_{-1}$ energy levels will be raised and lowered in energy, respectively, while 2s and $2p_0$ will be unaffected (see Fig. 4-14). This, in turn, will affect the atomic spectrum for absorption or emission. The splitting of spectral lines due to the imposition of an external magnetic field is known as *Zeeman splitting*. Because the splitting of levels depicted in Fig. 4-14 is proportional to the z component of orbital angular momentum, given by $m\hbar$, it is conventional to refer to m as the *magnetic* quantum number.

In the absence of external fields, eigenfunctions having the same n but different l and m are degenerate. We have seen that this allows us to take linear combinations of eigenfunctions, thereby arriving at completely real eigenfunctions like $2p_x$ and $2p_y$, instead of $2p_{+1}$ and $2p_{-1}$. When a magnetic field is imposed, the degeneracy no longer exists, and we are unable to perform such mixing. Under these conditions, $2p_x$, $2p_y$, $3d_{xy}$, etc. are *not* eigenfunctions, and we are restricted to the pure $m = 0, \pm 1, \pm 2 \ldots$ type solutions.

Thus far we have indicated that the stationary state functions for the hydrogenlike ions are eigenfunctions for $\hat{L}^2$ and $\hat{L}_z$, and we have compared this to the fact that $|\mathbf{L}|$ and L_z are constants of motion for a frictionless gyroscope precessing about an external field axis. But how about atoms with several electrons? And how about molecules? Are their stationary state functions also eigenfunctions for $\hat{L}^2$ and $\hat{L}_z$? A general approach to this kind of question is discussed in Chapter 6. For now we simply note that the spherical harmonics are eigenfunctions of $\hat{L}^2$ and $\hat{L}_z$ [Eqs. (4-55) and (4-57)] and that any state function of the form $\psi(r, \theta, \phi) = R(r)Y_{l,m}(\theta, \phi)$ will necessarily be an eigenfunction of these operators. But the spherical harmonics are solutions associated with spherically symmetric potentials. Therefore, it turns out that eigenfunctions of the time-independent hamiltonian operator are also eigenfunctions for $\hat{L}^2$ and $\hat{L}_z$ *only if the potential is spherically symmetric*. In the more restricted case in which ψ has

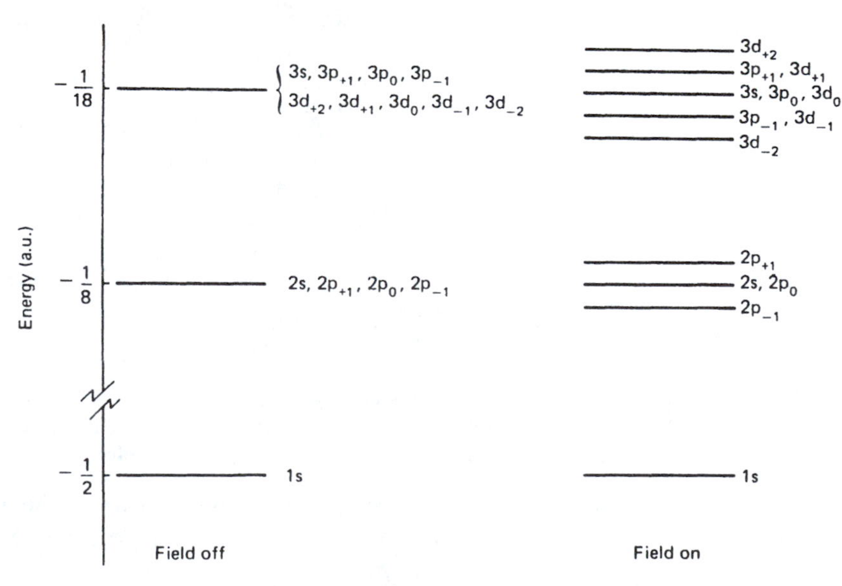

Figure 4-14 ▶ Energy levels of a hydrogenlike ion in the absence and presence of a z-directed magnetic field.

the form $\psi(r,\theta,\phi) = f(r,\theta)\exp(im\phi)$, ψ will still be an eigenfunction of $\hat{L}_z$ but not of $\hat{L}^2$. This situation applies in systems having cylindrically symmetric potentials, dependent on r and θ but not ϕ (e.g., H_2^+, N_2). We discuss such cases in more detail in Chapter 7.

4-7 Angular Momentum in Molecular Rotation—The Rigid Rotor

We have seen that the two-particle system of an electron and a nucleus rotating about a center of mass (COM) can be transformed to the one-particle system of a reduced mass rotating about a fixed point. However, this transformation can be made for *any* two-mass system, and so it applies also to the case of the *nuclei* of a rotating diatomic molecule. As we now show, the mathematical outcome for the rotating diatomic molecule is strikingly similar to that for the hydrogenlike ion.

The simplest treatment of molecular rotation ignores vibrational motion by assuming that the distance between the nuclei is fixed. The resulting model is therefore called the *rigid-rotor* model. Let there be two nuclear masses, m_1 and m_2, separated from the COM by distances r_1 and r_2, respectively. Then, because of the way that the COM is defined, we have that

$$m_1 r_1 = m_2 r_2 \tag{4-64}$$

The moment of inertia, I, is

$$I = m_1 r_1^2 + m_2 r_2^2 \tag{4-65}$$

It is not difficult to show (Problem 4-35) that the same moment of inertia results from a *reduced* mass μ rotating about a fixed point at a distance $r = r_1 + r_2$. That is, if

$$\mu = \frac{m_1 m_2}{m_1 + m_2} \tag{4-66}$$

then

$$I = \mu r^2 \tag{4-67}$$

Therefore, solving the problem of a reduced mass μ rotating about a fixed point at the fixed distance $r = r_1 + r_2$ is equivalent to solving the two-mass rigid-rotor problem. In effect, the rotating-diatomic problem is transformed to a particle-on-the-surface-of-a-sphere problem.

As usual, we write the Schrödinger equation by starting with the general prescription $[(-\hbar^2/2\mu)\nabla^2 + V]\psi = E\psi$. Then we recognize that V is constant over the spherical surface (corresponding to the diatomic molecule having no preferred orientation), so we can set $V = 0$. Since r is a constant, the first term in ∇^2 [Eq. (4-7)] vanishes due to the $\partial/\partial r$ operators. The resulting Schrödinger equation is [using Eq. (4-67)]

$$[(1/\sin\theta)(\partial/\partial\theta)\sin\theta(\partial/\partial\theta) + (1/\sin^2\theta)(\partial^2/\partial\phi^2)]\psi(\theta,\phi) = (-2IE/\hbar^2)\psi(\theta,\phi) \tag{4-68}$$

Equation (4-68) is the same as the Θ, Φ equations seen earlier [(4-36) and (4-39)] with $\beta = 2IE/\hbar$, and so we already know the eigenvalues and eigenfunctions. We noted earlier that $\beta = l(l+1)$. For molecular rotation it is conventional to symbolize the Θ quantum number as J, rather than l. This leads to

$$J(J+1) = 2IE/\hbar^2 \tag{4-69}$$

or

$$E = J(J+1)\hbar^2/2I \quad J = 0, 1, 2, \ldots \tag{4-70}$$

Since $V = 0$, E is entirely kinetic energy. Because r is not a variable, there is no analog to the principal quantum number n, and J is not limited in its highest value. (Note also that atomic units, which are designed to simplify *electronic* problems, are not normally used for molecular rotation or vibration.)

The eigenfunctions are, as before, the spherical harmonics $Y_{J,m_J}(\theta,\phi)$, with $m_J = 0, \pm1, \pm2, \ldots, \pm J$. Thus we have an s-type solution ($J = 0, m_J = 0$) that has constant value over the spherical surface, three p-type solutions ($J = 1, m_J = +1, 0, -1$), five d-type solutions, etc. For each value of J, there are $2J + 1$ eigenfunctions.

The s-type solution has zero energy. One can imagine that the reduced mass is motionless on the surface of the sphere and has equal probability for being found anywhere. This transforms back to a picture where the diatomic molecule is not rotating and where there is no preferred orientation. Since $E = 0$ when $J = 0$, we conclude that there is no zero-point energy for free rotation. (However, if rotation is *restricted* so that some orientations become *preferred*, the zero-point energy becomes finite.)

The three p-type solutions are degenerate, hence can be mixed to give real functions analogous to p_x, p_y, p_z. We could represent the p_z function by taking a globe and marking a circular region of positive phase around the northern polar region, with a matching region of negative phase around the southern pole. Clearly, these rigid-rotor wavefunctions have the same symmetries as their hydrogenlike counterparts. A p_z rotational state corresponds to a molecule (or ensemble of molecules) rotating with kinetic energy $2\hbar^2/2I$ (since $J = 1$) with little likelihood of finding the reduced mass near the equator (i.e., with little probability of finding the diatomic molecule oriented nearly perpendicular to the z axis.)

Angular momentum for the rigid rotor also follows the hydrogenlike system rules. The square of the total angular momentum equals $J(J+1)\hbar^2$, and the z component equals $m_J\hbar$.

Transitions between rotational energy levels can be detected spectroscopically. Once such energy differences are identified with specific changes in J, it becomes a simple matter to solve for r, the internuclear distance in the rigid rotor.

For example, suppose an absorption peak for $H^{81}Br$ seen at $101.58\,\text{cm}^{-1}$ is assigned to the $J = 6 \leftarrow 5$ transition. Since the energies of these levels are $42\hbar^2/2I$ and $30\hbar^2/2I$, respectively, their difference is $12\hbar^2/2I$. This energy is equal to that of the photons of light at 101.58 cm^{-1}. Solving this relation gives $I = 3.3069 \times 10^{-47}$ kg m^2. We know that this equals μr^2, and we know how to get μ from m_{H} and m_{Br} [Eq. (4-46)]. So we can solve for r, finding $r = 141.44$ pm (Problem 4-37).

4-8 Summary

1. The motion of two masses moving about a center of mass can be transformed to motion of a single *reduced* mass moving about a fixed point. The radius of rotation for the reduced mass is identical to the distance of separation of the original two masses. For hydrogenlike ions, the nuclear mass is so much greater than the electron mass that the reduced mass is almost identical to the electron mass.

2. The bound-state energies for time-independent states of the hydrogenlike ion depend on only the atomic number Z and the quantum number n (a positive integer) and vary as $-Z^2/2n^2$. This means that the energies get closer together as n increases and that there is an infinite number of such negative energy levels. Each such energy level has degeneracy n^2. A continuum of energies exists for unbound ($E > 0$) states.

3. Each stationary state wavefunction is characterized by three quantum numbers, n, l, and m, all integers, with l ranging from 0 to $n-1$ and m ranging from $-l$ to $+l$. If $l=0$, we have an s state and ψ is spherically symmetric with a cusp at the nucleus. If $l=1,2,\ldots$ we have a p, d, ... state, and ψ vanishes at the nucleus and is not spherically symmetric. In all states there is a finite probability for finding the electron beyond the classical turning point.

4. Eigenfunctions $R_{n,l}(r)\Theta_{l,m}(\theta)\Phi_m(\phi)$ with $m \neq 0$ are complex but can be mixed to form real eigenfunctions. However, if an external field causes states of different m to be nondegenerate, such mixing will produce noneigenfunctions.

5. All the stationary state eigenfunctions are orthogonal to each other, and radial and/or angular (usually planar) nodes are instrumental in this. The effect of a radial node on energy is the same as that of an angular node, so that all eigenfunctions with, say, three nodes (all radial, all angular, or a combination) are degenerate. This is peculiar to the $-r^{-1}$ potential.

6. Separation of variables is not "perfectly clean" since the differential equations for R and Θ (Eqs. (4-38) and (4-39)), are linked through β and those for Θ and Φ (Eq. 4-40) are linked through m. This leads to interdependencies in the values of n, l, and m.

7. Spherical harmonics are the angular parts of solutions to Schrödinger equations for systems having spherically symmetric potentials. These functions are eigenfunctions of $\hat{L}_z$ and $\hat{L}^2$ as well as $\hat{H}$, so such states have sharp values of L_z, $\mathbf{L}^2$, and E. The value of L_z is $m\hbar$, and for $\mathbf{L}^2$ it is $l(l+1)\hbar^2$, where l and m must be integers. In atomic units the quantity $\hbar$ does not appear in these formulas.

8. The z component of the magnetic moment due to orbital motion of a charged particle is proportional to $m\hbar$, and so m is called the magnetic quantum number. This magnetic moment contributes to the Zeeman splitting seen in spectra of hydrogenlike ions in magnetic fields.

9. Eigenfunctions other than spherical harmonics exist for $\hat{L}^2$ and $\hat{L}_z$, but these are not separable into θ- and ϕ-dependent functions. In these cases, l and m can be half-integers. These cases do not arise in orbital motion, but do arise in spin problems.

10. If V is cylindrically symmetric [i.e., $V = V(r, \theta)$], the eigenfunctions of the hamiltonian are still eigenfunctions for $\hat{L}_z$ but not for $\hat{L}^2$. Hence, $m\hbar$ still equals the z component of angular momentum for such a system. Here, z is the direction of the axis of rotational symmetry, i.e., the internuclear axis of the molecule.

11. The rules for allowed angular momentum magnitudes and orientations are the same for the rigid rotor as for the hydrogenlike ion. This leads to the following relationships in terms of the rotational quantum numbers $J = 0, 1, 2, \ldots, m_J = 0, \pm 1, \ldots, \pm J$:

Length of angular momentum vector $= \sqrt{J(J+1)}\hbar$.

Component of angular momentum perpendicular to internuclear axis $= m_J\hbar$.

Kinetic energy of rotation: $T_J = J(J+1)\hbar^2/2I$.

Degeneracy of level: $g_J = 2J + 1$.

4-8.A Problems

4-1. An observed spectroscopic transition in the hydrogen atom involves the $2p \leftarrow 1s$ transition. Using Eq. (4-8), evaluate this energy difference in units of hertz (Hz). ($1\ \mathrm{Hz} = 1\ \mathrm{s}^{-1}$.) Do the calculation using both m_e and μ. (See Appendix 10 for constants and conversion factors.) How much error in this calculation, in parts per million, is introduced by ignoring the finite mass of the nucleus (i.e., using m_e instead of μ)?

4-2. Sketch *qualitatively*, on the same r axis, $\psi^2(r)$ for a 1s state and $4\pi r^2$ for the variation of dv with r. Sketch the radial distribution function, which is the *product* of these two functions, and explain why it vanishes at $r = 0, \infty$.

4-3. For a hydrogen atom in the 1s state, $\psi = (1/\sqrt{\pi})\exp(-r)$, in atomic units.

a) Calculate the value of r (in a.u.) at the classical turning point.
b) Calculate the percentage of the electronic charge that is predicted to be beyond the classical turning point. (See Appendix 1 for useful integrals.)

4-4. Using atomic units, compute for a 1s electron of the hydrogenlike ion [$\psi = (\sqrt{Z^3/\pi})\exp(-Zr)$]. (See Appendix 1 for useful integrals.)

a) the *most probable distance* of the electron from the nucleus,
b) the *average distance* of the electron from the nucleus,
c) the distance from the nucleus of *maximum probability density*,
d) the *average value* of the potential energy ($V = -Z/r$). Note how these quantities depend on atomic number Z. Also, why do you think that, when $Z = 1$, (d) is not minus the reciprocal of (b)? Why is (d) *lower* than minus the reciprocal of (b)?

4-5. Demonstrate by integration that the 1s and 2s orbitals of the hydrogen atom are orthogonal. (See Appendix 1 for useful integrals.)

4-6. Normalize the function $r\exp(-r)\cos\theta$.

4-7. Obtain the average value of position $\bar{x}$, for a particle moving in a one-dimensional harmonic oscillator potential in a state with the normalized wavefunction

$$\psi = (\beta/48^2\pi)^{1/4}\left[(2\sqrt{\beta}x)^3 - 12\sqrt{\beta}x\right]\exp(-\beta x^2/2)$$

[There is an easy way to do this problem.]

4-8. For a particle in a one-dimensional box with boundaries at $x = 0$ and L and for any quantum number n:

a) Show how you would set up the calculation for the mean square deviation of the particle from its average position.
b) Explain qualitatively how you would expect the value of (a) to vary with quantum number n.
c) Evaluate the expression from part (a) in terms of n and L. Calculate the value for $n = 1, 2$. Discuss the relative values of these numbers for reasonableness.

4-9. Sketch the $2p_z$ and the $3d_{xy}$ wavefunctions. Demonstrate, without explicitly integrating, that these are orthogonal.

4-10. Repeat Problem 4-4, but for the $2p_0$ wavefunction.

4-11. Find an expression for the classical turning radius for a hydrogenlike ion in terms of n, l, m, and Z.

4-12. Show that the sum of the charge distributions of all five 3d orbitals is spherically symmetric.

4-13. Try to answer the following questions without looking up formulas or using pencil and paper. Use atomic units.

a) What is the energy of the hydrogen atom in the 1s state?
b) What is the energy of He^+ when $n = 1$? $n = 2$?
c) What is the degeneracy of the $n = 5$ energy level of hydrogen?
d) How many planar nodes does the $4d_{xz}$ orbital have? How many radial nodes?
e) What is the potential energy in a hydrogen atom when the electron is 0.50 a.u. from the nucleus?

4-14. You should be able to answer the following questions (use a.u.) in your head or with trivial calculations.
An unnormalized eigenfunction for the hydrogen atom is

$$\psi = (27 - 18r + 2r^2)\exp(-r/3)$$

a) What are the l and m quantum numbers for this state?
b) How many radial nodes does this function possess?
c) What is the energy of this state?
d) What is the classical turning radius for this state?

4-15. $\psi = N(6r - r^2)\exp(-r/3)\sin\theta\sin\phi$ is an eigenfunction, in a.u., for the hydrogen atom hamiltonian. N is the normalizing constant. Without looking at formulas in the text, answer the following questions by inspection:

a) Is there a node in the r coordinate? If so, where?
b) Which state is this? (Give orbital symbol—e.g., 1s, $2p_z$, etc.)

4-16. Without comparing to tabulated formulas, state whether each of the following could reasonably be expected to be an eigenfunction (unnormalized) of the hamiltonian for a hydrogenlike ion. Explain why.

a) $(27 - 18Zr + 2Z^2r^2)\exp(-2Zr/3)$
b) $r\exp(Zr/2)\sin\theta\cos\phi$
c) $r\sin\theta\exp(-i\phi)$.

4-17. Calculate the average value of x for the 1s state of the hydrogen atom. Explain why your result is physically reasonable.

4-18. a) Calculate the most probable value of θ in the $2p_z$ state of the hydrogen atom.
b) Calculate the values of θ corresponding to nodal cones in the $3d_{z^2}$ orbital.

4-19. Demonstrate that Eq. (4-30) for $3d_{xy}$ has the same angular dependence as the function xy.

4-20. Verify Eq. (4-45) using $P_1(x)$ with $P_1(x)$ and with $P_3(x)$ [from Eq. (4-42)]. What does Eq. (4-45) imply for the integral over all space of $\psi_{3p_0}\psi_{3d_{-1}}$?

4-21. Using Eqs. (4-51) and (4-43), write the normalized spherical harmonic function $Y_{3,-2}(\theta,\phi)$. For which type of hydrogenlike AO does this function give the angular dependence?

4-22. Verify Eq. (4-55).

4-23. Test the $2p_0$ eigenfunction to see if it is an eigenfunction for $\hat{L}_x$ or $\hat{L}_y$, [Eqs. (4-58) and (4-59)]. Show that the 1s function is an eigenfunction of $\hat{L}_x$, $\hat{L}_y$, $\hat{L}_z$, and $\hat{L}^2$. Explain, in terms of the vector model, this seeming violation of the discussion in the text.

4-24. Work out the value of $\hat{L}^2\psi_{2p_0}$ by brute force and show that the result agrees with Eq. (4-57).

4-25. Sketch the vector diagram (as in Fig. 4-13) for the 4f orbitals of hydrogen. How does this compare to the diagram for the 6f orbitals of He^+?

4-26. Sometimes eigenfunctions for an operator can be mixed together to produce new functions that are still eigenfunctions. Listed below are some operators with pairs of their eigenfunctions. Indicate in each case whether mixtures of these pairs will or will not continue to be eigenfunctions for the operator shown. (You should be able to do this by inspection, using your knowledge of these systems.)

Operator	Eigenfunctions	
a) $\hat{H}$(1 dim. box)	ψ_1	ψ_3
b) $\hat{H}$ (ring)	$\sin 3\phi$	$\cos 3\phi$
c) $\hat{L}_z$(ang. mom.)	$\exp(3i\phi)$	$\exp(-3i\phi)$
d) $\hat{H}$ (cubical box)	$\psi_{1,2,3}$	$\psi_{2,2,2}$
e) $\hat{H}$ (H atom)	ψ_{2s}	ψ_{2p_0}
f) $\hat{H}$ (H atom)	ψ_{3s}	ψ_{4s}

4-27. A hydrogenlike ion is in a state having a z-component of angular momentum equal to -2 a.u.

a) What is the smallest possible value of the *length* of the angular momentum vector for this state?
b) What symbol describes the state corresponding to your answer to part (a)?

4-28. Calculate the average angular momentum, $\bar{L}_z$, for a particle in a ring of constant potential having wavefunction

a) $(1/\sqrt{\pi})\sin 3\phi$
b) $(1/\sqrt{2\pi})\exp(-3i\phi)$

4-29. Evaluate each of the following integrals. Look for labor-saving approaches. Integrals are over all space unless otherwise indicated.

a) $\int \psi_{2p_x}\hat{L}_z\psi_{2p_x}dv$
b) $\int \psi_{2p_x}\hat{L}^2\psi_{2p_x}dv$
c) $\int \psi_{2p_x}\hat{L}_x\psi_{2p_x}dv$
d) $\int \psi_{3d_{x^2-y^2}}\psi_{2p_x}dv$
e) $\int_0^{2\pi} \exp(2i\phi)\exp(-3i\phi)\,d\phi$

4-30. Evaluate each of the following in a.u. (You should be able to answer these by inspection.) Note: $\psi_{n,l,m}$ stands for a hydrogen atom eigenfunction of $\hat{H}$.

a) $\hat{L}^2\psi_{3,2,1}$
b) $\hat{L}^2\psi_{2p_x}$
c) $\hat{H}\psi_{3p_x}$
d) $(1/i)(\partial/\partial\phi)\psi_{2p_{-1}}$

4-31. For each of the following operators, indicate by "yes" or "no" whether ψ_{3p_x} (with $Z=1$) is an eigenfunction. If it is, then also give the eigenvalue in a.u.

a) $-\frac{1}{2}\nabla^2 - 1/r$
b) $-\frac{1}{2}\nabla^2 - 3/r$
c) $\hat{L}_z$
d) $-\frac{1}{2}\nabla^2$
e) $\hat{L}_x$
f) $\hat{L}^2$

g) r
h) $1/r$

4-32. For the 3s state of the hydrogen atom, estimate the amount of electronic charge between 211 pm and 213 pm.

4-33. Calculate in hertz the splitting between $2p_0$ and $2p_{+1}$ of a hydrogen atom by a magnetic field of 2 tesla. Compare this with the $2p \longleftarrow 1s$ transition energy (in parts per million).

4-34. Try to answer (by inspection) the following questions about the $n = 4$ states of the hydrogen atom.

a) What is the energy of the level, in a.u.?
b) What is the degeneracy of the level?
c) What values for the length of the orbital angular momentum vector (in a.u.) are possible?
d) Into how many sublevels (of energy) is the $n = 4$ level split by imposition of a magnetic field?
e) What is the degeneracy of the *unshifted* portion of the sublevels referred to in (d)?

4-35. Show that the moment of inertia for two masses, m_1 and m_2, moving on a rigid massless bar about the center of mass at distances r_1 and r_2, respectively, is identical to that of a reduced mass $\mu = m_1 m_2/(m_1 + m_2)$ moving about a point at a distance of $r = r_1 + r_2$.

4-36. Show that Eq. (4-68) can be written, using $\hat{L}^2$, in a form that is analogous to the classical relation for a freely rotating mass, $\mathbf{L}^2/2I = T$.

4-37. Using that $m_{\mathrm{H}} = 1.0078$ a.m.u. and $m_{\mathrm{Br}} = 80.9163$ a.m.u., calculate μ and verify the value of r given at the end of Section 4-7.

4-38. Assuming an internuclear distance of 127.5 pm in $D^{35}Cl$, compute the expected positions in cm^{-1} for the absorption peaks corresponding to $J = 1 \leftarrow 0$, $2 \leftarrow 1$, $3 \leftarrow 2$ ($D = 2.0141$ a.m.u., $^{35}Cl = 34.9688$ a.m.u.).

4-39. The $J = 1 \leftarrow 0$ transition in $^{12}C^{16}O$ occurs at $3.86\ \mathrm{cm}^{-1}$. Calculate the internuclear distance. ($^{12}C = 12$ a.m.u. by definition, $^{16}O = 15.9949$ a.m.u.)

4-40. HCl has a permanent electric dipole moment, which means that the *reduced* mass has a partial electric charge in the transformed version of the system. This in turn means that there is a magnetic moment vector parallel to the total angular momentum vector. Describe qualitatively what happens to the energies of the $J = 3$ rotational states when HCl is subjected to a uniform magnetic field.

4-41. Uncertainty in position in one dimension, Δx, is defined as $\left[\overline{x^2} - \bar{x}^2\right]^{1/2}$ That is, it is the square root of the difference between the average squared position and the square of the average position. Calculate Δr for the 1s state of the hydrogen atom.

Multiple Choice Questions

(Try to answer these without referring to the text or using pencil and paper.)

1. A particle on the surface of a sphere has quantum number $J = 7$. The energy level to which this state belongs has a degeneracy of

a) 56
b) 49
c) 42
d) 14
e) None of these.

2. A particle on the surface of a sphere in the state having $J = 4$, $m_J = 4$

a) has $E = 16\hbar^2/2I$.
b) has a z-component of angular momentum of $4\hbar$.
c) doesn't exist because this state violates quantum number rules.
d) has a degeneracy of 20.
e) None of the above is a true statement.

3. HI and DI are made to undergo the same transition (say $J = 11 \longleftarrow J = 10$, but the particular J values are not important). The light frequency inducing the transition for HI is equal to ν. Approximately which frequency would you expect to induce the same transition for DI?

a) 2ν
b) $\nu/2$
c) $\sqrt{2}\nu$
d) $\nu/\sqrt{2}$
e) None of these.

4. The electronic energy of Li^{2+} in the 2s state is

a) the same as that of H in the 1s state.
b) nine times that of H in the 1s state.
c) one-fourth that of H in the 1s state.
d) four-ninths that of H in the 1s state.
e) nine-fourths that of H in the 1s state.

5. Consider the following expressions, where ψ is a hydrogen-atom wavefunction.

1) $\int^{allspace} \psi^* r \psi \, dv$
2) $\frac{d}{dr}\psi^*\psi r^2 \sin\theta = 0$
3) $\frac{d}{dr}\psi^*\psi = 0$
4) $\frac{d}{dr}\psi = 0$

Which one of the following is a true statement?

a) Expression 1 is equal to unity if ψ is normalized.
b) Expression 4 is true when r is the position of a radial node.
c) Expression 3 is true everywhere because $\psi^*\psi$ is a constant.
d) Expression 2 is true when r is at its most probable value.
e) Expression 3 is true when r is at its most probable value.

6. A $5d_{xy}$ atomic orbital has

a) 2 planar and 1 radial nodes.
b) 2 planar and 3 radial nodes.
c) 3 planar and 1 radial nodes.
d) 5 nodes all together.
e) None of the above is correct.

7. For a hydrogen atom in an $n = 4$ state, the maximum possible z-component of orbital angular momentum is

a) $2\hbar$
b) $3\hbar$
c) $\sqrt{12}\hbar$
d) $\sqrt{6}\hbar$
e) None of the above is correct.

8. The following is an eigenfunction for the hydrogen atom:
$\psi = \left[1/\sqrt{32\pi a_0^3}\right](r/a_0)\exp(-r/2a_0)\cos\theta$
Which one of the following statements about ψ is true?

a) The term on the left, up to the exponential, is the normalizing constant.
b) This ψ has a nonzero value at the nucleus.
c) For this state, $l = 1$ and $m_l = 0$.
d) This state has a spherical electron cloud distribution.
e) None of the above statements is true.

9. The radial distribution function for a 1s state, $4\pi r^2\psi_{1s}^2$, indicates that

a) the most probable value of the distance from the nucleus is zero.
b) the average value of r is zero.
c) the average value of r is greater than the most probable value.
d) the average value of r is equal to the most probable value.
e) the electron cloud density per cubic picometer is greatest at a radius other than zero.

References

[1] H. Eyring, J. Walter, and E. D. Kimball, *Quantum Chemistry*, Chapter VI. Wiley, New York, 1944.

[2] I. N. Levine, *Quantum Chemistry*, 5th ed. Prentice Hall, Upper Saddle River, New Jersey, 2000.

[3] L. Pauling and E. B. Wilson, Jr., *Introduction to Quantum Mechanics*. McGraw-Hill, New York, 1935.

[4] R. N. Zare, *Angular Momentum*. Wiley, New York, 1988.

Chapter 5

Many-Electron Atoms

5-1 The Independent Electron Approximation

In previous chapters we have dealt with the motion of a single particle in various potential fields. When we deal with more than one particle, new problems arise and new techniques are needed. Some of these are discussed in this chapter.

In constructing the hamiltonian operator for a many electron atom, we shall assume a fixed nucleus and ignore the minor error introduced by using electron mass rather than reduced mass. There will be a kinetic energy operator for each electron and potential terms for the various electrostatic attractions and repulsions in the system. Assuming n electrons and an atomic number of Z, the hamiltonian operator is (in atomic units)

$$H(1,2,3,\ldots,n) = -\frac{1}{2}\sum_{i=1}^{n}\nabla_i^2 - \sum_{i=1}^{n}(Z/r_i) + \sum_{i=1}^{n-1}\sum_{j=i+1}^{n}\frac{1}{r_{ij}} \tag{5-1}$$

The numbers in parentheses on the left-hand side of Eq. (5-1) symbolize the spatial coordinates of each of the n electrons. Thus, 1 stands for x_1, y_1, z_1, or r_1, θ_1, ϕ_1, etc. We shall use this notation frequently throughout this book. Since we are not here concerned with the quantum-mechanical description of the translational motion of the atom, there is no kinetic energy operator for the nucleus in Eq. (5-1). The index i refers to the electrons, so we see that Eq. (5-1) provides us with the desired kinetic energy operator for each electron, a nuclear electronic attraction term for each electron, and an interelectronic repulsion term for each *distinct* electron pair. (The summation indices guarantee that $1/r_{12}$ and $1/r_{21}$ will not *both* appear in H. This prevents counting the same physical interaction twice. The indices also prevent nonphysical self-repulsion terms, such as $1/r_{22}$, from occurring.) Frequently used alternative notations for the double summation in Eq. (5-1) are $\frac{1}{2}\sum_{i\neq j}^{n} 1/r_{ij}$, which counts each interaction twice and divides by two, and $\sum_{i<j}^{n}$ or $\sum_{i,j}'$ which is merely a shorthand symbol for the expression in Eq. (5-1). In each of these alternative notations, the summation is still over two indices, but the second $\sum$ symbol is "understood."

For the helium atom, Eq. (5-1) becomes (see Figure 5-1)

$$H(1,2) = -\frac{1}{2}\nabla_1^2 - \frac{1}{2}\nabla_2^2 - (2/r_1) - (2/r_2) + (1/r_{12}) \tag{5-2}$$

The helium hamiltonian (5-2) can be rewritten as

$$H(1,2) = h(1) + h(2) + 1/r_{12} \tag{5-3}$$

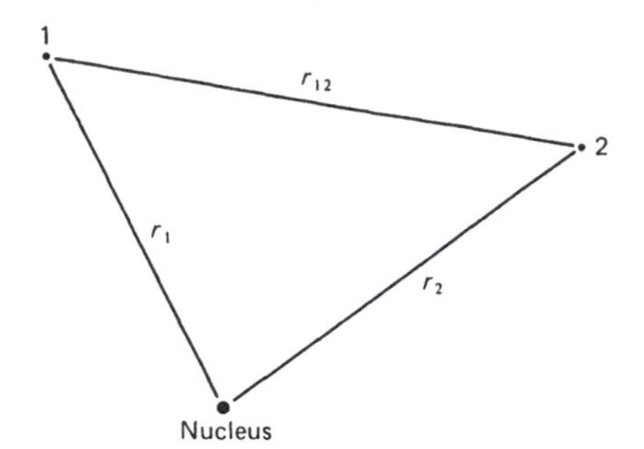

Figure 5-1 ▶ Interparticle coordinates for a three-particle system consisting of two electrons and a nucleus.

where

$$h(i) = -\frac{1}{2}\nabla_i^2 - 2/r_i \tag{5-4}$$

In Eq. (5-3) we have merely grouped H into two one-electron operators and one two-electron operator. There is no way to separate this hamiltonian *completely* into a sum of one-electron operators without loss of rigor. However, if we wish to *approximate* the hamiltonian for helium in such a way that it becomes separable, we might try simply ignoring the interelectronic repulsion term:

$$H_{\text{approx}} = h(1) + h(2) \tag{5-5}$$

If we do this, our approximate hamiltonian H_{approx} treats the kinetic and potential energies of each electron *completely independently* of the motion or position of the other. For this reason, such a treatment falls within the category of "independent electron approximations."

Notice that each individual one-electron hamiltonian (5-4) is just the hamiltonian for a hydrogenlike ion, so it has as eigenfunctions the 1s, 2s, 2p, etc., functions of Chapter 4 with $Z = 2$. Such one-electron functions are referred to as *atomic orbitals.*[1] Representing them with the symbol ϕ_i (e.g., ϕ_1 = 1s, ϕ_2 = 2s, $\phi_3 = 2p_x$, $\phi_4 = 2p_y$, etc.) we have, then,

$$h(1)\phi_i(1) = \epsilon_i\phi_i(1) \tag{5-6}$$

where ϵ_i is referred to as the *orbital energy*, or *one-electron energy* for atomic orbital ϕ_i. As we saw in Chapter 4, ϵ_i is given in atomic units by

$$\epsilon_i = -\frac{1}{2}Z^2/n^2 \tag{5-7}$$

where n is the principal quantum number for ϕ_i, and Z is the nuclear charge. The "1" in Eq. (5-6) indicates that $\phi_i(1)$ is a function whose variable is the position of electron 1.

We will now show that *products of the atomic orbitals ϕ are eigenfunctions of H_{approx}*. Let the general product of atomic orbitals for helium be written $\phi_i(1)\phi_j(2)$. Then

$$H_{\text{approx}}\phi_i(1)\phi_j(2) = (h(1) + h(2))\phi_i(1)\phi_j(2) \tag{5-8}$$

$$= h\,(1)\,\phi_i(1)\phi_j(2) + h\,(2)\,\phi_i(1)\phi_j(2) \tag{5-9}$$

[1]The term "atomic orbital" is used for any one-electron function used to describe the electronic distribution about an atom.

But $h(1)$ does not contain any of the variables in $\phi_j(2)$, and so they commute. Similarly, $h(2)$ and $\phi_i(1)$ commute, and

$$\begin{aligned} H_{\text{approx}}\phi_i(1)\phi_j(2) &= \phi_j(2)h(1)\phi_i(1) + \phi_i(1)h(2)\phi_j(2) \\ &= \phi_j(2)\epsilon_i\phi_i(1) + \phi_i(1)\epsilon_j\phi_j(2) \quad \text{[from Eq. (5-6)]} \\ &= (\epsilon_i + \epsilon_j)\,\phi_i(1)\phi_j(2) = E\phi_i(1)\phi_j(2). \end{aligned} \tag{5-10}$$

Thus, $\phi_i(1)\phi_j(2)$ is an eigenfunction of H_{approx}, and the eigenvalue E is equal to the sum of the orbital energies. These results are yet another example of the general rules stated in Section 2-7 for separable hamiltonians. Indeed, once we recognized that H_{approx} is separable, we could have written these results down at once.

Since the above terminology and results are so important for understanding many quantum-chemical calculations, we will summarize them here:

1. The hamiltonian for a multielectron system cannot be separated into one-electron parts without making some approximation.
2. Ignoring interelectron repulsion operators is one way to allow separability.
3. The one-electron operators in the resulting approximate hamiltonian for an atom are hydrogenlike ion hamiltonians. Their eigenfunctions are called *atomic orbitals*.
4. Simple products of atomic orbitals are eigenfunctions for the approximate hamiltonian.
5. *In this approximation* the total energy is equal to the sum of the one-electron energies.

EXAMPLE 5-1 What electronic energy is predicted by the above approximation for the lithium atom in its ground state? What is the experimental value for the total electronic energy, given that the first and second ionization energies are 0.198 a.u. and 2.778 a.u.?

SOLUTION ► The ground state configuration for lithium is $1s^2 2s$, so $E_{approx} = 2\varepsilon_{1s} + \varepsilon_{2s} = 2(-\frac{1}{2}\cdot\frac{3^2}{1^2}\text{ a.u.}) + (-\frac{1}{2}\cdot\frac{3^2}{2^2}\text{ a.u.}) = -10.125\text{ a.u.}$ The experimental value of E equals minus the sum of all three ionization energies. The first two values are given, and the third can be calculated using the formula for one-electron ions: $IE_3 = -E_{Li^{2+}} = -(-\frac{1}{2}\cdot\frac{3^2}{1^2}) = 4.500\text{ a.u.}$ Therefore, $E_{\text{exp}} = -(0.198 + 2.778 + 4.500)\text{ a.u.} = -7.476\text{ a.u.}$ Clearly, the approximate hamiltonian predicts an electronic energy that is much lower than the experimental value. ◄

5-2 Simple Products and Electron Exchange Symmetry

In the independent particle model just described, the wavefunction for the lowest-energy state for helium is 1s(1)1s(2) since this has the lowest possible sum of one-electron energies. The *electronic configuration* for this state is symbolized $1s^2$, the superscript telling us how many electrons are in 1s orbitals. What might we expect for the electronic configuration of the lowest excited state? The answer is 1s2s (superscript "ones" are implicit). (At this point there is no reason for preferring this configuration to, say,

$1s2p_x$, but we shall show later that, in multielectronic systems, the 2s orbital has a lower energy than does a 2p orbital, even though they have the same principal quantum number.) Thus, we might write

$$\psi(1,2) = 1s(1)2s(2) \equiv \underbrace{\sqrt{8/\pi}\exp(-2r_1)}_{\text{He}^+\ 1s}\underbrace{\sqrt{1/\pi}(1-r_2)\exp(-r_2)}_{\text{He}^+\ 2s} \tag{5-11}$$

If one were to calculate $\bar{r}_1$, the average distance from the nucleus for electron 1, using this wavefunction, one would obtain a value of $\frac{3}{4}$ a.u., consistent with the 1s state of a helium ion. For electron 2 one would find an average value, $\bar{r}_2$, of 3 a.u., characteristic of the 2s state (Problem 5-2). How does this correspond to what we would find experimentally?

Before answering this question, we must recall that there are special problems associated with measuring the properties of an atomic system. The process of "seeing" electrons in atoms well enough to pinpoint their positions perturbs an atom so strongly that it cannot be assumed to be in the same state after the measurement. To get around this problem, we can assume that we have a very large number of identically prepared helium atoms, and that a single measurement of electronic positions will be made on each atom. It is assumed that the average of the instantaneous r values for a billion systems is identical to the average r value for a billion instants in a single undisturbed system.

When we consider the measurement of average values for r_1 and r_2 in helium, we immediately encounter another problem. Say we can effect a simultaneous measurement of the two electronic distances in the first He atom. We call these r_1 and r_2 and tabulate them for future averaging. Then we move on to a new helium atom and measure r_1 and r_2 for it. But we clearly have no way of identifying a particular one of these electrons with a particular one of the earlier pair. There is no connection between r_1 for one atom and r_1 for the next since all electrons are identical. If we want to know $\bar{r}$, we can only average them all together and leave it at that.

Thus, the wavefunction (5-11) does not seem to be entirely satisfactory since it enables us to calculate $\bar{r}_1 \neq \bar{r}_2$, something that is *in principle* impossible to measure. We need to modify the wavefunction so that it yields an average value for r_1 and r_2 (or for any quantity) that is independent of our choice of electron labels. This means that the electron density itself, given by $\psi(1,2)^2$, must be independent of our electron labeling scheme.

In a two-electron system like helium, there are only two ways to arrange the labels "1" and "2" in a single product function. For example, the product 1s2s can be written

$$1s(1)2s(2) \quad \text{or} \quad 2s(1)1s(2) \tag{5-12}$$

Squaring these gives two different functions, namely,

$$\begin{aligned} 1s^2(1)2s^2(2) &= (8/\pi)\exp(-4r_1)(1/\pi)(1-2r_2+r_2^2)\exp(-2r_2) \\ 2s^2(1)1s^2(2) &= (8/\pi)\exp(-4r_2)(1/\pi)(1-2r_1+r_1^2)\exp(-2r_1) \end{aligned} \tag{5-13}$$

These are different since they predict, for instance, different distributions for electron 1. The functions (5-12) are said to differ by an interchange of electron indices, or coordinates. (Since electron labels denote position coordinates, interchange of labels in the mathematical formula corresponds to interchanging positions of electrons in the physical model.) For ψ^2 to be invariant under such an interchange, it is necessary that

ψ itself be either symmetric or antisymmetric under the interchange. That is, if P is an interchange operator such that $Pf(1,2) = f(2,1)$ then we need a ψ such that

$$P\psi = \pm\psi \tag{5-14}$$

since then

$$P(\psi^2) = (P\psi)^2 = (\pm\psi)^2 = \psi^2$$

One such wavefunction is given by the *sum* of eigenfunctions (5-12),

$$\psi_s = (1/\sqrt{2})[1s(1)2s(2) + 2s(1)1s(2)] \tag{5-15}$$

since

$$P\psi_s = (1/\sqrt{2})[1s(2)2s(1) + 2s(2)1s(1)] = \psi_s$$

(the factor $1/\sqrt{2}$ keeps the wavefunction normalized). Wavefunction (5-15) is thus symmetric under electron interchange. Is Eq. (5-15) still an eigenfunction for H_{approx}? Yes, because the eigenfunctions (5-12) are degenerate (both have $E = \epsilon_{1s} + \epsilon_{2s}$) and can therefore be mixed together in any way we choose and still be eigenfunctions. The antisymmetric combination is

$$\psi_a = (1/\sqrt{2})[1s(1)2s(2) - 2s(1)1s(2)] \tag{5-16}$$

Thus far we have shown that simple products of atomic orbitals give us two degenerate eigenfunctions of H_{approx} for the configuration 1s2s and that these eigenfunctions fail to have the required symmetry properties for interchange of electron coordinates. But we have shown that, by taking the sum and difference of these simple products, we can form new eigenfunctions of H_{approx} that are respectively symmetric and antisymmetric with respect to the interchange of electron coordinates, so that ψ^2 is invariant to electron interchange.

There is another way we can demonstrate that the helium atom eigenfunctions ought to be symmetric or antisymmetric for electron exchange: We can examine the hamiltonian operator. We have shown in Chapter 2 that nondegenerate eigenfunctions must be symmetric or antisymmetric for any operation that leaves the hamiltonian unchanged and that degenerate eigenfunctions may always be mixed together in some combination so that they too are symmetric or antisymmetric. This suggests that, for the case under discussion (the helium atom), the hamiltonian operator might be unchanged by an exchange of electrons. First we examine H_{approx}:

$$PH_{\text{approx}} = P[h(1) + h(2)] = h(2) + h(1) = H_{\text{approx}} \tag{5-17}$$

Our approximate hamiltonian is invariant to electron exchange, so any *nondegenerate* eigenfunctions must be symmetric or antisymmetric for interchange of electron labels (or positions). Only because the 1s2s configuration leads to *degenerate* eigenfunctions were we able to find unsymmetric[2] eigenfunctions like Eq. (5-12). This situation is reminiscent of the particle-in-a-ring system discussed in Chapter 2, where degenerate,

---

[2]A function is unsymmetric for any operation that produces neither plus nor minus that function; i.e., if $Pf = y$ and $y \neq \pm f$, then f is unsymmetric under the operation P.

symmetric exponential eigenfunctions could be mixed to form degenerate unsymmetric trigonometric eigenfunctions. Let us now examine the *full* hamiltonian H:

$$PH(1,2) = P[h(1) + h(2) + 1/r_{12}] = h(2) + h(1) + 1/r_{21} = H(1,2) \qquad (5\text{-}18)$$

Since r_{12} and r_{21} are the same distance, it is evident that the exact H is likewise invariant to interchange of electron labels. Thus, we see that appeal either to physical argument or to the invariance of H and of H_{approx} to exchange of electrons leads us to recognize the need to impose symmetry conditions on the wavefunctions.

We now summarize the points we have tried to convey in this section.

1. A simple product function of the type 1s(1)2s(2) enables one to calculate different values of $\bar{r}$ for electrons 1 and 2. This makes no physical sense since the electrons are *identical particles* and hence are not physically distinguishable.

2. Wavefunctions that overcome this difficulty must be either symmetric or antisymmetric with respect to exchange of electron labels (coordinates).

3. The fact that this kind of "exchange symmetry" must be present is also (or alternatively) seen from the fact that H (and also H_{approx}) is invariant under such an exchange operation.

EXAMPLE 5-2 Given the functions $f(x_1) = x_1^2$ and $g(x_2) = \exp(x_2)$, show that, for $x_1 = 1, x_2 = 2$, $f(x_1)g(x_2)$ is unsymmetric for exchange of the two x positions, $f(x_1)g(x_2) + g(x_1)f(x_2)$ is symmetric, and $f(x_1)g(x_2) - g(x_1)f(x_2)$ is antisymmetric.

SOLUTION ► For fg, we are examining what happens when $x_1^2 \exp(x_2)$ turns into $x_2^2 \exp(x_1)$. That is, we are comparing $1^2 \exp(2)$ to $2^2 \exp(1)$. The resulting values are 7.389 and 10.873—obviously neither plus or minus times each other. $fg + gf$ equals $1^2 \exp(2) + 2^2 \exp(1)$, or $7.389 + 10.873$. After switching positions, we get $10.873 + 7.389$, which is obviously the same thing. $fg - gf$ gives $7.389 - 10.873$. After switching, it gives $10.873 - 7.389$, which is obviously minus one times the first value. ◄

5-3 Electron Spin and the Exclusion Principle

Chemical and spectral evidence indicates that metals in Groups IA and IB of the periodic table are reasonably well represented by an electron configuration wherein one loosely held "valence" electron occupies an s orbital and all other electrons occur in pairs in orbitals of lower principal quantum number. Thus, lithium has a ground-state electronic structure approximated by the configuration $1s^2 2s$, sodium by $1s^2 2s^2 2p^6 3s$, copper by $1s^2 2s^2 2p^6 3s^2 3p^6 3d^{10} 4s$, etc. (A configuration indicating that all orbitals of given n and l are doubly occupied, leaving no other electrons, is often called a *closed shell*. Thus, the above-cited examples each consist of a closed shell plus one s valence electron.) The observation that each atomic orbital in such configurations is occupied by no more than two electrons was without a theoretical explanation for some time.

When an atom like sodium is placed in an external magnetic field, what should be the magnetic moment of the atom due to orbital motion of the electrons? The s electrons should contribute nothing since, by definition of s, $l=0$ and hence the magnetic quantum number $m=0$ for such electrons. An electron in a p orbital may have an orbital magnetic moment, but if all p levels ($l=1, m=+1, 0, -1$) are equally occupied, the net magnetic moment should be zero. It is clear, then, that we might expect atoms in Groups IA and IB to possess no magnetic moment due to electron orbital motion. Nevertheless, Gerlach and Stern [1, 2] found that, when a beam of unexcited silver atoms is passed through an inhomogeneous magnetic field, it splits into two components as though each silver atom possesses a small magnetic moment capable of taking on either of two orientations in the applied field. (In a *homogeneous* magnetic field, the north and south poles of a magnetic dipole experience equal but oppositely directed forces, causing the dipole to become *oriented*. An example is a magnetic compass in the magnetic field of the earth. In an *inhomogeneous* magnetic field the poles experience opposite but unequal forces, causing the entire dipole to be *accelerated through space* in addition to being oriented.) Uhlenbeck and Goudsmit [3] and Bichowsky and Urey [4] independently suggested that the electron behaves as though it were a particle of finite radius spinning about its center of mass. Such a spinning particle would classically have angular momentum and, since it is charged, an accompanying magnetic moment.[3]

If we accept the model of electron spin, then we can rationalize our experimental facts if we assume each electron is capable of being in one of but two possible states of opposite spin. This is done in the following way. If we attribute *opposite* spins to the two 1s electrons in, say, silver, their spin moments should cancel. Similarly, all other orbital-sharing electrons would contribute nothing if their spins were opposed. Only the outermost (5s) electron would have an uncanceled spin moment. Its two possible orientations would cause the beam to split into two components as is observed.[4]

The evident need for the introduction of the concept of electron spin means that our wavefunctions of earlier sections are incomplete. We need a wavefunction that tells us not only the probability that an electron will be found at given r, θ, ϕ coordinates in three-dimensional space, but also the probability that it will be in one or the other spin state. Rather than seeking detailed mathematical descriptions of spin state functions, we will simply symbolize them α and β. Then the symbol $\phi(1)\alpha(1)$ will mean that electron number 1 is in a spatial distribution corresponding to space orbital ϕ, and that it has spin α. In the independent electron scheme, then, we could write the *spin orbital* (includes space and spin parts) for the valence electron of silver either as $5s(1)\alpha(1)$ or $5s(1)\beta(1)$. These two possibilities both occur in the atomic beam and interact differently with the inhomogeneous magnetic field.

We now focus on the manner in which spin considerations affect wavefunction symmetry. The electrons are still identical particles, so our particle distribution must be

[3]This classical model, developed in the 1920s, is pedagogically useful and is responsible for the term *spin*, which is still employed to describe the fourth quantum number. However, it was not until 1948 and 1967 that mathematical studies of the properties of linearized equivalents of the Schödinger equation revealed the mathematical connection to this quantum number. For an entry to the literature, see Roman [10].

[4]Actually, other experimental evidence, such as splitting of atomic spectral lines due to applied magnetic fields, was also available. Furthermore, experience with the quantum theory of orbital angular momentum played a role in the treatment of electron spin. The reader should not think that the historical development of quantum theory of spin was as naive or simple as we make it appear here.

insensitive to our choice of labels. This last statement is equivalent to saying that ψ must be symmetric or antisymmetric for interchange of electron *space and spin coordinates*. Let us examine this situation in the case of ground state helium and lithium atoms.

In the independent electron approximation, the lowest-energy configuration for helium is $1s^2$. Let us write the various conceivable spin combinations for this configuration. They are

$$\left.\begin{array}{l} 1s(1)\alpha(1)1s(2)\alpha(2) \\ 1s(1)\alpha(1)1s(2)\beta(2) \\ 1s(1)\beta(1)1s(2)\alpha(2) \\ 1s(1)\beta(1)1s(2)\beta(2) \end{array}\right\} = 1s(1)1s(2)\left\{\begin{array}{ll} \alpha(1)\alpha(2) & \quad (5\text{-}19) \\ \alpha(1)\beta(2) & \quad (5\text{-}20) \\ \beta(1)\alpha(2) & \quad (5\text{-}21) \\ \beta(1)\beta(2) & \quad (5\text{-}22) \end{array}\right.$$

It is easy to see that the common space term 1s(1)1s(2) is symmetric for electron interchange. Likewise, $\alpha(1)\alpha(2)$ and $\beta(1)\beta(2)$ are each symmetric, so Eqs. (5-19) and (5-22) are totally symmetric wavefunctions. The spin parts of Eqs. (5-20) and (5-21) are unsymmetric (not antisymmetric) for interchange, so these wavefunctions are not satisfactory. However, we can take the sum and difference of Eqs. (5-20) and (5-21) to obtain

$$1s(1)1s(2)\left\{\begin{array}{ll} (1/\sqrt{2})[\alpha(1)\beta(2)+\beta(1)\alpha(2)] & \quad (5\text{-}23) \\ (1/\sqrt{2})[\alpha(1)\beta(2)-\beta(1)\alpha(2)] & \quad (5\text{-}24) \end{array}\right.$$

The $2^{-1/2}$ serves to maintain normality if we assume α and β to be orthonormal:

$$\int \alpha^*(1)\alpha(1)d\omega(1) = \int \beta^*(1)\beta(1)\,d\omega(1) = 1 \qquad (5\text{-}25)$$

$$\int \alpha^*(1)\beta(1)d\omega(1) = \int \beta^*(1)\alpha(1)\,d\omega(1) = 0 \qquad (5\text{-}26)$$

Here we use integrals and a differential element $d\omega$ in a "spin coordinate ω." This is notationally convenient but not, for our purposes, worth delving into. We can interpret integration over ω to be in effect equivalent to summing over the possible electron indices. If, for a particular electron index, the spins agree, then the integral equals unity. If they disagree, the integral vanishes. Wavefunction (5-23) consists of symmetric space and spin parts, so it is overall symmetric. Wavefunction (5-24) contains a symmetric space part times an antisymmetric spin part, so it is overall antisymmetric. We have succeeded, then, in writing down four wavefunctions for the configuration $1s^2$ having proper symmetry for electron interchange. Three of these, Eqs. (5-19), (5-22), (5-23), are symmetric and one, Eq. (5-24), is antisymmetric. Experimentally, we know that the ground state of helium is a *singlet*, that is, there is but one such state. *This suggests that the wavefunction must be antisymmetric for exchange of electron space and spin coordinates.*

EXAMPLE 5-3 We have just shown four wavefunctions resulting from four spin functions times a symmetric space part ($1s^2$). Can we manipulate the $1s^2$ configuration to obtain an **antisymmetric** space part, as we did for the 1s2s configuration?

SOLUTION ▶ We can try to produce an antisymmetric space function by taking the product difference 1s(1)1s(2)-1s(2)1s(1). Since these two products are really the same, this combination equals zero. Thus, whereas two different orbitals can be arranged in two product combinations, one symmetric and the other antisymmetric, we find that a single orbital, doubly occupied, can appear only in one simple product, which must be symmetric for electron exchange. ◀

Now let us try lithium. The lowest-energy configuration should be $1s^3$, and we can write eight unique space-spin orbital products:

$$1s(1)1s(2)1s(3)\begin{cases}\alpha(1)\alpha(2)\alpha(3) & (5\text{-}27)\\ \alpha(1)\alpha(2)\beta(3) & (5\text{-}28)\\ \alpha(1)\beta(2)\alpha(3) & (5\text{-}29)\\ \beta(1)\alpha(2)\alpha(3) & (5\text{-}30)\\ \alpha(1)\beta(2)\beta(3) & (5\text{-}31)\\ \beta(1)\alpha(2)\beta(3) & (5\text{-}32)\\ \beta(1)\beta(2)\alpha(3) & (5\text{-}33)\\ \beta(1)\beta(2)\beta(3) & (5\text{-}34)\end{cases}$$

Of these, the first and last are totally symmetric for all electron interchanges. The remaining six are unsymmetric for two out of three possible interchanges. Can we make appropriate linear combinations of these as we did for helium? Let us try. The problem is simplified by recognizing that, if we start with, say, two α's and one β, we still have that number of α's and β's after interchange of electron labels. Hence, we mix together only functions that agree in total numbers of α's and β's, i.e., (5-28), (5-29), (5-30) with each other, or (5-31), (5-32), (5-33) with each other. Let us try the sum of (5-28), (5-29), and (5-30). Ignoring normalization, this gives the spin function

$$\alpha(1)\alpha(2)\beta(3)+\alpha(1)\beta(2)\alpha(3)+\beta(1)\alpha(2)\alpha(3) \tag{5-35}$$

Interchanging electron spin coordinates 1 and 2 gives

$$\alpha(2)\alpha(1)\beta(3)+\alpha(2)\beta(1)\alpha(3)+\beta(2)\alpha(1)\alpha(3)$$

which, upon reordering each product, is easily seen to be identical to (5-35). The same result arises from interchanging 1 and 3 or 2 and 3, and so (5-35) is symmetric for all interchanges. The sum of (5-31), (5-32), and (5-33) is likewise symmetric. Can we find any combinations that are totally antisymmetric? A few attempts with pencil and paper should convince one that it is impossible to find a combination that is antisymmetric for all interchanges. Experimentally, we know that no state of lithium corresponds to a $1s^3$ configuration.

To summarize, we have found that for the configuration $1s^2$ we can write three wavefunctions that are symmetric and one that is antisymmetric under exchange of electron space and spin coordinates, while for the configuration $1s^3$ we can construct symmetric or unsymmetric wavefunctions, but no antisymmetric ones. The physical observation is that atoms exist in only *one* state having an electronic structure approximately represented by the configuration $1s^2$, but that there are no atoms having any state represented by $1s^3$. This and other physical evidence has led to the recognition of the *exclusion principle*: Wavefunctions must be *antisymmetric* with respect to simultaneous interchange of space *and* spin coordinates of electrons.[5] In invoking the exclusion principle, we exclude all of the $1s^3$ wavefunctions and three out of the four wavefunctions we were able to construct for the ground state of helium, leaving (5-24) as the only acceptable wave-function.

We have seen that the ground state configuration of lithium cannot be $1s^3$. Can we satisfy the exclusion principle with the next-lowest energy configuration $1s^2 2s$? We will try to find a satisfactory solution, but our manipulations will be simplified if we streamline our notation. We will write a function such as $1s(1)1s(2)2s(3)\alpha(1)\beta(2)\alpha(3)$ as $1s1s2s\alpha\beta\alpha$, allowing *position* in the sequence to stand for the electron label. Interchanging electrons 1 and 2 is then represented by switching the order of space functions in positions 1 and 2 and spin functions in positions 1 and 2 thus

$$1s1s2s\alpha\beta\alpha \xrightarrow{1\rightleftharpoons 2} 1s1s2s\beta\alpha\alpha \tag{5-36}$$

This interchange produced a new function rather than merely reversing the sign of our starting function. But if we take the difference between the two products in Eq. (5-36), we will have a function that is antisymmetric to 1, 2 interchange: $1s1s2s(\alpha\beta\alpha - \beta\alpha\alpha)$. Now we subject this to a 1, 3 interchange and the new products produced are subtracted to give a function that is antisymmetric to 1, 3 interchange: $1s1s2s(\alpha\beta\alpha - \beta\alpha\alpha) - 2s1s1s(\alpha\beta\alpha - \alpha\alpha\beta)$. The first pair of terms is still not antisymmetric to 2, 3 interchange, and the second pair is not antisymmetric to 1, 2 interchange. We can use either one of these interchanges to produce two new terms to subtract. Either way, the resulting wavefunction, antisymmetric for all interchanges, is

$$\frac{1}{\sqrt{6}}[1s1s2s(\alpha\beta\alpha - \beta\alpha\alpha) + 1s2s1s(\beta\alpha\alpha - \alpha\alpha\beta) + 2s1s1s(\alpha\alpha\beta - \alpha\beta\alpha)] \tag{5-37}$$

The factor $6^{-1/2}$ normalizes (5-37) since all of the six space-spin products are normalized and orthogonal to each other product by virtue of either space-orbital or spin-orbital disagreement, or both. Note that, whereas the two-electron wavefunction for helium was separable into a single space function times a spin function, the lithium wavefunction must be written as a linear combination of such products. This is usually true when we deal with more than two electrons.

Since $1s^2 2s$ is the lowest-energy configuration for which we can write an antisymmetrized wavefunction, this is the ground state configuration for lithium in this independent-electron approximation.

In summary, phenomenological evidence suggests that an electron can exist in either of two "spin states" in the presence of a magnetic field. Writing wavefunctions including spin functions and comparing these with experimental facts indicates that states exist only for wavefunctions that satisfy the exclusion principle.

[5]A broader statement is: Wavefunctions must be antisymmetric (symmetric) with respect to simultaneous interchange of space and spin coordinates of fermions (bosons). A fermion is characterized by half-integral spin quantum number; a boson is characterized by integral spin quantum number. Electrons have spin quantum number $\frac{1}{2}$ and are therefore fermions.

5-4 Slater Determinants and the Pauli Principle

It was pointed out by Slater [5] that there is a simple way to write wavefunctions guaranteeing that they will be antisymmetric for interchange of electronic space and spin coordinates: one writes the wavefunction as a determinant. For the $1s^2 2s$ configuration of lithium, one would write

$$\psi = \frac{1}{\sqrt{6}} \begin{vmatrix} 1s(1)\alpha(1) & 1s(2)\alpha(2) & 1s(3)\alpha(3) \\ 1s(1)\beta(1) & 1s(2)\beta(2) & 1s(3)\beta(3) \\ 2s(1)\alpha(1) & 2s(2)\alpha(2) & 2s(3)\alpha(3) \end{vmatrix} \tag{5-38}$$

Expanding this according to the usual rules governing determinants (see Appendix 2) gives

$$\begin{aligned} \psi = \frac{1}{\sqrt{6}}[&1s(1)\alpha(1)1s(2)\beta(2)2s(3)\alpha(3) + 2s(1)\alpha(1)1s(2)\alpha(2)1s(3)\beta(3) \\ &+1s(1)\beta(1)2s(2)\alpha(2)1s(3)\alpha(3) - 2s(1)\alpha(1)1s(2)\beta(2)1s(3)\alpha(3) \\ &-1s(1)\beta(1)1s(2)\alpha(2)2s(3)\alpha(3) - 1s(1)\alpha(1)2s(2)\alpha(2)1s(3)\beta(3)] \end{aligned} \tag{5-39}$$

This can be factored and shown to be identical to wavefunction (5-37) of the preceding section.

A simplifying notation in common usage is to delete the α, β symbols of the spin-orbitals and to let a bar over the space orbital signify β spin, absence of a bar being understood to signify α spin. In this notation, Eq. (5-38) would be written

$$\psi = \frac{1}{\sqrt{6}} \begin{vmatrix} 1s(1) & 1s(2) & 1s(3) \\ 1\bar{s}(1) & 1\bar{s}(2) & 1\bar{s}(3) \\ 2s(1) & 2s(2) & 2s(3) \end{vmatrix} \tag{5-40}$$

The general prescription to follow in writing a Slater determinantal wavefunction is very simple:

1. Choose the configuration to be represented. $1s1\bar{s}2s$ was used above. (Here we write $1s1\bar{s}2s$ rather than $1s^2 2s$ to emphasize that the two 1s electrons occupy different spin-orbitals.) For our general example, we will let U_i stand for a general spin-orbital and take a four-electron example of configuration $U_1 U_2 U_3 U_4$.

2. For n electrons, set up an $n \times n$ determinant with $(n!)^{-1/2}$ as normalizing factor. Every position in the first row should be occupied by the first spin-orbital of the configuration; every position in the second row by the second spin-orbital, etc. Now put in electron indices so that all positions in *column* 1 are occupied by electron 1, column 2 by electron 2, etc.

In the case of our four-electron configuration, the recipe gives

$$\psi = \frac{1}{\sqrt{4!}} \begin{vmatrix} U_1(1) & U_1(2) & U_1(3) & U_1(4) \\ U_2(1) & U_2(2) & U_2(3) & U_2(4) \\ U_3(1) & U_3(2) & U_3(3) & U_3(4) \\ U_4(1) & U_4(2) & U_4(3) & U_4(4) \end{vmatrix} \tag{5-41}$$

Notice that the principal diagonal (top left to bottom right) contains our original configuration $U_1U_2U_3U_4$. Often, the Slater determinant is represented in a space-saving way by simply writing the principal diagonal between short vertical bars. The normalizing factor is deleted. Thus, Eq. (5-41) would be symbolized as $|U_1(1)U_2(2) U_3(3)U_4(4)|$.

We have indicated the general recipe for writing down a Slater determinant, and we have seen that, for the configuration $1s1\bar{s}2s$, this gives an antisymmetric wavefunction. Now we will give a general proof of the antisymmetry of such wavefunctions for exchange of electrons. We have already seen that interchanging the space and spin coordinates of electrons 1 and 2 corresponds to going through the wavefunction and changing all the 1s to 2s and vice versa; i.e., electron *labels* denote coordinates. In a Slater determinant, interchanging electron labels 1 and 2 is the same thing as interchanging columns 1 and 2 of the determinant. [See Eq. (5-41) and note that columns 1 and 2 differ only in electron index.] But a determinant reverses sign upon interchange of two rows or columns. (See Appendix 2 for a summary of the properties of determinants.) Hence, any Slater determinant reverses sign (i.e., is antisymmetric) upon the interchange of space and spin coordinates of any two electrons.

Suppose we tried to put two electrons into the same space-orbital with the same spin. This would require that the same spin-orbital be written twice in the configuration, causing two rows of the Slater determinant to be identical. [If both 1s electrons in Eq. (5-40) had α spin, the bars would be absent from row 2.] We just stated that the determinant must reverse sign upon interchange of two rows. If we interchange two *identical* rows, we change nothing yet the sign must reverse: the determinant must be equal to zero. Thus, the determinantal wavefunction vanishes when we try to put more than one electron into the same spin-orbital, indicating that this is not a physically allowed situation. This is a generalization of our earlier discovery that no $1s^3$ configuration is allowed by the exclusion principle, such a configuration requiring at least two electrons to have the same space and spin functions.

This restriction on allowable electronic configurations is more familiar to chemists as the Pauli principle: *In assigning electrons to atomic orbitals in the independent electron scheme, no two electrons are allowed to have all four quantum numbers $(n, l, m, spin)$ the same*. The Pauli principle is a restatement of the exclusion principle as it applies in the special case of an orbital approximation to the wavefunction.

5-5 Singlet and Triplet States for the 1s2s Configuration of Helium

We showed in Section 5-2 that two *space* functions having proper space symmetry could be written for the configuration 1s2s. One was symmetric (Eq. 5-15) and one was antisymmetric (Eq. 5-16). Now we find that spin functions must be included in our wavefunctions, and in a way that makes the final result antisymmetric when space and spin coordinates are interchanged. We can accomplish this by multiplying the symmetric space function by an antisymmetric spin function, calling the result $\psi_{s,a}$. Thus,

$$\psi_{s,a}(1,2) = (1/\sqrt{2})\left[1s(1)2s(2) + 2s(1)1s(2)\right](1/\sqrt{2})\left[\alpha(1)\beta(2) - \beta(1)\alpha(2)\right] \quad (5\text{-}42)$$

Alternatively, we can multiply the antisymmetric space term by any one of the three possible symmetric spin terms:

$$\psi_{\mathrm{a,s}}(1,2) = (1/\sqrt{2})\left[1\mathrm{s}(1)2\mathrm{s}(2) - 2\mathrm{s}(1)1\mathrm{s}(2)\right] \begin{cases} \alpha(1)\alpha(2) & \text{(5-43a)} \\ (1/\sqrt{2})\left[\alpha(1)\beta(2) + \beta(1)\alpha(2)\right] & \text{(5-43b)} \\ \beta(1)\beta(2) & \text{(5-43c)} \end{cases}$$

All four of these wavefunctions satisfy the exclusion principle and each is linearly independent of the others, indicating that four distinct physical states arise from the configuration 1s2s.

There are a number of important points that can be illustrated using these wavefunctions. The first has to do with Slater determinants. Let us write down a Slater determinantal expression corresponding to wavefunction (5-43a). The configuration is $1\mathrm{s}(1)\alpha(1)2\mathrm{s}(2)\alpha(2)$, giving the Slater determinant (where absence of a bar indicates α spin)

$$\psi_{\mathrm{a,s}}(1,2) = \frac{1}{\sqrt{2}} \begin{vmatrix} 1\mathrm{s}(1) & 1\mathrm{s}(2) \\ 2\mathrm{s}(1) & 2\mathrm{s}(2) \end{vmatrix} \tag{5-44}$$

which, upon expansion, gives us Eq. (5-43a). If we attempt the same process to obtain Eq. (5-43b), we encounter a difficulty. The configuration $1\mathrm{s}(1)\alpha(1)2\mathrm{s}(2)\beta(2)$ leads to a 2×2 determinant, which, upon expansion, gives two product terms, whereas Eq. (5-43b) involves four product terms. The Slater determinantal functions corresponding to Eqs. (5-42) and (5-43b) are, in fact,

$$\psi_{\substack{\mathrm{s,a} \\ \mathrm{a,s}}}(1,2) = \frac{1}{\sqrt{2}} \left\{ \frac{1}{\sqrt{2}} \begin{vmatrix} 1\mathrm{s}(1) & 1\mathrm{s}(2) \\ 2\bar{\mathrm{s}}(1) & 2\bar{\mathrm{s}}(2) \end{vmatrix} \mp \frac{1}{\sqrt{2}} \begin{vmatrix} 1\bar{\mathrm{s}}(1) & 1\bar{\mathrm{s}}(2) \\ 2\mathrm{s}(1) & 2\mathrm{s}(2) \end{vmatrix} \right\} \tag{5-45}$$

The lesson to be gained from this is that a *single Slater determinant does not always display all of the symmetry possessed by the correct wavefunction.* (In this particular case, a single determinant restricts one of the AOs to α spin and the other to β, which is an artificial limitation.)

Next we will investigate the energies of the states as they are described by these wavefunctions. We have already pointed out that they are degenerate eigenfunctions of H_{approx}, but we will now examine their interactions with the full hamiltonian (5-2). Since our wavefunctions are not eigenfunctions of this hamiltonian, we cannot compare eigenvalues. Instead we must calculate the average values of the energy for each wavefunction, using the formula

$$\bar{E} = \frac{\int \psi^* H \psi \, d\tau}{\int \psi^* \psi \, d\tau} \tag{5-46}$$

The symbol "$d\tau$" stands for integration over space and spin coordinates of the electrons: $d\tau = dv\, d\omega$. Since both space and spin parts of our wavefunctions are normalized [cf. Eqs. (5-25) and (5-26)], the denominator of Eq. (5-46) is unity and may be ignored. The energy thus is given by the expression

$$\bar{E} = \int \psi^* \left[-\frac{1}{2}\nabla_1^2 - \frac{1}{2}\nabla_2^2 - (2/r_1) - (2/r_2) + (1/r_{12}) \right] \psi \, d\tau \tag{5-47}$$

Notice that the energy operator H contains no terms that would interact with spin functions α and β. (Such terms do arise at higher levels of refinement, but we ignore them for now.) Hence, the spin terms of ψ can be integrated separately, and, since all spin factors in Eqs. (5-42) and (5-43) are normalized, this gives a factor of unity in all four cases. This means that the average energies will be entirely determined by the space parts of the wavefunctions. This, in turn, means that all three states (5-43), which have the same space term, will have the same energy but that the state approximated by the function (5-42) may have a different energy. If our approximate representation of the exact eigenfunctions is physically realistic, we expect helium to display two excited state energies in the energy range consistent with a 1s2s configuration. Furthermore, we expect one of these state energies to be triply degenerate.

Which of these two state energies should be higher? To determine this requires that we expand our energy expression (5-47) for each of the two space functions (5-42) and (5-43).

$$\bar{E}_{\substack{1\\3}} = \frac{1}{2}\iint [1s^*(1)2s^*(2) \pm 2s^*(1)1s^*(2)]\left[-\frac{1}{2}\nabla_1^2 - \frac{1}{2}\nabla_2^2 - (2/r_1) - (2/r_2) + (1/r_{12})\right][1s(1)2s(2) \pm 2s(1)1s(2)]\,dv(1)\,dv(2) \qquad (5\text{-}48)$$

(The subscript on $\bar{E}$ refers to the degeneracy of whichever energy level we are considering.) This expands into a large number of terms. Integrals over one-electron operators may be written as products of two integrals, each over a different electron.[6] Thus, the expansion over the kinetic energy operators gives

$$\begin{aligned}
\tfrac{1}{2}\Bigg\{ & \int 1s^*(1)[-\tfrac{1}{2}\nabla_1{}^2]1s(1)\,dv(1) \int \cancelto{1}{2s^*(2)2s(2)\,dv(2)} \\
& + \int 2s^*(2)[-\tfrac{1}{2}\nabla_2{}^2]2s(2)\,dv(2) \int \cancelto{1}{1s^*(1)1s(1)\,dv(1)} \\
& + \int 2s^*(1)[-\tfrac{1}{2}\nabla_1{}^2]2s(1)\,dv(1) \int \cancelto{1}{1s^*(2)1s(2)\,dv(2)} \\
& + \int 1s^*(2)[-\tfrac{1}{2}\nabla_2{}^2]1s(2)\,dv(2) \int \cancelto{1}{2s^*(1)2s(1)\,dv(1)} \\
& \pm \int 1s^*(1)[-\tfrac{1}{2}\nabla_1{}^2]2s(1)\,dv(1) \int \cancelto{0}{2s^*(2)1s(2)\,dv(2)} \\
& \pm \int 2s^*(2)[-\tfrac{1}{2}\nabla_2{}^2]1s(2)\,dv(2) \int \cancelto{0}{1s^*(1)2s(1)\,dv(1)} \\
& \pm \int 2s^*(1)[-\tfrac{1}{2}\nabla_1{}^2]1s(1)\,dv(1) \int \cancelto{0}{1s^*(2)2s(2)\,dv(2)} \\
& \pm \int 1s^*(2)[-\tfrac{1}{2}\nabla_2{}^2]2s(2)\,dv(2) \int \cancelto{0}{2s^*(1)1s(1)\,dv(1)} \Bigg\}
\end{aligned} \qquad (5\text{-}49)$$

[6] We have already shown that, if the $1/r_{12}$ term is absent, the energy is equal to $E_{1s} + E_{2s}$, for He^+, which is equal to -2.5 a.u. Therefore, the detailed breakdown leading to Eqs. (5-49)–(5-51) is not necessary. However, we will present it in detail in the belief that some students will benefit from another specific example of integration of two-electron products over one-electron operators.

The orthogonality of the 1s and 2s orbitals causes the terms preceded by $\pm$ to vanish. Furthermore, integrals that differ only in the variable label [such as those in the second and third terms of (5-49)] are equal, so that this expansion becomes

$$\int 1s^*(1)\left[-\frac{1}{2}\nabla_1^2\right]1s(1)dv(1) + \int 2s^*(1)\left[-\frac{1}{2}\nabla_1^2\right]2s(1)\,dv(1) \tag{5-50}$$

Expansion of Eq. (5-48) over $(-2/r_1 - 2/r_2)$ proceeds analogously to give

$$\int 1s^*(1)(-2/r_1)1s(1)\,dv(1) + \int 2s^*(1)\,(-2/r_1)\,2s(1)\,dv(1) \tag{5-51}$$

The final term in the hamiltonian, $1/r_{12}$, occurs in four two-electron integrals:

$$\begin{aligned}\frac{1}{2}\bigg\{&\iint 1s^*(1)2s^*(2)(1/r_{12})1s(1)2s(2)\,dv(1)\,dv(2)\\ +&\iint 2s^*(1)1s^*(2)(1/r_{12})2s(1)1s(2)\,dv(1)\,dv(2)\\ \pm&\iint 1s^*(1)2s^*(2)(1/r_{12})2s(1)1s(2)\,dv(1)\,dv(2)\\ \pm&\iint 2s^*(1)1s^*(2)(1/r_{12})1s(1)2s(2)\,dv(1)\,dv(2)\bigg\}\end{aligned} \tag{5-52}$$

The first two integrals of (5-52) differ only by an interchange of labels "1" and "2 ," and so they are equal to each other. The same is true of the second pair. Thus, the average value of the energy is

$$\begin{aligned}\bar{E}_{\frac{1}{3}} = \bigg\{&\int 1s^*(1)\left[-\frac{1}{2}\nabla_1^2\right]1s(1)\,dv(1) + \int 1s^*(1)\,[-2/r_1]\,1s(1)\,dv(1)\\ +&\int 2s^*(1)\left[-\frac{1}{2}\nabla_1^2\right]2s(1)\,dv(1) + \int 2s^*(1)\,[-2/r_1]\,2s(1)\,dv(1)\\ +&\iint 1s^*(1)2s^*(2)(1/r_{12})1s(1)2s(2)\,dv(1)\,dv(2)\\ \pm&\iint 1s^*(1)2s^*(2)(1/r_{12})2s(1)1s(2)\,dv(1)\,dv(2)\bigg\}\end{aligned} \tag{5-53}$$

Notice that, since $-\frac{1}{2}\nabla^2 - 2/r$ is the hamiltonian for He^+, the first two integrals of Eq. (5-33) combine to give the average energy of He^+ in its 1s state. The second pair gives the energy for He^+ in the 2s state. Thus, Eq. (5-53) can be written

$$\bar{E}_{\frac{1}{3}} = E_{1s} + E_{2s} + J \pm K \tag{5-54}$$

where J and K represent the last two integrals in Eq. (5-53). No bars appear on E_{1s} or E_{2s} because these "average energies" are identical to the eigenvalues for the He^+ hamiltonian (Problem 5-15).

The integral J denotes electrons 1 and 2 as being in "charge clouds" described by 1s*1s and 2s*2s, respectively. The operator $1/r_{12}$ gives the electrostatic repulsion energy between these two charge clouds. Since these charge clouds are everywhere negatively charged, all the interactions are repulsive, and it is necessary that this "coulomb

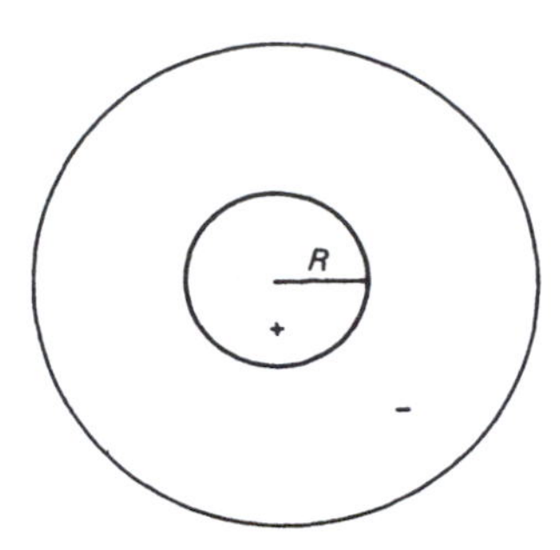

Figure 5-2 ▶ The function produced by multiplying together hydrogenlike 1s and 2s orbitals. R is the radius of the spherical nodal surface.

integral" J be positive. Alternatively, we can argue that the functions 1s*1s, 2s*2s, and $1/r_{12}$ are everywhere positive, so the integrand of J is everywhere positive and J must be positive.

The integral K is called an "exchange integral" because the two product functions in the integrand differ by an exchange of electrons. This integral gives the net interaction between an electron "distribution" described by 1s*2s, and another electron in the same distribution. (These distributions are mathematical functions, not physically realizable electron distributions.) The 1s2s function is sketched in Fig. 5-2. Because the 2s orbital has a radial node, the 1s2s function (which is the same as 1s*2s since the 1s function is real) also has a radial node. Now the function 1s(1)2s(1)1s(2)2s(2) will be positive whenever r_1 and r_2 are either both smaller or both larger than the radial node distance (R in Figure 5-2). But when one r value is smaller than R and the other is greater, corresponding to the electrons being on opposite sides of the nodal surface, the product 1s(1)2s(1)1s(2)2s(2) is negative. These positive and negative contributions to K are weighted by the function $1/r_{12}$, which is always positive and hence unable to affect the sign of the integrand. But $1/r_{12}$ is smallest when the electrons are far apart. This means that $1/r_{12}$ tends to reduce the contributions where the electrons are on opposite sides of the node (i.e., the negative contributions), and so the value of K turns out to be positive (although not as large in magnitude as J, which has no negative contribution at all).

Since the integral K is positive, we can see from Eq. (5-54) that the *triply degenerate energy level lies below the singly degenerate one, the separation between them being* $2K$. (We note in passing that these independent-electron wavefunction energies are not simply the sum of one-electron energies as was the case when we used H_{approx}, thereby ignoring interelectronic repulsion.)

The experimental observation agrees qualitatively with these results. There are two state energies associated with the 1s2s configuration. When the atom is placed in an external magnetic field, the lower-state-energy-level splits into three levels. The state having the higher energy has a "multiplicity" of one and is called a singlet. The lower-energy with multiplicity three is called a triplet. (The reference to a "triplet state" should not be construed to mean that this is one state. It is a triplet of states.)

It is possible to use vector arguments similar to those presented in Chapter 4 to understand why the triply degenerate level splits into three different levels in the presence of a homogeneous magnetic field. Let us first consider the case of a single electron. We have already indicated that two spin states are possible, which we have labeled α and β.

In a magnetic field the angular momentum vectors precess about the field axis z, as depicted in Figure 5-3. The z components of the angular momentum vectors are constant but the x and y components are not. Because the *allowed* z components must in general differ by one atomic unit (stated but not proved in Chapter 4), and because there are but two allowed values (inferred from observations such as the splitting of a beam of silver atoms into two components), and because the two kinds of state are *oppositely* affected by magnetic fields, it is concluded that the z components of angular momentum (labeled m_s) are equal to $+\frac{1}{2}$ and $-\frac{1}{2}$ a.u. for α and β, respectively. Following through using orbital angular momentum relations as a model, we postulate an electron spin quantum number s equal to the maximum z-component of spin angular momentum in a.u., $\frac{1}{2}$, and a spin angular momentum vector $\mathbf{s}$ having length $\sqrt{s(s+1)}$ a.u. The degeneracy, $g_s = 2s + 1$, equals 2, in agreement with the two orientations in Fig. 5-3.

As noted in Chapter 4, half-integer quantum numbers correspond to eigenfunctions that cannot be expressed as spherical harmonics. We will not pursue the question of detailed expressions for α and β here. (However, see the problems at the end of Chapter 9.)

Now let us turn to the two-electron system. We will assume that the magnetic moments of the two electrons interact independently with the external field. This ignores the fact that each electron senses a small contribution to the magnetic field resulting from the magnetic moment of the other electron.

Another factor that could affect the magnetic field sensed by the spin moment is the magnetic moment resulting from the *orbital* motions of the electrons, although this is not present if both electrons are taken to be in s atomic orbitals (AOs). For two electrons, we can imagine four situations: $\alpha\alpha$, $\alpha\beta$, $\beta\alpha$, and $\beta\beta$. We pointed out earlier (Section 4-5) that, for a system composed of several moving parts, the total angular momentum is the sum of the individual angular momenta, and the z component is the sum of the individual z components. For the four spin situations listed above, this means that the net z components of spin angular momentum (labeled M_s) are $+1, 0, 0$, and -1 a.u., respectively. The spin combinations $\alpha\beta + \beta\alpha$ [from the triplet (5-43b)] and $\alpha\beta - \beta\alpha$

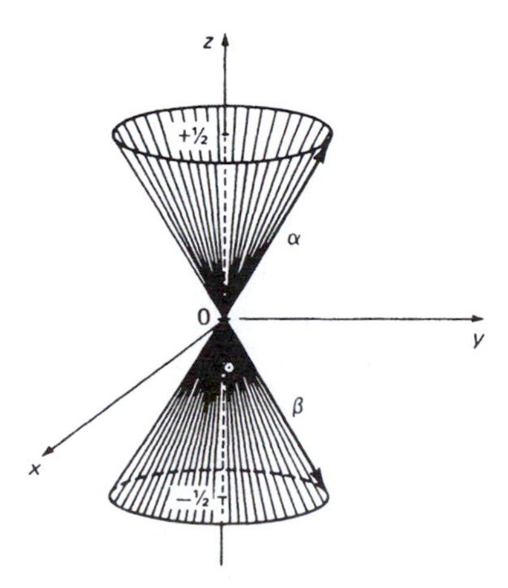

Figure 5-3 ▶ The angular momentum vectors for α and β precess around the magnetic field axis z. The z components of these vectors are constant and have values of $+\frac{1}{2}$ and $-\frac{1}{2}$ a.u. respectively.

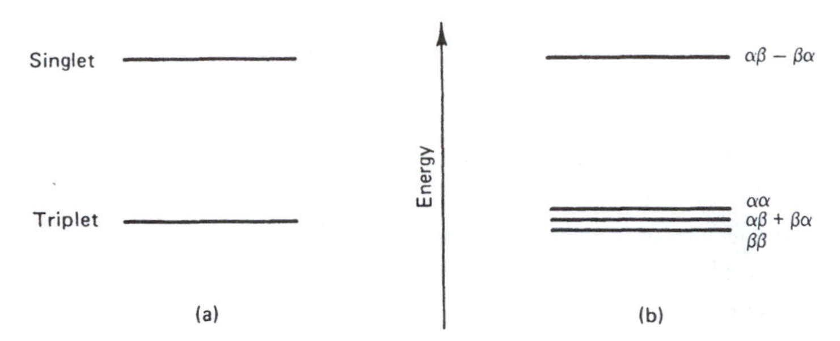

Figure 5-4 ▶ Energy levels for singlet and triplet levels of 1s2s helium in (a) absence, and (b) presence of an external magnetic field.

[from the singlet (5-42)] are linear combinations of $\alpha\beta$ and $\beta\alpha$. However, since these two functions have the same value for the z component of angular momentum (zero), their linear combinations will also have that value. It follows that the z components of the spin angular momenta of the triplet of states (5-43) are $+1, 0$, and -1 a.u., and for the singlet (5-42) it is zero. Because the electrons are charged, these spin angular momenta correspond to spin magnetic moments, which interact differently with the applied magnetic field to give splitting of the triplet (see Figure 5-4). It is customary to refer to all three spin states in (5-43) as having *parallel* spins even though the vector diagram for the (5-43b) state is not particularly in accord with this terminology. For the singlet state, the spins are said to be *opposed*, or *antiparallel*.

5-6 The Self-Consistent Field, Slater-Type Orbitals, and the Aufbau Principle

Up to now we have used wavefunctions that, while not being eigenfunctions of the hamiltonian, are eigenfunctions of an "effective hamiltonian" obtained by ignoring the interelectronic repulsion operator $1/r_{ij}$. That is, these wavefunctions would be exactly correct if the electrons in helium were attracted by the nucleus, but did not repel each other. For this reason, we have referred to this as an independent electron approximation. Because we have neglected interelectronic repulsion, we cannot expect such a wavefunction to give very good numerical predictions of charge density or energy. We can compare the energy of He in the $1s^2$ (ground) state as predicted by our independent electron wavefunction and H_{approx} (-108.84 eV) with the experimental value (-79.0143 eV). (See Table 5-1.) This shows that the predicted energy is much too low, which is understandable since we have neglected an important repulsive (hence positive) interaction energy. But we can account for much of this neglected energy by calculating the *average* value of the energy using H (with $1/r_{12}$ included) instead of H_{approx}. This gives a value of -74.83 eV—much better, though now too high by more than 4 eV. Even though we have now accounted for interelectronic repulsion, there is still a problem: Because we ignored interelectronic repulsion in arriving at these wavefunctions, they predict electron densities that are too large near the nucleus. In reality, interelectronic repulsion prevents so much build-up of charge density. Methods have been devised that partially overcome this problem by retaining the convenient form of orbital products but modifying the formulas for the orbitals themselves to make them more diffuse.

TABLE 5-1 ► Average Values for Energy Calculated from Helium Atom Ground State Approximate Wavefunctions[a]

Wavefunction description	$\bar{E}$(eV)
1) Product of He^+ orbitals	−74.83
2) Product of hyrdrogenlike orbitals with ζ fixed by SCF method	−77.48
3) Best product-type wavefunction	−77.870917
4) Nonorbital wavefunction of Pekeris [9]. This wavefunction uses functions of r_1, r_2 and r_{12} as coordinates and has the form of an exponential times a linear combination of 1078 terms	−79.00946912

[a] $\bar{E} = \int \psi^* H \psi d\tau / \int \psi^* \psi d\tau$, where H is given by Eq. (5-2).

Let the ground state of helium be our example. We take the ordinary independent-electron wavefunction as our initial approximation:

$$1s(1)1s(2) \equiv \sqrt{8/\pi}\exp(-2r_1)\sqrt{8/\pi}\exp(-2r_2) \tag{5-55}$$

These atomic orbitals are correct only if electrons 1 and 2 do not "see" each other *via* a repulsive interaction. They really do repel each other, and we can *approximate* this repulsion by saying that electron 2 "sees" electron 1 as a smeared out, time-averaged charge cloud rather than the rapidly moving point charge that is actually present. The initial description for this charge cloud is just the absolute square of the initial atomic orbital occupied by electron 2: $[1s(2)]^2$. Our approximation now has electron 1 moving in the field of a positive nucleus embedded in a spherical cloud of negative charge. Thus, for electron 1, the positive nuclear charge is "shielded" or "screened" by electron 2. Hence, electron 1 should occupy an orbital that is less contracted about the nucleus. Let us write this new orbital in the form

$$1s'(1) = \sqrt{\zeta^3/\pi}\exp(-\zeta r_1) \tag{5-56}$$

where ζ is related to the screened nuclear charge seen by electron 1. The mathematical methods used to evaluate ζ will be described later in this book.

Next we turn to electron 2, which we now take to be moving in the field of the nucleus shielded by the charge cloud *due to electron* 1, *now in its expanded orbital.* Just as before, we find a new orbital of form (5-56) for electron 2. The value of ζ that we calculate for electron 2, however, will be different from what we found for electron 1 because the shielding of the nucleus by electron 1 is different from what it was by electron 2 in our previous step. We now have a new distribution for electron 2, but this means that we must recalculate the orbital for electron 1 since this orbital was appropriate for the screening due to electron 2 in its old orbital. After revising the orbital for electron 1, we must revise the orbital for electron 2. This procedure is continued back and forth between electrons 1 and 2 until the value of ζ converges to an unchanging value (under the constraint that electrons 1 and 2 ultimately occupy orbitals having the same value of ζ). Then the orbital for each electron is consistent with the potential due to the nucleus and the charge cloud for the other electron: the electrons move in a "self-consistent field" (SCF).

The result of such a calculation is a wavefunction in much closer accord with the actual charge density distributions of atoms than that given by the complete neglect of interelectron repulsion.[7] A plot of the electron density distribution in helium as given by wavefunction (5-55) and by a similar wavefunction with optimized ζ is given in Fig. 5-5. Because each electron senses only the time-averaged charge cloud of the other in this approximation, it is still an independent-electron treatment. The hallmark of the independent electron treatment is a wavefunction containing only a product of one-electron functions. It will not contain functions of, say, r_{12}, which would make ψ depend on the instantaneous distance between electrons 1 and 2.

Atomic orbitals that are eigenfunctions for the one-electron hydrogenlike ion (for integral or nonintegral Z) are called *hydrogenlike orbitals*. In Chapter 4 we noted that many hydrogenlike orbitals have radial nodes. In actual practice, this mathematical aspect causes increased complexity in solving integrals in quantum chemical calculations. Much more convenient are a class of modified orbitals called *Slater-type orbitals* (STOs). These differ from their hydrogenlike counterparts in that they have no radial nodes. Angular terms are identical in the two types of orbital. The unnormalized radial term for an STO is

$$R(n, Z, s) = r^{(n-1)} \exp[-(Z - s)r/n] \tag{5-57}$$

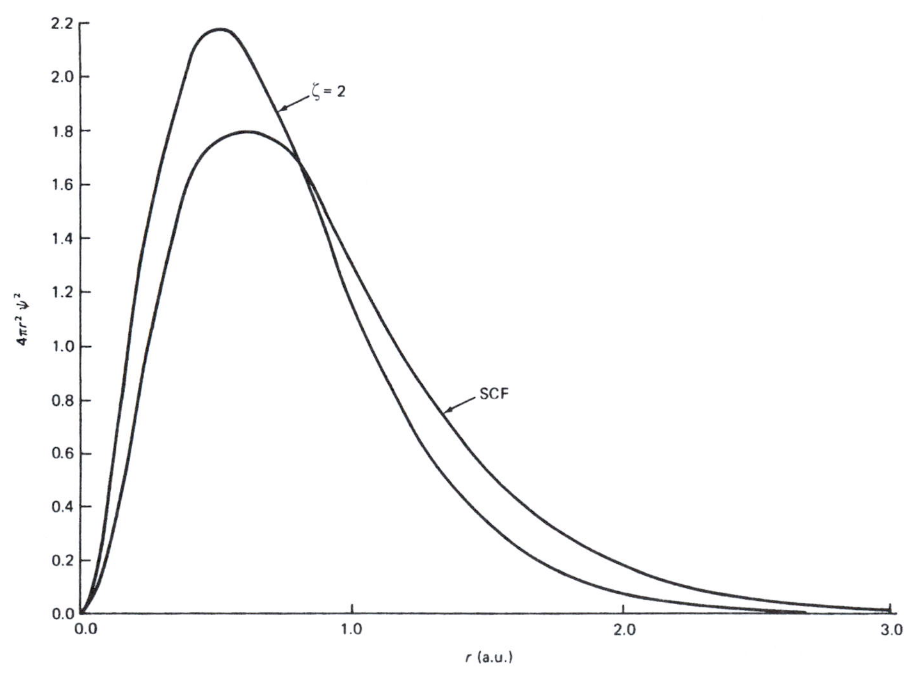

Figure 5-5 ▶ Electron distribution in helium as given by SCF and unshielded independent electron approximations.

[7] In practice, mathematical techniques have been found that lead to a self-consistent solution without explicit iteration between evolving AOs that converge to some final optimized ζ. Examples are described in Chapters 7 and 12. A thorough discussion of the SCF and related methods is given in Chapter 11.

where Z is the nuclear charge in atomic units, n is the principal quantum number, and s is a "screening constant" which has the function of reducing the nuclear charge Z "seen" by an electron. Slater [7] constructed rules for determining the values of s that will produce STOs in close agreement with those one would obtain by an SCF calculation. These rules, appropriate for electrons up to the 3d level, are

1. The electrons in the atom are divided up into the following groups: 1s|2s,2p|3s, 3p|3d.

2. The shielding constant s for an orbital associated with any of the above groups is the sum of the following contributions:

   a) Nothing from any electrons in groups to the right (in the above list) of the group under consideration
   b) 0.35 from each other electron in the group under consideration (except 0.30 in the 1s group)
   c) If the orbital under consideration is an s or p orbital, 0.85 for each electron with principal quantum number less by 1, and 1.00 for each electron still "farther in"; for a d orbital, 1.00 for all electrons farther in

For example, nitrogen, with ground state configuration $1s^2 2s^2\ 2p^3$, would have the same radial part for the 2s and 2p STOs. This would be given by the formula ($n = 2$, $Z = 7$, $s = 4 \times 0.35 + 2 \times 0.85 = 3.1$)

$$R_{2s,2p}(2, 7, 3.1) = r^{(2-1)} \exp[-(7 - 3.1)r/2] = r \exp(-1.95r)$$

For the 1s level, $n = 1$, $Z = 7$, $s = 0.30$, and

$$R_{1s} = \exp(-6.7r)$$

Comparing orbital exponents, we see that the 1s charge cloud is compressed much more tightly around the nucleus than are the 2s and 2p "valence orbital" charge clouds. Slater-type orbitals are very frequently used in quantum chemistry because they provide us with very good approximations to self-consistent field atomic orbitals (SCF–AOs) with almost no effort.

Clementi and Raimondi [8] have published a refined list of rules for the shielding constant, which extends to the 4p level. Their rules include contributions to shielding due to the presence of electrons in shells outside the orbital under consideration. Such contributions are not large, and, up to the 3d level, there is reasonably good agreement between these two sets of rules.

The fact that STOs have no radial nodes results in some loss of orthogonality. Angular terms still give orthogonality between orbitals having different l or m quantum numbers, but STOs differing only in their n quantum number are nonorthogonal. Thus, 1s, 2s, 3s, ... are nonorthogonal. Similarly $2p_z, 3p_z, \ldots$ or $3d_{xz}, 4d_{xy}, \ldots$ are nonorthogonal. In practice, this feature is handled easily. The only real problem arises if one *forgets* about this nonorthogonality when making certain calculations.

When carrying out SCF calculations on multielectronic atoms, one finds that the orbital energies for 2s and 2p functions are not the same. Similarly, 3s 3p, and 3d orbitals are nondegenerate. Yet these orbitals were degenerate in the one-electron hydrogenlike system in which energy was a function of n but not of l or m. Why are these orbital energies nondegenerate in the many-electron calculation? A reasonable explanation can be found by considering the comparative effectiveness with which a pair of 1s electrons screen the nucleus from a 3s or a 3p electron. Comparing the 3s, 3p, and 3d hydrogenlike orbital formulas in Table 4-2 shows that the 3s orbital is finite at the nucleus, decreasing proportionally to r for small r. The 3p orbitals vanish at the nucleus but grow as r for small r. The 3d orbitals vanish at the nucleus but grow as r^2 for small r. The result of all this is that an s electron spends a larger amount of its time near the nucleus than a p electron of the same principal quantum number, the p electron spending more time near the nucleus than the d, etc. Hence, the s electron penetrates the "underlying" charge clouds more effectively and is therefore less effectively shielded from the nucleus. Since the s electron sees a greater effective nuclear charge, its energy is lower than that of the p electron. (This effect is not obvious in STOs since the 3s and 3p STOs have the same radial function which vanishes at the nucleus. However, the STO for 3d does reflect the nondegeneracy since Slater's rules give it a different screening constant from 3s or 3p.)

The tendency for higher l values to be associated with higher orbital energies leads to the following orbital ordering:

$$1s\ 2s\ 2p\ 3s\ 3p\ 4s\ 3d\ 4p\ 5s\ 4d\ 5p\ 6s\ 5d\ 4f\ 6p\ 7s\ 6d\ 5f \ldots \tag{5-58}$$

When we get to principal quantum numbers of 3 and higher, the energy differences between different l values for the same n become comparable to the differences between different n levels. Thus, in some atoms, the 4s level is almost the same as the 3d level, etc.

In compiling data on ground states of atoms, Hund noticed that *greatest stability results if the AOs in a degenerate set are half-filled with electrons before any of them are filled.* This generalization, called *Hund's rule*, is sometimes stated in an alternative form: *Of the states associated with the ground state configuration of an atom or ion, those with greatest spin multiplicity lie deepest in energy.* Chemists generally find the former version to be more convenient, spectroscopists the latter. The reason for the equivalence of these statements will emerge later in this chapter.

EXAMPLE 5-4 An unexcited Fe atom has an electronic configuration of $1s^2 2s^2 2p^6 3s^2 3p^6 4s^2 3d^6$. What is its spin multiplicity?

SOLUTION ▶ All electrons below $3d^6$ are spin-paired in orbitals, hence contribute nothing to M_S. In $3d^6$, we have 4 electrons that can each occupy a 3d AO alone. If we follow Hund's rule and seek maximum spin multiplicity, we make all their spins the same (α). Then maximum, $M_S = 4 \cdot \frac{1}{2} = 2$, so $S = 2$, and spin multiplicity is $2S + 1 = 5$. The atom has a quintet ground "state" (really five states). ◀

The energy-ordering scheme (5-58) coupled with the Pauli or exclusion principle and Hund's rule leads us to a simple prescription for "building up" the electronic configurations of atoms. This *aufbau* principle is familiar to chemists and leads naturally to a correlation between electronic structure and the periodic table. The procedure is to place all the electrons of the atom into atomic orbitals, two to an orbital, starting at the low-energy end of the list (5-58) and working up in energy. In addition, when filling a set of degenerate levels like the five 3d levels, one half-fills all the levels with electrons of parallel spin before filling any of them. This prescription enables one to guess the electronic configuration of any atom, once its atomic number is known, unless it happens to put us into a region of ambiguity, where different levels have almost the same energy. (Electronic configurations for such atoms are deduced from experimentally determined chemical, spectral, and physical properties.) The configuration for carbon (atomic number 6) would be $1s^2 2s^2 2p^2$, with the understanding that p electrons occupy *different* p orbitals and have parallel spins. (Recall that we expect the most stable of all the states arising from the configuration $1s^2 2s^2 2p^2$ to be the one of highest multiplicity. The 2p electrons can produce either a singlet or a triplet state just as could the two electrons in the 1s2s configuration of helium. The triplet should be the ground state and this corresponds to parallel spins, which *requires* different p orbitals by the exclusion principle.)

It is important to realize that the orbital ordering (5-58) used in the *aufbau* process is not fixed, but depends on the atomic number Z. The ordering in (5-58) cannot be blindly followed in all cases. For instance, the ordering shows that 5s fills before 4d. It is true that element 38, strontium, has a $\cdots 4p^6 5s^2 4d^0$ configuration. But a later element, palladium, number 46, has $\cdots 4p^6 4d^{10} 5s^0$ as its ground state configuration. The effect of adding more protons and electrons has been to depress the 4d level more than the 5s level.

5-7 Electron Angular Momentum in Atoms

Most of our attention thus far has been with wavefunction symmetry and energy. However, understanding atomic spectroscopy or interatomic interactions (in reactions or scattering) requires close attention to angular momentum due to electronic orbital motion and "spin." In this section we will see what possibilities exist for the total electronic angular momenta of atoms and how these various states are distinguished symbolically.

We encountered earlier (Section 4-5) the notion that the total angular momentum for a classical system is the vector sum of the angular momenta of its parts. If the system interacts with a z-directed field, the total angular momentum vector precesses about the z axis, so the z component continues to be conserved and continues to be equal to the sum of z components of the system's parts. Since quantum hydrogenlike systems obey angular momentum relations analogous to a *precessing* classical system, it is this z-axis behavior that we focus on as we seek to construct the nature of the total angular momentum from the orbital and spin parts we already understand.

Because it is the *total* angular momentum that is conserved in a multicomponent classical system, it is the total angular momentum that obeys the quantum rules we have previously described for separate spin and orbital components. If we consider a one-electron system, the combined spin-orbital angular momentum can be associated with a quantum number symbolized by j (analogous to s and l). Then we can immediately say that the allowed z components of total angular momentum are, in a.u., $m_j = \pm j$, $\pm(j-1), \ldots$ and that the length of the vector is $\sqrt{j(j+1)}$ a.u.

The implication of accepting total angular momentum as the fundamental quantized quantity is that the spin and orbital angular momenta do not individually obey the quantum rules we have so far applied to them—s and l are not "good" quantum numbers. However, for atoms of low atomic number they are in fact quite good, especially for low-energy states, and we can continue to refer to the s and l quantum numbers in such cases with some confidence. (Classically, this corresponds to cases where there is little transfer of angular momentum between modes.)

5-7.A Combined Spin-Orbital Angular Momentum for One-Electron Ions

The key to understanding the following discussion is to remember that a quantum number l, s, or j really tells us *three* things:

1. It equals the *maximum* value of m_l, m_s, or m_j. If $l = 2$, the maximum allowed value of m_l is 2, and the maximum z component of orbital angular momentum is 2 a.u.

2. It allows us to know the *length* of the related angular momentum vector, **l**, **s**, or **j**, in a.u. For j, this is given by $\sqrt{j(j+1)}$. If $j = 2$, the length of the total angular momentum vector **j** is $\sqrt{6}$ a.u.

3. It allows us to know the *degeneracy*, g, of the energy level due to states having this angular momentum. For s, this is $2s + 1$. If $s = 1/2$, $g_s = 2$. The corresponding l degeneracies produce the s, p, d, f degeneracies of 1, 3, 5, 7.

Using the hydrogen atom as our example, let us consider what the *total* electronic angular momentum is in the ground (1s) state. For an s AO, $l = 0$, and so there is no orbital angular momentum. This means that the *total* angular momentum is the same as the spin angular momentum, so $j = s = 1/2$, $m_j = \pm 1/2$. The diagram for the vector **j**, then, looks just like that for **s** (Fig. 5-3).

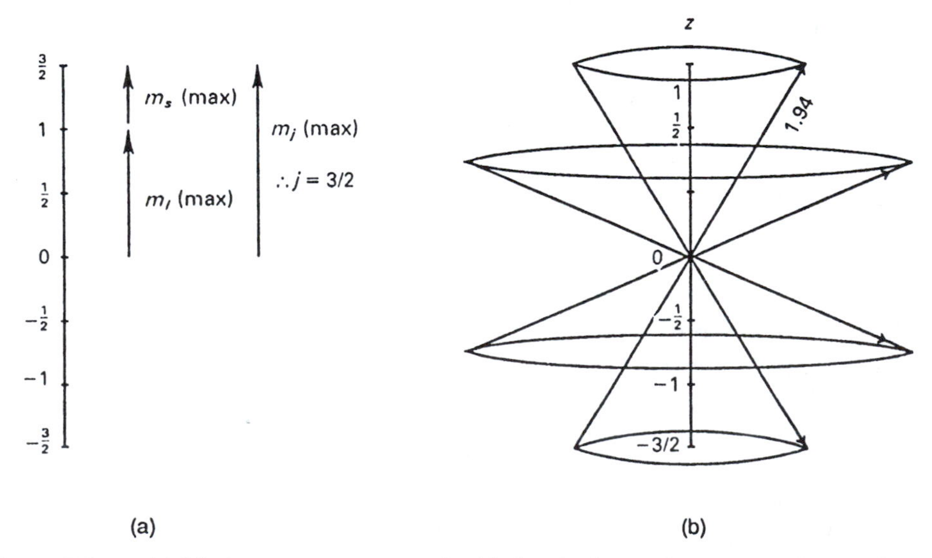

Figure 5-6 ▶ (a) Maximum z components of orbital and spin angular momenta for a p electron leading to a *total* z component of 3/2. (b) The four states corresponding to the $j = 3/2$ vector assuming its possible z intercepts (3/2, 1/2, −1/2, −3/2).

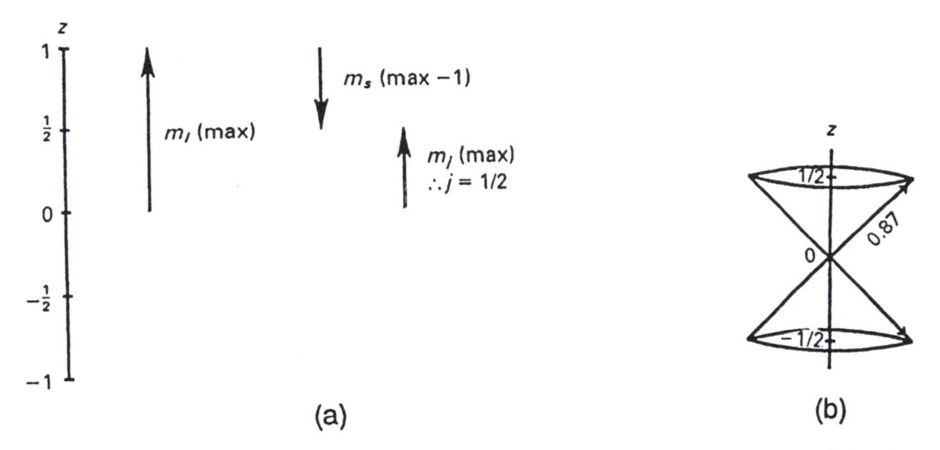

Figure 5-7 ▶ (a) $j = 1/2$, resulting when **l** is oriented with its maximum z intercept and **s** is oriented in its other orientation (other than as in Fig. 5-6). (b) The two states corresponding to the $J = 1/2$ vector assuming its possible z intercepts $(1/2, -1/2)$.

More interesting is an excited state, say 2p. Now $l = 1$ and $s = 1/2$. From $l = 1$ we can say that the maximum orbital z component of angular momentum is 1 a.u. $s = 1/2$ tells us that there is an additional maximum spin z component of 1/2. The maximum sum, then, is 3/2 for the z component of **j**. But, if this is the maximum m_j, then j itself must be 3/2 and the length of **j** must be $\sqrt{(3/2)(5/2)} = 1.94$ a.u. There must be $2j + 1 = 4$ allowed orientations of **j**, with z intercepts at 3/2, 1/2, −1/2, −3/2 in a.u. (Fig. 5-6).

We are not yet finished with the 2p possibilities. The total angular momentum is the sum of its orbital and spin parts, and we have so far found the way they combine to give the *maximum z* component. But this is not the *only* way they can combine. It is possible to have $m_l = 1$ and $m_s = -1/2$. Then the maximum $m_j = 1/2$, so $j = 1/2$, giving us a vector **j** of length $\sqrt{(1/2)(3/2)}$ a.u. and two orientations (Fig. 5-7).

So far we have identified six states, four with $j = 3/2$ and two with $j = 1/2$. This is all we should expect since we have three 2p AOs and two spins, giving a total of six combinations. It seems, though, that we could generate some more states by now letting $m_l = 0$ or -1 and combining these with $m_s = \pm 1/2$. However, these possibilities are already implicitly accounted for in the multiplicity of states we recognize to be associated with the $j = 3/2$, $1/2$ cases already found. This illustrates the general approach to be taken when combining two vectors: Orient the larger vector to give maximum z projection, and combine this projection with *each* of the allowed z components of the smaller vector. This gives all of the possible m_j(max) values, hence all of the j values. In other words, it gives us all of the allowed vectors, **j**, each oriented with maximum z component, and it remains only to recognize that these can have certain other orientations corresponding to z intercepts of $m_j - 1$, $m_j - 2, \ldots, -m_j$.

States can be *labeled* to reflect all of the angular momentum parts they possess. The main symbol is simply s, p, d, f, g, etc. depending on the l value as usual. A superscript at left gives spin multiplicity $(2s + 1)$ for the states. A subscript at right tells the j quantum number for the states. If an individual member of the group of states having the same j value is to be cited, it is identified by placing its m_j value at upper right.

Thus, all six of the states discussed above can be referred to as ^{2}p states. The two groups having different total angular momentum are distinguished as $^2p_{3/2}$ and $^2p_{1/2}$. One of the four states in the former group is the $^2p_{3/2}^{-1/2}$ state.

The general form of the symbol is $^{2s+1}l_j^{m_j}$. Such symbols are normally called *term symbols*, and the collection of states they refer to is called a *term* (except when an *individual* state is denoted by inclusion of the m_j value).

The reason for distinguishing between the $^2p_{3/2}$ and $^2p_{1/2}$ terms is that they occur at slightly different energies. This results from the different energies of interaction between the magnetic moments due to spin and orbital motions. For instance, if l and s are coupled so as to give the maximum j, their associated magnetic moments are oriented like two bar magnets side by side with north poles adjacent. This is a higher-energy arrangement than the other extreme, where l and s couple to give minimum j, acting as a pair of parallel bar magnets with the north pole of each next to the south pole of the other. So $^2p_{1/2}$ should be lower in energy than $^2p_{3/2}$ for hydrogen.

EXAMPLE 5-5 For a hydrogen atom having $n = 3, l = 2$, what are the possible j values, and how many states are possible? Indicate the lengths of the j vectors in a.u. What term symbols apply?

SOLUTION ▶ If $l = 2$ (d states), there are five AOs and two possible spins, so we expect a total of ten possible states. The maximum possible values of the z-component of angular moment for orbital and spin respectively are 2 and 1/2. So the maximum value is 5/2 giving a vector length of $\sqrt{35/4}$ a.u. and six possible z projections, hence six states. The term symbol is $^2d_{5/2}$. The remaining possible j value is $2 - 1/2 = 3/2$, accounting for four more states and giving a vector length of $\sqrt{15/4}$ a.u. and a term symbol of $^2d_{3/2}$. ◀

5-7.B Spin-Orbital Angular Momentum for Many-Electron Atoms

Much of what we have seen for one-electron ions continues to hold for many-electron atoms or ions. All the symbolism is the same, except that capital letters replace lowercase: The quantum numbers are L, S, and J, and the main symbol becomes S, P, D, F, G, etc. There is a total orbital angular momentum vector **L** with quantum number L that equals the maximum value of M_L. The length of **L** is $\sqrt{L(L+1)}$ a.u., and it has $2L + 1$ orientations. Vectors **S** and **J** behave analogously. When constructing the vectors **J**, we continue to place the larger of **L** and **S** to give maximum z intercept, and add to this the possible z intercepts of the smaller vector. The situation, then, is just as before except that we need to figure out the possible values for M_L and M_S by combining the allowed values of $m_l(1), m_l(2), \ldots$ and $m_s(1), m_s(2), \ldots$.[8]

[8]This procedure of first combining individual orbital contributions to find **L** and spin contributions to find **S** and then combining these to get **J** is referred to as "L–S coupling," or "Russell–Saunders coupling." The other extreme is to first combine **l** and **s** for the first electron to give **j**(1), **l** and **s** for the second electron to give **j**(2), ... and then combine these individual-electron **j**'s to give **J**. This is more appropriate for atoms having high atomic number (in which electrons move at relativistic speeds in the vicinity of the nucleus), and is referred to as "j–j coupling." We will not describe j–j coupling in this text. The reader should consult Herzberg [6] for a fuller treatment.

For example, if we have found that $M_L(\text{max}) = 2$ (which means $L = 2$) and $M_s(\text{max}) = 1$ (which means $S = 1$), we have that $M_J(\text{max})$ can be $2+1, 2+0$, and $2+(-1)$, or $3, 2$, and 1. This means that the possible values of J are $3, 2, 1$, giving three different **J** vectors. Since $L = 2$ and $S = 1$, the term symbols for these three J cases are $^3D_3$, $^3D_2$, and $^3D_1$. Notice that the multiplicities of these three terms—7, 5, and 3, respectively, obtained from $2J + 1$—total 15 states, which is just what we should expect for the ^3D symbol (spin multiplicity of 3, orbital multiplicity of 5). The 15 triplet-D states are found in three closely spaced levels, differing in energy because of different spin-orbital magnetic interactions.

The problem remains, how do we find the M_L and M_s values that allow us to construct term symbols? There are two situations to distinguish in this context, and a different approach is taken for each.

1. Nonequivalent Electrons. The first situation is exemplified by carbon in its $1s^22s^22p3p$ configuration. It is not difficult to show that the electrons in the 1s and 2s AOs contribute no net angular momentum and can be ignored: The spins of *paired* electrons are opposed, hence cancel, and the s-type AOs have no angular momentum, hence cannot contribute. However, even p, d, etc. sets of AOs cannot contribute *if they are filled* because then any orbital momentum having z intercept m_l is canceled by one with $-m_l$. The important result is that *filled subshells do not contribute to orbital or spin angular momentum*. The remaining 2p and 3p electrons occupy different sets of AOs, hence are called *nonequivalent* electrons.

Since these electrons are never in the same AO, they are not restricted to have opposite spins at any time—their AO and spin assignments are independent. There are three AO choices (p_1, p_0, p_{-1}) and two spin choices—six possibilities—for *each* electron, hence 36 unique possibilities. We should expect, therefore, 36 states to be included in our final set of terms.

We first find the possible L values. m_l for each electron is $1, 0$, or -1. We orient the larger of the **l** vectors to give the maximum $m_l(1) = +1$ and orient the second **l** in all possible ways, giving $m_l(2) = +1, 0, -1$. (Since the vectors have equal length in this case there is no "larger–smaller" choice to make.) The net M_L values are $+2, +1, 0$, and this tells us that the possible L values are $2, 1, 0$.

Treating m_s values similarly gives $M_s = 1, 0$, so $S = 1, 0$.

Thus, we have three **L** vectors and two **S** vectors. We now combine every one of the **L**, **S** pairs. In each case, we again take the longer in its position of greatest z overlap and combine it with the shorter in *all* of its orientations. This gives the J values shown in Table 5-2. The appropriate term symbols follow from L, S, and J in each case. Thus, our term symbols are $^3D_3$, $^3D_2$, $^3D_1$, $^1D_2$, $^3P_2$, $^3P_1$, $^3P_0$, $^1P_1$, $^3S_1$ and $^1S_0$ for a total of $7+5+3+5+5+3+1+3+3+1 = 36$ states.

In the absence of external fields, these 36 states occur in 10 energy levels, one for each term. These lie at different energies for several reasons. We have already seen, in our discussion of 1s2s helium states, that different spin multiplicities are associated with different symmetries of the spin wavefunction, meaning that the space part of the wavefunctions also differ in symmetry. This has a significant effect on energy, so ^1S and ^3S, for example, have rather different energies. Different L values amount to different occupancies of AOs, which also has an effect on the spatial wavefunctions, so ^3P and ^3S have different energies. Finally, we have already seen that different J values correspond to different relative orientations of orbital and spin angular momentum

TABLE 5-2 ► L and S Values for Two Nonequivalent Electrons and Resulting J Values and Term Symbols

L	S	J	Term
		3	$^3D_3$
2	1	2	$^3D_2$
		1	$^3D_1$
2	0	2	$^1D_2$
		2	$^3P_2$
1	1	1	$^3P_1$
		0	$^3P_0$
1	0	1	$^1P_1$
0	1	1	$^3S_1$
0	0	0	$^1S_0$

vectors, hence of magnetic moments. For light atoms, this is a relatively small effect, so $^3P_2$, $^3P_1$, and $^3P_0$ have only slightly different energies. The resulting energies for states of carbon in $1s^22s^22p^2$, $1s^22s^22p3p$, and $1s^22s^22p4p$ configurations are shown in Fig. 5-8. Only the major term-energy differences are distinguishable on the scale of the figure. The line for ^3D is really three very closely spaced lines corresponding to $^3D_3$, $^3D_2$, and $^3D_1$ terms.

2. Zeeman Effect. It was pointed out in Section 4-6 that the orbital energies of a hydrogen atom corresponding to the same n but different m_l undergo splitting when a magnetic field is imposed. Now we have seen that spin angular momentum is also present. Therefore, a proper treatment of the Zeeman effect requires that we focus on *total* angular momentum, not just the orbital component. Since there are $2J+1$ states with different M_J values in a given term, we expect each term to split into $2J+1$ evenly separated energies in the presence of a magnetic field, and this is indeed what is seen to happen (through its effects on lines in the spectrum). For example, a $^3P_2$ term splits into five closely spaced energies, corresponding to $M_J = 2, 1, 0, -1, -2$, and a $^1P_1$ term splits into three energies.

A surprising feature of this phenomenon is that the amount of splitting is not the same for all terms, despite the fact that adjacent members of any term always differ by ± 1 unit of angular momentum on the z axis. For instance, the spacing between adjacent members of the $^3P_2$ term mentioned above is 1.50 times greater than that in the $^1P_1$ term. It was recognized that terms wherein $S = 0$, so that J is entirely due to orbital angular momentum ($J = L$), undergo "normal splitting"—i.e., equal to what classical physics would predict for the amount of angular momentum and charge involved. On the other hand, terms wherein J is entirely due to spin ($L = 0$, so $J = S$) undergo *twice* the splitting predicted from classical considerations. [This extra factor of two (actually 2.0023) was without theoretical explanation until Dirac's relativistic treatment of quantum mechanics.]

Terms wherein J contains contributions from both L and S have Zeeman splittings other than one or two times the normal value, depending on the details of the way L and

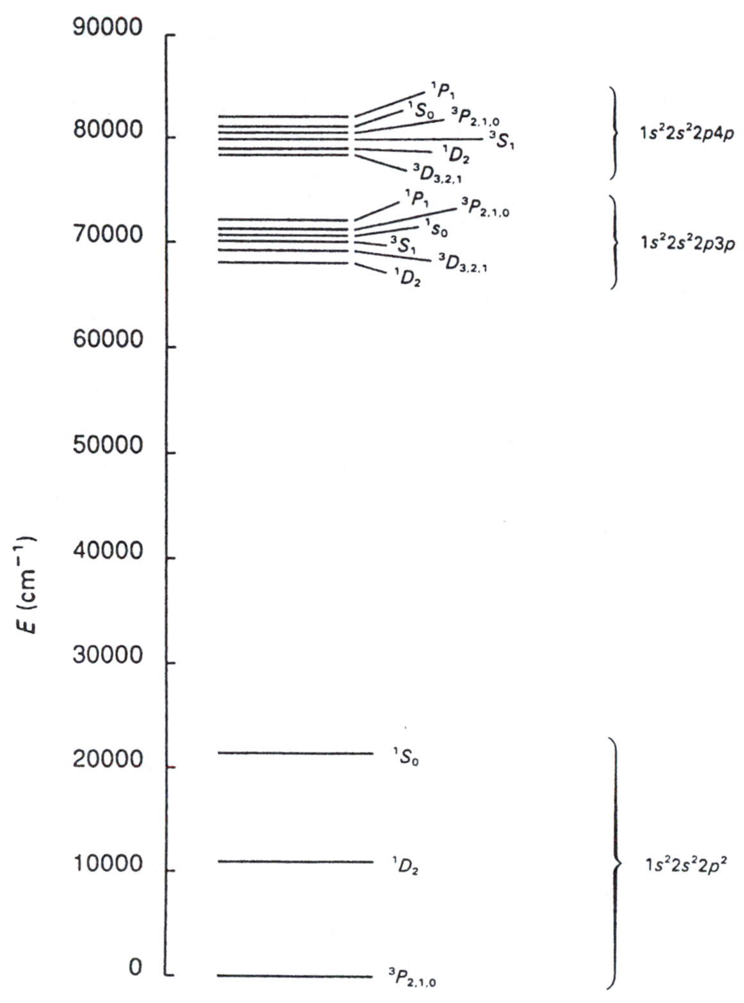

Figure 5-8 ▶ Energy levels for carbon atom terms resulting from configurations $1s^22s^22p^2$, $1s^22s^22p3p$, and $1s^22s^22p4p$.

S are combined. The extent to which a term member's energy is shifted by a magnetic field of strength B is

$$\Delta E = g\beta_e M_j B \tag{5-59}$$

where β_e is the Bohr magneton (Appendix 10) and g is the Landé g factor, which accounts for the different effects of L and S on magnetic moment that we have been discussing:

$$g = 1 + \frac{J(J+1) + S(S+1) - L(L+1)}{2J(J+1)} \tag{5-60}$$

It is not difficult to see that this formula equals one when $S = 0$, $J = L$, and equals two when $L = 0$, $J = S$. For the $^3P_2$ term, $S = 1$, $L = 1$, $J = 2$, and g equals 1.5, indicating that, in this state, half of the z-component of angular momentum is due to orbital motion, and half is due to spin (which is double-weighted in its effect on magnetic moment).

3. Equivalent Electrons. Observe that the energy-level diagram for carbon (Fig. 5-8) shows the 10 expected terms for the excited 2p3p and 2p4p configurations, but not for the ground $2p^2$ configuration. There are no new terms for the latter case, but some of the terms present for 2p3p or 2p4p are gone, namely ^3D, ^3S, and ^1P. The remaining terms account for 15 states. Evidently 21 states that are possible for a pair of nonequivalent p electrons are not allowed for a pair of *equivalent* electrons in a p^2 configuration. We will see that some of the states that are different for nonequivalent electrons become one and the same for equivalent electrons and must be excluded. Others are excluded by the Pauli exclusion principle because they would require two electrons to be in the same AO with the same spin.

We now demonstrate the method for discovering the terms that exist for equivalent electrons. This is more difficult than for nonequivalent electrons, even though there are fewer terms. We first list all the orbital-spin combinations (called microstates), strike out those that are redundant or that violate the Pauli exclusion principle, and then infer from the remaining microstates what terms exist.

Taking the p^2 case for illustration, we begin with the 36 microstates listed in Table 5-3. Some of these microstates are equivalent to others. For instance, $2p_1(1)2p_1(2)\alpha(1)\beta(2)$ is not a different state from $2p_1(1)2p_1(2)\beta(1)\alpha(2)$. These both correspond to a pair of electrons in the same pair of spin orbitals, $2p_1$ and $\overline{2p_1}$. Since electrons are indistinguishable, we cannot expect wavefunctions differing only in the order of electron labels to correspond to different physical states. [The single state that does exist would be accurately represented by $2p_1(1)2p_1(2)(\alpha(1)\beta(2) - \beta(1)\alpha(2))$, which is a linear combination of the microstates. But we do not need to go to this level of detail when finding terms. We only need to recognize that there is but one state here and omit one of the microstates as superfluous.] Accordingly, we strike out rows 3, 19, and 35 from Table 5-3, labeling them "R" for redundant.

Another way to recognize this equivalence is to observe that the microstates deemed redundant differ only by an interchange of a pair of electrons. This reveals that the set of four microstates with $2p_1(1)2p_0(2)$ is equivalent to the set with $2p_0(1)2p_1(2)$. Therefore, we can strike out rows 13–16. A similar argument removes rows 25–32. Already we have removed 15 microstates.

Next we look for violations of the Pauli exclusion principle. This leads us to strike out rows 1, 4, 17, 20, 33, and 36, labeling them "P" for Pauli. Our remaining microstates number 15 and are reassembled in Table 5-4, along with values of the quantum numbers for *z* components of the relevant angular momentum vectors for individual electrons as well as for their sum.

At this stage of the argument, the final column of Table 5-4 (the term symbols) is not yet known. We are about to fill out this column by making use of a simple rule that is based on the diagrammatic device described earlier—placing the larger vector so that it has the maximum *z* extension and then placing the shorter vector in all its allowed orientations. It is not difficult to see that the maximum resultant *z* component (M_J) can be achieved in one and only one way, namely when *both* vectors give their maximum *z* projection. This means that the maximum-M_J-member of a given set of states in the same term should be recognized as corresponding to one and only one microstate, because there is only one way to achieve this orientation. So we look for this maximum

TABLE 5-3 ► Unrestricted List of Space–Spin Combinations for a Pair of Electrons (Same Subshell). R = "Redundant," P = "Pauli"

	Electron number				
Row number	1	2	1	2	Comment
1	p_1	p_1	α	α	P
2	p_1	p_1	α	β	
3	p_1	p_1	β	α	R
4	p_1	p_1	β	β	P
5	p_1	p_0	α	α	
6	p_1	p_0	α	β	
7	p_1	p_0	β	α	
8	p_1	p_0	β	β	
9	p_1	p_{-1}	α	α	
10	p_1	p_{-1}	α	β	
11	p_1	p_{-1}	β	α	
12	p_1	p_{-1}	β	β	
13	p_0	p_1	α	α	R
14	p_0	p_1	α	β	R
15	p_0	p_1	β	α	R
16	p_0	p_1	β	β	R
17	p_0	p_0	α	α	P
18	p_0	p_0	α	β	
19	p_0	p_0	β	α	R
20	p_0	p_0	β	β	P
21	p_0	p_{-1}	α	α	
22	p_0	p_{-1}	α	β	
23	p_0	p_{-1}	β	α	
24	p_0	p_{-1}	β	β	
25	p_{-1}	p_1	α	α	R
26	p_{-1}	p_1	α	β	R
27	p_{-1}	p_1	β	α	R
28	p_{-1}	p_1	β	β	R
29	p_{-1}	p_0	α	α	R
30	p_{-1}	p_0	α	β	R
31	p_{-1}	p_0	β	α	R
32	p_{-1}	p_0	β	β	R
33	p_{-1}	p_{-1}	α	α	P
34	p_{-1}	p_{-1}	α	β	
35	p_{-1}	p_{-1}	β	α	R
36	p_{-1}	p_{-1}	β	β	P

TABLE 5-4 ► Allowed Space-Spin Combinations and M Quantum Numbers for a Pair of p Electrons (Same Subshell)

Microstate	$m_l(1)$	$m_l(2)$	$m_s(1)$	$m_s(2)$	M_L	M_S	M_J	State term
$p_1p_1\alpha\beta$	1	1	1/2	−1/2	2	0	2	${}^1D_2{}^2$
$p_1p_0\alpha\alpha$	1	0	1/2	1/2	1	1	2	${}^3P_2{}^2$
$p_1p_0\alpha\beta$	1	0	1/2	−1/2	1	0	1	$({}^1D_2{}^1)$
$p_1p_0\beta\alpha$	1	0	−1/2	1/2	1	0	1	$({}^3P_2{}^1)$
$p_1p_0\beta\beta$	1	0	−1/2	−1/2	1	−1	0	$({}^3P_2{}^0)$
$p_1p_{-1}\alpha\alpha$	1	−1	1/2	1/2	0	1	1	$({}^3P_1{}^1)$
$p_1p_{-1}\alpha\beta$	1	−1	1/2	−1/2	0	0	0	$({}^1D_2{}^0)$
$p_1p_{-1}\beta\alpha$	1	−1	−1/2	1/2	0	0	0	$({}^3P_1{}^0)$
$p_1p_{-1}\beta\beta$	1	−1	−1/2	−1/2	0	−1	−1	$({}^3P_2{}^{-1})$
$p_0p_0\alpha\beta$	0	0	1/2	−1/2	0	0	0	${}^1S_0{}^0$
$p_0p_{-1}\alpha\alpha$	0	−1	1/2	1/2	−1	1	0	$({}^3P_0{}^0)$
$p_0p_{-1}\alpha\beta$	0	−1	1/2	−1/2	−1	0	−1	$({}^1D_2{}^{-1})$
$p_0p_{-1}\beta\alpha$	0	−1	−1/2	1/2	−1	0	−1	$({}^3P_1{}^{-1})$
$p_0p_{-1}\beta\beta$	0	−1	−1/2	−1/2	−1	−1	−2	$({}^3P_2{}^{-2})$
$p_{-1}p_{-1}\alpha\beta$	−1	−1	1/2	−1/2	−2	0	−2	$({}^1D_2{}^{-2})$

M_J and, from its microstate, get the L and S values that go with it. That gives us the information we need to establish the term symbol.

We start, then, by seeking the maximum M_J value in Table 5-4. This is $M_J = 2$, and it occurs twice (in the first two rows). The first of these goes with $M_L = 2$, $M_S = 0$. Since these result when **L** and **S** are giving their maximum z component, we conclude that $L = 2$, $S = 0$. This, then, is a member of the ${}^1D_2$ term. (It is the ${}^1D_2^2$ member of that term, since $M_J = 2$.) We label this row ${}^1D_2^2$ and proceed to select microstates that can account for the other four members of this term. Our choice is controlled by the requirements that (1) the M_J values for the other members must be $1, 0, -1, -2$, and (2) we cannot have an $|M_s|$ value larger than zero or an $|M_L|$ value larger than 2. (That would be impossible for states resulting from vectors having $L = 2$ and $S = 0$.) Our selections are indicated in Table 5-4, with parentheses to indicate that these assignments follow from recognition of the leading member ${}^1D_2^2$. (All are symbolized as ${}^1D_2$.)

There is some arbitrariness in selecting the "inner" members, for which $|M_J| < J$: The parenthetical term ${}^1D_2^2$ could just as easily be assigned to $p_1p_0\beta\alpha$ as $p_1p_0\alpha\beta$. (Actually, neither of these microstates is a correct wavefunction for ${}^1D_2^2$. A linear combination of them is. But, if we only wish to designate term symbols, we need not worry about this.)

We have accomplished already the identification of a term ${}^1D_2$ and the elimination of five microstates from our list. The other microstate having $M_J = 2$ has $M_L = M_J = 1$, so we know this goes with $L = 1$, $S = 1$ and has the symbol ${}^3P_2^2$. Again, four other members exist down the table, and we select them, being careful that $|M_L|$ and $|M_s|$ do not exceed 1, while $M_J = 1, 0, -1, -2$.

At this point, we must recognize that we are not through with the ${}^3P$ family. The existence of the ${}^3P$ part of the symbol implies the existence of nine states, but ${}^3P_2$

accounts for only five of them. The others come from $^3P_1$ and $^3P_0$, resulting from $M_L = 1$ with $M_s = 0, -1$. (We did not worry about this for the $^1D_2$ term because only five states are implied by ^1D.) So we seek the microstates associated with these terms and label them as shown in the table.

Only one microstate remains. For this $M_L = M_s = 0$, so this is a state labeled $^1S_0^0$.

The term symbols for the p^2 configuration, then, are $^1D_2$, $^3P_2$, $^3P_1$, $^3P_0$, $^1S_0$, for a total of 15 states. The energies for these terms are shown in Fig. 5-8. The five terms fall into three energy groups, since the ^3P terms are found to be very close in energy. High resolution spectra can be used to see the slight energy differences between terms that appear to be at the same level at the energy scale used in Fig. 5-8. Delving further, the degenerate energies of microstates in the same term can be made to separate by imposing a magnetic field (Zeeman effect).

Based on spectroscopic assignments of energy levels for large numbers of atoms, Hund proposed a set of rules enabling one to predict the energy ordering for terms associated with *equivalent* electrons. These rules are, in order of decreasing influence:

1. Terms having greater spin multiplicity lie lower in energy.
2. Within each spin multiplicity, terms having greater L lie lower.
3. Within the same L and S, levels of different J behave oppositely, according to whether the subshell is more or less than half-filled: If less than half-filled, terms with lower J lie lower.

According to these rules, the five levels for carbon in its $1s^22s^22p^2$ configuration should be, in order of increasing energy, $^3P_0$, $^3P_1$, $^3P_2$ (closely spaced) followed by $^1D_0$, followed by $^1S_0$. The actual energies (in cm^{-1}) are, respectively, 0, 16.4, 43.5, 10194.8, 21647.7 (Fig. 5-8). The order of states for the excited 2p3p and 2p4p configurations is different. This is not a breakdown of Hund's rules because these are not equivalent-electron cases.

Hund's first rule is the source of the aufbau rule, cited earlier, that each AO of a subshell becomes half-filled before any of them become filled with electrons. The equivalence of these statements is easily demonstrated (Problem 5-30).

One can use Hund's rules to find the *lowest*-energy state term symbol without going through the tedious microstate process just described. For an atom having an outer subshell configuration of p^2 we would first recognize that we seek maximum S, so the electrons must have parallel spin, giving $S = 1$. (We use Hund's most influential rule first.) Subject to this constraint, we seek maximum L. Since the electrons cannot both be in p_1 with the same spin, p_1p_0 is next best, giving maximum $M_L = 1$, so $L = 1$. $S = 1$, $L = 1$ gives $J = 2, 1, 0$, so we know the corresponding terms are $^3P_2$, $^3P_1$, $^3P_0$. Since the 2p subshell is less than half- filled, $^3P_0$ is the ground term.

5-8 Overview

This chapter describes the new features that appear when we deal with systems having more than one electron. One of these features, interelectron repulsion, is easy to understand in its manifestation as operators in the hamiltonian and as coulomb repulsion integrals, J, in the average energy expression. Another feature, antisymmetry for electron exchange and the resulting existence of exchange integrals, K, is unfamiliar and unintuitive, without a classical counterpart, yet is enormously important in its effect on electronic structure.

In addition to these features, we have noted the importance of recognizing that atomic states conserve magnitude and z component of total angular momentum. Using this permits us to characterize states in terms of J, L, and S (even though the latter two are not "good" quantum numbers) for ground and excited configurations. This is essential in atomic spectroscopy (a topic we do not pursue in this book) and also allows one, with the assistance of Hund's rules, to predict the energy order for states associated with the ground configuration of any atom.

In closing this chapter, we should emphasize again a point frequently forgotten by chemist. In the orbital approach to many-electron systems we have a convenient *approximation*. This is an imperfect but useful way to describe atomic structure. There are more accurate ways to approximate eigenfunctions of many-electron hamiltonians, but this usually involves more difficulty in interpretation. The orbital representation of ψ appears to be the best compromise between accuracy and convenience for most chemical purposes.

5-8.A Problems

5-1. Write down the hamiltonian operator for the lithium atom, in a.u.

5-2. Calculate the values of $\bar{r}_1$and $\bar{r}_2$ consistent with the He wavefunction $\psi(1,2) = 1s(1)2s(2)\ldots$ (Eq. 5-11).

5-3. Calculate the energy in electron volts of a photon with associated wavelength 0.1 a.u. Compare this result with the ionization energy in electron volts of the hydrogen atom in its ground state. Why is this comparison relevant?

5-4. Show that the wavefunction (5-15) is normalized if the 1s and 2s orbitals are orthonormal.

5-5. Show that the wavefunction (5-16) is antisymmetric with respect to exchange of electron coordinates.

5-6. Show that the wavefunction (5-37) would vanish if 2s were replaced throughout by 1s, giving a $1s^3$configuration.

5-7. Produce a totally antisymmetric wavefunction starting from the configuration $1s(1)\alpha(1)2p(2)\beta(2)1s(3)\beta(3)$. Use the method described for Eq. (5-37) and use a determinantal function as a check.

5-8. Set up the integral of the product between $(1s1s2s\alpha\beta\alpha)^*$ and $2s1s1s\alpha\alpha\beta$. (Use symbols rather than explicit atomic orbital formulas.) Factor the integral into a product of integrals over one-electron space functions and one-electron spin functions. Indicate the value of each of the resulting six integrals and of their product.

5-9. a) Write down the Slater determinantal wavefunction for the configuration $1s1\bar{s}2p_z$.

b) Expand this determinant into a linear combination of products.

c) Write down the nonzero part of expansion (b) when $r_3 = 0$, $r_1 = 1$ a.u., $r_2 = 2$ a.u. [Do not evaluate the expression; just use symbols like 1s ($r = 1$).] Also write down the nonzero part of (b) when $r_2 = 0, r_1 = 1, r_3 = 2$, and when $r_1 = 0, r_2 = 2, r_3 = 1$. Is there any physical difference between saying "electron 3 is at the nucleus" and saying "an electron is at the nucleus?" Explain.

5-10. Wavefunction (5-38) describes a member of a *doublet*. Write the wavefunction for the other member.

5-11. A particle is capable of being in any one of *three* spin states. Call them α, β, and γ. Suppose you have two such particles in a molecule.

a) Write down all the spin functions you can that are symmetric for exchange of these two particles. (Do not worry about normalization.)

b) Write down all the antisymmetric cases.

5-12. The following wavefunction is proposed for an excited state of the lithium atom

$$\psi = \frac{1}{\sqrt{6}} \begin{vmatrix} \overline{1s}(1) & \overline{1s}(2) & \overline{1s}(3) \\ \overline{2s}(1) & \overline{2s}(2) & \overline{2s}(3) \\ \overline{3s}(1) & \overline{3s}(2) & \overline{3s}(3) \end{vmatrix}$$

Here 1s, 2s, and 3s are eigenfunctions for the Li^{2+} hamiltonian.

a) Does this wavefunction satisfy the Pauli exclusion principle? Explain.

b) Write the exact H for the lithium *atom* in atomic units.

c) Is ψ an eigenfunction for the exact hamiltonian?

d) If interelectronic repulsion terms are neglected in H, what energy, in a.u., is associated with ψ?

e) What z component of spin and orbital angular momentum (in atomic units) would you expect for the atom in this state, ignoring any nuclear contribution?

5-13. Write the normalized Slater determinantal wavefunction for beryllium in the $1s^2 2s^2$ configuration. Do *not* expand the determinant.

5-14. Write down the ground state configuration for the fluorine atom. Use Slater's rules to find the orbital exponents $\zeta = (Z - s)/n$ for 1s and 2s, 2p orbitals.

5-15. Show that the average value of an operator for a state described by an *eigenfunction* for that operator is identical to the eigenvalue associated with that eigenfunction.

5-16. Explain briefly the observation that the energy difference between the $1s^2 2s^1 (^2S_{1/2})$ state and the $1s^2 2p^1 (^2P_{1/2})$ state for Li is 14,904 cm^{-1}, whereas for Li^{2+} the $2s^1 (^2S_{1/2})$ and $2p^1 (^2P_{1/2})$ states are essentially degenerate. (They differ by only 2.4 cm^{-1}.)

5-17. In Chapter 4 it was stated that the magnitude of the square of the angular momentum is given by $l(l + 1)$ a.u., and that z components can be any of the values $-l, -l + 1, \ldots l - 1, l$ a.u. Similar relations hold for spin. From this fact plus the knowledge that the possible z components of spin angular momentum are $\pm\frac{1}{2}$ a.u., calculate the length of the spin angular momentum vector.

5-18. It has been shown that, for a single spinning electron, two spin states are possible having z components of spin angular momentum of $+1/2$ and $-1/2$ a.u. For two unpaired electrons, the state of greatest multiplicity is a triplet ($M_s = +1, 0, -1$). Show that, in general, the maximum spin multiplicity resulting from n unpaired electrons equals $n + 1$.

5-19. You have been shown symbolically that $1s(1)2s(2) \pm 2s(1)1s(2)$ and $\alpha(1)\beta(2) \pm \beta(1)\alpha(2)$ are symmetric or antisymmetric for exchange of electron labels (electron coordinates). For a more concrete and familiar example, take two functions: $f(x) = \exp(x)$ and $g(y) = y^3$. Construct combinations of these functions that are symmetric and antisymmetric for exchange of x and y coordinates. Set $x = 1$ and $y = 2$ and evaluate each function. Now set $x = 2$ and $y = 1$ and evaluate again. Compare results.

5-20. Give all the allowed term symbols for a hydrogen atom (a) in the $n = 1$ level, (b) in the $n = 2$ level. In each case, total up the states to see whether you have the expected number.

5-21. Consider the following helium atom wavefunction:

$$\psi = 1s(1)3d_{+2}(2)\alpha(1)\alpha(2)$$

a) Is this a satisfactory wavefunction in the sense of meeting general symmetry conditions resulting from particle indistinguishability and the exclusion principle? If not, how would you modify it to make it satisfactory?
b) Identify the term to which this state (modified if necessary) belongs.

5-22. How many *states* exist for the configuration spd?

5-23. A group of related terms has the common symbol ^2P. (This is called a term multiplet.)

a) What are the full term symbols for this multiplet?
b) How many energy levels exist (in the absence of a magnetic field) for this multiplet?
c) Indicate into how many levels each member of the multiplet splits in the presence of an external magnetic field.

5-24. Given the following space part of an approximate wavefunction for a Li^+ ion: $(1/\sqrt{2})[1s(1)2p_1(2) + 2p_1(1)1s(2)]$,

a) Write a physically possible spin part for this wavefunction.
b) What energy would this state have (in a.u.) if the $1/r_{12}$ term in H did not exist?
c) What *average* energy (expressed in terms of symbols like J) would this state have using the correct H (including $1/r_{12}$)?
d) You have not been shown the rules for operating with S^2, the operator for the square of total spin angular momentum, but you can nevertheless guess what the result would be if S^2 operates on this state function. What is your guess?

5-25. A state in the term $^3D_3$ is described by the wavefunction ψ. What is the value of x in each of the expressions Op $\psi = x\psi$, where Op is as given below? (Assume L–S coupling to be valid. If more than one x is possible, list them all.) (a) L^2 (b) S^2 (c) J^2 (d) L_z (e) S_z (f) J_z.

5-26. Carbon ($1s^22s^22p^2$) and oxygen ($1s^22s^22p^4$) have a "symmetrical" relation in their 2p occupancy: C has one electron *less* than a half-filled subshell, O has one electron *more*. Another way of stating this is to note that C has 2 electrons and 4 holes in its 2p shell, while O has 2 holes and 4 electrons.

a) Show that this leads to the same lowest-energy family (or "multiplet") of term symbols, $^3P_{2,0,1}$.
b) How do these atoms *differ* in the energy-ordering of these three terms?
c) Show that this agreement in lowest-energy multiplet terms holds in general for atoms having this symmetrical occupation relation.

5-27. Predict the ground state term symbol for each of the following atoms.

a) Na ($1s^22s^22p^63s$)
b) P ($1s^22s^22p^63s^23p^3$)
c) Ne ($1s^22s^22p^6$)
d) Ti ($1s^22s^22p^63s^23p^64s^23d^2$)

5-28. Calcium atoms are excited to the [Ar]4s4p configuration.

a) How many states are there?
b) What are the term symbols?

5-29. a) Find all the terms for boron in its ground configuration, $1s^22s^22p$, and order these terms according to energy.
b) Repeat for phosphorus, $[Ne]3s^23p^3$.

5-30. Explain how Hund's first rule is equivalent to the aufbau rule that degenerate AOs half-fill with electrons before any are filled, when forming the lowest-energy state(s).

5-31. How many states exist for each of the following term multiplets?

a) ^3D
b) ^5F

5-32. For a given electron configuration, are all of the following terms possible? Explain your reasoning. $^2P_{3/2}$, $^2P_{1/2}$, $^1S_0$.

5-33. How many states exist for each of the following configurations? (a) sd (b) sp (c) s^2p (d) pd (e) dd (nonequivalent)

5-34. a) How many states are associated with the ^4F term multiplet?
b) Write down the term symbols included in this multiplet.

5-35. By inspection, what is the term symbol with the maximum J value we can have for the configuration sd? What other terms would be included in the same multiplet?

5-36. Derive a formula for the number of states that exist for two *equivalent* electrons in a subshell having degeneracy g. How many states does this predict for p^2? for d^2?

5-37. Evaluate the splitting between adjacent lines in Zeeman-split terms $^3D_3$, $^3D_2$, $^3D_1$, $^1D_2$, when B equals 1 tesla.

Multiple Choice Questions

(Try to answer these without referring to the text.)

1. Which one of the following is an acceptable (unnormalized) approximate wavefunction for a state of the helium atom?

a) $[1s(1)1s(2) - 1s(1)1s(2)]\alpha(1)\alpha(2)$
b) $1s(1)1s(2)[\alpha(1)\beta(2) + \beta(1)\alpha(2)]$
c) $[1s(1)2s(2) + 2s(1)1s(2)]\alpha(1)\alpha(2)$
d) $[1s(1)2s(2) + 2s(1)1s(2)][\alpha(1)\beta(2) - \beta(1)\alpha(2)]$
e) None of the above is acceptable.

2. The spin multiplicity of an atom in its ground state and having the outer-shell configuration $4s^2 3d^7$ is

a) 19
b) 15
c) 7
d) 5
e) None of the above.

3. Which one of the following statements is NOT true for the ground state of the helium atom?

a) The atom's size (measured by r_{av}) is larger than the size of He^+ in its 1s state.
b) The ground state is a singlet.
c) The energy of the $2p_0$ orbital is above that of the 2s orbital.
d) The effective nuclear charge seen by both electrons is less than 2.
e) The atom's electronic energy is equal to -108.8 eV.

4. The crudest orbital model for the ground state of He uses the 1s atomic orbitals for He^+, for which $Z = 2$. Which of the following statements describes correctly the situation that pertains to a change to a more appropriate value?

a) The improved Z value is larger than 2, and the orbitals become more contracted.
b) The improved Z value is larger than 2, and the orbitals become more expanded.
c) The improved Z value is smaller than 2, and the orbitals become more contracted.
d) The improved Z value is smaller than 2, and the orbitals become more expanded.
e) The improved Z value is smaller than 2, but this only affects the computed energy, and not orbital size.

References

[1] O. Stern, Z. *Physik* **7**, 249 (1921).

[2] W. Gerlach and O. Stern, Z. *Physik* **8**, 110 (1922).

[3] E. G. Uhlenbeck and S. Goudsmit, *Naturwissenschaften* 13, 953 (1925); *Nature* **117**, 264 (1926).

[4] R. Bichowsky and H. C. Urey, *Proc. Natl. Acad. Sci.* **12**, 80 (1926).

[5] J. C. Slater, *Phys. Rev.* **34**, 1293 (1929).

[6] G. Herzberg, *Atomic Spectra and Atomic Structure*. Dover, New York, 1944.

[7] J. C. Slater, *Phys. Rev.* **36**, 57 (1930).

[8] E. Clementi and D. L. Raimondi, *J. Chem. Phys.* **38**, 2686 (1963).

[9] C. L. Pekeris, *Phys. Rev.* **115**, 1216 (1959).

[10] P. Roman, Origins of nonrelativistic spin, *Physics Today*, Jan. 1985, p.126.

Chapter 6

Postulates and Theorems of Quantum Mechanics

6-1 Introduction

The first part of this book has treated a number of systems from a fairly physical viewpoint, using intuition as much as possible. Now, armed with the concepts already developed, the reader should be in a better position to understand the more formal foundation to be described in this chapter. This foundation is presented as a set of postulates. From these follow proofs of various theorems. The ultimate test of the validity of the postulates comes in comparing the theoretical predictions with experimental data. The extra effort required to master the postulates and theorems is repaid many times over when we seek to solve problems of chemical interest.

6-2 The Wavefunction Postulate

We have already described most of the requirements that a wavefunction must satisfy: ψ must be acceptable (i.e., single-valued, nowhere infinite, continuous, with a piecewise continuous first derivative). For bound states (i.e., states in which the particles lack the energy to achieve infinite separation classically) we require that ψ be square integrable. So far we have considered only cases where the state of the system does not vary with time. For much of quantum chemistry, these are the cases of interest, but, in general, a state may change with time, and ψ will be a function of t in order to follow the evolution of the system.

Gathering all this together, we arrive at

Postulate I *Any bound state of a dynamical system of n particles is described as completely as possible by an acceptable, square-integrable function* $\Psi(q_1, q_2, \ldots q_{3n}, \omega_1, \omega_2, \ldots, \omega_n, t)$, *where the q's are spatial coordinates, ω's are spin coordinates, and t is the time coordinate.* $\Psi^*\Psi\, d\tau$ *is the probability that the space-spin coordinates lie in the volume element* $d\tau (\equiv d\tau_1 d\tau_2 \cdots d\tau_n)$ *at time t, if* Ψ *is normalized.*

For example, suppose we have a two-electron system in a time-dependent state described by the wavefunction $\Psi(x_1, y_1, z_1, \omega_1, x_2, y_2, z_2, \omega_2, t)$. The spin coordinates ω would each be some combination of spin functions α and β. If we integrate $\Psi^*\Psi$ over the spin coordinates of both electrons, we are left with a spin-free density function. Call it $\rho(x_1, y_1, z_1, x_2, y_2, z_2, t) \equiv \rho(v_1, v_2, t)$. We interpret $\rho(v_1, v_2, t)\, dv_1\, dv_2$ as the probability that electron 1 is in dv_1 (i.e., between x_1 and $x_1 + dx$, y_1 and $y_1 + dy$, and z_1 and $z_1 + dz$) and electron two is in dv_2 at time t. If we now integrate over the coordinates of electron 2, we obtain a new density function, $\rho'(v_1, t)$, which describes the probability of finding electron 1 in various volume elements at various times regardless of the position of electron 2.

6-3 The Postulate for Constructing Operators

Much of the substance of the second postulate is already familiar. We earlier used arguments based on de Broglie waves to construct hamiltonian operators. We then noted that the kinetic energy part of the operators can be identified with a classical term like $p_x^2/2m$ through the relation $p_x \leftrightarrow (\hbar/i)\partial/\partial x$. The potential energy terms in the hamiltonian operators are completely classical, however. Thus, we could have constructed the quantum mechanical hamiltonians by writing down the *classical* energy expressions in terms of momenta and position, and then replacing every momentum term by the appropriate partial differential operator. This is an example of the use of part *c* of:

Postulate II *To every observable dynamical variable M there can be assigned a linear hermitian operator $\hat{M}$. One begins by writing the classical expression, as fully as possible in terms of momenta and positions. Then:*

a) *If M is q or t, $\hat{M}$ is q or t. (q and t are space and time coordinates.)*

b) *If M is a momentum, p_j, for the jth particle, the operator is $(\hbar/i)\partial/\partial q_j$, where q_j is conjugate to p_j (e.g., x_j is conjugate to p_{xj}).*

c) *If M is expressible in terms of the q's, p's and t, $\hat{M}$ is found by substituting the above operators in the expression for M in such a way that $\hat{M}$ is hermitian.*

The reason for specifying that $\hat{M}$ must be hermitian is that the eigenvalues of a hermitian operator must be real numbers.[1] We shall discuss this and other aspects of hermiticity (including its definition) later in this chapter.

As an explicit example of this procedure, we reconsider the hydrogen atom. Assuming a fixed nucleus (infinite inertia), the classical expression for the total energy of the system is

$$E_{\text{classical}} = (1/2m_e)(p_x^2 + p_y^2 + p_z^2) - e^2/\left[4\pi\epsilon_0(x^2 + y^2 + z^2)^{1/2}\right]$$

where the first term is just the kinetic energy of the electron and the second term is the electrostatic potential energy. The coordinate origin is on the nucleus. Application of postulate II retains the position variables x, y, and z of the potential term unchanged, but replaces p_x by $(\hbar/i)\,\partial/\partial x$, etc.:

$$\frac{1}{2m_e}\left(p_x^2 + p_y^2 + p_z^2\right) \Rightarrow \frac{1}{2m_e}\left\{\left(\frac{h}{2\pi i}\frac{\partial}{\partial x}\right)^2 + \left(\frac{h}{2\pi i}\frac{\partial}{\partial y}\right)^2 + \left(\frac{h}{2\pi i}\frac{\partial}{\partial z}\right)^2\right\}$$
$$= \frac{-h^2}{8\pi^2 m_e}\nabla^2$$

[1] In this text (and in quantum chemistry in general) a caret indicates an operator and not a unit vector quantity as in classical physics.

Thus, we arrive at

$$\hat{H} = \frac{-h^2}{8\pi^2 m_e}\nabla^2 - \frac{e^2}{4\pi\epsilon_0(x^2+y^2+z^2)^{1/2}}$$

and we are now free to transform $\hat{H}$ to other coordinate systems (such as r, θ, ϕ) if we wish.

6-4 The Time-Dependent Schrödinger Equation Postulate

We have discussed only cases where neither the hamiltonian $\hat{H}$ nor ψ is time dependent. In those cases we required that ψ be an eigenfunction of $\hat{H}$. In the more general case in which Ψ and $\hat{\mathscr{H}}$ are time dependent,[2] a different requirement is imposed by

Postulate III *The state functions (or wavefunctions) satisfy the equation*

$$\hat{\mathscr{H}}\Psi(q,t) = \frac{-\hbar}{i}\frac{\partial}{\partial t}\Psi(q,t) \tag{6-1}$$

where $\hat{\mathscr{H}}$ is the hamiltonian operator for the system.

We should check to see if this is consistent with the time-independent Schrödinger equation we have been using. Suppose that the hamiltonian is time independent. Let us see if a solution to Eq. (6-1) exists when $\Psi(q,t)$ is separated into a product of space- and time-dependent functions: $\Psi(q,t) = \psi(q)f(t)$. Inserting this into Eq. (6-1) gives

$$\hat{H}\psi(q)f(t) = \frac{-\hbar}{i}\frac{\partial}{\partial t}\psi(q)f(t) \tag{6-2}$$

Dividing by $\psi(q)f(t)$ gives

$$\frac{\hat{H}\psi(q)}{\psi(q)} = \frac{(-\hbar/i)(\partial/\partial t)f(t)}{f(t)} \tag{6-3}$$

Since each side of Eq. (6-3) depends on a different variable, the two sides must equal the same constant, which we call E. This gives

$$\hat{H}\psi(q) = E\psi(q) \tag{6-4}$$

and

$$\frac{-\hbar}{i}\frac{d}{dt}f(t) = Ef(t) \tag{6-5}$$

The first of these equations is just the time-independent Schrödinger equation we have been using. The second equation has the solution $f(t) = A\exp(-iEt/\hbar)$. Hence, f^*f equals a constant, and so $\Psi^*\Psi = \psi^*\psi f^* f \propto \psi^*\psi$. Since f has no effect on energy or particle distribution, we can ignore it in dealing with stationary states. The situation is analogous to the case of standing waves discussed in Chapter 1.

[2] $\hat{\mathscr{H}}$ and Ψ symbolize time dependence; $\hat{H}$ and ψ symbolize time independence.

Note that while we have shown that solutions may exist in which Ψ is separable, this does not mean that every solution of Eq. (6-1) with $\hat{\mathscr{H}} = \hat{H}$ is separable (i.e., stationary). We can imagine a situation where a system in a stationary state is suddenly perturbed to produce a new time-independent hamiltonian. Ψ will change as the system adjusts to this new situation, giving us a case where the hamiltonian is time-independent (after the perturbation, at least) but Ψ is not a stationary state function. The way in which Ψ evolves in time is governed by Eq. (6-1).

EXAMPLE 6-1 Show that the average energy for a nonstationary state of the hydrogen atom is conserved as the system evolves, if $\hat{H}$ is not time-dependent.

SOLUTION ▶ We choose a simple example:

$$\psi = \frac{1}{\sqrt{2}}\{\psi_{1s}\exp(it/2\hbar) + \psi_{2s}\exp(it/8\hbar)\}$$

where each exponential equals $\exp(-iEt/\hbar)$. Then

$$\langle E\rangle = \int \psi^* \hat{H}\psi\, dv\, dt = \frac{1}{2}\int \psi_{1s}^* \exp(-it/2\hbar)\hat{H}\psi_{1s}\exp(it/2\hbar)\, dv\, dt$$

$+$ the analogous $2s2s$ term $+1s2s$ and $2s1s$ cross terms. Since $\hat{H}$ does not operate on functions of t, the exponentials in each of the first two integrals can join together, giving $\exp(0) = 1$. So time-dependence disappears from the first two integrals, and they become respectively equal to $-\frac{1}{2}$ a.u., and $-\frac{1}{8}$ a.u. Dependence on time does not disappear from the cross-term integrals, but that doesn't matter because the integration over space gives zero in each case, due to orbital orthogonality. Thus, $< E >= \frac{1}{2}(-\frac{1}{2}$ a.u.$) + \frac{1}{2}(-\frac{1}{8}$ a.u.$) = -\frac{5}{16}$ a.u., which has no time-dependence. ◀

6-5 The Postulate Relating Measured Values to Eigenvalues

The second postulate indicated that every observable variable of a system (such as position, momentum, velocity, energy, dipole moment) was associated with a hermitian operator. The connection between the observed value of a variable and the operator is given by

Postulate IV *Any result of a measurement of a dynamical variable is one of the eigenvalues of the corresponding operator.*

Any measurement always gives a real number, and so this postulate requires that eigenvalues of the appropriate operators be real. We will prove later that hermitian operators satisfy this requirement.

If we measured the electronic energy of a hydrogen atom (the negative of its ionization energy), we could get any of the allowed eigenvalues ($-1/2n^2$ a.u.) but no intermediate value. What if, instead, we measured the distance of the electron from the nucleus. By postulate II, the operator for this property is just the variable r itself. That is, $\hat{r} = r$. Hence, we need to consider the eigenvalues of r in the equation

$$r\,\delta(r,\theta,\phi) = \lambda\,\delta(r,\theta,\phi) \tag{6-6}$$

where δ is an eigenfunction and λ is a real number (corresponding to the distance of the electron from the nucleus). We can rewrite this equation as

$$(r-\lambda)\delta(r,\theta,\phi)=0 \tag{6-7}$$

This form makes it more apparent that the function δ must vanish at all points in space except those where $r=\lambda$. But λ is an eigenvalue of r and hence is a possible result of a measurement. Thus, we see that postulate IV implies some connection between a measurement of, say, $r=2$ a.u. and an eigenfunction of r that is finite only at $r=2$ a.u. We symbolize this eigenfunction $\delta(r-2\text{ a.u.})$, this "delta function" being zero whenever the argument is *not* zero. If we measured the electron's position to be at $r=5.3$ a.u., the corresponding eigenfunction would be $\delta(r-5.3\text{ a.u.})$—a function that is zero everywhere except in a shell of infinitesimal thickness at $r=5.3$ a.u. If instead we measured the *point* in space of the electron, rather than just the distance from the nucleus, and found it to be r_0, θ_0, ϕ_0, then the corresponding eigenfunction of the position operator would be $\delta(r-r_0)\delta(\theta-\theta_0)\delta(\phi-\phi_0)$. This function vanishes everywhere except at r_0, θ_0, ϕ_0.

It is evident that any value of λ from zero to infinity in Eq. (6-7) may be chosen without spoiling the ability of δ to serve as an eigenfunction of r. This means that, unlike the energy measurement, the measurement of the distance of the electron from the nucleus can have any value.

The eigenfunctions of the position operator are called *Dirac delta functions*. They are "spike" functions having infinitesimal width. They are normalized through the equation

$$\int \delta\,(x-x_0)\,dx=1 \tag{6-8}$$

where the integration range includes x_0.[3] On first acquaintance, these functions seem mathematically peculiar, but they make physical sense in the following way. One can interpret the actual measurement of position as a process that forces the particle to acquire a certain position at some instant. At that instant, ψ^2 for the system (now perturbed by the measuring process) ought to give unit probability for finding the particle at that point (where it definitely is) and zero probability elsewhere, and this is just what the Dirac delta function does.[4]

Postulate IV, then, is in accord with a picture wherein the process of measurement forces the measured system into an eigenstate for the appropriate operator, giving the corresponding eigenvalue as the measurement. This definition of "measurement" is somewhat restrictive and can be deceptive. Often scientists refer to measurements that are really measurements of average values rather than eigenvalues. This point is discussed further below.

---

[3]The reader should avoid confusing the Dirac delta function $\delta(x-x_0)$ with the Kronecker delta $\delta_{i,j}$ encountered earlier. They are similar in that both vanish unless $x=x_0$ in the former and $i=j$ in the latter. But they differ in that the *value* of $\delta_{i,j}$ is definite (unity) while the value of $\delta(x_0-x_0)$ is not defined. The Dirac delta function has definite value only in integrated expressions like Eq. (6-8). The spin functions α and β may be thought of as Dirac delta functions in the spin "coordinate" ω. The Dirac delta function is admittedly unusual, and one tends to be uneasy with it at first. This function is important and useful in quantum mechanics. However, since we will make almost no use of it in this text, we will not develop the topic further.

[4]Notice that Eq. (6-8) does not involve $\delta^*\delta$, but merely δ. Because δ is nonzero only at one point, $\delta^*\delta$ is likewise nonzero only at the same point. δ and $\delta^*\delta$ are therefore not independent functions. It is convenient to view the δ function as both the eigenfunction for the position operator and also as the probability distribution function for the particle.

6-6 The Postulate for Average Values

Suppose that we had somehow prepared a large number of hydrogen atoms so that they were all in the same, known, stationary state. Then we could measure the distance of the electron from the nucleus once in each atom and average these measurements to obtain an average value. We have already indicated that this average would be given by the sum of all the r values, each multiplied by its frequency of occurrence, which is given by $\psi^2 dv$ if ψ is normalized. Since r is a continuous variable, the sum becomes an integral. This is the content of

Postulate V *When a large number of identical systems have the same state function ψ, the expected average of measurements on the variable M (one measurement per system) is given by*

$$M_{\text{av}} = \int \psi^* \hat{M} \psi \, d\tau \bigg/ \int \psi^* \psi \, d\tau \tag{6-9}$$

The denominator is unity if ψ is normalized.

It is important to understand the distinction between average value and eigenvalue as they relate to measurements. A good example is the dipole moment. The dipole moment operator for a system of n charged particles is $\mu = \sum_{i=1}^{n} z_i \mathbf{r}_i$ where z_i is the charge on the ith particle and $\mathbf{r}_i$ is its position vector with respect to an arbitrary origin. (We get this by writing the classical formula and observing that momentum terms do not appear. Hence, the quantum-mechanical operator is the same as the classical expression.) What will the eigenfunctions and eigenvalues of $\hat{\mu}$ be like?

The charge z_i is only a number, while $\mathbf{r}_i$ is a position operator, which has Dirac delta functions as eigenfunctions. For a hydrogen atom, one eigenfunction of r_i would be a delta function at $r = 1$ a.u., $\theta = 0$, $\phi = 0$. The corresponding eigenvalue for $\hat{\mu}$ would be the dipole moment obtained when a proton and an electron are separated by 1 a.u., clearly a finite number. But "everybody knows" that an unperturbed atom in a stationary state has zero dipole moment. The difficulty is resolved when we recognize that measurement of a variable in postulates IV and V means measuring the value of a variable at a given instant. Hence, we must distinguish between the *instantaneous dipole moment* of an atom, which can have any value from among the eigenvalues of $\hat{\mu}$ and the *average dipole moment*, which is zero for the atom. In everyday scientific discussion, the term "dipole moment" is usually understood to refer to the *average* dipole moment. Indeed, the usual measurements of dipole moment are measurements that effectively average over many molecules or long times (in atomic terms) or both.

6-7 Hermitian Operators

Let ϕ and ψ be any square-integrable functions and $\hat{A}$ be an operator, all having the same domain. $\hat{A}$ is defined to be hermitian if

$$\int \psi^* \hat{A} \phi dv = \int \phi \hat{A}^* \psi^* dv \tag{6-10}$$

The integration is over the entire range of each spatial coordinate. Recall that the asterisk signifies reversal of the sign of i in a complex or imaginary term. The hermitian property has important consequences in quantum chemistry.

As an example of a test of an operator by Eq. (6-10), let us take the ψ and ϕ to be square-integrable functions of x and $\hat{A}$ to be $i(d/dx)$. Then the left-hand side of Eq. (6-10) becomes, upon integration by parts,

$$\int_{-\infty}^{+\infty} \psi^*(i\, d\phi/dx)\, dx = i\psi^*\phi \Big|_{-\infty}^{+\infty} {}^{\nearrow 0} - i\int_{-\infty}^{+\infty} (d\psi^*/dx)\phi\, dx = -i\int_{-\infty}^{+\infty} \phi(d\psi^*/dx)\, dx \tag{6-11}$$

Since ψ and ϕ are square integrable, they (and their product) must vanish at infinity, giving the zero term in Eq. (6-11). We now write out the right-hand side of Eq. (6-10):

$$\int_{-\infty}^{+\infty} \phi(i\, d/dx)^* \psi^* dx = -i\int_{-\infty}^{+\infty} \phi(d\psi^*/dx)\, dx \tag{6-12}$$

where the minus sign comes from carrying out the operation indicated by the asterisk. Equation (6-12) is equal to Eq. (6-11), and so the operator $i(d/dx)$ is hermitian. Since the effect of i was to introduce a necessary sign reversal, it is apparent that the equality would not result for $\hat{A} = d/dx$. Clearly, any hermitian operator involving a first derivative in any Cartesian coordinate must contain the factor i. The operators for linear momenta (Chapter 2) are examples of this.

It is important to realize that Eq. (6-10) does not imply that $\psi^* \hat{A}\phi = \phi \hat{A}^* \psi^*$. A simple example will make this clearer. Let $\hat{A}$ be the hydrogen atom hamiltonian, $\hat{H} = -\frac{1}{2}\nabla^2 - 1/r$, and let ϕ be the 1s eigenfunction: $\phi = (1/\sqrt{\pi})\exp(-r)$. Also, let $\psi = \sqrt{8/\pi}\exp(-2r)$ which is not an eigenfunction of $\hat{H}$. Then, since $\hat{H}\phi = -\frac{1}{2}\phi$,

$$\psi^* \hat{H}\phi = -\frac{1}{2}\psi^*\phi \tag{6-13}$$

But

$$\phi H^* \psi^* = \phi\left[-\frac{1}{2}(1/r^2)(d/dr)r^2(d/dr) - 1/r\right]\sqrt{8/\pi}\exp(-2r) \tag{6-14}$$

$$= \phi\,[(1/r) - 2]\sqrt{8/\pi}\exp(-2r) = [(1/r) - 2]\,\psi^*\phi \tag{6-15}$$

(Since ψ has no θ or ϕ dependence, the parts of $\hat{H}^*$ that include $\partial/\partial\theta$ and $\partial/\partial\phi$ have been omitted in Eq. (6-14).) Here we have two functions, $-\frac{1}{2}\psi^*\phi$ and $[(1/r) - 2]\psi^*\phi$. They are obviously different. However, by Eq. (6-10), their integrals are equal since $\hat{H}$ is hermitian.

6-8 Proof That Eigenvalues of Hermitian Operators Are Real

Let $\hat{A}$ be a hermitian operator with a square-integrable eigenfunction ψ. Then

$$\hat{A}\psi = a\psi \tag{6-16}$$

Each side of Eq. (6-16) must be expressible as a real and an imaginary part. The real parts must be equal to each other and so must the imaginary parts. Taking the complex conjugate of Eq. (6-16) causes the imaginary parts to reverse sign, but they remain equal. Therefore, we may write

$$\hat{A}^*\psi^* = a^*\psi^* \tag{6-17}$$

We multiply Eq. (6-16) from the left by ψ^* and integrate over all spatial variables:

$$\int \psi^*\hat{A}\psi \, dv = a\int \psi^*\psi \, dv \tag{6-18}$$

Similarly, we multiply Eq. (6-17) from the left by ψ and integrate:

$$\int \psi\hat{A}^*\psi^* dv = a^*\int \psi\psi^* dv \tag{6-19}$$

Since $\hat{A}$ is hermitian, the left-hand sides of Eqs. (6-18) and (6-19) are equal by definition (Eq. 6-10). Therefore, the right-hand sides are equal, and their difference is zero:

$$(a - a^*)\int \psi^*\psi \, dv = 0 \tag{6-20}$$

Since ψ is square integrable the integral cannot be zero. Therefore, $a - a^*$ is zero, which requires that a be real.

6-9 Proof That Nondegenerate Eigenfunctions of a Hermitian Operator Form an Orthogonal Set

Let ψ and ϕ be two square-integrable eigenfunctions of the hermitian operator $\hat{A}$:

$$\hat{A}\psi = a_1\psi \tag{6-21}$$
$$\hat{A}^*\phi^* = a_2\phi^* \tag{6-22}$$

Multiplying Eq. (6-21) from the left by ϕ^* and Eq. (6-22) from the left by ψ, and integrating gives

$$\int \phi^*\hat{A}\psi \, dv = a_1\int \phi^*\psi \, dv \tag{6-23}$$
$$\int \psi\hat{A}^*\phi^* \, dv = a_2\int \psi\phi^* \, dv \tag{6-24}$$

The left sides of Eqs. (6-23) and (6-24) are equal by (6-10), and

$$(a_1 - a_2)\int \phi^*\psi \, dv = 0. \tag{6-25}$$

If $a_1 \neq a_2$, the integral vanishes. This proves that nondegenerate eigenfunctions are orthogonal.

EXAMPLE 6-2 It has been shown (Section 6-7) that $i(d/dx)$ is a hermitian operator. We know that it has eigenfunctions $\exp(\pm ikx)$ with eigenvalues $\pm k$, which are real. So far, so good. However, this operator also has eigenfunctions $\exp(\pm kx)$, with eigenvalues $\pm ik$, which are imaginary. This appears to violate the proof that eigenvalues of hermitian operators are real. Explain why neither of these eigenfunction sets is covered by the proof of section 6-8, and how one of them manages to obey the rule anyway.

SOLUTION ▶ The test for Hermiticity requires that $i\psi^*\phi|_{-\infty}^{+\infty} = 0$. If ϕ is ψ, and if ψ is square-integrable, this condition is satisfied, because $\psi^*\psi$ vanishes at $\pm\infty$, giving $0 - 0 = 0$. But neither of the exponential functions given above is square-integrable: They are both unequal to zero at $\pm\infty$, so they both fall outside of the proof as given. Despite this, $\exp(\pm ikx)$ does have real eigenvalues, leading us to look more closely. Is it the case that $i\psi^*\psi|_{-\infty}^{+\infty} = 0$ for this set of functions, even though they do not vanish at infinity? It is indeed, since $\psi^*\psi = 1$, giving $i - i = 0$ for this term. Thus we see that our requirement that ψ be square integrable is more restrictive than what is necessary, namely that $i\psi^*\psi|_{-\infty}^{+\infty} = 0$. Note that the other set of exponentials, $\exp(\pm kx)$, leads to $i\psi^*\psi = i\exp(\pm 2kx)$, which does not produce a value of zero when values at $x = \infty$ and $x = -\infty$ are subtracted. Note also that the functions $\exp(\pm ikx)$ are orthogonal for different values of k, whereas the functions $\exp(\pm kx)$ are not. ◀

The point of the above example is that all of our proofs about eigenvalues or eigenfunctions of hermitian operators refer to cases where the eigenfunctions satisfy the requirement that $i\psi^*\psi|_{-\infty}^{+\infty} = 0$. Square-integrability guarantees this, but some nonsquare-integrable sets of functions can satisfy it too. A hermitian operator can have eigenfunctions that are associated with complex or imaginary eigenvalues, but these must result from eigenfunctions that do not satisfy the requirement.

6-10 Demonstration That All Eigenfunctions of a Hermitian Operator May Be Expressed as an Orthonormal Set

If $a_1 = a_2$, Eq. (6-25) is satisfied even when the integral is finite. Therefore, degenerate eigenfunctions need not be orthogonal. But they must be linearly independent or else they are the self-same function (to within a multiplicative constant), and if they are linearly independent, they can be converted to an orthogonal pair. Hence, it is always possible to express the *degenerate* eigenfunctions of a hermitian operator as an orthogonal set (and, as we have just proved, it is *necessary* that *nondegenerate* eigenfunctions be orthogonal). Furthermore, the functions must be square integrable, hence normalizable. In general, then, we are able to assume that all of the eigenfunctions of a hermitian operator can be expressed as an orthonormal set.

One way to orthogonalize two nonorthogonal, linearly independent functions (which may or may not be eigenfunctions) will now be demonstrated. Let the functions be ψ and ϕ (assumed normalized) and the integral of their product have the value S:

$$\int \psi^*\phi\, dv = S \tag{6-26}$$

We keep one of the functions unchanged, say ψ, and let $\phi' \equiv \phi - S\psi$ be our new second function. ψ and ϕ' are orthogonal since

$$\int \psi^* \phi' \, dv = \int \psi^* (\phi - S\psi) dv = \underbrace{\int \psi^* \phi \, dv}_{S} - S \underbrace{\int \psi^* \psi \, dv}_{1} = 0 \qquad (6\text{-}27)$$

(The new function ϕ' needs to be renormalized.) This process, known as *Schmidt orthogonalization*, may be generalized and applied sequentially to any number of linearly independent functions.

EXAMPLE 6-3 Two normalized $1s$ AOs are located on nearby nuclei, A and B, and overlap each other enough so that $\int 1s_A 1s_B \, dv = 0.500$. Construct a function from these two that is orthogonal to $1s_A$ and is normalized.

SOLUTION ▶ $1s'_B = 1s_B - 0.5 \cdot 1s_A$ is orthogonal to $1s_A$. It is not yet normalized because

$$\int (1s'_B)^2 \, dv = \int (1s_B^2 + 0.25 \cdot 1s_A^2 - 2 \cdot 0.5 \cdot 1s_A 1s_B) \, dv$$
$$= 1 + 0.25 - 2 \cdot 0.5 \cdot 0.5 = 0.75 = \frac{3}{4}.$$

So the normalized function we seek is $\frac{2}{\sqrt{3}}(1s_B - 0.5 \cdot 1s_A)$. ◀

6-11 Proof That Commuting Operators Have Simultaneous Eigenfunctions

$\hat{A}$ and $\hat{B}$ are commuting operators if, for the general square-integrable function f, $\hat{A}\hat{B}f = \hat{B}\hat{A}f$. This can be written $(\hat{A}\hat{B} - \hat{B}\hat{A})f = 0$, which requires that $\hat{A}\hat{B} - \hat{B}\hat{A} = \hat{0}$. ($\hat{0}$ is called the null operator. It satisfies the equation, $\hat{0}f = 0$.) This difference of operator products is called the *commutator* of $\hat{A}$ and $\hat{B}$ and is usually symbolized[5] by $[\hat{A}, \hat{B}]$. If the commutator $[\hat{A}, \hat{B}]$ vanishes, then $\hat{A}$ and $\hat{B}$ commute.

We will now prove an important property of commuting operators, namely, that they have "simultaneous" eigenfunctions (i.e., that a set of eigenfunctions can be found for one of the operators that is also an eigenfunction set for the other operator). Let β_i be the eigenfunctions for $\hat{B}: \hat{B}\beta_i = b_i\beta_i$. For the moment, assume all the numbers b_i are different (i.e., the eigenfunctions β_i are nondegenerate). Let $[\hat{A}, \hat{B}] = \hat{0}$. Then

$$\hat{B}(\hat{A}\beta_i) = \hat{A}\hat{B}\beta_i = \hat{A}b_i\beta_i = b_i(\hat{A}\beta_i) \qquad (6\text{-}28)$$

The parentheses emphasize that the function obtained by operating on β_i with $\hat{A}$ is an eigenfunction of $\hat{B}$ with eigenvalue b_i. But that function can only be a constant times β_i itself. Hence, for nondegenerate β_i we have that $\hat{A}\beta_i = c\beta_i$, and so β_i is an eigenfunction of $\hat{A}$. This proves that the nondegenerate eigenfunctions for one operator will also be eigenfunctions for any other operators that commute with it.

[5]Other less common conventions are $[\hat{A}, \hat{B}]_-$ and $(\hat{A}, \hat{B})$.

If β_i is degenerate with other functions $\beta_{i,k}$, then we can only go so far as to say that $\hat{A}\beta_i = \sum_k c_k\beta_{i,k}$, for this general linear combination is an eigenfunction of $\hat{B}$ having eigenvalue b_i. But if this is so, then β_i is evidently not necessarily an eigenfunction of $\hat{A}$. We shall not prove it here, but it is possible to show that one can always find *some* linear combinations of $\beta_{i,k}$ to produce a set of new functions, β_i', that *are* eigenfunctions of $\hat{A}$ (and remain eigenfunctions of $\hat{B}$ as well). Therefore we can state that, if $\hat{A}$ and $\hat{B}$ commute, there exists a set of functions that are eigenfunctions for $\hat{A}$ and $\hat{B}$ simultaneously.

An example of this property occurred in the particle-in-a-ring system described in Chapter 2. The hamiltonian and angular momentum operators commute for that system. There we found one set of functions, the trigonometric functions, that are eigenfunctions for $\hat{H}$ but not for $\hat{L}_z$. But by mixing the energy-degenerate sines and cosines we produced exponential functions that are eigenfunctions for both of these operators.

Another example concerns the familiar symmetry operations for reflection, rotation, etc. If one of these operations, symbolized $\hat{R}$, commutes with the hamiltonian, then we should expect there to be a set of eigenfunctions for $\hat{H}$ that are simultaneously eigenfunctions for $\hat{R}$. It was proved in Chapter 2 that this means that nondegenerate eigenfunctions must be symmetric or antisymmetric with respect to $\hat{R}$.

A symmetry operator that leaves $\hat{H}$ unchanged can be shown to commute with $\hat{H}$. That is, if $\hat{R}\hat{H} = \hat{H}$, then $\hat{R}\hat{H}f = \hat{H}\hat{R}f$, where f is any function. To show this, let $\hat{R}$ be, say, a reflection operator. Then $\hat{R}$ operates on functions and operators to its right by reflecting the appropriate coordinates: $\hat{R}f(q) = f(Rq)$. If $\hat{H}$ is invariant under reflection $\hat{R}$, then $H(q) = H(Rq)$, and it follows that $\hat{R}\hat{H}(q)f(q) = \hat{H}(Rq)f(Rq) = \hat{H}(q)f(Rq) = \hat{H}(q)\hat{R}f(q)$, and so $\hat{R}\hat{H}f = \hat{H}\hat{R}f$. We shall formally develop the ramifications of symmetry in quantum chemistry in Chapter 13.

The existence of simultaneous eigenfunctions for various operators has important ramifications for the measurement of a system's properties. This is discussed in Section 6-15.

6-12 Completeness of Eigenfunctions of a Hermitian Operator

In Chapter 3 we discussed the concept of completeness in connection with the power series expansion of a function. Briefly, a series of functions[6] $\{\phi\}$ having certain restrictions (e.g., all derivatives vary smoothly) is said to be complete if an arbitrary function f having the same restrictions can be expressed in terms of the series[7]

$$f = \sum_i c_i\phi_i \tag{6-29}$$

Proofs exist that certain hermitian operators corresponding to observable properties have eigenfunctions forming a complete set in the space of well-behaved (continuous,

[6]A symbol in braces is frequently employed to represent an entire set of functions.

[7]Equation (6-29) is overly restrictive in that it requires that the function and the series have identical values at every point, whereas it is possible for them to disagree at points of zero measure. However, at the level of this book, we can ignore this distinction and use Eq. (6-29) without encountering difficulty.

single-valued, square-integrable) functions. These proofs are difficult and will not be given here.[8] Instead we shall introduce

Postulate VI *The eigenfunctions for any quantum mechanical operator corresponding to an observable variable constitute a complete set.* (Furthermore, we have seen in Section 6-10 that we can assume that this set has been made orthonormal.)

We will now use this property to investigate further the nature of the average value of an operator. Let the system be in some state ψ (normalized), not an eigenfunction of $\hat{M}$. However, $\hat{M}$ possesses eigenfunctions $\{\mu\}$ that must form a complete set. Therefore, we can express ψ in terms of μ's:

$$\psi = \sum_i c_i \mu_i \tag{6-30}$$

Now we calculate the average value of M for the state ψ:

$$\begin{aligned} M_{\text{av}} &= \int \psi^* \hat{M} \psi \, dv = \int \sum_i c_i^* \mu_i^* \hat{M} \sum_j c_j \mu_j \, dv \\ &= \sum_i \sum_j c_i^* c_j \int \mu_i^* \hat{M} \mu_j \, dv \end{aligned} \tag{6-31}$$

But $\hat{M}\mu_i = m_i \mu_i$, and so

$$M_{\text{av}} = \sum_i \sum_j c_i^* c_j \int \mu_i^* m_j \mu_j \, dv = \sum_i \sum_j c_i^* c_j m_j \int \mu_i^* \mu_j \, dv \tag{6-32}$$

But we are assuming that $\{\mu\}$ is an orthonormal set, and so

$$M_{\text{av}} = \sum_i \sum_j c_i^* c_j m_j \delta_{ij} = \sum_i c_i^* c_i m_i \tag{6-33}$$

What does this expression mean? Each measurement of the property corresponding to $\hat{M}$ must give one of the eigenvalues m_i (postulate IV) and the average of many such measurements must be M_{av}. Equation (6-33) states how the individual measurements must be weighted to give the average, so it follows that each $c_i{}^* c_i$ is a measure of the relative frequency for observing the corresponding m_i. Putting it another way, the absolute squares of the mixing coefficients in Eq. (6-30) give the probabilities that a measurement of the variable M will give the corresponding eigenvalue. For example, if ψ happens to be equal to $(1/\sqrt{2})\mu_1 + (1/\sqrt{2})\mu_3$, it follows that $M_{\text{av}} = (1/2)m_1 + (1/2)m_3$.

EXAMPLE 6-4 What is the average value for the z-component of orbital angular momentum for the normalized function $\phi = (1/\sqrt{5})(\psi_{2s} + 2 \cdot \psi_{2p_{+1}})$?

SOLUTION ► Since we know that the $2s$ and $2p_{+1}$ eigenfunctions have z-components of angular momentum of 0 and +1 respectively, we can say at once that $\bar{p}_z = \frac{1}{5} \cdot 0 + \frac{4}{5} \cdot 1 = 0.8$ a.u. ◄

[8] See, e.g., Kemble [1, Section 25].

6-13 The Variation Principle

Many of the calculations of quantum chemistry are based on the Rayleigh-Ritz variation principle which states: *For any normalized, acceptable function* ϕ,

$$H_{\text{av}} \equiv \int \phi^* \hat{H} \phi \, d\tau \geq E_0 \tag{6-34}$$

where E_0 *is the lowest eigenvalue of* $\hat{H}$.

This statement is easily proved. We expand ϕ in terms of $\{\psi_i\}$, the complete, orthonormal set of eigenfunctions of $\hat{H}$:

$$\phi = \sum_i c_i \psi_i \tag{6-35}$$

As in the preceding section, this leads to

$$\int \phi^* \hat{H} \phi d\tau = \sum_i c_i^* c_i E_i \tag{6-36}$$

Now $c_i^* c_i$ is never negative, and so Eq. (6-36) is merely a weighted average of the eigenvalues E_i. Such an average can never be lower than the lowest contributing member and the principle is proved.

The variation principle is sometimes stated in an equivalent way by saying that the average value of $\hat{H}$ over ϕ is an *upper bound* for the lowest eigenvalue of $\hat{H}$. Following the approach of the example at the end of the previous section, if ϕ for a hydrogen atom happens to be a function equal to $(1/\sqrt{2})\psi_{1s} + (1/\sqrt{2})\psi_{2s}$, the average energy for ϕ is $(1/2)E_{1s} + (1/2)E_{2s}$, which obviously lies above the lowest eigenvalue E_{1s}.

6-14 The Pauli Exclusion Principle

We have already discussed the Pauli exclusion principle in Chapter 5. In its most general form, this is:

Postulate VII ψ *must be antisymmetric (symmetric) for the exchange of identical fermions (bosons).*

6-15 Measurement, Commutators, and Uncertainty

If we measure the exact position of the electron in a hydrogen atom, we force it into a state having a Dirac delta function as its wavefunction. Since this function is also an eigenfunction for the dipole moment operator, it follows that we also know the (instantaneous) dipole moment for the atom at that instant. In effect, measuring position measures dipole moment too. But the delta function is *not* an eigenfunction for the hamiltonian operator of the atom, and so we have *not* simultaneously measured the electronic energy of the atom.

We have earlier seen that an eigenfunction for one operator can serve also as eigenfunction for another operator when the operators commute. In the above example, the operators for position and dipole moment commute with each other but not with the hamiltonian operator. This leads us to recognize that we can simultaneously measure values for two variables only if their operators commute.

Let us consider this situation more deeply by imagining two successive measurements on a hydrogen atom, one immediately following the other. If we first measure position and find $r = 2.0$ a.u., and then measure dipole moment, we will get the value ($\mu = 2.0$ a.u.) corresponding to the electron being at $r = 2.0$ a.u. That is where we found it in the first measurement, and it has not had time to move elsewhere before the second measurement. If we immediately follow with yet another position measurement, the electron will still be found at $r = 2$ a.u. (We are imagining that no time elapses between measurements, which is a limit we cannot actually achieve. In the present case, though, since measurement of r is also a measure of μ, both measurements are done at once, so this is really not a problem.) Hence, it makes sense to say that we know these two values "simultaneously." However, if we first measure position and then measure energy, we find something very different. Suppose that we find $r = 2$ a.u. and then, in a subsequent measurement, $E = -1/2$ a.u. (E must, after all, be an eigenvalue of $\hat{H}$, according to postulate IV.) We know that the eigenfunction during the first measurement was $\delta(r - 2 \text{ a.u.})$, and that during the second measurement was a 1s AO. If we immediately do yet another position measurement, we can find any value of r (with probabilities given by $4\pi r^2 \psi_{1s}^2\, dr$). The processes of measuring position and energy are incompatible in the sense that there is no single function that can describe the situation that exists during both measurements. The energy-measuring process can be pictured as forcing a reconstruction of the wavefunction in such a manner that it no longer corresponds to a particular position, while measurement of position forces a state function that does not correspond to a particular energy. (In this case, separate measurements would really be necessary, so the impossibility of doing a second measurement truly immediately after the first must be recognized. Indeed, one has to allow for the fact that finding an electron in one place and then at some other place must imply a lapse of time permitting the electron to travel.)

The reader may suspect that there is some connection between commutators and the uncertainty principle, and this is indeed the case. It can be shown[9] that the product of widths of simultaneous measurements (i.e., the uncertainty in their values) of two variables satisfies the relation

$$\Delta a \cdot \Delta b \geq \frac{1}{2} \left| \int \psi^* \left[\hat{A}, \hat{B} \right] \psi \, d\tau \right| \tag{6-37}$$

where ψ is normalized, and the absolute value $|X|$ is defined as the positive square root of $X^* X$. If A and B are conjugate variables, such as position and momentum, Eq. (6-37) becomes $\Delta a \cdot \Delta b \geq \hbar/2$, which is Heisenberg's uncertainty relation. If $\hat{A}$ and $\hat{B}$ commute, the right-hand side of Eq. (6-37) vanishes, and the values of both variables may, in theory, be simultaneously known exactly.

Among the properties of greatest interest in molecular quantum mechanics are energy, symmetry, and electron orbital angular momentum because, for many molecules, some of these operators commute. Thus, if we know that an oxygen molecule is in a nondegenerate stationary electronic state, we know that it is possible to characterize that state by a definite value of the orbital angular momentum along

[9] See Merzbacher [2, Section 10-5].

the internuclear axis. Also, we know that the wavefunction must be symmetric or antisymmetric for inversion through the molecular midpoint.

6-16 Time-Dependent States

Much of quantum chemistry is concerned with *stationary* states, for which Ψ is a product of a space term ψ (an eigenfunction of $\hat{H}$) and a time-dependent factor $\exp(-iEt/\hbar)$, which we usually ignore because it has no effect on particle probability distribution. Sometimes, however, it becomes necessary to consider time-dependent states. In this section we illustrate how some of these may be treated.

There are two types of situation to distinguish. One is situations where the potential is changing as a function of time, and hence the hamiltonian operator is time dependent. An example is a molecule or atom in a time-varying electromagnetic field. The other is situations where the potential and hamiltonian operator do not change with time, but the particle is nonetheless in a nonstationary state. An example is a particle that is known to have been forced into a nonstationary state by a measurement of its position. We deal here with the second category.

As our first example, consider a particle in a one-dimensional box with infinite walls. Suppose that we measure the particle's position and find it in the left side of the box (i.e., between $x = 0$ and $L/2$; we will be more specific shortly) at some instant that we take to be $t = 0$. We are interested in knowing what this implies about a future measurement of the particle's position.

Our knowing that the particle is on the left side at $t = 0$ means that the wavefunction for this state is *not* one of the time-independent box eigenfunctions we saw in Chapter 2, because those all predict equal probabilities for finding the particle on the two sides of the box. If the state function is not stationary, it must be time dependent, and it must satisfy Schrödinger's time-dependent equation (6-1). We have, then, that the state function is time dependent, and that $\Psi^*\Psi \equiv |\Psi|^2$ is zero everywhere on the right side of the box when $t = 0$. (We have not yet been specific enough to describe $|\Psi|^2$ in detail on the left side of the box.)

Schrödinger's time-dependent equation (6-1) is not an eigenvalue equation. However, Eq. (6-2) shows that Schrödinger's time-dependent equation is satisfied by time-independent eigenfunctions of $\hat{H}$ if they are multiplied by their time-dependent factors $f(t) = \exp(-iEt/\hbar)$. Furthermore, Eq. (6-2) continues to be satisfied if the term $\psi(q)f(t)$ is replaced by a sum of such terms. (See Problem 6-9.) This means that we can seek to express the time-dependent state function, $\Psi(x,t)$, as a sum of time-independent box eigenfunctions as long as each of these is accompanied by its time factor $f(t)$. When $t = 0$, all the factors $f(t)$ equal unity, so at that point in time Ψ becomes the same as the sum of box eigenfunctions without their time factors.

Our strategy, then, is to find a linear combination of time-independent box eigenfunctions, ψ_n, that describe Ψ when $t = 0$. This is easy to do because the time factors are all equal to unity. Once we have found the proper mixture of ψ_n, we multiply each by its time factor and then observe the behavior of $|\Psi|^2$ as t increases.

In the case of our particle-in-a-box example, we can start with a very simple approximation to ψ at $t = 0$ by taking a 50–50 mixture of ψ_1 and ψ_2:

$$\Psi(x,t) = (1/\sqrt{2})\left[\psi_1 \exp(-iE_1t/\hbar) + \psi_2 \exp(-iE_2t/\hbar)\right] \tag{6-38}$$

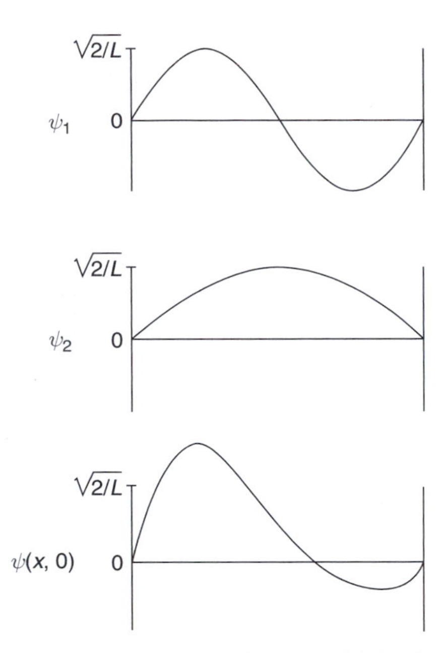

Figure 6-1 ▶ Stationary eigenfunctions ($n = 1, 2$) for the particle in a box and their normalized sum.

We have included the functions $f(t)$, even though they equal unity when $t = 0$, because they are needed to make $\Psi(x, t)$ a solution to Schrödinger's equation (6-1) and because they will inform us of the nature of Ψ at later times.

We choose this pair of functions because, when $t = 0$, both are positive on the left, but they differ in sign on the right, giving us some cancellation there. (See Fig. 6-1.) Obviously, we have not succeeded in describing a function that has no probability density on the right, but we already have a definite imbalance in that direction. (It is not difficult to see that some ψ_3 with a positive coefficient should help remove much of the remaining probability density on the right.)

Now we are in a position to examine this $|\Psi|^2$ as time progresses—the *time evolution of the square of a wavepacket* that describes the probability distribution for a particle that is known to have been in the left half of the box at $t = 0$. This is mathematically straightforward (Problem 6-20) and leads to the probability distributions sketched in Fig. 6-2 after time steps of Δt. The figure shows a changing distribution suggestive of the particle bouncing back and forth in the box with a cycle time of $8\Delta t$. It is not difficult to see why this happens. Ψ_1 and Ψ_2 have different "frequency factors" $\exp(-iE_n t/\hbar)$, so they behave like two waves oscillating at different frequencies. Since $E_2 = 4E_1$ (recall $E \propto n^2$ in the box), Ψ_2 oscillates four times faster than Ψ_1. This means that, by the time Ψ_1 has made half a cycle (and is equal to -1 times its starting coordinates), Ψ_2 has made two cycles and is just as it was at $t = 0$. It is easy to see from Fig. 6-1 that this will give a Ψ that is skewed to the *right*, leading to the distribution shown in Fig. 6-2(e). (This allows us to conclude that $4\Delta t$ equals $1/2$ of the cycle time of Ψ_1. See Problem 6-21.)

If we want a more accurate starting representation for the localized particle, we must mix together a larger number of stationary-state wavefunctions. In order to decide how much of each is needed, we must have a better-defined description of Ψ at $t = 0$.

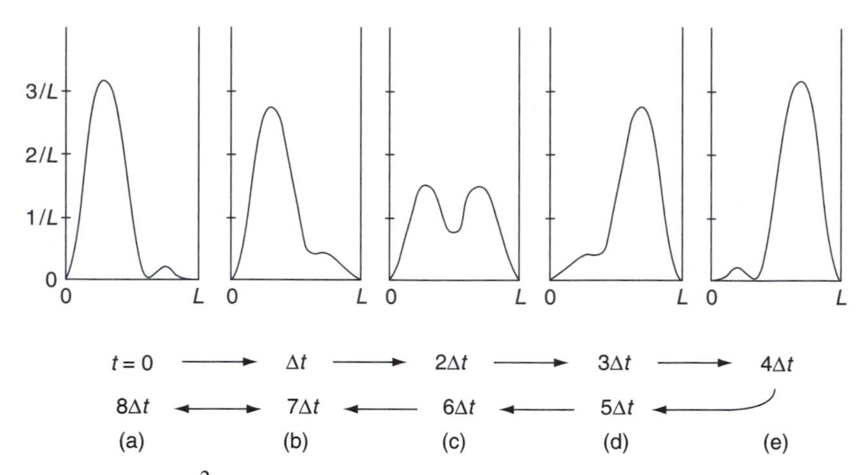

Figure 6-2 ▶ $|\Psi(x,t)|^2$ from Eq. (6-38) as it appears at various times.

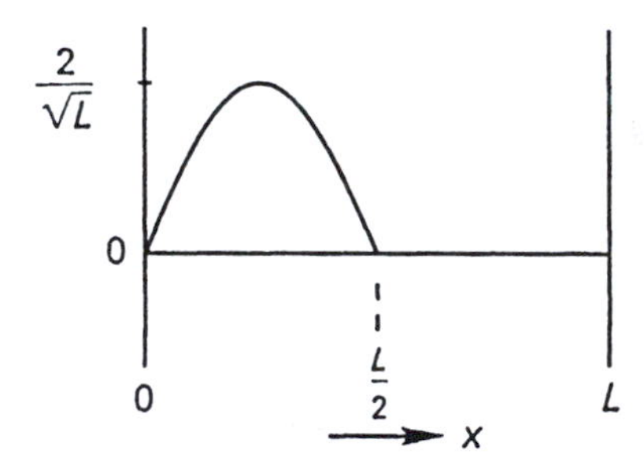

Figure 6-3 ▶ A normalized half sine wave in the left half of a "box." The numbers at left are values of Ψ, not of E.

Suppose, for instance, we choose to describe the starting wavefunction $\Psi(x,0)$ as a normalized half sine wave in the left side of the box and zero at the right (Fig. 6-3). Then we can calculate the amount (c_n) of each of the stationary-state functions ψ_n present in this function as follows:

$$c_n = \int_0^L \psi_n \Psi(x,0)dx \tag{6-39}$$

This follows from the completeness[10] and orthonormality of $\{\psi_n\}$. (See Problem 6-4.) Evaluation of Eq. (6-39) for the first few terms gives (Problem 6-22)

$$\Psi(x,t) = 0.600\psi_1 + 0.707\psi_2 + 0.360\psi_3 + 0.000\psi_4 + 0.086\psi_5 + \cdots \tag{6-40}$$

This modifies slightly our earlier commonsense combination and also verifies our prediction that ψ_3 times a positive coefficient would be beneficial.

[10]Because $\Psi(x,0)$ has a discontinuous derivative at the midpoint of the box, it falls outside the class for which $\{\psi_n\}$ is complete. However, because this problem is restricted to dx around one point, it should have little effect.

This example illustrates the basic approach to such problems:

1. Find a function that represents the initial particle distribution $\Psi(x,0)$.
2. Expand that function as a series of eigenfunctions for the hamiltonian, and include the time-dependent factor for each term.
3. Evaluate the probability distribution at other times t by examining $|\Psi(x,t)|^2$.

As a second example, suppose one were considering the behavior of the electronic state immediately after a tritium atom emits a beta particle to become a helium ion: ${}^3_1\mathrm{H} \rightarrow {}^3_2\mathrm{He} + {}^0_{-1}e$. A crude analysis could be attempted by imagining that the nuclear charge suddenly changes from 1 to 2 a.u., and the orbiting electron (not the beta particle) suddenly finds itself in a state (the original 1s state) that is not an eigenfunction for the new hamiltonian. We would accordingly set $\Psi(t=0)$ to be the 1s AO of hydrogen and expand this in terms of He^+ eigenfunctions. Only s-type AOs could contribute because of symmetry. The coefficients are given by

$$c_n = \int^{\text{allspace}} \psi_{1\mathrm{s}}(Z=1)\psi_{\mathrm{ns}}(Z=2)dv \tag{6-41}$$

and the time-dependent wavefunction is (in a.u.)

$$\begin{aligned}\Psi(r,\theta,\phi,t) = {} & c_1\psi_{1\mathrm{s}}(Z=2)\exp(-2it) + c_2\psi_{2\mathrm{s}}(Z=2)\exp(-it/2) \\ & + c_3\psi_{3s}(Z=2)\exp(-2it/9) + \cdots \end{aligned} \tag{6-42}$$

This function could be evaluated at various times t and would be found to give an oscillating spherical distribution, as though the electron cloud were shrinking, then rebounding to its original distance, then shrinking again, etc.

Our next example is perhaps the most important: It is a particle initially localized in some region of space, say by measurement of its position, and free to move anywhere thereafter. Taking the one-dimensional case, we imagine that the particle has been detected around $x=0$ at $t=0$ (the measurement caused it to be "unfree" for an instant). We assume that the average momentum of the particle is zero. We seek to know how the probability distribution function for the particle will evolve in time.

As before, we need a functional description of the wavefunction at $t=0$, $\Psi(x,0)$. We will then expand that in terms of eigenfunctions of the free-particle hamiltonian. As we have seen in Section 2-5, the free-particle eigenfunctions may be written $\exp(\pm i\sqrt{2mE}x/\hbar)$, where E is any nonnegative number. These are also eigenfunctions for the momentum operator, with eigenvalues $\sqrt{2mE}\hbar$.

The function usually selected to describe $\Psi(x,0)$ is a gaussian function:

$$\Psi(x,0) = \left(\sqrt[4]{2\alpha/\pi}\right)\exp(-\alpha x^2) \tag{6-43}$$

The constant α affects the width of the gaussian and reflects our degree of certainty in our knowledge of position. Large α gives a tight function and small uncertainty Δx. The relationship between the gaussian function in x and the coefficients of the eigenfunctions as a function of $\sqrt{2mE}/\hbar$ is depicted in Fig. 6-4. Remarkably, the coefficient values are also described by a gaussian function (in $\sqrt{2mE}/\hbar$). Furthermore, the tighter the gaussian function is in x, the broader the corresponding gaussian function

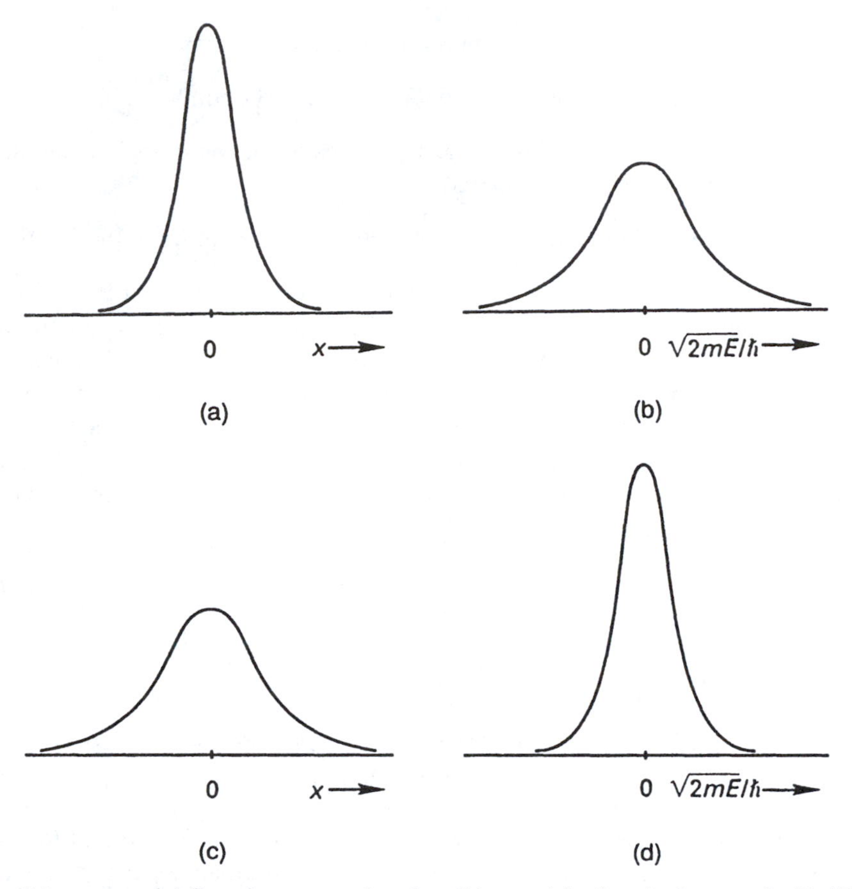

Figure 6-4 ▶ (a and c) Gaussian wave packets describing particles found to be at $x = 0$ with differing degrees of certainty. (b and d) Values of c_k (where $k = \sqrt{2mE}/\hbar$) for momentum eigenfunctions that combine to express the gaussian wave packets to their left. (a, b) corresponds to relatively certain position and relatively uncertain momentum, whereas (c, d) corresponds to the opposite situation.

is in $\sqrt{2mE}/\hbar$ (Problem 6-24). That is, we need to combine free-particle eigenfunction contributions from a wider range of momenta to create a tighter position function. This means that greater certainty in position goes with greater uncertainty in momentum, in accord with the uncertainty principle.

Once we have the appropriate mixture of momentum eigenfunctions, each with its time-dependent term, we can follow the time evolution of the particle wave packet. We find that the packet spreads out more and more about $x = 0$ as time passes, which means that our knowledge of position is decreasing as time passes. Even though the average position is not changing, the probability for finding the particle at a distance from $x = 0$ is increasing.

We can interpret this by remembering that the square of the wavefunction predicts the results of many experiments. In each of many position measurements finding the particle near $x = 0$, we impart some degree of momentum to the particle. Then, in a second measurement, we find that some of the particles have moved away from $x = 0$. The longer we wait before taking the second measurement, the greater this spread in x values. (Our assumption of zero *average* momentum amounts to saying that large deviations from zero momentum are equally likely for motion toward $x = +\infty$

and $-\infty$.) The more precise our first position measurement is, the greater the likelihood of introducing momenta quite different from zero and the more rapidly the wave packet spreads out as time passes.

In the first example, a packet located in half of a one-dimensional box, we saw $|\Psi|^2$ oscillate back and forth, changing shape in the process so that the motion cannot be described with a single frequency. A related important case is that of an oscillator moving in a harmonic potential. Let us assume the oscillator's position at $t = 0$ to be described by a gaussian wavefunction. If this function is *not* centered at the oscillator's equilibrium position, we have a time dependent situation analogous to the above particle-in-a-box case. For a harmonic potential, it can be shown[11] that $|\Psi|^2$ remains a gaussian function as time passes (i.e., does *not* change shape), and that the center of this gaussian oscillates back-and-forth about the equilibrium position with the classical frequency. This is a situation of interest to spectroscopists because it bears on the process of electronically exciting a sample of diatomic molecules. Suppose the molecules are initially in their ground electronic and ground vibrational states. Then their vibrational wavefunction is a simple gaussian function (the lowest-energy harmonic oscillator wavefunction) centered at R_e (i.e., *not* off-center). If the molecules are excited by a laser pulse to a new electronic state having an equilibrium internuclear distance of R_e', the vibrational wavefunction at $t = 0$ is still the simple gaussian centered at R_e, which means that it is now off-center. As time passes, the center of this function oscillates back and forth about R_e' in a coherent manner (i.e., describable with a single frequency). Thus, we have gone from an initial state describing an ensemble of molecules vibrating about R_e with zero-point energy $h\upsilon/2$ and with *random* phases (i.e., a time-independent state wherein such measurable properties as average molecular dipole moment appear to be constant) to a final state where the molecules are vibrating *in phase* about R_e' (a time-dependent state, wherein one might expect to see time variation of such properties). If R_e were quite a bit smaller than R_e', for instance, then almost all the molecules would find themselves to be "too short" at $t = 0$, "too long" a short time later, etc., as they swing in phase about R_e'. As a result of this simple behavior, it is possible to take advantage of it and time a second laser pulse to strike the molecules when they are almost all at their shortest, or all at their longest, extension. Of course, in real molecules the potential is not exactly harmonic. Furthermore, phase coherence is eventually lost due to collisions. So the second pulse must come very soon after the first one.

6-17 Summary

Some of the postulates and proofs described in this chapter are most important for what follows in this book, and we list these points here.

1. ψ describes a state as completely as possible and must meet certain mathematical requirements (single-valued, etc.). $\psi^*\psi$ is the probability density distribution function for the system.

2. For any observable variable, there is an operator (hermitian) which is constructed from the classical expression according to a simple recipe. (Operators related to

[11] See Schiff [3, pp. 67, 68], and Tanner [4].

"spin" are the exception because the classical analog does not exist.) The eigenvalues for such an operator are the possible values we can measure for that quantity. The act of measuring the quantity forces the system into a state described by an eigenfunction of the operator. Once in that state, we may know exact values for other quantities only if their operators commute with the operator associated with our measurement.

3. If the hamiltonian operator for a system is time independent, stationary eigenfunctions exist of the form $\psi(q,\omega)\exp(-iEt/\hbar)$. The time-dependent exponential does not affect the measurable properties of a system in this state and is almost always completely ignored in any time-independent problem.

4. The formula for the quantum-mechanical average value [Eq. (6-9)] is equivalent to the arithmetic average of all the possible measured values of a property times their frequency of occurrence [Eq. (6-33)]. This means that it is impossible to devise a function that satisfies the general conditions on ψ and leads to an average energy lower than the lowest eigenvalue of $\hat{H}$.

5. The square-integrable eigenfunctions for an operator corresponding to an observable quantity form a complete set, which may be assumed orthonormal. The eigenvalues are all real.

6. Any operation that leaves $\hat{H}$ unchanged also commutes with $\hat{H}$.

7. Wavefunctions describing time-dependent states are solutions to Schrödinger's time-dependent equation. The absolute square of such a wavefunction gives a particle distribution function that depends on time. The time evolution of this particle distribution function is the quantum-mechanical equivalent of the classical concept of a trajectory. It is often convenient to express the time-dependent wave packet as a linear combination of eigenfunctions of the time-independent hamiltonian multiplied by their time-dependent phase factors.

6-17.A Problems

6-1. Prove that d^2/dx^2 is hermitian.

6-2. Integrate the expressions in Eqs. (6-13) and (6-15) to show that their integrals are equal.

6-3. Prove that, if a normalized function is expanded in terms of an orthonormal set of functions, the sum of the absolute squares of the expansion coefficients is unity.

6-4. Show that a particular coefficient c_k in Eq. (6-30) is given by $c_k = \int \mu_k^* \psi \, dv$.

6-5. A particle in a ring is in a state with wavefunction $\psi = 1/\sqrt{\pi}\cos(2\phi)$.

a) Calculate the average value for the angular momentum by evaluating $\int \psi^* \hat{L}_z \psi \, d\phi$, where $\hat{L}_z = (\hbar/i)\, d/d\phi$. (Use symmetry arguments to evaluate the integral.)

b) Express ψ as a linear combination of exponentials and evaluate the average value of the angular momentum using the formula $L_{z,\text{av}} = \sum_i c_i^* c_i L_{zi}$ where L_{zi} is the eigenvalue for the ith exponential function.

6-6. Using Eq. (6-37), show that $\Delta x \cdot \Delta p_x \geq \hbar/2$.

6-7. What condition must the function ϕ satisfy for the *equality* part of $\geq$ to hold in Eq. (6-34)?

6-8. Suppose you had an operator and a set of eigenfunctions for it that were associated with real eigenvalues. Does it necessarily follow that the operator is hermitian as defined by Eq. (6-10)? [*Hint*: Consider d/dr and the set of all functions $\exp(-ar)$, where a is real and positive definite.]

6-9. a) Show that the nonstationary state having wavefunction

$$\Psi = (1/\sqrt{2})\psi_{1s}\exp(it/2) + (1/\sqrt{2})\psi_{2p_0}\exp(it/8)$$

is a solution to Schrödinger's time-dependent equation when $\hat{\mathscr{H}}$ is the time-independent $\hat{H}$ of the hydrogen atom. Use atomic units (i.e., $\hbar = 1$).

b) This time-dependent state is dipolar and oscillates with a characteristic frequency ν. Show that ν satisfies the relation $E_2 - E_1 \equiv \Delta E = 2\pi\nu$ in a.u. (The dipole oscillates at the same frequency as that of light required to drive the 1s $\longleftrightarrow$ 2p transition. This is central to the subject of spectroscopy.)

6-10. From the definition that $\phi' = \phi - S\psi$ [see the discussion following Eq. (6-26)], evaluate the normalizing constant for ϕ', assuming that ϕ and ψ are normalized.

6-11. Given the two normalized *nonorthogonal* functions $(1/\sqrt{\pi})\exp(-r)$ and $\sqrt{1/3\pi}\,r\exp(-r)$, construct a new function ϕ that is orthogonal to the first function and lies within the function space spanned by these two functions, *and is normalized.*

6-12. If the hydrogen atom 1s AO is expanded in terms of the He^+ AOs, what is the coefficient for (a) the He^+ 1s AO? (b) the He^+ $2p_0$ AO?

6-13. The lowest-energy eigenfunction for the one-dimensional harmonic oscillator is $\psi_{n=0} = (\beta/\pi)^{1/4}\exp(-\beta x^2/2)$.

a) Demonstrate whether or not momentum is a constant of motion (i.e., is a "sharp" quantity) for this state.

b) Calculate the *average* momentum for this state.

6-14. Demonstrate whether $x^2 d/dx$ and xd^2/dx^2 commute. What about xd/dx and x^2d^2/dx^2?

6-15. Evaluate the following integrals over all space. In neither case should you need to do this by brute force.

a) $\int (3d_{xy})\hat{L}^2(3d_{xy})\,dv$

b) $\int (3d_{xy})\,\hat{L}_z(3d_{xy})\,dv$

6-16. The operators for energy and angular momentum for an electron constrained to move in a ring of constant potential are, respectively, in a.u. $-(1/2)d^2/d\phi^2$ and $(1/i)d/d\phi$.

a) Discuss whether or not there should be a set of functions that are simultaneously eigenfunctions for both operators.
b) Discuss whether or not there is a set of functions that are eigenfunctions for one of these operators but not the other.
c) Discuss whether it is reasonable to expect these two physical quantities to be exactly knowable simultaneously or whether the uncertainty principle makes this impossible.

6-17. Suppose a hydrogen atom state was approximated by the function $\phi = (1/\sqrt{3})1s + (1/\sqrt{3})2s + (1/\sqrt{3})3s$, where 1s, 2s, and 3s are normalized eigenfunctions for the hydrogen atom hamiltonian. What would be the average value of energy associated with this function, in a.u.?

6-18. A function f is defined as follows: $f = 0.1 \cdot 1s + 0.2 \cdot 2p_1 + 0.3 \cdot 3d_2$, where 1s is the normalized eigenfunction for the 1s state of the hydrogen atom, etc. Evaluate the average value of the z component of angular momentum in a.u. for this function.

6-19. Without looking back at the text, prove that (a) eigenvalues of hermitian operators are real, (b) nondegenerate eigenfunctions of hermitian operators are orthogonal, (c) nondegenerate eigenfunctions of $\hat{A}$ must be eigenfunctions of $\hat{B}$ if $\hat{A}$ and $\hat{B}$ commute.

6-20. Using Eq. (6-38), obtain an expression for $|\Psi|^2$ as a function of x and t.

6-21. Evaluate Δt of Fig. 6-2 in terms of m, h, and L.

6-22. Verify the values of the coefficients in Eq. (6-40). How can you tell from simple inspection that (a) c_2 will be largest and positive, (b) c_4 will be zero, (c) c_i will tend toward small values at large i?

6-23. a) Evaluate the first *two* coefficients in Eq. (6-42).
b) What *qualitative* difference would you expect between Ψ of Eq. (6-42) and one that takes explicit account of the changing potential resulting as the beta particle travels away from the nucleus?

6-24. Show by qualitative arguments based on mathematical functions why coefficients c_k should drop off more rapidly with k if the position wave packet is broader. $c_k = \int \left(\sqrt[4]{2\alpha/\pi}\right) \exp(-\alpha x^2) \exp(ikx)\, dx$.

6-25. a) Prove that, if V is real and $\Psi(x, y, z, t)$ satisfies Schrödinger's time-*dependent* equation, then $\Psi(x, y, z, -t)^*$ is also a solution. (This is called "invariance under time reversal.")
b) Show that, for *stationary* states, invariance under time reversal means that $\hat{H}\psi^* = E\psi^*$ if $\hat{H}\psi = E\psi$ and if V is real.
c) Show from (b) that nondegenerate eigenfunctions of $\hat{H}$ (with real V) must be real.

d) What becomes of the $2p_{-1}$ AO (with time dependence included) upon complex conjugation and time reversal? the $2p_0$ AO?
e) Can the statement in (c) be reworded to say that all degenerate eigenfunctions of $\hat{H}$ (with real V) must be complex?

Multiple Choice Questions

(Try to answer these without referring to the text.)

1. Which one of the following statements about the eigenfunctions of a time-independent Hamiltonian operator is true?

a) Any linear combination of these eigenfunctions is also an eigenfunction for $\hat{H}$.
b) The state function for this system must be one of these eigenfunctions.
c) The eigenvalues associated with these eigenfunctions must all be real.
d) These eigenfunctions must all be orthogonal to one another.
e) These eigenfunctions have no time dependence.

2. A hydrogen atom is in a nonstationary state having the wavefunction $\frac{1}{2}\{\psi_{1s} \exp(it/2\hbar) + \psi_{2s}\exp(it/8\hbar) + \psi_{2p_0}\exp(it/8\hbar) + \psi_{2p_{+1}}\exp(it/8\hbar)\}$

Which statement is true at any time t?

a) A measurement of the energy has a 25% chance of giving -0.5 a.u.
b) The average value of the z component of angular momentum is 0.5 a.u.
c) The average energy is $-\frac{7}{8}$ a.u.
d) A measurement of the total angular momentum has a 75% chance of giving $\sqrt{2}$ a.u.
e) None of the above statements is true.

3. The function $r\exp(-0.3r^2)\cos\theta$ is expanded in terms of hydrogen atom wavefunctions. This series may have finite contributions from bound-state eigenfunctions

a) of all types: s, p_x, p_y, p_z, d_{xy}, d_{yz}, etc.
b) of all types except s.
c) of types p_x, p_y, p_z only.
d) of type p_z only.
e) of types p_z and d_{z^2} only.

References

[1] E. C. Kemble, *The Fundamental Principles of Quantum Mechanics with Elementary Applications*. Dover, New York, 1958.

[2] E. Merzbacher, *Quantum Mechanics*, 3rd ed. Wiley, New York, 1998.

[3] L. I. Schiff, *Quantum Mechanics*, 3rd ed. McGraw-Hill, New York, 1968.

[4] J. J. Tanner, *J. Chem. Ed.*, 67, 917 (1990).

Chapter 7

The Variation Method

7-1 The Spirit of the Method

The proof of the Rayleigh-Ritz variation principle (Section 6-12) involves essentially two ideas. The first is that any function can be expanded into a linear combination of other functions that span the same function space. Thus, for example, $\exp(ikx)$ can be expressed as $\cos(kx) + i\sin(kx)$. An exponential can also be written as a linear combination of powers of the argument:

$$\exp(x) = 1 + x + x^2/2! + x^3/3! + \cdots + x^n/n! + \cdots \tag{7-1}$$

The second idea is that, if a function is expressed as a linear combination of eigenfunctions for the energy operator, then the average energy associated with the function is a weighted average of the energy eigenvalues. For example, if

$$\phi = \left(\frac{1}{\sqrt{2}}\right)\psi_1 + \left(\frac{1}{\sqrt{2}}\right)\psi_2 \tag{7-2}$$

where

$$\hat{H}\psi_1 = E_1\psi_1, \quad \hat{H}\psi_2 = E_2\psi_2, \quad E_1 \neq E_2 \tag{7-3}$$

then measuring the energy of many systems in states described by ϕ would give the result E_1 half of the time and E_2 the other half. The average value, $\frac{1}{2}E_1 + \frac{1}{2}E_2$ must lie between E_1 and E_2. Alternatively, if

$$\phi' = \sqrt{\frac{1}{3}}\psi_1 + \sqrt{\frac{2}{3}}\psi_2 \tag{7-4}$$

measurements would give E_1 one-third of the time, and E_2 the rest of the time, for an average that still must lie between E_1 and E_2. It should be evident that, even when ϕ is a linear combination of many eigenfunctions ψ_i, the average value of E can never lie below the lowest or above the highest eigenvalue.

The variation *method* is based on the idea that, by varying a function to give the lowest average energy, we tend to maximize the amount of the lowest-energy eigenfunction ψ_0 present in the linear combination already discussed. Thus, if we minimize

$$\bar{E} = \frac{\int \phi^* \hat{H}\phi \, dv}{\int \phi^* \phi \, dv} \tag{7-5}$$

the resulting ϕ should tend to resemble ψ_0 since we have maximized (in a sense) the amount of ψ_0 in ϕ by this procedure.

7-2 Nonlinear Variation: The Hydrogen Atom

We have already seen (Chapter 4) that the lowest-energy eigenfunction for the hydrogen atom is (in atomic units)

$$\psi_{1s} = \left(\frac{1}{\sqrt{\pi}}\right)\exp(-r) \tag{7-6}$$

Suppose we did not know this and used the variation method to optimize the normalized trial function

$$\phi = \left(\sqrt{\frac{\zeta^3}{\pi}}\right)\exp(-\zeta r) \tag{7-7}$$

In this example, when $\zeta = 1$, ϕ becomes identical to ψ_{1s}, but in more complicated systems the trial function never becomes identical to an eigenfunction of the hamiltonian. Nevertheless, this is a good example to start with since there are few mathematical complexities to obscure the philosophy of the approach.

The variation method requires that we minimize

$$\bar{E} = \int \phi^* \hat{H}\phi \, dv \tag{7-8}$$

by varying ζ. [ϕ is normalized, so no denominator is required in Eq. (7-8).] Since the trial function ϕ has no θ- or ϕ-dependent terms for ∇^2 to operate on, only the radial part of ∇^2 is needed in $\hat{H}$. Thus [from Eq. (4-7)]

$$\hat{H} = -\frac{1}{2}\frac{1}{r^2}\frac{d}{dr}r^2\frac{d}{dr} - \frac{1}{r} \tag{7-9}$$

According to Eq. (7-8), we need first to evaluate the quantity $\hat{H}\phi$:

$$\hat{H}\phi = \left[-\frac{1}{2}\frac{1}{r^2}\frac{d}{dr}r^2\frac{d}{dr} - \frac{1}{r}\right]\sqrt{\frac{\zeta^3}{\pi}}\exp(-\zeta r) \tag{7-10}$$

$$\vdots$$

$$= \left[\frac{(\zeta - 1)}{r} - \frac{\zeta^2}{2}\right]\sqrt{\frac{\zeta^3}{\pi}}\exp(-\zeta r) \tag{7-11}$$

Incorporating this into Eq. (7-8) gives (after integrating θ and ϕ in dv to give 4π)

$$\bar{E} = 4\pi\left(\frac{\zeta^3}{\pi}\right)\int_0^\infty \left[\frac{(\zeta - 1)}{r} - \frac{\zeta^2}{2}\right]\exp(-2\zeta r)\, r^2\, dr \tag{7-12}$$

$$= 4\zeta^3\left\{(\zeta - 1)\int_0^\infty r\exp(-2\zeta r)\, dr - \left(\frac{\zeta^2}{2}\right)\int_0^\infty r^2\exp(-2\zeta r)\, dr\right\} \tag{7-13}$$

Using the integral table in Appendix 1, we obtain

$$\bar{E} = \left(\frac{\zeta^2}{2}\right) - \zeta \tag{7-14}$$

Now we have a simple expression for $\bar{E}$ as a function of ζ. To obtain the minimum, we set the derivative of $\bar{E}$ to zero:

$$\frac{d\bar{E}}{d\zeta}=0=\zeta-1 \tag{7-15}$$

As we expected, $\zeta=1$. Inserting this value for ζ into Eq. (7-14) gives $\bar{E}=-\frac{1}{2}$ a.u., which is identical with the lowest eigenvalue for the hydrogen atom.

This example demonstrates that minimizing $\bar{E}$ for a trial function causes the function to become like the lowest eigenfunction for the system. But it is more realistic to examine a case where the trial function is incapable of becoming exactly identical with the lowest eigenfunction. Suppose we assumed a trial form (normalized) of

$$\phi=\sqrt{\frac{\zeta^5}{3\pi}}r\exp(-\zeta r) \tag{7-16}$$

Proceeding as before, we first evaluate $\hat{H}\phi$:

$$\hat{H}\phi=\left[\left(-\frac{1}{r^2}\right)+\frac{2\zeta-1}{r}-\frac{\zeta^2}{2}\right]\phi \tag{7-17}$$

This leads to

$$\bar{E}(\zeta)=\frac{4}{3}\left[\frac{\zeta^2}{8}-\frac{3\zeta}{8}\right] \tag{7-18}$$

so that

$$\frac{d\bar{E}}{d\zeta}=0=\frac{4}{3}\left[\frac{\zeta}{4}-\frac{3}{8}\right] \tag{7-19}$$

and $\bar{E}$ is a minimum when $\zeta=\frac{3}{2}$. Thus, our energy-optimized ϕ is

$$\phi=\sqrt{\frac{3^5}{96\pi}}r\exp\left(-\frac{3r}{2}\right) \tag{7-20}$$

This is obviously not identical to the eigenfunction given by Eq. (7-6), but it must be expressible as a linear combination of hydrogen atom eigenfunctions, and the amount of ψ_{1s} present should be quite large unless the trial form was unwisely chosen. Since ϕ also must contain contributions from higher energy eigenfunctions, it follows that $\bar{E}$ must be higher in energy than $-\frac{1}{2}$ a.u. We test this by inserting $\zeta=\frac{3}{2}$ into Eq. (7-18), obtaining an $\bar{E}$ of $-\frac{3}{8}$ a.u., (-0.375 a.u.). This value is above the lowest eigenvalue, but it is well below the second-lowest eigenvalue ($-\frac{1}{8}$ a.u.) associated with 2s, 2p eigenfunctions, and so we know that ϕ does indeed contain much 1s character. We can find out exactly how much 1s eigenfunction is contained in ϕ by calculating the overlap between ϕ and the 1s eigenfunction. That is, since the 1s function is orthogonal to all the other hydrogen atom eigenfunctions;

$$\int \psi_{1s}\psi_j\,dv=\begin{cases}0 & j\neq 1s\\ 1 & j=1s\end{cases} \tag{7-21}$$

and since

$$\phi = c_{1s}\psi_{1s} + \sum_{j \neq 1s} c_j \psi_j \tag{7-22}$$

it follows that

$$\int \psi_{1s}\phi \, dv = c_{1s} \int \psi_{1s}\psi_{1s} \, dv + \sum_{j \neq 1s} c_j \int \psi_{1s}\psi_j \, dv = c_{1s} \tag{7-23}$$

Integrating $\int \psi_{1s}\phi \, dv$, where ϕ is given by Eq. (7-20), gives $c_{1s} = 0.9775$, so ϕ does indeed "contain" a large amount of ψ_{1s}, (If $c_{1s} = 1$, then ψ_{1s} and ϕ are identical.)

This suggests another way of trying to get a "best" approximate wavefunction. We could find the value of ζ that maximized the overlap between ϕ and ψ_{1s}. This maximizes c_{1s} in Eq. (7-22). If one does this (Problem 7-6), one obtains $\zeta = \frac{5}{3}$, which corresponds to an overlap of 0.9826 and an $\bar{E}$ of -0.370 a.u. At first sight, this seems puzzling. $\phi(\zeta = \frac{5}{3})$ has a larger amount of ψ_{1s} in it, but $\phi(\zeta = \frac{3}{2})$ has a lower average energy. But it is really not so unreasonable. Maximizing the value of c_{1s} causes ζ to take on a certain value that may cause the coefficient for some high-energy state (say, 6s) to become relatively larger, producing a tendency to raise the average energy. On the other hand, minimizing $\bar{E}$ is a process that is implicitly concerned with what all the coefficients are doing. This allows for a different sort of compromise wherein c_{1s} may be allowed to be a bit smaller if the associated energy loss is more than compensated for by a favorable shifting in values of higher coefficients (e.g., if c_{2s} increases and c_{6s} decreases). The purpose of this discussion is to emphasize that the variation method optimizes the trial function in a certain sense (best energy), but that other kinds of optimization are conceivable. The optimization that gives best overlap is not generally useful because it requires that we know the exact solution to begin with. If we vary ζ to obtain optimal agreement for $\bar{r}$, or $\bar{V}$, or $\overline{r^2}$, we would find a different value of ζ appropriate for each property. In each case, however, we would need to know the correct value of $\bar{r}$, etc., before starting. A great virtue of the energy variation method is that it does not require foreknowledge of the exact eigenvalue or eigenfunction. However, the function that gives the lowest value for $\bar{E}$ might not be especially good in describing other properties. This is demonstrated in Fig. 7-1 and Table 7-1. The figure indicates that the trial function differs from the exact function chiefly near the nucleus, where $r < 1$. This discrepancy shows up when we compare average values for various powers of r. The operators r, r^2, and r^3 become large when r is large. Thus, these operators magnify $\psi^2 \, dv$ at large r. Since the two functions are fairly similar at large r, the average values show fair agreement. But r^{-1} and r^{-2} become large when r is small. Thus, the fact that the approximate function is too small at small r shows up as a marked disagreement in the average value of r^{-2}, this average being much larger for the exact function (see Table 7-1). *Any trial function that is known to be especially inaccurate in some region of space (e.g., at small r) will give unreliable average values for operators that are largest in that region of space (e.g.,* r^{-2}*).*

Choosing a trial form such as Eq. (7-16), which must vanish at $r = 0$, might seem foolish since we know that ψ_{1s} does not vanish at $r = 0$. And if we were interested in electron density at the nucleus for, say, calculating the Fermi contact interaction, this would indeed be a self-defeating choice. But if our interest is in energies or other properties having operators that are large in regions where the trial function is not too deficient,

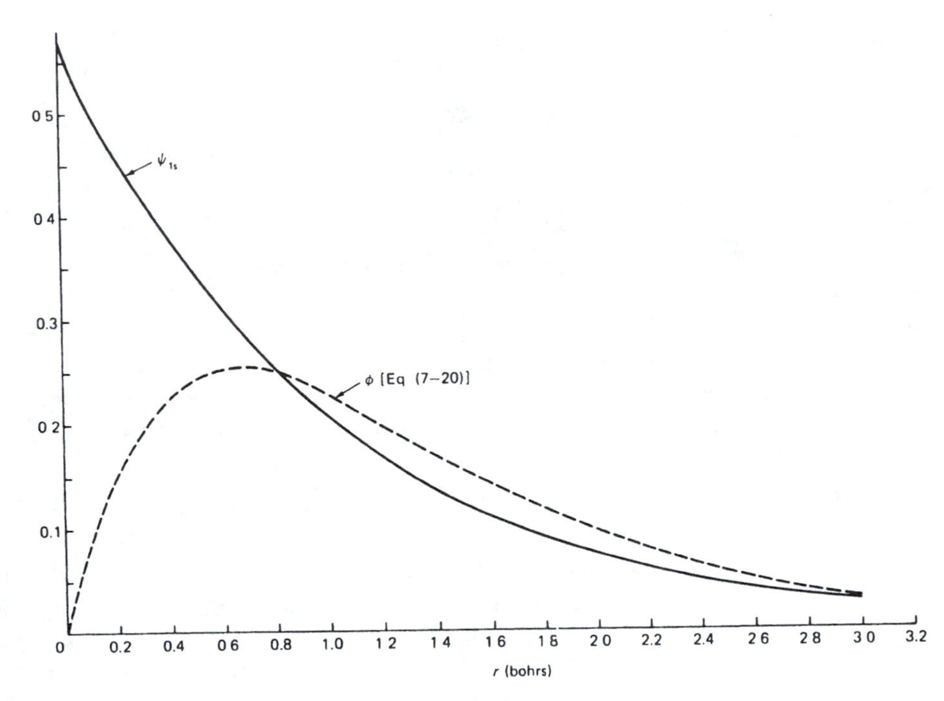

Figure 7-1 ▶ Plots of ψ_{1s} and ϕ [Eq. (7-20)] versus r.

TABLE 7-1 ▶ Comparison between Exact Values for Some Properties of (1s) Hydrogen and Values Calculated from the Function Eq. (7-20)

Quantity	Exact (1s) value (a.u.)	Trial function value (a.u.)
$\bar{E}$	$-\frac{1}{2}$	$-\frac{3}{8}$
Electron density at nucleus	$\frac{1}{\pi}$	0
$\bar{r}$	1.5	1.67
$\overline{r^2}$	3.0	3.33
$\overline{r^3}$	7.5	7.78
$\overline{r^{-1}}$	1.0	0.75
$\overline{r^{-2}}$	2.0	0.75

this choice would serve. One often settles for a mathematically convenient form even though it is known to be inadequate in some way. Care must then be exercised, however, to avoid using that trial wavefunction in ways which emphasize its inadequacies.

7-3 Nonlinear Variation: The Helium Atom

We mentioned in Chapter 5 that the ground-state wavefunction 1s(1)1s(2) for helium was much too contracted if the 1s functions were taken from the He^+ ion without modification. Physically, this arises because, in He^+, the single electron sees only a doubly positive nucleus, whereas in He each electron sees a doubly positive nucleus *and* another electron, so that in He the repulsion between electrons prevents them from spending as much time near the nucleus as in He^+. Somehow, the 1s functions should be modified to reflect this behavior. We will now show how the variation method may be used to accomplish this.

The form of the hydrogenlike ion 1s solution is

$$1s = \sqrt{\frac{Z^3}{\pi}} \exp(-Zr) \tag{7-24}$$

For He^+, $Z = 2$, but we have just seen that this gives a function that is too contracted. Smaller values of Z would cause the function to die away more slowly with r. Therefore, it is reasonable to replace the atomic number Z with a variable parameter ζ and find the value of ζ that gives the lowest average energy. Hence, we let

$$1s'(1) = \sqrt{\frac{\zeta^3}{\pi}} \exp(-\zeta r_1) \tag{7-25}$$

and our trial wavefunction is[1]

$$\phi(1,2) = 1s'(1)1s'(2)\left(1/\sqrt{2}\right)\left[\alpha(1)\beta(2) - \beta(1)\alpha(2)\right] \tag{7-26}$$

The average energy is [since $\phi(1,2)$ is normalized]

$$\bar{E} = \int\int \phi^*(1,2)\hat{H}(1,2)\phi(1,2)d\tau(1)d\tau(2) \tag{7-27}$$

Since $\hat{H}(1,2)$ contains no spin operators at our level of approximation, the integral separates into an integral over the space coordinates of both electrons and an integral over the spin coordinates of both electrons. The integration over spins gives a factor of unity. There remains

$$\bar{E} = \int\int 1s'(1)1s'(2)\hat{H}(1,2)1s'(1)1s'(2)dv(1)dv(2) \tag{7-28}$$

where

$$\hat{H}(1,2) = -\frac{1}{2}\nabla_1^2 - \frac{1}{2}\nabla_2^2 - \left(\frac{2}{r_1}\right) - \left(\frac{2}{r_2}\right) + \left(\frac{1}{r_{12}}\right) \tag{7-29}$$

and where the θ and ϕ parts of ∇^2 can be ignored since $\phi(1,2)$ is independent of these variables. The calculation is easier if we recognize that

$$\hat{H}(1,2) = \hat{H}_{He^+}(1) + \hat{H}_{He^+}(2) + \frac{1}{r_{12}} \tag{7-30}$$

[1] In this trial function, ζ has the same value in each atomic orbital. This is not a necessary restriction. There is no physical reason for not choosing the more general trial function where orbitals with different ζ are used. Symmetry requires that such a function be written $2^{-1/2}[1s'(1)1s''(2) + 1s''(1)1s'(2)]2^{-1/2}[\alpha(1)\beta(2) - \beta(1)\alpha(2)]$. This type of function is called a "split shell" wavefunction. It gives a lower energy for He than does the function (7-26). However, for most quantum-chemical calculations split shells are not used, the gain in accuracy usually not being commensurate with the increased computational effort.

This allows us to express Eq. (7-28) as the sum of three integrals, the first of these being

$$\iint 1s'(1)1s'(2)\hat{H}_{He^+}(1)\, 1s'(1)1s'(2)dv(1)dv(2) \tag{7-31}$$

Since the operator in the integrand operates only on coordinates of electron 1, we can separate this into a product of two integrals:

$$\int 1s'(2)1s'(2)dv(2)\int 1s'(1)\hat{H}_{He^+}(1)\, 1s'(1)dv(1) \tag{7-32}$$

The $1s'$ functions are normalized, and so the first integral is unity. The second integral is almost identical to the integral in Eq. (7-8) and has the value $(\zeta^2/2) - 2\zeta$. Therefore, the first of the three integrals mentioned above equals $(\zeta^2/2) - 2\zeta$. The second of the three integrals is identical with Eq. (7-31) except that the operator acts on electron 2 instead of 1. This integral is evaluated in the same manner and gives the same result. The third integral, in which the operator is $1/r_{12}$, is more difficult to evaluate. This interesting and instructive problem constitutes a detour from the main sequence of ideas in this chapter and is therefore discussed in Appendix 3. We here simply take the result, $\frac{5}{8}\zeta$, and proceed with the variation calculation.

We now have an expression for $\bar{E}$ as a function of ζ:

$$\bar{E} = 2\left[(\zeta^2/2) - 2\zeta\right] + \frac{5}{8}\zeta = \zeta^2 - \frac{27}{8}\zeta \tag{7-33}$$

Minimizing $\bar{E}$ with respect to ζ gives

$$\frac{d\bar{E}}{d\zeta} = 0 = 2\zeta - \frac{27}{8} \tag{7-34}$$

so that

$$\zeta = \frac{27}{16} \tag{7-35}$$

This value of ζ is smaller than the unmodified He^+ value of 2, as we anticipated. Let us see how much the average energy has been improved. According to Eq. (7-33), when ζ is 2, $\bar{E}$ is equal to -2.75 a.u. When $\zeta = \frac{27}{16}$, $\bar{E}$ equals -2.848 a.u., so the average energy has been lowered by approximately 0.1 a.u., or 2.7 eV, or 62 kcal/mole. (The exact nonrelativistic energy for He is -2.903724377 a.u.) Further analysis would show that, by decreasing ζ, we have decreased the average kinetic energy (the less compressed wavefunction changes slope less rapidly), raised the nuclear-electron attraction energy from a large negative to a smaller negative value (the decreased attraction resulting from the electrons being farther from the nucleus, on the average), and decreased the interelectronic repulsion energy from a higher positive value to a lower one. The variational procedure has allowed the wavefunction to adjust to the best compromise it can achieve among these three factors. If ζ becomes less than $\frac{27}{16}$ the loss of nuclear–electron attraction is too great to be offset by the loss of interelectronic repulsion and kinetic energy.

Varying a parameter in the argument of an exponential produces a nonlinear change in the function, and so calculations of the type described above are referred to as *nonlinear* variation calculations. Such calculations tend to become mathematically complicated and are not frequently used except for fairly simple systems. The fact that the hamiltonian operator is a *linear* operator [i.e., $\hat{H}(c_1\phi_1 + c_2\phi_2) = (c_1\hat{H}\phi_1 + c_2\hat{H}\phi_2)$] makes a linear variation procedure more convenient for most purposes.

7-4 Linear Variation: The Polarizability of the Hydrogen Atom

Suppose we wish to express a wavefunction ψ (which may be approximate) as a linear combination of two known functions ϕ_1 and ϕ_2:

$$\psi(c_1, c_2) = c_1\phi_1 + c_2\phi_2 \tag{7-36}$$

The question is, what values of c_1 and c_2 give a ψ that best approximates the exact wavefunction for a particular system? The usual approach is to determine which values of c_1 and c_2 give the ψ associated with the minimum average energy attainable. The technique for achieving this, called the *linear variation method*, is by far the most common type of quantum chemical calculation performed.

An example of a problem that can be treated by this method is the polarizability of the hydrogen atom. The wavefunction for the unperturbed hydrogen atom in its ground state is spherically symmetrical. But, when a uniform z-directed external electric field of strength F is imposed, the positive nucleus and the negative electron are attracted in opposite directions, which leads to an electronic distribution that is skewed with respect to the nucleus. The wavefunction describing this skewed distribution may be *approximated* by mixing with the unperturbed 1s function some $2p_z$ function: $\psi = c_1 1s + c_2 2p_z$. As indicated in Fig. 7-2, this produces a skewed wavefunction because the $2p_z$ function is of the same sign as the 1s function on one side of the nucleus and of the opposite sign on the other. We will work out the details of this example after developing the method for the general case.

Let the *generalized* trial function ψ be a linear combination of known functions $\phi_1, \phi_2, \ldots, \phi_n$. (This set of functions is called the *basis set* for the calculation.)

$$\psi = c_1\phi_1 + c_2\phi_2 + \cdots + c_n\phi_n \tag{7-37}$$

where the coefficients c are to be determined so that

$$\frac{\int \psi^* \hat{H} \psi \, d\tau}{\int \psi^* \psi \, d\tau} = \bar{E} \tag{7-38}$$

is minimized. Substituting Eq. (7-37) into Eq. (7-38) gives

$$\begin{aligned} \bar{E} &= \frac{\int \left(c_1^*\phi_1^* + c_2^*\phi_2^* + \cdots + c_n^*\phi_n^*\right) \hat{H} \left(c_1\phi_1 + c_2\phi_2 + \cdots + c_n\phi_n\right) d\tau}{\int \left(c_1^*\phi_1^* + c_2^*\phi_2^* + \cdots + c_n^*\phi_n^*\right)\left(c_1\phi_1 + c_2\phi_2 + \cdots + c_n\phi_n\right) d\tau} \\ &= \frac{\text{num}}{\text{denom}} \end{aligned} \tag{7-39}$$

Since we will be dealing with cases in which the c's and ϕ's are real, we will temporarily omit the complex conjugate notation to simplify the derivation. At the minimum value of $\bar{E}$,

$$\frac{\partial \bar{E}}{\partial c_1} = \frac{\partial \bar{E}}{\partial c_2} = \cdots = \frac{\partial \bar{E}}{\partial c_n} = 0 \tag{7-40}$$

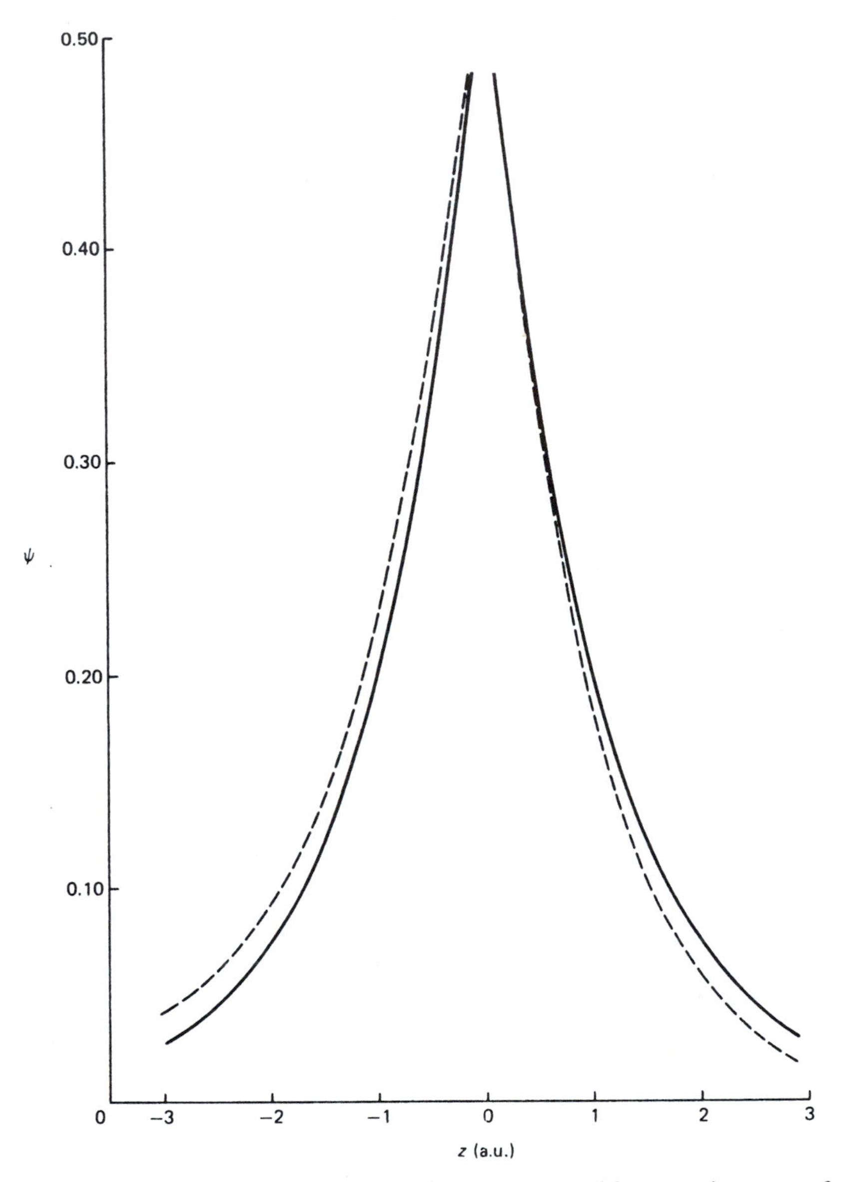

Figure 7-2 ► Values of ψ versus z for 1s state of H atom (—) and for approximate wavefunction given by 0.982 1s −0.188 $2p_z$ (- - -). The nucleus is at $z=0$ for each case.

The partial derivative of Eq. (7-39) with respect to c_1 is

$$\frac{\partial \bar{E}}{\partial c_1} = \frac{\int \phi_1 \hat{H} (c_1\phi_1 + \cdots + c_n\phi_n)\, d\tau}{\text{denom}} + \frac{\int (c_1\phi_1 + \cdots + c_n\phi_n)\, \hat{H}\phi_1\, d\tau}{\text{denom}}$$
$$- (\text{num})\,(\text{denom})^{-2}\left[\int \phi_1\,(c_1\phi_1 + \cdots + c_n\phi_n)\, d\tau + \int (c_1\phi_1 + \cdots + c_n\phi_n)\,\phi_1 d\tau\right]$$
$$= 0 \tag{7-41}$$

Multiplying through by denom, recalling that num/denom equals $\bar{E}$, and rearranging, gives

$$\begin{aligned} & c_1\left[\int \phi_1 \hat{H}\phi_1\, d\tau - \bar{E}\int \phi_1\phi_1\, d\tau\right] \\ & +c_2\left[\int \phi_1 \hat{H}\phi_2\, d\tau - \bar{E}\int \phi_1\phi_2\, d\tau\right] \\ & +\cdots + c_n\left[\int \phi_1 \hat{H}\phi_n\, d\tau - \bar{E}\int \phi_1\phi_n\, d\tau\right] \\ & =0 \end{aligned} \tag{7-42}$$

At this point, it is convenient to switch to an abbreviated notation:

$$\int \phi_i \hat{H}\phi_j\, d\tau \equiv H_{ij} \tag{7-43}$$

$$\int \phi_i \phi_j\, d\tau \equiv S_{ij} \tag{7-44}$$

The integral S_{ij} is normally called an *overlap integral* since its value is, in certain cases, an indication of the extent to which the two functions ϕ_i and ϕ_j occupy the same space. Use of this abbreviated notation produces, for Eq. (7-42),

$$c_1\left(H_{11}-\bar{E}S_{11}\right)+c_2\left(H_{12}-\bar{E}S_{12}\right)+\cdots+c_n\left(H_{1n}-\bar{E}S_{1n}\right)=0 \tag{7-45}$$

A similar treatment for $\partial\bar{E}/\partial c_i$ gives a similar equation:

$$c_1\left(H_{i1}-\bar{E}S_{i1}\right)+c_2\left(H_{i2}-\bar{E}S_{i2}\right)+\cdots+c_n\left(H_{in}-\bar{E}S_{in}\right)=0 \tag{7-46}$$

Thus, requiring that $\partial\bar{E}/\partial c_i$ vanish for all coefficients produces n homogeneous linear equations (homogeneous, all equal zero; linear, all c_i's to first power). If one chooses a value for $\bar{E}$, there remain n unknowns—the coefficients c_i. (The integrals H_{ij} and S_{ij} are presumably knowable since $\hat{H}$ and the functions ϕ_i are known.) Of course, one trivial solution for Eqs. (7-46) is always possible, namely, $c_1=c_2=\cdots=c_n=0$. But this corresponds to $\psi=0$, a case of no physical interest. Are there nontrivial solutions as well? *In quantum chemical calculations, nontrivial solutions usually exist only for certain discrete values of* $\bar{E}$. This provides the approach for solving the problem. First, find those values of $\bar{E}$ for which nontrivial coefficients exist. Second, substitute into Eqs. (7-46) whichever of these values of $\bar{E}$ one is interested in and solve for the coefficients. (Each value of $\bar{E}$ has its own associated set of coefficients.) But how do we find these particular values of $\bar{E}$ that yield nontrivial solutions to Eqs. (7-46)? The answer is given in Appendix 2, where it is shown that *the condition which must be met by the coefficients of a set of linear homogeneous equations in order that nontrivial solutions exist is that their determinant vanish.* Notice that, in the standard treatment given in Appendix 2, the coefficients are *known* and x, y, and z are unknown. Here, however, the coefficients c_i are unknown, and H_{ij} and S_{ij} are known. Therefore, it is the determinant of the H's, S's and $\bar{E}$ in Eqs. (7-46) that must equal zero:

$$\begin{vmatrix} H_{11}-\bar{E}S_{11} & H_{12}-\bar{E}S_{12} & \cdots & H_{1n}-\bar{E}S_{1n} \\ H_{21}-\bar{E}S_{21} & H_{22}-\bar{E}S_{22} & \cdots & H_{2n}-\bar{E}S_{2n} \\ \vdots & \vdots & & \vdots \\ H_{n1}-\bar{E}S_{n1} & H_{n2}-\bar{E}S_{n2} & \cdots & H_{nn}-\bar{E}S_{nn} \end{vmatrix}=0 \tag{7-47}$$

Expansion of this determinant gives a single equation containing the unknown $\bar{E}$. Any value of $\bar{E}$ satisfying this equation is associated with a nontrivial set of coefficients. The lowest of these values of $\bar{E}$ is the minimum average energy achievable by variation of the coefficients. Substitution of this value of $\bar{E}$ back into Eqs. (7-46) produces n simultaneous equations for the n coefficients. Equation (7-47) is referred to as the *secular equation*, and the determinant on the left-hand side is called the *secular determinant*.

This method is best illustrated by example, and we will now proceed with the problem of a hydrogen atom in a z-directed uniform electric field of strength F a.u. As mentioned earlier, a suitable choice of functions to mix together to approximate the accurate wavefunction is the 1s and 2p, hydrogenlike functions. The choice of two basis functions leads to a 2×2 secular determinantal equation:

$$\begin{vmatrix} H_{11} - \bar{E}S_{11} & H_{12} - \bar{E}S_{12} \\ H_{21} - \bar{E}S_{21} & H_{22} - \bar{E}S_{22} \end{vmatrix} = 0 \tag{7-48}$$

If we arbitrarily associate the 1s function with the index 1 and the $2p_z$ function with index 2 (consistent with $\psi = c_1 1s + c_2 2p_z$), then the terms in the determinant are (returning to general complex conjugate notation):

$$H_{11} = \int 1s^* \hat{H} 1s \, d\tau \qquad H_{12} = \int 1s^* \hat{H} 2p_z \, d\tau$$

$$H_{21} = \int 2p_z^* \hat{H} 1s \, d\tau \qquad H_{22} = \int 2p_z^* \hat{H} 2p_z \, d\tau \tag{7-49}$$

(The electron label has been omitted since there is only one electron.) The corresponding S integrals are obtained from these if $\hat{H}$ is omitted in each case. The hamiltonian operator is just that for a hydrogen atom with an additional term to account for the z-directed field:

$$\hat{H} = -\frac{1}{2}\nabla^2 - (1/r) - Fr\cos\theta \tag{7-50}$$

[The energy of a charge $-e$ in a uniform electric field of strength F and direction z is $-eFz$. In atomic units, one unit of field strength is $e/a_0^2 = 5.142 \times 10^{11}\ V/m$. The unit of charge in atomic units is e, so this symbol does not appear explicitly in Eq. (7-50). Also, the identity $z = r\cos\theta$ has been used.] This may also be written

$$\hat{H} = \hat{H}_{\text{hyd}} - Fr\cos\theta \tag{7-51}$$

where $\hat{H}_{\text{hyd}}$ is the hamiltonian for the unperturbed hydrogen atom.

The secular determinant contains four H-type and four S-type terms. However, evaluating these eight integrals turns out to be much easier than one might expect. In the first place, $S_{12} = S_{21}$ since these integrals differ only in the order of the two functions in the integrand, and the functions commute. Also, because $\hat{H}$ is hermitian, it follows immediately that $H_{21} = H_{12}^*$. This leaves us with three S terms and three H terms to evaluate. The three S terms are simple. Because the hydrogenlike functions are orthonormal, S_{11} and S_{22} equal unity, and S_{12} vanishes. These points have already reduced the secular determinantal equation to

$$\begin{vmatrix} H_{11} - \bar{E} & H_{12} \\ H_{12}^* & H_{22} - \bar{E} \end{vmatrix} = 0 \tag{7-52}$$

Consider next the term H_{11}. This may be written as

$$H_{11} = \int 1s^* \hat{H}_{\text{hyd}} 1s\, d\tau - \int 1s^* (Fr\cos\theta)\, 1s\, d\tau \tag{7-53}$$

But the 1s function is an eigenfunction of $\hat{H}_{\text{hyd}}$ with eigenvalue $-\frac{1}{2}$ a.u. Therefore, the first integral on the right-hand side of Eq. (7-53) is

$$\int 1s^* \hat{H}_{\text{hyd}} 1s\, d\tau = -\frac{1}{2}\int 1s^* 1s\, d\tau = -\frac{1}{2}\text{a.u.} \tag{7-54}$$

The second term on the right-hand side of Eq. (7-53) is zero by symmetry since $1s^*1s$ is symmetric for reflection in the xy plane while $r\cos\theta (= z)$ is antisymmetric. Thus, $H_{11} = -\frac{1}{2}$ a.u. Similarly, $H_{22} = -\frac{1}{8}$ a.u. [Recall that the 8 eigenvalues of hydrogen are equal to $-1/(2n^2)$, and here $n = 2$.] All that remains is H_{12}:

$$H_{12} = \int 1s^* \hat{H}_{\text{hyd}} 2p_z\, d\tau - F\int 1s^* (r\cos\theta)\, 2p_z\, d\tau \tag{7-55}$$

The first term on the right-hand side is easily shown to be zero:

$$\int 1s^* \hat{H}_{\text{hyd}} 2p_z\, d\tau = \int 1s^* \left(-\frac{1}{8}\right) 2p_z\, d\tau = 0 \tag{7-56}$$

Here we employ the fact that $2p_z$ is an eigenfunction of $\hat{H}_{\text{hyd}}$ and then the fact that 1s and $2p_z$ AOs are orthogonal. The second term on the right-hand side must be written out in full and integrated by "brute force." It is remarkable that, of the eight terms originally considered, only one needs to be done by detailed integration. Proceeding, we substitute formulas for 1s and $2p_z$ in this last integral to obtain

$$-F\iiint (\pi)^{-1/2}\exp(-r)\left[r\cos\theta\right](32\pi)^{-1/2} r\exp(-r/2)\cos\theta\,(r^2\sin\theta)\, dr\, d\theta\, d\phi \tag{7-57}$$

We consider the integration over spin to have been carried out already, giving a factor of unity. Integrating over ϕ to obtain 2π, and regrouping terms gives

$$-2\pi F/(4\sqrt{2}\pi)\int_0^\infty r^4\exp(-3r/2)\, dr \int_0^\pi \cos^2\theta\sin\theta\, d\theta \tag{7-58}$$

$$= -\left(F/2\sqrt{2}\right)\left[\frac{4!}{\left(\frac{3}{2}\right)^5}\right]\left[\frac{2}{3}\right] = -\frac{2^{15/2}F}{3^5}\ \text{a.u.} \tag{7-59}$$

This completes the task of evaluating the terms in the secular determinant.[2] The final result is (in atomic units)

$$\begin{vmatrix} -\frac{1}{2} - \bar{E} & -2^{15/2}(F/3^5) \\ -2^{15/2}(F/3^5) & -\frac{1}{8} - \bar{E} \end{vmatrix} = 0 \tag{7-60}$$

[2]Since H_{12} is real, it is clear that $H_{21} = H_{12}$.

which expands to

$$\frac{1}{16}+(5\bar{E}/8)+\bar{E}^2-2^{15}F^2/3^{10}=0 \tag{7-61}$$

This is quadratic in $\bar{E}$ having roots

$$\bar{E}=-\frac{5}{16}\pm\frac{\left(\frac{9}{64}+2^{17}F^2/3^{10}\right)^{1/2}}{2} \tag{7-62}$$

When no external field is present, $F=0$ and the roots are just $-\frac{1}{2}$ and $-\frac{1}{8}$ a.u., the 1s and $2p_z$ eigenvalues for the unperturbed hydrogen atom. As F increases from zero, the roots change, as indicated in Fig. 7-3. We see that, for a given field strength, there are only two values of $\bar{E}$ that will cause the determinant to vanish. If either of these two values of $\bar{E}$ is substituted into the simultaneous equations related to Eq. (7-52), then nontrivial values for c_1 and c_2 can be found. Thus, at $F=0.1$ a.u., $\bar{E}=-0.51425$ a.u., and $\bar{E}=-0.1107$ a.u. are values of $\bar{E}$ for which $\partial\bar{E}/\partial c_1$, and $\partial\bar{E}/\partial c_2$ both vanish. The former is the minimum, the latter the maximum in the curve of $\bar{E}$ versus c_1 (Fig. 7-4). (The normality requirement results in the two c's being dependent, and so $\bar{E}$ may be plotted against either of them.) Since the variation principle tells us that $\bar{E}\geq E_{\text{lowest exact}}$, we can say immediately that the energy of the hydrogen atom in a uniform electric field of 0.1 a.u. is -0.51425 a.u. or lower. That is, -0.51425 a.u. is an *upper bound* to the true energy.

Now that we have the value of the lowest $\bar{E}_1$ we can solve for c_1 and c_2 and obtain the approximate ground state wavefunction. The homogeneous equations related to the determinant in Eq. (7-52) are

$$c_1(H_{11}-\bar{E})+c_2H_{12}=0 \tag{7-63}$$

$$c_1H_{12}+c_2\left(H_{22}-\bar{E}\right)=0 \tag{7-64}$$

Substituting -0.51425 for $\bar{E}$, and inserting the values for H_{11}, H_{22}, and H_{12} found earlier gives (when $F=0.1$ a.u.)

$$0.01425c_1-0.074493c_2=0 \tag{7-65}$$

$$-0.074493c_1+0.38925c_2=0 \tag{7-66}$$

Equation (7-65) gives

$$c_1=5.2275c_2 \tag{7-67}$$

If we substitute this expression for c_1 into Eq. (7-66) we get

$$-0.3892c_2+0.3892c_2=0 \tag{7-68}$$

This is useless for evaluating c_2. It is one of the properties of such a set of homogeneous equations that the last unused equation is useless for determining coefficients. This arises because an equation like (7-63) still equals zero when c_1 and c_2 are both multiplied by the same arbitrary constant. Therefore, these equations are inherently capable of telling us the ratio of c_1 and c_2 only, and not their absolute values. We shall determine

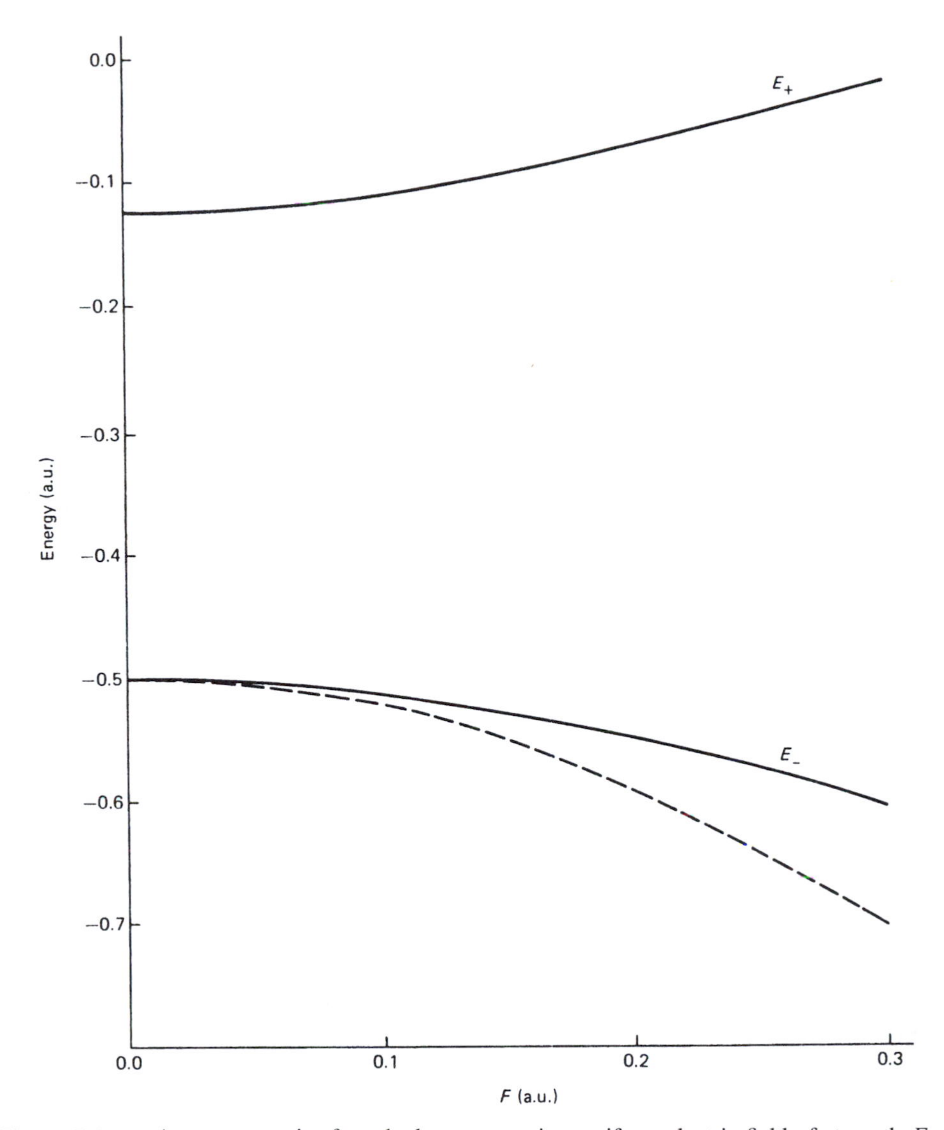

Figure 7-3 ▶ Average energies for a hydrogen atom in a uniform electric field of strength F as given by a linear variation calculation using a 1s, $2p_z$ basis. (- - -) Results from accurate calculations.

absolute values by invoking the requirement that ψ be normalized. In this case, this means that

$$c_1^2 + c_2^2 = 1 \tag{7-69}$$

or

$$(5.2275c_2)^2 + c_2^2 = 1 \tag{7-70}$$

which gives

$$c_2 = \pm 0.18789 \tag{7-71}$$

If we arbitrarily choose the positive root for c_2, it follows from Eq. (7-67) that $c_1 = 0.98219$.

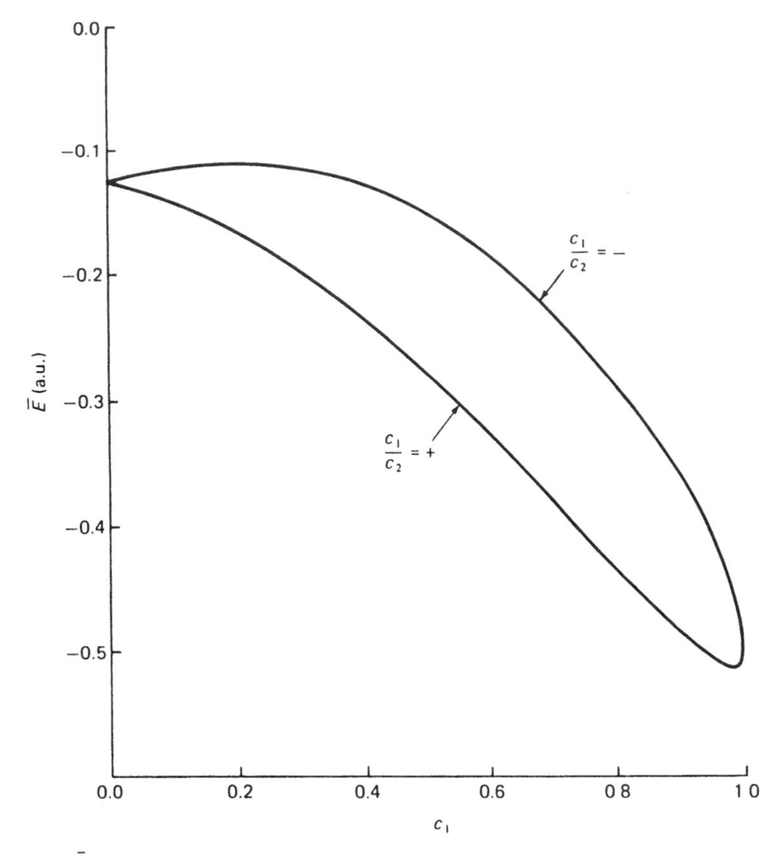

Figure 7-4 ▶ $\bar{E}$ versus c_1 for a hydrogen atom in a uniform electric field of strength 0.1 a.u.

Thus, when $F = 0.1$ a.u., the linear variation method using a 1s, $2p_z$ basis set gives an upper bound to the energy of −0.51425 a.u. and a corresponding approximate wavefunction of

$$\psi = 0.98219\ 1s + 0.18789\ 2p_z \tag{7-72}$$

As mentioned earlier, the admixture of $2p_z$ with 1s produces the skewed charge distribution shown in Fig. 7-2.

It is important to note that the extent of mixing between 1s and $2p_z$ depends partly on the size of the off-diagonal determinantal element H_{12}. When H_{12} is zero (no external field), no mixing occurs. As H_{12} increases, mixing increases. Generally speaking, the larger the size of the off-diagonal element connecting two basis functions in the secular determinant, the greater the degree of mixing of these basis functions in the final solution, other factors being equal. H_{ij} is sometimes referred to as the *interaction element* between basis functions i and j.

If we carried through the same procedure using the *maximum* $\bar{E}$ of −0.1107 a.u., we would obtain the approximate wavefunction

$$\psi' = 0.98219\ 2p_z - 0.18789\ 1s \tag{7-73}$$

This wavefunction is orthogonal to ψ. It may be proved that the *second root*, $\bar{E} = -0.1107$ is an upper bound for the energy of the *second-lowest state* of the hydrogen atom in the field. However, ψ' is probably not too good an approximation to the exact wavefunction for that state. This is partly because the wavefunction ψ' is one that *maximizes* $\bar{E}$. Hence, there is no particular tendency for the procedure to isolate the second-lowest state from the infinite manifold of states. Also, our basis set was chosen with an eye toward its appropriateness for approximating the lowest state. The true second-lowest state might be expected to contain significant amounts of the 2s AO, which is not included in this basis.

The values of $\bar{E}$ versus c_1 are plotted in Fig. 7-4. The values of $\bar{E}$ that we obtained by the variation procedure correspond to the extrema in this figure. The low-energy extreme corresponds to a wavefunction that shifts negative charge in the direction it is attracted by the field. The high-energy extreme corresponds to a wavefunction that shifts charge in the opposite direction. (When a calculation is performed over a more extensive basis to produce more than two roots, the highest and lowest roots correspond, respectively, to the maximum and minimum, the other roots to saddle points, on the energy hypersurface.)

The detailed treatment just completed is rather involved, and so we now summarize the main points. Step 1 involved selection of a basis set of functions which is capable of approximating the exact solution. Step 2 was the construction of the secular determinant, including evaluation of all the H_{ij}- and S_{ij}-type integrals. Step 3 was the conversion of the determinantal equation into its equivalent equation in powers of $\bar{E}$ and solution for the roots $\bar{E}$. Step 4 was the substitution of an $\bar{E}$ of interest into the simultaneous equations that are related to the secular determinant and solution for c_1/c_2. Finally, we used the normality requirement to arrive at convenient values for c_1 and c_2.

There are many ways one could increase the flexibility of the trial function in an effort to increase the accuracy of the calculation. By adding additional basis functions, one would stay within the linear variation framework, merely increasing the size of the secular determinant. If these additional basis functions are of appropriate symmetries, they will cause the minimum energy root to be lowered further and will mix into the corresponding wavefunction to make it a better approximation to the lowest-energy eigenfunction for the system. Also, the additional functions will increase the number of roots $\bar{E}$, thereby providing upper bounds for the energies of the third-, fourth-, etc. lowest states of the system. Another possibility is to allow nonlinear variation of the 1s and $2p_z$ orbital exponents, in combination with linear variation. This would be more involved than the calculation we have shown here, but could easily be accomplished with the aid of a computer.

EXAMPLE 7-1 Suppose the variational process just described were performed using $\psi(c_1, c_2, c_3) = c_1 1s + c_2 2p_0 + c_3 3d_0$. Would all three of these AOs be present in the lowest-energy solution?

SOLUTION ▶ We already know that 1s and $2p_0$ will be mixed. They differ in reflection symmetry in just the right manner to provide a wavefunction skewed in the z direction, and this is manifested as a nonzero interaction element H_{12}. $3d_0$, however, is, like 1s, symmetric for reflection through the x, y plane. Therefore, it cannot skew 1s in the z direction, and its interaction element with 1s, H_{13}, equals zero (by symmetry). Therefore, it is tempting to think that that $3d_0$ will not contribute. However, because $3d_0$ and $2p_0$ have opposite reflection symmetries through the x, y

plane, they do interact, H_{23} is not zero, and so all three basis functions show up in the lowest-energy solution. 1s and $3d_0$ are indirectly linked because they are each directly linked to $2p_0$. Physically, one can argue that a $2p_0$ AO that has been polarized by admixture with $3d_0$ can better polarize the 1s AO than can the pure $2p_0$ AO. (One could say that $3d_0$ is brought in "on the coat-tails" of $2p_0$.) Note also that 1s has zero angular momentum along the z axis before the electric field is turned on. The z-directed field distorts the ground state, but has no effect on the z component of angular momentum. Therefore, only eigenfunctions having $m_l = 0$ can contribute to the polarized ground state. ◀

7-5 Linear Combination of Atomic Orbitals: The H_2^+ Molecule–Ion

We are now ready to consider using the linear variation method on molecular systems. We begin with the simplest case, H_2^+. This molecule–ion has enough symmetry so that we could guess many important features of the solution without calculation. However, to demonstrate the method, we shall first simply plunge ahead mathematically, and discuss symmetry later.

The H_2^+ system consists of two protons separated by a variable distance R, and a single electron (see Fig. 7-5). The hamiltonian for this molecule is, in atomic units

$$\hat{H}(\mathbf{r}_A, \mathbf{r}_B, \mathbf{r}_1) = -\frac{1}{2}\left[\nabla_1^2 + \frac{\nabla_A^2}{1836} + \frac{\nabla_B^2}{1836}\right] - \left(\frac{1}{r_{A1}}\right) - \left(\frac{1}{r_{B1}}\right) + \left(\frac{1}{R}\right) \quad (7\text{-}74)$$

Since all the particles in the system are capable of motion, the exact eigenfunction of $\hat{H}$ will be a function of the coordinates of the electron and the protons. However, the protons are each 1836 times as heavy as the electron, and in states of chemical interest their velocity is much smaller than that of the electron. This means that, to a very good degree of approximation, the electron can respond instantly to changes in internuclear separation. In other words, whenever the nuclei are separated by a given distance R, no matter how they got there, the motion of the electron will always be described in the same way by ψ. This means that we can separate the electronic and nuclear motions with little loss of accuracy.[3] Given an internuclear separation, we can solve for the electronic wavefunction by ignoring nuclear motion [i.e., omitting ∇_A^2 and

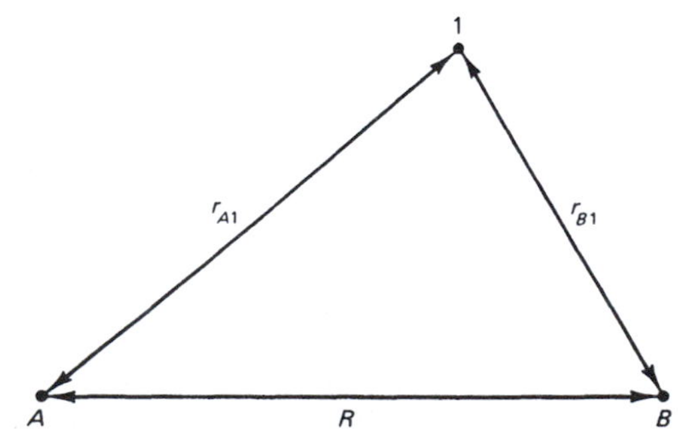

Figure 7-5 ▶ The H_2^+ molecule–ion. A and B are protons. The electron is numbered "1."

[3]However, there are times when coupling of electronic and nuclear motions becomes important.

∇_B^2 in Eq. (7-74)]. This gives us a hamiltonian for the electronic energy and nuclear repulsion energy of the system:

$$\hat{H}(\mathbf{r}_A, \mathbf{r}_B, \mathbf{r}_1) = -\frac{1}{2}\nabla_1^2 - \left(\frac{1}{r_{A1}}\right) - \left(\frac{1}{r_{B1}}\right) + \left(\frac{1}{R}\right) \tag{7-75}$$

For a given internuclear separation R, the internuclear repulsion $1/R$ is a constant, and we can omit it and merely add it on again after we have found the electronic energy. If we let $\hat{H}_{\text{elec}}$ stand for the first three terms on the right-hand side of Eq. (7-75), we can write

$$\hat{H}_{\text{elec}}\psi_{\text{elec}}(\mathbf{r}_1) = E_{\text{elec}}\psi_{\text{elec}}(\mathbf{r}_1) \tag{7-76}$$

and

$$E_{\text{elec}} + E_{\text{nuc rep}} = E_{\text{elec}} + 1/R \tag{7-77}$$

Solving Eq. (7-76) for E_{elec} for every value of R allows us to plot the electronic and also the total energy of the system as a function of R. But $E_{\text{elec}}(R) + 1/R$ is just the *potential energy for nuclear motion*. Therefore, this quantity can be inserted as the potential in the hamiltonian operator for *nuclear* motion:

$$\hat{H}_{\text{nuc}}(\mathbf{r}_A, \mathbf{r}_B) = -\frac{1}{2}\left[\frac{\nabla_A^2}{1836} + \frac{\nabla_B^2}{1836}\right] + E_{\text{elec}}(R) + 1/R \tag{7-78}$$

$$\hat{H}_{\text{nuc}}(\mathbf{r}_A, \mathbf{r}_B)\psi_{\text{nuc}}(\mathbf{r}_A, \mathbf{r}_B) = E_{\text{nuc}}\psi_{\text{nuc}}(\mathbf{r}_A, \mathbf{r}_B) \tag{7-79}$$

The eigenfunctions of Eq. (7-79) describe the translational, vibrational, and rotational states of the molecule. Note that the eigenvalues of the hamiltonian for *nuclear* motion are *total* energies for the system because they contain the electronic energy in their potential parts. Hence,

$$E_{\text{tot}} = E_{\text{nuc}} \tag{7-80}$$

But

$$\psi_{\text{tot}}(\mathbf{r}_A, \mathbf{r}_B, \mathbf{r}_1) = \psi_{\text{elec}}(\mathbf{r}_1)\psi_{\text{nuc}}(\mathbf{r}_A, \mathbf{r}_B) \tag{7-81}$$

This approximation—that the electronic wavefunction depends only on the positions of nuclei and not on their momenta—is called the *Born–Oppenheimer* approximation.[4] Only to the extent that this approximation holds true is it valid, for example, to separate electronic and vibrational wavefunctions and treat various vibrational states as a subset existing in conjunction with a given electronic state. We will assume the Born–Oppenheimer approximation to be valid in all cases treated in this book.

Making the Born–Oppenheimer approximation for H_2^+, we seek to solve for the electronic eigenfunctions and eigenvalues with the nuclei fixed at various separation distances R. We already know these solutions for the two extremes of R. When the two

[4]It is analogous to the concept of reversibility in thermodynamics: The piston moves so slowly in the cylinder that the gas can always maintain equilibrium, so pressure depends only on piston position; the nuclei move so slowly in a molecule that the electrons can always maintain their optimum motion at each R, so electronic energy depends only on nuclear position.

nuclei are widely separated, the lowest-energy state is a 1s hydrogen atom and a distant proton. (Since there is a choice about which nucleus is "the distant proton," there are really two degenerate lowest-energy states.) When $R=0$, the system becomes He^+ These two extremes are commonly referred to as the *separated-atom* and *united-atom* limits, respectively.

In carrying out a linear variation calculation on H_2^+, our first problem is choice of basis. In the separated-atom limit, the appropriate basis for the ground state would be a 1s atomic orbital (AO) on each proton. Then, regardless of which nucleus the electron resided at, the basis could describe the wavefunction correctly. The appropriate basis at the united-atom limit is a hydrogenlike 1s wavefunction with $Z=2$. At intermediate values of R, choice of an appropriate basis is less obvious. One possible choice is a large number of hydrogenlike orbitals or, alternatively, Stater-type orbitals (STOs), all centered at the molecular midpoint. Such a basis is capable of approximating the exact wavefunction to a high degree of accuracy, provided a sufficiently large number of basis functions is used.[5] Calculations using such a basis are called single-center expansions. A different basis, and one that is much more popular among chemists, is the *separated-atom basis*—a 1s hydrogen AO centered on each nucleus. At finite values of R, this basis can produce only an approximation to the true wavefunction. One way to improve this approximation is to allow additional AOs on each nucleus, 2s, 2p, 3s, etc., thereby increasing the mathematical flexibility of the basis. If we restrict our basis to AOs that are occupied in the separated-atom limit ground state (the 1s AOs in this case), then we are performing what is called a *minimal basis set* calculation. For now, we will use a minimal basis set.

The wavefunction that we produce by linear variation will extend over the whole H_2^+ molecule, and its square will tell us how the electron density is distributed in the molecule. Hence, the one-electron molecular wavefunction is referred to as a *molecular orbital* (MO) just as the one-electron atomic wavefunction is referred to as an *atomic orbital* (AO). With a basis set of the type we have selected, the MOs are expressed as linear combinations of AOs. For this reason, this kind of calculation is referred to as a *minimal basis set linear combination of atomic orbitals–molecular orbital* (LCAO–MO) calculation.

Our second problem, now that we have selected a basis, is construction of the secular determinant. Since we have only two basis functions ($1s_A$, $1s_B$), we expect a 2×2 determinant:

$$\begin{vmatrix} H_{AA}-\bar{E}S_{AA} & H_{AB}-\bar{E}S_{AB} \\ H_{BA}-\bar{E}S_{BA} & H_{BB}-\bar{E}S_{BB} \end{vmatrix} = 0 \tag{7-82}$$

Here we have used the notation developed earlier, where

$$H_{AB} = \int 1s_A^*(1)\,\hat{H}_{\text{elec}}(1)\,1s_B(1)\,d\tau(1) \tag{7-83}$$

etc. $\bar{E}$ of Eq. (7-82) is the *average* value of the energy. Henceforth, the bar is omitted.

If we take our basis functions to be normalized, $S_{AA}=S_{BB}=1$. Since our basis functions and hamiltonian are all real, their integrals will be real. Therefore, $S_{AB}=S_{BA}$

[5]The hydrogenlike orbitals are a complete set if the continuum functions are included. Hence, this set can allow one to approach arbitrarily close to the exact eigenfunction and eigenvalue. The Slater-type orbitals do not constitute a complete set.

and $H_{AB} = H_{BA}^* = H_{BA}$. Since the hamiltonian is invariant to an interchange of the labels A and B, it follows that $H_{AA} = H_{BB}$. (H_{AA} is the energy of an electron when it is in a 1s AO on one side of the molecule, H_{BB} when it is on the other side.) This leaves us with

$$\begin{vmatrix} H_{AA} - E & H_{AB} - ES_{AB} \\ H_{AB} - ES_{AB} & H_{AA} - E \end{vmatrix} = 0 \tag{7-84}$$

For Eq. (7-84) to be satisfied, the product of the diagonal terms must equal that of the off-diagonal terms, which means that

$$H_{AA} - E = \pm(H_{AB} - ES_{AB}) \tag{7-85}$$

This gives two values for E:

$$E_{\pm} = \frac{H_{AA} \pm H_{AB}}{1 \pm S_{AB}} \tag{7-86}$$

To arrive at numerical values for E_+ and E_- requires that we choose a value for R and explicitly evaluate H_{AA}, H_{AB}, and S_{AB}. We know in advance that S_{AB} increases monotonically from zero at $R = \infty$ to unity at $R = 0$ because $1s_A$ and $1s_B$ are each normalized and everywhere positive. H_{AA} is the average energy of an electron in a 1s AO on nucleus A, subject also to an attraction by nucleus B. Hence, H_{AA} should be lower than the energy of the isolated H atom ($-\frac{1}{2}$ a.u.) whenever R is finite. H_{AB} is easily expanded to

$$H_{AB} = \int 1s_A \left(-\frac{1}{2}\nabla^2 - 1/r_B\right) 1s_B dv + \int 1s_A(-1/r_A) 1s_B \, dv \tag{7-87}$$

The operator in the first integrand is simply the hamiltonian operator for a hydrogen atom centered at nucleus B. This operator operates on $1s_B$ to give $-\frac{1}{2}1s_B$. Hence, the first integral becomes simply $-\frac{1}{2}S_{AB}$. The second integral in Eq. (7-87) gives the attraction between a nucleus and the "overlap charge." Thus, H_{AB} is zero at $R = \infty$ and negative for finite R. The formulas for these terms are (after nontrivial mathematical evaluation)

$$S_{AB} = \int 1s_A 1s_B \, dv = \exp(-R)\left[1 + R + R^2/3\right] \tag{7-88}$$

$$H_{AA} = \int 1s_A \hat{H}_{\text{elec}} 1s_A \, dv = -\frac{1}{2} - (1/R)\left[1 - e^{-2R}(1 + R)\right] \tag{7-89}$$

$$H_{AB} = \int 1s_A \hat{H}_{\text{elec}} 1s_B \, dv = -S_{AB}/2 - e^{-R}(1 + R) \tag{7-90}$$

When $R = 2$ a.u., $S_{AB} = 0.586$, $H_{AA} = -0.972$ a.u. and $H_{AB} = -0.699$ a.u. Inserting these values into Eq. (7-86) gives $E_+ = -1.054$ a.u. and $E_- = -0.661$ a.u. These are electronic energies. Internuclear repulsion energy ($+\frac{1}{2}$ a.u.) can be added to these values to give -0.554 a.u. and -0.161 a.u., respectively.

The ways in which H_{AA}, H_{AB}, S_{AB}, and $1/R$ contribute to the energy are illustrated for $R = 2$ a.u. in Fig. 7-6. H_{AA} is lower than the energy of an isolated H atom because

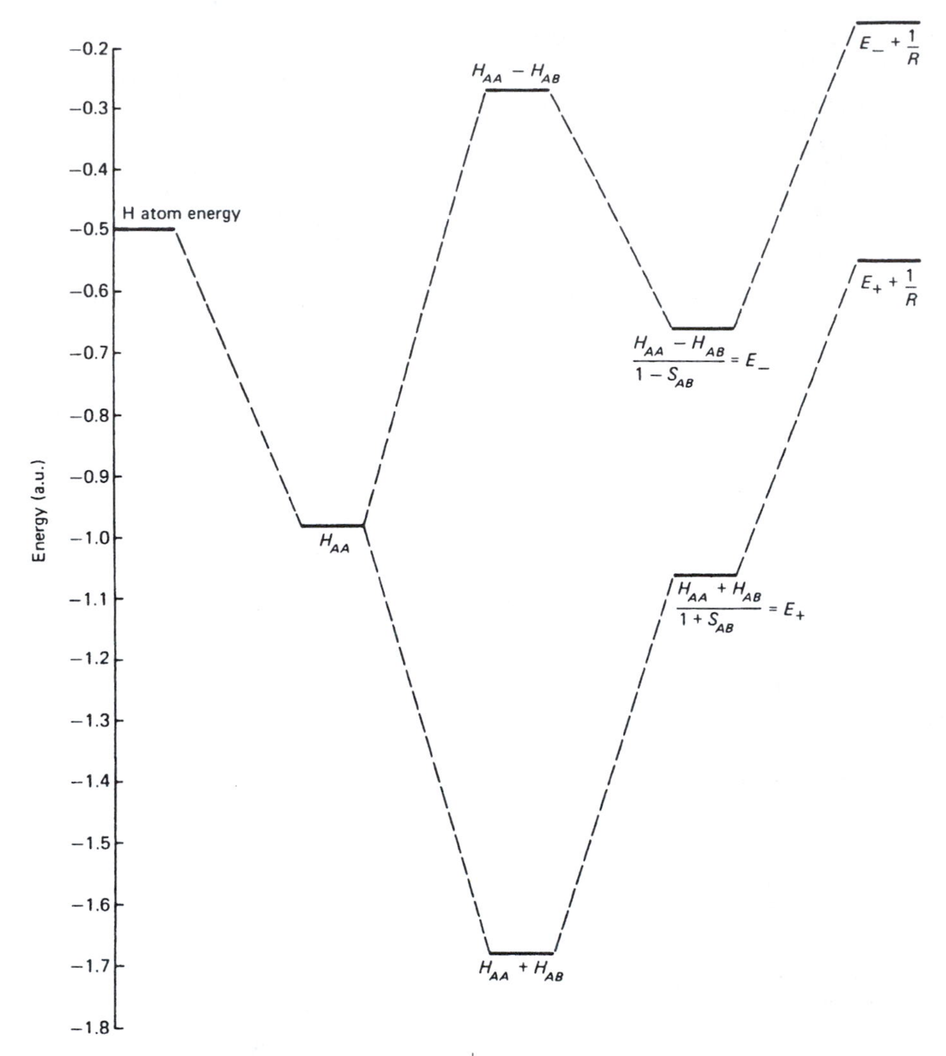

Figure 7-6 ▶ Contributions to the energy of H_2^+ at $R = 2$ a.u. in a minimal basis LCAO–MO calculation.

the electron experiences additional nuclear attraction at $R = 2$. The effect of the H_{AB} interaction element is to split the energy into two levels equally spaced above and below H_{AA}. The S_{AB} term has the effect of partially negating this splitting. The internuclear repulsion energy $1/R$ merely raises each level by $\frac{1}{2}$ a.u. The lower energy, $E_+ + 1/R$, has a final value that is lower than the separated-atom energy of $-\frac{1}{2}$ a.u. Since the *exact* energy at $R = 2$ must be as low or lower than our value of -0.554 a.u., we can conclude that the H_2^+ molecule–ion has a state that is stable, with respect to dissociation into $H + H^+$, by at least 0.054 a.u., or 1.47 eV, or 33.9 kcal/mole, neglecting vibrational energy effects.

The data depicted in Fig. 7-6 are sometimes presented in the abbreviated form of Fig. 7-7. The energy levels for the pertinent AOs of the separated atoms are indicated on the left and right, and the final energies (either electronic or electronic plus internuclear) are shown in the center.

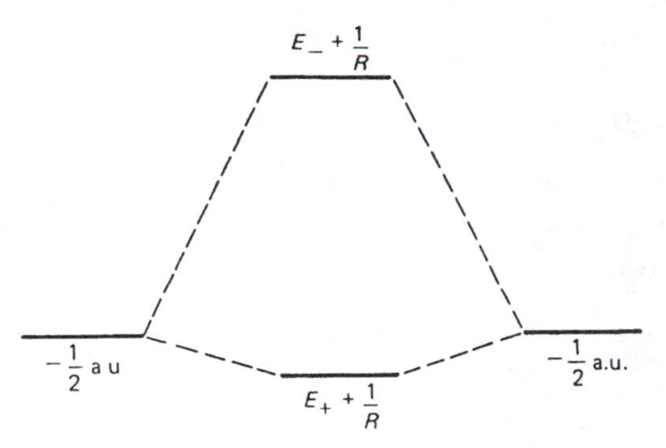

Figure 7-7 ▶ Separated atom energies and energies at an intermediate R for H_2^+.

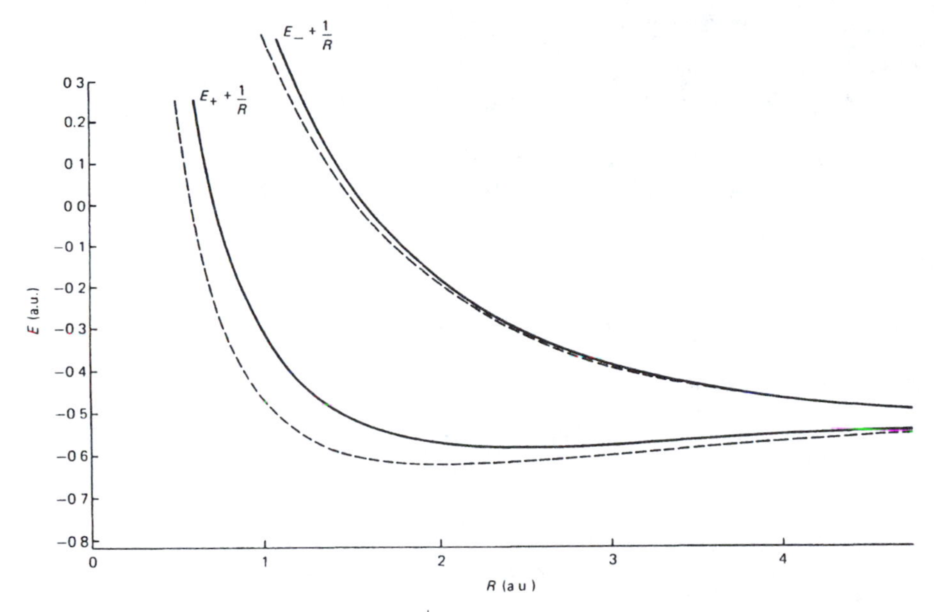

Figure 7-8 ▶ $E_{\pm} + 1/R$ versus R for H_2^+. (—) Calculation described in text. (- - -) Exact calculation.

The behavior of these energies as a function of R is plotted in Fig. 7-8. Included for comparison are the exact energies for the two lowest-energy states of H_2^+. Only the lower of these shows stability with respect to molecular dissociation. Both energy levels approach infinity asymptotically as R approaches zero because of internuclear repulsion. (The zero of energy corresponds to complete separation of the protons and electron.)

Having found the roots $E_{\pm}$ for the secular determinant, we can now solve for the coefficients which describe the approximate wavefunctions in terms of our basis set. Let us first find the approximate wavefunction corresponding to the lower energy. To do this, we substitute the expression for E_+ [Eq. (7-86)] into the simultaneous equations associated with the secular determinant (7-84):

$$c_A(H_{AA} - E_+) + c_B(H_{AB} - E_+S_{AB}) = 0 \tag{7-91}$$

$$c_A(H_{AB} - E_+S_{AB}) + c_B(H_{AA} - E_+) = 0 \tag{7-92}$$

Equation (7-91) leads to

$$c_A\left[H_{AA}-(H_{AA}+H_{AB})/\left(1+S_{AB}\right)\right]=-c_B\left[H_{AB}-(H_{AA}+H_{AB})S_{AB}/\left(1+S_{AB}\right)\right] \tag{7-93}$$

which ultimately gives

$$c_A=c_B \tag{7-94}$$

The same procedure for E_- produces the result

$$c_A=-c_B \tag{7-95}$$

The normality requirement is

$$\int \psi^*\psi\, dv = 1 = \int (c_A 1s_A + c_B 1s_B)^2\, dv$$

$$= c_A{}^2 \int \cancelto{1}{1s_A^2\, dv} + c_B{}^2 \int \cancelto{1}{1s_B^2\, dv} + 2c_A c_B \int \cancelto{S_{AB}}{1s_A 1s_B\, dv} \tag{7-96}$$

so that

$$c_A^2+c_B^2+2c_Ac_BS_{AB}=1 \tag{7-97}$$

For $c_A=c_B$, this gives

$$c_A=1/\left[2\left(1+S_{AB}\right)\right]^{1/2}=c_B \tag{7-98}$$

For $c_A=-c_B$,

$$c_A=1/\left[2\left(1-S_{AB}\right)\right]^{1/2}=-c_B \tag{7-99}$$

Therefore, the LCAO–MO wavefunction corresponding to the lower energy E_+ is

$$\psi_+=\frac{1}{\sqrt{2(1+S_{AB})}}(1s_A+1s_B) \tag{7-100}$$

The higher-energy solution is

$$\psi_-=\frac{1}{\sqrt{2(1-S_{AB})}}(1s_A-1s_B) \tag{7-101}$$

Just as was true for AOs, there are a number of ways to display these MOs pictorially. One possibility is to plot the value of $\psi_\pm$ or $\psi_\pm^2$ along a ray passing through both nuclei. (Other rays could also be chosen if we were especially interested in other regions.) Another approach is to plot contours of ψ or ψ^2 on a plane containing the internuclear axis, Still another way is to sketch a three dimensional view of a surface of constant value of ψ or ψ^2 containing about 90–95% of the wavefunction or the electron charge. Finally, one can plot the value of ψ as distance above or below the plane containing the internuclear axis. All of these schemes are shown in Fig. 7-9 for ψ_+ and ψ_-.

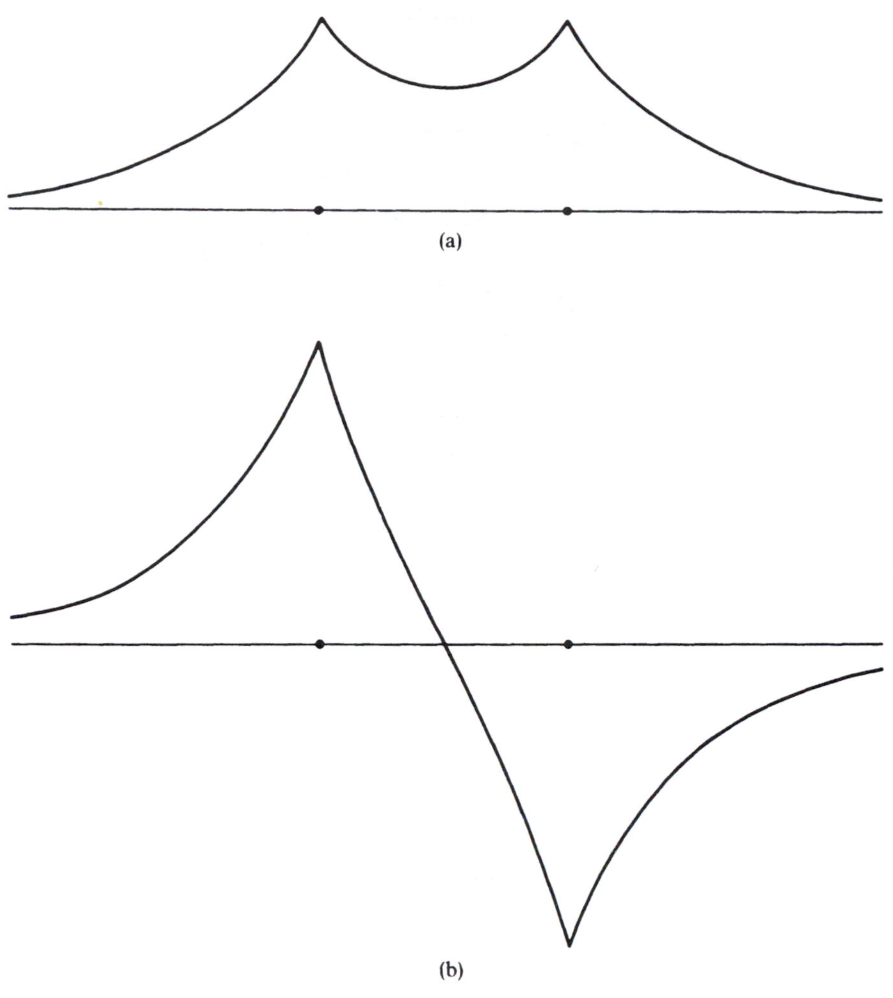

Figure 7-9 ▶ (a) Plot of ψ_+, along the z axis. [Eq. (7-100)]. (b) Plot of ψ_- along the z axis [Eq. (7-101)].

The wavefunctions ψ_+ and ψ_- may be seen from Fig. 7-9 to be, respectively, symmetric and antisymmetric for inversion through the molecular midpoint. [They are commonly called *gerade* (German for *even*) and *ungerade*, respectively.] This would be expected for nondegenerate *eigenfunctions* of the hamiltonian, since it is invariant to inversion. But ψ_+ and ψ_- are not eigenfunctions. They are only approximations to eigenfunctions. How is it that they show the proper symmetry characteristics of eigenfunctions? The reason is that the symmetry of the H_2^+ molecule is manifested as a symmetry in our secular determinant of Eq. (7-84). Note that the determinant is symmetric for reflection across either diagonal. The symmetry for reflection through the principal diagonal (which runs from upper left to lower right) is due to the hermiticity of $\hat{H}$, and is always present in the secular determinant for any molecule regardless of symmetry.[6] Symmetry for reflection through the other diagonal is due to the fact that the hamiltonian is invariant to inversion and also to the fact that the basis functions at

[6]Hermiticity requires that $H_{ij} = H_{ji}{}^*$. If H_{ij} is imaginary, the determinant will be antisymmetric for reflection through the principal diagonal.

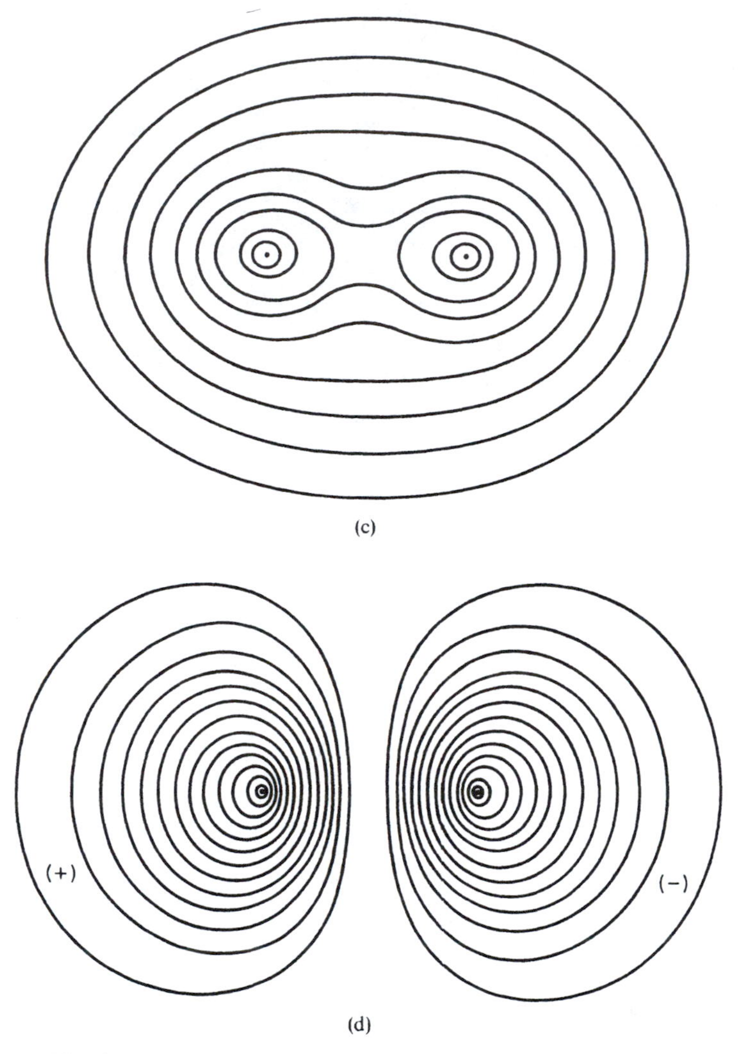

Figure 7-9 ▶ (Continued) (c) Contour diagram of ψ_+. (d) Contour diagram of ψ_-.

the two ends of the molecule are identical. When the AOs in a basis set are interchanged (times ± 1) by a symmetry operation of the molecule, the basis set is said to be *balanced* for that operation. Thus, a $1s_A$ and $1s_B$ basis is balanced for inversion in H_2^+, but a $1s_A$ and $2s_B$ basis is not. Whenever a symmetry-balanced basis is used for a molecule, the symmetry of the molecule is manifested in the secular determinant and ultimately leads to approximate MOs that show the proper symmetry characteristics.

Since the eigenfunctions for H_2^+ must be gerade or ungerade, and since we started with the simple balanced basis $1s_A$ and $1s_B$, it should be evident that it is unnecessary to go through the linear variation procedure for this case. With such a simple basis set, there is only one possible gerade linear combination of AOs, namely $1s_A + 1s_B$. Similarly, $1s_A - 1s_B$ is the only possible ungerade combination. Therefore, we could have used symmetry to guess our solutions at the outset. Usually, however, we are not so limited in our basis. We shall see that, while symmetry is useful in such circumstances, it does not suffice to produce the variationally best solution.

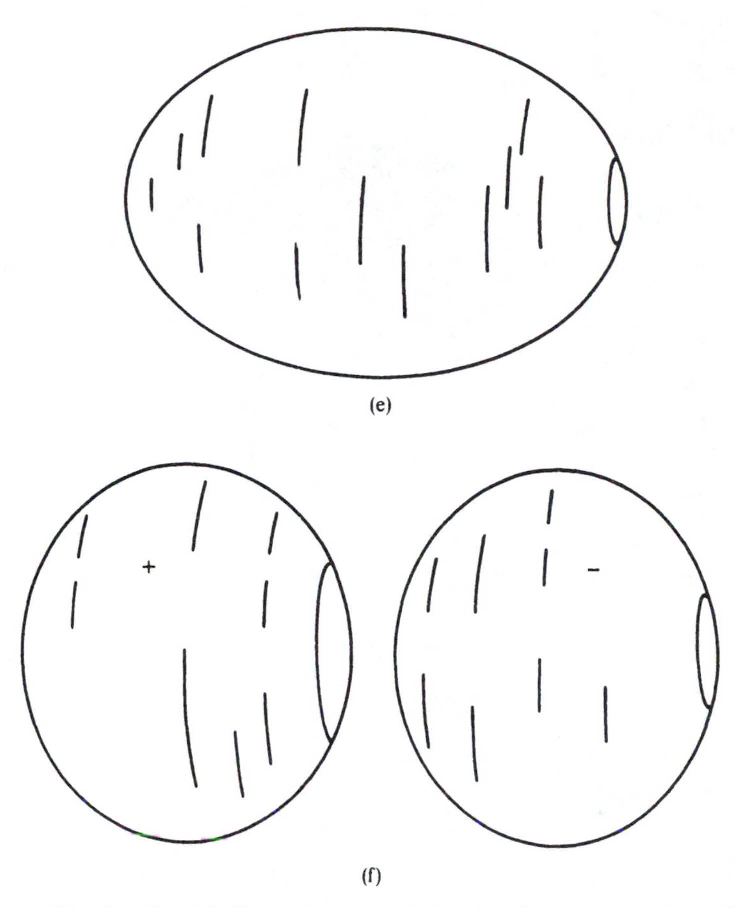

Figure 7-9 ▶ (Continued) (e) Three-dimensional sketch of contour envelope for ψ_+ and (f) for ψ_-.

According to the theorem mentioned earlier (but not proved), the nth lowest root of a linear variation calculation for a state function must lie above the nth lowest exact eigenvalue for the system. However, the two states we are dealing with have different symmetries. In such a case, a more powerful boundedness theorem holds—one that holds even if we are not using a linear variation procedure. To prove this, we first recognize that every H_2^+ *eigenfunction* is either *gerade* (i.e., symmetric for inversion) or *ungerade*. Since the lowest-energy approximate wavefunction ψ_+ is *gerade*, it must be expressible as a linear combination of these *gerade* eigenfunctions. Hence, its average energy E_+ cannot be lower than the lowest eigenvalue for the *gerade* eigenfunctions. Similarly, E_- cannot be lower than the lowest eigenvalue for the *ungerade* eigenfunctions, and so we have a separate lower bound for the average energy of trial functions of each symmetry type. For this reason, our approximate energies in Fig. 7-8 must lie above the exact energies for both states. If we were to make further efforts to lower the average energy of the ungerade function, even by going outside the linear variation procedure, we could never fall below the exact energy for the lowest-energy *ungerade* state unless we somehow allowed our trial function to change symmetry. This means that a lowest average energy criterion can be used in attempting to find the lowest-energy state *of each symmetry type* for a system, by either linear or nonlinear variation methods.

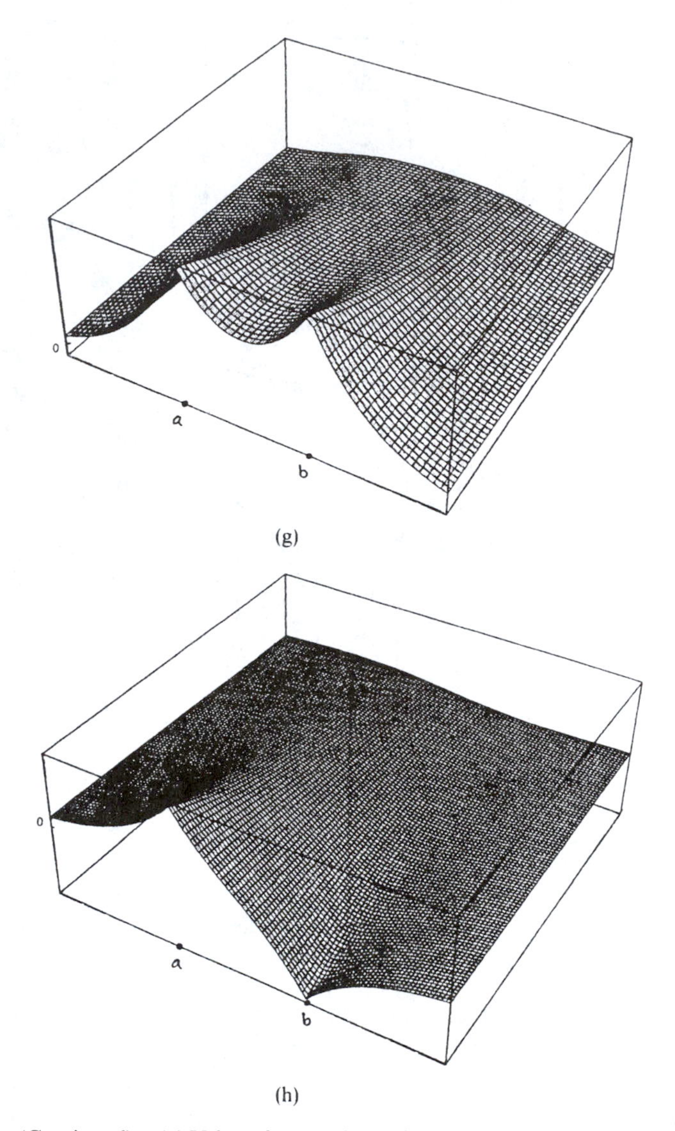

Figure 7-9 ▶ (Continued) (g) Value of ψ_+ and (h) of ψ_- versus position in plane containing the nuclei at points a and b.

> **EXAMPLE 7-2** Suppose one were to do a variational calculation on the hydrogen atom using the unnormalized trial wavefunction $\phi = \exp(-\alpha r^2)\cos\theta$, with α being varied. What lower bound could we expect for the average energy?

SOLUTION ▶ This function has a node in the x, y plane (where $\theta = \pi/2$), and this nodal plane exists regardless of the value of α. Therefore, our trial function is like the $2p_0$ AO in symmetry. It will be represented by a linear combination of p_0 AOs–a combination that changes as α is varied. The average value of energy cannot be lower than the lowest eigenvalue in the set, which is $-1/8$ a.u. (for the $2p_0$ AO). ◀

Inspection of Fig. 7-9 shows that the charge distribution for the state described by ψ_+ is augmented at the molecular midpoint compared to the charge due to unperturbed atoms. This state is also the one that gives H_2^+ stability at finite R. Because this MO puts electronic charge into the bond region and stabilizes the molecule, it is commonly called a *bonding* MO. In the state described by ψ_-, charge is shifted out of the bond region and the molecule is unstable at finite R, and so ψ_- is called an *antibonding* MO.

Because the potential in H_2^+ (or in any linear molecule) is independent of ϕ, the angle about the internuclear axis, the ϕ dependence of the wavefunctions is always of the form

$$\Phi(\phi) = (1/\sqrt{2\pi})\exp(im\phi), \quad m = 0, \pm 1, \pm 2, \ldots \tag{7-102}$$

This fact may be arrived at in two ways. One way is to write down the Schrödinger equation for H_2^+ using spherical polar coordinates or elliptical coordinates. (ϕ is a coordinate in each of these coordinate systems.) Then one attempts to separate coordinates and finds that the ϕ coordinate is indeed separable from the others and yields the equation

$$\frac{d^2\Phi(\phi)}{d\phi^2} = -m^2\Phi(\phi) \tag{7-103}$$

The acceptable solutions of this are the functions (7-102). The other approach is to note that, since the potential in $\hat{H}$ has no ϕ dependence, $\hat{H}$ commutes with $\hat{L}_z$, the angular momentum operator, so that the eigenfunctions of $\hat{H}$ are simultaneously eigenfunctions of $\hat{L}_z$. We know the eigenfunctions of $\hat{L}_z$ are the functions (7-102), and thus, we know that these functions must also give the ϕ dependence of the eigenfunctions of $\hat{H}$.

Two important conclusions emerge. First, each nondegenerate H_2^+ wavefunction must have a ϕ dependence given by one of the functions (7-102). This tells us something about the shapes of the wavefunctions. Second, the nondegenerate H_2^+ wavefunctions are eigenfunctions of $\hat{L}_z$, which means that an electron in any one of these states has a definite, unvarying (i.e., sharp) component of angular momentum of value $m\hbar$ mks units (m a.u.) along the internuclear axis. It is thus useful to know the m value associated with a given one-electron wavefunction. A standard notation is used, which is analogous to the atomic orbital notation wherein s, p, d, f, correspond to l values of 0, 1, 2, 3, respectively. The corresponding Greek lower-case letters σ, π, δ, ϕ indicate values of $|m|$ of 0, 1, 2, 3, respectively, in one-electron orbitals of linear molecules. Thus, for the case at hand, ψ_+ and ψ_- are both σ MOs because they are cylindrically symmetrical (i.e., no ϕ dependence), which requires that m be zero. Because ψ_+ is a *gerade* function, it is symbolized σ_g. ψ_-, then, is a σ_u MO. (A simple way to determine whether an MO is σ, π, or δ is to imagine viewing it (in its real form) end-on (i.e., along a projection of the internuclear axis.) If the MO from this view looks like an s AO, it is a σ MO. If it looks like a p AO, it is a π MO. A δ MO has the four-leaf-clover appearance of a d AO.)

Let us now examine the dependence of our LCAO–MO results on our original choice of basis. The 1s AOs we have used are capable of giving the exact energy when $R = \infty$, but become increasingly inadequate as R decreases. As a result, Fig. 7-8 shows that, for both states, the approximate energy deviates more and more from the exact energy as R decreases. At $R = 0$, our σ_g function becomes a single 1s H atom ($Z = 1$) AO centered on a doubly positive nucleus. Yet we know that the lowest-energy *eigenfunction* for that situation is a single 1s $He^+(Z = 2)$ AO. An obvious way to improve our wavefunction, then, is to allow the 1s basis functions to change their orbital exponents as R changes. This adds a nonlinear variation, and the calculation is more complicated. It is very easily performed with the aid of a computer, however, and we summarize the results in Figs. 7-10 and 7-11. The internuclear repulsion has been omitted from the energies in Fig. 7-10. The lowest approximate energy curve is now in perfect agreement with the exact electronic energy both at $R = 0$ and $R = \infty$, and shows improved, though still not perfect, agreement at intermediate R. The R dependence of the orbital exponent for this wavefunction (Fig. 7-11) varies smoothly from 1 at $R = \infty$ to 2 at $R = 0$, as expected. In contrast, the σ_u function fails to reach a well-defined energy at $R = 0$ because the function becomes indeterminate at that point. [At $R = 0$, $1s_A - 1s_B$ becomes $(1s_A - 1s_A) = 0$.] However, Figs. 7-10 and 7-11 indicate that the exact energy E_- is $-\frac{1}{2}$ a.u. at $R = 0$, corresponding to the $n = 2$ level of He^+, and that the orbital exponent in our 1s basis functions approaches 0.4 in the effort to approximate this state function at small R. To understand this behavior, we must once again consider the symmetries of these functions.

We have already seen that the nondegenerate eigenfunctions of H_2^+ must be either *gerade* or *ungerade*, and we note in Fig. 7-10 that the energy curve for the gerade state is continuous as is the one for the ungerade state. In other words, as we move along a given curve, we are always referring to a wavefunction of the *same* symmetry. This continuity of symmetry along an energy curve is central to many applications of quantum chemistry. The reason for continuity of symmetry can be seen by considering a molecule having some element of symmetry, and having a nondegenerate wavefunction or molecular orbital (which must be symmetric or antisymmetric with respect to the symmetry operations of the molecule). If we change the molecule infinitesimally (without destroying its symmetry), we expect the wavefunction to change infinitesimally also.

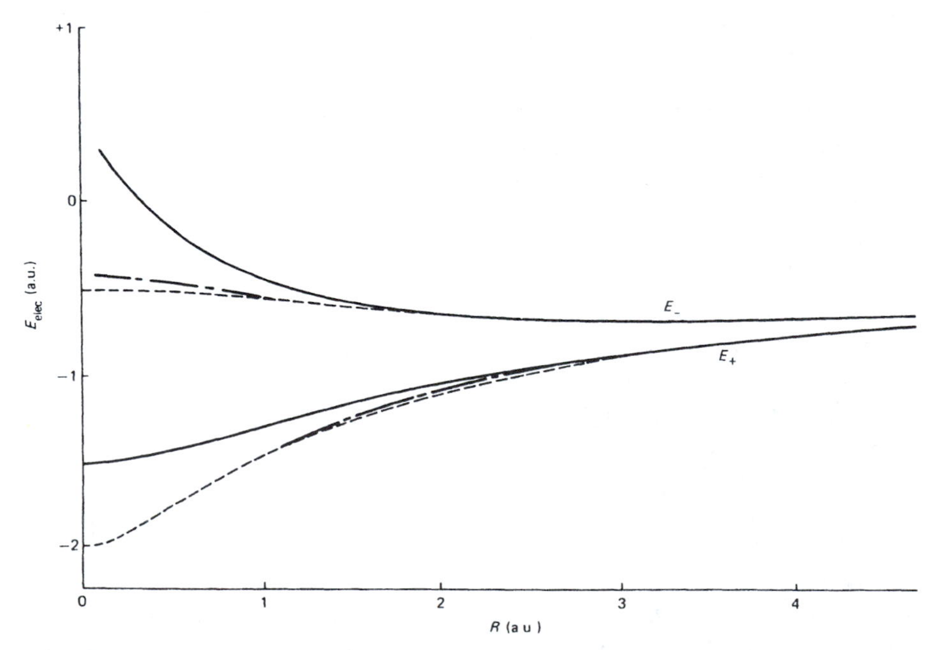

Figure 7-10 ▶ E_+ and E_- for H_2^+ from (- - -) exact, ($\cdots$) variable ζ, and (—) fixed ζ ($\zeta = 1$) treatments.

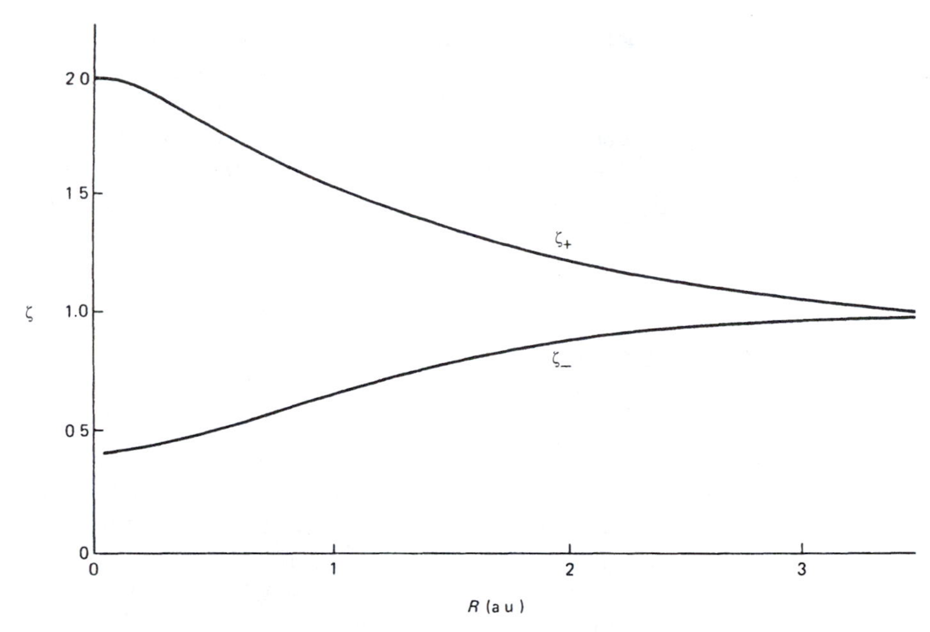

Figure 7-11 ▶ Values of ζ minimizing E_+ and E_- as a function of R.

In particular it should not change symmetry because this is generally not an infinitesimal change. (To change symmetry requires adding or removing nodes, changing signs in parts of the function. Such changes have more than infinitesimal effects on ψ and on kinetic and potential parts of the energy.) The entire curve can be traversed by an infinite number of such infinitesimal but symmetry conserving steps.

The continuity of symmetry enables the σ_g state of H_2^+ to correlate with an s-type AO of He^+ as R goes to zero. This correlation is *symmetry allowed* because the s-type AOs of He^+ have the proper symmetry characteristics—they are *gerade* and have no dependence on angle about the axis that is the internuclear axis when $R > 0$ (see Fig. 7-12). In contrast, the σ_u MO cannot correlate with an s-type AO. It must correlate with an AO that is cylindrically symmetrical about the old internuclear axis but is antisymmetric for inversion. A p-type AO pointing along the old internuclear axis (called a p_σ AO) satisfies these requirements.

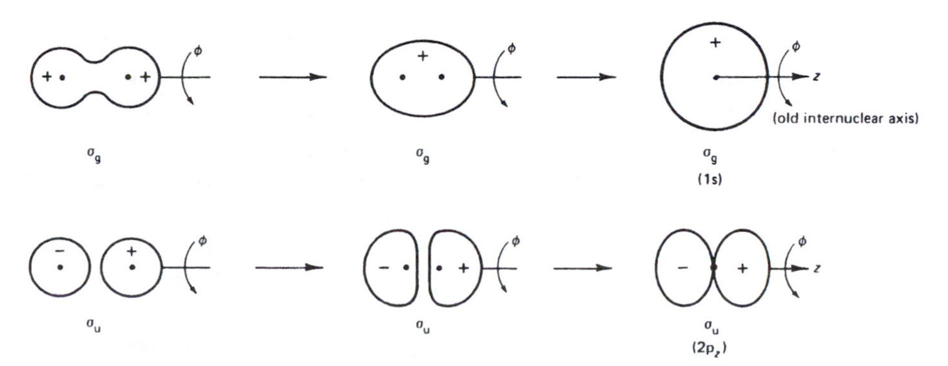

Figure 7-12 ▶ Sketches demonstrating how separated atom functions can be related to united-atom functions through symmetry invariance. These functions are not drawn to a common scale.

These symmetry requirements help us understand the σ_u curves of Figs. 7-10 and 7-11. The exact energy for E_- goes to $-\frac{1}{2}$ a.u. at $R=0$ because the σ_u state of H_2^+ correlates with a 2p AO of He^+. Our basis set is incapable of reproducing a 2p AO at $R=0$, and so the calculated energy curve fails to rejoin the exact curve at $R=0$. At small R, our basis set is attempting to approximate the two lobes of an evolving 2p AO. Apparently the orbital exponent that best enables the basis to accomplish this is around 0.4.

7-6 Molecular Orbitals of Homonuclear Diatomic Molecules

We have already seen how one produces the ground configurations for many-electron *atoms* by placing pairs of electrons of opposite spin into AOs, starting with the lowest-energy AO and working up. Subsequent manipulation of this product function to produce proper space and spin symmetry yields the approximate wavefunction. Precisely the same procedure is used for molecules. Thus, the electronic configurations for H_2^+, H_2, and H_2^- are $1\sigma_g$, $(1\sigma_g)^2$, and $(1\sigma_g)^2 1\sigma_u$, respectively, and the approximate wavefunction for H_2 is provided by the Slater determinant $\left|1\sigma_g(1)\,1\bar{\sigma}_g(2)\right|$. When we come to consider heavier homonuclear diatomic molecules, such as O_2, we must place electrons in higher energy MOs. Such MOs are still provided by the minimal basis set, which now includes 1s, 2s, and three 2p AOs on each atom since these AOs are occupied in the separated atoms. We now consider the natures of the additional MOs produced by this larger basis.

We begin by making a change to a basis set that is mathematically equivalent to the starting set but is more convenient for discussing and analyzing the problem. This new set is the set of *symmetry orbitals* (SOs) $(1s_A \pm 1s_B)$, $(2s_A \pm 2s_B)$, etc. From our original ten AOs, we thus produce ten SOs. These may be normalized, if desired. Each of these SOs has definite symmetry. The SOs built from 2s AOs must be of σ_g and σ_u symmetry since the 2s AOs act like 1s AOs for all the symmetry operations of the molecule. The 2p AOs pointing along the internuclear axis have cylindrical symmetry and hence also give rise to a σ_g and a σ_u SO. (We will take the internuclear axis to be coincident with the z axis, and so these SOs are constructed from $2p_0$ (or $2p_z$) AOs.) The functions $(2p_{+1_A} \pm 2p_{+1_B})$ are π SOs because $|m|=1$. Since it is difficult to visualize complex functions, however, the usual practice is to take linear combinations of the complex π functions to produce a corresponding set of real functions. (This is completely analogous to forming real p_x and p_y AOs from complex p_{+1} and p_{-1} AOs.) Thus, we obtain $(2p_{x_A} \pm 2p_{x_B})$ and $(2p_{y_A} \pm 2p_{y_B})$, which are not eigenfunctions for the $\hat{L}_z$ operator anymore, but are still given the symbol π. From Fig. 7-13, we see that the *positive* combinations give *ungerade* π SOs. This is just the opposite of the case for σ-type SOs. We conclude from Fig. 7-13 that σ_g and π_u SOs will tend to place charge *into* the bond and hence contribute to bonding, whereas σ_u and π_g SOs will contribute antibonding character.

Even though these SOs are only a basis set, the reader may nevertheless recognize that conversion to this symmetrized basis goes a long way toward producing our ultimate MOs. Indeed, our $(1s_A \pm 1s_B)$ SOs, if normalized, are the same as the MOs we obtained

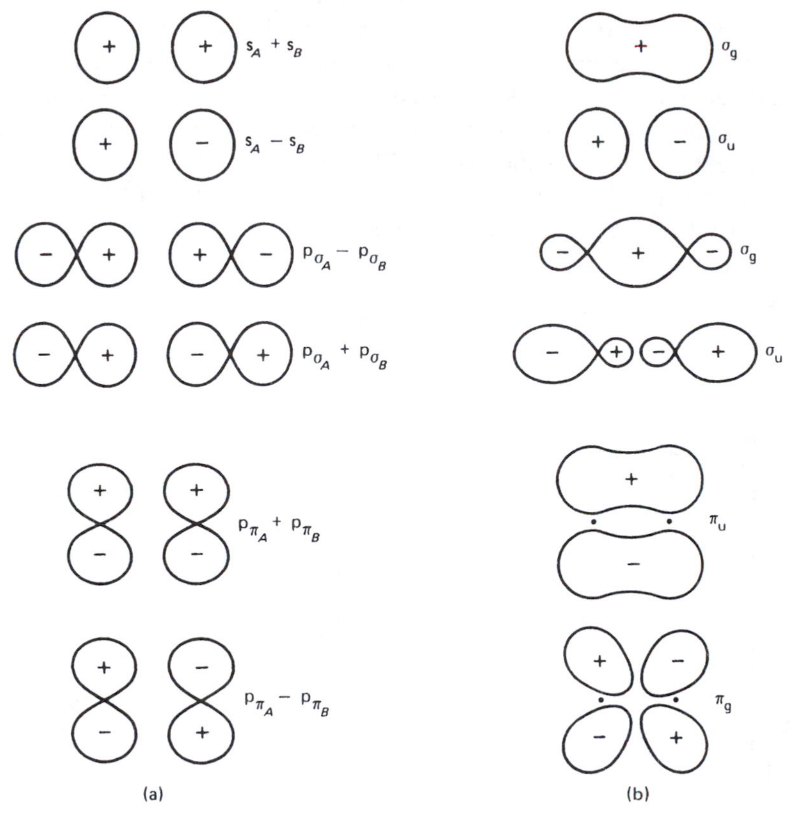

Figure 7-13 ▶ Symmetry orbitals constructed from s- and p-type AOs. (a) Sketches according to an idealized convention that ignores overlap between AOs on A and B. (b) Effects of overlap. Note that the $p\sigma_A - p\sigma_B$ combination is bonding. This depends on our having chosen a common z axis for both atoms. Sometimes the z axes are chosen to point from each atom toward the other. In that case, $p\sigma_A - p\sigma_B$ becomes antibonding.

for H_2^+. The essential advantage of a symmetrized basis set is that it simplifies the secular determinant and makes it easier to understand and describe the mixing of the basis functions by the hamiltonian. For instance, since our MOs must have pure σ, π, δ, ... and also g or u symmetry, and since our SOs are already of pure symmetry, we expect no further mixing to occur between SOs of *different* symmetry in forming MOs. This suggests that the interaction element H_{ij} between SOs ϕ_i and ϕ_j of different symmetry should vanish. This is easily proved by noting that $\hat{H}$ is symmetric for all symmetry operations of the molecule, and, if ϕ_i and ϕ_j differ in symmetry for some operation, their product is antisymmetric for that operation, and therefore $\phi_i^* H \phi_j$ is antisymmetric and its integral vanishes. Similarly, S_{ij} vanishes, and $H_{ij} - ES_{ij}$ vanishes except in positions connecting basis functions of identical symmetry. As a result, our secular determinant over SOs has the form (7-104), where the notation $\sigma_g[1s]$ indicates a σ_g SO made from $1s_A$, $1s_B$ AOs, etc. By placing basis functions of like symmetry together, we have emphasized the block-diagonal form of our determinant, all elements in the nonshaded areas being zero by symmetry. (The $\pi_g[2p_x]$ and $\pi_g[2p_y]$ do not interact because of symmetry disagreements for reflection in the xz and yz planes.) Each of the nonzero blocks of (7-104) is a separate determinant (which is just a number), and the value of determinant (7-104) is simply the product of these six smaller determinants. Hence, if any *one* of these small determinants is zero, the large determinant is zero, thereby satisfying our determinantal equation. Therefore, each of these small determinants may be employed in a separate determinantal equation, and the problem is said to be *partitioned* into six smaller problems. It follows immediately that we can get mixing among the three σ_g SOs to produce three σ_g MOs and likewise for the σ_u set, and that the π SOs can undergo no further mixing and hence are already MOs. (Notice that SOs in one block are not linked by off-diagonal elements *either directly or indirectly*, unlike the case described in Example (7-1).)

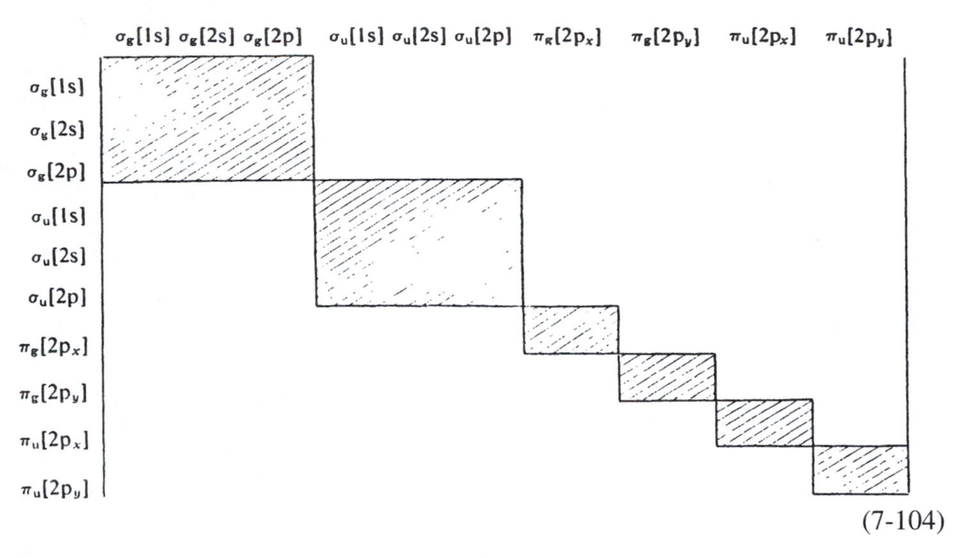

(7-104)

What will be the nature of the lowest-energy σ_g MO? It will not be pure σ_g[1s] because admixture of σ_g[2s] and σ_g[2p] SOs can produce charge shifts that will lower the energy. But the σ_g[2s] and σ_g[2p] SO energies are much higher than the σ_g[1s] (mainly because the 2s and 2p AOs are higher in energy in the atoms, and the atomic contributions still dominate in the molecule). Any energy decrease to be gained by charge shifting must be weighed against the energy increase due to the mixing in of such high energy components. The former very quickly become overbalanced by the latter, so the lowest energy σ_g MO is almost pure σ_g[1s] SO, the σ_g[2s] and σ_g[2p] SOs coming in only very slightly. This exemplifies an important general feature of quantum chemical calculations: mixing between basis orbitals tends to be small if they have widely different energies in the system. *Thus, we now have two factors governing the extent of mixing of functions ϕ_i and ϕ_j—the size of the interaction element H_{ij}, and the difference in energy between them in the system $(H_{ii} - H_{jj})$.*

A label we can use for this lowest-energy MO that avoids implying that it is identical to the σ_g[1s] SO is $1\sigma_g$. This stands for "the lowest-energy σ_g MO." Because of the low energy of the 1s AOs, the next-lowest MO is almost pure σ_u[1s], and we label it $1\sigma_u$.

When considering the remaining two σ_g MOs, we can expect substantial mixing between σ_g[2s] and σ_g[2p] SOs because these functions are not very different in energy. In the hydrogen atom, the 2s and 2p AOs are degenerate, and, as we move along the periodic table, they become split farther and farther apart in energy. Therefore, we might expect to find the greatest mixing for B_2 and C_2, and to find less mixing for O_2 and F_2. Figure 7-14 is a schematic diagram of the MO energy levels we should expect for F_2. Here the $2\sigma_g$ MO is primarily the σ_g[2s] SO and the $3\sigma_g$ MO is mainly the σ_g[2p]

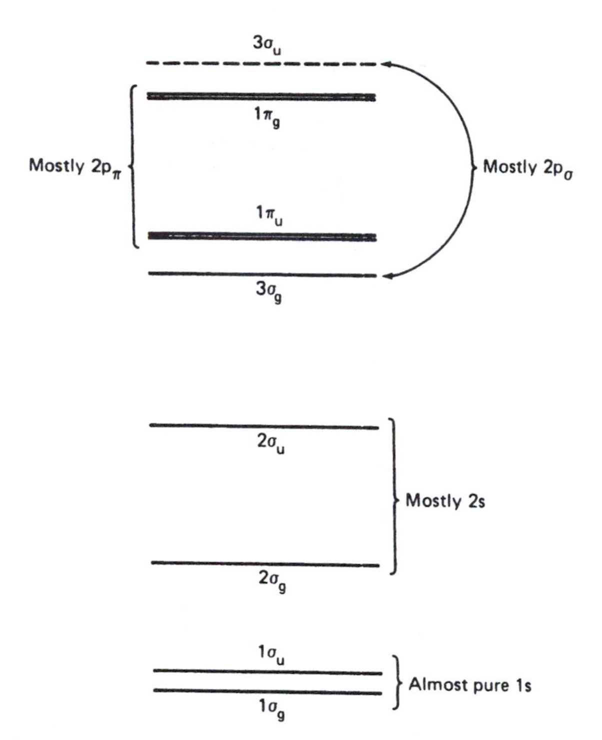

Figure 7-14 ▶ Schematic of MO energy level order for F_2. The vertical axis is *not* an accurate energy scale. For instance, the 1σ levels are very much lower in energy relative to the other levels than is suggested by the drawing. (- - -) indicates that the orbital is unoccupied in the ground state.

SO. Similarly $2\sigma_u$ and $3\sigma_u$ are mainly $\sigma_u[2s]$ and $\sigma_u[2p]$, respectively. The 1π and 3σ MOs are degenerate at the separated atom limit, where they are all 2p AOs. As the atoms come together and interact, the π levels split apart less than the σ levels because the $2p_x$ and $2p_y$ AOs approach each other side to side, whereas the $2p_\sigma$ AOs approach end to end. The latter mode produces larger overlap and leads to larger interaction elements and greater splitting. (Because of the symmetry of the molecule, the $1\pi_u$ MOs are always degenerate, and they may be mixed together in any way. In particular, we can regard them as being $1\pi_{ux}$ and $1\pi_{uy}$ or $1\pi_{u+1}$ and $1\pi_{u-1}$ with equal validity. The same situation holds for the $1\pi_g$ pair.) From the ordering of energy levels in Fig. 7-14 we obtain for F_2 the configuration

$$F_2 : (1\sigma_g)^2(1\sigma_u)^2(2\sigma_g)^2(2\sigma_u)^2(3\sigma_g)^2(1\pi_u)^4(1\pi_g)^4 \tag{7-105}$$

Now we will consider what happens for lighter molecules. Recall that here the 2s and 2p AOs are closer together in energy. This allows greater mixing between the SOs containing these AOs and produces increased energy level splitting. Thus, the $\sigma_g[2s]$ and $\sigma_g[2p]$ SOs mix together more, and the resulting splitting causes an additional lowering in energy of the $2\sigma_g$ level and an increase for the $3\sigma_g$ level, compared to the F_2 case. In a similar way the $2\sigma_u$ and $3\sigma_u$ levels are lowered and raised, respectively, by increased mixing between $\sigma_u[2s]$ and $\sigma_u[2p]$ SOs. The resulting energy level pattern for C_2 is shown in Fig. 7-15. Note that the extra splitting has pushed the $3\sigma_g$ level *above* the $1\pi_u$ level. No other change in the orbital energy ordering has occurred. As a result, C_2 has, the configuration

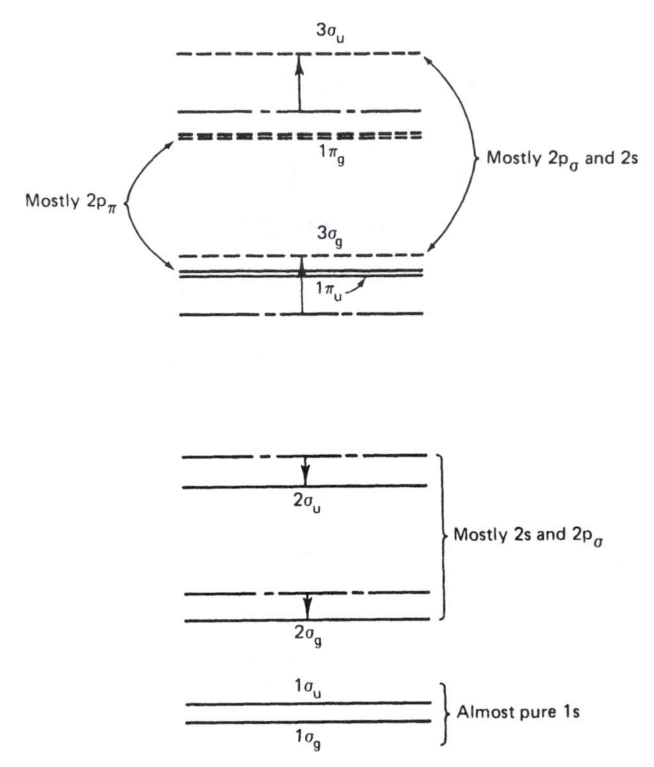

Figure 7-15 ▶ A schematic diagram of the MO energy level order for C_2. The 2s and $2p\sigma$ symmetry functions mix more strongly in producing 2σ and 3σ MOs than is the case in F_2. The level shifts discussed in the text are indicated by dashed lines and arrows. (——-——) represent F_2 levels. Note the inversion of the order of the $1\pi_u$ and $3\pi_g$ levels. The orbital ordering is deduced from spectra. (See Mulliken [1].) (– – –) represent energies of orbitals not occupied by electrons. Note that there is no well-defined energy ordinate for this figure, and no accurate relationship between absolute values of orbital energies within a molecule or between molecules is implied.

$$C_2 : (1\sigma_g)^2(1\sigma_u)^2(2\sigma_g)^2(2\sigma_u)^2(1\pi_u)^4 \tag{7-106}$$

The above discussion is an effort to *rationalize* orbital energies obtained from calculations or deduced from molecular spectra. We have not yet described the details of how one goes about carrying out MO calculations on these molecules, nor will we in this chapter. However, even in the absence of precise numerical results, it is possible for such qualitative arguments to be very useful in understanding and predicting features of a wide range of chemical reactions.

The molecular configurations obtained from the patterns of Figs. 7-14 and 7-15 predict molecular properties that are in strikingly good qualitative agreement with experimental observations. The data in Table 7-2 show that, when we go from H_2^+ to H_2, adding a second electron to a bonding MO, the bond length decreases and the dissociation energy increases. Adding a third electron (He_2^+) causes partial occupation of the antibonding $1\sigma_u$ MO and causes the bond length to increase and the dissociation energy to decrease. The four electron molecule He_2 is not observed as is consistent with its configuration. The relative inertness of N_2 as a chemical reactant becomes understandable from the fact that it has six more "bonding electrons" than antibonding electrons, giving a net of three bonds—one σ and two π bonds. For N_2 to react, it

TABLE 7-2 ► Some Properties of Homonuclear Diatomic Molecules and Ions in their Ground Electronic States

Molecule	MO configuration	Net number of bonding electrons	Binding energy, D_e (eV)	Equilibrium internuclear separation, R_e(Å)	Term[c]
H_2^+	$1\sigma_g$	1	2.7928	1.06	$^2\Sigma_g$
H_2	$1\sigma_g{}^2$	2	4.747745	0.7414	$^1\Sigma_g{}^+$
H_2^-	$1\sigma_g{}^2 1\sigma_u$	1	1.7[a]	0.8	$^2\Sigma_u{}^+$
$He_2{}^+$	$1\sigma_g{}^2 1\sigma_u$	1	2.5	1.08	$^2\Sigma_u{}^+$
He_2	$1\sigma_g{}^2 1\sigma_u{}^2$	0	0.001[b]	2.88	$(^1\Sigma_g{}^+)$
$He_2{}^-$	$[He_2]2\sigma_g$	1		No data	$^2\Sigma_g{}^+$
$Li_2{}^+$	$[He_2]2\sigma_g$	1	1.29	3.14	$^2\Sigma_g{}^+$
Li_2	$[He_2]2\sigma_g{}^2$	2	1.05	2.673	$^1\Sigma_g{}^+$
$Li_2{}^-$	$[He_2]2\sigma_g{}^2 2\sigma_u$	1	~1.3(?)	3.2	$^2\Sigma_u{}^+$
$Be_2{}^+$	$[He_2]2\sigma_g{}^2 2\sigma_u$	1		No definitive data	$^2\Sigma_u{}^+$
Be_2	$[He_2]2\sigma_g{}^2 2\sigma_u{}^2$	0	0.1	2.49	$^1\Sigma_g{}^+$
$Be_2{}^-$	$[Be_2]1\pi_u$	1	~0.3	2.4	$^2\Pi_u$
$B_2{}^+$	$[Be_2]1\pi_u$	1	1.8	—	$^2\Pi_u$
B_2	$[Be_2]1\pi_u{}^2$(?)	2	~3	1.589	$^3\Sigma_g{}^-$
$B_2{}^-$	$[Be_2]1\pi_u{}^3$	3		No data	$^2\Pi_u$
$C_2{}^+$	$[Be_2]1\pi_u{}^3$	3	5.3	1.301	$^2\Pi_u$
C_2	$[Be_2]1\pi_u{}^4$	4	6.36	1.2425	$^1\Sigma_g{}^+$
$C_2{}^-$	$[Be_2]1\pi_u{}^4 3\sigma_g$	5	8.6	—	$^2\Sigma_g{}^+$
$N_2{}^+$	$[Be_2]1\pi_u{}^4 3\sigma_g$	5	8.86	1.116	$^2\Sigma_g{}^+$
N_2	$[Be_2]1\pi_u{}^4 3\sigma_g{}^2$	6	9.90	1.098	$^1\Sigma_g{}^+$
$N_2{}^-$	$[Be_2]1\pi_u{}^4 3\sigma_g{}^2 1\pi_g$	5	~8.3	—	$^2\Pi_g$
$O_2{}^+$	$[Be_2]1\pi_u{}^4 3\sigma_g{}^2 1\pi_g$	5	6.7796	1.1171	$^2\Pi_g$
O_2	$[Be_2]3\sigma_g{}^2 1\pi_u{}^4 1\pi_g{}^2$	4	5.2132	1.2075	$^3\Sigma_g{}^-$
$O_2{}^-$	$[Be_2]3\sigma_g{}^2 1\pi_u{}^4 1\pi_g{}^3$	3	4.14	1.32	$^2\Pi_g$
$F_2{}^+$	$[Be_2]3\sigma_g{}^2 1\pi_u{}^4 1\pi_g{}^3$	3	3.39	1.32	$^2\Pi_g$
F_2	$[Be_2]3\sigma_g{}^2 1\pi_u{}^4 1\pi_g{}^4$	2	1.65	1.42	$^1\Sigma_g{}^+$
$F_2{}^-$	$[Be_2]3\sigma_g{}^2 1\pi_u{}^4 1\pi_g{}^4 3\sigma_u$	1	~1.3	1.9	$^2\Sigma_u{}^+$
$Ne_2{}^+$	$[Be_2]3\sigma_g{}^2 1\pi_u{}^4 1\pi_g{}^4 3\sigma_u$	1	~1.1	1.7	$^2\Sigma_u{}^+$
Ne_2	$[Be_2]3\sigma_g{}^2 1\pi_u{}^4 1\pi_g{}^4 3\sigma_u{}^2$	0	0.003[b]	3.09	$(^1\Sigma_g{}^+)$

[a] This state is unstable with respect to loss of an electron, but is stable with respect to dissociation into an atom and a negative ion.

[b] From Hirschfelder et al. [2]. It may be shown that any two neutral atoms will have some range of R where the attractive part of the van der Waals' interaction dominates. For He_2, this minimum is so shallow and the nuclei so light that a stable state (including vibrations) probably cannot exist. For Ne_2, a stable state should exist. The data for He_2 and Ne_2 are *calculated* from considerations of intermolecular forces.

[c] The term symbol corresponds to the configuration of column 2.

is necessary to supply enough energy to at least partially break these bonds prior to forming new bonds. The reversal of the orbital energy order between $1\pi_u$ and $3\sigma_g$ can be seen to occur between N_2 and O_2.

The configurations of Table 7-2 indicate that some molecules will have closed shells, whereas others will have one or more unpaired electrons. For example, O_2 has a configuration in which the degenerate $1\pi_g$ level "contains" two electrons. Several states can be produced from such a configuration, just as several states could be produced from the 1s2s configuration for He. Hund's first rule has been found to hold for molecules, and so the state of highest multiplicity is lowest in energy. For n unpaired electrons, the highest multiplicity achievable is $n+1$. For O_2, this is three, and we expect the ground state of the O_2 molecule to be a triplet. This is conveniently demonstrated to be so by observing that liquid oxygen is attracted into the gap between poles of a magnet, consistent with the existence of uncanceled electron spin magnetic moments.

Close perusal of Table 7-2 indicates that some of the data are not in accord with the simple qualitative ideas just presented. For example, Li_2^+ is more strongly bonded than is Li_2, even though the former has fewer bonding electrons. However, the Li_2^+ ion-molecule is longer than Li_2. H_2^- is less strongly bound than isoelectronic He_2^+, yet it has a shorter equilibrium internuclear separation. Irregularities such as these require more detailed treatment. However, one of the useful characteristics of a *qualitative* approach is that it enables us to recognize cases that deviate from our expectations and therefore warrant further study.

In Figs. 7-14 and 7-15, we saw how MOs are related to SOs for the separated atoms. Let us now consider how the separated atom SOs correlate with the united atom AOs. Recall that these orbitals are correlated by requiring them to be of identical symmetry. In Fig. 7-16 some of the possible SOs and united-atom AOs, together with their symmetry labels, are shown. Note that the σ_g SOs can correlate with s or d_σ AOs, σ_u SOs with p_σ AOs, π_u SOs with p_π AOs and π_g SOs with d_π AOs. This gives us all the information we need except for resolving the ambiguities *within* a given symmetry type. For instance, which of the 1s, 2s, 3s, $3d_\sigma, \ldots$ in the AOs correlates with which of the $\sigma_g[1s], \sigma_g[2s], \sigma_g[2p], \sigma_g[3s], \ldots$ in the SOs? This question is resolved by use of the noncrossing rule, which states that, *in correlation diagrams, energy levels associated with orbitals or states of the same symmetry will not cross*. This requires that we match up the lowest-energy united-atom AO of a given symmetry with the lowest energy SO of that symmetry, and so on up the ladder. This leads to the diagram in Fig. 7-17. The line interconnecting the 1s AO of the united atom with the 1s AOs of the separated atoms refers, at intermediate R, to the $1\sigma_g$ MO. We have already seen that this MO may contain contributions from $\sigma_g[2s]$ and $\sigma_g[2p]$ SOs. Thus, the correlation diagram tells us what orbitals the $1\sigma_g$ MO "turns into" at the limits of R, but does *not* imply that, at other R values, this MO is comprised totally of 1s AOs.

Study of this correlation diagram reveals that the antibonding MOs (σ_u and π_g) are the MOs that correlate with higher energy united-atom AOs and hence favor the separated atoms in terms of energy. This results from the fact that antibonding MOs have a nodal plane bisecting the bond which is preserved as we proceed to the united atom, yielding a united atom AO having one more node than the separated AOs we started with. Thus, $1s\sigma_u$ goes to p, $2p\pi_g$ goes to d, etc. The MO is said to be "promoted."

We have now seen several ways to explain the effects of an orbital. We may focus on energies, and note that bonding and antibonding MOs correlate with low-energy and high-energy united-atom orbitals, respectively. Or, as we saw earlier, we can focus

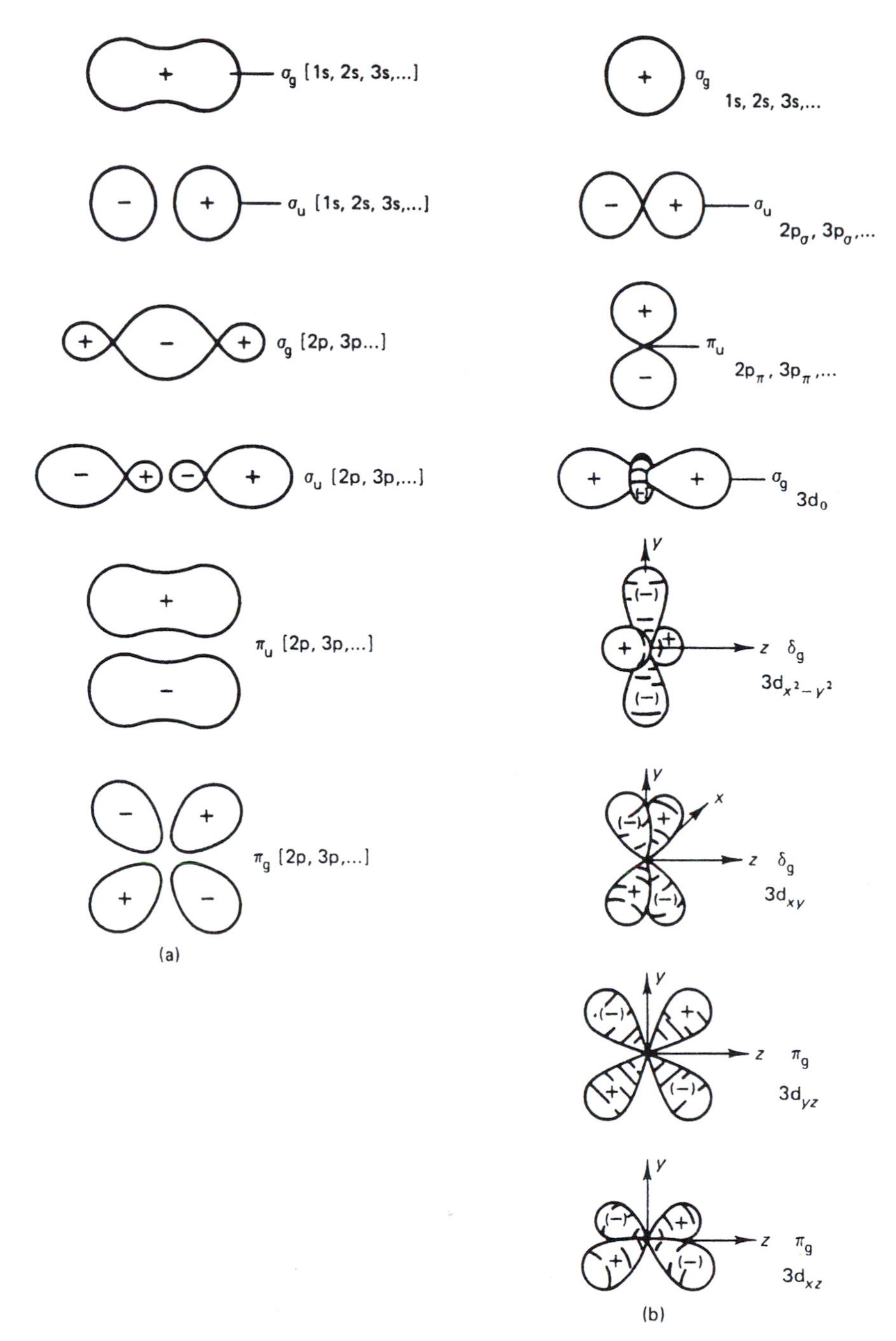

Figure 7-16 ▶ (a) Symmetry orbitals for homonuclear diatomic molecules. (b) United-atom AOs characterized by symmetry with respect to z axis.

on charge distributions and their attractions for nuclei, and note that bonding MOs concentrate charge in the bond region, attracting the nuclei together, whereas antibonding MOs shift charge *outside* the bond, attracting the nuclei apart. Alternatively, we can recognize that bonding MOs result from AOs on each atom coming together *in phase*, resulting in *positive overlap*, while antibonding MOs result from opposite phases coming together to give negative overlap.

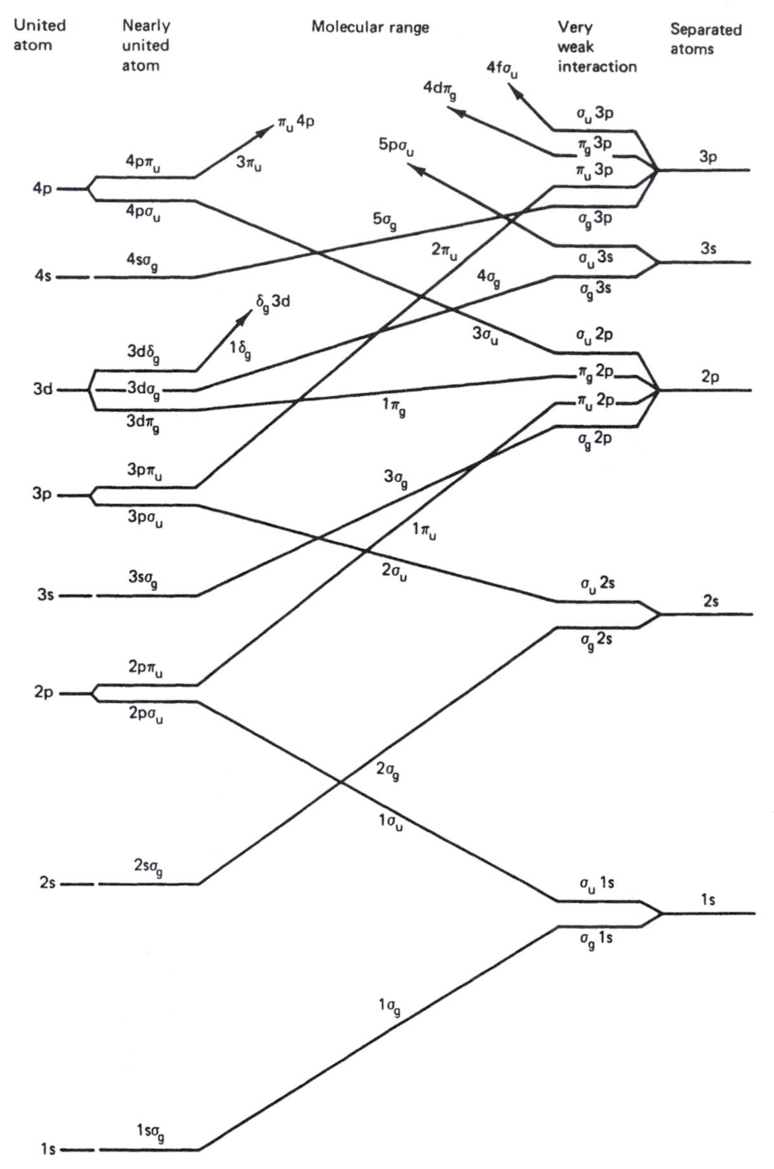

Figure 7-17 ▶ Correlation diagram between separated-atom orbitals and united-atom orbitals for homonuclear diatomic molecules. Energy ordinate and internuclear separation abscissa are only suggestive. No absolute values are implied by the sketch. [Note: For H_2^+ $\sigma_g 2p$ correlates with $3d\sigma_g$, $\sigma_g 3s$ with $3s\sigma_g$. This arises because the separability of the H_2^+ hamiltonian (in the Born–Oppenheimer approximation) leads to an additional quantum number for this molecule. In essence, the H_2^+ wavefunction in elliptical coordinates may be written $\psi = L(\lambda)\,M(\mu)\,e^{im\phi}$. The function L may have nodal surfaces of elliptical shape. M may have nodes of hyperbolic shape. In the correlation diagram for H_2^+, it is necessary that ellipsoidal nodes correlate with spherical nodes in the united atom, while hyperboloid nodes correlate with hyperboloid nodes (which may be planar). Sketching $\sigma_g 2p$ and $\sigma_g 3s$ and comparing them with $3s\sigma_g$ and $3d\sigma_g$ (i.e., $3d_{3z^2-r^2}$) makes clear how this "nodal control" results in what, at first sight, appears to be a violation of the noncrossing rule. For H_2^+, modifications for higher-energy states will also be required. For example, $4s\sigma_g$ will correlate with $\sigma_g 4s$, not $\sigma_g 3p$.]

Three common labeling conventions are used in Fig. 7-17. A level may be labeled with reference to the separated atom AOs to which it correlates. The separated atom AO symbol is placed *to the right* of the MO symmetry symbol (e.g., $\sigma_g 2s$). Note the absence of square brackets, which we used to symbolize the SO $(2s_A + 2s_B)$. The symbol $\sigma_g 2s$ means "the MO of σ_g symmetry that correlates with 2s AOs at $R = \infty$." An alternative label indicates the united-atom orbital with which the MO correlates. Here the AO label is placed *to the left* of the symmetry symbol (e.g., $3p\sigma_u$). The u and g subscripts in the united-atom notation are redundant and are often omitted. However, they are helpful in drawing correlation diagrams. Finally, the MOs may be simply numbered in their energy order within each symmetry type, as mentioned earlier (e.g., $2\sigma_g$).

EXAMPLE 7-3 A symmetry orbital is produced by taking $3d_{yz,\,a} - 3d_{yz,\,b}$, where a and b are points on the z axis. Sketch the situation. What is the sign of the overlap between these AOs? Is this SO bonding or antibonding? What is the symmetry symbol for this SO?

SOLUTION ▶ The sketch would show two four-leafed AOs, each like the $3d_{yz}$ AO shown in Fig. 7-16, except that the phase signs of one would be minus those in the matching lobes of the other. These AOs are positioned side by side, and their phases are such that the inner lobes pointing towards each other are in phase. Hence the overlap is positive and the SO is bonding. Looking at this SO along the z axis, it appears like a p_y AO, so it is π. Inversion causes interchange of lobes of opposite sign, so it has u symmetry. Its symbol, then, is π_u. ◀

Term symbols for electronic states of homonuclear diatomic molecules are much like those in atoms. The main symbol gives information about the component of electronic angular momentum along the z axis. If $M_L = 0, \pm 1, \pm 2$, etc., then the main symbol is Σ, Π, Δ, etc. M_L is the sum of m_l values for the electrons. When identifying m_l, we assume that the π MOs are complex, with $m_l = +1$ and -1 rather than being mixed to give real π_x and π_y MOs. The main symbol is decorated with a superscript at left, giving spin multiplicity, and a g or u subscript at right for overall inversion symmetry. As an example, C_2^+, with configuration $1\sigma_g^2 1\sigma_u^2 2\sigma_g^2 2\sigma_u^2 1\pi_u^3$ has $M_L = \pm 1$ (zero for all σ electrons and either $+1, +1, -1$ or $+1, -1, -1$ for the three π electrons) so its main symbol is Π. The inversion symmetry is u because there is an odd number of electrons in ungerade MOs. There is one unpaired electron, so the multiplicity is 2. The term, symbol is $^2\Pi_u$ ("doublet-pi-you").

Σ terms are given, in addition, a + or − superscript at the right, indicating whether the wavefunction is symmetric or antisymmetric for reflection through a plane containing the nuclei. Such a reflection has the effect of reversing the direction of ϕ: Clockwise motion around the internuclear axis becomes counterclockwise when viewed in a mirror containing that axis. This transforms $\exp(i\phi)$ and $\exp(-i\phi)$ into each other, so that π_+ and π_- turn into (minus) each other. To achieve a Σ MO in the first place requires that m_z values sum to 0. This happens for occupied equivalent π MOs only if π_+ and π_- are equally occupied, which means that both have one electron or both are full. The full case must yield $^1\Sigma_g^+$, i.e., must be symmetric for every operation, because each MO occurs twice in the electron list, and even antisymmetric functions give symmetric results when multiplied by themselves. For the half-filled case there are only two possibilities: $\pi_+(1)\pi_-(2) \pm \pi_-(1)\pi_+(2)$. The "plus" case is symmetric for electron exchange, hence goes with the antisymmetric spin function $\alpha\beta - \beta\alpha$, hence is a singlet state. Reflection causes $\pi_+ \leftrightarrow \pi_-$, but this returns the same function, so this case goes with a $^1\Sigma^+$ term. Similar reasoning shows that the "minus" case goes with a $^3\Sigma^-$ term. Hund's rule predicts the latter term to lie below the former in energy.

EXAMPLE 7-4 What term symbols represent possible excited states of N_2 produced by promoting an electron from the highest occupied MO to the lowest unoccupied MO?

SOLUTION ▶ When an electron is promoted, the possible resulting $1\pi_u$, $1\pi_g$ configurations can be pictured as follows:

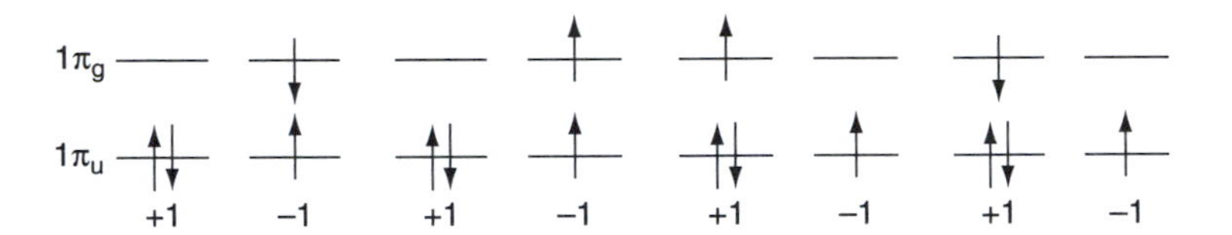

The excited molecule has two half-filled MOs, for which the electron spins can be opposite (singlet) or the same (triplet). The sum of electron orbital momenta will remain zero (giving Σ if the electron is excited from $m_l = -1$ to -1 or from $+1$ to $+1$, but will change to $+2$ or -2 (giving Δ) if changed from -1 to $+1$ or from $+1$ to -1. In all cases, the excited state has an odd number of electrons in the MOs of ungerade symmetry, so all terms will have a u subscript. The term symbols, then, are, respectively $^1\Sigma_u, {}^3\Sigma_u, {}^3\Delta_u, {}^1\Delta_u$. ◀

These conventions for term symbols apply for any linear molecule having inversion symmetry (e.g., HCCH, CO_2). For linear molecules lacking inversion symmetry (CO, HCN) all is the same except that there is no g–u symbol.

The noncrossing rule mentioned above is an important aid in constructing correlation diagrams for many processes. It is called a rule rather than a law because it can only be shown to be highly improbable, not impossible, for two levels of the same symmetry to cross. Thus, imagine that we have a molecule with some variable parameter λ and also with a symmetry operation R which is not lost as λ varies. For example, λ might be the H–O–H angle in water, and R could be reflection through the plane bisecting the H–O–H angle. Suppose that we have a complete set of basis functions and that, at each value of λ, we manage to express exactly all but two of the eigenfunctions for the molecule. This uses up all but two dimensions of our function space, leaving us, at each value of λ, with two eigenfunctions to determine and two functions in terms of which to express them. (These functions change with λ, but the above argument has nevertheless served to reduce our problem to two dimensions.) Now let the two functions remaining from our original basis be mixed to become orthonormal and also individually either symmetric or antisymmetric for R. We label these symmetrized basis functions χ_1 and χ_2. Because we began with a complete basis, it must be possible to express the as yet undetermined wavefunctions ψ_1 and ψ_2 exactly as linear combinations of χ_1 and χ_2 at each value of λ. Furthermore, if ψ_1 and ψ_2 are, say, both antisymmetric for R, it is necessary that χ_1 and χ_2 also both be antisymmetric. If ψ_1 and ψ_2 have opposite symmetries, however, χ_1 and χ_2 also have opposite symmetries. (In the latter case, χ_1 and χ_2 can only mix to produce unsymmetric functions, and so we know that χ_1 and χ_2

are already identical with ψ_1 and ψ_2.) To determine the mixing coefficients and state energies for ψ_1 and ψ_2, we solve the 2×2 secular equation over the basis χ_1, χ_2:

$$\begin{vmatrix} H_{11} - E & H_{12} \\ H_{12} & H_{22} - E \end{vmatrix} = 0 \tag{7-107}$$

The roots are

$$E_{\pm} = \frac{1}{2}\left\{H_{11} + H_{22} \pm \left[4H_{12}^2 + (H_{11} - H_{22})^2\right]^{1/2}\right\} \tag{7-108}$$

The crossing of energy levels for ψ_1 and ψ_2 requires that, at some value of λ, E_+ equals E_-. From Eq. (7-108), we see that this requires that the term in square brackets vanish, which requires that H_{12} and $H_{11} - H_{22}$ vanish. Now, if χ_1 and χ_2 (and hence ψ_1 and ψ_2) have opposite symmetries for R, H_{12} vanishes for all values of λ, and the curves will cross whenever H_{11} equals H_{22}. But if χ_1 and χ_2 (and hence ψ_1 and ψ_2) have the *same* symmetry for R, H_{12} is not generally zero. In this situation, the curve crossing requires that both H_{12} *and* $H_{11} - H_{22}$ *happen to pass through zero* at the same value of λ. This simultaneous occurrence of two functions passing through zero is so unlikely that it is safe to assume it will not happen.

If the molecule possesses several elements of symmetry, H_{12} will vanish at all λ if ψ_1 and ψ_2 disagree in symmetry for *any one* of them, so the noncrossing rule applies only to states having wavefunctions of identical symmetry for all symmetry operations of the molecule.

A similar treatment for orbitals and orbital energy levels is possible, and the noncrossing rule applies for orbital energies as well as for state energies.

7-7 Basis Set Choice and the Variational Wavefunction

One of the places where human decision can effect the outcome of a variational calculation is in the choice of basis. Some insight into the ways this choice effects the ultimate results is necessary if one is to make a wise choice of basis, or recognize which calculated results are "physically real" and which are artifacts of basis choice.

One question we can ask is this: Is a minimal basis set equally appropriate for calculating an MO wavefunction for, say, B_2 as F_2? In each case we use 10 AOs and 2 spin functions producing a total of 20 spin MOs. With B_2, however, we have 10 electrons to go into these spin MOs, and in F_2 we have 18 electrons. In all but the crudest MO calculations, the total energy is minimized in a manner that depends on the natures of only the occupied MOs. In effect, then, the calculation for B_2 produces the 10 "best" spin MOs from a basis set of 20 spin-AOs, whereas that for F_2 produces the 18 best MOs from a different basis set of 20 spin-AOs. In a sense, then, the basis for F_2 is less flexible than that for B_2. Of course, the use of separated atom orbitals is a conscious effort to choose that basis that best spans the same function space as the best MOs. To the extent that this strategy is successful, the above problem is obviated (i.e., if both sets are perfect, additional flexibility is useless). The strategy is not completely successful, however, and comparison of results of minimal basis set calculations down a series of molecules such as B_2, C_2, N_2, O_2, and F_2 may be partially hampered by this ill-defined inequivalence in basis set adequacy. In contrast, comparison of calculated results in a series of molecules such as C_nH_{2n+2} is much less likely to suffer from this particular problem because the minimal basis set grows with increasing n in a way to keep pace with the number of electrons.

Let us now briefly consider how basis sets might vary in adequacy for different states of a given molecule. We will compare the wavefunction for the ground state of a molecule with the wavefunction for a Rydberg state. Rydberg states are so named because their spectral lines progress toward the ionization limit in a manner similar to the spectral pattern for hydrogenlike ions (called a *Rydberg series*).[7] Hence, the Rydberg states of molecules are in some way like excited states of the hydrogen atom. This can be understood by visualizing an excited state for, say, N_2 wherein one electron is, on the average, very far away from the rest of the molecule, which is now an N_2^+ "core." As the excited electron moves to orbitals farther and farther out, the N_2^+ core becomes effectively almost like a point positive charge. As a result, the coupling between the angular momentum of this orbital and the internuclear axis grows progressively weaker, so that the motion of the Rydberg electron becomes more and more independent of orientation of the core. It is not surprising that a hydrogen-like AO centered in the bond becomes more and more appropriate as a basis for describing this orbital. In contrast, such a "single-center" basis normally requires many terms to accurately describe MOs in ground or non-Rydberg excited states. Thus, for a Rydberg state of N_2, one would do well to choose a basis set of AOs located on the nuclei to describe the MOs of the N_2^+ core, and to use an AO (or several AOs) centered between the nuclei to describe the orbital for the Rydberg electron.

Thus far we have kept the discussion within the framework of homonuclear diatomic molecules. When we come to heteronuclear diatomics, for example, CO, we lose inversion symmetry and we can no longer symmetry balance our basis. This means that a given basis may be more inadequate for representing the wavefunction on one end of the molecule than on the other. As a result, the electronic charge will be shifted toward the end where the basis set is best able to minimize the energy. This charge shift is an artifact of basis set imbalance, but, since we have no way to evaluate this imbalance, it is difficult to tell how much it affects our results. Mulliken has published some calculations on the HF molecule that illustrate this problem in a striking way. Table 7-3 is a list of total energies and dipole moments calculated for HF using a variety of basis sets.

The first column of data arises from a minimal basis set of STOs ($1s_H$, $1s_F$, $2s_F$, $2p_{\sigma_F}$, $2p_{\pi x_F}$, $2p_{\pi y_F}$) with orbital exponents evaluated from Slater's rules for atoms. The second column results if the orbital exponents are allowed to vary independently to minimize the molecular energy. The basis set for the third column is obtained by augmenting the previous basis with additional STOs centered on the H nucleus ($2s_H$, $2p_{\sigma_H}$, $2p_{\pi x_H}$, $2p_{\pi y_H}$). Finally, the fourth column results from use of a basis set that has been augmented (over the minimal basis) at *both* nuclei in a way thought to be appropriately balanced. As the basis set grows increasingly flexible, the average energy becomes lower, but the expectation value for the dipole moment does not converge uniformly toward the observed value. In particular, by augmenting the basis on hydrogen only, we create a very unbalanced basis, which causes charge to shift too much toward the hydrogen end of the molecule.

[7] See A. B. F. Duncan [3].

TABLE 7-3 ► Energies and Dipole Moments for Hydrogen Fluoride Calculated by the Variation Method Using Different Basis Sets[a]

	Min STO Slater ζ	Min STO best ζ	Min STO F; Aug. STO H (very unbalanced)	Aug. STO F and H (balanced)	Exp
E (a.u.)	−99.4785	−99.5361	−99.6576	−100.0580	−100.527
$\mu(H^+F^-)$	$0.85D$	$1.44D$	$0.92D$	$1.98D$	$1.82D$[b]

[a] See Mulliken [4].
[b] Data from Weiss [5].

These problems with basis set adequacy are difficult to overcome completely. Fortunately, with a certain amount of experience, insight, and caution, it is nevertheless possible to carry out variational calculations and interpret their results to obtain reliable and useful information.

7-8 Beyond the Orbital Approximation

Most of our discussion of the variation method has been restricted to calculations within the orbital approximation. To avoid leaving an inaccurate impression of the capabilities of the variation method, we shall briefly describe some calculations on some small (two-electron) systems where the method can be employed to its fullest capabilities. These calculations are listed in Table 7-4.

The calculation on He by Kinoshita expresses the spatial part of the wavefunction as

$$\psi(ks, kt, ku) = e^{-ks/2} \sum_{\substack{l,m,n=0 \\ n,\text{even}}}^{\infty} c_{l,m,n} (ks)^{l-m} (ku)^{m-n} (kt)^{n} \tag{7-109}$$

where

$$s = r_1 + r_2, \quad u = r_{12}, \quad t = -r_1 + r_2 \tag{7-110}$$

and k and $c_{l,m,n}$ are variable parameters. The exponential term causes the wavefunction to vanish as either electron goes to infinite r, and the terms in the sum build up a polynomial in one- and two-electron coordinates, reminiscent of the form of eigenfunctions for the harmonic oscillator and the hydrogenlike ion. Kinoshita carried out his calculation to as many as 39 terms, obtaining an energy that he estimated to differ from the exact result by no more than 1.2×10^{-6} a.u. A subsequent calculation by Pekeris, using a related approach, required solving a secular determinant of order 1078 and yielded an energy estimated to be accurate to 1.0×10^{-9} a.u. Applying corrections for coupling between electronic and nuclear motions, and also for relativistic effects, Pekeris arrived at a theoretical value for the ionization energy of He of 198310.687 cm^{-1} compared to the experimental value of $198310.8_2 \pm 0.15\,\text{cm}^{-1}$.

TABLE 7-4 ▶ Results of Some Very Accurate Variational Calculations on Two-Electron Systems

System	Minimized energy[a] (a.u.)	Estimated maximum energy error $\bar{E} - E_{\text{exact}}$ (a.u.)
He (ground singlet)[b]	−2.9037225	0.0000012
He (ground singlet)[c]	−2.903724375	0.000000001
He (lowest triplet)[c]	−2.17522937822	0.00000000001
H_2 (ground singlet)[d]	−1.17447498302[e]	0.000000001

[a] Uncorrected for nuclear motion and relativistic effects.
[b] From Kinoshita [6].
[c] From Pekeris [7].
[d] From Kolos and Wolniewicz [8].
[e] At $R = 1.401078$ a.u.

In the 35 years since Pekeris work was reported, even more extensive calculations have been reported. For example, Drake et al. [9] have calculated an upper bound for the ground state of helium having 22 significant figures using a wavefunction having 2358 terms.

Extremely accurate variational calculations have been carried out on H_2 by Kolos and Wolniewicz. They used elliptic coordinates and an r_{12} coordinate and expressed their wavefunction as an expansion in powers of these coordinates, analogous in spirit to the Kinoshita wavefunction described above. Their most accurate wavefunctions contain 100 terms and are calculated for a range of R values. After including corrections for relativistic effects and nuclear motion, Kolos and Wolniewicz arrived at a theoretical value for the *dissociation energy* in H_2 of $36117.4\,\text{cm}^{-1}$ compared to what was then the best experimental value $36113.6 \pm 0.5\,\text{cm}^{-1}$. Subsequent redetermination of the experimental value gave $36117.3 \pm 1.0\,\text{cm}^{-1}$.[8]

The *dissociation energy*, D_0, is the energy required to separate a molecule into its constituent atoms, starting with a molecule in its lowest vibrational state. The *binding energy*, D_e, is the energy for the corresponding process if we omit the vibrational energy of the molecule (see Fig. 7-18). These quantities are often much more sensitive measures of the accuracies of calculations than are total energies. The reason for this is easily understood when we recognize that the binding energy is a fairly small difference between two large numbers—the total energy of the molecule and the total energy of the separated atoms. Unless our errors in these two large energies are equal, the residual error is magnified (in terms of percentage) when we take the difference. Thus, the best total energy for H_2 in a certain orbital approximation is −1.133629 a.u., which is 96.7% of the total energy. However, the corresponding binding energy is −0.133629 a.u., which is 76.6% of the correct value. The need to calculate accurate binding energies is sometimes referred to as the need to achieve "chemical accuracy."

The variational calculations cited above are among the most accurate performed, and they give an indication of the capabilities of the method. Properties other than

[8] See Herzberg [10].

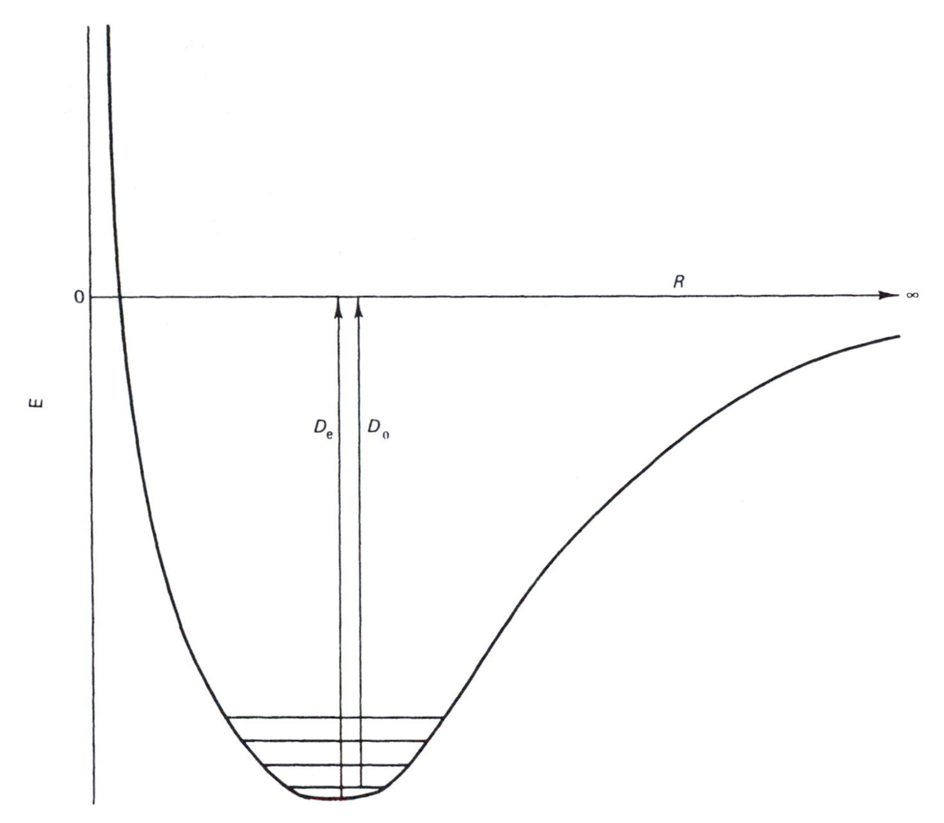

Figure 7-18 ▶ Schematic drawing showing the distinction between D_0, the dissociation energy from the lowest vibrational level, and D_e, the binding energy, which does not take vibrational energy into account. The zero of energy is the energy of the separated atoms.

energy predicted from such wavefunctions are also very accurate. For example, Pekeris' best wavefunction for the first triplet state of helium gives an electron density at the nucleus of 33.18416 electrons per cubic bohr compared with the experimental value 33.18388 ± 0.00023 deduced from hyperfine splitting. For most systems of chemical interest, calculations of this sort become much too impractical to be considered. For this reason the orbital approximation, with all its limitations, is used in most quantum-chemical calculations on systems having more than two electrons.

7-8.A Problems

7-1. Given the following two functions, $f(x)$ and $g(x)$, for the range $0 \geq x \geq L$:

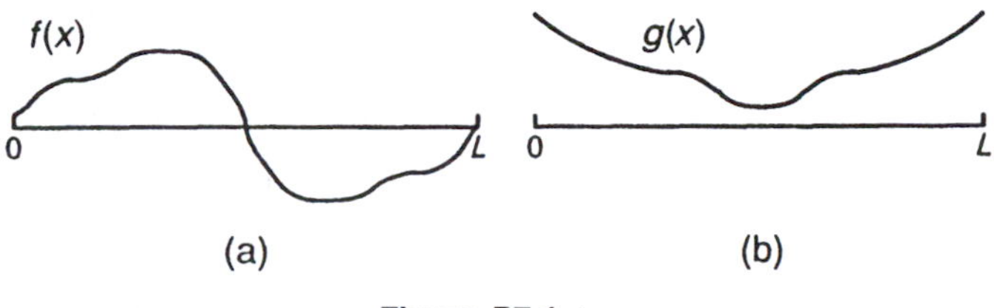

Figure P7-1 ▶

Can these functions be expressed accurately as linear combinations of particle-in-a-box eigenfunctions (box walls at $x = 0, L$)? Indicate your reasoning. If yes, what is the expression for the first two coefficients in the expansion? Can you evaluate any of these by inspection?

7-2. Consider a particle in a box with a biased potential that is higher at $x = L$ than at $x = 0$. An approximate solution for the ground state could be $\phi = \sqrt{0.9}\psi_1 + \sqrt{0.1}\psi_2$, where ψ_1 and ψ_2 are the first and second eigenfunctions for the unbiased box. (a) Make a rough sketch of ϕ, showing how it skews the particle distribution. (b) What is the *average* kinetic energy for ϕ, in terms of h, m, and L?

7-3. The normalized function $\phi = (2/45\pi)^{1/2} r^2 \exp(-r)$ can be expanded in terms of hydrogen atom eigenfunctions:

$$\phi = c_1\psi_{1s} + c_2\psi_{2s} + c_3\psi_{2p_0} + \cdots$$

where $\psi_{1s} = (1/\sqrt{\pi})\exp(-r)$ and $\psi_{2p_0} = (1/\sqrt{32\pi}) r \exp(-r/2)\cos\theta$. Evaluate c_1 and c_3.

7-4. Given the approximate wavefunction for the lowest state of a particle in a one-dimensional box (Fig. P7-4):

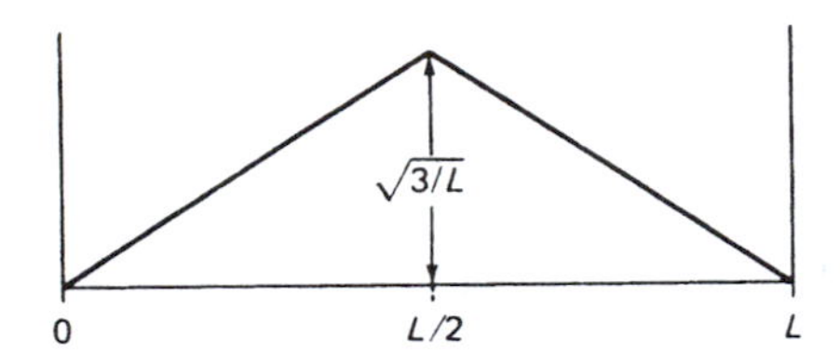

Figure P7-4 ▶

$$\phi = \sqrt{3/L}(2x/L), \qquad 0 \le x \le L/2$$
$$\phi = \sqrt{3/L}[2(L-x)/L], \qquad L/2 \le x \le L$$
$$\phi = 0, \qquad x < 0,\ x > L$$

a) Resolve ϕ into the box eigenfunctions. That is, evaluate c_n in the expression

$$\phi = \sum_{n=1}^{\infty} c_n \psi_n,$$

where

$$\psi_n = \sqrt{2/L}\sin(n\pi x/L), \quad 0 \le x \le L, \quad \psi_n = 0, \quad x < 0, x > L$$

b) Using the coefficients from part (a) compare the value of ϕ at $x = L/2$ with the values one obtains from the

$$\phi_{\text{approx}} = \sum_{n=1}^{m} c_n \psi_n, \quad \text{with } m = 1, 3, 5, 7, \text{ and } 9$$

c) Use the coefficients from part (a) to obtain an expression for $\bar{E}$ appropriate for ϕ. Estimate the value of the infinite series and thereby estimate $\bar{E}$. Compare this value to E_{exact}.

7-5. Let $\phi = \exp(-\alpha r^2)$ be a trial function (not normalized) for the ground state of the hydrogen atom. Use the variation method to determine the minimum energy attainable from this form by variation of α. Find the average value of r and the most probable value of r for this wavefunction. Compare these r values and the average energy with the exact values.

7-6. Let $\phi(\alpha) = (\alpha^5/3\pi)^{1/2} r \exp(-\alpha r)$ be a trial function for the ground state of the hydrogen atom:

a) Verify that the variation method gives $\alpha = \frac{3}{2}$, $\bar{E} = -\frac{3}{8}$ a.u.
b) Verify that $\phi\left(\frac{3}{2}\right)$ has an overlap of 0.9775 with the 1s function.
c) Find the value of α that maximizes the overlap of ϕ with the 1s function and determine the average energy of this new ϕ.

7-7. A *normalized* approximate wavefunction ϕ for the hydrogen atom is projected into its components and found to be $\phi = (1/\sqrt{2})\psi_{1s} + (1/\sqrt{4})\psi_{2s} + (1/\sqrt{8})\psi_{3s} + c_{4s}\psi_{4s}$. There are no higher terms.

a) What is the value of c_{4s}?
b) What is the value of the average energy?

7-8. A normalized, *spherically symmetric* variationally optimized function for the hydrogen atom is analyzed by projecting out coefficients c_{1s}, c_{2s}, c_{2p_0}, etc. It is found that c_{1s} is equal to $\sqrt{0.80}$ and $c_{2s} = \sqrt{0.15}$.

a) What is the maximum possible value for c_{2p_0}? Explain your reasoning.
b) What is the minimum possible value for $\bar{E}$ that could correspond to this function, based on the above data? Explain your reasoning.

7-9. A normalized trial wavefunction of the form (in a.u.)

$$\phi = \left[(2\zeta)^7/(4\pi 6!)\right]^{1/2} r^2 \exp(-\zeta r)$$

is variationally optimized to give the lowest possible energy for the hydrogen atom. The results part way along this process are $\int \phi \hat{H} \phi \, dv = \zeta^2/10 - \zeta/3$.

a) Complete the variational process to obtain the optimum ζ *and* the minimized average energy.
b) Calculate the value of c_1 in the expression $\phi_{\text{opt}} = c_1\psi_{1s} + c_2\psi_{2s} + \cdots$.
c) Produce a new, *normalized* function χ that is orthogonal to ψ_{1s} but is otherwise as similar as possible to ϕ_{opt}. (Use the unexpanded symbolic terms ψ_{1s} and ϕ in your answer.)
d) Suppose you are told that the average energy associated with your new function χ is -0.133 a.u. Do you find this reasonable? Explain your answer.

7-10. $\phi = (\alpha^5/3\pi)^{1/2} r \exp(-\alpha r)$ is a normalized trial function for the hydrogen atom. The energy is minimized when $\alpha = 3/2$. Calculate the average value of the *potential* energy predicted by this function if $\alpha = 3/2$.

7-11. Prove that optimized trial function (7-20) *must* contain contributions from continuum wavefunctions.

7-12. Compare the orbital exponent for a 1s AO in He as found by the variation method [Eq. (7-35)] with that given by Slater's rules (Chapter 5).

7-13. A different trial function for calculating the polarizability of the hydrogen atom in a uniform electric field of strength F is

$$\psi_{\text{trial}} = \psi_{1s}(c_1 + c_2 z)$$

This is somewhat similar to the example in the text, since $z\psi_{1s}$ gives a p-like function, but not exactly the $2p_z$ eigenfunction.

a) Use this form to find an expression for the minimum $\bar{E}$ as a function of F. What value of $\bar{E}$ does this give for $F = 0.1$ a.u.? Can you suggest why this trial function is superior to the one used in the text?
b) The polarizability is defined to be α in the expression

$$E = -\frac{1}{2} - \frac{1}{2}\alpha F^2$$

What value of α do you obtain? [Exact $\alpha = 4.5$ a.u.] 1 a.u. of field strength is equal to e/a_0^2. Deduce the value of 1 a.u. of polarizability.

7-14. Which hydrogen atom state should be more polarizable, the 1s or 2s? [Consider the factors that determine the extent of mixing between basis functions.] Explain your reasoning.

7-15. ϕ_a and ϕ_b are chosen to be a normalized set of basis functions for an LCAO wavefunction for a one-electron homonuclear diatomic system. It is found that the values for the integrals involving these functions are

$$\int \phi_a^* \hat{H} \phi_a \, dv = -2 \text{ a.u.}, \quad \int \phi_b^* \hat{H} \phi_b \, dv = -2 \text{ a.u.},$$
$$\int \phi_a^* \hat{H} \phi_b \, dv = -1 \text{ a.u.}, \quad \int \phi_a^* \phi_b \, dv = \frac{1}{4}.$$

Find an upper bound for the exact lowest electronic energy for this system. Find the corresponding LCAO *normalized* approximate wavefunction.

7-16. ϕ_a and ϕ_b are chosen as the normalized basis functions for an LCAO wavefunction for a one-electron, heteronuclear, diatomic molecule. It is found that the values for some integrals involving these functions are

$$\int \phi_a \hat{H} \phi_a \, dv = -2 \text{ a.u.}, \quad \int \phi_b \hat{H} \phi_b \, dv = -1 \text{ a.u.},$$
$$\int \phi_a \hat{H} \phi_b \, dv = -\frac{1}{2} \text{ a.u.}, \quad \int \phi_a \phi_b \, dv = \frac{1}{3}.$$

where $\hat{H}$ is the molecular hamiltonian. Set up the secular determinantal equation and find the lowest electronic energy that can be computed from an LCAO wavefunction $c_a\phi_a + c_b\phi_b$. Find c_a and c_b such that $\bar{E}$ is minimized and the wavefunction is normalized.

7-17. A possible basis function for representing the $1\sigma_g$ wavefunction of H_2^+ is a 1s-like AO $(\zeta^3/\pi)^{1/2}\exp(-\zeta r)$ located at the bond center. Assuming an internuclear separation of 2 a.u., find the ζ value that minimizes $\bar{E}$. Is this basis function adequate to predict a bound H_2^+ molecule? [Use Appendix 3 to help you develop your formulas.]

7-18. Without referring to the text, and by inspection, what is the united atom limit for the $1\sigma_u$ molecular orbital's energy (in a.u.) for H_2^+?

7-19. Show that, at $R=\infty$, the ψ_+ and ψ_- wavefunctions for H_2^+ are capable of describing a state wherein the electron is in a 1s orbital on atom A.

7-20. Evaluate Eqs. (7-89) and (7-90) at $R=0$ to show that $H_{AA}=H_{AB}$ at this point.

7-21. Examining Eq. (7-86), and letting $H_{AB}=kH_{AA}$ what relationship between k and S_{AB} is necessary if the σ_g MO is to be lower in energy than the σ_u MO? [Assume that H_{AA} is negative, and that k and S_{AB} are positive.]

7-22. Consider the one-electron molecule–ion HeH^{2+}:

a) Write down the hamiltonian operator (nonrelativistic, Born–Oppenheimer approximation) for the electronic energy in atomic units for this system.
b) Calculate the electronic energies for the lowest energy state of this system in the separated atom and united atom limits.

7-23. For a *homo*nuclear diatomic molecule aligned as shown in Fig. P7-23, characterize each of the following MOs as σ, π, δ, and g or u, and bonding or antibonding.

a) $2p_{y_a}+2p_{y_b}$
b) $2p_{z_a}+2p_{z_b}$
c) $3d_{z_a^2}+3d_{z_b^2}$
d) $3d_{xy_a}+3d_{xy_b}$
e) $3d_{xz_a}-3d_{xz_b}$

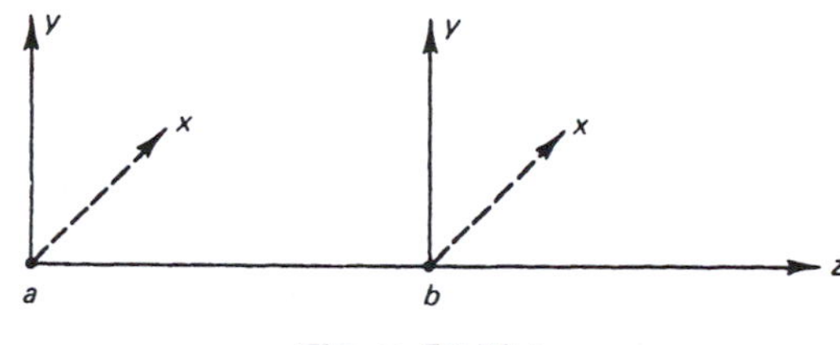

Figure P7-23 ▶

7-24. Characterize each of the following *atomic* orbitals with the symbols σ, π, δ, and also g or u. Let the z axis be the reference axis for angular momentum.

1s	$2p_z$	$3p_y$	$3d_{xy}$
2s	$2p_x$	$3d_{z^2}$	$3d_{xz}$

7-25. Indicate whether you expect each of the following homonuclear diatomic MOs to be bonding or antibonding. Sketch the MO in each case: (a) σ_u (b) π_u (c) δ_g

7-26. A homonuclear diatomic molecule MO of π_u symmetry is to be expressed as a linear combination of AOs centered on the nuclei, which lie on the z axis. Which of the AOs in the following list can contribute to the MO?

$$1s \quad 2s \quad 2p_x \quad 2p_y \quad 2p_z \quad 3s \quad 3p_x$$
$$3p_y \quad 3p_z \quad 3d_{xy} \quad 3d_{xz} \quad 3d_{yz} \quad 3d_{z^2} \quad 3d_{x^2-y^2}$$

7-27. Use sketches and symmetry arguments to decide which of the following integrals vanish for diatomic molecules (the x, y, and z axes are shown in Fig. P7-23):

a) $\int 2p_{za} 1s_b \, dv$
b) $\int 2p_{ya} 1s_b \, dv$
c) $\int 2p_{za} 2p_{yb} \, dv$
d) $\int 2p_{ya} 3d_{yzb} \, dv$
e) $\int 2p_{za} 3d_{yzb} \, dv$
f) $\int 1s_a \hat{H} 2p_{xa} \, dv$
g) $\int 1s_a \hat{H} 2p_{za} \, dv$

7-28. Show that, for reflection through a plane containing the nuclei, if the MO $\delta_{x^2-y^2}$ is symmetric, then the MO δ_{xy} is antisymmetric. Show that the same is true for π MOs constructed from d_{xz} and d_{yz} AOs. (The internuclear axis is assumed to lie along the z coordinate.)

7-29. Assuming the internuclear axis to lie along the z coordinate, what are the possible M_L quantum numbers for an MO constructed from $3d_{z_a^2} - 3_{z_b^2}$?

7-30. In a homonuclear diatomic correlation diagram, what MO symmetry symbols (σ, π, δ, g, u) could correlate with each of the united atom AOs listed below? Assume z to be the "old" internuclear axis. Indicate for each case whether this united atom orbital is the terminus for a *bonding* or an *antibonding* MO. (a) $2p_z$ (b) $2p_x$ (c) $3d_{xz}$ (d) $3d_{xy}$

7-31. A homonuclear diatomic system has the ground-state MO configuration $1\sigma_g^2 1\sigma_u^2 2\sigma_g^2 2\sigma_u^2 3\sigma_g^2 1\pi_u^4 1\pi_g^2$:

a) What is the *net* number of bonding electrons?
b) What spin multiplicity would you expect for the ground state?
c) What would you expect the effect to be on the dissociation energy of this molecule of ionization (1) from the $1\pi_g$ MO? (2) from the $3\sigma_g$ MO?
d) Upon ionization (one-electron) from the $1\pi_g$ level, what would be the spin multiplicity of the resulting ion?
e) To what type of united atom AO does the π_u MO correlate?

7-32. a) *Without referring to the text*, write out the ground state configuration for O_2^+ using MO symmetry symbols ($1\sigma_g^2$ etc.)
b) What is the net number of bonding electrons?
c) How does the dissociation energy for this ion compare to that for O_2?
d) What is the ground state term symbol for this ion?
e) Which occupied MOs may contain contributions from $2p_z$ AOs, assuming z to be the internuclear axis?

7-33. The reduced symmetry of heteronuclear (compared to homonuclear) diatomic molecules results in their having a different correlation diagram. Set up a correlation diagram for heteronuclear diatomics. Be sure to indicate that the energy levels of each type of AO are not identical for the separated atoms. Comparing correlation diagrams for homonuclear and heteronuclear molecules, does it seem reasonable that He_2 is unstable, whereas the isoelectronic LiH and $LiHe^+$ are stable molecules?

7-34. Following are some Slater orbital coefficients for some MO's of F_2 calculated by Ransil [11] (the $2p_\sigma$ STOs are defined according to z axes pointing from each atom toward the other):

$1\sigma_g$	$c_{1s,A} = c_{1s,B} = 0.70483$	$2\sigma_g$	$c_{1s,A} = c_{1s,B} = 0.17327$
	$c_{2s,A} = c_{2s,B} = 0.00912$		$c_{2s,A} = c_{2s,B} = -0.67160$
	$c_{2p_\sigma,A} = c_{2p_\sigma,B} = -0.00022$		$c_{2p_\sigma,A} = c_{2p_\sigma,B} = -0.08540$

We see that the $1\sigma_g$ MO is almost entirely made from 1s AOs on A and B. However, the $2\sigma_g$ MO contains what appears to be an anomalously large amount of 1s AO. This turns out to be an artifact of the fact that Slater-type 2s orbitals are not orthogonal to 1s AOs on the same center. For F_2, the STO 1s, 2s overlap is 0.2377. Use this fact to construct a new orbital, 2s′, that is orthogonal to 1s. Express the $2\sigma_g$ MO of Ransil in terms of the basis functions 1s, 2s′, and $2p_\sigma$ on centers A and B. You should find the 1s coefficients much reduced.

7-35. R_e for H_2^+ equals 2.00 a.u. At this distance, $E_{elec} = -1.1026$ a.u. What is the value of D_e for H_2^+?

Multiple Choice Questions

(Try to answer these without referring to the text.)

1. A homonuclear diatomic MO is given by $\phi = 2p_{z,a} + 2p_{z,b}$, where the z axis is the same as the internuclear axis. Which one of the following statements about ϕ is correct?

a) ϕ is an antibonding MO, symbolized σ_u.
b) ϕ is a bonding MO, symbolized π_u.
c) ϕ is an antibonding MO, symbolized π_g.
d) ϕ is a bonding MO, symbolized σ_g.
e) ϕ is an antibonding MO, symbolized π_u.

2. For the three species N_2, N_2^+, N_2^-, which one of the following orders for the bond energy (i.e., bond strength) is most reasonable?

a) $N_2 > N_2^+ > N_2^-$
b) $N_2^+ > N_2 > N_2^-$
c) $N_2^- > N_2 > N_2^+$
d) $N_2^- > N_2^+ > N_2$
e) Only N_2 forms a bond; N_2^+ and N_2^- do not.

3. According to the LCAO-MO model, which one of the following second period diatomic molecules has a double bond in the ground electronic state?

a) Li_2
b) Be_2
c) B_2
d) C_2
e) N_2

4. Which one of the following statements concerning H_2^+ is *false*?

a) The nondegenerate LCAO-MOs (without spin) must be either symmetric or antisymmetric for inversion.
b) The lowest energy MO (without spin) of the molecule is antisymmetric for inversion.
c) The MOs transform into the AOs of the helium ion as the two nuclei are fused together.
d) The ground state has a multiplicity of two.
e) The Born-Oppenheimer approximation permits finding the solution for the purely electronic wave function at each value of the internuclear distance.

5. Which one of the following statements is *false* for bonding MOs formed from linear combinations of AOs on atoms a and b?

a) Only AOs that have nonzero overlap can form bonding MOs.
b) Only AOs that have similar energies can form strongly bonding MOs.
c) Bonding MOs cannot have a nodal plane perpendicular to the internuclear axis and midway between a and b.
d) A p AO on b can combine with a p AO on a to form σ, π, or δ MOs.
e) A maximum of three bonding MOs can be formed from 2p AOs on a and b.

References

[1] R. S. Mulliken, *Rev. Mod. Phys.* **2**, 60, 506 (1930); **3**, 90 (1931); **4**, 1 (1932).

[2] J. O. Hirschfelder, C. F. Curtiss, and R. B. Bird, *Intermolecular Forces*. Wiley, New York, 1964.

[3] A. B. F. Duncan, *Rydberg Series in Atoms and Molecules*. Academic Press, New York, 1971.

[4] R. S. Mulliken, *J. Chem Phys.* **36**, 3428 (1962).

[5] R. Weiss, *Phys. Rev.* **131**, 659 (1963).

[6] T. Kinoshita, *Phys. Rev.* **150**, 1490 (1957).

[7] C. L. Pekeris, *Phys. Rev.* **115**, 1216 (1959).

[8] W. Kolos and L. Wolniewicz, *J. Chem. Phys.* **49**, 404 (1968).

[9] G. W. F. Drake, M. M. Cassar, and R. A. Nistor, *Phys. Rev. A* **65**, 54501 (2002).

[10] G. Herzberg, *J. Mol. Spectrosc.* **33**, 147 (1970).

[11] B. J. Ransil, *Rev. Mod. Phys.* **32**, 245 (1960).

Chapter 8

The Simple Hückel Method and Applications

8-1 The Importance of Symmetry

Our discussions of the particle in a box, the harmonic oscillator, the hydrogen atom, and homonuclear diatomic molecules have all included emphasis on the role that symmetry plays in determining the qualitative nature of the eigenfunctions. When we encounter larger systems, detailed and accurate solutions become much more difficult to perform and interpret, but symmetry continues to exert strong control over the solutions.

In this chapter, we will describe a rather simple quantum chemical method that was formulated in the early 1930s by E. Hückel. One of the strengths of this method is that, by virtue of its crudeness and simplicity, the effects of symmetry and topology on molecular characteristics are easily seen. Also, the simplicity of the model makes it an excellent pedagogical tool for illustrating many quantum chemical concepts, such as bond order, electron densities, and orbital energies. Finally, the method and some of its variants continue to be useful for certain research applications. Indeed, it is difficult to argue against the proposition that every graduate student of organic and inorganic chemistry should be acquainted with the Hückel molecular orbital (HMO) method.

8-2 The Assumption of σ–π Separability

The simple Hückel method was devised to treat electrons in unsaturated molecules like ethylene and benzene. By 1930 it was recognized that unsaturated hydrocarbons are chemically more reactive than are alkanes, and that their spectroscopic and thermodynamic properties are different too. The available evidence suggested the existence of loosely held electrons in unsaturated molecules.

We have already seen that, when atoms combine to form a linear molecule, we can distinguish between MOs of type $\sigma, \pi, \delta, \ldots$ depending on whether the MOs are associated with an m quantum number of $0, \pm 1, \pm 2, \ldots$ Thus, in acetylene (C_2H_2), the minimal basis set of AOs on carbon and hydrogen lead to σ and π MOs. Let us imagine that our acetylene molecule is aligned along the z Cartesian axis. Then the $p_x\,\pi$-type AOs on the carbons are antisymmetric for reflection through a plane containing the molecular axis and the y axis. Similarly, the $p_y\,\pi$-type AOs are antisymmetric for reflection through a plane containing the molecular axis and the x axis. The p_z AOs, which are σ-type functions, are symmetric for reflection through any plane containing the molecular axis. It has become standard practice to carry over the $\sigma-\pi$ terminology to planar (but nonlinear) molecules, where m is no longer a "good" quantum number. In this expanded usage, *a π orbital is one that is antisymmetric for reflection through the plane of the molecule*, a σ orbital being symmetric for that reflection.

Hückel found that, by treating only the π electrons explicitly, it is possible to reproduce theoretically many of the observed properties of unsaturated molecules such as the uniform C–C bond lengths of benzene, the high-energy barrier to internal rotation about double bonds, and the unusual chemical stability of benzene. Subsequent work by a large number of investigators has revealed many other useful correlations between experiment and this simple HMO method for π electrons.

Treating only the π electrons explicitly and ignoring the σ electrons is clearly an approximation, yet it appears to work surprisingly well. Physically, Hückel's approximation may be viewed as one that has the π electrons moving in a potential field due to the nuclei and a "σ core," which is assumed to be frozen as the π electrons move about. Mathematically, the $\sigma-\pi$ separability approximation is

$$E_{\text{tot}} = E_\sigma + E_\pi \tag{8-1}$$

where E_{tot} is taken to be the electronic energy E_{el} plus the internuclear repulsion energy V_{nn}.

Let us consider the implications of Eq. (8-1). We have already seen (Chapter 5), that a sum of energies is consistent with a sum of hamiltonians and a product-type wavefunction. This means that, if Eq. (8-1) is true, the wavefunction of our planar molecule should be of the form (see Problem 8-1)

$$\psi(1,\ldots,n) = \psi_\pi(1,\ldots,k)\psi_\sigma(k+1,\ldots,n) \tag{8-2}$$

and our hamiltonian should be separable into π and σ parts:

$$\hat{\mathscr{H}}(1,2,\ldots,n) = \hat{\mathscr{H}}_\pi(1,2,\ldots,k) + \hat{\mathscr{H}}_\sigma(k+1,\ldots,n) \tag{8-3}$$

Equations (8-2) and (8-3) lead immediately to Eq. (8-1):

$$\begin{aligned}\bar{E} &= \frac{\int \psi_\pi^*\psi_\sigma^*\left(\hat{\mathscr{H}}_\pi + \hat{\mathscr{H}}_\sigma\right)\psi_\pi\psi_\sigma d\tau(1,\ldots,n)}{\int \psi_\pi^*\psi_\sigma^*\psi_\pi\psi_\sigma d\tau(1,\ldots,n)} \\ &= \frac{\int \psi_\pi^*\hat{\mathscr{H}}_\pi\psi_\pi d\tau(1,\ldots,k)}{\int \psi_\pi^*\psi_\pi d\tau(1,\ldots,k)} + \frac{\int \psi_\sigma^*\hat{\mathscr{H}}_\sigma\psi_\sigma d\tau(k+1,\ldots,n)}{\int \psi_\sigma^*\psi_\sigma d\tau(k+1,\ldots,n)} \\ &= E_\pi + E_\sigma \end{aligned} \tag{8-4}$$

If these equations were valid, one could ignore ψ_σ and legitimately minimize E_π by varying ψ_π, But the equations are *not* valid because it is impossible to rigorously satisfy Eq. (8-3). We cannot define $\hat{\mathscr{H}}_\pi$ and $\hat{\mathscr{H}}_\sigma$ so that they individually depend completely on separate groups of electrons and still sum to the correct total hamiltonian. Writing these operators explicitly gives

$$\hat{\mathscr{H}}_\pi(1,\ldots,k) = -\frac{1}{2}\sum_{i=1}^{k}\nabla_i^2 + \sum_{i=1}^{k}V_{\text{ne}}(i) + \frac{1}{2}\sum_{i=1}^{k}\sum_{j=1,j\neq i}^{k}\frac{1}{r_{ij}} \tag{8-5}$$

$$\hat{\mathscr{H}}_\sigma(k+1,\ldots,n) = -\frac{1}{2}\sum_{i=k+1}^{n}\nabla_i^2 + \sum_{i=k+1}^{n}V_{\text{ne}}(i) + \frac{1}{2}\sum_{i=k+1}^{n}\sum_{j=k+1,j\neq i}^{n}\frac{1}{r_{ij}} + V_{\text{nn}} \tag{8-6}$$

where $V_{\text{ne}}(i)$ represents the attraction between electron i and all the nuclei. These hamiltonians do indeed depend on the separate groups of electrons, but they leave out the operators for repulsion between σ and π electrons:

$$\hat{\mathscr{H}} - \hat{\mathscr{H}}_\pi - \hat{\mathscr{H}}_\sigma = \sum_{i=1}^{k} \sum_{j=k+1}^{n} \frac{1}{r_{ij}} \tag{8-7}$$

In short, the σ and π electrons really do interact with each other, and the fact that the HMO method does not *explicitly* include such interactions must be kept in mind when we consider the applicability of the method to certain problems. Some account of σ–π interactions is included *implicitly* in the method, as we shall see shortly.

8-3 The Independent π-Electron Assumption

The HMO method assumes further that the wavefunction ψ_π is a product of one-electron functions and that the hamiltonian $\hat{\mathscr{H}}_\pi$ is a sum of one-electron operators. Thus, for $n\pi$ electrons,

$$\psi_\pi(1, 2, \ldots, n) = \phi_i(1)\phi_j(2)\ldots\phi_l(n) \tag{8-8}$$

$$\hat{\mathscr{H}}_\pi(1, 2, \ldots, n) = \hat{H}_\pi(1) + \hat{H}_\pi(2) + \cdots + \hat{H}_\pi(n) \tag{8-9}$$

and

$$\frac{\int \phi^*{}_i(1)\hat{H}_\pi(1)\phi_i(1)d\tau(1)}{\int \phi^*{}_i(1)\phi_i(1)d\tau(1)} \equiv E_i \tag{8-10}$$

It follows that the total π energy E_π is a sum of one-electron energies:

$$E_\pi = E_i + E_j + \cdots + E_l \tag{8-11}$$

This means that the π electrons are being treated as though they are independent of each other, since E_i depends only on ϕ_i and is not influenced by the presence or absence of an electron in ϕ_j. However, this cannot be correct because π electrons in fact interact strongly with each other. Once again, such interactions will be roughly accounted for in an implicit way by the HMO method.

The implicit inclusion of interelectronic interactions is possible because we never actually write down a detailed expression for the π one-electron hamiltonian operator $\hat{H}_\pi(i)$. (We *cannot* write it down because it results from a π–σ separability assumption and an independent π-electron assumption, and both assumptions are incorrect.) $\hat{H}_\pi(\mathrm{i})$ is considered to be an "effective" one-electron operator—an operator that somehow includes the important physical interactions of the problem so that it can lead to a reasonably correct energy value E_i. A key point is that the HMO method ultimately evaluates E_i *via* parameters that are evaluated by appeal to experiment. Hence, it is a *semiempirical* method. Since the experimental numbers must include effects resulting from all the interelectronic interactions, it follows that these effects are implicitly included to some extent in the HMO method through its parameters.

It was pointed out in Chapter 5 that, when the independent electron approximation [Eqs. (8-8)–(8-11)] is taken, all states belonging to the same configuration become degenerate. In other words, considerations of space-spin symmetry do not affect the energy in that approximation. Therefore, the HMO method can make no explicit use of spin orbitals or Slater determinants, and so ψ_π is normally taken to be a single product function as in Eq. (8-8). The Pauli principle is provided for by assigning no more than two electrons to a single MO.

EXAMPLE 8-1 If O_2 were treated by the HMO method, what would be the form of the wavefunction and energy for the ground state?

SOLUTION ▶ The ground state configuration for O_2 is $1\sigma_g^2 1\sigma_u^2 2\sigma_g^2 2\sigma_u^2 3\sigma_g^2 1\pi_{u,x}^2 1\pi_{u,y}^2 \times 1\pi_{g,x} 1\pi_{g,y}$, where we have shown the degenerate members of π levels explicitly and in their real forms. The HMO wavefunction is simply a product of the pi MOs, one for each of the six pi electrons: $1\pi_{u,x}(1)1\pi_{u,x}(2)1\pi_{u,y}(3)1\pi_{u,y}(4)1\pi_{g,x}(5)1\pi_{g,y}(6)$. The HMO energy is $2E_{\pi,u,x} + 2E_{\pi,u,y} + E_{\pi,g,x} + E_{\pi,g,y}$, which reduces to $4E_{\pi,u} + 2E_{\pi,g}$. Note that, because O_2 is linear, there is no unique molecular plane containing the internuclear axis. Therefore this molecule has two sets of π MOs, one pair pointing in the x direction, the other pair pointing along y. For a planar molecule, only one of these pairs would qualify as π MOs, as will be seen in the next section. ◀

8-4 Setting up the Hückel Determinant

8-4.A Identifying the Basis Atomic Orbitals and Constructing a Determinant

The allyl radical, C_3H_5, is a planar molecule[1] with three unsaturated carbon centers (see Fig. 8-1). The minimal basis set of AOs for this molecule consists of a 1s AO on each hydrogen and 1s, 2s, $2p_x$, $2p_y$, and $2p_z$ AOs on each carbon. Of all these AOs only the $2p_z$ AOs at the three carbons are antisymmetric for reflection through the molecular plane.

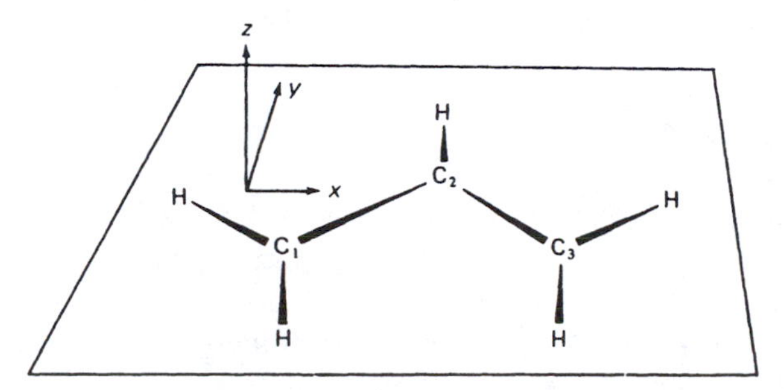

Figure 8-1 ▶ Sketch of the nuclear framework for the allyl radical. All the nuclei are coplanar. The z axis is taken to be perpendicular to the plane containing the nuclei.

[1]The minimum energy conformation of the allyl system is planar. We will ignore the deviations from planarity resulting from vibrational bending of the system.

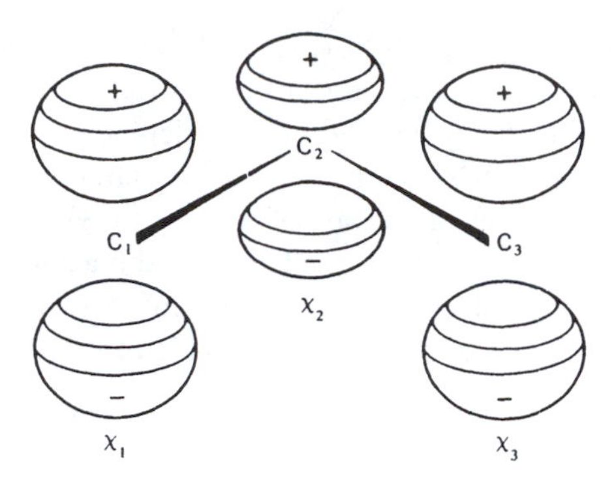

Figure 8-2 ▶ The three π-type AOs in the minimal basis set of the allyl radical.

Following Hückel, we ignore all the σ-type AOs and take the three $2p_z$ AOs as our set of basis functions. Notice that this restricts us to the carbon atoms: the hydrogens are not treated explicitly in the simple HMO method. We label our three basis functions χ_1, χ_2, χ_3 as indicated in Fig. 8-2. We will assume these AOs to be normalized.

Suppose that we now perform a linear variation calculation using this basis set. We know this will lead to a 3×3 determinant having roots that are MO energies which can be used to obtain MO coefficients. The determinantal equation is

$$\begin{vmatrix} H_{11} - ES_{11} & H_{12} - ES_{12} & H_{13} - ES_{13} \\ H_{21} - ES_{21} & H_{22} - ES_{22} & H_{23} - ES_{23} \\ H_{31} - ES_{31} & H_{32} - ES_{32} & H_{33} - ES_{33} \end{vmatrix} = 0 \tag{8-12}$$

where

$$H_{ij} = \int \chi_i \hat{H}_\pi \chi_j \, dv \tag{8-13}$$

$$S_{ij} = \int \chi_i \chi_j \, dv \tag{8-14}$$

Since H_{ij} and S_{ij} are integrals over the space coordinates of a single electron, the electron index is suppressed in Eqs. (8-13) and (8-14).

8-4.B The Quantity α

We have already indicated that there is no way to write an explicit expression for $\hat{H}_\pi$ that is both consistent with our separability assumptions and physically correct. But, without an expression for $\hat{H}_\pi$, how can we evaluate the integrals H_{ij}? The HMO method sidesteps this problem by carrying certain of the H_{ij} integrals along as symbols until they can be evaluated empirically by matching theory with experiment.

Let us first consider the integrals H_{11}, H_{22} and H_{33}. The interpretation consistent with these integrals is that H_{11}, for instance, is the average energy of an electron in AO χ_1 experiencing a potential field due to the entire molecule. Symmetry requires that $H_{11} = H_{33}$. H_{22} should be different since an electron in AO χ_2 experiences a different environment than it does when in χ_1 or χ_3. It seems likely, however, that H_{22} is not *very* different from H_{11}. In each case, we expect the dominant part of the potential to arise from interactions with the local carbon atom, with more distant atoms playing a secondary role. Hence, one of the approximations made in the HMO method is that all H_{ii} are identical if χ_i is on a carbon atom. The symbol α is used for such integrals. Thus, for the example at hand, $H_{11} = H_{22} = H_{33} = \alpha$. The quantity α is often called the *coulomb integral*.[2]

8-4.C The Quantity β

Next, we consider the *resonance integrals* or *bond integrals* H_{12}, H_{23}, and H_{13}. (The requirement that $\hat{H}_\pi$ be hermitian plus the fact that the χ's and $\hat{H}_\pi$ are real suffices to make these equal to H_{21}, H_{32}, and H_{31}, respectively.) The interpretation consistent with these integrals is that H_{12}, for instance, is the energy of the overlap charge between χ_1 and χ_2. Symmetry requires that $H_{12} = H_{23}$ in the allyl system. However, even when symmetry does not require it, the assumption is made that all H_{ij} are equal to the same quantity (called β) when i and j refer to "neighbors" (i.e., atoms connected by a σ bond). It is further assumed that $H_{ij} = 0$ when i and j are not neighbors. Therefore, in the allyl case,

$$H_{12} = H_{23} \equiv \beta, \; H_{13} = 0.$$

8-4.D Overlap Integrals

Since the χ's are normalized, $S_{ii} = 1$. The overlaps between neighbors are typically around 0.3. Nevertheless, in the HMO method, all S_{ij} $(i \neq j)$ are taken to be zero. Although this seems a fairly drastic approximation, it has been shown to have little effect on the qualitative nature of the solutions.

8-4.E Further Manipulation of the Determinant

Our determinantal equation for the allyl system is now much simplified. It is

$$\begin{vmatrix} \alpha - E & \beta & 0 \\ \beta & \alpha - E & \beta \\ 0 & \beta & \alpha - E \end{vmatrix} = 0 \tag{8-15}$$

Dividing each row of the determinant by β corresponds to dividing the whole determinant by β^3. This will not affect the equality. Letting $(\alpha - E)/\beta \equiv x$, we obtain the result

$$\begin{vmatrix} x & 1 & 0 \\ 1 & x & 1 \\ 0 & 1 & x \end{vmatrix} = 0 \tag{8-16}$$

---

[2]The term "coulomb integral" for α is unfortunate since the same name is used for repulsion integrals of the form $\int \chi_1(1)\chi_2(2)(1/r_{12})\chi_1(1)\chi_2(2)\,dv$. The quantity α also contains kinetic energy and nuclear–electronic attraction energy.

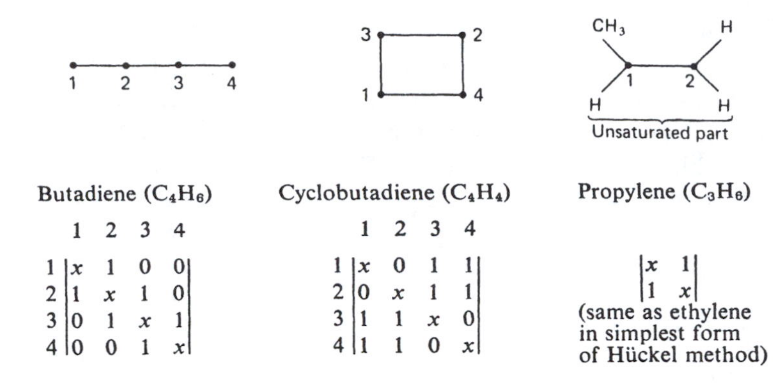

Figure 8-3 ▶ HMO determinants for some small systems.

which is the form we will refer to as the *HMO determinantal equation.* Notice that x occurs on the principal diagonal, 1 appears in positions where the indices correspond to a bond, 0 appears in positions (e.g., 1,3) corresponding to no bond. This gives us a simple prescription for writing the HMO determinant for any unsaturated hydrocarbon system directly from a sketch of the molecular structure. The rules are (1) sketch the framework defined by the n *unsaturated* carbons; (2) number the atoms $1, \ldots, n$ (the ordering of numbers is arbitrary); (3) fill in the $n \times n$ determinant with x's on the diagonal, 1's in positions where row column indices correspond to bonds, 0's elsewhere. See Fig. (8-3) for examples. As a check, it is useful to be sure that the determinant is symmetric for reflection through the diagonal of x's. This is necessary since, if atoms i and j are neighbors, 1's must appear in positions i, j *and* j, i of the determinant.

Since the Hückel determinant contains only information about the number of unsaturated carbons and how they are connected together, it is sometimes referred to as a *topological determinant.* (Topology refers to properties that are due to the *connectedness* of a figure, but are unaffected by twisting, bending, etc.)

8-5 Solving the HMO Determinantal Equation for Orbital Energies

The HMO determinantal equation for the allyl system (8-16) can be expanded to give

$$x^3 - 2x = 0 \tag{8-17}$$

or

$$x(x^2 - 2) = 0 \tag{8-18}$$

Thus, the roots are $x = 0$, $x = \sqrt{2}$, and $x = -\sqrt{2}$. Recalling the definition of x, these roots correspond respectively to the energies $E = \alpha$, $E = \alpha - \sqrt{2}\beta$, $E = \alpha + \sqrt{2}\beta$.

How should we interpret these results? Since α is supposed to be the energy of a pi electron in a carbon 2p AO in the molecule, we expect this quantity to be negative (corresponding to a bound electron). Since β refers to an electron in a bond region, it too should be negative. Therefore, the lowest-energy root should be $E_1 = \alpha + \sqrt{2}\beta$, followed by $E_2 = \alpha$, with $E_3 = \alpha - \sqrt{2}\beta$ being the highest-energy root.

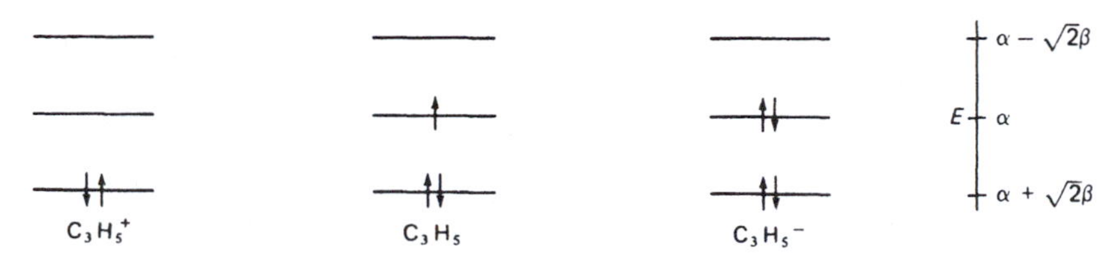

Figure 8-4 ▶ π-Electron configurations and total energies for the ground states of the allyl cation, radical, and anion.

(It is convenient to number the orbital energies sequentially, starting with the lowest, as we have done here.)

We have just seen that bringing three $2p_\pi$ AOs together in a linear arrangement causes a splitting into three MO energy levels. This is similar to the splitting into two energy levels produced when two 1s AOs interact, discussed in connection with H_2^+. In general, n linearly independent separated AOs will lead to n linearly independent MOs.

The ground-state π-electron configuration of the allyl system is built up by putting electrons in pairs into the MOs, starting with those of lowest energy. Thus far, we have been describing our system as the allyl radical. However, since we have as yet made no use of the number of π electrons in the system, our results so far apply equally well for the allyl cation, radical, or anion.

Configurations and total π energies for these systems in their ground states are depicted in Fig. 8-4. The total π-electron energies are obtained by summing the one-electron energies, as indicated earlier.

EXAMPLE 8-2 For a planar, unsaturated hydrocarbon having formula C_xH_y, where all the carbons are part of the unsaturated framework, how many pi MOs are there?

SOLUTION ▶ Each carbon atom brings one $2p_\pi$ AO into the basis set, so there are x basis AOs. These x independent AOs mix to form x independent MOs. ◀

8-6 Solving for the Molecular Orbitals

We still have to find the coefficients that describe the MOs as linear combinations of AOs. Recall from Chapter 7 that this is done by substituting energy roots of the secular determinant back into the simultaneous equations. For the allyl system, the simultaneous equations corresponding to the secular determinant (8-16) are

$$c_1x + c_2 = 0 \tag{8-19}$$

$$c_1 + c_2x + c_3 = 0 \tag{8-20}$$

$$c_2 + c_3x = 0 \tag{8-21}$$

(Compare these equations with the secular determinant in Eq. (8-16) and note the obvious relation.) As we noted in Chapter 7, homogeneous equations like these can give us only ratios between c_1, c_2, and c_3, not their absolute values. So we anticipate using only two of these equations and obtaining absolute values by satisfying the normality condition. Because we are neglecting overlap between AOs, the latter step corresponds to requiring

$$c_1^2 + c_2^2 + c_3^2 = 1 \tag{8-22}$$

The roots x are, in order of increasing energy, $-\sqrt{2}, 0, +\sqrt{2}$. Let us take $x = -\sqrt{2}$ first. Then

$$-\sqrt{2}c_1 + \quad c_2 \qquad = 0 \tag{8-23a}$$

$$c_1 - \sqrt{2}c_2 + \quad c_3 = 0 \tag{8-23b}$$

$$c_2 - \sqrt{2}c_3 = 0 \tag{8-23c}$$

Comparing Eqs. (8-23a) and (8-23c) gives $c_1 = c_3$. Equation (8-23a) gives $c_2 = \sqrt{2}c_1$. Inserting these relations into the normality equation (8-22) gives

$$c_1^2 + \left(\sqrt{2}c_1\right)^2 + c_1^2 = 1 \tag{8-24}$$

$$4c_1^2 = 1, \qquad c_1 = \pm\frac{1}{2} \tag{8-25}$$

It makes no difference which sign we choose for c_1 since any wavefunction is equivalent to its negative. (Both give the same ψ^2.) Choosing $c_1 = +\frac{1}{2}$ gives

$$c_1 = \frac{1}{2}, \quad c_2 = \frac{1}{\sqrt{2}}, \quad c_3 = \frac{1}{2} \tag{8-26}$$

These coefficients define our lowest-energy MO, ϕ_1:

$$\phi_1 = \frac{1}{2}\chi_1 + \frac{1}{\sqrt{2}}\chi_2 + \frac{1}{2}\chi_3 \tag{8-27}$$

A similar approach may be taken for $x = 0$ and $x = +\sqrt{2}$. The results are

$$(x = 0): \quad \phi_2 = \frac{1}{\sqrt{2}}\chi_1 - \frac{1}{\sqrt{2}}\chi_3 \tag{8-28}$$

$$\left(x = +\sqrt{2}\right): \quad \phi_3 = \frac{1}{2}\chi_1 - \frac{1}{\sqrt{2}}\chi_2 + \frac{1}{2}\chi_3 \tag{8-29}$$

The allyl system MOs are sketched in Fig. 8-5.

The lowest-energy MO, ϕ_1, has no nodes (other than the molecular-plane node common to all π MOs) and is said to be bonding in the $C_1 - C_2$ and $C_2 - C_3$ regions. It is reasonable that such a bonding MO should have an energy wherein the bond-related term β acts to lower the energy, as is true here. The second-lowest energy MO, ϕ_2, has a nodal plane at the central carbon. Because there are no π AOs on *neighboring* carbons in this MO, there are no interactions at all, and β is absent from the energy expression. This MO is said to be *nonbonding*. The high-energy MO, ϕ_3 has nodal planes intersecting both bonds. Because the π AOs show sign disagreement across both bonds, this MO is everywhere antibonding and β terms act to raise the orbital energy above α.

EXAMPLE 8-3 According to HMO theory, do the π electrons favor a linear, or a bent allyl radical?

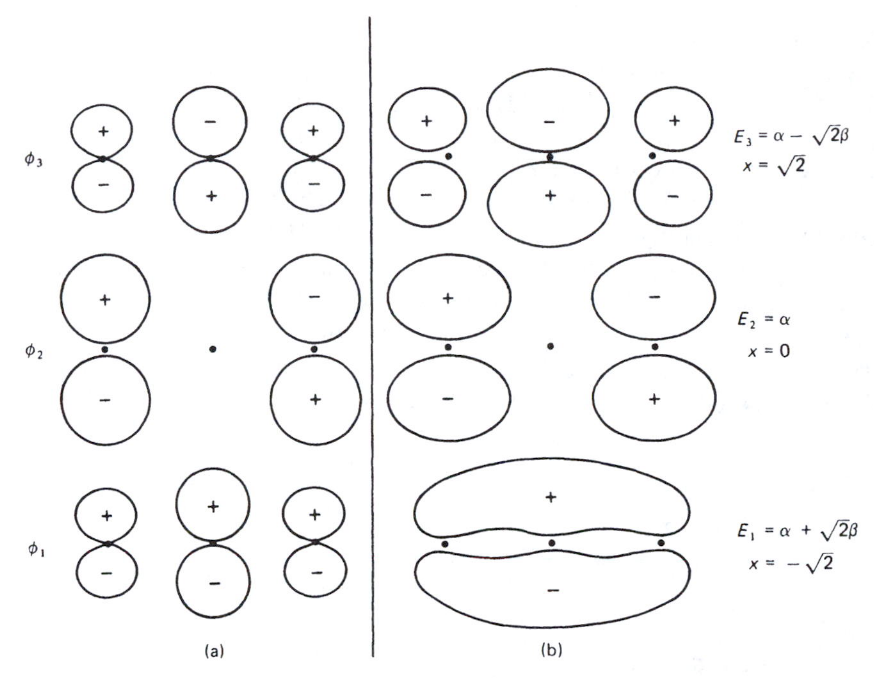

Figure 8-5 ▶ Sketches of the allyl system MOs. (a) emphasizes AO signs and magnitudes. (b) resembles more closely the actual contours of the MOs.

SOLUTION ▶ HMO theory favors neither. The difference between linear and bent allyl shows up as a difference in $C_1 - C_2 - C_3$ angle and in C_1 to C_3 distance. The HMO method has no angular-dependent features and explicitly omits interactions between non-neighbor carbons, like C_1 and C_3. ◀

8-7 The Cyclopropenyl System: Handling Degeneracies

The allyl system results when three π AOs interact in a linear arrangement wherein $H_{12} = H_{23} = \beta$, but $H_{13} = 0$. We can also treat the situation where the three π AOs approach each other on vertices of an ever-shrinking equilateral triangle. In this case, each AO interacts equally with the other two. This triangular system is the cyclopropenyl system C_3H_3 shown in Fig. 8-6.

The HMO determinantal equation for this system is

$$\begin{vmatrix} x & 1 & 1 \\ 1 & x & 1 \\ 1 & 1 & x \end{vmatrix} = 0, \quad x^3 + 2 - 3x = 0 \tag{8-30}$$

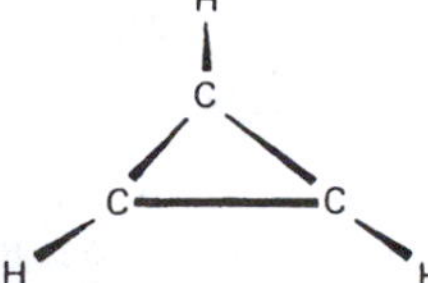

Figure 8-6 ▶ The cyclopropenyl system (all nuclei are coplanar).

This equation can be factored as

$$(x+2)(x-1)(x-1)=0 \tag{8-31}$$

Therefore, the roots are $x=-2, +1, +1$.

Since the root $x=1$ occurs twice, we can expect there to be two independent HMOs having the same energy–a doubly degenerate level. The energy scheme and ground state electron configuration for the cyclopropenyl radical (three π electrons) (**I**) gives a total E_π of $3\alpha+3\beta$. We can surmise from these orbital energies that ϕ_1 is a bonding MO, whereas ϕ_2 and ϕ_3 are predominantly antibonding. To see if this is reflected in the nodal properties of the MOs, let us solve for the coefficients. The equations consistent with the HMO determinant and with orbital normality are

$$\begin{aligned} c_1x + c_2 \ \ + c_3 \ \ &= 0 \\ c_1 + c_2x + c_3 \ \ &= 0 \\ c_1 + c_2 \ \ + c_3x &= 0 \\ c_1^2 + c_2^2 \ \ + c_3^2 \ \ &= 1 \end{aligned} \tag{8-32}$$

Setting $x=-2$ and solving gives

$$\phi_1 = \frac{1}{\sqrt{3}}\chi_1 + \frac{1}{\sqrt{3}}\chi_2 + \frac{1}{\sqrt{3}}\chi_3 \tag{8-33}$$

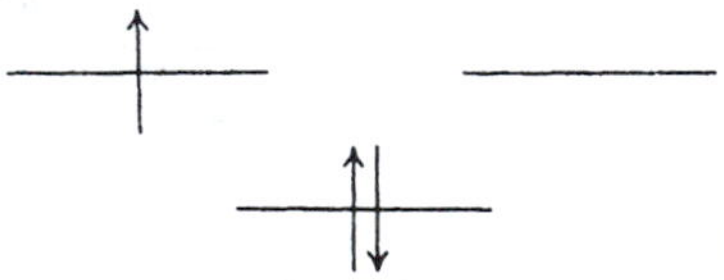

$E_2 = E_3 = \alpha - \beta \quad (x = +1)$

$E_1 = \alpha + 2\beta \quad (x = -2)$

(**I**)

For this MO, the coefficients are all of the same sign, and so the AOs show phase agreement across all bonds and all interactions are bonding.

To find ϕ_2 and ϕ_3 is trickier. We begin by inserting $x=\pm 1$ into our simultaneous equations. This gives

$$c_1 + c_2 + c_3 = 0 \quad \text{(three times)} \tag{8-34}$$

$$c_1^2 + c_2^2 + c_3^2 = 1 \tag{8-35}$$

With three unknowns and two equations, an infinite number of solutions is possible. Let us pick a convenient one: $c_1 = -c_2, c_3 = 0$. The normalization requirement then gives $c_1 = 1/\sqrt{2}$, $c_2 = -1/\sqrt{2}$, $c_3 = 0$. Let us call this solution ϕ_2:

$$\phi_2 = \frac{1}{\sqrt{2}}\chi_1 - \frac{1}{\sqrt{2}}\chi_2 \tag{8-36}$$

We still need to find ϕ_3. There remain an infinite number of possibilities, so let us pick one: $c_1 = 1/\sqrt{2}$, $c_2 = 0$, $c_3 = -1/\sqrt{2}$. We have used our experience with ϕ_2 to choose c's that guarantee a normalized ϕ_3. Also, it is clear that ϕ_3 is linearly independent of

ϕ_2 since they contain different AOs. But it is desirable to have ϕ_3 *orthogonal* to ϕ_2. Let us test ϕ_2 and ϕ_3 to see if they are orthogonal:

$$S = \int \phi_2\phi_3\, dv = \tfrac{1}{2}\int (\chi_1 - \chi_2)(\chi_1 - \chi_3)\, dv$$

$$= \tfrac{1}{2}\left\{\int \cancelto{1}{\chi_1^2\, dv} - \int \cancelto{0}{\chi_1\chi_3\, dv} - \int \cancelto{0}{\chi_1\chi_2\, dv} + \int \cancelto{0}{\chi_2\chi_3\, dv}\right\} = \tfrac{1}{2} \qquad (8\text{-}37)$$

Since $S \neq 0$, ϕ_2 and ϕ_3 are nonorthogonal. We can project out that part of ϕ_3 that is orthogonal to ϕ_2 by using the Schmidt orthogonalization procedure described in Section 6-10. We seek a new function ϕ_3' given by

$$\phi_3' = \phi_3 - S\phi_2 \qquad (8\text{-}38)$$

where

$$S = \int \phi_2\phi_3 dv = \frac{1}{2} \qquad (8\text{-}39)$$

Therefore,

$$\phi_3' = \phi_3 - \frac{1}{2}\phi_2 = \frac{1}{2\sqrt{2}}(\chi_1 + \chi_2 - 2\chi_3) \qquad (8\text{-}40)$$

This function is orthogonal to ϕ_2 but is not normalized. Renormalizing gives

$$\phi_3'' = \frac{1}{\sqrt{6}}(\chi_1 + \chi_2 - 2\chi_3) \qquad (8\text{-}41)$$

In summary, to produce HMO coefficients for degenerate MOs, pick any two independent solutions from the infinite choice available, and orthogonalize one of them to the other using the Schmidt (or any other) orthogonalization procedure.

The MOs for the cyclopropenyl system *as seen from above the molecular plane* are sketched in Fig. 8-7. The MO ϕ_2 can be seen to have both antibonding (C_1–C_2) and nonbonding (C_1–C_3, C_2–C_3) interactions. ϕ_3'' has antibonding (C_1–C_3, C_2–C_3) and bonding (C_1–C_2) interactions. The interactions are of such size and number as to give an equal net energy value ($\alpha - \beta$) in each case. Since nodal planes produce antibonding or nonbonding situations, it is not surprising that higher and higher-energy HMOs in a

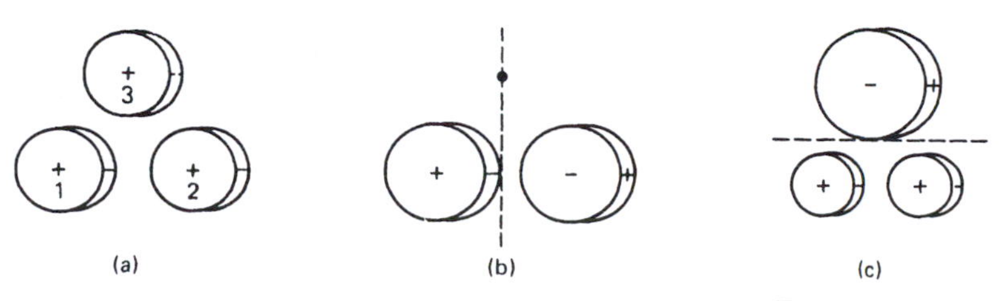

Figure 8-7 ▶ The HMOs for the cyclopropenyl system: (a) $\phi_1 = (1/\sqrt{3})(\chi_1 + \chi_2 + \chi_3)$; (b) $\phi_2 = (1/\sqrt{2})(\chi_1 - \chi_2)$ (c) $\phi_3'' = (1/\sqrt{6})(\chi_1 + \chi_2 - 2\chi_3)$. The nodal planes intersect the molecular plane at the dashed lines.

system display more and more nodal planes. Notice that the MOs ϕ_2 and ϕ_3'' have the same number of nodal planes (one, not counting the one in the molecular plane) but that these planes are perpendicular to each other. This is a common feature of some degenerate, orthogonal MOs in cyclic molecules.

It is important to notice the symmetry characteristics of these MOs. ϕ_1 is either symmetric or antisymmetric for every symmetry operation of the molecule. (It is antisymmetric for reflection through the molecular plane, symmetric for rotation about the threefold axis, etc.) This must be so for any nondegenerate MO. But the degenerate MOs ϕ_2 and ϕ_3'' are neither symmetric nor antisymmetric for certain operations. (ϕ_2 is antisymmetric for reflection through the plane indicated by the dashed line in Fig. 8-7, but is neither symmetric nor antisymmetric for rotation about the threefold axis by 120°.) In fact, one can easily show that, given a cycle with an odd number of centers, each with one AO of a common type, there is but *one* way to combine the AOs (to form a *real* MO) so that the result is symmetric or antisymmetric for all rotations and reflections of the cycle. Hence, an HMO calculation for a three-, five, seven-, ... membered ring can give only *one* nondegenerate MO. However, for a cycle containing an *even* number of centers, the analogous argument shows that *two* nondegenerate MOs exist.

8-8 Charge Distributions from HMOs

Now that we have a method that provides us with orbitals and orbital energies, it should be possible to get information about the way the π-electron charge is distributed in the system by squaring the total wavefunction ψ_π. In the case of the neutral allyl radical, we have (taking ψ_π to be a simple product of MOs)

$$\psi_\pi = \phi_1(1)\phi_1(2)\phi_2(3) \tag{8-42}$$

Hence, the probability for simultaneously finding electron 1 in $dv(1)$, electron 2 in $dv(2)$ and electron 3 in $dv(3)$ is

$$\psi_\pi^2(1,2,3)dv(1)dv(2)dv(3) = \phi_1^2(1)\phi_1^2(2)\phi_2^2(3)dv(1)dv(2)dv(3) \tag{8-43}$$

For most physical properties of interest, we need to know the probability for finding *an* electron in a three-dimensional volume element dv. Since the probability for finding an electron in dv is the *sum* of the probabilities for finding each electron there, the *one-electron density function* ρ for the allyl radical is

$$\rho = 2\phi_1^2 + \phi_2^2 \tag{8-44}$$

where we have suppressed the index for *the* electron. If we integrate ρ over all space, we obtain a value of three. This means we are certain of finding a total π charge corresponding to three π electrons in the system.

To find out how the π charge is distributed in the molecule, let us express ρ in terms of AOs. First, we write ϕ_1^2 and ϕ_2^2 separately:

$$\phi_1^2 = \frac{1}{4}\chi_1^2 + \frac{1}{2}\chi_2^2 + \frac{1}{4}\chi_3^2 + \frac{1}{\sqrt{2}}\chi_1\chi_2 + \frac{1}{\sqrt{2}}\chi_2\chi_3 + \frac{1}{2}\chi_1\chi_3$$

$$\phi_2^2 = \frac{1}{2}\chi_1^2 + \frac{1}{2}\chi_3^2 - \chi_1\chi_3 \tag{8-45}$$

If we were to integrate ϕ_1^2 we would obtain

$$\int \phi_1^2\,dv = \frac{1}{4}\overbrace{\int \chi_1^2\,dv}^{\longrightarrow 1} + \frac{1}{2}\overbrace{\int \chi_2^2\,dv}^{\longrightarrow 1} + \frac{1}{4}\overbrace{\int \chi_3^2\,dv}^{\longrightarrow 1} + \frac{1}{\sqrt{2}}\overbrace{\int \chi_1\chi_2\,dv}^{\longrightarrow 0}$$

$$+\frac{1}{\sqrt{2}}\overbrace{\int \chi_2\chi_3\,dv}^{\longrightarrow 0} + \frac{1}{2}\overbrace{\int \chi_1\chi_3\,dv}^{\longrightarrow 0}$$

$$= \frac{1}{4} + \frac{1}{2} + \frac{1}{4} = 1 \tag{8-46}$$

Thus, one electron in ϕ_1 shows up, upon integration, as being "distributed" $\frac{1}{4}$ at carbon 1, $\frac{1}{2}$ at carbon 2, and $\frac{1}{4}$ at carbon 3. We say that the *atomic π-electron densities* due to an electron in ϕ_1 are $\frac{1}{4}$, $\frac{1}{2}$, $\frac{1}{4}$ at C_1, C_2, and C_3, respectively, If we accumulate these figures for all the electrons, we arrive at a total π-electron density for each carbon. For the allyl radical, Table 8-1 shows that each atom has a π-electron density of unity.

Generalizing this approach gives for the total π-electron density q_i on atom i

$$q_i = \sum_{k}^{\text{all MOs}} n_k c_{ik}^2 \tag{8-47}$$

Here k is the MO index, c_{ik} is the coefficient for an AO on atom i in MO k, and n_k, the "occupation number," is the number of electrons (0, 1, or 2) in MO k. (In those rare cases where c_{ik} is complex, c_{ik}^2 in Eq. (8-47) must be replaced by $c_{ik}^* c_{ik}$.)

If we apply Eq. (8-47) to the cyclopropenyl radical, we encounter an ambiguity. If the unpaired electron is assumed to be in MO ϕ_2 of Fig. 8-7, we obtain $q_1 = q_2 = \frac{7}{6}$, $q_3 = \frac{4}{6}$. On the other hand, if the unpaired electron is taken to be in ϕ_3'', $q_1 = q_2 = \frac{5}{6}$, $q_3 = \frac{8}{6}$. The HMO method resolves this ambiguity by assuming that each of the degenerate MOs is occupied by half an electron. This has the effect of forcing the charge distribution to show the overall symmetry of the molecule. In this example, it follows that $q_1 = q_2 = q_3 = 1$. The general rule is that, for purposes of calculating electron distributions, the electron occupation is averaged in any set of partially occupied, degenerate MOs.

TABLE 8-1 ▶ HMO π Electron Densities in the Allyl Radical

	Carbon atom		
Electron	1	2	3
1 in ϕ_1	$\frac{1}{4}$	$\frac{1}{2}$	$\frac{1}{4}$
2 in ϕ_1	$\frac{1}{4}$	$\frac{1}{2}$	$\frac{1}{4}$
3 in ϕ_2	$\frac{1}{2}$	0	$\frac{1}{2}$
	—	—	—
Sum	1	1	1

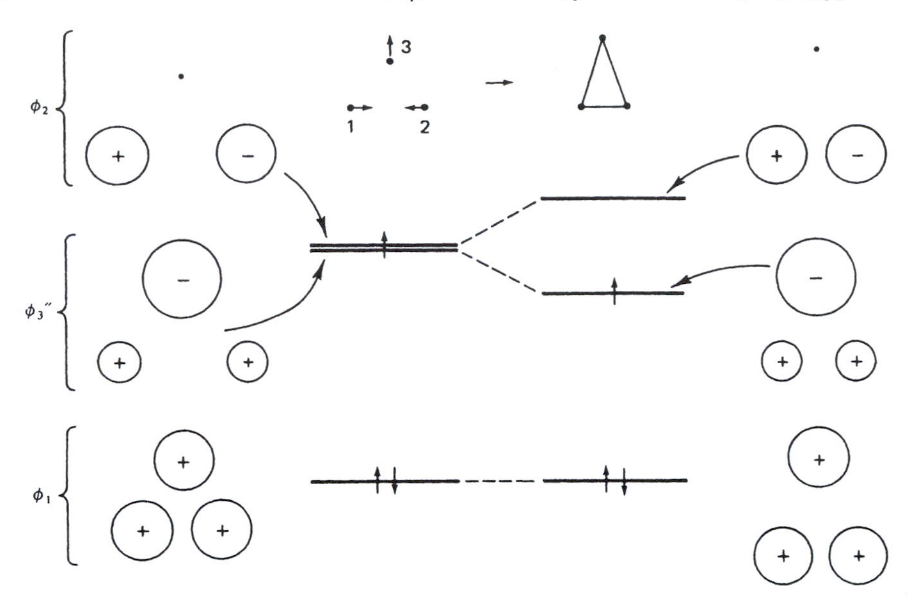

Figure 8-8 ▶ When the equilateral structure is distorted by decreasing R_{12} and increasing R_{13}, R_{23}, the energies associated with ϕ_1, ϕ_2, ϕ_3'' shift as shown.

In actuality, the equilateral triangular structure for the cyclopropenyl radical is unstable, and therefore the above-described averaging process is only a theoretical idealization. It is fairly easy to see that a distortion from equilateral to isosceles form will affect the MO energies E_1, E_2, and E_3'' differently. In particular, a distortion of the sort depicted in Fig. 8-8 would have little effect on E_1 but would raise E_2 (increased antibonding) and lower E_3'' (decreased antibonding and increased bonding). Thus, there is good reason for the cyclopropenyl radical to be more stable in an isosceles rather than equilateral triangular form. This is an example of the Jahn–Teller theorem, which states, in effect, that *a system having an odd number of electrons in degenerate MOs will change its nuclear configuration in a way to remove the degeneracy.*[3] The preference of the cyclopropenyl radical for a shape less symmetrical than what we might have anticipated is frequently called *Jahn–Teller distortion.*[4]

Many times we are interested in comparing the π-electron distribution *in the bonds* instead of on the atoms. In the integrated expression (8-46) are cross terms that vanish under the HMO assumption of zero overlap. But the overlaps are not actually zero, especially between AOs on nearest neighbors. Hence, we might view the factors $1/\sqrt{2}$ as indicating how much overlap charge is being placed in the C_1–C_2 and C_2–C_3 bonds by an electron in ϕ_1. The C_1–C_3 bond is usually ignored because these atoms are not nearest neighbors and therefore have much smaller AO overlap. Since $S_{12} = S_{23} = S_{ij}$ for neighbors i and j in any π system (assuming equal bond distances), we need not include S_{ij}, explicitly in our bond index. If we proceed in this manner, two electrons in

[3]Linear systems are exceptions to this rule. Problems are also encountered if there is an odd number of electrons and spin-orbit coupling is substantial. The reader should realize that the above statement of the theorem is a little misleading inasmuch as it makes it sound like the molecule finds itself in a symmetric geometry that produces denerate MOs and then "distorts" to a lower-energy geometry. It is actually we who have guessed a geometry that is too symmetric. When our calculations reveal that this results in degenerate orbital energies containing an odd number of electrons, we are alerted that we have erred in our assumption, and that the molecule is really in a less symmetric, lower energy geometry.

[4]See Salem [1, Chapter 8].

ϕ would then give us a "bond order" of $2/\sqrt{2} = 1.414$. It is more convenient in practice to divide this number in half, because then the calculated π-bond order for ethylene turns out to be unity rather than two. Since ethylene has one π-bond, this can be seen to be a more sensible index.

As a result of these considerations, the π-bond order (sometimes called *mobile bond order*) of the allyl radical is $1/\sqrt{2} = 0.707$ in each bond. (Electrons in ϕ_2 make no contribution to bond order since c_2 vanishes. This is consistent with the *nonbonding* label for ϕ_2.)

Generalizing the argument gives, for p_{ij}, the *π-bond order* between *nearest-neighbor* atoms i and j:

$$p_{ij} = \sum_{k}^{\text{all MOs}} n_k c_{ik} c_{jk} \tag{8-48}$$

where the symbols have the same meanings as in Eq. (8-46). In cases in which partially filled degenerate MOs are encountered, the averaging procedure described in connection with electron densities must be employed for bond orders as well.

EXAMPLE 8-4 Calculate p_{13} for the cyclopropenyl radical, using data in Fig. 8-7.

SOLUTION ▸ There are 2 electrons in ϕ_1 and the coefficients on atoms 1 and 3 are $\frac{1}{\sqrt{3}}$, so this MO contributes $2 \times (\frac{1}{\sqrt{3}})^2 = 2/3$. We allocate $\frac{1}{2}$ electron to ϕ_2. Since $c_3 = 0$ in this MO, the contribution to p_{13} is zero. The remaining $\frac{1}{2}$ electron goes to ϕ_3, yielding a contribution of $\frac{1}{2} \times \frac{1}{\sqrt{6}} \times \frac{-2}{\sqrt{6}} = \frac{-1}{6}$. So $p_{13} = \frac{2}{3} - \frac{1}{6} = \frac{1}{2}$. ◂

8-9 Some Simplifying Generalizations

Thus far we have presented the bare bones of the HMO method using fairly small systems as examples. If we try to apply this method directly to larger molecules, it is very cumbersome. A ten-carbon-atom system leads to a 10×10 HMO determinant. Expanding and solving this for roots and coefficients is tedious. However, there are some short cuts available for certain cases. In the event that the system is too complicated to yield to these, one can use computer programs which are readily available.

For straight chain and monocyclic planar, conjugated hydrocarbon systems, simple formulas exist for HMO energy roots and coefficients. These are derivable from the very simple forms of the HMO determinants for such systems.[5] We state the results without proof.

For a straight chain of n unsaturated carbons numbered sequentially,

$$x = -2\cos[k\pi/(n+1)], \quad k = 1, 2, \ldots, n \tag{8-49}$$

$$c_{lk} = [2/(n+1)]^{1/2} \sin[kl\pi/(n+1)] \tag{8-50}$$

where l is the atom index and k the MO index.

[5] See Coulson [2].

For a cyclic polyene of n carbons,

$$x = -2\cos(2xk/n), \quad k=0, 1, \ldots, n-1 \tag{8-51}$$

$$c_{lk} = n^{-1/2}\exp[2\pi ik(l-1)/n], \quad i=\sqrt{-1} \tag{8-52}$$

The coefficients derived from Eq. (8-52) for monocyclic polyenes will be complex when the MO is one of a degenerate pair. In such cases one may take linear combinations of these degenerate MOs to produce MOs with real coefficients, if one desires.

There is also a diagrammatic way to find the energy levels for linear and monocyclic systems.[6] Let us consider monocycles first. One begins by drawing a circle of radius $2|\beta|$. Into this circle inscribe the cycle, point down, as shown in Fig. 8-9 for benzene. Project sideways the points where the polygon intersects the circle. The positions of these projections correspond to the HMO energy levels if the circle center is assumed to be at $E=\alpha$ (see Fig. 8-9). The number of intersections at a given energy is identical to the degeneracy. The numerical values for E are often obtainable from such a sketch by inspection or simple trigonometry.

For straight chains, a modified version of the above method may be used: For an n-carbon chain, inscribe a cycle with $2n+2$ carbons into the circle as before. Projecting out all intersections *except the highest and lowest*, and *ignoring degeneracies* gives the proper roots. This is exemplified for the allyl system in Fig. 8-10.

Examination of the energy levels in Figs. 8-9 and 8-10 reveals that the orbital energies are symmetrically disposed about $E=\alpha$. Why is this so? Consider the allyl system. The lowest-energy MO has two bonding interactions. The highest-energy MO differs *only* in that these interactions are now antibonding. [See Fig. 8-5 and note that the coefficients in ϕ_1 and ϕ_3 are identical except for sign in Eqs. (8-27) and (8-29).] The role of the β terms is thus reversed and so they act to raise the orbital energy for ϕ_3 just as much as they lower it for ϕ_1. A similar situation holds for benzene. As we will see shortly, the lowest energy corresponds to an MO without nodes between atoms, so this is a totally bonding MO. The highest-energy MO has nodal planes between all neighbor carbons, and so every interaction is antibonding. An analogous argument holds for the degenerate pairs of benzene MOs. These observations suggest that the energy of an

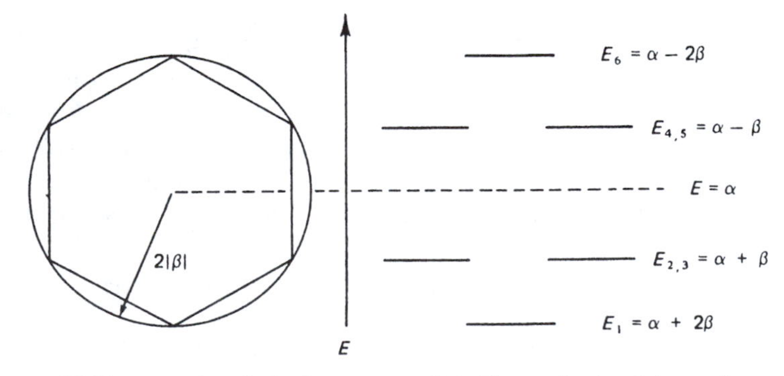

Figure 8-9 ▶ HMO energy levels for benzene produced by projecting intersections of a hexagon with a circle of radius $2|\beta|$.

[6]See Frost and Musulin [3].

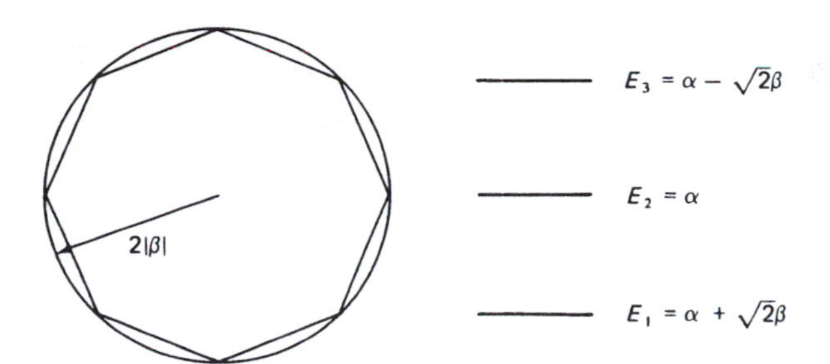

Figure 8-10 ▶ HMO energy levels for the allyl system ($n = 3$) produced by projecting the intersections of an octagon ($n = 2 \times 3 + 2$) with a circle of radius $2\,|\beta|$.

MO should be expressible as a function of the net bond order associated with it, and this is indeed the case. The energy of the ith MO is given by the expression

$$E_i = \int \phi_i \hat{H}_\pi \phi_i \, dv = \int \sum_k c_{ki} \chi_k \hat{H}_\pi \sum_l c_{li} \chi_l \, dv \tag{8-53}$$

$$= \sum_k \sum_l c_{ki} c_{li} \int \chi_k \hat{H}_\pi \chi_l \, dv \tag{8-54}$$

When the atom indices k and l are identical, the integral is equal to α; when k and l are neighbors, it equals β. Otherwise it vanishes. Hence, we may write

$$E_i = \sum_k c_{ki}^2 \alpha + \sum_{k,l}^{\text{neighbors}} c_{ki} c_{li} \beta \tag{8-55}$$

However, c_{ki}^2 is $q_{k,i}$, the electron density at atom k due to one electron in MO ϕ_i, and $c_{ki} c_{li}$ is $p_{kl,i}$, the bond order between atoms k and l due to an electron in ϕ_i. Therefore,

$$E_i = \sum_k q_{k,i} \alpha + 2 \sum_{k<l}^{\text{neighbors}} p_{kl,i} \beta \tag{8-56}$$

We have seen that the sum of electron densities must equal the total number of electrons present. For one electron in ϕ_i, this gives additional simplification.

$$E_i = \alpha + 2\beta \sum_{k<l}^{\text{bonds}} p_{kl,i} \tag{8-57}$$

The *total* π-electron energy is the sum of one-electron energies. For $n\pi$ electrons

$$E_\pi = n\alpha + 2\beta \sum_{k<l}^{\text{bonds}} p_{kl} \tag{8-58}$$

where p_{kl} is the total π-bond order between neighbors k and l. Hence, the individual orbital energies directly reflect the amount of bonding or antibonding described by the MOs, and the total energy reflects the net bonding or antibonding due to all the π electrons together.

EXAMPLE 8-5 The total π energy of cycloheptatrienyl radical (C_7H_7) is $7\alpha + 8.5429\beta$. What is the bond order for any bond in this molecule, assuming it to be heptagonal?

SOLUTION ▶ The total π-bond order must be 8.5429/2 or 4.2714. This results from seven identical bonds, so each bond order is $4.2714/7 = 0.6102$. ◀

Does this pairing of energy levels observed for allyl and benzene always occur? It is easy to show that it cannot in rings with an odd number of carbon centers. Consider the cyclopropenyl system. The lowest-energy MO is nodeless, totally bonding and has an energy of $2\alpha + 2\beta$. [Note from Eq. (8-51) and also from the diagram method that every monocyclic system has a totally bonding MO at this energy.] To transform these three bonding interactions into antibonding interactions of equal magnitude requires that we cause a sign reversal across every bond. This is impossible, for, if c_1 disagrees in sign with c_2 and c_3, then c_2 and c_3 must agree in sign and cannot yield an antibonding interaction.

Not surprisingly, this has all been considered in a rigorous mathematical fashion. Systems containing a ring with an odd number of atoms are "nonalternant" systems. All other homonuclear unsaturated systems are "alternant" systems. An alternant system can always have asterisks placed on some of the centers so that no two neighbors are both asterisked or unasterisked. For nonalternants, this is not possible (see Fig. 8-11). It is convenient to subdivide alternant systems into even alternants or odd alternants according to whether the number of centers is even or odd. With this terminology defined, we can now state the *pairing theorem* and some of its immediate consequences.

The theorem states that, for alternant systems, (1) energy levels are paired such that, for each level at $E = \alpha + k\beta$ there is a level at $E = \alpha - k\beta$; (2) MOs that are paired in energy differ only in the signs of the coefficients for one of the sets (asterisked or unasterisked) of AOs.

It is easy to see that an immediate result of this theorem is that an odd-alternant system, which must have an odd number of MOs, must have a nonbonding ($E = \alpha$) MO that is not paired with another MO. It is also possible to show that the electron density is unity at every carbon for the neutral ground state of an alternant system. The proofs of the pairing theorem and some of its consequences are given in Appendix 5.

Another useful short cut exists that enables one to sketch qualitatively the MOs for any linear polyene. The HMOs for the allyl and butadiene systems are given in Fig. 8-12. Notice that the envelopes of positive (or negative) phase in these MOs are similar in appearance to the particle in a one-dimensional "box" solutions described

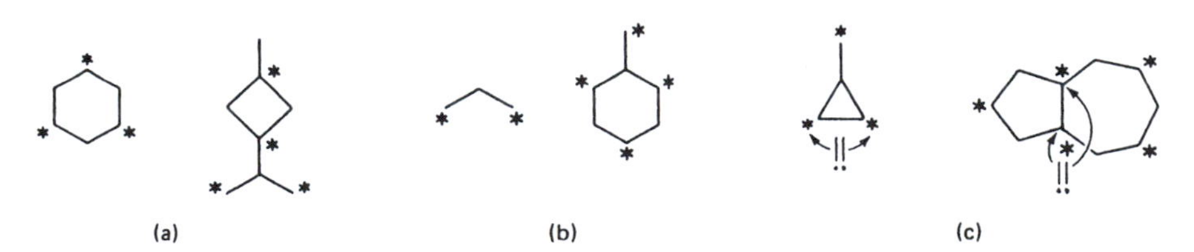

Figure 8-11 ▶ (a) Even and (b) odd alternants have no two neighbors identical in terms of an asterisk label. (c) Nonalternants have neighbors that are identical.

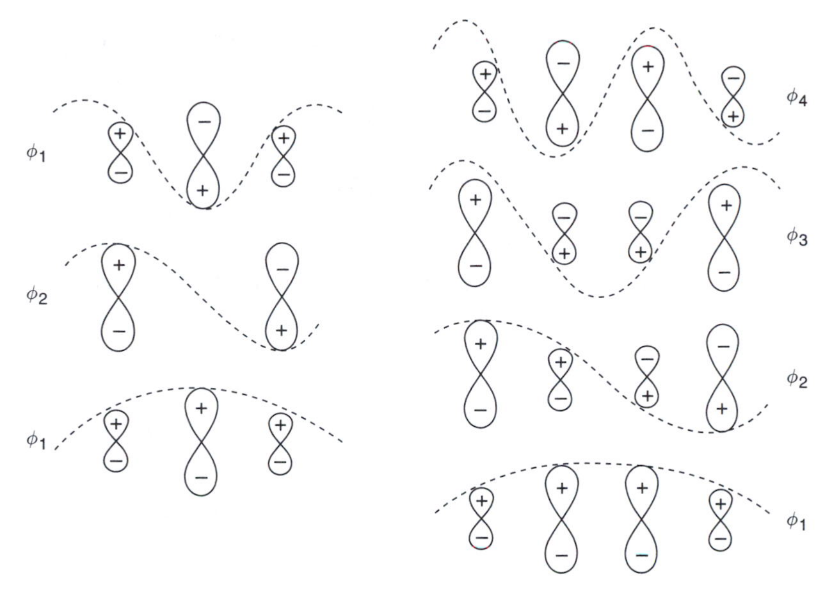

Figure 8-12 ► MOs for the allyl and butadiene systems. The dashed lines emphasize the similarity between an envelope, or contour, of positive ϕ_π for these systems and the particle in a one-dimensional box solutions.

in Chapter 2. This similarity makes it fairly easy to guess the first few MOs for pentadienyl, hexatriene, etc. Also, if one knows the lowest-energy half of the MOs for such molecules, one can generate the remaining MOs by appeal to the second part of the pairing theorem. (The edges of the one-dimensional box should extend one C–C bond length beyond the terminal atoms.)

As we consider larger systems, brute-force solution of determinantal equations becomes too labor-intensive, and so such cases are always either solved on a computer[7] or else by appeal to HMO tabulations in print.[8] However, the above generalizations continue to be useful in understanding the MO results.

8-10 HMO Calculations on Some Simple Molecules

Thus far, we have used the allyl and cyclopropenyl systems as examples. We will now describe the results of HMO calculations on some other simple but important systems.

8-10.A Ethylene (Even Alternant)

The Hückel determinantal equation is

$$\begin{vmatrix} x & 1 \\ 1 & x \end{vmatrix} = 0$$

[7]Many types of quantum-chemical computer programs are available from: Quantum Chemistry Program Exchange, Chemistry Department. Indiana Üniversity, Bloomington, Indiana 47401. On the Internet at http://www.QCPE.Indiana.edu/

[8]See Coulson and Streitwieser [4], Streitwieser and Brauman [5], and Heilbronner and Straub [6]. See also Appendix 6 of this text.

and so $x^2 - 1 = 0$; $x = +1, -1$. The resulting orbital energies and coefficients are

$$E_1 = \alpha + \beta, \phi_1 = \frac{1}{\sqrt{2}}\chi_1 + \frac{1}{\sqrt{2}}\chi_2$$
$$E_2 = \alpha - \beta, \phi_2 = \frac{1}{\sqrt{2}}\chi_1 - \frac{1}{\sqrt{2}}\chi_2 \tag{8-59}$$

These, with the ground state electronic configuration indicated, are shown in (**II**). π-Electron densities and π-bond order are indicated in the diagram beneath the MO sketches. Ethylene is an even alternant, so it has paired energies, unit electron densities, and coefficients related by a sign change.

$E_2 = \alpha - \beta$

$E_1 = \alpha + \beta$

$E_\pi = 2\alpha + 2.00\beta$

(II)

8-10.B Butadiene (Even Alternant)

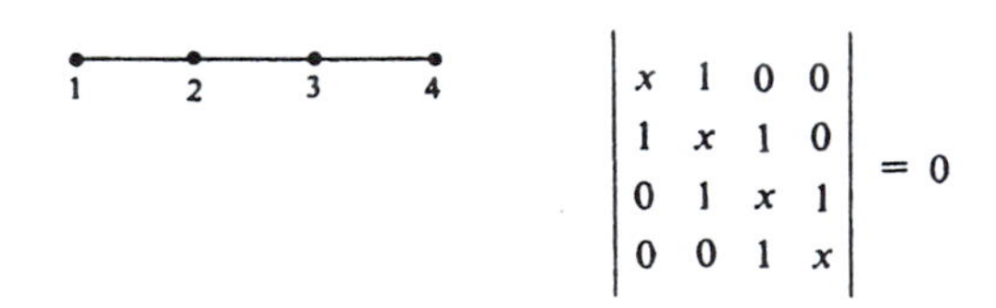

$$\begin{vmatrix} x & 1 & 0 & 0 \\ 1 & x & 1 & 0 \\ 0 & 1 & x & 1 \\ 0 & 0 & 1 & x \end{vmatrix} = 0$$

This problem can be solved by expansion to a polynomial in x and factoring, but it is simpler to use Eq. (8-49) or the decagon in a circle of radius $2|\beta|$. The coefficients are obtainable from Eq. (8-50). The results are

$$E_4 = \alpha - 1.618\beta, \qquad E_3 = \alpha - 0.618\beta, \qquad E_2 = \alpha + 0.618\beta,$$
$$E_1 = \alpha + 1.618\beta, \qquad E_\pi = 4\alpha + 4.472\beta$$
$$\underset{(4)}{\phi_1} = 0.372\chi_1 \underset{(-)}{+} 0.602\chi_2 + 0.602\chi_3 \underset{(-)}{+} 0.372\chi_4$$
$$\underset{(3)}{\phi_2} = 0.602\chi_1 \underset{(-)}{+} 0.372\chi_2 - 0.372\chi_3 \underset{(+)}{-} 0.602\chi_4 \tag{8-60}$$

0.894 0.447 0.894

1 1 1 1

The MOs are given in Fig. 8-12. ϕ_1 is bonding in all bonds, ϕ_2 is bonding in the outer bonds, antibonding in the central bond. As a result, butadiene has a lower π-bond order in the central bond than in the outer bonds. This is in pleasing accord with the experimental observation that the central bond in butadiene is significantly longer than the outer bonds.

The formal structural formula for butadiene (**III**), indicating two pure double bonds and one pure single bond, is clearly not an adequate description since we have just

(**III**)

found the central bond to have some π-bonding order (0.447) and the outer bonds to be less π bonding than ethylene (0.894 as opposed to 1.000). The MO parlance is that the π electrons in butadiene are *delocalized* over the entire carbon system rather than being restricted to the formal double bonds only.

Because the HMO method allows for no interaction between terminal carbons (i.e., $H_{1,4} = 0$), there is no distinction between *cis*- and *trans*-butadiene in this calculation. However, if a weak interaction were postulated, it is not difficult to see that ϕ_1 would give a bonding end-to-end contribution, ϕ_2 a substantially larger antibonding interaction, leading to a prediction (in agreement with experiment) that *trans*-butadiene is the more stable form.

8-10.C Cyclobutadiene (Even Alternant)

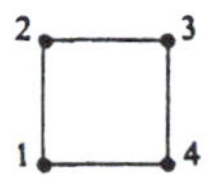

If we assume that this system is perfectly square, we can inscribe the square in a circle, giving us the orbital energies immediately (**IV**).

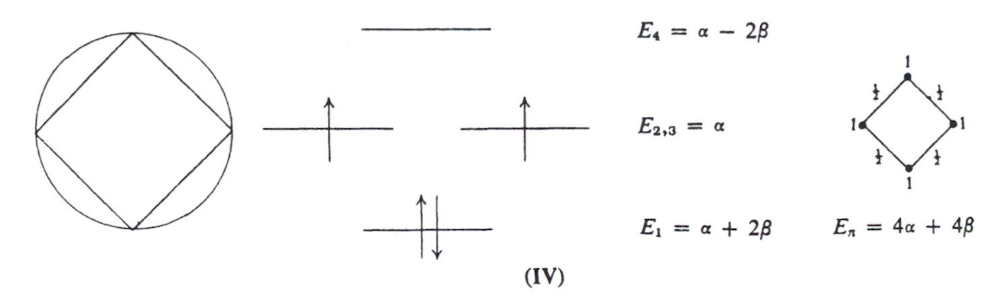

(**IV**)

Since ϕ_1 and ϕ_4 are nondegenerate, they must be symmetric or antisymmetric for the various rotations and reflections of the molecule. Also, ϕ_1 and ϕ_4 must have the same coefficients, except for sign changes.

It follows at once that

$$\phi_1 = \frac{1}{2}\chi_1 + \frac{1}{2}\chi_2 + \frac{1}{2}\chi_3 + \frac{1}{2}\chi_4, \quad \phi_4 = \frac{1}{2}\chi_1 - \frac{1}{2}\chi_2 + \frac{1}{2}\chi_3 - \frac{1}{2}\chi_4$$

ϕ_2 and ϕ_3 are degenerate, so there is some arbitrariness here. However, we expect each of these MOs to have a nodal plane, and these planes should be perpendicular to each other if ϕ_2 and ϕ_3 are to be orthogonal. Therefore, we choose the pair having nodal planes indicated by dashed lines (**V**). The four MOs for cyclobutadiene *as seen from* *above* are shown in **(VI)**. It is clear that ϕ_2 and ϕ_3 are nonbonding because the nodal planes prevent interactions between neighbors, but it looks like ϕ_2 and ϕ_3 violate the pairing theorem since they cannot be interchanged by changing signs of coefficients. This is only an apparent violation, because it is easy to find an equivalent pair of MOs that follow the rule. We need only choose nodal planes that are rotated 45° **(VII)** from those we selected previously. These MOs are linear combinations of ϕ_2 and ϕ_3 and are equivalent to them for purposes of calculating electron densities and bond orders. They are still nonbonding MOs (E still equals α), but now it is not because of a nodal plane preventing nearest-neighbor interactions, but because bonding and antibonding interactions occur in equal number and magnitude.

Notice that this is an example of an even alternant system with nonbonding MOs. Thus, whereas an odd alternant system *must* have *an* unpaired nonbonding MO, even alternants *may* have nonbonding MOs in pairs.

$\phi_2 = (1/\sqrt{2})\chi_1 - (1/\sqrt{2})\chi_3$ $\quad$ $\phi_3 = (1/\sqrt{2})\chi_2 - (1/\sqrt{2})\chi_4$

(V)

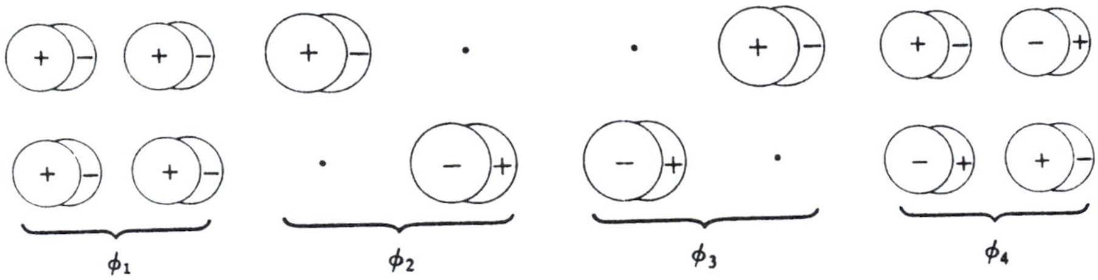

ϕ_1 $\quad$ ϕ_2 $\quad$ ϕ_3 $\quad$ ϕ_4

(VI)

$\phi_2' = \frac{1}{2}\chi_1 - \frac{1}{2}\chi_2 - \frac{1}{2}\chi_3 + \frac{1}{2}\chi_4$ $\quad$ $\phi_3' = \frac{1}{2}\chi_1 + \frac{1}{2}\chi_2 - \frac{1}{2}\chi_3 - \frac{1}{2}\chi_4$

(VII)

Since this system is alternant, it must have π-electron densities of unity in its ground neutral state. This would be necessary however, even if the molecule were not alternant, due to the fact that all carbons are equivalent by symmetry. Hence, they must all have the same electron density. Since it must sum to four electrons, the density of each atom must be unity. The same argument applies to the cyclopropenyl radical, a nonalternant that, nevertheless, has all electron densities equal to unity (if the unpaired electron is divided between degenerate MOs).

Our assumed square symmetry also requires all four bonds to be identical. Since the total energy, $4\alpha + 4\beta$, is related to bond order through Eq. (8-58), it follows at once that the total bond order is 2, and so each bond has order $\frac{1}{2}$.

EXAMPLE 8-6 We have assumed that cyclobutadiene is square. Could it be otherwise?

SOLUTION ▶ Square cyclobutadiene leads to degenerate, nonbonding pi MOs, each of which we are assuming to contain one electron. (See **IV**.) We have further assumed that these two electrons have the same spin. This gives a triplet state (assumed to be the lower-energy choice) which forces us to put each electron into a different MO. If cyclobutadiene were distorted into a rectangle by shortening bonds 1–2 and 3–4, and lengthening 2–3 and 1–4, we would expect ϕ_2' of **VII** to rise in energy and ϕ_3' to drop. If we continue having one electron in each MO, these changes would tend to cancel but, if we put both electrons into ϕ_3', the π energy would drop. However, this would require paired spins—a singlet. We have here, then, a competition between two factors: the energy rise of forming a singlet and the energy lowering from being distorted to a rectangle. HMO theory doesn't tell us which factor wins, but it does alert us to a possible extension to the Jahn-Teller effect: An odd number of electrons (one or three) in a doubly-degenerate level **must** cause distortion. An even number (two) **may** cause distortion if the destabilization caused by going from triplet to singlet is more than counterbalanced by the stabilization resulting from "distortion."[9] ◀

8-10.D Benzene (Even Alternant)

Benzene is another molecule whose high symmetry enables one to use shortcuts. The orbital energies have already been found from the hexagon-in-a-circle diagram (see Fig. 8-9). The lowest- and highest-energy MOs must show all the symmetry of the molecule, as they are nondegenerate. Therefore,

$$\underset{(6)}{\phi_1} = \frac{1}{\sqrt{6}}\left(\chi_1 \underset{(-)}{+} \chi_2 + \chi_3 \underset{(-)}{+} \chi_4 + \chi_5 \underset{(-)}{+} \chi_6\right)$$

where the carbon atoms are numbered sequentially around the ring.

The degenerate MOs ϕ_2 and ϕ_3 should have one nodal plane each, and these should be perpendicular to each other.[10] If we take one plane as shown in (**VIII**), we can immediately write down ϕ_2. The node for ϕ_3 is given in (**IX**). It is obvious that χ_6, χ_1, χ_2 have coefficients of the same sign, and that $c_2 = c_6 = -c_5 = -c_3$, and also $c_1 = -c_4$. However, c_1 need not equal c_2 as these atoms are differently placed with respect to the nodal plane. To determine these coefficients, we will use the fact that the neutral ground-state π densities are all unity in this system. We consider first atom number 1. Its electron density due to two electrons in ϕ_1 and two electrons in ϕ_2 is $\frac{1}{6} + \frac{1}{6} + 0 + 0 = \frac{1}{3}$.

$$\underset{(4)}{\phi_2} = \tfrac{1}{2}\left(\chi_2 \underset{(-)}{+} \chi_3 \underset{(+)}{-} \chi_5 - \chi_6\right)$$

(**VIII**)

(**IX**)

[9]Experiments and high-accuracy calculations indicate that the lowest-energy state for cyclobutadiene is a rectangular singlet. Surprisingly, calculations of high accuracy also indicate that, even in the square geometry, the $^1B_{1g}$ *singlet* state (corresponding to each degenerate π MO containing one electron, but with spins paired) lies below the triplet in energy—the opposite of what Hund's rule predicts. This is thought to result from energy lowering due to spin polarization. For a detailed review of studies on cyclobutadiene, see Minkin et al. [19].

[10]This statement applies to the real forms of ϕ_2 and ϕ_3. The complex forms [derivable from Eq. (8-52)] do not have a planar node. The situation is analogous to the $2p_{+1}$, $2p_{-1}$ versus $2p_x$, $2p_y$ orbitals for the H atom.

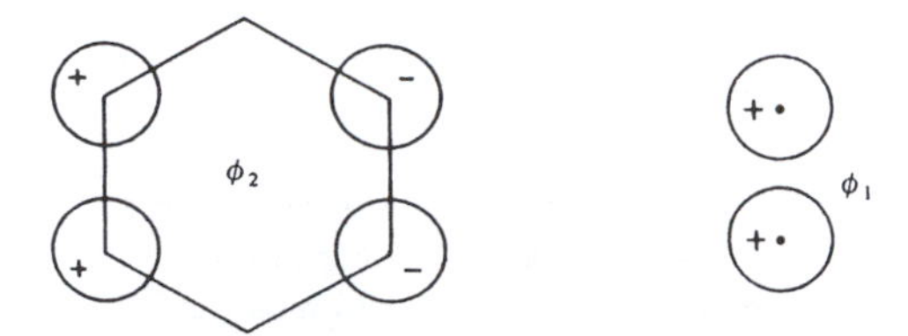

Figure 8-13 ▶ ϕ_2 for benzene and ϕ_1 for ethylene have the same HMO energy. The MOs are sketched as seen from above.

Therefore, two electrons in ϕ_3 must produce a contribution of $\frac{2}{3}$. Hence the coefficient for this atom in ϕ_3 must be $1/\sqrt{3}$. A similar argument for atom 2 gives a coefficient of $1/\sqrt{12}$ (or, one can use the normality condition for ϕ_3). As a result,

$$\underset{(5)}{\phi_3} = \frac{1}{\sqrt{3}}\left(\chi_1 \underset{(-)}{+} \frac{1}{2}\chi_2 - \frac{1}{2}\chi_3 \underset{(+)}{-} \chi_4 - \frac{1}{2}\chi_5 \underset{(-)}{+} \frac{1}{2}\chi_6\right)$$

Appeal to the pairing theorem generates ϕ_4 and ϕ_5 as indicated.

By symmetry, all bond orders must be identical, and their sum must be 4, since $E = 6\alpha + 8\beta$. Therefore, $p_{12} = p_{23} = \cdots = p_{61} = \frac{4}{6} = 0.667$.

The hexagon in a circle applies to ethylene ($n = 2$, $2n + 2 = 6$) as well as to benzene. As a result, the orbital energies for ϕ_2, ϕ_3 of benzene are identical to E_1 for ethylene. Examination of these MOs (Fig. 8-13) makes the reason for their energy agreement clear. Molecular orbital ϕ_2 of benzene is more revealing than is ϕ_3 in this context. The nodal plane produces an MO corresponding to two *noninteracting* ethylene MOs. Hence, an electron in this MO is always in a situation that is indistinguishable (under HMO approximations) from that in ϕ_1 of ethylene.

In this section, we have tried to illustrate some of the properties of HMO solutions for simple systems and to indicate how symmetry and other relations are useful in producing and understanding HMO results. It is often convenient to have HMO results for various simple systems readily available in a condensed form. Therefore, a summary of results for a number of molecules is provided in Appendix 6.

8-11 Summary: The Simple HMO Method for Hydrocarbons

1. The assumption is made that the π-electron energy can be minimized independently of σ electrons. This is an approximation.

2. The assumption is made that each π electron sees the same field (the repulsion due to the other π electrons is presumably included "in effect," in a time averaged way) so that the π electrons are treated as independent particles. This approximation leads to a total wavefunction that is a simple product of one-electron MOs and a total π energy that is a sum of one-electron energies. Except for use of the Pauli principle to build up configurations, no explicit treatment is made of electron spin.

3. The basis set is chosen to be a $2p_\pi$ AO from each carbon atom in the unsaturated system. Choosing a basis set of AOs means our MOs will be linear combinations of AOs, and so this is an LCAO–MO method.

4. The Hückel determinant summarizes the connectedness of the unsaturated system, and is independent of *cis–trans* isomerism or bond length variation.

5. The energy of each MO is expressed in terms of atomic terms, α, and bond terms, β. The amount of α in each MO energy is always unity because the sum of π-electron densities for one electron in the MO is always unity. The amount of β present is related to the net bonding or antibonding character of the MO.

6. Alternant systems display paired energy levels and corresponding MOs having coefficients related by simple sign reversals. For ground-state neutral alternants, the π electron densities are all unity.

7. A caveat: One can perform HMO calculations on very large systems such as pentahelicene (**X**) thereby making the implicit assumption that this is a planar molecule. But repulsion between protons on the terminal rings is sufficient to cause this molecule to deviate from planarity. Similarly, cyclooctatetraene is tub-shaped, probably due to considerations of angle strain that would occur in the planar molecule. Hence, one must recognize that, in certain cases, an HMO calculation refers to a planar "ideal" not actually achieved by the molecule.

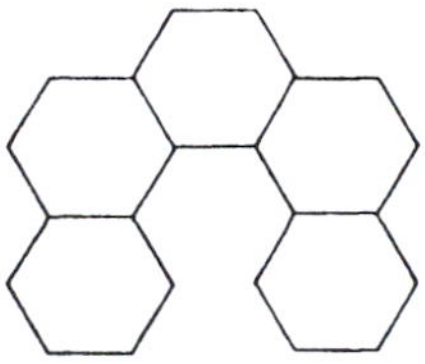

(X)

8-12 Relation Between Bond Order and Bond Length

In this and following sections we will describe some of the relations between HMO theoretical quantities and experimental observations.[11]

It is natural to look for a correlation between calculated π-bond orders and experimentally determined bond lengths. A high bond order should correspond to a large π charge in the bond region, which should yield a shorter, stronger bond. The bond-order–bond-length results for certain simple systems, given as a graph in Fig. 8-14, do indeed show the anticipated behavior. However, as more and more data are added (Fig. 8-15) it becomes clear that an exact linear relation between these quantities does not exist at this level of refinement.

Nevertheless, the correlation between bond order and bond length is good enough to make it useful for rough predictions of bond length *variations*. An example of this is given in Fig. 8-16, where calculated and observed bond lengths for phenanthrene are plotted. Even though the predicted *absolute* values of bond lengths are imperfect, the theoretical values show a rough parallelism with observed *changes* in values as we go from bond to bond. The relation used here to calculate the theoretical bond lengths from HMO bond orders is due to Coulson [8] and has the form

$$R = s - \frac{s-d}{1 + k(1-p)/p} \tag{8-61}$$

[11] For a more complete discussion of these phenomena as well as many others, see Streitwieser [7].

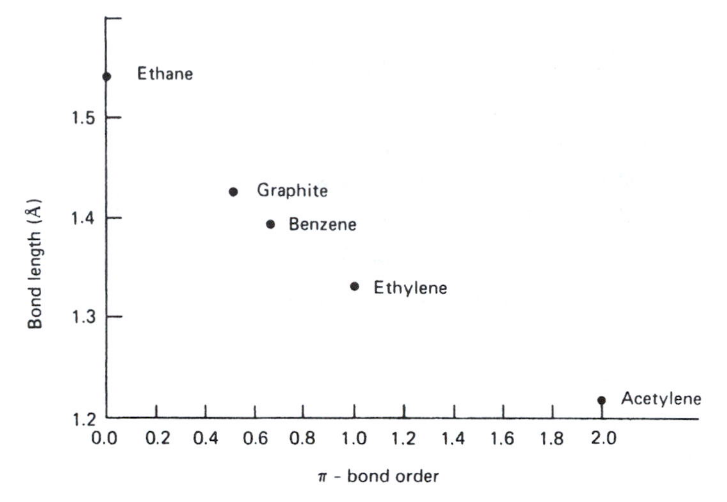

Figure 8-14 ▶ π-Bond order versus bond length for some simple unsaturated hydrocarbons.

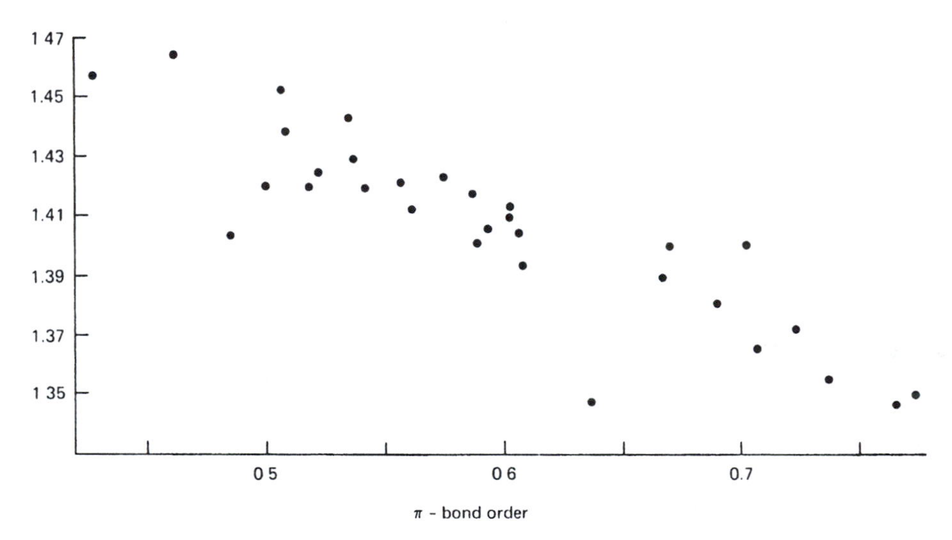

Figure 8-15 ▶ Bond lengths versus π-bond orders for benzene, graphite, naphthalene, anthracene, phenanthrene, triphenylene, and pyrene.

Here, s is the single bond length, d the double bond length, p the π-bond order, k an adjustable parameter, and R the predicted length. The double bond length d is taken to be 1.337 Å, the bond length of ethylene. The single bond length may be taken to be the length of the C–C bond in ethane, 1.54 Å, or it may be set by fitting to data points (such as those in Fig. 8-15) on the assumption that the single bond between two CH_2 groups (i.e., ethylene with its π bond "turned off") is not necessarily the same length as the single bond between two CH_3 groups. Both of these alternatives have been used in Fig. 8-16. Other mathematical forms have also been suggested. Because of the scatter in the data, there is little basis for preferring one formula over another. It seems generally true that all the proposed relationships work best for bonds in condensed ring systems, and most poorly for bonds in acyclic polyenes (e.g., butadiene) or between rings (e.g., biphenyl).

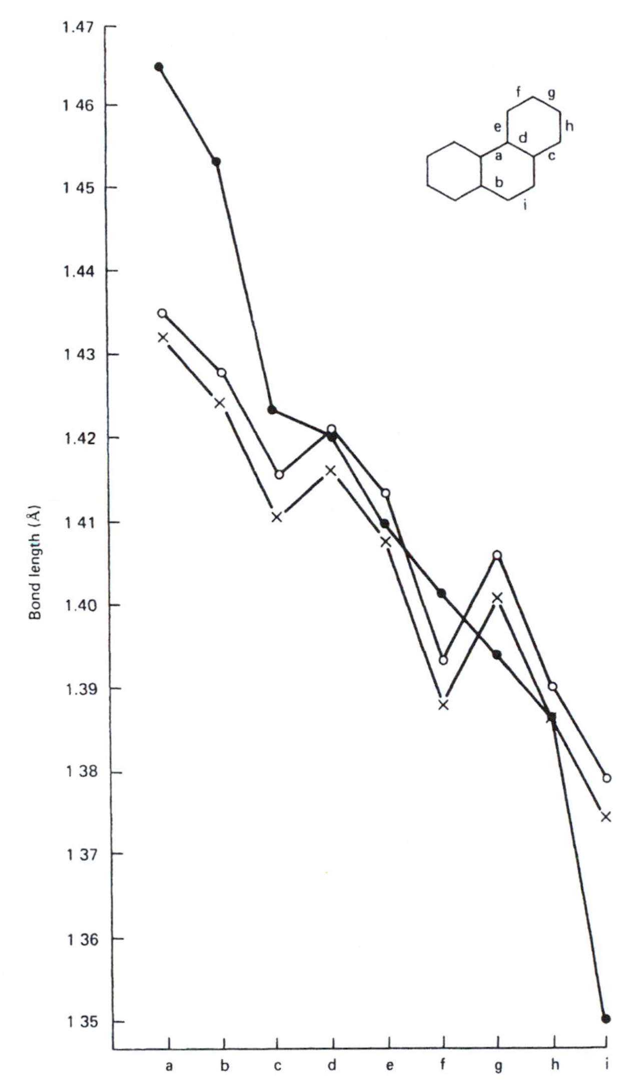

Figure 8-16 ▶ Theoretical versus experimental bond lengths for phenanthrene in order of decreasing observed length: (•) experimental; (×) theoretical with $s = 1.54\,\text{Å}$, $k = 0.765$; (○) theoretical with $s = 1.515\,\text{Å}$, $k = 1.05$.

8-13 π-Electron Densities and Electron Spin Resonance Hyperfine Splitting Constants[12]

If a molecule with one or more unpaired electron spins is placed in an external magnetic field, the interaction energy of an unpaired electron's spin dipole with the external field will be different depending on whether the electron's spin is α or β. This produces two slightly different energy levels for an unpaired electron. Suppose microwave radiation is supplied in an effort to excite such electrons from the lower to the higher of these two levels. If the radiation's energy is larger than required, no absorption occurs. But, if we gradually increase the strength of the magnetic field, thereby increasing the energy-level

[12]A number of reviews on this subject have been published. See, for example, Gerson and Hammons [9].

gap, we eventually achieve a magnetic field where absorption of the microwave energy occurs. This would give us a spectrum (absorption vs. magnetic field) with just one line.

Now suppose that our radical has a hydrogen atom attached to it. The proton nucleus of the hydrogen can have a spin of α or β, and this can have a small effect on the energies of the electron whose spin we are trying to flip. We find now that half of the radicals absorb at a slightly different applied magnetic field than the other half giving us a two-line spectrum: The original single line now exhibits *hyperfine splitting* due to the presence of a hydrogen atom, and the distance between the two peaks on the magnetic axis is called the *hyperfine splitting constant*, usually given in millitesla (mT), or in gauss (G).

If the unpaired electron is a π electron, HMO theory can be applied to the value of the ESR splitting constant. The interaction between an *unpaired π electron* and a proton (or other nucleus with nonzero nuclear spin) falls off rapidly with increasing separation. Therefore, the hyperfine structure is generally ascribed to interactions involving protons directly bonded to carbons in the π system (α protons) or else separated from the π system by two σ bonds (β protons). The equilibrium position of an α proton is in the nodal plane of the π system, so it is clear that any net spin density at the proton must be only indirectly due to the presence of an unpaired π electron. This indirect effect arises because the unpaired π electron interacts slightly differently with α- and β-spin σ electrons on carbon, and so the spatial distributions of these become slightly different in the σ MOs, ultimately producing net spin density at the proton; the σ electrons are said to be *spin polarized* by the π electron (see Fig. 8-17).

The extent of spin polarization at a given hydrogen should depend on the percentage of time the unpaired π electron spends on the carbon to which that hydrogen is bonded. It is therefore reasonable to look for relationships between the distribution of *π-spin* density in a radical and the hyperfine coupling constants characterizing its ESR spectrum. The simplest assumption one can make in this regard, called the McConnell relation, is that the hyperfine splitting constant, a_{H_μ}, for a proton directly bonded to the μth carbon, is proportional to the net π-spin density ρ_μ on that carbon:

$$a_{H_\mu} = Q\rho_\mu \tag{8-62}$$

Thus, if the unpaired π electron density is twice as great on carbon 1 as on carbon 2, the ESR splitting constant for a proton attached to carbon 1 should be twice as great as that for a proton attached to carbon 2.

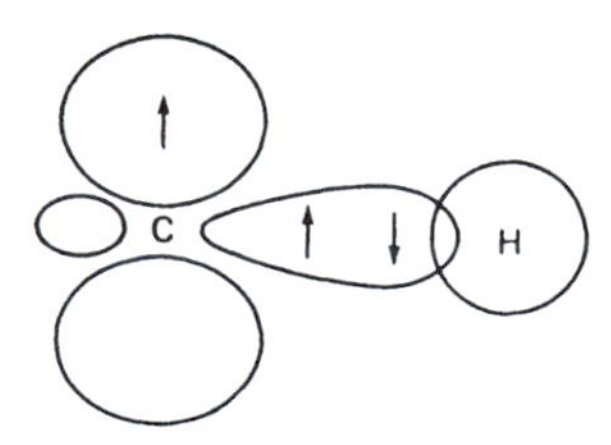

Figure 8-17 ▶ The unpaired π-spin density at carbon repels both σ electrons in the C–H bond region, but does not repel them equally. As a result, slight spin imbalance due to σ electrons occurs at the proton.

The HMO method predicts that the spin density at the μth carbon due to an unpaired electron in the mth MO is simply $|c_{\mu m}|^2$. This is only a first approximation to ρ_μ, but a graph of observed splitting constants plotted against ρ_μ calculated in this way shows a respectable correlation (Fig. 8-18). The proportionality factor Q, given by the slope of the line of best fit, varies somewhat depending on the type of system and the charge of the radical. Thus, in Fig. 8-18, where all the data are from aromatic fused ring hydrocarbon radical anions, the correlation is quite good. As data for nonalternants and radical cations are added (Fig. 8-19) the scatter increases.

Some ESR spectra are interpreted to be consistent with the presence of some *negative* spin density. That is, if the extra π electron is taken to have α spin, some of the carbons appear to have an excess of β-spin π density. This has the effect of giving us a total spin density that is greater than unity, even though we have only one electron with unpaired spin. (It sounds absurd, but read on.) Now the amount of splitting seen in ESR spectra depends only on the *magnitudes* of spin densities at the protons, not on whether these spin densities are α or β, and the presence of negative spin density does not produce a *qualitative* change in an ESR spectrum (such as a "negative splitting," whatever that might be). Rather it leads to an increase in the total *amount* of spin density in the system, and this, in turn, leads to an increase in the sum of splitting constants (over

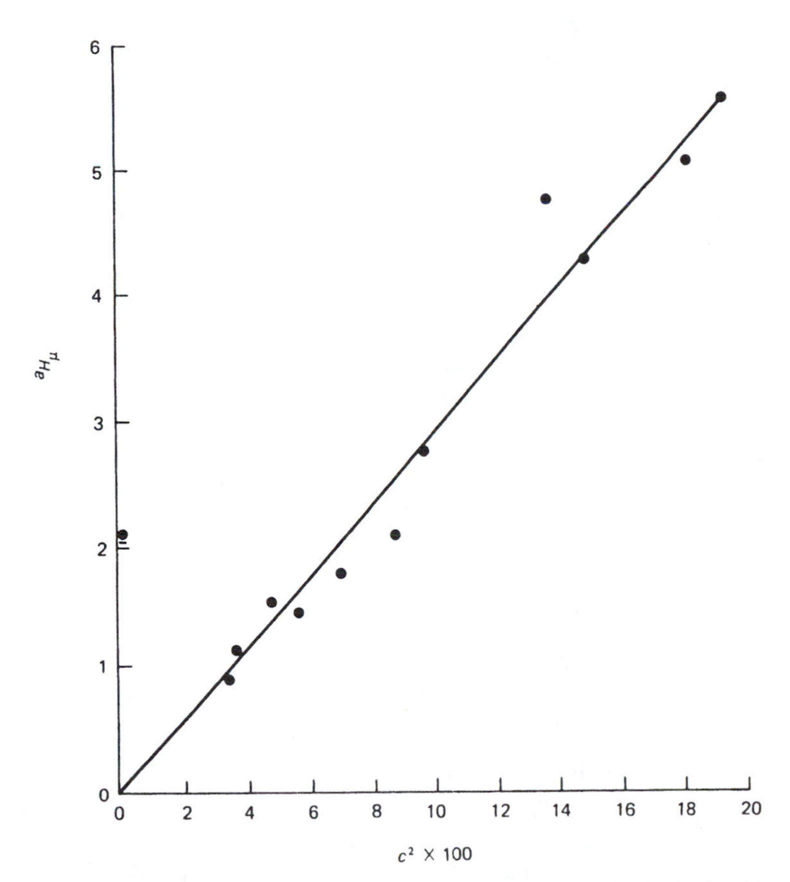

Figure 8-18 ▸ ESR splitting constants a_{H_μ} in gauss versus HMO unpaired spin densities. The systems are fused ring alternant hydrocarbon radical anions (naphthalene, anthracene, tetracene, pyrene). The underlined point is thought to result from negative spin density. (Data from Streitwieser [7].)

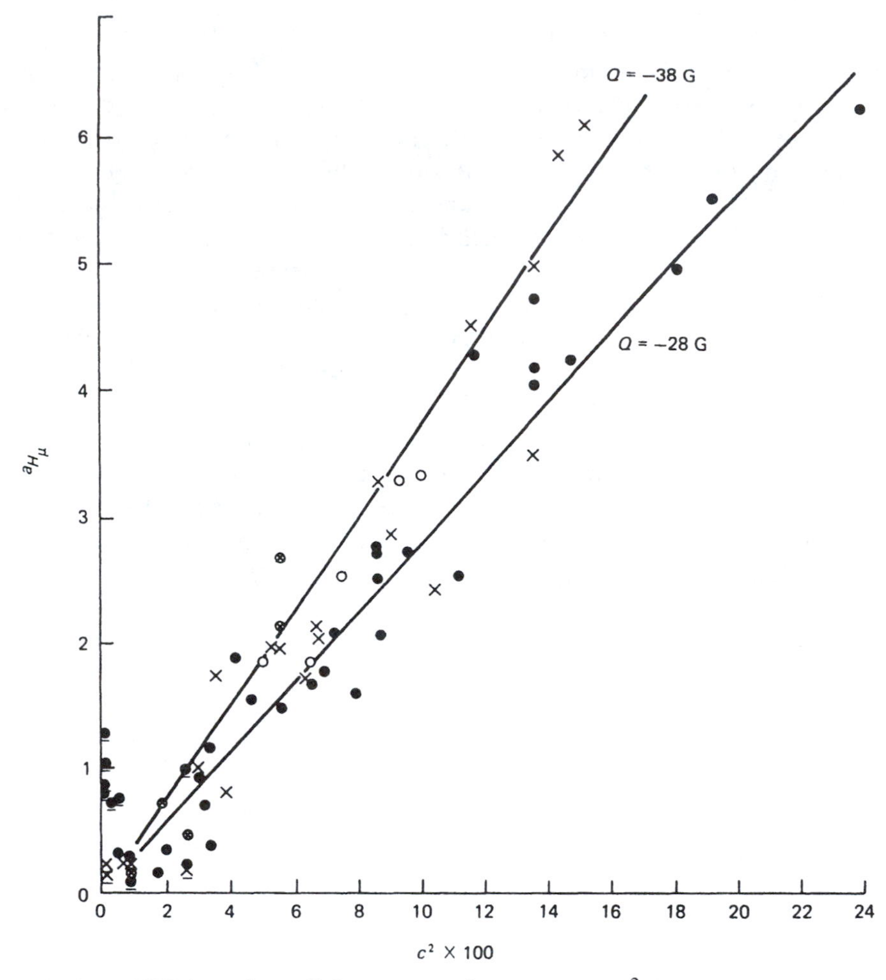

Figure 8-19 ▶ ESR hyperfine splitting constants in gauss versus c^2 for the highest occupied MO of the radical. All data are from hydrocarbon radical anions or cations of alternant or nonalternant type. (•) Anion radical; (×) cation radical. Uncertain assignments: (○) anion; (⊗) cation. The two correlation lines are merely sight fitted to the anion and cation points separately and suggest that Q for cations should be larger than for anions. Points thought to result from negative spin density are underlined. [See Tables I, III, VIII, XII, XIII, XIV, XV, XVI, XVII, XVIII of Gerson and Hammons [9], and Table 6.2 of Streitwieser [7] for data plotted here.]

what would be predicted in the absence of negative spin density). For example, the HMO prediction for allyl radical would give a net spin density of $\frac{1}{2}$ at each terminal carbon and zero at the central carbon. [The unpaired electron is in the nonbonding MO $\frac{1}{\sqrt{2}}\chi_1 - \frac{1}{\sqrt{2}}\chi_3$.] However, we might imagine the situation wherein there is a net spin of $\frac{2}{3}\alpha$ at each terminal carbon and $\frac{1}{3}\beta$ at the central atom. This retains a net spin value of 1α yet results in a larger splitting constant a_{H_μ} for every proton in the molecule. It is possible to think of a physical explanation for this kind of spin distribution. The π electrons in lower-energy, filled MOs are being spin polarized similarly to the σ electrons mentioned earlier. If an α-spin electron in effect repels electrons of β spin more strongly than those of α spin, a buildup of β spin on the central carbon of allyl radical could be expected. Because this secondary effect is expected to be fairly small, a net negative spin density is likely only on carbons where the primary effect ($c_{\mu m}^2$) is zero or quite small. Methods for calculating this spin polarization (which will not be described here) indicate that certain of the splitting constants plotted in Figs. 8-18 and 8-19 do in fact result from negative spin densities. These data points are indicated in the figures, and it can be seen that they do indeed occur where $c_{\mu m}^2$ and a_{H_μ} are small. This secondary effect is masked at higher values of $c_{\mu m}^2$ and a_{H_μ}. but presumably accounts for some of the scatter in the data. (Some scatter also results from solvent dependence of a_{H_μ}. Several solvents were used in experiments yielding the data in Fig. 8-19.)

Because of the pairing-theorem relation between coefficients of MOs, the ESR spectra of the radical cation and anion of an alternant system should appear very similar. Hückel molecular orbital theory would predict them to be identical except for a slight change of scale due to a change in the factor Q. In practice, this similarity has been observed to hold fairly well.

8-14 Orbital Energies and Oxidation-Reduction Potentials

Many conjugated hydrocarbons can be oxidized or reduced in solution using standard electrochemical techniques. Since oxidation involves removing an electron from the highest occupied (π) MO (HOMO), it is reasonable to expect molecules with lower-lying HOMOs to have larger oxidation potentials. Similarly, we might expect reduction to be easier for compounds wherein the lowest unoccupied MO (LUMO) is lower in energy [see (**XI**)]. Thus, compound A should have a lower oxidation potential than compound B since $E_m^A > E_m^B$, but compound B should have the lower reduction potential. A plot of oxidation potential versus E_m in units of β for a series of aromatic hydrocarbons (Fig. 8-20) yields a correlation that is remarkably free of scatter. Figure 8-21 indicates a similar correlation for reduction potential versus E_{m+1}.

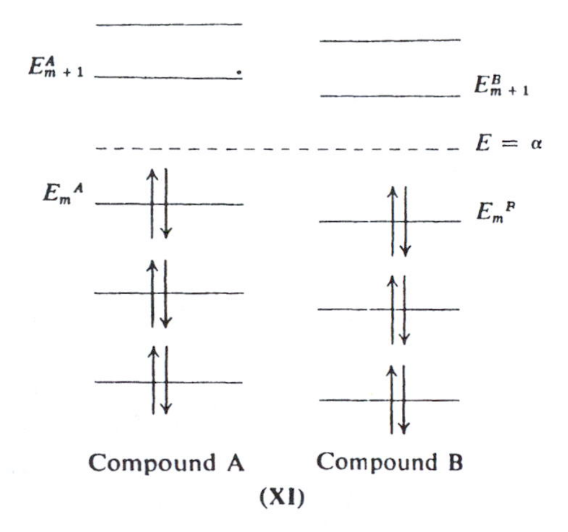

(**XI**)

Since oxidation-reduction potentials correlate with HOMO–LUMO energies, and since these energies are paired in even alternant systems, we should expect a plot of oxidation vs. reduction potential for such compounds to be linear also. (Problem 8-14.)

The data in Figs. 8-20 and 8-21 provide a connection between theoretical energy differences in units of β, and experimental energies. From the slope in Fig. 8-20 we

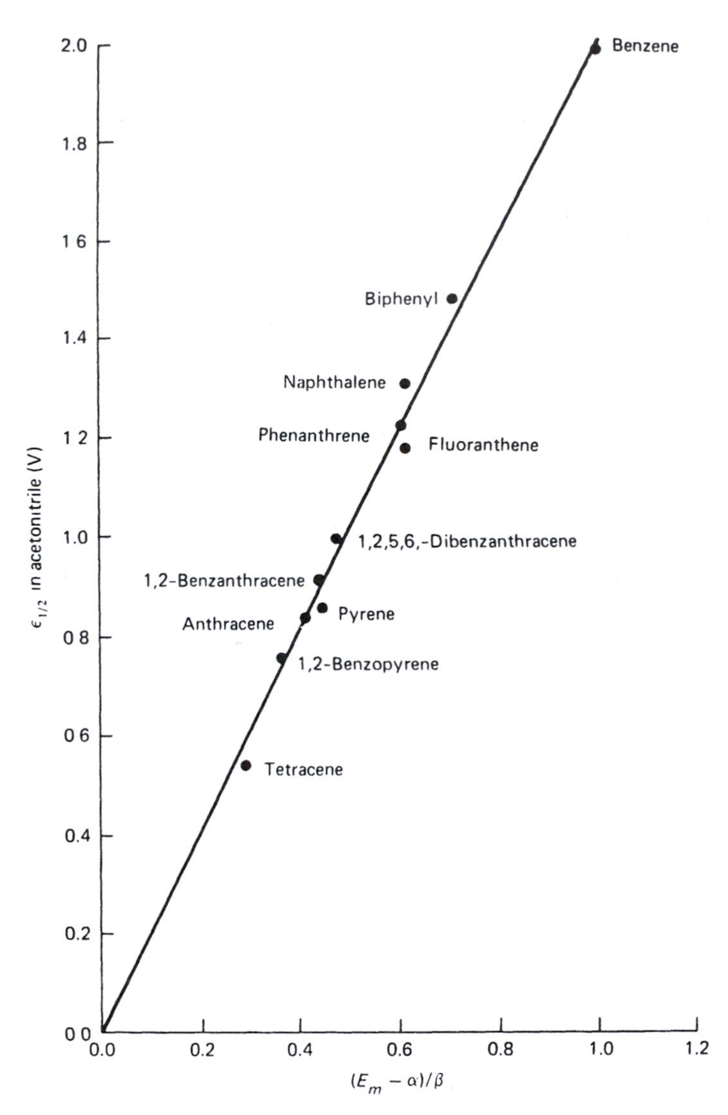

Figure 8-20 ▶ Oxidation potentials in acetonitrile solution versus energy of HOMO (in units of β). ($\epsilon_{1/2}$ from Lund [10].)

obtain that $\beta \cong -2.03\,\text{eV}$, or $-46.8\,\text{kcal/mole}$. Similarly, the reduction potential data of Fig. 8-21 give $\beta = -2.44\,\text{eV} = -56.3\,\text{kcal/mole}$.

One must be cautious in interpreting the above values of β. The problem is that the experimental numbers include effects of physical processes not included in the theory. For example, when a neutral molecule in solvent becomes oxidized or reduced, solvation energy changes occur. One might argue that the fit of the data to straight lines in Figs. 8-20 and 8-21 implies this sort of contribution to be small, but the sensitivity of redox potentials to solvent nature indicates that this is not the case. However, in larger molecules, solvation energy change upon ionization tends to be smaller, and larger molecules also tend to have E_m and E_{m+1} closer to the $E = \alpha$ level. In other words, we expect both solvation energy change *and* redox potential to be proportional to E_m or E_{m+1}. Therefore, they can be combined in a single linear relation. There

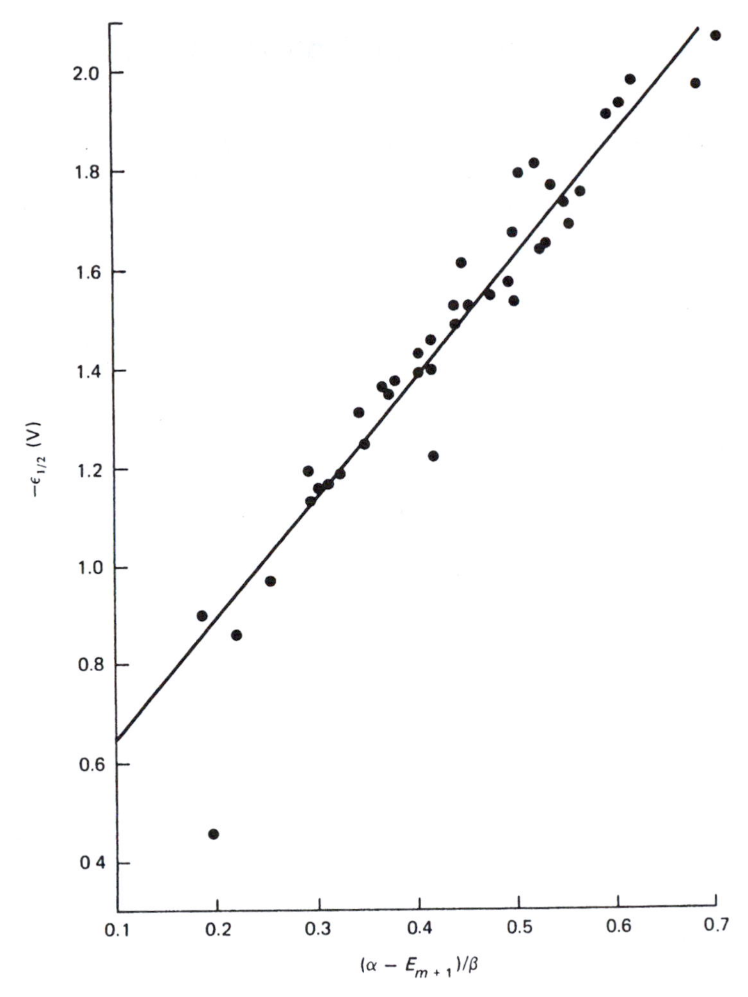

Figure 8-21 ▶ First reduction potential in 2-methoxyethanol versus energy of LUMO (in units of β). ($\epsilon_{1/2}$ from Bergman [11].)

is, in general, no guarantee that "extra" contributing effects will be of a nature to be correctly assimilated into the theoretical formula, but, for this effect, it happens that this is at least partially the case.

Another extra effect to consider is the π-electron repulsion energy. Because the energies of the HOMO and LUMO are both associated with the neutral molecule, their energies fail to reflect the decrease or increase of π-electron repulsion energy resulting from loss or gain of a π electron. Here again, however, we expect the magnitude of the effect to be larger for smaller molecules, where the change in π densities is greatest, and also where E_m and E_{m+1} deviate most from α. Thus, as before, this extra effect will not necessarily upset the linear relation expected from simpler considerations.

In view of the crudity of the HMO method together with the fact that the empirical value of β includes the effects of several extra processes, the β values cited above for oxidation and reduction are considered to be in fairly good agreement, especially considering that the electron repulsion change operates to make oxidation easier and reduction harder.

8-15 Orbital Energies and Ionization Energies

Suppose that monochromatic light is beamed into a gaseous sample of a compound. If the light is of sufficient energy, electrons will be "knocked out" of the molecules. The kinetic energy of such a photoelectron will be equal to the kinetic energy of the incident photon ($h\nu$) minus the energy needed to remove the electron from the molecule, that is, the ionization energy. Measurement of the kinetic energies of photoelectrons emitted in this manner is known as "photo-electron spectroscopy" [12].

In measuring the kinetic energies of photoelectrons from, say, benzene, it is found that a large number are near a particular energy value, another large number of electrons are near a different value, and so on for several kinetic energy values. Because the photoelectrons tend to clump near several kinetic energy values, it follows that we are, in effect, measuring several ionization energies. It is reasonable to associate these with removal of electrons from different MOs of the molecule.[13] A correlation plot (Fig. 8-22) between HMO orbital energies and experimentally measured ionization energies for a number of alternant and nonalternant hydrocarbons has been produced by Brogli and Heilbronner [13]. Their best fit was achieved using $\alpha = -6.553 \pm 0.340\,\text{eV}$, $\beta = -2.734 \pm 0.333\,\text{eV}$, where the limits define a range for the predicted ionization energy (IE) that will include the experimental value nine

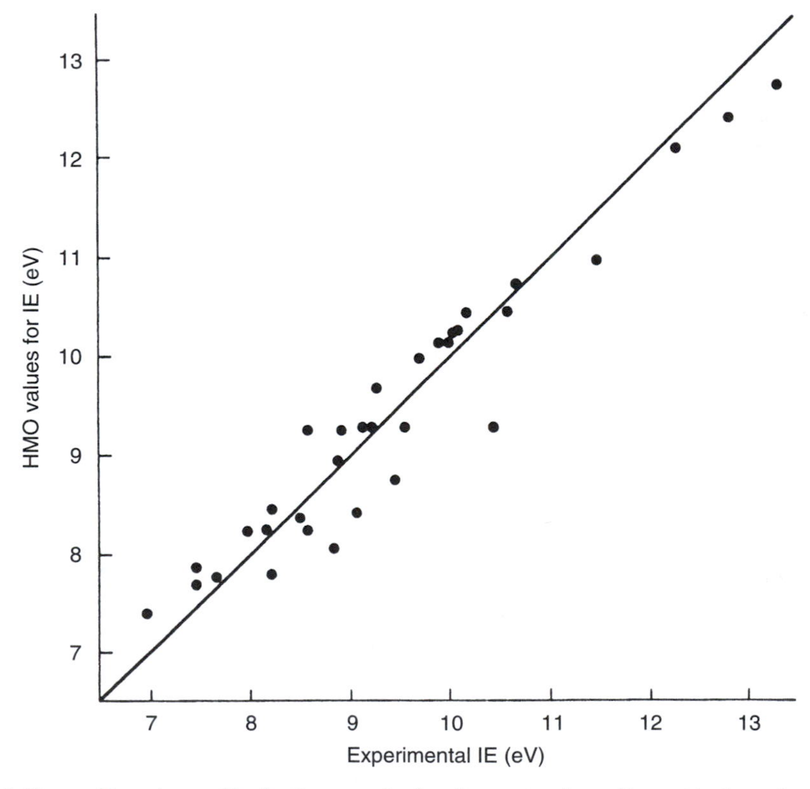

Figure 8-22 ▶ Experimental ionization energies for alternant and nonalternant hydrocarbons versus HMO orbital energies using $\alpha = -6.553\,\text{eV}$, $\beta = -2.734\,\text{eV}$.

[13] Note that each ionization energy is associated with removal of an electron from a different MO of the *neutral* molecule. This differs from the first, second, etc. ionization energies produced by *successive* ionization.

times out of ten (a 90% confidence level). There is a fair degree of scatter in the correlation plot. By making some additional refinements in the theory, Brogli and Heilbronner succeeded in substantially improving the correlation. We will describe these refinements in Section 8-18.

In this section, we have dealt only with ionization energies assigned to removal of a π electron. Removal of electrons from σ orbitals is also observed. If the impinging photons are of X-radiation frequencies, the inner shell electron ionization energies are seen. Quantum-chemical treatments of these ionization energies are possible using theoretical methods described later.

8-16 π-Electron Energy and Aromaticity

When propene is hydrogenated to form propane, the increase in heat content ΔH°_{298} (called the *heat of hydrogenation*) is -30.1 kcal/mole. The heat of hydrogenation for 1-butene is -30.3 kcal/mole. In general, the heat of hydrogenation of an isolated double bond is about -30 kcal/mole. A similar constancy holds for the contribution of a double bond to the heat of formation of a molecule. Therefore, isolated double bonds fit easily into the usual chemical device of estimating the energy of a molecule by adding together contributions from the substituent parts. This additivity appears to be violated when double bonds are conjugated. Thus, the heat of hydrogenation for *trans*-1,3-butadiene is -57.1 kcal/mole compared with the value of -60.6 kcal/mole for a pair of butene double bonds. Since the hydrogenation product in each case is butane, the energy difference of 3.5 kcal/mole must be due to the greater stability of conjugated double bonds.

There is a theoretical parallel to this. The HMO energy for butadiene is $4\alpha + 4.472\beta$. For a pair of *isolated* double bonds, we double the energy of ethylene to obtain $4\alpha + 4\beta$. Therefore, the HMO method indicates that the conjugated double bonds are stabilized by 0.472β.

Because the π electrons in butadiene are delocalized over all three C–C bonds, this 0.472β has often been referred to as *delocalization energy*. It was common practice for many years to equate this theoretical delocalization energy for a molecule to its experimentally measured "extra" stability (e.g., for butadiene, $0.472\,|\beta| = 3.5$ kcal/mole).

In 1969, Dewar and co-workers [14, Chapter 5; 15; 16] demonstrated that conjugated double bonds may be successfully included in an additivity scheme. Hess and Schaad [17, 18] have considered this idea in the context of the HMO method. They distinguish between several kinds of C–C single and double bond energy as indicated in Table 8-2. Using these values of bond energy, the π energy of butadiene is calculated to be $2 \times 2.000\beta + 0.4660\beta + 4\alpha = 4\alpha + 4.4660\beta$ compared with the HMO result of $4\alpha + 4.472\beta$. The difference is 0.006β, less than 0.002β per π electron. Hess and Schaad show that this level of agreement holds for acyclic polyenes in general, even when there is much branching. Thus, it seems that conjugated double bonds in acyclic molecules fit into an additivity scheme after all.

There is an important difference between the energy additivity scheme for conjugated systems described above and the familiar additivity scheme used for C–H, C–C, and isolated C–C bonds. In the latter cases the energy "contributed" by the bond is generally thought of as being the same as the energy *of that bond in the molecule* (at least in an averaged way–some care must be exercised with definitions), and, furthermore, the bond is thought to be fairly independent of the identity of the molecule. For instance,

TABLE 8-2 ► π-Bond Types and Effective Binding Energies for Carbon–Carbon Double and Single Bonds

Bond type		Effective binding energy in units of β
(C)(H)C=C(H)(H)		2.0000
(C)(H)C=C(C)(H)	(*cis* or *trans*)	2.0699
(C)(C)C=C(H)(H)		2.0000
(C)(C)C=C(C)(H)		2.1083
(C)(C)C=C(C)(C)		2.1716
(H)(C=)C–C(H)(=C)	(*cis* or *trans*)	0.4660
(C=)(C)C–C(H)(=C)	(*cis* or *trans*)	0.4362
(C=)(C)C–C(=C)(C)	(*cis* or *trans*)	0.4358

[a]The eight numbers associated with these bonds are not unique. They satisfy six simultaneous equations. Hence, any two of them may be given arbitrary values and the remaining six found by solving the simultaneous equations. Different arbitrary assignments lead to different sets of numbers, all equally valid and all giving identical π-electron energies. These numbers can only be applied when a bond has an unambiguous formal identity, which means we must deal with acyclic polyenes (even number of centers and electrons).

a C–H bond in butane is very similar to one in heptane. But this is not the case for conjugated molecules. Inspection of Fig. 8-23 shows that a single C–C bond between (formal) double bonds varies significantly in bond order from molecule to molecule. Since the total π-electron energy depends on bond order [Eq. (8-58)], we can tell at once that the *actual* theoretical energy contribution due to such a "single" bond varies from $2 \times 0.4472 = 0.8944\beta$ in butadiene to 1.088β in decapentaene. However, examination of Fig. 8-23 reveals that, in going from molecule to molecule, as the "single" bonds increase in π-bond order, the "double" bonds decrease in order. Thus, adding an additional C–C–C group to a chain adds a constant amount of bond energy (about 2.54β) to the total π energy, but this energy increment contains contributions from bond order changes over the whole molecule. We have, then, an *additive* scheme for

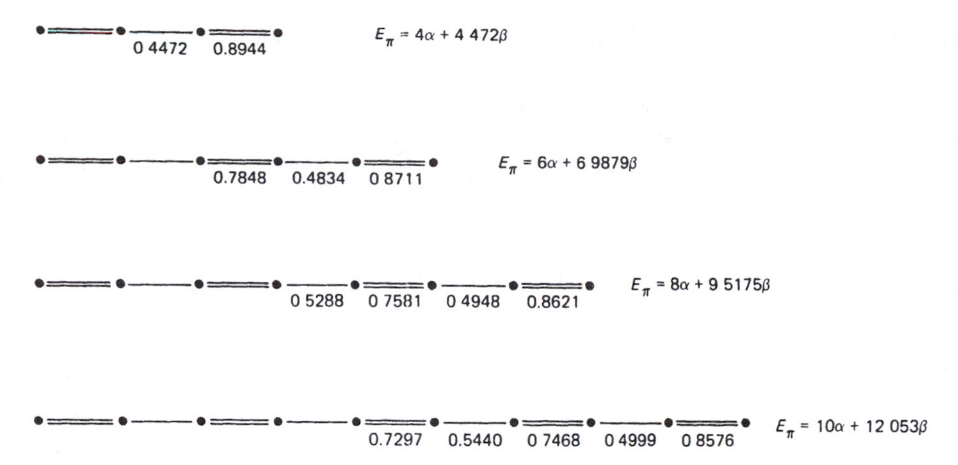

Figure 8-23 ▶ HMO bond orders for butadiene, hexatriene, octatetraene, and decapentaene.

a *delocalized* effect. For this reason, the bond energy contributions of Table 8-2 are called *effective* bond energies.

We are now in a position to consider the concept of aromaticity. The term "aromatic" originally referred to organic molecules having pleasant odors. Later it referred to a class of molecules having a high degree of unsaturation. Benzene was recognized as the parent compound for many such molecules, and the term *aromatic* has come to mean *having chemical properties peculiar to benzene and some of its relatives*. The chemical stability of these molecules, their relatively low heats of combustion or of hydrogenation, and their tendency to prefer substitution rather than addition (thereby preserving their π systems intact) distinguish these molecules from ordinary polyenes and have come to be called *aromatic properties*, or manifestations of *aromaticity*.[14] These properties suggest that the π electrons in aromatic systems are unusually low in energy, contributing to both the thermodynamic and the kinetic stability of the systems. We can test whether this is the case by calculating an expected π energy for benzene using the bond energies of Table 8-2 for an alternating single-double bond, or Kekulé, structure. The result, $6\alpha + 7.61\beta$, is significantly less stable (by 0.39β) than the HMO energy of $6\alpha + 8\beta$. We shall refer to this difference as the *resonance energy* (RE) of the system.

$$RE = E_n(HMO) - E_n \quad \text{(from Table 8-2)} \qquad (8\text{-}63)$$

By this definition, a positive RE (in units of β) corresponds to extra molecular stabilization. If we divide the RE by the number of π electrons, we obtain the RE per electron (REPE). Hence, the REPE for benzene is 0.065β. Following Dewar we shall refer to a system having significantly positive REPE as *aromatic*, significantly negative REPE as *antiaromatic*, and negligible REPE as *nonaromatic*, or polyolefinic. (Recall that β is a negative quantity. When REPE is tabulated in eV, it is conventional to use the absolute value of β in eV so as to retain positive REPE for *aromatic* molecules.)

Cyclic polyenes differ from acyclic polyenes in that many of them show significant values for REPE. Some results of Schaad and Hess [18] are reproduced in Fig. 8-24.

[14] The definition of the term "aromatic" is not generally agreed upon. Various criteria based on structural, magnetic, or reactivity properties have been proposed. See Minkin et al. [19] for a general review, and Schleyer et al. [20] for consideration of ring-current-induced nmr chemical shifts as an aromaticity probe.

It is evident that a strong correlation exists between REPE and the presence or absence of aromatic properties.

The change from very aromatic to very antiaromatic nature as we go from a six-membered to a four-membered system is striking, If we examine the HMO energy levels, it is not difficult to see why these monocycles are so different. All six π electrons in benzene occupy bonding MOs, whereas cyclobutadiene has two bonding and two nonbonding electrons (**XII**). The situation is optimal for benzene not only because all of its π electrons are bonding, but also because all of the bonding levels are fully occupied. Because all monocycles have a nondegenerate lowest level followed by higher-energy pairs of degenerate levels, it is not hard to describe the conditions that should produce maximum stability. There should be $4n+2$ electrons (2 for the lowest level and 4 for each of the n higher bonding levels), and there should be more than $4n$ centers (either $4n+1$, $4n+2$, or $4n+3$) to force the n doubly degenerate, occupied levels to all be bonding. The stability of cyclopentadienyl anion (6 electrons, 5 centers) and cycloheptatrienyl cation (6 electrons, 7 centers) was correctly predicted from these simple considerations (**XIII**). Extensive research has gone on in efforts to find examples

Molecule	REPE (eV) ($\lvert\beta\rvert = 1.4199$ eV)	*Chemical properties*
	0.092	Very stable; Undergoes substitution only under forcing conditions
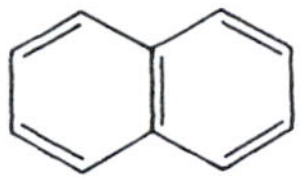	0.078	
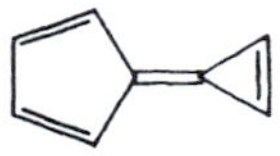	0.061	Simple derivatives isolated; undergo electrophilic substitution
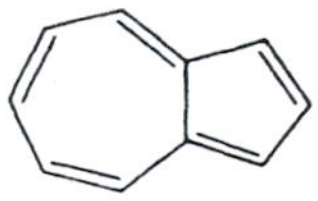	0.032	Known, stable compound
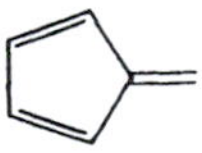	−0.003	Isolated, but reactive
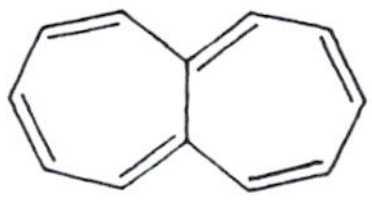	−0.006	Isolated, but reactive
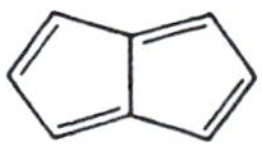	−0.026	Me derivative observed spectroscopically at −196°C
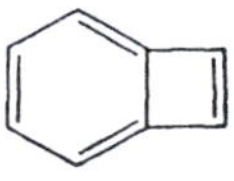	−0.038	Never prepared
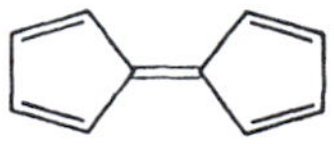	−0.047	Prepared in very dilute solution; extremely reactive
	−0.381	Prepared at 8°K, vanished at 35°K

Figure 8-24 ▶ Resonance energy per electron for a number of molecules. (From Schaad and Hess [18].)

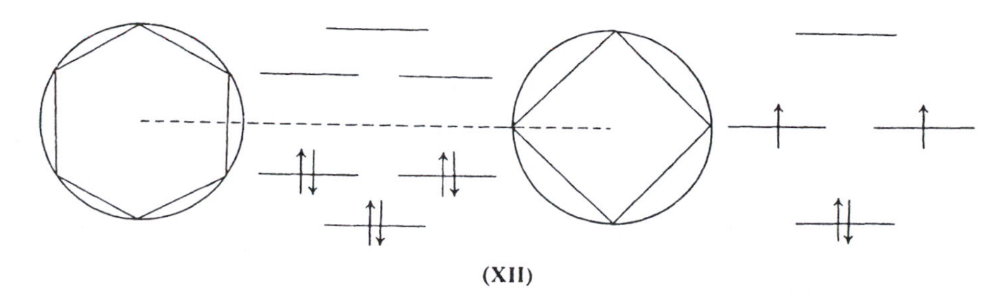

(XII)

where $n \neq 1$, but results are often complicated by angle strain and deviation from planarity in the molecule.[15]

EXAMPLE 8-7 What would the above approach predict for the relative stabilities of cyclic ions $C_5H_5^+$ and $C_5H_5^-$?

SOLUTION ▶ $C_5H_5^-$, the cyclopentadienyl anion shown in XIII, satisfies the $4n+2$ rule and should be unusually stable. $C_5H_5^+$ has only four π electrons, hence should be less stable. (It has two fewer bonding electrons than $C_5H_5^-$.) ◀

In a molecule containing both cyclic and acyclic parts, does the RE arise only from the cyclic part? For certain cases, the answer is yes. Dewar has recognized that side or connecting chains contribute nothing to the RE of a system when the chain in question is the same in all formal structures for the molecule. For example, stilbene can be written in four equivalent formal ways, as shown below. The linking chain is identical in all cases, and it should not contribute to the RE. Calculations confirm that the RE for stilbene is just double that for benzene.

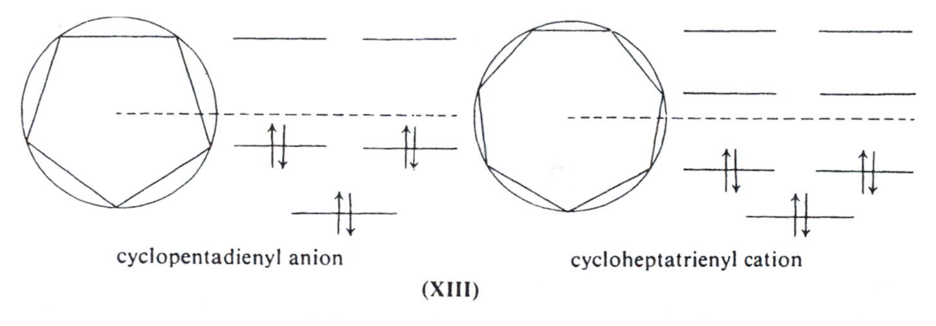

(XIII)

The intimate relation between aromaticity and the possibility for more than one equivalent formal structure for a molecule has long been recognized. These "mobile" bonds tend to favor equal bond lengths in contrast to the strong alternation characteristic of acyclic polyenes. In benzene, the extra stabilization may be viewed as resulting from the fact that all six bonds are identical and have a higher bond order (0.667) than the average of 0.634 for double and single acyclic bonds. In antiaromatic cyclobutadiene, the four identical bonds have a bond order of 0.5, which is significantly below the

[15]Research on aromaticity has been extended to Mobius conjugated hydrocarbons, where aromatic behavior should be associated with 4n electrons (see Herges [21]), and also to clusters of boron or silicon atoms (see Ritter [22]), as well as metal atoms (see Chen et al. [23]).

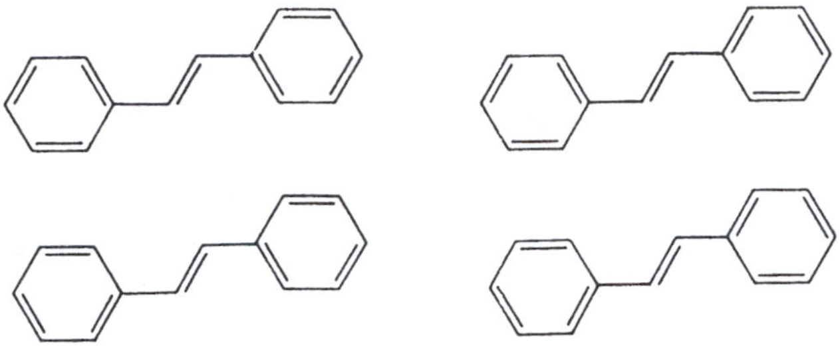

(XIV)

average for acyclics. The very different bond orders in these two molecules is fully consistent with the energy argument based on the $4n+2$ rule described earlier.[16]

To summarize, the π bonds in acyclic polyenes exhibit *delocalization* in the sense that bond orders are not transferable from one molecule to another, *additivity* in *effective* bond energies, *immobility* in formal bonds. Cyclic molecules do not exhibit additivity of effective energies or immobility of formal bonds. Their energy deviations from energy calculated assuming additivity and immobility are good indicators of kinetic and thermodynamic stability. Aromatic molecules possess extra stability because their π electrons[17] are more bonding than those in acyclic polyenes. Antiaromatics are unstable because their π electrons are less bonding.

From a consideration of experimental heats of atomization, Schaad and Hess have evaluated β to be $-1.4199\,\text{eV}$. The physical processes involved in dissociating a gas-phase molecule into constituent atoms are quite different from those involved in adding or removing a π electron from a molecule in a solvent. Therefore, it is not surprising that the β value obtained from heats of atomization differs substantially from values obtained from redox experiments. Indeed, it is this variability of β as we compare HMO theory with different types of experiment that compensates for many of the oversights and simplifications of the approach. It is remarkable that, with but one such parameter, HMO theory does as well as it does.

8-17 Extension to Heteroatomic Molecules

The range of application of the HMO method could be greatly extended if atoms other than carbon could be treated. Consider pyridine as an example (**XV**). A π electron at a carbon atom contributes an energy α to E_π. The contribution due to a π electron at nitrogen is presumably something different. Let us take it to be $\alpha' = \alpha + h\beta$, where h is a parameter that will be fixed by fitting theoretical results to experiment. If the π electron is attracted more strongly to nitrogen than to carbon, h will be a positive

[16]It has been noted that bond length equalization associated with bond mobility results in it energy lowering when the C–C–C angles are near 120°. When the angle is very different from this, π energies are higher than expected. This has led to suggestions that "strain energy" may be an important factor in aromaticity. Because of the present lack of a quantum mechanical quantity equivalent to strain energy, and because the HMO method may include effects of σ electrons in an implicit but poorly understood way, it is very hard to know whether such suggestions are at variance with other statements or are simply equivalent to them but stated from a different viewpoint.

[17]It is not necessarily true that all the "extra" stability of aromatic molecules is attributable to π-electron effects; σ-electron energies also depend on bond lengths and bond angles. Hence, we may be seeing, once again, a situation where the π-electron treatment includes other effects implicitly. Schaad and Hess [18] indicate that σ energies and π energies are indeed simply related over the bond-length range of interest.

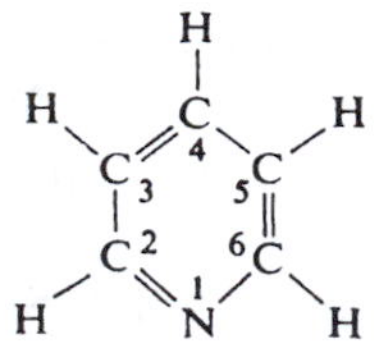

(XV)

number. In a similar spirit, we will take the energy of a π electron in a C–N bond to be $\beta' = k\beta$ and evaluate k empirically. Not surprisingly, the values of h and k appropriate for various heteroatoms depend somewhat on which molecules and properties are used in the evaluation procedure. A set of values compiled and critically discussed by Streitwieser [7] is given in Table 8-3. Other sets have been published.[18]

The dots over each symbol indicate the number of π electrons contributed by the atom. In pyridine, the formal bond diagram indicates a six π-electron system, implying that the nitrogen atom contributes one π electron. We also can argue that, of the five

TABLE 8-3 ▶ Parameters for Heteroatoms in the Hückel Method[a]

Heteroatom	h	Heteroatomic bond	k
$\dot{\mathrm{N}}$	0.5	$\mathrm{C}\cdots\ddot{\mathrm{N}}$	1.0
$\ddot{\mathrm{N}}$	1.5	$\mathrm{C}—\ddot{\mathrm{N}}$	0.8
$\overset{+}{\mathrm{N}}$	2.0		
$\dot{\mathrm{O}}$	1.0	$\mathrm{C}=\dot{\mathrm{O}}$	1.0
$\ddot{\mathrm{O}}$	2.0	$\mathrm{C}—\ddot{\mathrm{O}}$	0.8
$\overset{+}{\mathrm{O}}$	2.5	$\mathrm{N}—\ddot{\mathrm{O}}$	0.7
$\ddot{\mathrm{F}}$	3.0	$\mathrm{C}—\ddot{\mathrm{F}}$	0.7
$\ddot{\mathrm{Cl}}$	2.0	$\mathrm{C}—\ddot{\mathrm{Cl}}$	0.4
$\ddot{\mathrm{Br}}$	1.5	$\mathrm{C}—\ddot{\mathrm{Br}}$	0.3
S'^{b}	0.0	$\mathrm{C}—\mathrm{S}'$	0.8
S''	0.0	$\mathrm{C}—\mathrm{S}''$	0.8
		$\mathrm{S}'—\mathrm{S}'$	1.0
Methyl (inductive $\dot{\mathrm{C}}_\alpha$—Me)−0.5	$h_{\mathrm{C}_\alpha} = -0.5$	none	—
Methyl (heteroatom $\dot{\mathrm{C}}_\alpha—\ddot{\mathrm{M}}\mathrm{e}$)0.2	$h_{\mathrm{Me}} = 0.2$	C_α—Me	0.7
Methyl (conjugative $\dot{\mathrm{C}}_\alpha—\mathrm{C}\overset{..}{\cdots}\mathrm{H}_3$)	$h_{\mathrm{C}_\alpha} = -0.1$	C_α—C	0.8
	$h_c = -0.1$	$C—\mathrm{H}_3$	3.0
	$h_{\mathrm{H}_3} = -0.5$		

[a] Consistent with the philosophy of this approach is a distinction between single, double, and intermediate C—C bonds. Streitwieser recommends $k_{\mathrm{C-C}} = 0.9$, $k_{\mathrm{C\cdots C}} = 1.0$, $k_{\mathrm{C=C}} = 1.1$.

[b] Sulfur is treated as a pair of AOs with a total of two π electrons, i.e., a sulfur in an aromatic ring is *formally* treated as *two* adjacent atoms S′ and S″ with the indicated parameters.

[18] See McGlynn et al. [24, p. 87].

valence electrons of nitrogen, two are involved in σ covalent bonds with neighboring carbons, two more are in a σ lone pair, leaving one for the π system. Therefore, the atom parameter to use for this molecule is $h = 0.5$. The pyridine ring, like benzene, admits two equivalent structural formulas, and so the C–N bonds should be intermediate between double and single, symbolized $\text{C} \mathrel{\overset{\cdots}{-}} \dot{\text{N}}$ in Table 8-3. Since $k = 1.0$ in this case, $\beta' = \beta$, and pyridine will have an HMO determinant differing from the benzene determinant only in the diagonal position corresponding to the nitrogen atom–the 1, 1 position according to our (arbitrary) numbering scheme. For this position, instead of x, we will have

$$x' = (\alpha' - E)/\beta = (\alpha + 0.5\beta - E)/\beta = (\alpha - E)/\beta + 0.5\beta/\beta = x + 0.5 \qquad (8\text{-}64)$$

The pyrrole molecule has a nitrogen atom of the type $\ddot{\text{N}}$(**XVI**). Since three valence electrons of nitrogen are in covalent σ bonds, two remain for inclusion in the π system.

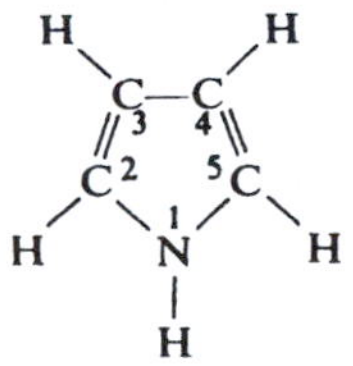

(XVI)

Therefore, pyrrole has a total of six π electrons. The unique structural formula indicates that the C–$\ddot{\text{N}}$ bond is formally single, and $k = 0.8$, $h = 1.5$ are the appropriate parameters here. Also, the carbon–carbon bonds are now formally single or double. If we choose to distinguish among these bonds using the parameters in note a of Table 8-3, the resulting HMO determinant is

$$\begin{vmatrix} x+1.5 & 0.8 & 0 & 0 & 0.8 \\ 0.8 & x & 1.1 & 0 & 0 \\ 0 & 1.1 & x & 0.9 & 0 \\ 0 & 0 & 0.9 & x & 1.1 \\ 0.8 & 0 & 0 & 1.1 & x \end{vmatrix}$$

The methyl group can also be incorporated into the HMO method. Several approaches have been suggested. One is simply to modify the coulomb integral α for the carbon to which the methyl group is attached. A methyl group is thought to release sigma electrons to the rest of the molecule as compared to a substituent hydrogen. This suggests that an atom having a methyl group attached to it will be a bit electron rich and hence will be less attractive to π electrons. Use of a negative h parameter for this carbon is appropriate. This method is called the *inductive* model. Use of the inductive model does not add any new centers or any more π electrons to the conjugated system to which the methyl group is attached: The carbon to which the methyl is bonded is merely treated as a less attractive atom. A second approach is to treat the methyl group itself as a heteroatom. As we shall see shortly, the methyl group has two electrons that

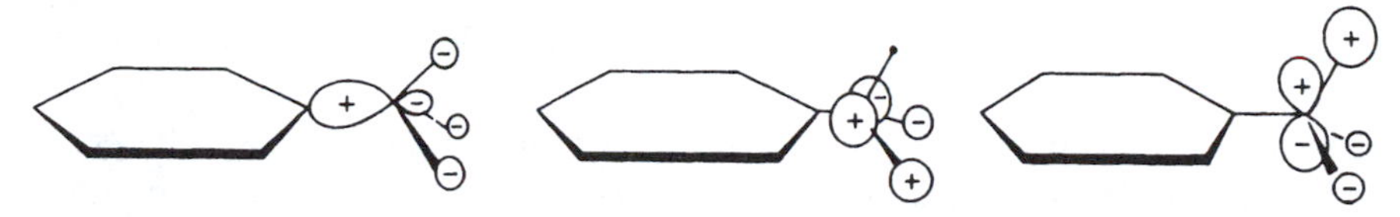

Figure 8-25 ▶ Three MO symmetry solutions for a methyl group attached to a benzene ring.

can participate (to a slight extent) in the π system, and so use of this *heteroatom* model adds one more center and two more π electrons for each methyl group included in this way. A third approach is the *conjugative* model. Because the methyl group has *local* threefold symmetry, one of the σ MOs for the methyl group resembles the π MOs of cyclopropenyl in symmetry characteristics. The three types of symmetry solutions are given in Fig. 8-25 for a methyl group on a benzene ring (compare with Fig. 8-7). Notice that the MO at the right of the figure has its phases arranged so that it has nonzero overlap with a π AO on the benzene ring. This means that the two electrons in this "methyl group MO" can participate in the π system of the molecule. To emulate this picture in our HMO determinant, we must add two π electrons and two more centers to our system (one for C and one for H_3) and find h and k values for the two new centers and bonds. The inductive effect of the methyl group on the neighboring ring carbon is often included in this model. In most cases, it is probably safe to say that the conjugative model is superior to the heteroatom model, which is, in turn, better than the inductive model, but no extensive critical comparison of these three models has been made. Parameters for all three approaches are included in Table 8-3. (C_α is the ring carbon.)

8-18 Self-Consistent Variations of α and β

Efforts have been made to improve the HMO method by taking account of molecular π charge distribution. Suppose that we carry out an HMO calculation on a nonalternant molecule and find an electron density of 1.2 at one carbon and 0.8 at another. It is reasonable to argue that a π electron at the latter carbon is more strongly bound because it experiences less repulsion from other π electrons there. We can try to account for this by making α at that atom more negative. Thus, we could take

$$\alpha_i' = \alpha_i + \omega(1 - q_i)\beta \tag{8-65}$$

where q_i is the π-electron density at atom i and ω is a parameter (assumed positive) to be fixed empirically. If $q < 1$, then α' is more negative than α. If $q > 1$, α' is less negative. Having now modified α (using a trial value for ω), we must set up our new HMO determinant and solve it again. This yields new MOs, new values of q_i, and therefore new values of α'. We repeat this process over and over until electron densities remain essentially unchanged for two successive iterations. At this point, the electron densities *leading to* the HMO determinant are the same as those *produced* by the determinant, and the solution is said to be self-consistent with respect to electron densities. This procedure, often referred to as the "ω technique," discourages extreme deviations of electronic densities from the "norm" of unity at each carbon and thereby helps to compensate for the lack of explicit inclusion of π electron repulsion in the HMO method. Streitwieser's calculations have led him to favor a value for ω of 1.4.

A similar idea has been applied to variations of the bond integral β. Suppose that we carry out an HMO calculation and find a π-bond order of 0.5 in one bond and 0.9 in another. We expect that the latter bond is in fact shorter than the former. We could roughly predict how much shorter it is by using the bond-order–bond-length relation described earlier. It is reasonable to modify β in these bonds on the basis of predicted length differences, set up a new HMO determinant, solve again, find new bond orders, and iterate until self-consistency is achieved with respect to bond orders.

These modifications to the simple HMO method improve predictions of some properties but not others. For example, Brogli and Heilbronner [13] have found that orbital energy correlation with ionization energy, determined by photoelectron spectroscopy, is significantly improved through inclusion of the effects of bond length variations in the *neutral molecule and the cation*. This improved correlation, shown in Fig. 8-26, showed no additional improvement upon subsequent variation of α as a function of electron density. This is reasonable, since β variation affects primarily bond order, hence MO energy, and that is the property measured by photoelectron spectroscopy. Variation of α shifts charge from atom to atom, but has smaller energy effects. On the other hand, π-electron contributions to dipole moments, calculated from electronic excess or deficiency at each center, are very sensitive to variation of α, and quite insensitive

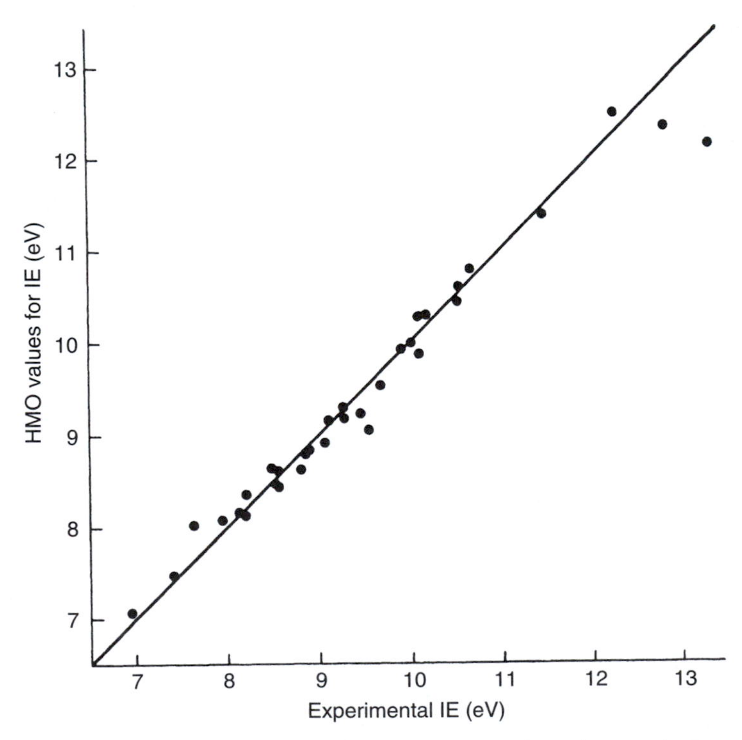

Figure 8-26 ▸ Experimental ionization energy for alternant and nonalternant hydrocarbons versus orbital energy using a modified HMO technique which includes provisions for bond length variation in molecule and cation. (Compare with Fig. 8-22.)

to variation of β. Again, this is reasonable because dipole moments are sensitive to electronic distribution rather than MO energy.

8-19 HMO Reaction Indices

In this section, we discuss some applications of the HMO method to *reactivities* of conjugated molecules. The reactions of conjugated molecules that have received most of this theoretical treatment are:

1. electrophilic aromatic substitution (**XVII**)

+ Cl_2, $FeCl_3$ $\xrightarrow{k_1}$ 1-chloronaphthalene + H^+; $\xrightarrow{k_2}$ 2-chloronaphthalene + H^+

(XVII)

2. nucleophilic aromatic substitution (**XVIII**)

+ CH_3Li $\xrightarrow{k_1}$ 1-methylnaphthalene + LiH; $\xrightarrow{k_2}$ 2-methylnaphthalene + LiH

(XVIII)

3. radical addition (**XIX**)

+ $CCl_3^{\bullet}$ $\xrightarrow{k_1}$ (CCl₃, H at position 1, radical); $\xrightarrow{k_2}$ (CCl₃, H at position 2, radical)

(XIX)

For any of these reactions, we imagine there to be a path of least energy connecting reactants with products. For the two distinct reaction positions, 1 and 2 on naphthalene, the activation energies ϵ may differ, as indicated in Fig. 8-27.

The problem is somehow to relate the differences in ϵ (inferred from relative rate data) to a number based on quantum chemical calculations. To do this in a sensible way requires that we have some idea of the detailed way in which the reaction proceeds–we have to know what the reaction coordinate is. In some cases, this is fairly well known. For electrophilic aromatic substitution reactions, evidence suggests that a

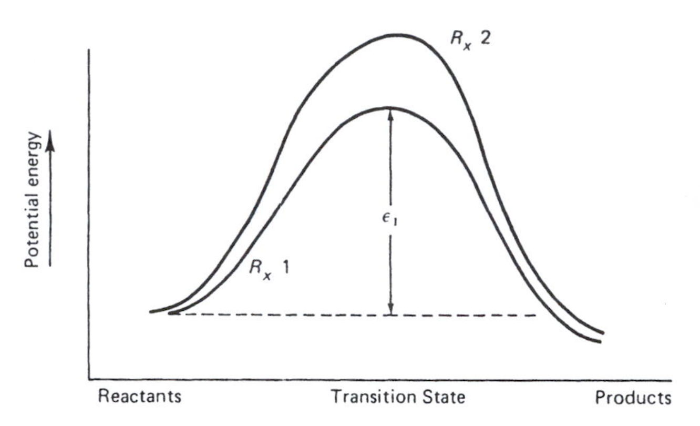

Figure 8-27 ▶ Generalized energy versus reaction coordinate for reaction at two positions in naphthalene.

positive electrophile (e.g., Cl^+) approaches the substrate (say, naphthalene). As it draws closer, it causes a significant polarization of the π-electron charge distribution, drawing it toward the site of attack. Ultimately, it forms a partial bond with the carbon. At this stage, the carbon has already begun to loosen its bond to hydrogen, but it is at least partially bonded to four atoms (**XX**). This means that the ability of the carbon to participate in the aromatic system is temporarily hampered. At this point, the system is at or near the transition state. Thereafter, as H^+ leaves, the potential energy decreases and Cl moves into the molecular plane. A similar detailed mechanism is thought to apply for nucleophilic reactions except that the attacking group is negative.

(XX)

For attack by a neutral radical, electrostatic attraction and charge polarization should not be significant factors. The radical bonds to the site of attack to produce a more-or-less tetrahedrally bonded carbon. This again leads to an interruption of the π system, but now it is not temporary as it was in the substitution reactions.

Based on these simple pictures, a number of MO quantities, often referred to as *reaction indices*, have been proposed as indicators of preferred sites for reaction. It is useful to divide these into two categories–those purporting to relate to early stages of the reaction, and those specifically related to the intermediate stage.

Perhaps the most obvious reaction index to use for the earliest stages of electrophilic or nucleophilic reactions is the *π-electron density*. If Cl^+ is attracted to π charge, it should be attracted most to those sites where π density is greatest. (Such an ion should be attracted to sites having excessive σ charge density also, but our basic HMO assumptions ignore any variations in σ density.) For an alternant hydrocarbon like naphthalene, all π densities are unity, so this index is of no use. For nonalternant molecules, however, it can be quite helpful. Azulene has varying HMO π densities (**XXI**). (More sophisticated calculations described in future chapters are in qualitative agreement with these π-electron density variations.) Experimentally, it is found that electrophilic substitution by Cl occurs almost entirely at position 1 (or 3). Nucleophilic substitution by CH_3 (from CH_3Li) occurs at the position of least π-electron density, namely 4 (or 8).

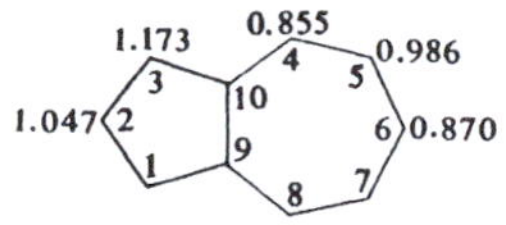

1.173 0.855 0.986
1.047 0.870

(XXI)

The charge density index refers to the nature of the molecule before allowance is made for perturbing effects due to the approaching reactant. Such a method is often called a "first-order" method, a terminology that is discussed more fully in Chapter 12. For alternant molecules, it is necessary to proceed to a high-order method, one that reflects the ease with which molecular charge is drawn toward some atom, or pushed away from it, as approach by a charged chemical reactant makes that atom more or less attractive for electrons. An index which measures this is called *atom self-polarizability*, symbolized $\pi_{r,r}$. The formulas for this and related polarizabilities are derived in Chapter 12. For now, we simply note that the formula is

$$\pi_{r,r} \equiv \frac{\partial q_r}{\partial \alpha_r} = 4 \sum_{j}^{\text{occ}} \sum_{k}^{\text{unocc}} \frac{c_{rj}^2 c_{rk}^2}{E_j - E_k} \tag{8-66}$$

A larger absolute value of π_{rr} means that a larger change in π density q_r occurs as a result of making atom r more or less attractive for electrons. ($\pi_{r,r}$ is negative since $E_j - E_k$ is negative. This makes physical sense because it means that if $\delta\alpha_r$ is negative, making atom r more attractive, δq_r is positive, indicating that charge accumulates there.) Since the most polarizable site should most easily accommodate either a positive- or a negative-approaching reactant, this index should apply for both electrophilic and nucleophilic reactions. For naphthalene, the values are $\pi_{11} = -0.433/|\beta|$, $\pi_{22} = -0.405/|\beta|$. This agrees with the experimentally observed fact that the 1 position of naphthalene is more reactive for both types of reaction.

Examination of Eq. (8-66) indicates that the MOs near the energy gap between filled and empty MOs will tend to contribute most heavily to $\pi_{r,r}$ because, for these, $E_j - E_k$ is smallest. For this reason, the highest occupied and lowest unfilled MO (HOMO and LUMO) are often the determining factor in relative values of π_{rr}. Fukui[19] named these the *frontier orbitals* and suggested that electrophilic substitution would occur preferentially at the site where the HOMO had the largest squared coefficient. In nucleophilic substitution, the approaching reagent seeks to *donate* electronic charge to the substrate, so here the largest squared coefficient for the LUMO should determine the preferred site. For even alternants like naphthalene, the pairing theorem forces these two MOs to have their absolute maxima at the same atom. The HOMO–LUMO coefficients for naphthalene are 0.425 and 0.263 for atoms 1 and 2, respectively, in accord with our expectations. For the nonalternant molecule azulene, discussed above, the largest HOMO coefficient occurs at atoms 1 and 3, which have already been mentioned to be the preferred sites for electrophilic attack. The largest LUMO coefficient occurs at

[19] See Fujimoto and Fukui [25].

atom 6, with atoms 4 and 8 having the second-largest value (see Appendix 6). Atoms 4 and 8 are the preferred sites for nucleophilic attack. Here, then, is a case where the charge density and frontier MO indices are not in agreement. The results suggest that, when significant π-density variations occur, this factor should be favored over higher-order indices. However, the whole approach is so crude that no ironclad rule can be formulated.

For radical attack, some other index should be used, for we do not expect electrostatic or polarization effects to be important in such reactions. An index called the *free valence*[20] has been proposed for free radical reactions. One assumes that the free radical begins bonding to a carbon atom in early stages of the reaction and that the ease with which this occurs depends on how much residual bonding capacity the carbon has after accounting for its regular π bonds. Thus, free valence is taken to be the difference between the maximum π bonding a carbon atom is capable of and the amount of π bonding it actually exhibits in the unreacted substrate molecule. The extent of π bonding is taken as the sum of all the orders of π bonds involving the atom in question. A common choice of reference for a maximally bonded carbon is the central atom in trimethylenemethane in its planar conformation (which is not the most stable). (**XXII**). Each bond in this neutral system has a π-bond order of $1/\sqrt{3}$, and so the total π-bond

(XXII)

order associated with the central carbon is $\sqrt{3}$.[21] Using this as reference, the free valence for some atom r in *any* unsaturated hydrocarbon is defined as

$$F_r = \sqrt{3} - \sum_{z}^{\text{neighbors of } r} p_{rs} \qquad (8\text{-}67)$$

A common way of representing the situation schematically is indicated in (**XXIII**) for butadiene. Bond orders are indicated on the bonds and free valences by arrows. It is clear that butadiene has a good deal more "residual bonding capacity" on its terminal atoms, and this is consistent with the fact that free radical attack on butadiene occurs predominantly on the end atoms. Other examples of correlation between free valence and rate of free radical addition have been reported.[22] A plot of rate data for methyl

C —0.894— C —0.447— C —— C
↓ 0.838 ↓ 0.391

(XXIII)

[20] See Coulson [26].

[21] Sometimes the three sigma bonds are included in this calculation, giving $3 + \sqrt{3}$. This has no effect on the question of *relative* values of F_r.

[22] See Streitwieser [7] and Salem [1].

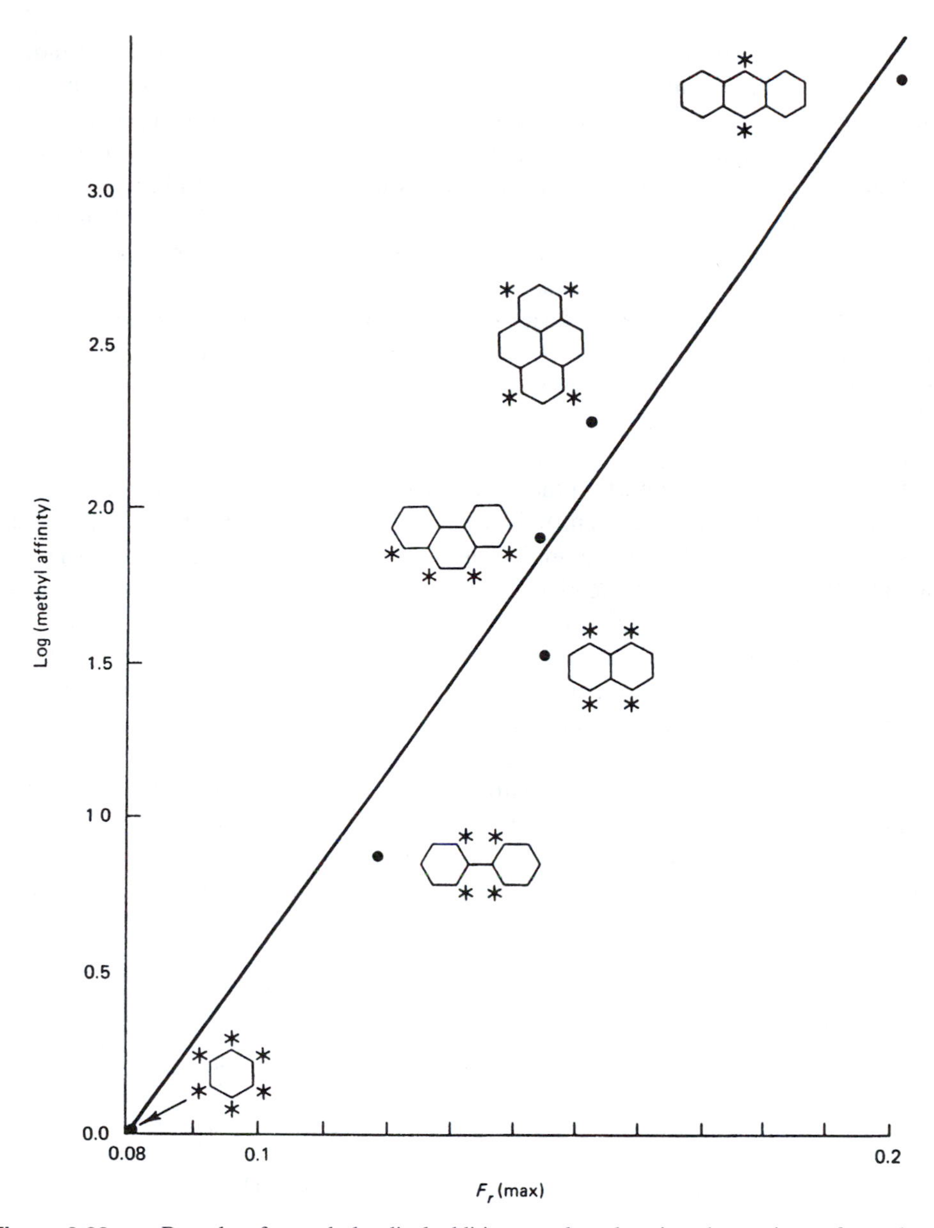

Figure 8-28 ▶ Rate data for methyl radical addition are plotted against the maximum free valence found in each molecule. The original "methyl affinity" (Levy and Szwarc [27]) has been multiplied by $6/m$, where m is the number of sites having maximum free valence. (The asterisks in the figure identify these sites.)

radical addition to conjugated molecules versus the largest free valence of the molecule is shown in Fig. 8-28. (We assume that the kinetics is dominated by the atom(s) having the maximum free valence.)

The indices described above are most appropriate for indicating the relative ease of reaction *in the early stages*. By the time the reactants have reached the transition state, the substrate is quite far from its starting condition, so charge densities, polarizabilities, free valences calculated from the wavefunction of the unperturbed molecule may no longer be very appropriate. If the energy curves being compared through our indices behave in the simple manner described in Fig. 8-27, so that the higher-energy curve in early stages is also the higher-energy curve in the region of the transition state, such indices can be useful. Also, such simple behavior is very likely to occur when we compare a single type of reaction down a series of molecules of similar type, as in Fig. 8-28 (see also Problem 8-25). Experience, however, has indicated that indices more closely linked to the nature of the transition state are more generally reliable. We now describe one such reactivity index.

We mentioned earlier that addition and substitution reactions are expected to interrupt the π system at the site of attack. For substitution reactions, this interruption is only temporary and is presumably most severe in the transition state. For addition reactions, it is permanent. The *localization energy* is defined as the π energy lost in this process of interrupting the π system.

As an example, let us return to the naphthalene molecule. The situations resulting from interruption of the π system by *neutral radical* attack at positions 1 and 2 are illustrated in (**XXIV**). The remaining unsaturated fragment is, in each case, a neutral radical. [If attack were by a negative ion (nucleophilic), the fragment would be topologically the same but would be negatively charged. Likewise, attack by a positive electrophilic reagent leads to a positively charged fragment.] No matter where attack occurs, our π energy must go from $10\alpha + \cdots \beta$ to $9\alpha + \cdots \beta$. This decrease by α is thus not expected to differ from case to case and hence is ignored in our localization energy calculation. The decrease in π energy is thus $2.299\,|\beta|$ for attack at position 1 and $2.480\,|\beta|$ for position 2. These localization energies indicate that attack at position 1 should be favored since the energy cost is smaller there, and this is in accord with observation. (Localization energies are symbolized $L_r^{\cdot}$, L_r^{+}, L_r^{-} depending, respectively, on whether attack is by a free radical, an electrophilic cation, or a nucleophilic anion.)

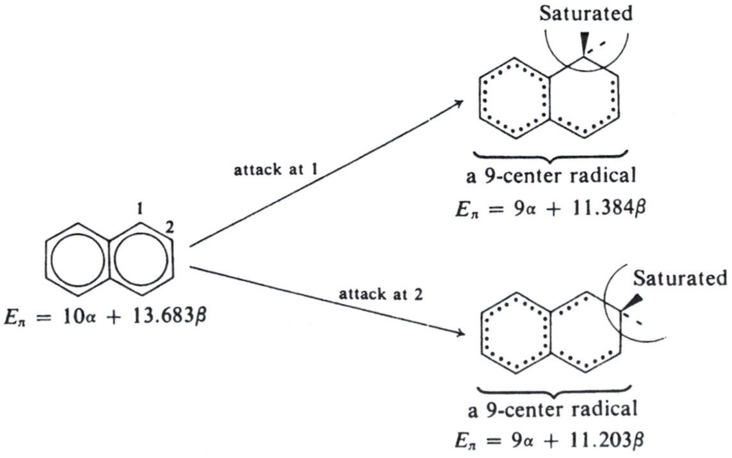

(**XXIV**)

It is interesting to compare the various indices we have discussed for a single molecule to see how well they agree. Data for azulene are collected in Table 8-4. Experimentally, azulene is known to preferentially undergo electrophilic substitution at positions 1 (and 3), nucleophilic substitution at positions 4 (and 8), and radical addi-

TABLE 8-4 ► Reactivity Indices for Azulene

r (atom no)	q_r	$-\lvert\beta\rvert\pi_{r,r}$	HOMO c_r^2	LUMO c_r^2	Fr	$Lr^{+}(\lvert\beta\rvert)$	$Lr^{\bullet}(\lvert\beta\rvert)$	$Lr^{-}(\lvert\beta\rvert)$
1	*1.173*	0.425	*0.2946*	0.0040	0.480	*1.924*	2.262	2.600
2	1.047	0.419	0.0000	0.0997	0.420	2.362	2.362	2.362
4	*0.855*	*0.438*	0.0256	0.2208	*0.482*	2.551	*2.240*	*1.929*
5	0.986	0.429	0.1126	0.0104	0.429	2.341	2.341	2.341
6	0.870	0.424	0.0000	*0.2610*	0.454	2.730	2.359	1.988

tion in positions 1 (and 3). Consider first electrophilic reaction. Examining the table indicates that position 1 is heavily favored by q_r, HOMO distribution, and L_r^+. The only other index relevant for this process, π_{rr}, favors position 4. For nucleophilic reaction, q_r, π_{rr}, L_r^- all favor position 4. The LUMO index favors position 6, but not decisively over position 4. For radical addition, LUMO favors position 6, whereas F_r and $L_r^{\cdot}$ favor position 4. The latter two indices, however, favor 4 over 1 by only a slight margin. Thus, for a nonalternant molecule like azulene, these numbers are not completely trustworthy and must be interpreted with caution. One difference between a molecule like anthracene and one like azulene is that all the C–C–C angles in the former molecule are similar ($\sim 120°$) whereas in azulene they differ. One might anticipate that the smaller angles in the five-membered subunit, being already closer to the tetrahedral angle characteristic of saturated carbons, would allow easier substitution or addition than would be the case in the seven-membered subunit. This factor is ignored in our calculations of $L_r^{\cdot}$ and might easily tip the balance to favor position 1 over position 4 for radical attack since $L_1^{\cdot}$ and $L_4^{\cdot}$ are so close in value. In short, these HMO reactivity index approaches are once again techniques that ignore many aspects of the physical processes being followed. It seems likely that many of these will cancel out of comparisons among similar molecules, but dissimilarities between or within molecules (most often encountered in nonalternant systems) will cause such cancellations to be less complete. For more discussion of these and other HMO reaction indices, the reader is referred to more specialized discussions.[23] Application of some of these indices to carcinogenicities of polycyclic aromatic hydrocarbons has been reviewed.[24]

8-20 Conclusions

In this chapter, we have seen how certain basic features of molecular structure manifest themselves in molecular properties. The *connectedness*, or σ bond network, defines the bond positions where π electrons can congregate to lower the energy of the system.

[23] See Streitwieser [7], Salem [1], Dewar [14], and Klopman [28].

[24] See Lowe and Silverman [29]. For a dialogue for for nonspecialists by the same authors, see [30].

The extent of congregation is a useful measure of bond length and also is directly contributory to the total energy of the system.

The symmetry restrictions for MOs have been emphasized. The fairly successful correlation of HMO results with certain experimental measurements suggests that the method effectively accounts for the controlling factors in some molecular properties.

However, we have omitted discussion of other properties that correlate only poorly with HMO theory. Notable in this regard are spectral energies. The natural idea of relating a spectral transition energy to a difference between orbital energies has not been very successful in HMO theory, except when one restricts attention to a particular band in a series of related molecules. This is at least partly due to improper handling of electron exchange symmetry in the HMO method leading to an inability to distinguish between different states of a given configuration.

The HMO method can be an extremely instructive way to approach a problem since it can describe the manner in which certain important factors are operating. Also, for some situations, its predictive power is rather good. However, the limitation to conjugated systems, the reliance on an increasing number of parameters as extensions are made, the inability to conform to some kinds of experimental measurement, and the conceptual slipperiness of the quantities used in the method have all contributed to a decline of interest in further development of HMO theory. More powerful computers have made it possible for more complicated but better defined methods to be used.

8-20.A Problems

8-1. Show that, if $\hat{H} = \hat{H}_1 + \hat{H}_2 + \hat{H}_3$, and $\hat{H}_i\phi_j(i) = E_j\phi_j(i)$, then $\psi_{\text{prod}} = \phi_1(1)\phi_2(2)\phi_3(3)$ and $\psi_{\text{det}} = |\phi_1(1)\phi_2(2)\phi_3(3)|$ are both eigenfunctions of $\hat{H}$ and have the same eigenvalue. Show also that $\psi = (1/\sqrt{2})[\phi_1(1)\phi_2(2)\phi_3(3) + \phi_1(1)\phi_2(2)\phi_4(3)]$ is an eigenfunction of $\hat{H}$ if and only if $E_3 = E_4$.

8-2. Set up the HMO determinant for each of the following molecules:

(a)

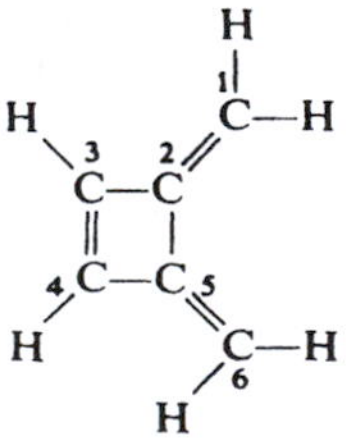

(b)

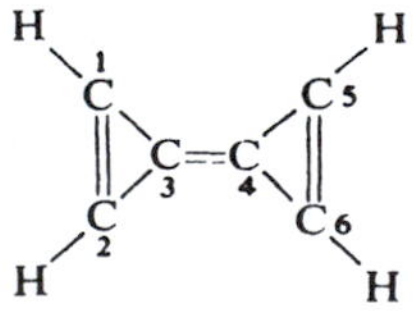

(c) $H_3C{-}CH_2{-}CH_2{-}CH{=}CH_2$

(d) $H_2C{=}CH{-}CH_2{-}CH_2{-}CH{=}CH_2$

(e) $H{-}C{\equiv}C{-}H$

8-3. Set up and solve the Hückel determinantal equation for 2-allylmethyl (also called trimethylenemethane) (**XXV**). Display the orbital energy levels and indicate the

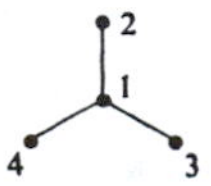

(XXV)

electron configuration for the neutral ground state. Calculate E_n. Find the coefficients for all MOs. (Be sure that degenerate MOs are orthogonal.) Calculate the charge densities and bond orders.

8-4. Suppose that two MOs of a molecule are given by the formulas

$$\phi_1 = (1/\sqrt{3})\chi_1 + (1/\sqrt{3})\chi_2 + (1/\sqrt{3})\chi_3,$$
$$\phi_2 = (1/\sqrt{3})\chi_3 + (1/\sqrt{3})\chi_4 + (1/\sqrt{3})\chi_5,$$

where the χ's are AOs which are assumed to be orthonormal. By inspection, what is the overlap between these MOs?

8-5. Given the following two degenerate MOs for cyclobutadiene (assumed square planar):

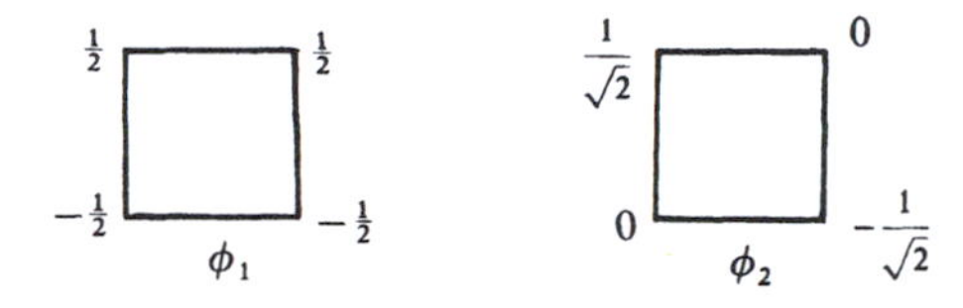

Use the Schmidt procedure to obtain a normalized MO that has the same energy as ϕ_2 but is orthogonal to ϕ_1.

8-6. For cyclooctatetraene (in its idealized, but incorrect, planar, octahedral form), see if you can answer the following without reference to tabulations:

a) What is the HMO energy of the *highest* energy π MO?
b) Sketch this MO (from above the molecule) showing signs and magnitudes (actual numbers) of the MO coefficients.

8-7. a) What is the Hückel *orbital* energy for the following MO? (Assume that all centers are carbons.) [*Hint*: Use Eq. (8-56).]

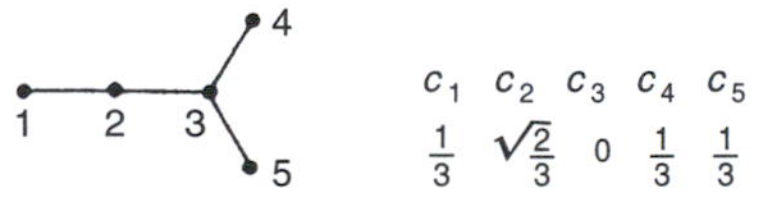

c_1	c_2	c_3	c_4	c_5
$\frac{1}{3}$	$\sqrt{\frac{2}{3}}$	0	$\frac{1}{3}$	$\frac{1}{3}$

b) Calculate the contributions to bond orders due to *one* electron in this MO.
c) Calculate the Hückel energy of the following MO. (Figure out the value of a if you need to use it.)

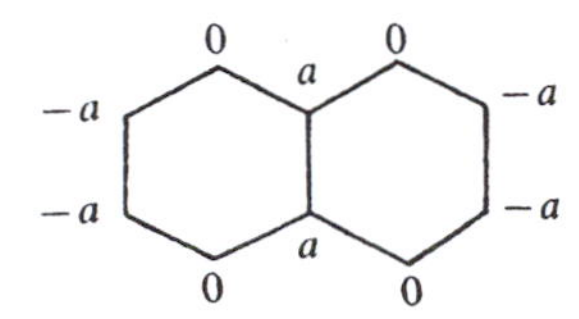

8-8. Without performing an HMO calculation, sketch the MOs for the pentadienyl radical. Use the particle-in-a-box solutions and the pairing theorem as a guide.

8-9. Which, if any, of the following systems would you expect to exhibit Jahn–Teller distortion? Indicate your thinking. (You should be able to answer without reference to HMO data tables.) (a) Benzyl radical, C_7H_7, (b) Cyclopentadienyl radical, C_5H_5. (c) Cyclobutadienyl radical cation, $C_4H_4^{+}$. (d) Benzene cation, $C_6H_6^{+}$.

8-10. The bond lengths given in Table P8-10 have been reported for ovalene (**XXVI**). Using a library source or a computer program, obtain HMO bond orders for ovalene and calculate theoretical bond lengths using a relation from Section 8-12. Make a comparison plot for ovalene of the type shown in Fig. 8-16. (See caption of Fig. 8-16 for values of k and s.)

TABLE P8-10 ►

Bond	Length (Å)	Bond	Length (Å)
1–2	1.445	1 – 10	1.401
2–3	1.354	6 – 7	1.419
3–4	1.432	4 – 25	1.411
4–5	1.429	24 – 25	1.366
5–6	1.429	5 – 22	1.424
6–1	1.425	7 – 20	1.435

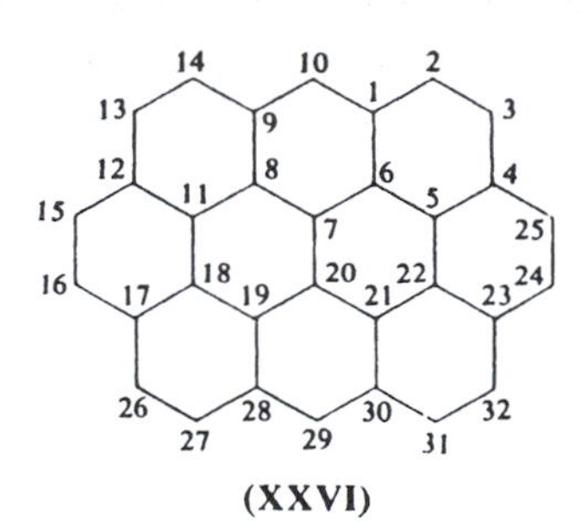

(XXVI)

8-11. Horrocks et al. [31] report experimental bond lengths for quinoline complexed to nickel. They display a comparison plot that uses theoretical data from an MO method more refined than the simple HMO method. Using the appropriate heteronuclear parameters, perform on the computer an HMO calculation for quinoline. Calculate theoretical C–C bond distances and compare them with the experimental and theoretical data of Horrocks et al.

8-12. When the molecule $CH_2{=}CH{-}CH{=}O$ absorbs light of a certain frequency, a lone-pair electron on oxygen (called "n" for nonbonding) is promoted to the lowest empty π MO of the molecule (called π^*; hence, an $n \longrightarrow \pi^*$ transition). Assuming that the π MOs of this molecule are identical to those in butadiene, which C–C bond would you expect to become longer and which shorter as a result of this transition? Calculate the expected bond length changes using butadiene data, using either set of k, s values in the caption of Fig. 8-16. (*Observed*: $\Delta CH_2{-}CH \cong 0.06$ Å, $\Delta CH{-}CH \cong -0.04$ Å.)

8-13. ESR coupling constants are shown in Table P8-13 for six hydrocarbon anion radicals. Use HMO tabulations in the literature (or a computer) to obtain π-electron MO coefficients for these systems. Construct a plot of coupling constant a_{H_μ} versus $c_{\mu i}^2$, where i is the MO containing the unpaired electron (a_{H_μ} values are in gauss). The numbered positions in Table P8-13 refer to hydrogen atoms.

TABLE P8-13 ►

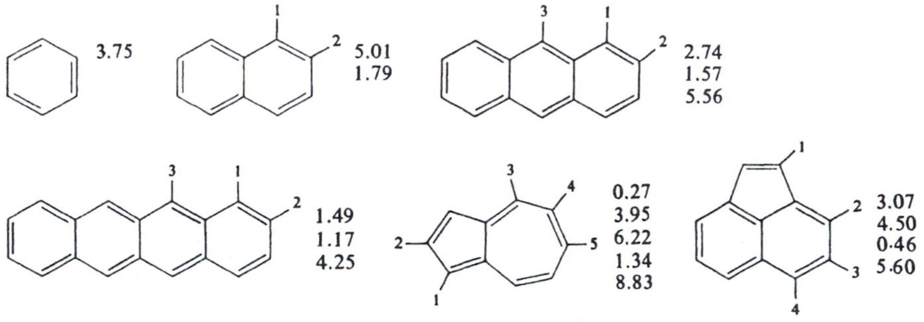

3.75

1 2 5.01 1.79

3 1 2 2.74 1.57 5.56

3 1 2 1.49 1.17 4.25

3 4 2 5 1 0.27 3.95 6.22 1.34 8.83

1 2 3 4 3.07 4.50 0.46 5.60

8-14. Polarographic half-wave potentials for oxidation and reduction of aromatic hydrocarbons are given in Table P8-14.

a) Make separate plots of these data against energy (in units of β) of the highest occupied and lowest empty MO respectively. (Use tabulations or a computer program.)

b) Now plot reduction versus oxidation half-wave potential for this series. Explain adherence to or deviation from linearity.

TABLE P8-14 ►

Compound	Structure	Reduction half-wave potential in 2-methyoxyethanol (V)	Oxidation half-wave potential in acetonitrile (V)
Tetracene	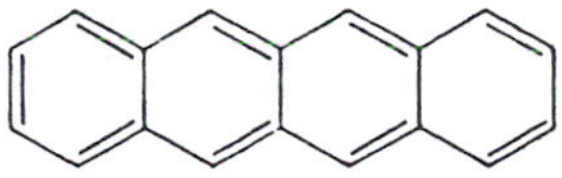	1.135	0.54
1,2-Benzpyrene	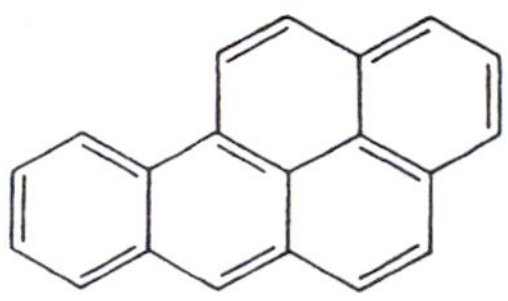	1.36	0.76
Anthracene	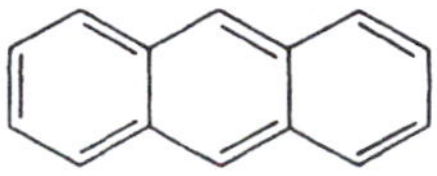	1.46	0.84

(Continued)

TABLE P8-14 ► (Continued)

Compound	Structure	Reduction half-wave potential in 2-methyoxyethanol (V)	Oxidation half-wave potential in acetonitrile (V)
Pyrene	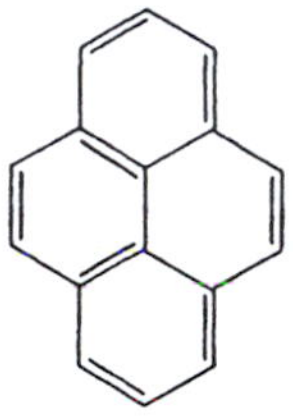	1.61	0.86
1,2-Benzanthracene	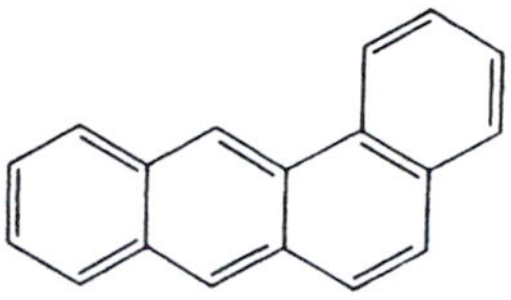	1.53	0.92
1,2,5,6-Dibenzanthracene	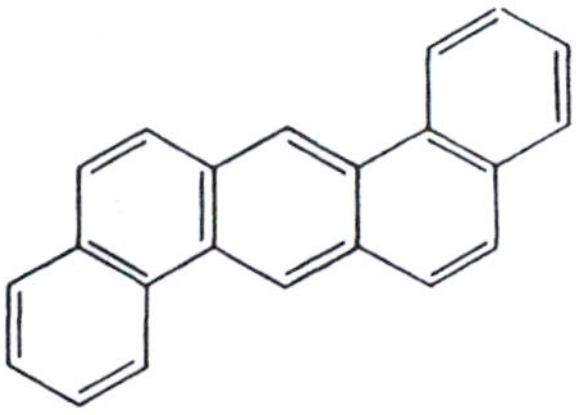	1.545	1.00
Phenanthrene	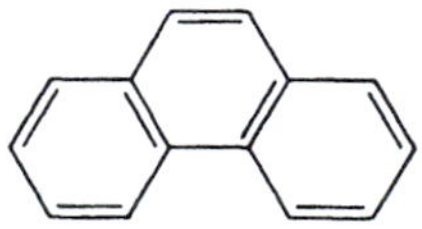	1.935	1.23
Fluoranthene	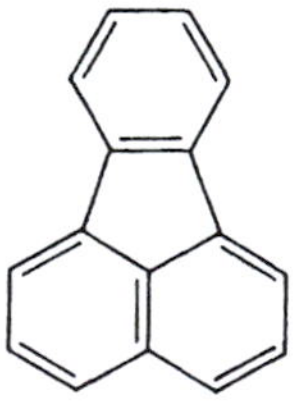	1.345	1.18
Naphthalene	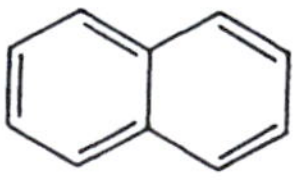	1.98	1.31
Biphenyl	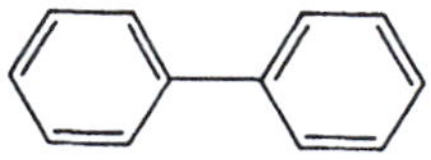	2.075	1.48

8-15. Use tabulated or computer-generated HMO data for *neutral* azulene (**XXVII**) to answer the following questions (Tabulated data may be found in Appendix 6.):

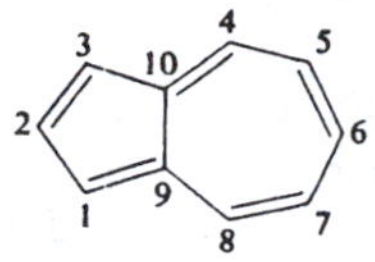

(XXVII)

a) What values would you expect for oxidation and reduction half-wave potentials for this molecule under conditions described in Problem 8-14?
b) If an electron were removed from the highest occupied MO to produce an ion, which bonds would you expect to lengthen, which to shorten?

8-16. Use the effective bond energies of Table 8-2 to calculate the expected π energy for (**XXVIII**). Compare this with the HMO energy of $18\alpha + 21.906\beta$.

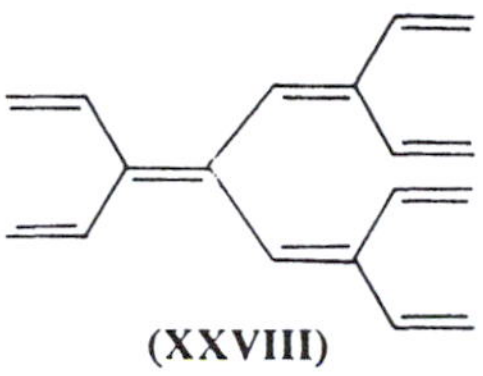

(XXVIII)

8-17. Obtain the HMO data for naphthalene (**XXIX**) and perylene (**XXX**):

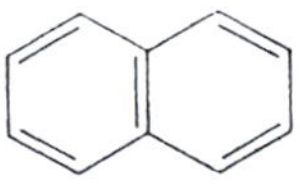

(XXIX)

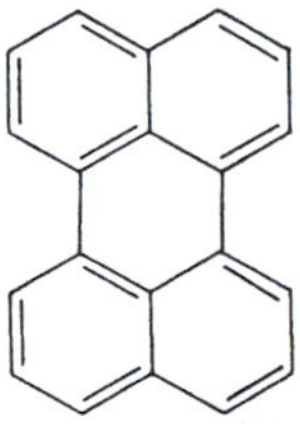

(XXX)

a) For each molecule, compare E to the energy predicted by use of Table 8-2. Categorize each molecule as aromatic, nonaromatic or antiaromatic.
b) Compare the RE for these two molecules. Does the central ring in perylene appear to be contributing?
c) Draw formal bond structures for perylene. What can you conclude about the two bonds connecting naphthalene units in perylene?
d) Use HMO bond orders to calculate a predicted length for these two bonds. How do they compare with the observed 1.471 Å value? Is this observed length consistent with your conclusion of part (c)?

8-18. The third, fourth, and fifth molecules in Fig. 8-24 have dipole moments. Assuming that the individual rings attract or repel charge in accordance with our expectations from the $4n + 2$ rule, predict the direction of the π-electronic contribution to the dipoles. (Dipoles are defined by chemists as being directed from positive toward negative ends of electric dipoles. Physicists use the opposite convention.) How would you expect the π-electron densities to vary in these molecules? Compare your expectations with tabulated densities.

8-19. How many π electrons are there in each of the following neutral molecules? Assume that each one is planar.

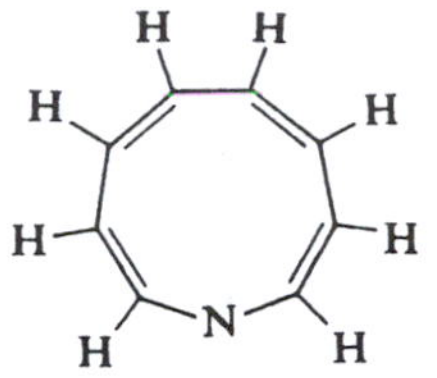

(a) 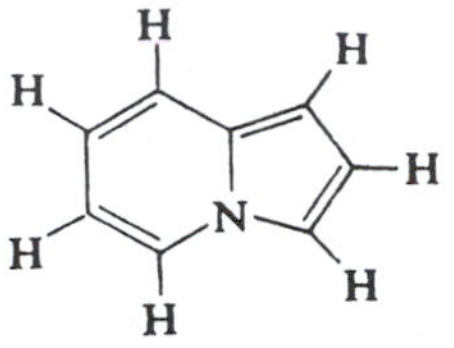 (b) 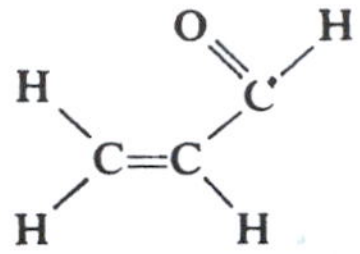 (c)

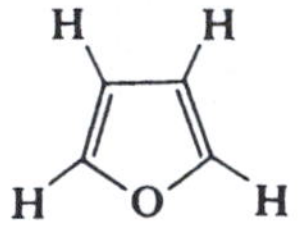

(d) 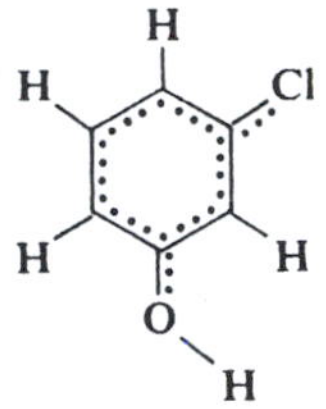 (e)

8-20. Use the parameters in Table 8-3 to construct the HMO determinant for molecule (**XXXI**). Use the conjugative model for the methyl group.

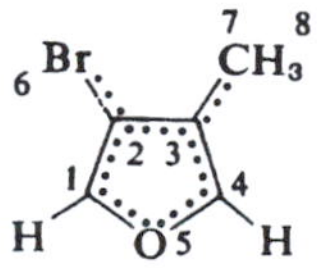

(XXXI)

8-21. Substitution of a nitrogen for a carbon in benzene changes the HMO energy levels, but not drastically. Hence the stability of the six π-electron molecule pyridine can still be rationalized by the $4n + 2$ rule. Which member of each of the following pairs of molecules would you expect to be stable on the basis of such arguments

(a)  or 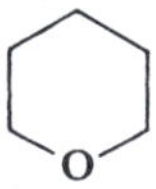 (b) 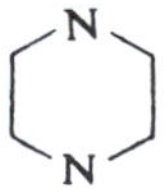 or 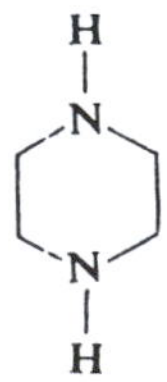 (c)  or

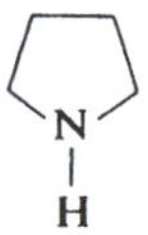

8-22. It is observed that many even alternant hydrocarbons tend to undergo nucleophilic substitution, electrophilic substitution, and radical addition at the same site(s). Rationalize this behavior in terms of the following indices (where appropriate): q_r, HOMO, LUMO, $L_r^{\cdot}$.

8-23. Can π_{rr} be a successful index for nucleophilic and electrophilic substitution if these are observed to occur at different sites?

8-24. For the methylene cyclopropene system C_4H_4 (see Appendix 6 for HMO data and atomic numbering scheme):

a) Calculate the free valences for the neutral molecule.
b) Decide, using three appropriate reactivity indices, which site is most susceptible to electrophilic attack.
c) Decide which protons would lead to hyperfine splitting of the ESR spectrum of the radical anion, according to the simple Hückel approach.
d) Decide whether the second-lowest MO is net bonding or net antibonding. Why?

8-25. Published data for free radical ($CCl_3^{\cdot}$) addition to hydrocarbons are shown in Table P8-25; see Kooyman and Farenhorst [32]. Assuming planarity in each case, use standard HMO tabulations or computer programs to obtain F_r values for each molecule. Choose the largest value for each molecule and plot this against the log of the *modified* rate constant. In each case, modify the rate constant by dividing k by the *number* of sites on the molecule having the maximum F_r value.

TABLE P8-24 ►

Molecule	Structure	Rate constant
Benzene		$<10^{-3}$ (use this limit)
Biphenyl	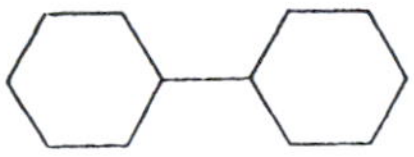	2.7×10^{-3}
Triphenylene	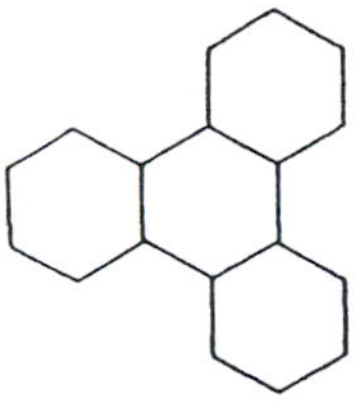	$<4 \times 10^{-2}$ (use this limit)
Phenanthrene	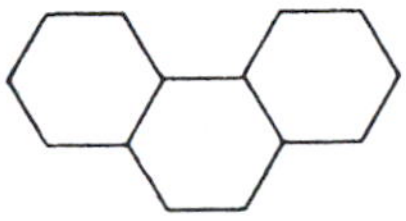	1.6×10^{-2}
Naphthalene	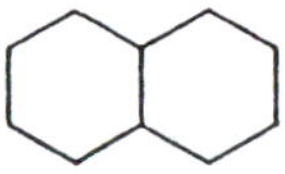	4×10^{-2}

(Continued)

TABLE P8-24 ▶ (Continued)

Molecule	Structure	Rate constant
Chrysene	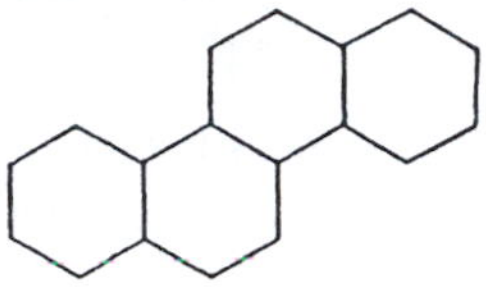	6.7×10^{-2}
Pyrene	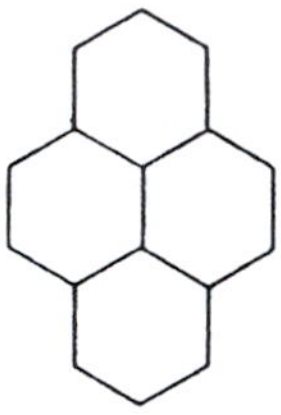	1.3
Stilbene	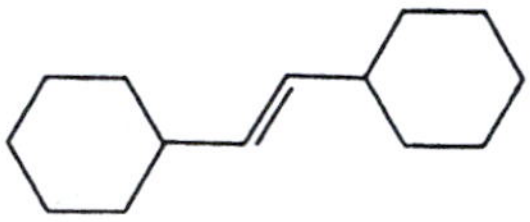	1.0
1,2,5,6-Dibenzanthracene	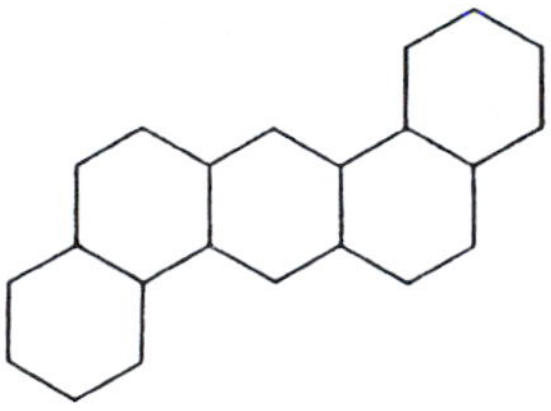	3.7
Styrene		12.5
Anthracene	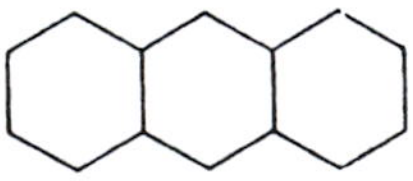	22
Benzanthracene	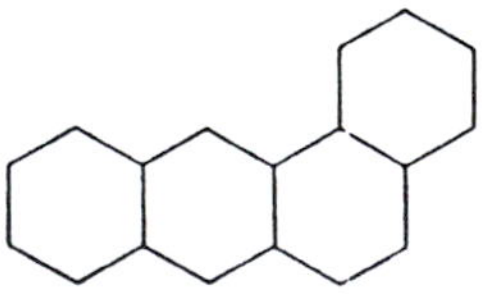	30
3,4-Benzopyrene	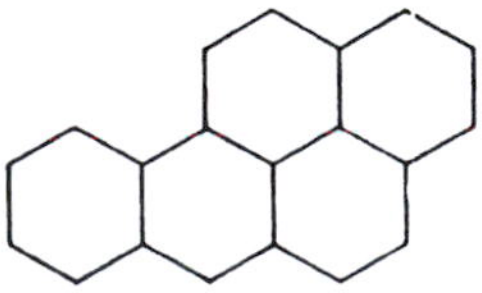	70
Naphthacene	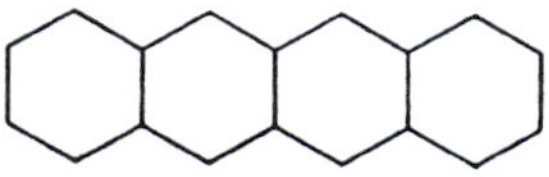	102

(If a molecule has F_r values differing by 0.002 or less, treat them as equal.) If any data points deviate greatly from the general trend, try to give an explanation.

8-26. Which ring of the fourth molecule in Fig. 8-24 would you predict an electrophilic reagent would be more likely to attack, and why?

8-27.* Which of the following neutral unsaturated planar hydrocarbons should experience greater changes in bond order when a π electron is added to form the anion, and why?

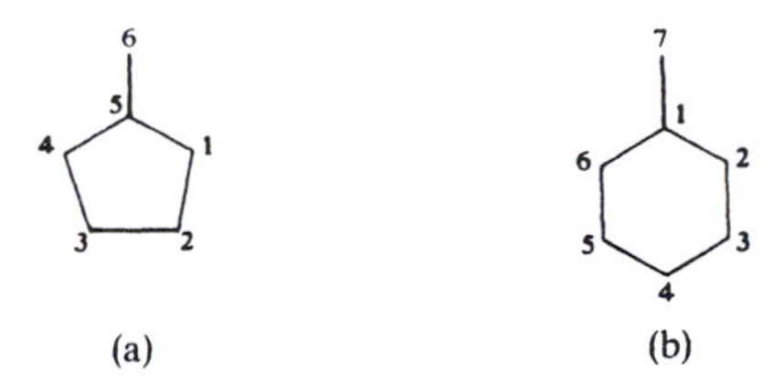

(a) (b)

8-28.* Consider an ESR experiment on the odd-alternant radical shown. Identify the site(s) that have attached hydrogen atoms that should give

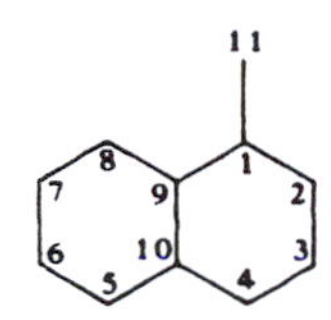

a) the largest coupling constant.
b) the second-largest coupling constant.
c) no coupling constant.

8-29.* Which of the following two seven-π-electron systems should be easier to ionize, and why?

(a) (b)

Multiple Choice Questions

(Try to answer the following questions from sketches that you generate without reference to the text.)

1. According to simple HMO theory, which one of the following statements about butadiene is true?

a) It has a nonbonding MO.
b) Exciting an electron from the HOMO to the LUMO will not change the π bond orders.

*This problem assumes prior study of Appendix 5.

c) Exciting an electron from the HOMO to the LUMO will not change the π charge densities.
d) Ionizing an electron from the HOMO results in all π charge densities becoming equal to 0.75.
e) It is most likely to undergo electrophilic substitution at one of the two inner carbons.

2. Which of the following is a prediction that would result from a simple HMO treatment of the butadienyl cation, $C_4H_6^+$?

a) The ESR coupling constant is larger for hydrogens attached to the two inner carbons.
b) The central C–C bond has a higher π-bond order than it has in the neutral molecule.
c) The positive charge resides mostly on the two central carbons.
d) There are only three π energy levels.
e) The MO coefficients on the central pair of carbons are larger in all of the π MOs than they are in the neutral molecule.

References

[1] L. Salem, *The Molecular Orbital Theory of Conjugated Systems*. Benjamin, New York, 1966.

[2] C. A. Coulson, *Proc. Roy. Soc. (London)* **A164**, 383 (1938).

[3] A. A. Frost and B. Musulin, *J. Chem. Phys.*, **21**, 572 (1953).

[4] C. A. Coulson and A. Streitwieser, Jr., *Dictionary of π-Electron Calculations*. Freeman, San Francisco, 1965.

[5] A. Streitwieser, Jr., and J. I. Brauman, *Supplemental Tables of Molecular Orbital Calculations*. Pergamon, Oxford, 1965.

[6] E. Heilbronner and P. A. Straub, *HMOs*. Springer-Verlag, Berlin and New York, 1966.

[7] A. Streitwieser, Jr., *Molecular Orbital Theory for Organic Chemists*. Wiley, New York, 1961.

[8] C. A. Coulson, *Proc. Roy. Soc. (London)* **A169**, 413 (1939).

[9] F. Gerson and J. H. Hammons, in *Nonbenzenoid Aromatics* (P. J. Snyder, ed.), Vol. II. Academic Press, New York, 1971.

[10] H. Lund, *Acta Chem. Scand.* **11**, 1323 (1957).

[11] I. Bergman, *Trans. Faraday Soc.* **50**, 829 (1954).

[12] D. W. Turner, *Molecular Photoelectron Spectroscopy*. Wiley (Interscience), New York, 1970.

[13] F. Brogli and E. Heilbronner, *Theoret. Chim. Acta* **26**, 289 (1972).

[14] M. J. S. Dewar, *The Molecular Orbital Theory of Organic Chemistry*. McGraw-Hill, New York, 1969.

[15] M. J. S. Dewar and C. de Llano, *J. Amer. Chem. Soc.* **91**, 789 (1969).

[16] M. J. S. Dewar, A. J. Harget, and N. Trinajstic, *J. Amer. Chem. Soc.* **91**, 6321 (1969)

[17] B. A. Hess, Jr., and L. J. Schaad, *J. Amer. Chem. Soc.* **93**, 305, 2413 (1971).

[18] L. J. Schaad and B. A. Hess, Jr., *J. Amer. Chem. Soc.* **94**, 3068 (1972).

[19] V. I. Minkin, M. N. Glukhovtsev, and B. Y. Simkin, *Aromaticity and Antiaromaticity*. Wiley (Interscience), New York, 1994.

[20] P. v. R. Shleyer, C. Maerker, A. Dransfeld, H. Jiao, and N. J. R. van Eikema Hommes, *J. Amer. Chem. Soc.* **118**, 6317, (1996).

[21] R. Herges, and D. Ajami, *Nature*, **426**, 819 (2003).

[22] S. K. Ritter, *Chem. Eng. News*, **82**, vol. 9, p. 28 (2004).

[23] Z. Chen, C. Corminboeuf, T. Heine, J. Bohmann, and P v. R. Schleyer, *J. Amer. Chem. Soc.*, **125**, 13, 930, (2003).

[24] S. P. McGlynn, L. G. Vanquickenborne, M. Kinoshita, and D. G. Carroll, *Introduction To Applied Quantum Chemistry*. Holt, New York, 1972.

[25] H. Fujimoto and K. Fukui, in *Chemical Reactivity and Reaction Paths* (G. Klopman, ed.). Wiley (Interscience), New York, 1974.

[26] C. A. Coulson, *Trans. Faraday Soc.* **42**, 265 (1946).

[27] M. Levy and M. Szwarc, *J. Chem. Phys.* **22**, 1621 (1954).

[28] G. Klopman, ed., *Chemical Reactivity and Reaction Paths.* Wiley (Interscience), New York, 1974.

[29] J. P. Lowe and B. D. Silverman, *Accounts Chem Res.*, **17**, 332, 1984.

[30] J. P. Lowe and B. D. Silverman, *J. Mol. Structure (Theochem.)* **179**, 47 (1988).

[31] W. de W. Horrocks, D. H. Templeton, and A. Zalkin, *Inorg. Chem.* **7**, 2303 (1968).

[32] E. C. Kooyman and E. Farenhorst, *Trans. Faraday Soc.* **49**, 58 (1953).

Chapter 9

Matrix Formulation of the Linear Variation Method

9-1 Introduction

In Chapter 7 we developed a method for performing linear variational calculations. The method requires solving a determinantal equation for its roots, and then solving a set of simultaneous homogeneous equations for coefficients. This procedure is not the most efficient for programmed solution by computer. In this chapter we describe the *matrix* formulation for the linear variation procedure. Not only is this the basis for many quantum-chemical computer programs, but it also provides a convenient framework for formulating the various quantum-chemical methods we shall encounter in future chapters.

Vectors and matrices may be defined in a formal, algebraic way, but they also may be given geometric interpretations. The formal definitions and rules suffice for quantum-chemical purposes. However, the terminology of matrix algebra is closely connected with the geometric ideas that influenced early development. Furthermore, most chemists are more comfortable if they have a physical or geometric model to carry along with mathematical discussion. Therefore, we append some discussion of geometrical interpretation to the algebraic treatment.[1]

9-2 Matrices and Vectors

9-2.A Definitions

A matrix is an ordered array of elements satisfying certain algebraic rules. We write our matrices with parentheses on the left and right of the array.[2] Unless otherwise stated, we restrict the elements to be numbers (which need not be real). In general, however, as long as the rules of matrix algebra can be observed, there is no restriction on what the elements may be. In expression (9-1), we have written a matrix in three ways:

$$\begin{pmatrix} 1.2 & 3.8 & -4.0 \\ 5.0 & 1.0 & 0.0 \\ 9.1 & 0.0 & -3+4i \\ 6.0 & -1.0 & -1.0 \end{pmatrix} \equiv \begin{pmatrix} c_{11} & c_{12} & c_{13} \\ c_{21} & c_{22} & c_{23} \\ c_{31} & c_{32} & c_{33} \\ c_{41} & c_{42} & c_{43} \end{pmatrix} \equiv \mathrm{C} \tag{9-1}$$

[1] More thorough discussions of matrix algebra at a level suitable for the nonspecialist are given by Aitken [1] and Birkhoff and MacLane [2].

[2] Some authors use brackets.

On the left, the numerical elements are written explicitly. In the center they are symbolized by a subscripted letter. On the right the entire matrix is indicated by a single symbol. We will use sans serif, upper-case symbols to represent matrices.

Individual matrix elements will be symbolized either by a subscripted lower-case symbol (e.g., c_{12}) or by a subscripted symbol for the matrix in parentheses (e.g., $(\mathsf{C})_{12}$).

It is useful to recognize rows and columns in a matrix. The sample matrix given above has four rows and three columns, so it is said to have *dimensions* 4×3. When subscripts are used to denote position in a matrix, the convention is that the first subscript indicates the row, the second indicates the column. (The order "row-column" is important to remember. The mnemonic "RC," or "Roman Catholic" is helpful.) Rows are numbered from top to bottom, columns from left to right. There is no limit on the dimensions for matrices, but we usually will be concerned in a practical way with finite-dimensional matrices in this book.

If a matrix has only one column or row, it is called a *column vector* or *row vector*, respectively. We will use sans serif, lower-case symbols to denote *column* vectors. Additional symbols, described shortly, will be used to denote row vectors. These two kinds of vector behave differently under the rules of matrix algebra, so it is important to avoid confusing them.

If a matrix has only one row and one column, its behavior under the rules of matrix algebra becomes identical to the familiar behavior of ordinary *scalars* (i.e., numbers), and so a 1×1 matrix is simply a number.

The similarity in appearance between a matrix and a determinant may be deceptive. A determinant is denoted by bounding with vertical straight lines, and is equal to a *number* that can be found by reducing the determinant according to a prescribed procedure (see Appendix 2). For this to be possible, the determinant must be square (i.e., have the same number of rows as columns). A matrix is not equal to a number and need not be square. (However, one can take the determinant of a square matrix A. This *number* is symbolized $|\mathsf{A}|$ and is *not* the same as A without the vertical bars.)

Two matrices are *equal* if all elements in corresponding positions are equal. Thus, $\mathsf{A} = \mathsf{B}$ means $a_{ij} = b_{ij}$ for all i and j.

9-2.B Complex Conjugate, Transpose, and Hermitian Adjoint of a Matrix

We define the *complex conjugate* of a matrix A to be the matrix A^*, formed by replacing every element of A by its complex conjugate. If $\mathsf{A} = \mathsf{A}^*$, A is a *real* matrix. (Every element is real.)

We define the *transpose* of a matrix A to be the matrix $\tilde{\mathsf{A}}$, formed by interchanging row 1 and column 1, row 2 and column 2, etc. The transpose of the 4×3 matrix in expression (9-1) is the 3×4 matrix given in

$$\tilde{\mathsf{C}} = \begin{pmatrix} 1.2 & 5.0 & 9.1 & 6.0 \\ 3.8 & 1.0 & 0.0 & -1.0 \\ -4.0 & 0.0 & -3+4i & -1.0 \end{pmatrix} \tag{9-2}$$

If we denote some column vector as p, we can symbolize the corresponding row vector as $\tilde{\mathsf{p}}$. Thus, the tilde symbol is one device we can use to indicate a row vector.

Transposing a square matrix corresponds to "reflecting" it through its *principal diagonal* (which runs from upper left to lower right) as indicated in

$$\mathsf{A} = \begin{pmatrix} 1 & 2 & 3 \\ 4 & 5 & 6 \\ 7 & 8 & 9 \end{pmatrix}, \quad \tilde{\mathsf{A}} = \begin{pmatrix} 1 & 4 & 7 \\ 2 & 5 & 8 \\ 3 & 6 & 9 \end{pmatrix} \tag{9-3}$$

If $\mathsf{A} = \tilde{\mathsf{A}}$, A is a *symmetric* matrix.

We define the *hermitian adjoint* of A to be the matrix $\mathsf{A}^\dagger$, formed by taking the transpose of the complex conjugate of A (or the complex conjugate of the transpose. The order of these operations is immaterial.) Hence, $\mathsf{A}^\dagger = (\tilde{\mathsf{A}})^* = (\tilde{\mathsf{A}}^*)$. If $\mathsf{A} = \mathsf{A}^\dagger$, A is a *hermitian* matrix.

9-2.C Addition and Multiplication of Matrices and Vectors

Multiplication of a matrix by a scalar is equivalent to multiplying every element in the matrix by the scalar. Addition of two matrices is accomplished by adding elements in corresponding positions in the matrices. Thus, for example,

$$\begin{pmatrix} 1 & 2 \\ 3 & 4 \end{pmatrix} + 2\begin{pmatrix} 5 & 6 \\ 7 & 8 \end{pmatrix} = \begin{pmatrix} 1+10 & 2+12 \\ 3+14 & 4+16 \end{pmatrix} = \begin{pmatrix} 11 & 14 \\ 17 & 20 \end{pmatrix}$$

and it is evident that the operation of matrix addition is possible only within sets of matrices of identical dimensions.

Matrix multiplication is a bit more involved. We start by considering multiplication of vectors. Two types of vector multiplication are possible. If we multiply a row vector on the left times a column vector on the right, we take the *product* of the leading element of each *plus* the product of the second element of each, plus ..., etc., thereby obtaining a *scalar* as a result. Hence, this is called *scalar* multiplication of vectors. For example,

$$\begin{pmatrix} 1 & 2 & 3 \end{pmatrix}\begin{pmatrix} 4 \\ 5 \\ 6 \end{pmatrix} = 1 \times 4 + 2 \times 5 + 3 \times 6 = 32$$

This kind of multiplication requires an equal number of elements in the two vectors. It is important to retain in mind the basic operation described here: summation of products taken by sweeping *across a row on the left* and *down a column on the right*. Since there exists but *one* row on the left and *one* column on the right, we obtain *one* number as the result.

The other possibility is to multiply a column vector on the left times a row vector on the right. Employing the same basic operation as above, we sweep across row 1 on the left and down column 1 on the right, obtaining a product that we will store in position (1, 1) of a matrix to keep track of its origin. The product of row 1 times column 2 gives us element (1, 2) and so on. In this way, we generate a whole matrix of numbers. For example,

$$\begin{pmatrix} 1 \\ 2 \end{pmatrix}\begin{pmatrix} 3 & 4 & 5 \end{pmatrix} = \begin{pmatrix} 1\times 3 & 1\times 4 & 1\times 5 \\ 2\times 3 & 2\times 4 & 2\times 5 \end{pmatrix} = \begin{pmatrix} 3 & 4 & 5 \\ 6 & 8 & 10 \end{pmatrix}$$

This is an example of *matrix* multiplication of vectors. Just as before, the number of columns on the left (one) equals the number of rows on the right. Now, however, the number of elements in the vectors may differ, and the dimensions of the matrix reflect the dimensions of the original vectors.

The two types of vector multiplication may be symbolized as follows, using our notation for scalars, row vectors, column vectors, and matrices:

$$\tilde{\mathsf{a}}\mathsf{b} = c \tag{9-4}$$

$$\mathsf{a}\tilde{\mathsf{b}} = \mathsf{C} \tag{9-5}$$

Multiplying two matrices together is most simply viewed as scalar multiplying all the rows in the left matrix by all the columns of the right matrix. Thus, in $\mathsf{AB} = \mathsf{C}$, the element c_{ij} is the (scalar) product of row i in A times column j in B. This process is possible only when the number of columns in A equals the number of rows in B. Thus, AB may exist as a matrix C, while BA may not exist due to a disagreement in number of rows and columns. If A and B are both square matrices and have equal dimension, then AB and BA both exist, but they still need not be equal. That is, *matrix multiplication is not commutative.*

In a triple product of matrices, ABC, one can multiply AB first (call the result D) and then multiply DC to get the final result (call it F). Or, if one takes $\mathsf{BC} = \mathsf{E}$, then one always finds $\mathsf{AE} = \mathsf{F}$. Thus, the result is invariant to the choice between $(\mathsf{AB})\mathsf{C}$ or $\mathsf{A}(\mathsf{BC})$, and so matrix multiplication is *associative*.

9-2.D Diagonal Matrices, Unit Matrices, and Inverse Matrices

A *diagonal matrix* is a square matrix having zeros everywhere except on the principal diagonal. Diagonal matrices of equal dimension commute with each other, but a diagonal matrix does not, in general, commute with a nondiagonal matrix.

A *unit* matrix $\mathsf{1}$ is a special diagonal matrix. Every diagonal element has a value of unity. A unit matrix times any matrix (of appropriate dimension) gives that same matrix as product. That is, $\mathsf{1A} = \mathsf{A1} = \mathsf{A}$. It follows immediately that the unit matrix commutes with any *square* matrix of the same dimension.

We define the *left inverse* of a matrix A to be A^{-1}, satisfying the matrix equation $\mathsf{A}^{-1}\mathsf{A} = \mathsf{1}$. The right inverse is defined to satisfy $\mathsf{AA}^{-1} = \mathsf{1}$.

In most of our quantum-chemical applications of matrix algebra, we will be concerned only with vectors and *square* matrices. For square matrices, the left and right inverses are identical, and so we refer simply to the inverse of the matrix.

9-2.E Complex Conjugate, Inverse, and Transpose of a Product of Matrices

If $\mathsf{AB} = \mathsf{C}$, then $\mathsf{C}^* = (\mathsf{AB})^* = \mathsf{A}^*\mathsf{B}^*$. In words, the complex conjugate of a product of matrices is equal to the product of the complex conjugate matrices. This is demonstrable from the observation that $(\mathsf{C})_{ij} = (\mathsf{A})_{i1}(\mathsf{B})_{1j} + (\mathsf{A})_{i2}(\mathsf{B})_{2j} + \cdots$ and $(\mathsf{C})_{ij}^* = (\mathsf{A})_{i1}^*(\mathsf{B})_{1j}^* + (\mathsf{A})_{i2}^*(\mathsf{B})_{2j}^* + \cdots$ and so the complex conjugate is produced by taking the complex conjugate of every element in A and B but not changing their order of combination.

If $\mathsf{AB} = \mathsf{C}$, then $\mathsf{C}^{-1} = (\mathsf{AB})^{-1} = \mathsf{B}^{-1}\mathsf{A}^{-1}$. In words, the inverse of a product of matrices is equal to the product of inverses, but *with the order reversed*. We can easily show that this satisfies the rules of matrix algebra. $\mathsf{C}^{-1}\mathsf{C} = (\mathsf{AB})^{-1}\mathsf{AB} = \mathsf{B}^{-1}\mathsf{A}^{-1}\mathsf{AB} = \mathsf{B}^{-1}\mathsf{1B} = \mathsf{B}^{-1}\mathsf{B} = \mathsf{1}$. If we failed to reverse the order, we would instead have $\mathsf{A}^{-1}\mathsf{B}^{-1}\mathsf{AB}$ and, because the matrices do not commute, we would be prevented from carrying through the reduction to $\mathsf{1}$.

If $\mathsf{AB} = \mathsf{C}$, then $\tilde{\mathsf{C}} = (\widetilde{\mathsf{AB}}) = \tilde{\mathsf{B}}\tilde{\mathsf{A}}$. The transpose of a product is the product of transposes, again in reverse order. Since $\tilde{\mathsf{C}}$ has c_{ij} and c_{ji} interchanged, it follows that, where we had row i of A times column j of B, we must now have row j of A times column i of B. But this is the same as column j of $\tilde{\mathsf{A}}$ times row i of $\tilde{\mathsf{B}}$. To obtain row on left and column on right for proper multiplication, we must have $\tilde{\mathsf{B}}\tilde{\mathsf{A}}$.

9-2.F A Geometric Model

Consider a vector in two-dimensional space emanating from the origin of a Cartesian system as indicated in Fig. 9-1a. We can summarize the information contained in this vector (magnitude and direction) by writing down the x and y components of the vector terminus, (3 2) in this case. It must be understood that the first number corresponds to the x component and not the y, and so the vector (3 2) carries its information through number *position* as well as number *value*.

If we multiply both numbers in the vector by 2, the result, (6 4), corresponds to a vector collinear to the original but twice as long. Therefore, multiplying a vector by a *number* results in a change of *scale* but no change in direction. Hence, the term "scalar" is often used in place of "number" in vector terminology.

Suppose that we rotated the Cartesian axes counterclockwise through an angle θ, maintaining them orthogonal to each other and not varying the distance scales. We imagine our original vector to remain unrotated during this *coordinate transformation*. (Equivalently, we can imagine rotating the vector *clockwise* by θ, keeping the axes fixed.) We wish to know how to express our vector in the new coordinate system. The situation is depicted in Fig. 9-1b. Inspection reveals that the new coordinates $(x'\,y')$ are related to the old $(x\;y)$ as follows:

$$x' = x\cos\theta + y\sin\theta, \quad y' = -x\sin\theta + y\cos\theta \tag{9-6}$$

If we make use of matrix algebra, we can express Eqs. (9-6) as a matrix equation:

$$\begin{pmatrix} x' \\ y' \end{pmatrix} = \begin{pmatrix} \cos\theta & \sin\theta \\ -\sin\theta & \cos\theta \end{pmatrix} \begin{pmatrix} x \\ y \end{pmatrix} \tag{9-7}$$

Multiplying the two-dimensional vector (call it v) by the 2×2 matrix (call it R) generates a new vector v′, that gives the coordinates of our vector in the new coordinate system:

$$\mathsf{v}' = \mathsf{Rv} \tag{9-8}$$

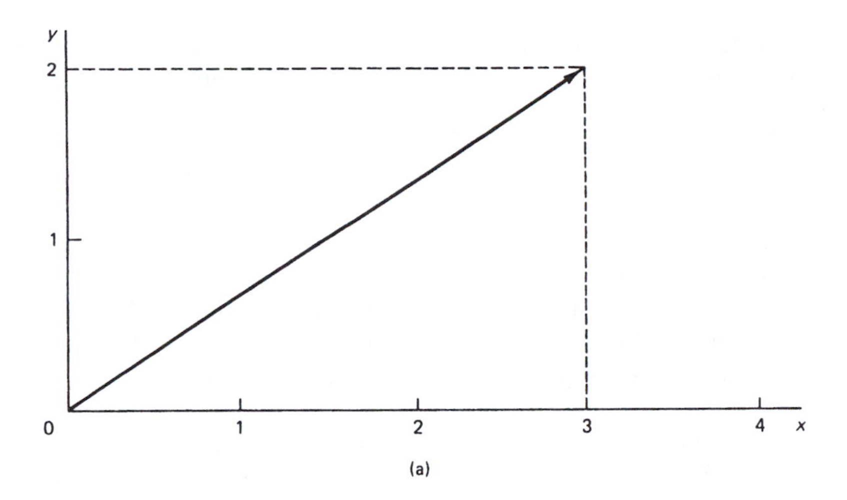

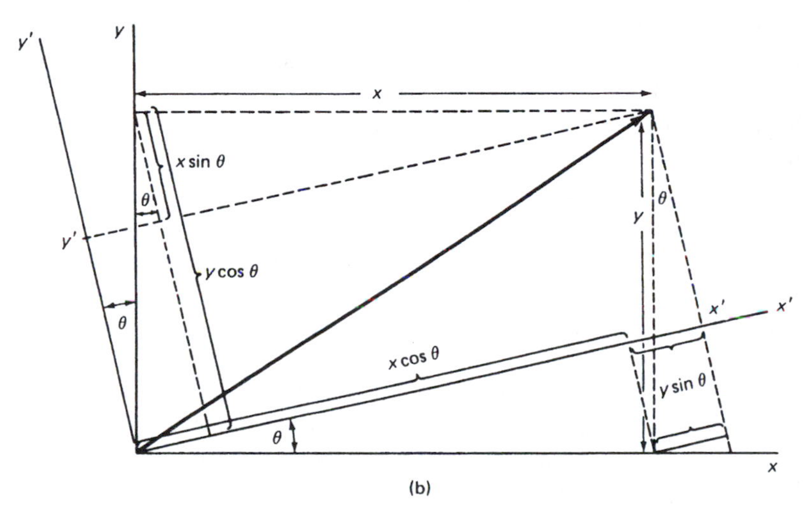

Figure 9-1 ▶ (a) The vector (3 2). (b) The same vector and its relationship to two Cartesian axis systems.

We have, then, a parallel between the vectors V' and V and matrix R on the one hand, and the two-dimensional "geometrical" vector and rotating coordinate system on the other. R *represents* the rotation and is often referred to as a *rotation* matrix.

If we were to perform the rotation in the opposite direction, the rotation matrix would be the same except for the $\sin\theta$ terms, which would reverse sign:

$$\left.\begin{array}{l}\text{rotation of coordinates}\\ \text{clockwise by }\theta\end{array}\right\} \longrightarrow \begin{pmatrix}\cos\theta & -\sin\theta\\ \sin\theta & \cos\theta\end{pmatrix} \tag{9-9}$$

Note that this is just $\tilde{\mathsf{R}}$, the transpose of R.

If we were to rotate counterclockwise by θ and then clockwise by θ, we should end up with our original coordinates for the vector. Thus, we should expect

$$\tilde{\mathsf{R}}\mathsf{R}\mathsf{v} = \mathsf{v} \tag{9-10}$$

or

$$\tilde{\mathsf{R}}\mathsf{R} = 1 \tag{9-11}$$

(Note that the *order* of operations is consistent with reading from right to left, just as with differential operators. Thus, $\tilde{\mathsf{R}}\mathsf{R}$ means that first R is performed, then $\tilde{\mathsf{R}}$.) Relation (9-11) is easily verified by explicit multiplication. Thus we see that, if we think of a matrix as representing some coordinate transformation in geometrical space, the inverse of the matrix represents the reverse transformation. In this *particular* example, the *transpose* of the transformation matrix turns out to be the inverse transformation matrix. When this is so, the matrix is said to be *orthogonal*. (Orthogonal transformations do not change the angles between coordinate axes; orthogonal axes remain orthogonal—hence the name "orthogonal.") The analogous transformation for matrices having complex or imaginary coefficients is called a *unitary* transformation. A unitary matrix has its hermitian adjoint as inverse: $\mathsf{A}^{\dagger}\mathsf{A} = 1$.

While it is easy to visualize a coordinate transformation in two or three dimensions, it is more difficult in higher-dimensional situations. Nevertheless, the mathematics and terminology carry forward to any desired dimension and are very useful. One may, if one wishes, talk of vectors and coordinate transformations in hyperspace, or one can eschew such mental constructs and simply follow the mathematical rules without a mental model.

One must be a bit cautious about inverses of matrices. In the rotation described above, we have a unique way of relating each x, y point in one coordinate system to a point at x', y' in the other. The transformation does not entail any loss of information and can therefore be "undone."

Such transformations (and their matrices) are called *nonsingular*. A nonsingular matrix is recognizable through the fact that its determinant must be nonzero. If we had a transformation which, for example, caused all or some points in one coordinate system to coalesce into a single point in the transformed system, we would lose our ability to back-transform in a unique way. Such a *singular* transformation has no inverse, and the determinant of a singular matrix equals zero.

9-2.G Similarity Transformations

A matrix product of the form $\mathsf{A}^{-1}\mathsf{H}\mathsf{A}$ is called a *similarity transformation* on H. If A is orthogonal, then $\tilde{\mathsf{A}}\mathsf{H}\mathsf{A}$ is a special kind of similarity transformation, called an *orthogonal transformation*. If A is unitary, then $\mathsf{A}^{\dagger}\mathsf{H}\mathsf{A}$ is a *unitary transformation* on H. There is a physical interpretation for a similarity transformation, which will be discussed in a later chapter. For the present, we are concerned only with the mathematical definition of such a transformation. The important feature is that the eigenvalues, or "latent roots," of H are preserved in such a transformation (see Problem 9-5).

In this section we have quickly presented the salient rules of matrix algebra and hinted at their connection with geometric operations. The results are summarized in Table 9-1 for ease of reference.

TABLE 9-1 ► Some Matrix Rules and Definitions for a Square Matrix A of Dimension n

$\mathsf{A} = \mathsf{B}$	Matrix equality; means $a_{ij} = b_{ij}$, $i, j = 1, n$
$\mathsf{A} + \mathsf{B} = \mathsf{C}$	Matrix addition; $c_{ij} = a_{ij} + b_{ij}$, $i, j = 1, n$
$c\mathsf{A} = \mathsf{B}$	Multiplication of A by scalar; $b_{ij} = c \cdot a_{ij}$, $i, j = 1, n$
$\mathsf{AB} = \mathsf{C}$	Matrix multiplication; $c_{ij} = \sum_{k=1}^{n} a_{ik} b_{kj}$, $i, j = 1, n$
$\lvert\mathsf{A}\rvert$	The determinant of the matrix A (see Appendix 2)
A^{-1}	The inverse of A; $\mathsf{A}^{-1}\mathsf{A} = \mathsf{A}\mathsf{A}^{-1} = 1$ If A^{-1} exists, A is *nonsingular* and $\lvert\mathsf{A}\rvert \neq 0$.
A^*	The complex conjugate of A; $a_{ij} \to a_{ij}^*$, $i, j = 1, n$ If $\mathsf{A}^* = \mathsf{A}$, A is *real*.
$\tilde{\mathsf{A}}$	The transpose of A; $(\tilde{\mathsf{A}})_{ij} = a_{ji}$ (rows and columns inter changed) If $\tilde{\mathsf{A}} = \mathsf{A}$, symmetric; if $\tilde{\mathsf{A}} = -\mathsf{A}$, antisymmetric; if $\tilde{\mathsf{A}} = \mathsf{A}^{-1}$, orthogonal.
$\mathsf{A}^\dagger$	The hermitian adjoint of A; $(\mathsf{A}^\dagger)_{ij} = a_{ji}^* (\mathsf{A}^\dagger = \tilde{\mathsf{A}}^*)$ If $\mathsf{A}^\dagger = \mathsf{A}$, hermitian. If $\mathsf{A}^\dagger = \mathsf{A}^{-1}$, unitary.
$(\mathsf{ABC})^* = \mathsf{A}^*\mathsf{B}^*\mathsf{C}^*$	Complex conjugate of product
$\widetilde{(\mathsf{ABC})} = \tilde{\mathsf{C}}\tilde{\mathsf{B}}\tilde{\mathsf{A}}$	Transpose of product
$(\mathsf{ABC})^\dagger = \mathsf{C}^\dagger\mathsf{B}^\dagger\mathsf{A}^\dagger$	Hermitian adjoint of product
$(\mathsf{ABC})^{-1} = \mathsf{C}^{-1}\mathsf{B}^{-1}\mathsf{A}^{-1}$	Inverse of product
$\lvert\mathsf{ABC}\rvert = \lvert\mathsf{A}\rvert \cdot \lvert\mathsf{B}\rvert \cdot \lvert\mathsf{C}\rvert$	Determinant of product (any order)
$\mathsf{T}^{-1}\mathsf{AT}$	A similarity transformation If $\mathsf{T}^{-1} = \mathsf{T}^\dagger$, this is a *unitary* transformation. If $\mathsf{T}^{-1} = \tilde{\mathsf{T}}$, this is an *orthogonal* transformation.

9-3 Matrix Formulation of the Linear Variation Method

We have seen that the independent-electron approximation leads to a series of MOs for a molecular system. If the MOs are expressed as a linear combination of n basis functions (which are often approximations to AOs, although this is not necessary), the variation method leads to a set of simultaneous equations:

$$
\begin{aligned}
(H_{11} - ES_{11})c_1 + (H_{12} - ES_{12})c_2 + \cdots + (H_{1n} - ES_{1n})c_n &= 0 \\
&\vdots \\
(H_{n1} - ES_{n1})c_1 \qquad\qquad + \cdots + (H_{nn} - ES_{nn})c_n &= 0
\end{aligned}
\tag{9-12}
$$

All terms have been defined in Chapter 7. Given a value for E that satisfies the associated *determinantal* equations, we can solve this set of simultaneous equations for ratios between the c_i's. Requiring MO normality establishes convenient numerical values for the c_i's.

A matrix equation equivalent to Eq. (9-12) is[3]

$$\begin{pmatrix} H_{11}-ES_{11} & H_{12}-ES_{12} & \cdots & H_{1n}-ES_{1n} \\ \vdots & & & \\ H_{n1}-ES_{n1} & & \cdots & H_{nn}-ES_{nn} \end{pmatrix} \begin{pmatrix} c_1 \\ c_2 \\ \vdots \\ c_n \end{pmatrix} = \begin{pmatrix} 0 \\ 0 \\ \vdots \\ 0 \end{pmatrix} \tag{9-13}$$

The matrix in Eq. (9-13) is clearly the difference between two matrices. This enables us to rewrite the equation in the form

$$\begin{pmatrix} H_{11} & H_{12} & \cdots & H_{1n} \\ \vdots & & & \vdots \\ H_{n1} & & \cdots & H_{nn} \end{pmatrix} \begin{pmatrix} c_1 \\ c_2 \\ \vdots \\ c_n \end{pmatrix} = E \begin{pmatrix} S_{11} & S_{12} & \cdots & S_{1n} \\ \vdots & & & \vdots \\ S_{n1} & & \cdots & S_{nn} \end{pmatrix} \begin{pmatrix} c_1 \\ c_2 \\ \vdots \\ c_n \end{pmatrix} \tag{9-14}$$

or

$$\mathsf{H}\mathsf{c}_i = E_i \mathsf{S}\mathsf{c}_i, \quad i = 1, 2, \ldots, n \tag{9-15}$$

where we have introduced the subscript i to account for the fact that there are many possible values for E and that each one has its own characteristic set of coefficients. Note that the "*eigenvector*" c_i is a *column* vector and that each element in c_i is (effectively) multiplied by the scalar E_i according to Eq. (9-15).

In general, there are as many MOs as there are basis functions, and so Eq. (9-15) represents n separate matrix equations. We can continue to use matrix notation to reduce these to a single matrix equation. We do this by stacking the n c vectors together, side by side, to produce an $n \times n$ matrix C. The numbers E must also be combined into an appropriate matrix form. We must be careful to do this in such a way that the scalar E_1 still multiplies only c_1 (now column 1 of C) E_2 multiplies only c_2, and so forth. This is accomplished in the following equation

$$\begin{pmatrix} H_{11} & \cdots & H_{1n} \\ \vdots & & \vdots \\ H_{n1} & \cdots & H_{nn} \end{pmatrix} \begin{pmatrix} c_{11} & \cdots & c_{1n} \\ \vdots & & \vdots \\ c_{n1} & \cdots & c_{nn} \end{pmatrix} = \begin{pmatrix} S_{11} & \cdots & S_{1n} \\ \vdots & & \vdots \\ S_{n1} & \cdots & S_{nn} \end{pmatrix} \begin{pmatrix} c_{11} & \cdots & c_{1n} \\ \vdots & & \vdots \\ c_{n1} & \cdots & c_{nn} \end{pmatrix} \begin{pmatrix} E_1 & 0 & 0 & \cdots & 0 \\ 0 & E_2 & 0 & \cdots & 0 \\ \vdots & \vdots & \vdots & & \vdots \\ 0 & 0 & 0 & \cdots & E_n \end{pmatrix} \tag{9-16}$$

or

$$\mathsf{HC} = \mathsf{SCE} \tag{9-17}$$

The matrix E is a diagonal matrix of orbital energies (often referred to as the *matrix* of *eigenvalues*). C is the matrix of coefficients (or *matrix* of *eigenvectors*), and each

[3]Quantum-chemical convention is to use upper case letters for individual elements of the matrices H, S, and E. This differs from the usual convention.

column in C refers to a different MO. The first column refers to the MO having energy E_1. In multiplying E by C from the left, each coefficient in column 1 becomes multiplied by E_1. This would not occur if we multiplied E by C_1 from the right. Therefore, HC = SCE is correct, whereas HC = ESC is incorrect.

9-4 Solving the Matrix Equation

Since we know the basis functions and the effective hamiltonian (in principle, at least), we are in a position to evaluate the elements in H and S. How do we then find C and E?

Let us first treat the simplified situation where our basis set of functions is orthonormal, either by assumption or design. Then all the off-diagonal elements of S (which correspond to overlap between *different* basis functions) are zero, and all the diagonal elements are unity because of normality. In short, S = 1. Therefore, Eq. (9-17) becomes

$$\mathsf{HC} = \mathsf{CE} \tag{9-18}$$

and our problem is, given H, find C and E.

Now, we want a set of coefficients that correspond to normalized MOs. We have seen earlier that, for an orthonormal basis set, this requires each MO to have coefficients satisfying the equation (assuming real coefficients)

$$c_{1i}^2 + c_{2i}^2 + \cdots + c_{ni}^2 = 1 \tag{9-19}$$

We can write this as a vector equation

$$\tilde{\mathsf{c}}_i \mathsf{c}_i \equiv \begin{pmatrix} c_{i1} & c_{i2} & \cdots & c_{in} \end{pmatrix} \begin{pmatrix} c_{1i} \\ c_{2i} \\ \vdots \\ c_{ni} \end{pmatrix} = 1 \tag{9-20}$$

Furthermore, we know that any two different MOs must be orthogonal to each other. That is $\tilde{\mathsf{c}}_i \mathsf{c}_j = 0$, $i \neq j$. All this may be summarized in the matrix equation

$$\tilde{\mathsf{C}}\mathsf{C} = \mathsf{1} \tag{9-21}$$

Hence, the coefficient matrix is *orthogonal*. In the more general case in which coefficients may be complex, C is *unitary*; i.e., $\mathsf{C}^\dagger \mathsf{C} = \mathsf{1}$. Our problem then is, given H, find a unitary matrix C such that HC = CE with E diagonal.

We can multiply both sides of a matrix equation by the same matrix and preserve the equality. However, because matrices do not necessarily commute, we must be careful to carry out the multiplication from the left on both sides, or from the right on both sides. Thus, multiplying Eq. (9-18) from the left by $\mathsf{C}^\dagger$, we obtain

$$\mathsf{C}^\dagger \mathsf{HC} = \mathsf{C}^\dagger \mathsf{CE} = \mathsf{1E} = \mathsf{E} \tag{9-22}$$

where we have used the fact that C is unitary. Now our problem may be stated as, given H, find a unitary matrix C such that $\mathsf{C}^\dagger \mathsf{HC}$ is diagonal.[4] Several techniques

[4]Not every matrix (not even every square matrix) can be diagonalized by a unitary transformation, but every *hermitian* matrix can be so diagonalized.

exist for finding such a matrix C. These are generally much more suitable for machine computation than are determinantal manipulations.

We can illustrate that the allyl radical energies and coefficients already found by the HMO method do in fact satisfy the relations $\mathsf{C}^{\dagger}\mathsf{C} = \mathsf{1}$ and $\mathsf{C}^{\dagger}\mathsf{HC} = \mathsf{E}$. The matrix C can be constructed from the HMO coefficients and is

$$\mathsf{C} = \begin{pmatrix} \frac{1}{2} & \frac{1}{\sqrt{2}} & \frac{1}{2} \\ \frac{1}{\sqrt{2}} & 0 & -\frac{1}{\sqrt{2}} \\ \frac{1}{2} & -\frac{1}{\sqrt{2}} & \frac{1}{2} \end{pmatrix} \tag{9-23}$$

Therefore

$$\mathsf{C}^{\dagger}\mathsf{C} = \begin{pmatrix} \frac{1}{2} & \frac{1}{\sqrt{2}} & \frac{1}{2} \\ \frac{1}{\sqrt{2}} & 0 & -\frac{1}{\sqrt{2}} \\ \frac{1}{2} & -\frac{1}{\sqrt{2}} & \frac{1}{2} \end{pmatrix} \begin{pmatrix} \frac{1}{2} & \frac{1}{\sqrt{2}} & \frac{1}{2} \\ \frac{1}{\sqrt{2}} & 0 & -\frac{1}{\sqrt{2}} \\ \frac{1}{2} & -\frac{1}{\sqrt{2}} & \frac{1}{2} \end{pmatrix}$$

$$= \begin{pmatrix} 1 & 0 & 0 \\ 0 & 1 & 0 \\ 0 & 0 & 1 \end{pmatrix} = \mathsf{1} \tag{9-24}$$

The matrix H for the allyl radical is, in HMO theory,

$$\mathsf{H} = \begin{pmatrix} \alpha & \beta & 0 \\ \beta & \alpha & \beta \\ 0 & \beta & \alpha \end{pmatrix} \tag{9-25}$$

The reader should verify that

$$\mathsf{C}^{\dagger}\mathsf{HC} = \begin{pmatrix} \alpha + \sqrt{2}\beta & 0 & 0 \\ 0 & \alpha & 0 \\ 0 & 0 & \alpha - \sqrt{2}\beta \end{pmatrix} = \mathsf{E} \tag{9-26}$$

The diagonal elements can be seen to correspond to the HMO energies. Note that the energy in the (1 1) position of E corresponds to the MO with coefficients appearing in *column* 1 of C, illustrating the positional correlation of eigenvalues and eigenvectors referred to earlier.

If the basis functions are not orthogonal, $\mathsf{S} \neq 1$ and the procedure is slightly more complicated. Basically, one first transforms to an *orthogonal* basis to obtain an equation of the form $\mathsf{H}'\mathsf{C}' = \mathsf{C}'\mathsf{E}$. One diagonalizes H' as indicated above to find E and C', where C' is the matrix of coefficients in the *orthogonalized* basis. Then one back transforms C' into the original basis set to obtain C. There are many choices available for the orthogonalizing transformation. The Schmidt transformation, based on the Schmidt orthogonalization procedure described in Chapter 6, is popular because it is very rapidly performed by a computer. Here we will simply indicate the matrix algebra involved. Let the matrix that transforms a nonorthonormal basis to an orthonormal one be symbolized A. This matrix satisfies the relation

$$\mathsf{A}^{\dagger}\mathsf{SA} = \mathsf{1} \tag{9-27}$$

Furthermore, $|\mathsf{A}| \neq 0$ and so A^{-1} exists. We can insert the unit matrix (in the form AA^{-1}) wherever we please in the matrix equation $\mathsf{HC} = \mathsf{SCE}$ without affecting the equality. Thus,

$$\mathsf{HAA}^{-1}\mathsf{C} = \mathsf{SAA}^{-1}\mathsf{CE} \tag{9-28}$$

Multiplying from the left by $\mathsf{A}^{\dagger}$ gives

$$\mathsf{A}^{\dagger}\mathsf{HAA}^{-1}\mathsf{C} = \mathsf{A}^{\dagger}\mathsf{SAA}^{-1}\mathsf{CE} \tag{9-29}$$

By Eq. (9-27), this reduces to

$$(\mathsf{A}^{\dagger}\mathsf{HA})(\mathsf{A}^{-1}\mathsf{C}) = (\mathsf{A}^{-1}\mathsf{C})\mathsf{E} \tag{9-30}$$

where the parentheses serve only to make the following discussion clearer. If we *define* $\mathsf{A}^{\dagger}\mathsf{HA}$ to be H', and $\mathsf{A}^{-1}\mathsf{C}$ to be C', Eq. (9-30) becomes

$$\mathsf{H}'\mathsf{C}' = \mathsf{C}'\mathsf{E} \tag{9-31}$$

Since we know the matrix H and can compute A from knowledge of S, it is possible to write down an *explicit* H' matrix for a given problem. Then, knowing H' (which is just the *hamiltonian matrix* for the problem in the orthonormal basis), we can seek the unitary matrix C' such that $\mathsf{C}'^{\dagger}\mathsf{H}'\mathsf{C}'$ is diagonal. These diagonal elements are our orbital energies. (Note that E in Eq. (9-31) is the same as E in $\mathsf{HC} = \mathsf{SCE}$.) To find the coefficients for the MOs in terms of the *original* basis (i.e., to find C), we use the relation

$$\mathsf{AC}' = \mathsf{A}(\mathsf{A}^{-1}\mathsf{C}) = \mathsf{1C} = \mathsf{C} \tag{9-32}$$

One nice feature of this procedure is that, even though we use the inverse matrix A^{-1} in our *formal* development, we never need to actually compute it. (A and $\mathsf{A}^{\dagger}$ are used to find H', and A is used to find C.) This is fortunate because calculating inverse matrices is a relatively slow process.

A few more words should be said about the process of diagonalizing a hermitian matrix H with a unitary transformation. Two methods are currently in wide use. The older, slower method, known as the *Jacobi* method, requires a series of steps on the starting matrix. In the first step, a matrix O_1 is constructed that causes the largest off-diagonal pair of elements of H to vanish in the transformation $\mathsf{H}_1 = \tilde{\mathsf{O}}_1\mathsf{HO}_1$. Now a second transformation matrix O_2 is constructed to force the largest off-diagonal pair of elements in H_1 to vanish in the transformation $\tilde{\mathsf{O}}_2\mathsf{H}_1\mathsf{O}_2 = \tilde{\mathsf{O}}_2\tilde{\mathsf{O}}_1\mathsf{HO}_1\mathsf{O}_2$. This procedure is continued. However, since each transformation affects more elements in the matrix than just the biggest pair, we eventually "unzero" the pair that was zeroed in forming H_1 or H_2, etc. This means that many more transformations are required than there are off-diagonal pairs. Eventually, however, the off-diagonal elements will have been nibbled away (while the diagonal elements have been building up) until they are all smaller in magnitude than some preselected value, and so we stop the process. The transformation matrix C corresponds to the accumulated product $\mathsf{O}_1\mathsf{O}_2\mathsf{O}_3\ldots$.

A more recently discovered, faster procedure is the Givens–Householder–Wilkinson method. Here, H is first *tridiagonalized*, which means that all elements are made to vanish except those on the main diagonal *as well as on the codiagonals above and below the main diagonal*. This similarity transformation can be done in a few steps, each step zeroing all the necessary elements in an entire row and column. The eigenvalues for the tridiagonal matrix (and hence for the original matrix) may be found one at a time as desired. If only the third lowest eigenvalue is of interest, that one alone can be computed. This is a useful degree of freedom which results in substantial savings of time. Once an eigenvalue is found, its corresponding eigenvector may be computed.

9-5 Summary

The steps to be performed in a matrix solution for a linear variation calculation are:

1. From the basis set, calculate the overlap matrix S.

2. From the basis set and hamiltonian operator, calculate the hamiltonian matrix H.

3. If $\mathsf{S} \neq \mathsf{1}$, find an orthogonalization procedure. In the Schmidt method, A is such that $\mathsf{A}^\dagger\mathsf{S}\mathsf{A} = \mathsf{1}$. The matrix equation may now be written in the form $\mathsf{H}'\mathsf{C}' = \mathsf{C}'\mathsf{E}$.

4. Find C' such that $\mathsf{C}'^\dagger\mathsf{H}'\mathsf{C}'$ is a diagonal matrix. The diagonal elements are the roots E.

5. If necessary, back transform: $\mathsf{A}\mathsf{C}' = \mathsf{C}$. The columns of C contain the MO coefficients appropriate for the original basis set.

9-5.A Problems

9-1. Evaluate the following according to the rules of matrix algebra:

a) $\begin{pmatrix} 6 & 7 & 8 \end{pmatrix} \begin{pmatrix} 9 \\ 10 \\ 11 \end{pmatrix}$

b) $\begin{pmatrix} 6 \\ 7 \end{pmatrix} \begin{pmatrix} a & b & c \end{pmatrix}$

c) $\begin{pmatrix} 4 & 6 \\ i & -3 \end{pmatrix} + 7 \begin{pmatrix} 3 & 1 \\ -1 & 3 \end{pmatrix}$

d) $\begin{pmatrix} \cos\theta & -\sin\theta \\ \sin\theta & \cos\theta \end{pmatrix} \begin{pmatrix} \cos\theta & \sin\theta \\ -\sin\theta & \cos\theta \end{pmatrix}$

e) $\begin{pmatrix} i & 4 \\ 1 & 7 \\ 0 & -3 \end{pmatrix} \begin{pmatrix} 3 & 2 \\ 4 & 7 \end{pmatrix}$

f) $\begin{pmatrix} 3 & 2 \\ 4 & 7 \end{pmatrix} \begin{pmatrix} i & 4 \\ 1 & 7 \\ 0 & -3 \end{pmatrix}$

g) $\begin{vmatrix} \cos\theta & -\sin\theta \\ \sin\theta & \cos\theta \end{vmatrix}$

9-2. If $H_{ij} = \int \chi_i^* \hat{H} \chi_j \, d\tau$ and $\hat{H}$ is hermitian, show that H is a hermitian matrix.

9-3. Let

$$\mathsf{A} = \begin{pmatrix} a_{11} & a_{12} \\ a_{21} & a_{22} \end{pmatrix}, \quad \mathsf{B} = \begin{pmatrix} b_{11} & b_{12} \\ b_{21} & b_{22} \end{pmatrix}$$

Show that, in general, $\mathsf{AB} \neq \mathsf{BA}$.

9-4. Let

$$\mathsf{A} = \begin{pmatrix} 1 & 0 & 0 \\ 0 & 2 & 0 \\ 0 & 0 & 3 \end{pmatrix}, \quad \mathsf{B} = \begin{pmatrix} 4 & 0 & 0 \\ 0 & 5 & 0 \\ 0 & 0 & 6 \end{pmatrix}, \quad \text{and } \mathsf{C} = \begin{pmatrix} 1 & 0 & 1 \\ 0 & 1 & 0 \\ 1 & 0 & 1 \end{pmatrix}$$

Show that $\mathsf{AB} = \mathsf{BA}$, but $\mathsf{AC} \neq \mathsf{CA}$. Compare the matrix AC with CA. Do these matrices show any simple relationship? Can you relate this to properties of A and C mathematically?

9-5. The "latent roots" λ_i of A are solutions to the equation $|\mathsf{A} - \lambda_i \mathsf{1}| = 0$, $i = 1, 2, \ldots, n$, where n is the dimension of A.

a) Show that, under a similarity transformation $\mathsf{B} = \mathsf{T}^{-1}\mathsf{AT}$, the latent roots are preserved.
b) Demonstrate that diagonalization of A *via* a similarity transformation produces the latent roots as the diagonal elements.

9-6. Show that, if a matrix has any latent roots equal to zero, it has no inverse.

9-7. The *trace* (or *spur*) of a matrix is the sum of the elements on the principal diagonal. Thus, tr $\mathsf{A} = \Sigma_{i=1}^{n} a_{ii}$.

a) Show that the trace of a triple product of matrices is invariant under cyclic permutation. That is, $\mathrm{tr}(\mathsf{ABC}) = \mathrm{tr}(\mathsf{CAB}) = \mathrm{tr}(\mathsf{BCA})$ but not $\mathrm{tr}(\mathsf{CBA})$.
b) Show that the trace of a matrix is invariant under a similarity transformation.

9-8. The *norm* of a matrix is the positive square root of the sum of the absolute squares of all the elements.

For a real matrix A,

$$\text{norm } \mathsf{A} = \left[\sum_{i,j=1}^{n} a_{ij}^2 \right]^{1/2} = \left[\sum_i \sum_j (\tilde{\mathsf{A}})_{i,j} (\mathsf{A})_{j,i} \right]^{1/2}$$

Prove that the norm of a real matrix is preserved in an orthogonal transformation (or, you may prefer to prove that the norm of any matrix is preserved in a unitary transformation).

9-9. Use the facts that the trace, the determinant, and the norm of a matrix are invariant under an orthogonal transformation to find the eigenvalues of the following matrices:

a) $\begin{pmatrix} 0 & 1 & 1 \\ 1 & 0 & 1 \\ 1 & 1 & 0 \end{pmatrix}$

b) $\begin{pmatrix} \frac{1}{2} & \frac{1}{\sqrt{2}} & \frac{1}{2} \\ \frac{1}{\sqrt{2}} & 0 & -\frac{1}{\sqrt{2}} \\ \frac{1}{2} & -\frac{1}{\sqrt{2}} & \frac{1}{2} \end{pmatrix}$

c) $\begin{pmatrix} 0 & 1 & 0 \\ 1 & 2 & 1 \\ 0 & 1 & 0 \end{pmatrix}$

9-10. Consider the matrix

$$\begin{pmatrix} \cos\theta & 0 \\ -\sin\theta & 0 \end{pmatrix}$$

What is the effect of this transformation on $\begin{pmatrix} 3 \\ 2 \end{pmatrix}$? On $\begin{pmatrix} 3 \\ 3 \end{pmatrix}$? Can the transformation be uniquely reversed? (That is, for, say, $\theta = 0$, and given a transformed vector $\begin{pmatrix} 3 \\ 0 \end{pmatrix}$, can one uniquely determine the vector this was transformed from?) Does the matrix have an inverse? Evaluate its determinant.

9-11. What are the eigenvectors for the matrix

$$\mathsf{H} = \begin{pmatrix} 1 & 0 & 0 \\ 0 & -3 & 0 \\ 0 & 0 & -2 \end{pmatrix}$$

9-12. Show that, if A and B have "simultaneous eigenvectors" (i.e., both diagonalized by the same similarity transformation), then A and B commute.

9-13. If $\mathsf{HC} = \mathsf{CE}$, and $\mathsf{C}^\dagger\mathsf{C} = \mathsf{1}$, then $\mathsf{C}^\dagger\mathsf{HC} = \mathsf{E}$, and we seek a unitary transformation that diagonalizes H. If $\mathsf{HC} = \mathsf{SCE}$, and $\mathsf{C}^\dagger\mathsf{SC} = \mathsf{1}$, then $\mathsf{C}^\dagger\mathsf{HC} = \mathsf{C}^\dagger\mathsf{SCE} = \mathsf{1E}$, and $\mathsf{C}^\dagger\mathsf{HC} = \mathsf{E}$. Since this is the same working equation as the one we found above, why do we not proceed in the same way? Why do we bother orthogonalizing our basis first?

9-14. We have mentioned that a matrix may be used to represent the rotation of coordinates by some angle θ. Such a rotation is a geometric *operation*, so we have, in effect, represented an *operator* with a matrix. It is possible to represent other operators in a similar way. Indeed, an alternative approach to quantum mechanics exists in which the whole formalism is based on matrices and their properties (matrix mechanics, as opposed to wave mechanics). A particularly interesting example is provided by the matrices constructed by Pauli to represent spin operators and functions. It was mentioned in Chapter 5 that spin functions α and β satisfy rules similar to those for orbital angular momentum. Two of these are

$$\hat{S}_z\alpha = \frac{1}{2}\alpha, \quad \hat{S}_z\beta = -\frac{1}{2}\beta$$

But it was pointed out that α and β could not be expressed in terms of spherical harmonics. Pauli represented this operator and functions by

$$\alpha = \begin{pmatrix} 1 \\ 0 \end{pmatrix}, \quad \beta = \begin{pmatrix} 0 \\ 1 \end{pmatrix}, \quad \hat{S}_z = \frac{1}{2}\begin{pmatrix} 1 & 0 \\ 0 & -1 \end{pmatrix}$$

Using these definitions, show that

$$\int \alpha^\dagger \beta \, d\omega = \int \beta^\dagger \alpha \, d\omega = 0, \quad \int \alpha^\dagger \alpha \, d\omega = \int \beta^\dagger \beta \, d\omega = 1,$$

$$\hat{S}_z\alpha = \frac{1}{2}\alpha, \quad \hat{S}_z\beta = -\frac{1}{2}\beta$$

[*Note*: since α and β are essentially the Dirac delta functions in the spin coordinate ω, the process of integration reduces here to scalar multiplication of vectors.]

References

[1] A. C. Aitken, *Determinants and Matrices*, 9th ed. Wiley (Interscience), New York, 1958.

[2] G. Birkhoff and S. MacLane, *A Survey of Modern Algebra*, 5th ed. Macmillan, New York, 1995.

Chapter 10

The Extended Hückel Method

10-1 The Extended Hückel Method

The extended Hückel (EH) method is much like the simple Hückel method in many of its assumptions and limitations. However, it is of more general applicability since it takes account of all valence electrons, σ *and* π, and it is of more recent vintage because it can only be carried out on a practical basis with the aid of a computer. The basic methods of extended Hückel calculations have been proposed at several times by various people. We will describe the method of Hoffmann [1], which, because of its systematic development and application, is the EHMO method in common use.

The method is described most easily by reference to an example. We will use methane (CH_4) for this purpose.

10-1.A Selecting Nuclear Coordinates

The first choice we must make is the molecular geometry to be used. For methane, we will take the H–C–H angles to be tetrahedral and C–H bond distances of 1.1 Å. We can try altering these dimensions later.

Cartesian coordinates for the five atoms are listed in Table 10-1, and the orientation of the nuclei in Cartesian space is indicated in Fig. 10-1. (Even though the eigenvalues and MOs one finally obtains are independent of how CH_4 is oriented in Cartesian space,[1] it is generally a good idea to choose an orientation that causes some Cartesian and symmetry axes to coincide. The resulting expressions for MOs in terms of AOs are generally much simpler to sketch and interpret.)

10-1.B The Basis Set

Next we must select the basis set of functions with which to express the MOs. The extended Hückel method uses the normalized valence AOs for this purpose. For CH_4, this means a 1s AO on each hydrogen and a 2s, $2p_x$, $2p_y$, and $2p_z$ AO on carbon. The inner-shell 1s AO on carbon is not included. The AOs are represented by Slater-type orbitals (STOs). Except for the 1s AOs on hydrogen, the exponential parameters of the STOs are determined from Slater's rules (Chapter 5). Various values for the hydrogen 1s AO exponent have been suggested. These have ranged from the 1.0 given by Slater's

[1] It sometimes happens that an approximation is made that causes the solution to depend on orientation. This is called "loss of rotational invariance."

TABLE 10-1 ► Cartesian Coordinates (in Angstroms) for Atoms of Methane Oriented as Shown in Fig. 10-1

Atom	x	y	z
C	0.0	0.0	0.0
H_a	0.0	0.0	1.1
H_b	1.03709	0.0	−0.366667
H_c	−0.518545	0.898146	−0.366667
H_d	−0.518545	−0.898146	−0.366667

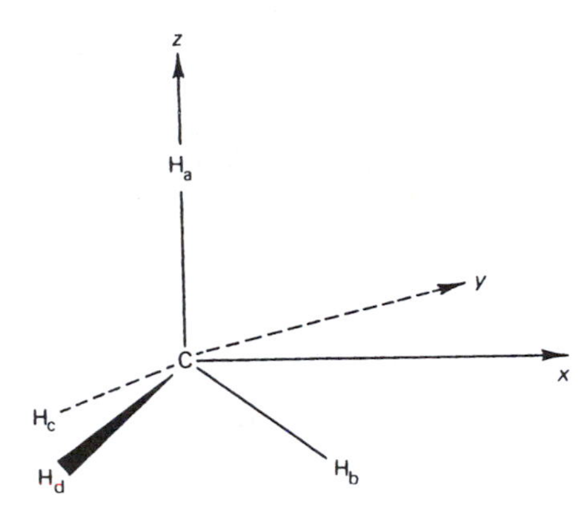

Figure 10-1 ► Orientation of methane in a Cartesian axis system.

rules to a value of $\sqrt{2}$. We will use a value of 1.2, which is near the optimal value for H_2 (see Chapter 7). The STOs for methane are listed in Table 10-2.

10-1.C The Overlap Matrix

Knowing the AO functions and their relative positions enables us to calculate all of their overlaps. This would be a tedious process with pencil and paper. However, the formulas have been programmed for automatic computation, so this step is included in any EH computer program.[2] The computed overlap matrix for the methane molecule is shown in Table 10-3.

This matrix is symmetric (since the overlap between two AOs is independent of their numbering order) and has diagonal elements of unity since the AOs are normalized. The zero values in the first four rows and columns reflect the orthogonality between the s and all p AOs on carbon. Other zero values result when hydrogen 1s AOs are centered in nodal planes of carbon p AOs. The geometry of the system is clearly reflected in the overlap matrix. For instance, the overlap of the $2p_z$ AO of carbon with the hydrogen 1s AO at H_a is large and positive, while its overlaps with AOs on H_b, H_c, and H_d are negative, equal, and of smaller magnitude. The 2s AO of carbon, on the other hand, overlaps all 1s AOs equally. Also, the overlap between every pair of hydrogen 1s AOs

[2] Several such programs are available from Quantum Chemistry Program Exchange, Chemistry Dept., Room 204, Indiana University, Bloomington, Indiana 47401. http://www.QCPE.Indiana.edu

TABLE 10-2 ► Basis AOs for Methane

AO no.	Atom	Type	n^a	l^a	m^a	exp
1	C	2s	2	0	0	1.625
2	C	$2p_z$	2	1	0	1.625
3	C	$2p_x$	2	1	$(1)^b$	1.625
4	C	$2p_y$	2	1	$(1)^b$	1.625
5	H_a	1s	1	0	0	1.200
6	H_b	1s	1	0	0	1.200
7	H_c	1s	1	0	0	1.200
8	H_d	1s	1	0	0	1.200

[a] n, l, m are the quantum numbers described in Chapter 4.
[b] $2p_x$ and $2p_y$ are formed from linear combinations of $m=+1$ and $m=-1$ STOs, and neither of these AOs can be associated with a particular value of m.

TABLE 10-3 ► Overlap Matrix for STOs of Table 10-2

	1	2	3	4	5	6	7	8
1	1.0000	0.0	0.0	0.0	0.5133	0.5133	0.5133	0.5133
2	0.0	1.0000	0.0	0.0	0.4855	−0.1618	−0.1618	−0.1618
3	0.0	0.0	1.0000	0.0	0.0	0.4577	−0.2289	−0.2289
4	0.0	0.0	0.0	1.0000	0.0	0.0	0.3964	−0.3964
5	0.5133	0.4855	0.0	0.0	1.0000	0.1805	0.1805	0.1805
6	0.5133	−0.1618	0.4577	0.0	0.1805	1.0000	0.1805	0.1805
7	0.5133	−0.1618	−0.2289	0.3964	0.1805	0.1805	1.0000	0.1805
8	0.5133	−0.1618	−0.2289	−0.3964	0.1805	0.1805	0.1805	1.000

is the same. Features such as these provide a useful check on the correctness of our initial Cartesian coordinates.

10-1.D The Hamiltonian Matrix

We have the overlap matrix S. Next we must find the hamiltonian matrix H. Then we will be in a position to solve the equation $\mathsf{HC}=\mathsf{SCE}$ for C and E. (See Chapter 9.) The matrix H is calculated from a very approximate but simple recipe. The basic ideas are similar in spirit to those described in connection with the interpretations of α and β in the simple Hückel method. The energy integral H_{ii} in the EH method is taken to be equal to the energy of an electron in the ith AO of the isolated atom in the appropriate state. The various ionization energies of atoms are known,[3] so this presents no great difficulty. However, one special problem must be dealt with, namely, finding the

[3] See Moore [2].

appropriate state. In the isolated carbon atom, the lowest-energy states are associated with the configuration $1s^2 2s^2 2p^2$. In a saturated molecule such as methane, however, carbon shares electrons with four hydrogens, and calculations indicate that the 2s and all three 2p AOs are about equally involved in forming occupied MOs. That is, in the molecule, carbon behaves as though it were in the $2s2p^3$ configuration. This configuration (shortened to sp^3) is referred to as the *valence state* of carbon in this molecule. Since this is an "open-shell" configuration (i.e., not all the electrons are spin-paired in filled orbitals), there are several actual physical states (corresponding to different spin and orbital angular momenta of an isolated carbon atom) that are associated with this configuration. Thus, there are several real physical states, with different ionization energies, associated with our mentally constructed sp^3 valence state for the atom in a molecule. The question is, what real ionization energies should we use to evaluate our valence state ionization energy (VSIE)? The approach that is used is simply to *average* the real IEs for loss of a 2p or a 2s electron, the average being taken over all states associated with the sp^3 configuration. Various authors recommend slightly different sets of VSIEs.[4] We use here the values tabulated by Pople and Segal [7]. Because of the rather crude nature of the EH method, the slight variations in VSIE resulting from different choices are of little consequence.[5] For methane, we have

$$(\mathrm{C_{2s}}):\ H_{11} = -19.44\,\mathrm{eV} = -0.7144\,\mathrm{a.u.} \tag{10-1}$$

$$(\mathrm{C_{2p}}):\ H_{22} = H_{33} = H_{44} = -10.67\,\mathrm{eV} = -0.3921\,\mathrm{a.u.} \tag{10-2}$$

$$(\mathrm{H_{1s}}):\ H_{55} = H_{66} = H_{77} = H_{88} = -13.60\,\mathrm{eV} = -0.50000\,\mathrm{a.u.} \tag{10-3}$$

The off-diagonal elements of H are evaluated according to[6]

$$H_{ij} = KS_{ij}\left(\frac{H_{ii} + H_{jj}}{2}\right) \tag{10-4}$$

where K is an adjustable parameter. The rationalization for such an expression is that the energy of interaction should be greater when the overlap between AOs is greater, and that an overlap interaction energy between low-energy AOs should be lower than that produced by an equal amount of overlap between higher-energy AOs. We will discuss the energy *versus* overlap relation in more detail in a later section. The value of K suggested by Hoffmann [1] is 1.75. The reasons for choosing this value will be discussed shortly. For now, we accept this value and arrive at the hamiltonian matrix given in Table 10-4.

By examining H we can guess in advance some of the qualitative features of the MOs that will be produced. For instance, the value of H_{25} (−0.3790 a.u.) indicates a strong energy-lowering interaction between the $2p_z$ AO and the 1s AO on H_a. This interaction refers to AOs with positive and negative lobes *as they are assigned in the basis set.* Therefore, we expect a low-energy (bonding) MO to occur where these AOs are mixed with coefficients of the same sign so as not to affect this AO sign relation. A high-energy MO should also exist where the mixing occurs through coefficients of opposite sign, producing an antibonding interaction. The values of H_{26}, H_{27}, and H_{28} are positive,

[4] See Skinner and Pritchard [3], Hinze and Jaffé [4], Basch et al. [5], Anno [6], and Pople and Segal [7].

[5] The proper valence state for carbon in methane differs from that in ethylene, which in turn differs from that in acetylene. Generally, this is ignored in EHMO calculations and a compromise set of VSIEs is selected for use over the whole range of molecules.

[6] This formula is often called the Wolfsberg–Helmholtz relation.

TABLE 10-4 ► The Extended Hückel Hamiltonian Matrix for CH_4[a]

	1	2	3	4	5	6	7	8
1	−0.7144	0.0	0.0	0.0	−0.5454	−0.5454	−0.5454	−0.5454
2	0.0	−0.3921	0.0	0.0	−0.3790	0.1263	0.1263	0.1263
3	0.0	0.0	−0.3921	0.0	0.0	−0.3573	0.1787	0.1787
4	0.0	0.0	0.0	−0.3921	0.0	0.0	−0.3094	0.3094
5	−0.5454	−0.3790	0.0	0.0	−0.5000	−0.1579	−0.1579	−0.1579
6	−0.5454	0.1263	−0.3573	0.0	−0.1579	−0.5000	−0.1579	−0.1579
7	−0.5454	0.1263	0.1787	−0.3094	−0.1579	−0.1579	−0.5000	−0.1579
8	−0.5454	0.1263	0.1787	0.3094	−0.1579	−0.1579	−0.1579	−0.5000

[a] All energies in a.u.

due to negative overlap in corresponding positions of S. In this case, energy lowering will be associated with mixing $2p_z$ with 1s AOs on hydrogens b, c, and d, *but now the mixing coefficients will have signs that reverse the AO sign relations from those pertaining in the original basis*. The AOs that are orthogonal have zero interaction, and so mixing between such AOs will not affect MO energies. (When such AOs are mixed in the same MO, it is often the result of arbitrary mixing between degenerate MOs. In such cases, one can find an orthogonal pair of MOs such that two noninteracting AOs do not appear in the same MO. Methane will be seen to provide an example of this.)

10-1.E The Eigenvalues and Eigenvectors

Having H and S, we now can use the appropriate matrix-handling programs to solve HC = SCE for the matrix eigenvalues on the diagonal of E and the coefficients for the MOs, which are given by the columns of C. The eigenvalues for methane, together with their occupation numbers, are given in Table 10-5. The corresponding coefficients are given in Table 10-6.

TABLE 10-5 ► Energies for Methane by the Extended Hückel Method

MO no.	Energy (a.u.)	Occ. no.
8	1.1904	0
7	0.2068	0
6	0.2068	0
5	0.2068	0
4	−0.5487	2
3	−0.5487	2
2	−0.5487	2
1	−0.8519	2

TABLE 10-6 ▶ Coefficients Defining MOs for Methane

	MO number							
	1	2	3	4	5	6	7	8
1(2s)	0.5842	0.0	0.0	0.0	0.0	0.0	0.0	1.6795
$2(2p_z)$	0.0	0.5313	−0.0021	−0.0007	−0.0112	−0.0137	1.1573	0.0
$3(2p_x)$	0.0	0.0021	0.5313	−0.0021	1.1573	−0.0178	0.0110	0.0
$4(2p_y)$	0.0	0.0007	0.0021	0.5313	0.0176	1.1572	0.0139	0.0
$5(1s_a)$	0.1858	0.5547	−0.0022	−0.0007	0.0105	0.0128	−1.0846	−0.6916
$6(1s_b)$	0.1858	−0.1828	0.5237	−0.0019	−1.0260	0.0114	0.3518	−0.6916
$7(1s_c)$	0.1858	−0.1853	−0.2589	0.4542	0.4943	−0.8977	0.3558	−0.6916
$8(1s_d)$	0.1858	−0.1865	−0.2626	−0.4516	0.5213	0.8734	0.3770	−0.6916

Only two of the eight MOs are nondegenerate. These two MOs must be symmetric or antisymmetric for every symmetry operation of the molecule. This is easily checked by sketching the MOs, referring to the coefficients in the appropriate columns of Table 10-6. The lowest nondegenerate energy occurs in position 1 of our eigenvalue list (Table 10-5), and so the coefficients for this MO are to be found in column 1 of Table 10-6. This column indicates that the MO ϕ_1 is equal to $0.5842\ 2s + 0.1858\ 1s_a + 0.1858\ 1s_b + 0.1858\ 1s_c + 0.1858\ 1s_d$, where the symbols 2s, $1s_a$, etc., stand for AOs on carbon, H_a, etc. A sketch of this MO appears in Fig. 10-2. It is obviously symmetric for all rotations and reflections of a tetrahedron. Notice that this MO is bonding in all four C–H bond regions since the 2s STO on carbon is in phase agreement with all the hydrogen 1s AOs. The higher-energy, nondegenerate MO ϕ_8 is qualitatively similar to ϕ_1 except that the signs are reversed on the 1s AOs (see Table 10-6). Hence, this MO has the same symmetry properties as ϕ_1, but is antibonding in the C–H regions.

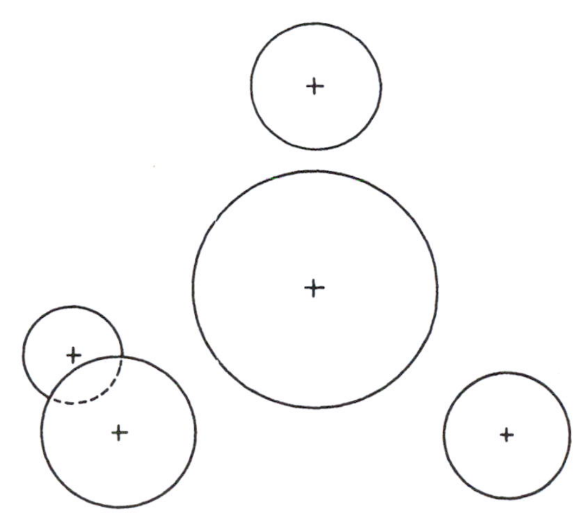

Figure 10-2 ▶ A drawing of the lowest-energy nondegenerate EHMO for methane. The AOs are drawn as though they do not overlap. This is done only to make the drawing simpler. Actually, the AOs overlap strongly.

Since these two MOs resemble the 2s STO in being symmetric for all the symmetry operations of a tetrahedron, we will refer to them as s-type MOs.

The remaining six MOs are grouped into two energy levels, each level being triply degenerate. (It is possible to predict from symmetry considerations alone that the energy levels resulting from this calculation will be nondegenerate and triply degenerate. This is discussed in a later chapter.) Because they are degenerate, the MOs cannot be expected to show *all* the symmetry of the molecule, but it should be possible for them to show *some* symmetry. Consider ϕ_2, as given by column 2 of Table 10-6. This is mainly constructed from the $2p_z$ AO on carbon and 1s AOs on the four hydrogens. Small contributions from $2p_x$ and $2p_y$ are also present, however. It would be nice to remove these small contributions and "clean up" the MO. We can do this, as mentioned earlier, by mixing ϕ_2 with appropriate amounts of ϕ_3 and ϕ_4 since these are all degenerate. The equations that our cleaned-up MO, ϕ_2', must satisfy are

$$\phi_2' = d_2\phi_2 + d_3\phi_3 + d_4\phi_4 \tag{10-5}$$

where (in order to cause all $2p_x$ and $2p_y$ contributions to vanish)

$$\begin{aligned} 0.0021d_2 + 0.5313d_3 + -0.0021d_4 &= 0.0 \\ 0.0007d_2 + 0.0021d_3 + \quad 0.5313d_4 &= 0.0 \\ d_2^2 + \quad d_3^2 + \quad d_4^2 &= 1 \end{aligned} \tag{10-6}$$

As a result,

$$\phi_2' = 0.9999\phi_2 - 0.0040\phi_3 - 0.0013\phi_4 \tag{10-7}$$

A similar procedure to produce an orbital with no p_z or p_y contribution (ϕ_3'), and one with no p_z or p_x contribution (ϕ_4') gives

$$\phi_3' = 0.9999\phi_3 + 0.0040\phi_2 - 0.0040\phi_4 \tag{10-8}$$

$$\phi_4' = 0.9999\phi_4 + 0.0013\phi_2 + 0.0039\phi_3 \tag{10-9}$$

The coefficients for these MOs appear in Table 10-7.

TABLE 10-7 ► Coefficients for MOs $\phi_2', \phi_3', \phi_4'$

	ϕ_2'	ϕ_3'	ϕ_4'
2s	0.0	0.0	0.0
$2p_z$	0.5313	0.0	0.0
$2p_x$	0.0	0.5313	0.0
$2p_y$	0.0	0.0	0.5313
$1s_a$	0.5547	0.0	0.0
$1s_b$	−0.1849	0.5228	0.0
$1s_c$	−0.1849	−0.2614	0.4529
$1s_d$	−0.1849	−0.2614	−0.4529

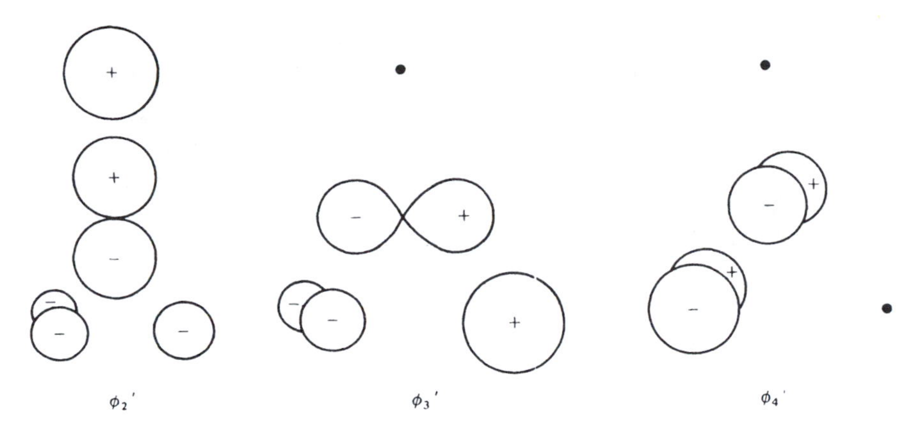

Figure 10-3 ▶ The three lowest-energy degenerate MOs of methane.

There is no fundamental change produced by intermixing degenerate MOs in this way. The total electronic density and the orbital energies are uninfluenced. The only advantage is that the cleaned-up MOs are easier to sketch and visualize. The MOs ϕ_2', ϕ_3', and ϕ_4' are sketched in Fig. 10-3.

Each of the MOs in Fig. 10-3 is symmetric or antisymmetric for some of the operations that apply to a tetrahedron. ϕ_2' is symmetric for rotations about the z axis by $2\pi/3$, and also for reflection through the xz plane. This same reflection plane is a symmetry plane for $\phi_3'$and ϕ_4', but neither of these MOs shows symmetry or antisymmetry for rotation about the z axis. Each MO contains one p AO and, perforce, has the symmetry of that AO. We shall refer to these as p-type MOs. Note that hydrogen 1s AOs lying in the nodal plane of a p AO do not mix with that p AO in formation of MOs. This results from zero interaction elements in H, which, in turn, results from zero overlap elements in S. Note also that the MO ϕ_2' is the MO that we anticipated earlier on the basis of inspection of the matrix H. Because of phase agreements between the hydrogen 1s AOs and the adjacent lobes of the p AOs, ϕ_1', ϕ_2', and ϕ_3' are C–H bonding MOs.

A similar "cleaning up" procedure can be performed on ϕ_5, ϕ_6, and ϕ_7. These turn out to be the C–H antibonding mates to the MOs in Fig. 10-3.

The broad results of this calculation are that there are four occupied C–H bonding MOs, one of s type and three of p type. At higher energies are four unoccupied C–H antibonding MOs, again one of s type and three of p type.

Note that the s- and the three p-type MOs fall into the same energy pattern as the s and p AOs of isolated carbon. Because of their highly symmetric tetrahedral geometry, the hydrogen atoms do not lift the degeneracy of the p AOs. There are many molecules and complexes in which a cluster of atoms or molecules surrounds a central atom in such a highly symmetric way that the degeneracies among certain AOs on the central atom are retained.

10-1.F The Total Energy

The total EH energy is taken as the sum of the one-electron energies. For methane, this is $2 \times (-0.8519) + 6 \times (-0.5487)$, or -4.9963 a.u. There is some ambiguity as to how this energy is to be interpreted. For instance, does it include any of the internuclear repulsion energy? Also, what problems will arise from our neglect of inner-shell electrons? By comparing EH total energy changes with experimental energy changes, it has been decided[7] that the *change* in EH total energy upon change of geometry is approximately the same as the actual change in total electronic plus nuclear repulsion energy for the system. Thus, our value of −4.9963 a.u. is not a realistic value for the total (nonrelativistic) energy of methane (the actual value is −40.52 a.u.), but it is meaningful when compared to EH energies for methane at other geometries. For example, if we uniformly lengthen or shorten all the C–H bonds in methane and repeat our EH calculation several times, we can generate an energy curve versus $R_{C–H}$ for the symmetrical stretch vibrational mode of methane. The resultant plot is given in Fig. 10-4. The EH total energy is minimized at about $R_{C–H} = 1$ Å, reasonably close to the experimentally observed 1.1- Å distance for the minimum *total* energy of methane.

The appearance of the curve in Fig. 10-4 does encourage us to equate EH total energy changes to changes in actual electronic-plus-nuclear-repulsion energies. As we shall see later, this procedure fails for some molecules (notably H_2) and for methane may be regarded as fortuitous.

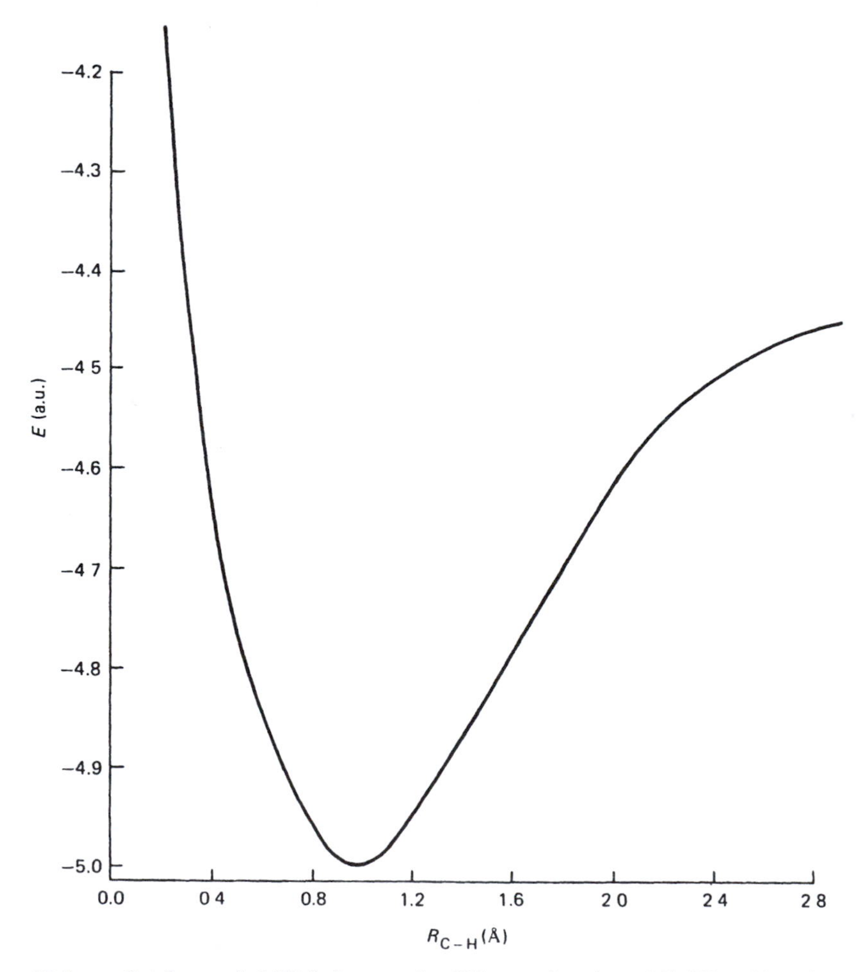

Figure 10-4 ▶ Total extended Hückel energy for CH_4 as a function of C–H bond length.

[7] See Hoffmann [1].

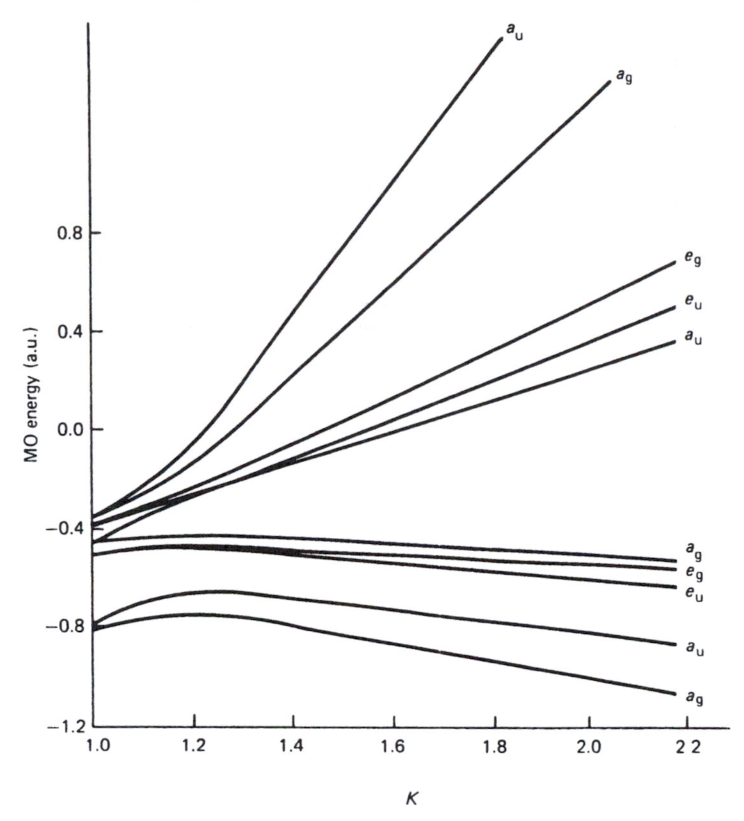

Figure 10-5 ▶ Staggered ethane MO energies versus K. C–H distances are 1.1 Å; C–C distance is 1.54 Å; all angles are tetrahedral.

10-1.G Fixing the Parameter *K*

We mentioned earlier that Hoffmann suggested a value of 1.75 for K. We will now indicate the considerations behind this suggestion.

Hoffmann used the ethane molecule C_2H_6 to evaluate K. A plot of the orbital energies of staggered ethane as a function of K is shown in Fig. 10-5. The energies are linearly dependent on K at values of K greater than about 1.5. At lower values, the lines curve and some crossing occurs. Hence, in order that the MO energy order not be highly sensitive to K, its value should exceed 1.5. A plot of the amount of electronic charge in a bond or at an atom in ethane, as calculated by the EHMO method, versus K is shown in Fig. 10-6. We will describe the details of such calculations shortly, but for now, we merely note that the disposition of charge in ethane becomes rather insensitive to K at values of K greater than about 1.5. In Fig. 10-7 a plot of the EH total energy difference between staggered and eclipsed ethanes versus K is given. This energy difference has an experimentally determined value of 2.875 ± 0.025 kcal/mole, the staggered form being more stable. To give reasonable agreement with this experimental value, K should be about 1.75. Thus, the value $K = 1.75$ is selected because the MO energy order and charge distribution are not sensitive to K in this region and because this value of K gives the correct total energy change for a known physical process. We have also seen earlier that this same value of K leads to a reasonable prediction for the equilibrium bond length in CH_4. Notice that the evaluation of K comes by matching the

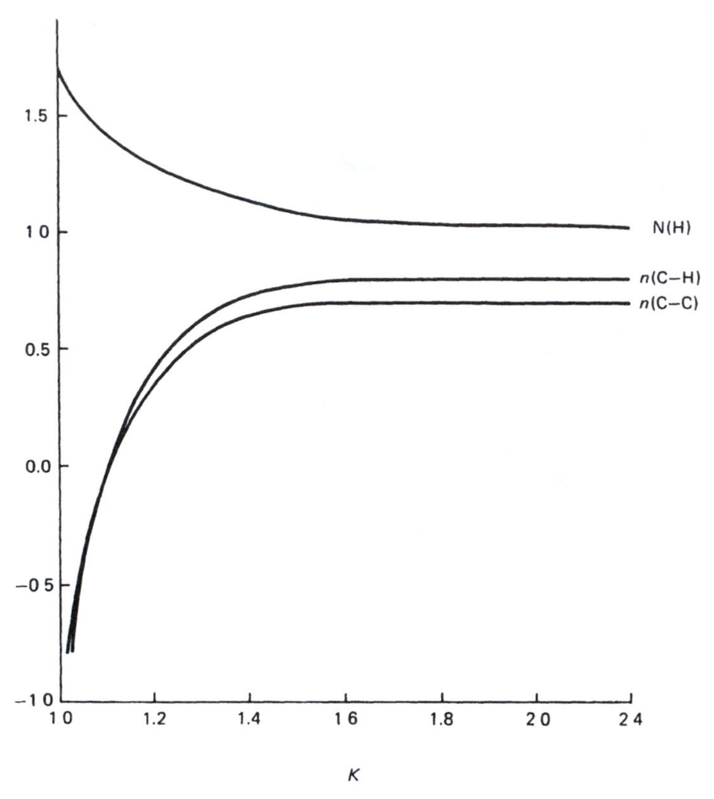

Figure 10-6 ▶ Mulliken gross population on H [N(H)] and Mulliken overlap populations in C–C and C–H bonds [n(C–C) and n(C–H)] of ethane as calculated by the EH method with various values of K.

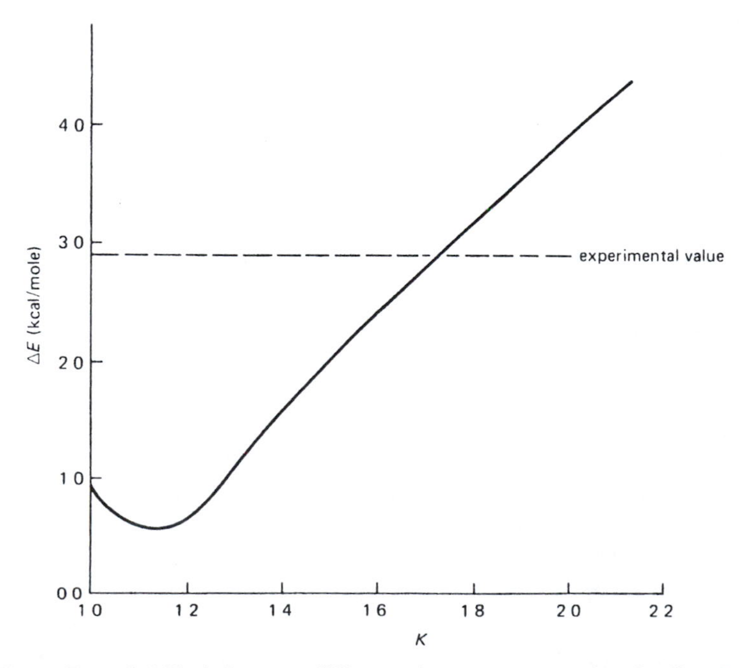

Figure 10-7 ▶ Extended Hückel energy difference between staggered and eclipsed ethanes as a function of K.

EH total energy *change* to the total (nuclear repulsion plus electronic) observed energy *change* for internal rotation in ethane. This is consistent with our earlier interpretation of EH total energy.

10-2 Mulliken Populations

We found that the electron densities and bond orders calculated in the simple Hückel method were extremely useful for relating theory to observable molecular properties such as electron spin-resonance splittings or bond lengths. Hence, it is desirable that we find analogous quantities to describe the distribution of electrons in an all-valence-electron method like the EH method. A number of suggestions have been made. The one we use is due to Mulliken [8]. It is the most widely used, and, as we shall see, it has an especially direct and useful connection with the EH method.

Consider a real, normalized MO, ϕ_i, made up from two normalized AOs, χ_j and χ_k:

$$\phi_i = c_{ji}\chi_j + c_{ki}\chi_k \tag{10-10}$$

We square this MO to obtain information about electronic distribution:

$$\phi_i^2 = c_{ji}^2\chi_j^2 + c_{ki}^2\chi_k^2 + 2c_{ji}c_{ki}\chi_j\chi_k \tag{10-11}$$

If we integrate Eq. (10-11) over the electronic coordinates, we obtain (since ϕ_i, χ_j, and χ_k are normalized)

$$1 = c_{ji}^2 + c_{ki}^2 + 2c_{ji}c_{ki}S_{jk} \tag{10-12}$$

where S_{jk} is the overlap integral between χ_j and χ_k. Mulliken suggested that one electron in ϕ_i should be considered to contribute c_{ji}^2 to the electron *net AO population* of χ_j, c_{ki}^2 to the population of χ_k, and $2c_{ji}c_{ki}S_{jk}$ to the *overlap population* between χ_j and χ_k. If there are two electrons in ϕ_i, then these populations should be doubled.

Let q_j^i symbolize the net AO population of χ_j due to one electron in $\text{MO}_{\phi i}$, and p_{jk}^i symbolize the overlap population between χ_j and χ_k due to this same electron. The above example leads to the following general definitions:

$$q_j^i = c_{ji}^2 \tag{10-13}$$

$$p_{jk}^i = 2c_{ji}c_{ki}S_{jk} \tag{10-14}$$

We can now sum the contributions due to all the electrons present in the model system, obtaining a *Mulliken net AO population* q_j for each AO χ_j, and a *Mulliken overlap population* p_{jk}, for each distinct AO pair χ_j and χ_k:

$$q_j = \sum_i^{\text{MOs}} n_i c_{ji}^2 \equiv \sum_i^{\text{MOs}} n_i q_j^i \tag{10-15}$$

$$P_{jk} = 2\sum_i^{\text{MOs}} n_i c_{ji}c_{ki}S_{jk} \equiv \sum_i^{\text{MOs}} n_i p_{jk}^i \tag{10-16}$$

Notice that the sum of all the net AO *and* overlap populations must be equal to the total number of electrons in the model system. (In the EH method, this is the total number of *valence* electrons.) This contrasts with the situation in the simple Hückel method where the sum of electron densities *alone*, exclusive of bond orders, equals the total number of π electrons.

The Mulliken populations are useful indices of the location of electronic charge in the molecule and its bonding or antibonding nature. The contributions to such populations from *one* electron in MO ϕ_4' of methane are given below (data taken from Tables 10-3 and 10-7):

$$q_{2p_y}^{4'} = (0.5313)^2 = 0.2823 \tag{10-17}$$

$$q_{1s_c}^{4'} = q_{1s_d}^{4'} = (0.4529)^2 = 0.2051 \tag{10-18}$$

$$p_{2p_y-1s_c}^{4'} = p_{2p_y-1s_d}^{4'} = 2(0.5313)(0.4529)(0.3964) = 0.1908 \tag{10-19}$$

$$p_{1s_c-1s_d}^{4'} = 2(0.4529)(-0.4529)(0.1805) = -0.0740 \tag{10-20}$$

All the other populations for $\phi_{4'}$ are zero. These numbers show that an electron in this MO contributes about 28%, of an electron to the carbon $2p_y$ net AO population, 20–21% to each of two hydrogen AOs, 19% to each of two C–H bonds, and a negative 7% to the region between H_c and H_d. The last corresponds to an antibonding interaction, as is obvious from an examination of the sketch of $\phi_{4'}$ in Fig. 10-3. (A negative overlap population is interpreted to mean that the amount of charge in the overlap region is less than what would exist if one squared the two AOs and then combined them.)

By summing over the contributions due to all eight valence electrons, we obtain the Mulliken populations shown in Table 10-8. These data are part of the normal output of an EH computer program. The matrix is symmetric, and only the unique elements are tabulated. Such a matrix is named a *Mulliken overlap population matrix*.

The data in Table 10-8 indicate that the hydrogens have positive overlap populations with the 2s and 2p AOs of carbon but have small negative overlap populations with each other. This corresponds to saying that the hydrogens interact with the carbon AOs in a bonding way and in a weakly antibonding way with each other. The symmetry equivalence of the four hydrogen atoms results in their having identical net AO populations.

Often we are interested in knowing how the hydrogen atoms interact with the carbon *in toto*, rather than with the 2s and 2p AOs separately. This can be obtained simply

TABLE 10-8 ▶ Mulliken Net AO and Overlap Populations for Methane as Computed by the Extended Hückel Method

	2s	$2p_z$	$2p_x$	$2p_y$	$1s_a$	$1s_b$	$1s_c$	$1s_d$
2s	0.6827	0.0	0.0	0.0	0.2229	0.2229	0.2229	0.2229
$2p_z$		0.5645	0.0	0.0	0.5723	0.0636	0.0636	0.0636
$2p_x$			0.5645	0.0	0.0	0.5087	0.1272	0.1272
$2p_y$				0.5645	0.0	0.0	0.3815	0.3815
$1s_a$					0.6844	−0.0491	−0.0491	−0.0491
$1s_b$						0.6844	−0.0491	−0.0491
$1s_c$							0.6844	−0.0491
$1s_d$								0.6844

TABLE 10-9 ► Reduced Net AO and Overlap Population Matrix for Methane

	C	H_a	H_b	H_c	H_d
C	2.3762	0.7952	0.7952	0.7952	0.7952
H_a		0.6844	−0.0491	−0.0491	−0.0491
H_b			0.6844	−0.0491	−0.0491
H_c				0.6844	−0.0491
H_d					0.6844

by summing all the carbon atom AO contributions together to give a reduced overlap population matrix, shown in Table 10-9. The reduced population matrix makes evident the equivalence of the four C–H bonds, all of which have a total overlap population of 0.7952 electrons.

The Mulliken population scheme described above assigns some electronic charge to AOs, the rest to overlap regions. An alternative scheme, which assigns all the charge to AOs, was also proposed by Mulliken. One simply divides each overlap population in half, assigning half of the charge to each of the two participating AOs. When all the overlap populations have been reassigned in this way, the electronic charge is all in the AOs, and the sum of these AO charges still equals the total number of electrons. Mulliken called the AO populations resulting from this procedure *gross AO populations*. We will use the symbol $N(X)$ for the gross population in X, where X can be an AO or an atom (i.e., the sum of all gross AO populations on one atom). Clearly

$$N(X) = q_x + \frac{1}{2}\sum_{j \neq x} p_{xj} \tag{10-21}$$

The gross AO populations, gross atomic populations, and the resultant atomic charges (obtained by combining electronic and nuclear charges) for methane are listed in Table 10-10. These data suggest that the carbon has lost a very small amount of charge to hydrogen upon formation of the molecule. It would be very risky, however, to place much faith in such an interpretation. It turns out that populations are rather sensitive to choice of VSIEs. For example, Hoffmann's original choice of VSIEs differs from that

TABLE 10-10 ► Gross AO Populations, Gross Atomic Populations, and Net Atomic Charges for Methane

	Gross AO population	Gross atom population	Net atomic charge
C_{2s}	1.128	3.966	+0.0334
C_{2p}[a]	0.946		
H[a]	1.008	1.008	−0.0083

[a] All 2p AOs and all H AOs have identical values because they are equivalent through symmetry.

used here, and he obtained gross populations for hydrogen atoms of around 0.9, giving net positive charges of 0.1. Thus, absolute values are not very useful, but *changes* in gross population as we go from one hydrogen to another in the same hydrocarbon molecule or in closely related molecules do appear to be rather insensitive to VSIE choice and are often in accord with results of more accurate calculations.

10-3 Extended Hückel Energies and Mulliken Populations

In Chapter 8 it was shown that a simple quantitative relation exists between the energy of a simple Hückel MO and the contributions of the MO to bond orders [Eq. (8-56)]. The more bonding such an MO is, the lower is its energy. A similar relationship will now be shown to hold for extended Hückel energies and Mulliken populations.

The orbital energy for the real MO ϕ_i is (ignoring the spin variable)

$$E_i = \frac{\int \phi_i \hat{H} \phi_i \, dv}{\int \phi_i^2 \, dv} \tag{10-22}$$

$$= \frac{\sum_{j,k}^{\text{AOs}} c_{ji} c_{ki} H_{jk}}{\sum_{j,k}^{\text{AOs}} c_{ji} c_{ki} S_{jk}} \tag{10-23}$$

Assume that ϕ_i is normalized: The denominator is unity. The numerator of Eq. (10-23) contains diagonal and off-diagonal terms. If we separate these, and substitute the relation (10-4) for the off-diagonal terms, we obtain

$$E_i = \sum_{j,k}^{\text{AOs}} c_{ji}^2 H_{jj} + 2 \sum_{j<k}^{\text{AOs}} c_{ji} c_{ki} K S_{jk} \frac{H_{jj} + H_{kk}}{2} \tag{10-24}$$

Comparison with Eqs. (10-13) and (10-14) shows that the first sum contains contributions to net AO populations and the second sum contains overlap population contributions. That is,

$$E_i = \sum_{j}^{\text{AOs}} q_j^i H_{jj} + \sum_{j<k}^{\text{AOs}} p_{jk}^i K \frac{H_{jj} + H_{kk}}{2} \tag{10-25}$$

This equation indicates that the energy of an extended Hückel MO is equal to its net contributions to AO populations times AO energy weighting factors plus its contributions to overlap populations times overlap energy weighting factors.

By summing over all MOs times occupation numbers, we arrive at a relation between *total* EH energy and net AO and overlap populations:

$$E = \sum_{j}^{\text{AOs}} q_j H_{jj} + \sum_{j<k}^{\text{AOs}} p_{jk} K \frac{H_{jj} + H_{kk}}{2} \tag{10-26}$$

Equations (10-25) and (10-26) are useful because they permit us to understand computed energies in terms of electron distributions. This is helpful when we seek to understand energy changes which occur as molecules are stretched, bent, or twisted.

There is an important difference between the extended Hückel formulas (10-25) and (10-26) and their simple Hückel counterparts. In a simple Hückel MO, the charge density contributions q^i always add up to unity, contributing α to the MO energy (omitting heteroatom cases). The deviation of the MO energy from α is thus due only to the bond-order contributions p^i. As a result, for hydrocarbons, the simple Hückel MO that is the more bonding of a pair is *always* lower in energy. This is not so, however, in extended Hückel MOs. Since the sum of AO and overlap populations must equal the number of electrons present, we can increase the total amount of overlap population *only* at the expense of net AO populations. Therefore, energy lowering due to the second sum of Eq. (10-25) is purchased at the expense of that due to the first sum. This complicates the situation and requires that we take a more detailed look before issuing any blanket statements.

The following simple example provides a convenient reference point for discussion. Consider two identical 1s AOs, χ_a and χ_b, on identical nuclei separated by a distance R. These AOs combine to form MOs of σ_g and σ_u symmetry. The σ_g MO has an associated positive overlap population and equal positive net AO populations for χ_a and χ_b. If R now is decreased slightly, the overlap population increases slightly (say by 2δ), so the net AO populations must each decrease by δ. The energy change for the MO, then, is

$$\Delta E_{\sigma_g} = -\delta\,(H_{aa} + H_{bb}) + 2\delta K \frac{H_{aa} + H_{bb}}{2} \tag{10-27}$$

Since the molecule is homonuclear, $H_{aa} = H_{bb}$, and

$$\Delta E_{\sigma_g} = 2\delta H_{aa}\,(K - 1) \tag{10-28}$$

If $K > 1$, ΔE_{σ_g} is negative (since H_{aa} is negative). This analysis shows that the choice of 1.75 as the value for K has the effect of making the increase in overlap population dominate the energy change. If K were less than unity, the net AO population changes would dominate.

The above example suggests that the EHMO method lowers the energy by maximizing weighted overlap populations at the expense of net AO populations. It also suggests that the EH energy should be lowered whenever a molecule is distorted in a way that enables overlap population to increase. These are useful rules of thumb, but some caution must be exercised since the existence of several different kinds of atom in a molecule leads to a more complicated relation than that in the above example.

Our methane example illustrates the above ideas. We have already seen that the total EH energy for methane goes through a minimum around $R_{\text{C–H}} = 1$ Å. Let us see how the individual MO energies change as a function of $R_{\text{C–H}}$ and try to rationalize their behaviors in terms of the overlap population changes. A plot of the MO energies is given in Fig. 10-8. The lowest-energy MO is s type and C–H bonding. The overlap between H 1s AOs and the C 2s STO increases as $R_{\text{C–H}}$ decreases. As a result, the energy of this MO decreases as $R_{\text{C–H}}$ decreases, favoring formation of the united atom. The second-lowest energy level is triply degenerate and belongs to p-type C–H bonding MOs. A glance at Fig. 10-3 indicates that overlap between AOs will first increase in magnitude as $R_{\text{C–H}}$ decreases. But ultimately, at small $R_{\text{C–H}}$ this behavior must be reversed because the 1s and 2p AOs are orthogonal when they are isocentric. Therefore, as $R_{\text{C–H}}$ decreases, the overlap population first increases, then decreases toward zero. Consequently, the EH energy of this level first decreases, then increases. Since there are six electrons in this

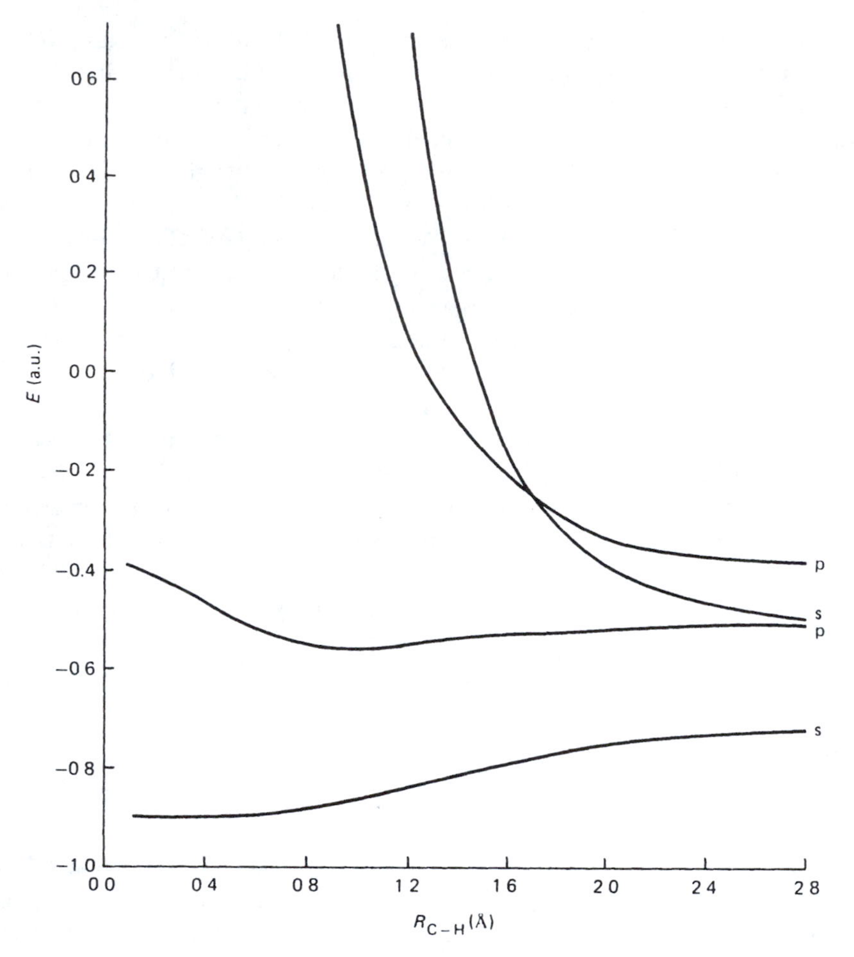

Figure 10-8 ▶ Extended Hückel MO energies for methane as a function of R_{C-H}.

level and only two in the lowest-energy MO, this one dominates the total energy and is responsible for the eventual rise in total energy at small R that creates the minimum in Fig. 10-4. The two remaining (unoccupied) levels belong to antibonding MOs, and the overlap population becomes more negative as R_{C-H} decreases. The p-type level rises less rapidly and eventually passes through a maximum (not shown), again because the overlap population must ultimately approach zero as R_{C-H} approaches zero.

It is important to remember that all these remarks apply to the EH method only. The relationships between the EH method and other methods or with experimental energies is yet to be discussed.

10-4 Extended Hückel Energies and Experimental Energies

We have seen that a change in EH energy reflects certain changes in calculated Mulliken populations. We now consider the circumstances that should exist in order that such EH energy changes agree roughly with actual total energy changes for various systems.

In essence, there are two requirements. First, the population changes calculated by the EH method ought to be in qualitative agreement with the charge shifts that actually occur in the real system. This condition is not always met. Some systems, especially those having unpaired electrons, are not accurately representable by a single-configuration wavefunction (that is, by a single product or a single Slater determinant). An example of this was seen for some 1s2s states of helium (see Chapter 5). Since the EHMO method is based on a single configuration, it is much less reliable for treating such systems. Special methods exist for handling such systems, but they normally are not applied at the Hückel level of approximation.

The second requirement is that, as the charge shifts in the real system, the total energy should change in the way postulated by the EH method. For instance, if the charge increases in bond regions, this should tend to lower the total energy. In actuality, this does not always happen. We can see this by returning to our example of two identical 1s AOs separated by R. The associated σ_g MO increases its overlap population monotonically as R decreases. Therefore, the EH postulate predicts that the total energy should decrease monotonically. It is clear, however, that this does not happen in any real system. If our system is H_2, we know that the experimental energy decreases with decreasing R until $R = R_e$, and increases thereafter (Fig. 7-18). If our system is He_2^{2+}, the experimental total energy *increases* as R decreases, due to the dominance of internuclear repulsion.[8] Consideration of many examples such as these indicates that the postulate is usually qualitatively correct when we are dealing with interactions between neutral, fairly nonpolar systems that are separated by a distance greater than a bond length typical for the atoms involved. These, then, are circumstances under which the EHMO method might be expected to be qualitatively correct.

In brief, we find two kinds of condition limiting the applicability of the EH method. The first is that the system must be reasonably well represented by a single configuration. Hence, closed-shell systems are safest. The second condition is that we apply the method to uncharged nonpolar systems where the nuclei that are undergoing relative motion are not too close to each other.

These conditions are often satisfied by molecules undergoing internal rotation about single bonds. Thus, it is indeed appropriate that the internal rotation barrier in ethane was used to help calibrate the method. A comparison of other EH calculated barriers with experimental values (Table 10-11) gives us some idea of the capabilities and limitations of the method. For all the molecules listed, the most stable conformation predicted by the EH method agrees with experiment. This suggests that one can place a fair amount of reliance on the conformational predictions of the method, at least for threefold symmetric rotors. Also, certain gross quantitative trends, such as the significant reduction as we proceed down the series ethane, methylamine, methanol, are displayed in the EH results, but it is evident that the quantitative predictions of barrier height are not very good. This is not too surprising since the method was calibrated on hydrocarbons. Introduction of halogens, nitrogen, or oxygen produces greater polarity and might be expected to require a different parametrization.

For reasons outlined earlier, we expect even less accuracy in EH calculations of molecular deformations involving bond-angle or bond-length changes. This expectation is generally reinforced by calculation. Molecular shape predictions become poorer

[8]Actually, He_2^{2+} has a minimum at short R in its energy curve due to an avoidance of curve crossing (discussed in Chapter 14), but this minimum is unstable with respect to the separated ions. See Pauling [9].

TABLE 10-11 ► Energy Barriers for Internal Rotation about Single Bonds[a]

	Barrier (kcal/mole)[b]	
Molecule	Calculated	Experiment
CH_3–CH_3	3.04	2.88
CH_3–NH_2	1.66	1.98
CH_3–OH	0.45	1.07
CH_3–CH_2F	2.76	3.33
CH_3–CHF_2	2.39	3.18
CH_3–CF_3	2.17	3.25
CH_3–CH_2Cl	4.58	3.68
CH_3–$CHCH_2$	1.20	1.99
cis–CH_3–CHCHCl	0.11	0.62
CH_3–CHO	0.32	1.16
CH_3–NCH_2	0.44	1.97

[a]Calculated barriers are for rigid rotation, where no bond length or angle changes occur except for the torsional angle change about the internal axis.

[b]The stable form for the first seven molecules has the methyl C–H bonds staggered with respect to bonds across the rotor axis. For the last four molecules, the stable form has a C–H methyl bond eclipsing the double bond.

for more polar molecules (water is calculated to be most stable when it is linear), and bond-length predictions are quite poor too.

Results such as these have tended to restrict use of the EH method to qualitative predictions of conformation in molecules too large to be conveniently treated by more accurate methods. However, just as the simple Hückel method underwent various refinements (such as the ω technique) to patch up certain inadequacies, so has the EH method been refined. Such refinements[9] have been shown to give marked improvement in numerical predictions of various properties. The EH method has been overtaken in popularity by a host of more sophisticated computational methods. (See Chapter 11.) However, it is still sometimes used as a first step in such methods as a way to produce a starting set of approximate MOs. The EHMO method also continues to be important as the computational equivalent of qualitative MO theory (Chapter 14), which continues to play an important role in theoretical treatments of inorganic and organic chemistries (as, for example, in Walsh's Rules and in Woodward-Hoffmann Rules).

10-4.A Problems

The following output is produced by an EH calculation on the formaldehyde molecule, and is referred to in Problems 10-1 to 10-10.

[9]See Kalman [10], Boyd [11], and the references cited in these papers. Also, see Anderson and Hoffmann [12] and Anderson [13].

Formaldehyde (Ground State) Orbital Numbering

Orbital	Atom	n	l	m	x	y	z	exp H_{ii}
1	H-1	1	0	0	−0.550000	0.952600	0.0	1.200–13.60
2	H-2	1	0	0	−0.550000	−0.952600	0.0	1.200–13.60
3	C-3	2	0	0	0.0	0.0	0.0	1.625–19.44
4	C-3	2	1	0	0.0	0.0	0.0	1.625–10.67
5	C-3	2	1	1	0.0	0.0	0.0	1.625–10.67
6	C-3	2	1	1	0.0	0.0	0.0	1.625–10.67
7	O-4	2	0	0	1.220000	0.0	0.0	2.275–32.38
8	O-4	2	1	0	1.220000	0.0	0.0	2.275–15.85
9	O-4	2	1	1	1.220000	0.0	0.0	2.275–15.85
10	O-4	2	1	1	1.220000	0.0	0.0	2.275–15.85

$H_{ij} = KS_{ij}(H_{ii} + H_{jj})/2$, with $K = 1.75$.

Distance Matrix (a.u.)

	1	2	3	4
1	0.0	3.6004	2.0787	3.7985
2	3.6004	0.0	2.0787	3.7985
3	2.0787	2.0787	0.0	2.3055
4	3.7985	3.7985	2.3055	0.0

Total effective nuclear repulsion = 17.69537317 a.u.

Formaldehyde Eigenvalues

Eigenvalues (a.u.)	Occ. no.	Eigenvalues (a.u.)	Occ. no.
$E(1) = 1.039011$	0	$E(6) = -0.587488$	2
$E(2) = 0.472053$	0	$E(7) = -0.597185$	2
$E(3) = 0.314551$	0	$E(8) = -0.611577$	2
$E(4) = -0.342162$	0	$E(9) = -0.755816$	2
$E(5) = -0.517925$	2	$E(10) = -1.242836$	2

Sum = −8.625654 a.u.

Total Overlap Matrix

	1	2	3	4	5	6	7	8	9	10
1	1.0000	0.1534	0.5133	0.0	−0.2428	0.4204	0.0813	0.0	−0.0729	0.0392
2	0.1534	1.0000	0.5133	0.0	−0.2428	−0.4204	0.0813	0.0	−0.0729	−0.0392
3	0.5133	0.5133	1.0000	0.0	0.0	0.0	0.3734	0.0	−0.3070	0.0
4	0.0	0.0	0.0	1.0000	0.0	0.0	0.0	0.2146	0.0	0.0
5	−0.2428	−0.2428	0.0	0.0	1.0000	0.0	0.4580	0.0	−0.3056	0.0
6	0.4204	−0.4204	0.0	0.0	0.0	1.0000	0.0	0.0	0.0	0.2146
7	0.0813	0.0813	0.3734	0.0	0.4580	0.0	1.0000	0.0	0.0	0.0
8	0.0	0.0	0.0	0.2146	0.0	0.0	0.0	1.0000	0.0	0.0
9	−0.0729	−0.0729	−0.3070	0.0	−0.3056	0.0	0.0	0.0	1.0000	0.0
10	0.0392	−0.0392	0.0	0.0	0.0	0.2146	0.0	0.0	0.0	1.0000

Eigenvectors

	1	2	3	4	5	6	7	8	9	10
1	0.5279	0.7683	0.8924	0.0	−0.4281	−0.2016	0.0	−0.2141	−0.2721	0.0011
2	0.5279	0.7683	−0.8924	0.0	0.4281	−0.2016	0.0	0.2141	−0.2721	0.0011
3	−1.3964	−0.5553	0.0000	0.0	−0.0000	−0.0460	0.0	−0.0000	−0.4875	0.2550
4	0.0	0.0	0.0	0.9940	0.0	0.0	0.2456	0.0	0.0	0.0
5	−0.6043	1.1727	−0.0000	0.0	0.0000	0.2768	0.0	0.0000	0.2245	0.0685
6	0.0000	−0.0000	−1.2519	0.0	−0.3813	0.0000	0.0	−0.3179	0.0000	0.0000
7	0.8367	−0.4799	0.0000	0.0	0.0000	−0.0884	0.0	−0.0000	0.3066	0.8481
8	0.0	0.0	0.0	−0.4532	0.0	0.0	0.9181	0.0	0.0	0.0
9	−0.6960	0.3412	−0.0000	0.0	0.0000	−0.8317	0.0	−0.0000	0.3327	0.0252
10	−0.0000	0.0000	0.2511	0.0	0.6475	0.0000	0.0	−0.7600	−0.0000	−0.0000

Mulliken Overlap Populations for 12 Electrons

	1	2	3	4	5	6	7	8	9	10
1	0.6876	−0.0702	0.2920	0.0	0.1134	0.3890	−0.0210	0.0	−0.0225	−0.0180
2	−0.0702	0.6876	0.2920	0.0	0.1134	0.3890	−0.0210	0.0	−0.0225	−0.0180
3	0.2920	0.2920	0.6097	0.0	0.0	0.0	0.1058	0.0	0.1443	0.0
4	0.0	0.0	0.0	0.1207	0.0	0.0	0.0	0.1936	0.0	0.0
5	0.1134	0.1134	0.0	0.0	0.2634	0.0	0.1876	0.0	0.1880	0.0
6	0.3890	0.3890	0.0	0.0	0.0	0.4929	0.0	0.0	0.0	−0.0046
7	−0.0210	−0.0210	0.1058	0.0	0.1876	0.0	1.6420	0.0	0.0	0.0
8	0.0	0.0	0.0	0.1936	0.0	0.0	0.0	1.6857	0.0	0.0
9	−0.0225	−0.0225	0.1443	0.0	0.1880	0.0	0.0	0.0	1.6061	0.0
10	−0.0180	−0.0180	0.0	0.0	0.0	−0.0046	0.0	0.0	0.0	1.9939

Charge Matrix for MOs with Two Electrons in Each

	1	2	3	4	5	6	7	8	9	10
1	0.1664	0.3881	0.4266	0.0	0.4258	0.1088	0.0	0.1476	0.3363	0.0004
2	0.1664	0.3881	0.4266	0.0	0.4258	0.1088	0.0	0.1476	0.3363	0.0004
3	0.9171	0.0561	0.0000	0.0	0.0000	0.0028	0.0	0.0000	0.7358	0.2882
4	0.0	0.0	0.0	1.7825	0.0	0.0	0.2175	0.0	0.0	0.0
5	0.3198	1.1155	0.0000	0.0	0.0000	0.3257	0.0	0.0000	0.1775	0.0614
6	0.0000	0.0000	1.1204	0.0	0.4593	0.0000	0.0	0.4203	0.0000	0.0000
7	0.2082	0.0241	0.0000	0.0	0.0000	0.0021	0.0	0.0000	0.1123	1.6534
8	0.0	0.0	0.0	0.2175	0.0	0.0	1.7825	0.0	0.0	0.0
9	0.2220	0.0282	0.0000	0.0	0.0000	1.4518	0.0	0.0000,	0.3017	−0.0037
10	0.0000	0.0000	0.0264	0.0	0.6891	0.0000	0.0	1.2845	0.0000	0.0000

Reduced Overlap Population Matrix Atom by Atom

	1	2	3	4
1	0.6876	−0.0702	0.7945	−0.0615
2	−0.0702	0.6876	0.7945	−0.0615
3	0.7945	0.7945	1.4866	0.8148
4	−0.0615	−0.0615	0.8148	6.9277

Orbital Charges

1	1.018973	6	0.879574
2	1.018973	7	1.767672
3	1.026764	8	1.782529
4	0.217471	9	1.749778
5	0.564637	10	1.973630

Net Charges

1	−0.018973	3	1.311555
2	−0.018973	4	−1.273609

Total charge = 0.000000.

10-1. Use the output to determine the orientation of the molecule with respect to Cartesian coordinates. Sketch the molecule in relation to these axes and number the *atoms* in accord with their numbering in the output.

10-2. Use the orbital numbering data together with the overlap matrix to figure out which of the labels 1s, 2s, $2p_x$, $2p_y$, $2p_z$ goes with each of the ten AOs (i.e., it is obvious that AO 1 is a 1s AO, but it is not so obvious what AO 5 is).

10-3. Use your conclusions from above, together with coefficients in the eigenvector matrix, to sketch the MOs having energies of −0.756, −0.611, and −0.597 a.u. Which of these are π MOs? Which are σ MOs?

10-4. Label each of the ten MOs "π" or "σ" by inspecting the coefficient matrix.

10-5. What is the Mulliken overlap population between C and O $2p_\pi$ AOs in this molecule? Should removal of an electron from MO 7 cause the C=O bond to shorten or to lengthen?

10-6. Demonstrate that MO 7 satisfies Eq. (10-24). (Note that H_{ii} values in the first table are in units of electron volts, while orbital energies are in atomic units.)

10-7. Use the reduced overlap population matrix to verify that the sum of AO and overlap populations is equal to the number of valence electrons.

10-8. Using MO 7 as your example, verify that the charge matrix table is a tabulation of the contributions of each MO to *gross* atom populations.

10-9. In the list of "orbital charges," are the "orbitals" MOs or AOs? Demonstrate how these numbers are derived from those in the charge matrix.

10-10. What is the physical meaning of the "net charges" in the data? Would you characterize these results as indicative of low polarity? Which end of the molecule should correspond to the negative end of the dipole moment?

10-11. How many MOs will be produced by an EHMO calculation on butadiene?

References

[1] R. Hoffmann, *J. Chem. Phys.* **39**, 1397 (1963).

[2] C. E. Moore, Atomic Energy Levels, *Natl. Bur. Std. (U.S.) Circ. 467*. Natl. Bur. Std., Washington D.C., 1949.

[3] H. A. Skinner and H. O. Pritchard, *Trans. Faraday Soc.* **49**, 1254 (1953).

[4] J. Hinze and H. H. Jaffé, *J. Amer. Chem. Soc.* **84**, 540 (1962).

[5] H. Basch, A. Viste, and H. B. Gray, *Theoret. Chim. Acta* 3, 458 (1965).

[6] T. Anno, *Theoret. Chim. Acta* **18**, 223 (1970).

[7] J. A. Pople and G. A. Segal, *J. Chem. Phys.* **43**, S136 (1965).

[8] R. S. Mulliken, *J. Chem. Phys.* **23**, 1833, 1841, 2338, 2343 (1955).

[9] L. Pauling, *J. Chem. Phys.* **1**, 56 (1933).

[10] B. L. Kalman, *J. Chem. Phys.* **60**, 974 (1974).

[11] D. B. Boyd, *Theoret. Chim. Acta* **30**, 137 (1973).

[12] A. B. Anderson and R. Hoffmann, *J. Chem. Phys.* **60**, 4271 (1974).

[13] A. B. Anderson, *J. Chem. Phys.* **62**, 1187 (1975).

Chapter 11

The SCF-LCAO-MO Method and Extensions

11-1 *Ab Initio* Calculations

A rigorous variational calculation on a system involves the following steps:

1. Write down the hamiltonian operator $\hat{H}$ for the system.
2. Select some mathematical functional form ψ as the trial wavefunction. This form should have variable parameters.
3. Minimize

$$\bar{E} = \frac{\int \psi^* \hat{H} \psi \, d\tau}{\int \psi^* \psi \, d\tau} \tag{11-1}$$

with respect to variations in the parameters.

The simple and extended Hückel methods are not rigorous variational calculations. Although they both make use of the secular determinant technique from linear variation theory, no hamiltonian operators are ever written out explicitly and the integrations in H_{ij} are not performed. These are *semiempirical* methods because they combine the theoretical form with parameters fitted from experimental data.

The term *ab initio* ("from the beginning") is used to describe calculations in which no use is made of experimental data. In an *ab initio* variational method, all three steps listed above are explicitly performed. In this chapter we describe a certain kind of *ab initio* calculation called the self-consistent field (SCF) method. This is one of the most commonly encountered types of *ab initio* calculation for atoms or molecules. We also describe a few popular methods for proceeding beyond the SCF level of approximation.

The SCF method and extensions to it are mathematically and physically considerably more complicated than the one-electron methods already discussed. Thus, one normally does not perform such calculations with pencil and paper, but rather with complicated computer programs. Therefore, in this chapter we are not concerned with how one does such calculations because, in most cases, they are done by acquiring a program written by a group of specialists. Rather we are concerned with a description of the mathematical and physical underpinnings of the method. Because the method is simultaneously complicated and rigorously defined, a special jargon has developed. Terms like "Hartree–Fock," or "correlation energy" have specific meanings and are pervasive in the literature. Hence, a good deal of emphasis in this chapter is put on defining some of these important terms.

11-2 The Molecular Hamiltonian

In practice, one usually does not use the complete hamiltonian for an isolated molecular system. The complete hamiltonian includes nuclear and electronic kinetic energy operators, electrostatic interactions between all charged particles, and interactions between all magnetic moments due to spin and orbital motions of nuclei and electrons. Also an accounting for the fact that a moving particle experiences a change in mass due to relativistic effects is included in the complete hamiltonian. The resulting hamiltonian is much too complicated to work with. Usually, relativistic mass effects are ignored, the Born–Oppenheimer approximation is made (to remove nuclear kinetic energy operators), and all magnetic interactions are ignored (except in special cases where we are interested in spin coupling). The resulting hamiltonian for the electronic energy is, in atomic units,

$$\hat{H} = -\frac{1}{2}\sum_{i=1}^{n}\nabla_i^2 - \sum_{\mu=1}^{N}\sum_{i=1}^{n} Z_\mu/r_{\mu i} + \sum_{i=1}^{n-1}\sum_{j=i+1}^{n} 1/r_{ij} \tag{11-2}$$

where i and j are indices for the n electrons and μ is an index for the N nuclei. The nuclear repulsion energy V_{nn} is

$$V_{\mathrm{nn}} = \sum_{\mu=1}^{N-1}\sum_{\nu=\mu+1}^{N} Z_\mu Z_\nu / r_{\mu\nu} \tag{11-3}$$

In choosing this hamiltonian, we are in effect electing to seek an energy of an idealized nonexistent system—a nonrelativistic system with clamped nuclei and no magnetic moments. If we wish to make a very accurate comparison of our computed results with experimentally measured energies, it is necessary to modify either the experimental or the theoretical numbers to compensate for the omissions in $\hat{H}$.

11-3 The Form of the Wavefunction

The wavefunction for an SCF calculation is one or more antisymmetrized products of one-electron spin-orbitals. We have already seen (Chapter 5) that a convenient way to produce an antisymmetrized product is to use a Slater determinant. Therefore, we take the trial function ψ to be made up of Slater determinants containing spin-orbitals ϕ. If we are dealing with an atom, then the ϕ's are atomic spin-orbitals. For a molecule, they are molecular spin-orbitals.

In our discussion of many-electron atoms (Chapter 5), we noted that certain atoms in their ground states are fairly well described by assigning two electrons, one of each spin, to each AO, starting with the lowest-energy AO and working up until all the electrons are assigned. If the last electron completes the filling of all the AOs having a given principal quantum number, n, we have a *closed shell* atomic system. Examples are $\mathrm{He}(1s^2)$ and $\mathrm{Ne}(1s^2 2s^2 2p^6)$. Atoms wherein the last electron completes the filling of all AOs having a given l quantum number are said to have a *closed subshell*. An example is $\mathrm{Be}(1s^2 2s^2)$. Both types of system tend to be well approximated by a single determinantal wavefunction if the highest filled level is not too close in energy to the lowest empty level. (Beryllium is the least successfully treated of these three at this level of approximation because the 2s level is fairly close in energy to the 2p level.) A similar situation holds for molecules; that is, the wavefunctions of many molecules in their ground states are well represented by single determinantal wavefunctions with electrons of paired spins occupying identical MOs. Such molecules are said to be *closed-shell systems*. We can represent a trial wavefunction for a $2n$-electron closed-shell system as

$$\psi_{\text{closed shell}} = \left|\phi_1(1)\bar{\phi}_1(2)\phi_2(3)\bar{\phi}_2(4)\cdots\phi_n(2n-1)\bar{\phi}_n(2n)\right| \tag{11-4}$$

where we have used the shorthand form for a Slater determinant described in Chapter 5. *For the present, we restrict our discussion to closed-shell single-determinantal wavefunctions.*

11-4 The Nature of the Basis Set

Some functional form must be chosen for the MOs ϕ. The usual choice is to approximate ϕ as a linear combination of "atomic orbitals" (LCAO), these AOs being located on the nuclei. The detailed nature of these AOs, as well as the number to be placed on each nucleus, is still open to choice. We consider these choices later. For now, we simply recognize that we are working within the familiar LCAO-MO level of approximation. If we represent the basis AOs by χ, we have, for the ith MO,

$$\phi_i = \sum_j c_{ji}\chi_j \tag{11-5}$$

where the constants c_{ji} are as yet undetermined.

11-5 The LCAO-MO-SCF Equation

Having a hamiltonian and a trial wavefunction, we are now in a position to use the linear variation method. The detailed derivation of the resulting equations is complicated and notationally clumsy, and it has been relegated to Appendix 7. Here we discuss the results of the derivation.

For our restricted case of a closed-shell single-determinantal wavefunction, the variation method leads to

$$\hat{F}\phi_i = \epsilon_i\phi_i \tag{11-6}$$

These equations are sometimes called the Hartree–Fock equations, and $\hat{F}$ is often called the *Fock operator*. The detailed formula for $\hat{F}$ is (from Appendix 7)

$$\hat{F}(1) = -\frac{1}{2}\nabla_1^2 - \sum_\mu Z_\mu/r_{\mu 1} + \sum_{j=1}^{n}(2\hat{J}_j - \hat{K}_j) \tag{11-7}$$

The symbols $\hat{J}_j$ and $\hat{K}_j$ stand for operators related to the $1/r_{ij}$ operators in $\hat{H}$. $\hat{J}_j$ is called a *coulomb operator* because it leads to energy terms corresponding to charge cloud repulsions. It is possible to write $\hat{J}_j$ explicitly:

$$\hat{J}_j = \int \phi_j^*(2)\left(1/r_{12}\right)\phi_j(2)d\tau(2) \tag{11-8}$$

$\hat{K}_j$ leads ultimately to the production of exchange integrals, and so it is called an *exchange operator*. It is written explicitly in conjunction with a function on which it is operating, viz.

$$\hat{K}_j \phi_i(1) = \int \phi_j^*(2)\,(1/r_{12})\phi_i(2)d\tau(2)\,\phi_j(1) \tag{11-9}$$

Notice that an index exchange has been performed. It is not difficult to see that the expression (see Appendix 9 for bra-ket notation)

$$\langle \phi_i | \hat{F} | \phi_i \rangle = \epsilon_i \tag{11-10}$$

will lead to integrals such as

$$\langle \phi_i | \hat{J}_j | \phi_i \rangle = \langle \phi_i(1)\,\phi_j(2)\,|1/r_{12}|\phi_i(1)\,\phi_j(2)\rangle = J_{ij} \tag{11-11}$$

$$\langle \phi_i | \hat{K}_j | \phi_i \rangle = \langle \phi_i(1)\,\phi_j(2)\,|1/r_{12}|\phi_i(2)\,\phi_j(1)\rangle = K_{ij} \tag{11-12}$$

which are formally the same as the coulomb and exchange terms encountered in Chapter 5 in connection with the helium atom. Notice that, if the spins associated with spin-orbitals ϕ_i and ϕ_j differ, K_{ij} must vanish. This arises because integrations over space and spin coordinates of electron 1 (or 2) in Eq. (11-12) lead to integration over two different (and orthogonal) spin functions. On the other hand, J_{ij} is not affected by such spin agreement or disagreement.

It would appear from Eq. (11-6) that the MOs ϕ are eigenfunctions of the Fock operator and that the Fock operator is, in effect, the hamiltonian operator. There is an important qualitative difference between $\hat{F}$ and $\hat{H}$, however. *The Fock operator is itself a function of the* MOs ϕ. Since the summation index j in Eq. (11-7) includes i, the operators $\hat{J}_i$ and $\hat{K}_i$ must be known in order to write down $\hat{F}$, but $\hat{J}_i$ and $\hat{K}_i$ involve ϕ_i, and ϕ_i is an eigenfunction of $\hat{F}$. Hence, we need $\hat{F}$ to find ϕ_i, and we need ϕ_i to know $\hat{F}$. To circumvent this problem, an iterative approach is used. One makes an initial guess at the MOs ϕ. (One could use a semiempirical method to produce this starting set.) Then these MOs are used to construct an operator $\hat{F}$, which is used to solve for the new MOs ϕ'. These are then used to construct a new Fock operator, which is in turn used to find new MOs, which are used for a new $\hat{F}$, etc., until at last no significant change is detected in two successive steps of this procedure. At this point, the ϕ's *produced by* $\hat{F}$ are the same as the ϕ's that produce the coulomb-and-exchange fields *in* $\hat{F}$. The solutions are said to be self-consistent, and the method is referred to as the self-consistent-field (SCF) method.

11-6 Interpretation of the LCAO-MO-SCF Eigenvalues

The physical meaning of an eigenvalue ϵ_i is best understood by expanding the integral

$$\epsilon_i = \langle \phi_i | \hat{F} | \phi_i \rangle \tag{11-13}$$

with $\hat{F}$ given by Eq. (11-7). We obtain

$$\epsilon_i = \left\langle \phi_i \left| -\frac{1}{2}\nabla_1^2 \right| \phi_i \right\rangle - \sum_{\mu} \langle \phi_i \,|Z_u/r_{\mu 1}|\, \phi_i \rangle + \sum_{j=1}^{n} (2J_{ij} - K_{ij}) \tag{11-14}$$

It is common practice to combine the first two terms of Eq. (11-14), which depend only on the nature of ϕ_i, into a single expectation value of the one-electron part of the hamiltonian, symbolized H_{ii}. Thus,

$$\epsilon_i = H_{ii} + \sum_{j=1}^{n}(2J_{ij} - K_{ij}) \tag{11-15}$$

The quantity H_{ii} is the average kinetic plus nuclear-electronic attraction energy for the electron in ϕ_i.

The sum of coulomb and exchange integrals in Eq. (11-15) contains all the electronic interaction energy. Observe that the index j runs over all the occupied MOs. For a particular value of j, say $j = k \neq i$, this gives $2J_{ij} - K_{ij}$ as an interaction energy. This means that an electron in ϕ_i, experiences an interaction energy with the *two* electrons in ϕ_k of

$$2\,\langle \phi_i(1)\phi_k(2)|1/r_{12}|\phi_i(1)\phi_k(2)\rangle - \langle \phi_i(1)\phi_k(2)\,|1/r_{12}|\,\phi_k(1)\phi_i(2)\rangle \tag{11-16}$$

The first part is the classical repulsion between the electron having an orbital charge cloud given by $|\phi_i|^2$ and the two electrons having charge cloud $|\phi_k|^2$. The second part is the exchange term which, as we saw in Chapter 5, arises from the antisymmetric nature of the wavefunction. It enters (11-16) only once because the electron in ϕ_i, agrees in spin with only one of the two electrons in ϕ_k, [Equation (11-15) applies because we have restricted our discussion to closed-shell systems.]

The summation over $2J_{ij} - K_{ij}$ includes the case $j = i$. Here we get $2J_{ii} - K_{ii}$. However, examination of Eqs. (11-11) and (11-12) shows that $J_{ii} = K_{ii}$, and so we are left with J_{ii}. This corresponds to the repulsion between the electron in ϕ_i (the energy of which we are calculating) and the other electron in ϕ_i. Because these electrons must occur with opposite spin, there is no exchange energy for this interaction.

In brief, then, the quantity ϵ_i, often referred to as an *orbital energy* or a *one-electron energy*, is to be interpreted as the energy of an electron in ϕ_i, resulting from its kinetic energy, its energy of attraction for the nuclei, and its repulsion and exchange energies due to all the other electrons in their charge clouds $|\phi_j|^2$.

11-7 The SCF Total Electronic Energy

It is natural to suppose that the total electronic energy is merely the sum of the one-electron energies, but this is not the case in SCF theory. Consider a two-electron system. The energy of electron 1 includes its kinetic and nuclear attraction energies and its repulsion and exchange energies for electron 2. The energy of electron 2 includes its kinetic and nuclear attraction energies and its repulsion and exchange energies for electron 1. If we sum these, we have accounted properly for kinetic and nuclear attraction energies, but we have included the interelectronic interactions *twice* as much as they actually occur. (The energy of repulsion, say, between two charged particles, 1 and 2, is given by the repulsion of 1 for 2 or of 2 for 1, but not by the sum of these.) Therefore, if we sum one-electron energies, we get the total electronic energy plus an extra measure of electron repulsion and exchange energy. We can correct this by subtracting this extra measure away. Thus, for our closed-shell system

$$E_{\text{elec}} = \sum_{i=1}^{n} \left[2\epsilon_i - \sum_{j=1}^{n} (2J_{ij} - K_{ij}) \right] \tag{11-17}$$

where the summation is over the *occupied orbitals*. Comparing Eq. (11-17) with (11-18) makes it evident that we can also write

$$E_{\text{elec}} = \sum_{i=1}^{n} \left[2H_{ii} + \sum_{j=1}^{n} (2J_{ij} - K_{ij}) \right] \tag{11-18}$$

or

$$E_{\text{elec}} = \sum_{i=1}^{n} (\epsilon_i + H_{ii}) \tag{11-19}$$

To obtain the *total* (electronic plus nuclear) energy, we add the internuclear repulsion energy for the N nuclei:

$$E_{\text{tot}} = E_{\text{elec}} + V_{\text{nn}} \tag{11-20}$$

$$V_{\text{nn}} = \sum_{\mu=1}^{N-1} \sum_{\nu=\mu+1}^{N} \frac{Z_\mu Z_\nu}{r_{\mu\nu}} \tag{11-21}$$

11-8 Basis Sets

A great deal of research effort has gone into devising and comparing basis sets for *ab initio* calculations. There are essentially two important criteria:

1. We want a basis set that is capable of describing the actual wavefunction well enough to give chemically useful results.
2. We want a basis set that leads to integrals F_{ij} and S_{ij} that we can evaluate reasonably accurately and quickly on a computer.

Many types of basis set have been examined and two of these have come to dominate the area of *ab initio* molecular calculations. These two, which we refer to as the gaussian and the Slater-type-orbital (STO) basis sets, are actually very similar in many important respects.

Let us consider the STO basis set first. The essence of this basis choice is to place on each nucleus one or more STOs. The number of STOs on a nucleus and the orbital exponent of each STO remain to be chosen. Generally, the larger the number of STOs and/or the greater the care taken in selecting orbital exponents, the more accurate the final wavefunction and energy will be.

At the least sophisticated end of the spectrum of choices is the *minimal basis set of STOs*, which we encountered in Chapter 7. This includes only those STOs that correspond to occupied AOs in the separated atom limit. If we choose a minimal basis set, then we must still decide how to evaluate the orbital exponents in the STOs. One way is to use Slater's rules, which are actually most appropriate for isolated atoms. Another way is to vary the orbital exponents until the energy of the molecular system is minimized. This amounts to performing a nonlinear variational calculation along with the linear variational calculation. For molecules of more than a few atoms, this procedure consumes much computer time, for reasons we will describe shortly, but for small molecules (two or three first-row atoms plus a few hydrogens) it is possible to accomplish this task. From this, one discovers what orbital exponent best suits an STO in a molecular environment. This leads us to the third way of choosing orbital exponents—choose the values that were found best for each type of atom in nonlinear variational calculations in smaller molecules.

One may improve the basis by adding additional STOs to various nuclei. Suppose, for example, each carbon 2p AO were represented as a linear combination of two p-type STOs, each having a different orbital exponent. An example of the basic principle involved is indicated in Fig. 11-1. If we treat these functions independently and do a linear variational calculation, they will both be mixed into the final wavefunction to some degree. If the *linear* coefficient for the "inner" STO is much larger, it means that the p-type charge cloud around this atom in the molecule is calculated to be fairly contracted around the nucleus. To describe a more diffuse charge cloud, the wavefunction would contain quite a lot of the "outer" STO, and not so much of the "inner" STO.

Thus, we have a linear variation procedure that, in effect, allows for AO expansion and contraction. It is akin to optimizing an orbital exponent, but it does not require nonlinear variation. Of course, one still has to choose the values of the "larger ζ" and "smaller ζ" of Fig. 11-1. This is normally done by optimizing the fit to very accurate atomic wavefunctions or by a nonlinear variation on atoms. A basis set in which every minimal basis AO is represented by an "inner-outer" pair of STOs is often referred to as a "double-zeta" basis set.

A further kind of extension is frequently made. In addition to the above types of STO, one includes STOs with symmetries different from those present in the minimal basis. This has the effect of allowing charge to be shifted in or out of bond regions in new ways. For example, one could add p-type STOs on hydrogen nuclei. By mixing this with the s-type STOs there, one can describe a skewed charge distribution in the regions of the protons. We have already seen (Chapter 7) that a hydrogen atom in a uniform

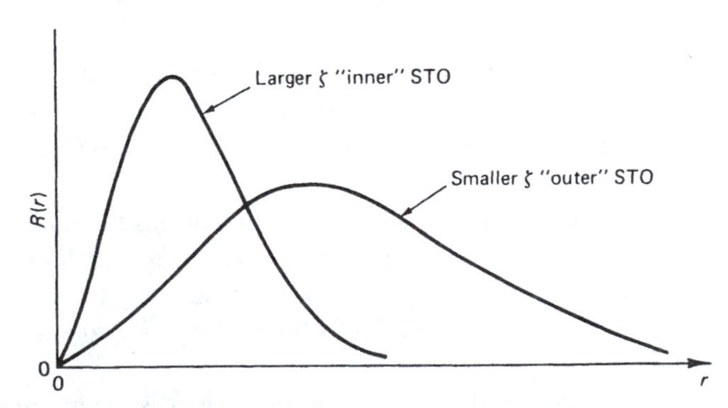

Figure 11-1 ► Radial functions $R(r) = r\exp(-\zeta r)$ for 2p-type STOs. The larger ζ value gives an STO more contracted around the nucleus. Hence, it is sometimes called the "inner" STO.

electric field is polarized in a way that is reasonably well described by an s-p linear combination. Since the hydrogen atom in a molecule experiences an electric field due to the remainder of the molecule, it is not surprising that such p functions are indeed mixed into the wavefunction by the variational procedure if we provide them in the basis set. Similarly, d-type STOs may be added to atoms that, in the minimal basis set, carried only s- and p-type STOs. Functions of this nature are often called *polarization functions* because they allow charge polarization to occur within the molecule as a result of the internally generated electric field.

It should be evident that one could go on indefinitely, adding more and more STOs to the basis, even placing some of them in bonds, rather than on nuclei. This is not normally done because the computing task goes up enormously as we add more basis functions. In fact, the number of integrals to be calculated eventually increases as N^4, where N is the number of basis functions. The evaluation of integrals can be a logistic bottleneck in *ab initio* calculations, and for this reason nonlinear variations (of orbital exponents) are impractical for any but smaller molecules. Each new orbital exponent value requires re-evaluation of all the integrals involving that orbital. In essence, a change of orbital exponent is a change of basis set. In linear variations, the basis functions are mixed together but they do not change. Once all the integrals between various basis functions have been evaluated, they are usable for the remainder of the calculation.

The STO basis would probably be the standard choice if it were not for the fact that the many integrals encountered in calculating F_{ij} elements are extremely time consuming to evaluate, even on a computer. This problem has led to the development of an alternative basis set class that is based on gaussian-type functions. Gaussian functions include an exponential term of the form $\exp(-\alpha r^2)$. The radial dependence of such a function is compared to that for a hydrogenlike 1s function (which is identical to a 1s STO) in Fig. 11-2. There are two obvious problems connected with using gaussian functions as basis functions:

1. They do not have cusps at $r = 0$ as s-type hydrogen-like AOs do.

2. They decay faster at larger r than do hydrogen-like AOs.

Both of these deficiencies are relevant in molecules because, at $r = 0$ (on a nucleus) and at $r = \infty$, the molecular potential is like that in an atom, so similar cusp and asymptotic behavior are expected for molecular and atomic wavefunctions. Balanced against these deficiencies is an advantage: gaussian functions have mathematical properties that make it extremely easy to compute the integrals they produce in F_{ij}. This has led to a practice of replacing each STO in a basis set by a *number* of gaussian functions.

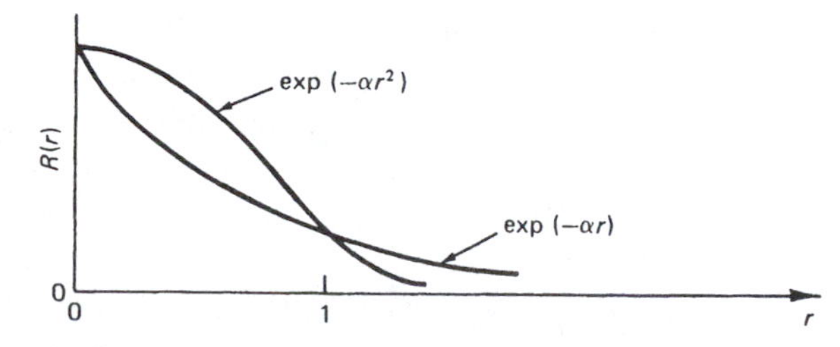

Figure 11-2 ▶ Radial dependence of hydrogen-like and gaussian functions.

By choosing several values of α in $\exp(-\alpha r^2)$, one can create a set of "primitive" gaussian functions ranging from very compact to very diffuse, and then take a linear combination of these to build up an approximation to the radial part of an STO function. Multiplication by the standard θ and ϕ dependences (spherical harmonics) generates p, d, etc. functions. Once this approximation is optimized, the linear combination of gaussian functions is "frozen," being treated thereafter as a single function insofar as the subsequent molecular variational calculation is concerned. This linear combination of primitive gaussian functions is called a *contracted gaussian function*.

Once we have a contracted gaussian function corresponding to each STO, we can go through the same hierarchy of approximations as before—minimal basis set, double-ζ basis set, double-ζ plus polarization functions-only now using contracted gaussian functions in place of STOs. Typically, *ab initio* calculations on systems involving only light elements, e.g., H–Ne, involve anywhere from 1 to 15 primitive gaussian functions for each contracted gaussian function. Basis sets for heavier elements, however, can contain more than 30 primitive gaussians for each contracted gaussian.

We have already described a certain amount of quantum-chemical jargon. Some of the basis set descriptions that one commonly encounters in the modern literature are as follows:

- DZP [double-ζ gaussian basis with polarization]
- STO-3G [each STO approximated as a linear combination of three gaussian primitives]
- 6-31G [each inner shell STO represented by a sum of six gaussians and each valence shell STO split into inner and outer parts (i.e., double-ζ) described by three and one gaussian primitives, respectively]
- 6-31G* [the 6-31G basis set augmented with six d-type gaussian primitives on each heavy ($Z > 2$) atom, to permit polarization]
- 6-31G** [same as 6-31G* but with a set of gaussian p-type functions on H and He atoms. Good for systems where hydrogen is a bridging atom, as in diborane or in hydrogen bonds]
- 6-31+G* [the 6-31G* basis set augmented with a set of diffuse s- and p-type gaussian functions on each heavy atom, to permit representation of diffuse electronic distribution, as in anions]
- cc-pVnZ, $n =$ D, T, Q, 5 [correlation consistent polarized valence n-ζ gaussian basis sets. The inner shell STOs are described by single contracted gaussian functions while the valence STOs are described by n contracted gaussian functions, $n =$ D for double-ζ, $n =$ T for triple-ζ, etc. Both the number and angular momentum symmetry type of the polarization functions are increased with each successive correlation consistent basis set in a systematic manner. For example, the cc-pVDZ basis set has a set of 5 d-type gaussian primitive functions on each heavy atom, while the cc-pVTZ basis set has 2 sets of d-type gaussian functions and one set of 7 f-type gaussian primitives. These families of basis sets are designed to converge the total energy to the complete basis set limit for the SCF method and its extensions.]

- aug-cc-pVnZ [cc-pVnZ basis sets augmented with one set of diffuse primitive gaussian functions for each angular momentum symmetry present in the cc-pVnZ basis set, to provide an accurate description of anions and weak interactions, e.g., van der Waals forces and hydrogen bonding.]

11-9 The Hartree–Fock Limit

It should be apparent that different choices of basis set will produce different SCF wavefunctions and energies. Suppose that we do an SCF calculation on some molecule, using a minimal basis set and obtain a total electronic energy E_1. If we now choose a double-ζ basis and do a new SCF calculation, we will obtain an energy E_2 that normally will be lower than E_1. (If one happens to choose the first basis wisely and the second unwisely, it is possible to find E_2 higher than E_1. We assume here that each improvement to the basis extends the mathematical flexibility while including the capabilities of all preceding bases.) If we now add polarization functions and repeat the SCF procedure, we will find E_3 to be lower than E_2. We can continue in this way, adding new functions in bonds and elsewhere, always increasing the capabilities of our basis set, but always requiring that the basis describe MOs in a single determinantal wavefunction. The electronic energy will decrease with each basis set improvement, but eventually this decrease will become very slight for any improvement; that is, the energy will approach a limiting value as the basis set approaches mathematical completeness. This limiting energy value is the lowest that can be achieved for a single determinantal wavefunction. It is called the *Hartree–Fock energy*. The MOs that correspond to this limit are called Hartree–Fock orbitals (HF orbitals), and the determinant is called the HF wavefunction.

Sometimes the term *restricted Hartree–Fock* (RHF) is used to emphasize that the wavefunction is restricted to be a single determinantal function for a configuration wherein electrons of α spin occupy the same space orbitals as do the electrons of β spin. When this restriction is relaxed, and different orbitals are allowed for electrons with different spins, we have an *unrestricted Hartree–Fock* (UHF) calculation. This refinement is most likely to be important when the numbers of α- and β-spin electrons differ. We encountered this concept in Section 8-13, where we noted that the unpaired electron in a radical causes spin polarization of other electrons, possibly leading to negative spin density.

11-10 Correlation Energy

The Hartree–Fock energy is not as low as the true energy of the system. The mathematical reason for this is that our requirement that ψ be a single determinant is restrictive and we can introduce additional mathematical flexibility by allowing ψ to contain many determinants. Such additional flexibility leads to further energy lowering.

There is a corresponding *physical* reason for the HF energy being too high. It is connected with the independence of the electrons in a single determinantal wavefunction. To understand this, consider the four-electron wavefunction

$$\psi = \left|\phi_1(1)\overline{\phi_1}(2)\phi_2(3)\overline{\phi_2}(4)\right| \tag{11-22}$$

Recall from Chapter 5 that the numbers in parentheses stand for the spatial coordinates of an electron; that is, $\phi_1(1)$ really means $\phi_1(x_1, y_1, z_1)\alpha(1)$ or $\phi_1(r_1, \theta_1, \phi_1)\alpha(1)$.[1] In other words, if we pick values of r, θ, and ϕ for each of the four electrons and insert them into Eq. (11-22) we will be able to evaluate each function and we will obtain a determinant of numbers which can be evaluated to give a numerical value for ψ and ψ^2. The latter number (times dv) can be taken as the probability for finding one electron in the volume element around r_1, θ_1, and ϕ_1, another electron simultaneously in dv_2 at r_2, θ_2, and ϕ_2, etc. The important point to notice is that the effect on ψ^2 of a particular choice of r_1, θ_1, and ϕ_1, is not dependent on choices of r, θ, ϕ for other electrons because the form of the wavefunction is *products of functions of independent coordinates*. Physically, this corresponds to saying that the probability for finding an electron in dv_1, at some instant is not influenced by the presence or absence of another electron in some nearby element dv_2, at the same instant. This is consistent with the fact that the Fock operator $\hat{F}$ [Eq. (11-7)] treats each electron as though it were moving in the *time-averaged* potential field due to the other electrons.

Because electrons repel each other, there is a tendency for them to keep out of each other's way. That is, in reality, their motions are *correlated*. The HF energy is higher than the true energy because the HF wavefunction is formally incapable of describing correlated motion. The energy difference between the HF and the "exact" (for a simplified nonrelativistic hamiltonian) energy for a system is referred to as the *correlation energy*.

11-11 Koopmans' Theorem

Despite the fact that the total electronic energy is not given by the sum of SCF one-electron energies, it is still possible to relate the ϵ_i's to physical measurements. If certain assumptions are made, it is possible to equate orbital energies with molecular ionization energies or electron affinities. This identification is related to a theorem due to Koopmans.

Koopmans [1] proved[2] that the wavefunction obtained by removing one electron from ϕ_k, or adding one electron to the virtual (i.e., unoccupied) MO ϕ_j in a Hartree–Fock wavefunction is stable with respect to any subsequent variation in ϕ_k, or ϕ_j. Notice that this ignores the question of subsequent variation of all of the MOs ϕ with unchanged occupations. It is not necessarily true that they remain optimized, since the potential they experience is changed by addition or removal of an electron. Nevertheless, Koopmans' theorem suggests a model. It suggests that we approximate the wavefunction for a positive ion by removing an electron from one of the occupied HF MOs for a neutral molecule without reoptimizing any of the MOs. Let us do this and compare the electronic energies for the two wavefunctions.

For the neutral molecule, which we assume is a closed-shell system,

$$E = \sum_i \left[2H_{ii} + \sum_j (2J_{ij} - K_{ij}) \right] \tag{11-23}$$

[1] Note that ϕ_1 in parentheses represents a coordinate of electron 1, whereas ϕ_1 outside the parentheses represents an MO.

[2] See also Smith and Day [2].

For the cation, produced by removing an electron from ϕ_k,

$$E_k^+ = \sum_{i \neq k} \left[2H_{ii} + \sum_{j \neq k} (2J_{ij} - K_{ij}) \right] + H_{kk} + \sum_{i \neq k} (2J_{ik} - K_{ik}) \tag{11-24}$$

The first sum in Eq. (11-24) gives the total electronic energy due to all but the unpaired electron in ϕ_k. H_{kk} gives the kinetic and nuclear attraction energies for the unpaired electron and the final sum gives the repulsion and exchange energy between this electron and all the others. Now we note that the last sum is exactly equal to the void produced in the first sum due to the restriction $j \neq k$. Therefore, we can combine these by removing the index restriction and deleting the last sum. This gives

$$E_k^+ = \sum_{i \neq k} \left[2H_{ii} + \sum_{j} (2J_{ij} - K_{ij}) \right] + H_{kk} \tag{11-25}$$

To compare this with E of (11-23) we should remove the remaining index restriction. We do this by allowing i to equal k in the sum and simultaneously subtracting the new terms thus produced:

$$E_k^+ = \sum_{i} \left[2H_{ii} + \sum_{j} (2J_{ij} - K_{ij}) \right] - H_{kk} - \sum_{j} (2J_{kj} - K_{kj}) \tag{11-26}$$

But, by virtue of Eqs. (11-15) and (11-23), this is

$$E_k^+ = E - \epsilon_k \tag{11-27}$$

Hence, the ionization energy I_k^0, for ionization from the ϕ_k is

$$I_k^0 = E_k^+ - E = -\epsilon_k \tag{11-28}$$

This illustrates that, within the context of this simplified *model*, the negative of the orbital energies for occupied HF MOs are to be interpreted as ionization energies.

Another way to see the relation between I_k^0 and $-\epsilon_k$, is to recognize that the physical interactions lost upon removal of an electron from φ_k, are precisely those that constitute ϵ_k, [See Eq. (11-15).]

A similar result holds for orbital energies of unoccupied HF MOs and electron affinities. (However, this is less successful in practice; see Problem 11-3.)

In actuality, the relation (11-28) is only approximately obeyed. One reason for this has to do with our assumption that doubly occupied SCF MOs produced by a variational procedure on the neutral molecule will be suitable for the doubly occupied MOs of the cation as well. These MOs minimize the energy of the neutral molecule but give an energy for the cation that is higher than what would be produced by an independent variational calculation. For this mathematical reason, we expect the Koopmans' theorem prediction for the ionization energy to be higher than the value predicted by taking the difference between separate SCF calculations on the molecule and cation (which we will symbolize ΔSCF). The corresponding *physical* argument is that use of Eq. (11-28) views ionization as removal of an electron without any reorganization of the remaining

TABLE 11-1 ► Ionization Energies (in electron volts) of Water as Measured Experimentally and as Predicted from SCF Calculations

		SCF (near HF limit)[b]	
Cation state	Observed[a]	Koopmans	ΔSCF
$^2B_2$	12.62	13.79	11.08
$^2A_1$	14.74	15.86	13.34
$^2B_2$	18.51	19.47	17.61

[a] From Potts and Price [3].
[b] From Dunning et al. [4].

electronic charge. This neglects a process that stabilizes the cation and lowers the ionization energy. Whichever argument we choose, we have here a reason for expecting $-\epsilon$ to be an *overestimate* of the value obtained by independent calculations, ΔSCF.

Another error results from the neglect of change in correlation energy. We have seen that the total SCF energy for the molecule is too high because the single determinantal form of the wavefunction cannot allow for correlated electronic motion. The SCF energy for the cation is too high for the same reason, but the error is different for the two cases because there are fewer electrons in the cation. We expect the neutral molecule to have the greater correlation energy (since it has more electrons)[3] so that proper inclusion of this feature would lower the energy of the neutral molecule more than the cation, making the true I_k^0 larger than that obtained by neglect of correlation. Hence, this leads us expect ΔSCF to *underestimate* I_k^0. Since $-\epsilon$ overestimates ΔSCF, and ΔSCF underestimates the ionization energy, we can expect some cancellation of errors in using Eq. (11-28).

An illustration of these relations is provided in Table 11-1, where observed *vertical* ionization energies (i.e., no nuclear relaxation), the appropriate values of $-\epsilon$, and the values of ΔSCF are compared.

11-12 Configuration Interaction

There are several techniques for going beyond the SCF method and thereby including some effects of electron correlation. Some extremely accurate calculations on small atoms and molecules, making explicit use of interparticle coordinates, were described in Section 7-8. There is one general technique, however, that has traditionally been used for including effects of correlation in many-electron systems. This technique is called *configuration interaction* (CI).

The mathematical idea of CI is quite obvious. Recall that we restricted our SCF wavefunction to be a single determinant for a closed-shell system. To go beyond the optimum (restricted Hartree–Fock) level, then, we allow the wavefunction to be a linear

[3] This reasoning is rather naive. Significant correlation energy contribution can result from a small energy-level separation between filled and empty MOs (rather than from merely the number of electrons), but production of a cation should normally increase this gap and lead to reduced correlation.

combination of determinants. Suppose we choose two determinants D_1 and D_2, each corresponding to a different orbital occupation scheme (i.e., different configurations). Then we can let

$$\psi = c_1 D_1 + c_2 D_2 \tag{11-29}$$

and minimize E as a function of the linear mixing coefficients c_1 and c_2.

If we go through the mathematical formalism and express $\bar{E}$ as $\langle\psi|\hat{H}|\psi\rangle/\langle\psi|\psi\rangle$, expand this as integrals over D_1 and D_2, and require $\partial\bar{E}/\partial c_i = 0$, we obtain the same sort of 2×2 determinantal equation that we find when minimizing an MO energy as a function of mixing of two AOs. That is, we obtain

$$\begin{vmatrix} H_{11} - \bar{E}S_{11} & H_{12} - \bar{E}S_{12} \\ H_{21} - \bar{E}S_{21} & H_{22} - \bar{E}S_{22} \end{vmatrix} = 0 \tag{11-30}$$

where now

$$H_{ij} = \left\langle D_i \middle| \hat{H} \middle| D_j \right\rangle \tag{11-31}$$

$$S_{ij} = \langle D_i | D_j \rangle \tag{11-32}$$

We see that, whereas before we might have had two AOs interacting to form two MOs, here we have two configurations (i.e., two determinantal functions) interacting to form two approximate wavefunctions. Our example involves only two configurations, but there is no limit to the number of configurations that can be mixed in this way.

Since each configuration D contains products of MOs, each of which is typically a sum of AOs, the integrals H_{ij} and S_{ij} can result in very large numbers of integrals over basis functions when they are expanded. This is the sort of situation where a computer is essential, and CI on atoms and molecules, while still expensive compared to SCF, have become routine on modern computers.

Our purpose in this chapter is not to describe how to carry out a CI calculation, but rather to convey what a CI calculation is and what its predictive capabilities are. Therefore, we will not concern ourselves with the mathematical complexities of evaluating H_{ij} and S_{ij}.[4] But we will consider one practical aspect of CI calculations, namely, how one goes about choosing which configurations should be mixed together, and which ones may be safely ignored.

We begin by considering the H_2 molecule. The LCAO-MO-SCF method expresses the ground state wavefunction for H_2 as

$$\psi(1,2) = \begin{vmatrix} 1\sigma_g(1)\alpha(1) & 1\sigma_g(2)\alpha(2) \\ 1\sigma_g(1)\beta(1) & 1\sigma_g(2)\beta(2) \end{vmatrix} \tag{11-33}$$

that is, as the configuration $1\sigma_g^2$. The SCF procedure mixes the AO basis functions together in the optimum way to produce the $1\sigma_g$ MO.

We have noted at several points in this book that, if one begins with a basis set of n linearly independent functions, one ultimately arrives at n independent MOs. Hence, the $1\sigma_g$ MO of Eq. (11-33) is but one of several MOs produced by the SCF procedure.

[4] In most actual calculations, the D's are orthonormal, and $S_{ij} = \delta_{ij}$.

It is called an *occupied* MO because it is occupied with electrons in this configuration. All the other MOs in this case are *unoccupied* or *virtual* MOs. The virtual MOs of H_2 have symmetry properties related to the molecular hamiltonian, just as does the occupied MO. Thus, we can refer to $1\sigma_u$, $2\sigma_g$, $2\sigma_u$, $1\pi_u$, $1\pi_g$, etc., virtual MOs of H_2. Which of these virtual MOs are produced by an SCF calculation depends on the number and nature of the AO basis set provided at the outset. If no π-type AOs are provided, no π-type MOs will be produced. If only a minimal basis ($1s_a$ and $1s_b$) is provided, $1\sigma_u$ will be the only virtual MO produced.

It is important to distinguish between the physical content of occupied versus virtual SCF MOs. The SCF procedure finds the set of *occupied* MOs for a system leading to the lowest SCF electronic energy. The virtual orbitals are the residue of this process. The virtual MOs span that part of the basis set function space that the SCF procedure found *least* suitable for describing ψ. The subspace is sometimes referred to as the *orthogonal complement* of the occupied orbital subspace. (Note that this situation differs from that pertaining to Hückel-type calculations, where MOs and energy levels are calculated without regard for electron occupancy. Only after the variational procedure are electrons added.)

Our concern with virtual MOs is due to the fact that they provide a ready means for constructing new configurations to mix with our $1\sigma_g^2$ configuration for H_2. Thus, using some of the above-mentioned virtual MOs, we could write determinantal functions corresponding to the excited configurations $1\sigma_g 1\sigma_u$, $1\sigma_g 2\sigma_g$, $1\sigma_g 2\sigma_u$, $1\sigma_g 1\pi_u$, etc.[5] These are commonly referred to as *singly excited* configurations because one electron has been promoted from a ground-state-occupied MO to a virtual MO. (This is *not* meant to imply that the orbital energy difference is equal to the expected spectroscopic energy of the transition.) It is also possible to construct *doubly excited* configurations, such as $1\sigma_u^2$, $1\sigma_u 2\sigma_g$, $2\sigma_g^2$, $1\sigma_u 2\sigma_u$, $1\sigma_u 1\pi_u$, etc. For systems having more electrons, one can write determinants corresponding to triple, quadruple, etc., excitations. If one has a reasonably large number, say 50, of virtual orbitals and, say, 10 electrons to distribute among them, then there is an enormous number of possible configurations. A major step in doing a CI calculation is deciding which configurations might be important in affecting the results and ought therefore to be included.

We can gain insight into this problem by considering our minimal basis set H_2 problem in more detail. We have

$$1\sigma_g = N_g(1s_A + 1s_B) \tag{11-34}$$

$$1\sigma_u = N_u(1s_A - 1s_B) \tag{11-35}$$

where N_g and N_u are normalization constants. The spatial part of the ground configuration is

$$\psi_{\text{space}} = 1\sigma_g(1)1\sigma_g(2) \tag{11-36}$$

[5]As was shown in Chapter 5, the symmetry requirements of the wavefunction require that each of these open shell configurations be expressed as a linear combination of two 2×2 determinants; for example, $1\sigma_g 2\sigma_u$ stands for the combination

$$\left(1/\sqrt{2}\right)\left\{\left|1\sigma_g(1)2\bar{\sigma}_u(2)\right| \pm \left|1\bar{\sigma}_g(1)2\sigma_u(2)\right|\right\}$$

which expands to

$$\psi_{\text{space}} = N_{\text{g}}^2 \left[1s_A(1)\,1s_A(2) + 1s_B(1)\,1s_B(2) + 1s_A(1)\,1s_B(2) + 1s_B(1)\,1s_A(2)\right]$$

If both electrons are near nucleus A, the first term is quite large. This may be rephrased to say that ψ^2 gives a sizable probability for finding both electrons near nucleus A. The second term gives a similar likelihood for finding both electrons near B. These two terms are referred to as *ionic* terms because they become large whenever the instantaneous electronic dispositions correspond to $H_A^- H_B^+$ and $H_A^+ H_B^-$, respectively. The last two terms cause ψ^2 to be sizable whenever an electron is near each nucleus. Hence, these are called *covalent terms*, and their presence means that ψ contains significant "covalent character." In fact, because all four terms have the same coefficient, the configuration $1\sigma_{\text{g}}^2$ is said to have 50% covalent and 50% ionic character.

Is this bad? It turns out to be no problem at all when the nuclei are close together. Indeed, in the united-atom (helium) limit, the ionic-covalent distinction vanishes. But at large internuclear separations it is very inaccurate to describe H_2 as 50% ionic. In reality, H_2 dissociates to two neutral ground state H atoms—that is, 100% "covalent," with an electron near each nucleus. In short, the SCF-MO description does not properly describe the molecule as it dissociates. This means that the calculation of $\bar{E}$ versus R_{AB} for H_2 will deviate from experiment more and more as R_{AB} increases. This defect in the SCF treatment of H_2 occurs for many other molecular species also.

Can we correct this defect through use of CI? We ask the question this way: "What configuration could we mix with $1\sigma_{\text{g}}^2$ in order to make the mixture of covalent and ionic character variable?" Since $1\sigma_{\text{g}}^2$ expands to give us covalent and ionic terms of the *same* sign, we need an additional configuration that will give them with *opposite* sign. Then admixture of the two configurations will affect the two kinds of term differently. The configuration that will accomplish this is $1\sigma_{\text{u}}^2$:

$$\begin{aligned} 1\sigma_{\text{u}}(1)\,1\sigma_{\text{u}}(2) = N_{\text{u}}^2 [&1s_A(1)\,1s_A(2) + 1s_B(1)\,1s_B(2) - 1s_A(1)\,1s_B(2) \\ &- 1s_B(1)\,1s_A(2)] \end{aligned} \tag{11-37}$$

Mixing these two configurations together gives

$$\begin{aligned} \psi(c_1/c_2) &= c_1 1\sigma_{\text{g}}(1)\,1\sigma_{\text{g}}(2) + c_2 1\sigma_{\text{u}}(1)\,1\sigma_{\text{u}}(2) \\ &= \left(c_1 N_{\text{g}}^2 + c_2 N_{\text{u}}^2\right)[1s_A(1)\,1s_A(2) + 1s_B(1)\,1s_B(2)] \\ &\quad + \left(c_1 N_{\text{g}}^2 - c_2 N_{\text{u}}^2\right)[1s_A(1)\,1s_B(2) + 1s_B(1)\,1s_A(2)] \end{aligned} \tag{11-38}$$

If c_1/c_2 is readjusted at each value of R_{AB} to minimize $\bar{E}$, it is evident that the relative weights of covalent and ionic character in Eq. (11-38) will change to suit the circumstances. Actual calculations on this system show that, as R_{AB} gets large, c_1/c_2 approaches a value such that $c_1 N_{\text{g}}^2 + c_2 N_{\text{u}}^2$ approaches zero, so that the ionic component of ψ vanishes.

This example illustrates that CI of this sort has an associated physical picture. It suggests that, in any CI calculation involving the dissociation (or extensive stretching) of a covalent bond, important configurations are likely to include double excitations into the antibonding virtual "mates" of occupied bonding MOs.

What about other configurations for H_2? What will $1\sigma_g 2\sigma_g$ do for the calculation, assuming now an extended basis set has produced a $2\sigma_g$ MO? Suppose we take as our trial function

$$\psi = c_1 1\sigma_g^2 + c_2 1\sigma_g 2\sigma_g \tag{11-39}$$

where the configurations are understood to stand for determinants. If the $1\sigma_g$ MO has been produced by an SCF calculation on the ground state, and $2\sigma_g$ is a virtual MO from that SCF calculation, then it is possible to show that the CI energy minimum occurs when c_2 in Eq. (11-39) is zero. In other words, these determinants will not mix when they are combined in this way. An equivalent statement is that the mixing element $H_{12} = \langle 1\sigma_g^2 \,|\hat{H}|1\sigma_g 2\sigma_g\rangle$ vanishes. Hence, the CI determinant (11-33) is already in diagonal form, and no variational mixing will occur. This is an example of *Brillouin's theorem*, which may be stated as follows:

EXAMPLE 11-1 If D_1 is an optimized single determinantal function and D_j is a determinant corresponding to any single excitation out of an orbital ϕ_j occupied in D_1 and into the virtual subspace (orthogonal complement) of D_1, then no improvement in energy is possible by taking $\psi = c_1 D_1 + c_2 D_j$.

The proof of Brillouin's theorem is very simple. We start with a basis set that spans a function space. An SCF calculation is performed, which produces the best single-determinantal wavefunction we can possibly get within this function space. This is D_1. D_j differs from D_1 in only one orbital, which means they differ in only one row. A general property of determinants is that, if two of them differ in only one row or column, any linear combination of the two can be written as a single determinant (see Problem 11-4). This means that any combination $c_1 D_1 + c_2 D_j$ is still expressible as a single determinant. Since D_j makes no use of functions outside our original basis set, $c_1 D_1 + c_2 D_j$ is a single determinant within our original function space. However, D_1 is already known to be the single determinant within this function space that gives the lowest energy, and $c_1 D_1 + c_2 D_j$ cannot do better. QED.

A doubly excited configuration differs from D_1 in two rows, and mixing such a configuration with D_1 produces a result that cannot be expressed as a single determinant.

Because of Brillouin's theorem, one might decide to omit all single excitations from CI calculations. But it is important to recognize that singly excited configurations can affect the results of CI calculations *in the presence of doubly excited configurations*. This comes about because nonzero mixing elements can occur between singly and doubly excited configurations in the CI determinant. To illustrate, let ψ_0 be an SCF single determinant, ψ_1 be a singly excited configuration, and ψ_2 be a "double." Then the CI determinant could be, assuming orthogonal determinants,

$$\begin{vmatrix} H_{00} - E & 0 & H_{02} \\ 0 & H_{11} - E & H_{12} \\ H_{02} & H_{12} & H_{22} - E \end{vmatrix} = 0 \tag{11-40}$$

The zeros result from Brillouin's theorem. However, H_{12} does not necessarily vanish, and solution of this 3×3 determinantal equation leads to a wavefunction of the form

$$\psi = c_0\psi_0 + c_1\psi_1 + c_2\psi_2 \tag{11-41}$$

with c_1 not zero. ψ_1 comes in on the coattails of ψ_2 and is referred to as a second-order correction. This is not a guarantee that it will be unimportant, however. (See Example 7-4 for similar behavior in a different context.)

Another rule that is useful for recognizing configurations that may be omitted is the rather obvious one that each configuration must share the same set of eigenvalues for operators commuting with the hamiltonian. That is, if ψ is to be associated with a particular symmetry, angular momentum, spin angular momentum, etc., then each configuration in ψ must have that same symmetry, angular momentum, etc. This means that, for the ground state of H_2, $1\sigma_g^2$ will not mix with $1\sigma_g 1\sigma_u$ because the latter has overall u symmetry. $1\sigma_u 2\sigma_u$ could contribute, but the symmetrized combination corresponding to the *singlet* state $(|1\sigma_u 2\bar{\sigma}_u| - |1\bar{\sigma}_u 2\sigma_u|)$ must be used rather than the (positive) triplet state combination. The configuration $1\sigma_u 1\pi_u$ will not contribute because it has the wrong total angular momentum.

Even with the aid of all these rules, a calculation on a molecule such as N_2 or O_2 using a reasonably extended basis set gives rise to an enormous number of possible configurations. Additional rules of thumb have been found to help choose the major configurations. It has been found, for example, that triply or higher excited configurations are usually of lesser importance than doubly excited configurations. [Since the hamiltonian contains only one- and two-electron operators, interaction elements must vanish between the ground-state configuration and all triply or higher-excited configurations. But, like singly excited configurations, these can, in principle, come in on the coattails of doubly (or other) excited configurations.] In addition, a study of the energy change in some process involving primarily the valence electrons (e.g., stretching N_2) really does not require calculation of the correlation energy of the 1s electrons since they are fairly unaffected by the change. Any correlation energy for these electrons tends to cancel itself when initial and final state energies are subtracted. Therefore, in a CI calculation of such a process, it is reasonable to omit configurations corresponding to excitation of a 1s electron unless high accuracy is desired.

The acronym CID refers to a CI calculation in which only all doubly excited configurations are included. Inclusion of all singly and doubly excited configurations is referred to as a CISD calculation. *Full* CI (FCI) means all excited configurations have been included, and this is the limit that gives all of the correlation energy within the chosen basis set. The combination of full CI and a complete basis gives the exact energy (generally nonrelativistic and within the Born–Oppenheimer approximation).

11-13 Size Consistency and the Møller–Plesset and Coupled Cluster Treatments of Correlation

Whenever certain parts of a well-defined procedure are omitted, as when full CI is truncated to CID or CISD, one must consider whether systematic errors are introduced. This is indeed the case in the above example. Suppose CID calculations are made for the energy of N_2 as a function of internuclear distance. At short distances, we treat the system as a 14-electron molecule, including configurations in which 12 of the electrons are in their HF-occupied MOs. At very large distances we have two nitrogen atoms, which we normally treat as having twice the energy of one atom. Now CID on atom A includes the HF configuration, D_0^A, as well as doubly excited configurations in which five of the seven electrons are in their HF-occupied AOs. Let D_2^A represent this class of configuration. Then $\psi^A = c_0 D_0^A + c_2 D_2^A$. Atom B has a similar CID wavefunction: $\psi^B = c_0 D_0^B + c_2 D_2^B$. The wavefunction for the overall, noninteracting system is the antisymmetrized product of these wavefunctions. It will contain terms like $D_0^A D_0^B$, $D_0^A D_2^B$, and $D_2^A D_2^B$. There are no terms present corresponding to a single excitation at each atom, $D_1^A D_1^B$, and such terms *would* be present in a CID treatment of the combined system. Also, there is a class of terms present corresponding to *four* promoted electrons, $D_2^A D_2^B$, and these terms would *not* be present in a CID treatment of the combined system. The dilemma is that, if we treat the system as a single 14-electron "molecule," which is appropriate at small R, we mix in different terms than if we treat it as two separate atoms, which is appropriate at large R. If we choose some fairly large R value to redefine N_2 as two separate atoms, we change the nature of the CI in a discontinuous way at an arbitrary point. This feature of truncated CI is called the problem of *size consistency*; CID and CISD methods are not size consistent. Doing separate calculations on each of two separated atoms and combining the energies yields a different result from doing a calculation on one system made up of two separated atoms.

A correlation method that is size consistent has been developed by Pople and co-workers.[6] It is based on perturbation theory that was introduced many years ago by Møller and Plesset.[7] This approach divides the process of treating correlation into a series of corrections to an unperturbed starting point. If one chooses to do such a calculation to, say, third order (MP3, standing for Møller–Plesset to third order), then the set of configurations to be included is determined by the perturbation formulas.[8] It does not require further decision by the person doing the calculation and can be wholly managed by a computer program. Møller–Plesset perturbation theory is different from standard CI in at least two important respects: It is size consistent, and it is not variationally bound. One cannot assume, therefore, that going to higher and higher orders of perturbation will cause the calculated energy to approach closer and closer to the true energy from above.

Because of the way MP theory defines the unperturbed system, the starting point energy (MP0) is the sum of HF one-electron energies. The first-order correction to the energy (MP1) brings in the appropriate electronic coulomb and exchange integrals, giving the correct HF energy. MP2 brings in contributions wherein doubly excited configurations "interact with" (i.e., occur in the same integral with) the ground configuration. MP3 adds contributions due to doubly excited configurations interacting with each other. MP4 brings in interactions involving also single, double, triple, and quadruple excitations. The selection of interaction terms by the perturbation formalism is what produces size consistency, but it leaves out certain terms at each level that would be included in a variational calculation.

In coupled cluster (CC) approaches, which are also size consistent and generally not variationally bound, instead of including all configurations to a particular order as

[6]See Binkley and Pople [5].

[7]See Møller and Plesset [6].

[8]Perturbation theory is presented in Chapter 12. The present discussion avoids mathematical details.

in MP theory, each class of excited configurations is included to *infinite* order. This is accomplished via an exponential excitation operator,

$$\Psi_{CC} = e^{\hat{T}}\psi_0 = \left[1 + \hat{T} + \frac{\hat{T}^2}{2!} + \frac{\hat{T}^3}{3!} + \ldots\right]\psi_0 \qquad (11\text{-}42)$$

where ψ_0 is the HF determinant for an N-electron system, and $\hat{T} = \hat{T}_1 + \hat{T}_2 + \hat{T}_3 + \cdots + \hat{T}_N$. $\hat{T}_1$ produces singly excited determinants, $\hat{T}_2$ doubly excited ones, and so on. Because of the exponential nature of the excitation operator, each class of excitations is included to all orders, e.g., terms in $\hat{T}_2$ would include products of double excitations ($\hat{T}_2^2$) that would be considered a subset of the possible quadruple excitations in CI. This is what makes CC theory size consistent. Usually coupled cluster theory is truncated to include just $\hat{T}_1$ and $\hat{T}_2$, i.e., CCSD. One of the most accurate post-HF methods has been shown to be the CCSD(T) method,in which a CCSD calculation is followed by a contribution due to triple excitations ($\hat{T}_3$) via perturbation theory.

11-14 Multideterminant Methods

Up to this point, the methods that have been presented for describing electron correlation effects have been constructed with the single determinant SCF wavefunction as a starting point. For most molecules near their equilibrium geometries, this is a very good zeroth-order approximation, but as we saw earlier for the H_2 molecule, as covalent bonds are stretched towards dissociation multiple determinants are required for even a qualitative description. This puts much stronger demands on these so-called single reference methods, and their accuracy can be much degraded or even unphysical in these regions. In a multiconfigurational SCF (MCSCF) calculation one writes the wavefunction as a linear combination of determinants exactly as in a CI calculation, and the energy is minimized as a function of the linear CI coefficients. However, in an MCSCF calculation one also *simultaneously* optimizes the MO coefficients of the orbitals that are used to construct the determinants, using methods analogous to SCF theory. Because this greatly adds to the complexity of the calculation, the number of determinants used in MCSCF is generally much smaller than in a standard HF-based CI calculation. In the simplest case, only the additional determinants that allow for a qualitative treatment of the process under study are included, e.g., one would include only the determinants corresponding to excitations of bonding electrons into their respective antibonding orbitals when stretching the triple bond of N_2. This procedure results in a set of MCSCF molecular orbitals (some strongly occupied, some weakly occupied) that smoothly changes in character from equilibrium to dissociation.

In multireference CISD (MRCISD) calculations, the wavefunction is written as

$$\psi = \sum_i c_i\psi_i + \sum_s c_s\psi_s + \sum_d c_d\psi_d \qquad (11\text{-}43)$$

where $\sum_i c_i\psi_i$ is the set of MCSCF *reference* determinants, ψ_s are new determinants formed by single excitations into the virtual orbitals relative to all of the reference determinants, and ψ_d are doubly excited determinants. An MRCISD calculation of this type can yield a very balanced and accurate description of a molecule's potential energy surface, but often at a relatively steep cost in terms of computational requirements.

11-15 Density Functional Theory Methods

The wavefunction ψ for an n-electron molecule is a function of $3n$ spatial coordinates and n spin coordinates. From ψ we can produce the molecule's spin-free electron density function, $\rho(1)$, by integrating $\psi^*\psi$ over all of the spin coordinates and all the the space coordinates except those for one of the electrons:[9]

$$\rho(1) = \int |\psi(1, 2, \ldots, N)|^2 d\omega_1 d\tau_2 \ldots d\tau_n \tag{11-44}$$

which is a function of only the three spatial coordinates.[10] We have seen that, in the early days of quantum chemistry, a major challenge was the evaluation of integrals over the interelectronic-repulsion term in the hamiltonian, as well as dealing with the related problem of electron correlation. Several methods were devised that attempted to approximate these quantities from the density function $\rho(1)$, with moderate success. However, the continuing progress in computer speed and the development of sophisticated *ab initio* methods gradually shifted attention away from approaches using the density function.

In 1964, proof by Hohenberg and Kohn [7] of a connection between the ground state energy, E_0, for a system and ρ_0, the ground state density function,[11] sparked new interest in finding a rigorous way to go from knowledge of the attractively simple three-dimensional density function to a value for E_0.

Recall that, for a system having n electrons and N nuclei, the hamiltonian operator for the electronic energy is

$$H = -\frac{1}{2}\sum_{i=1}^{n} \nabla_i^2 + \sum_{i=1}^{n}\sum_{\alpha=1}^{N} \frac{-Z_\alpha}{r_{i\alpha}} + \sum_{i=1}^{n-1}\sum_{j=i+1}^{n} \frac{1}{r_{ij}} \tag{11-45}$$

The first and last terms can be written down immediately if we know how many electrons are present, but the middle term depends on $\sum_{\alpha=1}^{N} \frac{-Z_\alpha}{r_{i\alpha}}$, which is a function of nuclear charges and locations. This quantity is called the *external potential*, symbolized $v_{ext}(\vec{r})$, because it results from the presence of fields produced by particles not included in the group of electrons.

Hohenberg and Kohn were able to prove that there is a uniqueness relation between ρ_0 and the external potential: No two external potentials could give the same ρ_0. This raises the possibility that one could work backwards from ρ_0 to find $v_{ext}(\vec{r})$ and then E_0. The following route comes first to mind: Integrate ρ_0 to get the number of electrons n. Figure out $v_{ext}(\vec{r})$ from ρ_0. This would allow one to write down the hamiltonian operator. Then, using *ab initio* methods, one could get to an accurate E_0 and ψ_0, and from ψ_0 one could calculate T_0, V_{ne_0}, V_{ee_0}, and all the other properties of interest for the system.

Two problems exist with this scenario. First, there is no generally applicable procedure known for getting from ρ_0 to $v_{ext}(\vec{r})$. We can posit that $v_{ext}(\vec{r})$ is a *functional* of ρ_0, which we symbolize $v_{ext}[\rho_0]$, but we don't know what the functional relationship is. Second, even if we could get back to the hamiltonian operator, it would simply

[9] Because ψ is antisymmetric for exchange of electrons, the density function is independent of our choice as to which electron's coordinates should be spared from integration.

[10] Recall that ω is the spin coordinate and τ is the coordinate for space *and* spin. $d\tau = dv\, d\omega$.

[11] We henceforth suppress the electron index in ρ.

land us back on square one: We would still have to solve the whole problem in the traditional way. Nevertheless, the hopes raised by this uniqueness theorem have led to the current goal of density functional theory, which is to find a procedure that takes us from ρ to E in a rigorous way that avoids the complexities of landing on square one and proceeding using standard *ab initio* methods. [8]

A subsequent relation proved by Hohenberg and Kohn [7] indicated a way to proceed. They proved that an *approximate* density function, $\rho_{0,approx}$, when subjected to the (unknown) procedure that relates the exact ρ_0 to the exact E_0, must yield an energy higher than the exact E_0 : $E_{0,approx} \geq E_0$, so a variational bound exists. Note that the unknown process referred to here is one that assumes $v_{ext}(\vec{r})$ to be the same for the analysis of ρ_0 and $\rho_{0,approx}$, which means that the same nuclear framework applies in both cases.

If a procedure were known for finding E from ρ, then the existence of a variational bound would allow a variational procedure analogous to what we have applied earlier. One would start with a trial ρ, calculate its energy, and vary ρ to locate the ρ that gives the lowest energy.

The barrier to proceeding is the lack of a way to get E from ρ. Hence, the development of approximate functionals that relate the energy to the electron density is an extremely active area of current research and probably will be for some time to come.

In analogy to wavefunction methods, the functional that connects E to ρ, $E[\rho]$, can be separated into an electronic kinetic energy contribution, $T[\rho]$, a contribution due to nuclear-electron attractions, $E_{ne}[\rho]$, and the electron-electron repulsions, $E_{ee}[\rho]$. The latter term can be further decomposed into Coulomb and exchange terms, $J[\rho]$ and $K[\rho]$. Both the nuclear-electron attraction and the interelectronic Coulomb terms can be easily written in terms of the density using their classical expressions as in wavefunction methods. For an accurate treatment of the electronic kinetic energy term, however, one must differentiate a wavefunction,[12] and this has led to the practice first proposed by Kohn and Sham [9] of expressing the density in terms of one-electron orbitals ϕ (constructed numerically or from a basis set of Slater or gaussian functions). These orbitals serve two purposes. They allow us to calculate a value of the kinetic energy within a single Slater determinant framework similar to Hartree–Fock theory,

$$T_S = \sum_{i=1}^{n} \langle \phi_i | -\frac{1}{2}\nabla^2 | \phi_i \rangle \tag{11-46}$$

and to obtain the electron density, defined in terms of these Kohn–Sham orbitals as

$$\rho_s = \sum_{i}^{n} |\phi_i|^2 \tag{11-47}$$

The final DFT energy expression is then written as

$$E_{DFT}[\rho] = T_S[\rho] + E_{ne}[\rho] + J[\rho] + E_{xc}[\rho] \tag{11-48}$$

where the *exchange correlation* functional $E_{xc}[\rho]$ contains the difference between the exact kinetic energy and T_S, the nonclassical (exchange) part of electron-electron repulsions, $K[\rho]$, and correlation contributions to both $K[\rho]$ and $J[\rho]$. The Kohn–Sham

[12] As far as we know, we must differentiate a wavefunction to get kinetic energy. If there is a functional that permits us to obtain kinetic energy directly from the density function, we might avoid having to use orbitals.

orbitals are eigenfunctions of an effective one-electron hamiltonian that is nearly identical in form to the Fock operator in the SCF equations. In the Kohn–Sham case, however, the HF exchange operators are replaced by the functional derivative of the exchange correlation energy. Assuming the existence of $E_{xc}[\rho]$ and an initial guess for the electron density, one then solves the Kohn–Sham eigenvalue equations for the orbitals, which can then be used to define a new electron density and effective hamiltonian. These iterations continue until the density is converged to within a specified threshold.

The exact form of $E_{xc}[\rho]$ is not currently known, however, and a rapidly growing list of approximate exchange correlation functionals have appeared in the literature. Because these are all estimates of a part of the overall energy, the total energy finally calculated is not an upper bound to the true energy. Also, DFT is not size-consistent.

Generally, most existing exchange correlation functionals are split into a pure exchange and correlation contribution, $E_x[\rho]$ and $E_c[\rho]$ and the current functional nomenclature often reflects this with two-part acronyms, e.g., the BLYP DFT method uses an exchange functional from Becke (B) [10] and a correlation functional by Lee, Yang, and Parr (LYP) [11]. In principle, the exchange contribution could be calculated exactly (for a single determinant) in the same manner as T_S, but this is generally not done since this disturbs the balance between $E_x[\rho]$ and $E_c[\rho]$. In hybrid DFT, a percentage of this exact exchange is included in $E_{xc}[\rho]$.

The great benefit of present day DFT methods is computational cost. With the exchange correlation functionals commonly used, the computational effort is similar to a SCF calculation, but since $E_{xc}[\rho]$ implicitly includes some amount of electron correlation, the accuracy of DFT (depending on the chosen functional) is often similar to that obtained with MP2 or better. The great weakness of DFT at the present time, however, is the inability to systematically improve upon $E_{xc}[\rho]$ and converge towards the exact Born–Oppenheimer energy like one might conceptually do in a wavefunction-based CI or CC calculation, e.g., SCF, CCSD, CCSDT, CCSDTQ, etc. with sequences of correlation consistent basis sets.

One of the simplest DFT methods is the local density approximation (LDA), which assumes the density behaves locally like a uniform electron gas. Generally this does not lead to an accurate description of molecular properties, but if one makes $E_x[\rho]$ and $E_c[\rho]$ depend also on the gradient of the density, yielding gradient corrected DFT or the generalized gradient approximation (GGA), the results are much more accurate. Finally, the definition of $E_{xc}[\rho]$ also lends itself to semiempirical contributions. One such parameterization that has been very successful is the B3LYP hybrid DFT method, which includes 20% exact exchange and involves three semiempirical parameters that were obtained by fits to experimental thermochemical data (heats of formation, etc.) of small molecules [12].

11-16 Examples of *Ab Initio* Calculations

Self-consistent-field and correlated calculations have now been made for a very large number of systems. The best way to judge the capabilities of these methods is to survey some of the results.[13]

[13] For extensive surveys, see Schaefer [13], Hehre et al. [14], and Raghavachari [15].

Table 11-2 provides information on energies for a number of atoms in their ground states. Self-consistent-field energies are presented for three levels of basis set complexity. In the STO single-ζ level, a minimal basis set of one STO per occupied AO is used, and the energy is minimized with respect to independent variation of every orbital exponent ζ. The STO double-ζ basis set is similar except that there are two STOs for each AO, the only restriction being that the STOs have the same spherical harmonics as the AOs to which they correspond.

The Hartree–Fock energies are estimated by extrapolating from more extensive basis sets, and represent the limit achievable for the SCF approach using a complete basis set. We can make the following observations:

1. The improvement in energy obtained when one goes from a single-ζ to a double-ζ STO basis set is substantial, especially for atoms of higher Z.
2. The agreement between the optimized double-ζ data and the HF energies is quite good. Even for neon, the error is only about 10^{-2} a.u. (0.27 eV). Thus, for atoms, the double-ζ basis is capable of almost exhausting the energy capabilities of a single-configuration wavefunction.
3. The disagreement between HF and "exact" energies (i.e., the correlation energy) grows progressively larger down the list. For neon it is almost 0.4 a.u. (10 eV), which is an unacceptable error in chemical measurements.

One might think that the magnitude of the correlation energy in these examples would make SCF calculations on heavy atoms useless for quantitative purposes, but this

TABLE 11-2 ▶ Ground-State Energies (in atomic units) for Atoms, as Computed by the SCF Method and from Experiment

	STO				
Atom	Single ζ^a	Double ζ^a	Hartree–Fock[a]	Exact[b]	Correlation[c] energy
He	−2.8476563	−2.8616726	−2.8616799	−2.9037	−0.0420
Li	−7.4184820	−7.4327213	−7.4327256	−7.4774	−0.0447
Be	−14.556740	−14.572369	−14.573021	−14.6663	−0.0933
B	−24.498369	−24.527920	−24.529057	−24.6519	−0.1228
C	−37.622389	−37.686749	−37.688612	−37.8420	−0.1534
N	−54.268900	−54.397951	−54.400924	−54.5849	−0.1840
O	−74.540363	−74.804323	−74.809370	−75.0607	−0.2513
F	−98.942113	−99.401309	−99.409300	−99.7224	−0.3131
Ne	−127.81218	−128.53511	−128.54705	−128.925	−0.378
Ar	−525.76525	−526.81511	−526.81739	−527.542	−0.725

[a]From Roetti and Clementi [16].
[b]"Exact" equals experimental with relativistic correction but without correction for Lamb shift. See Veillard and Clementi [17].
[c]Correlation energy is "exact" minus HF energy.

is not the case. Most frequently we are not concerned with the value of the *total* energy of a system so much as with *energy changes* (e.g., in spectroscopy or in reactions) or else with other properties such as transition moments (for spectroscopic intensities) or, in molecular systems, dipole moments.

Let us, therefore, see how well SCF calculations can predict atomic ionization energies. We have already indicated (Section 11-11) that there are two ways we can get ionization energies from SCF calculations. The first, and simplest, is to take the various $-\epsilon_i$, as suggested by Koopmans' theorem. Table 11-3 shows that this gives only rough agreement with experimental values for neon. Another way is to do separate SCF calculations for each excited state produced by removal of an electron from an orbital (i.e., for each "hole state") and equate the ionization energies to the energy differences between these and the neutral ground state (ΔSCF). This second method requires much more effort. As Table 11-3 indicates, however, the extra effort leads to great improvement in agreement between theoretical and experimental values. We conclude that SCF calculations on atoms and ions give quantitatively useful data on ionization energies, even for ionization out of deep-lying levels. The Koopmans' theorem approach is less accurate, although still qualitatively useful.

A related problem is the calculation of energies of excited states of atoms. Weiss [19] has reported calculations on some of the excited states of carbon, and his results are summarized in Fig. 11-3. Inspection of this figure reveals that near-HF calculations only roughly reproduce the energy spectrum, but CI (with four or five configurations) brings about marked improvement. Weiss has omitted configurations involving excitations of 1s electrons, and so these results ignore correlation energy for the inner-shell electrons. The agreement suggests that these electrons experience almost no change in correlation for transitions among these states. Weiss has also calculated oscillator strengths[14] associated with atomic transitions and he finds that CI is necessary before reasonable agreement with experiment is achieved.

TABLE 11-3 ► Ionization Energies of Neon[a]

	Ionization potential (a.u.)		
Ion configuration	Koopmans	ΔSCF	Experiment
$1s2s^22p^6$	32.7723	31.9214	31.98
$1s^22s2p^6$	1.9303	1.8123	1.7815
$1s^22s^22p^5$	0.8503	0.7293	0.7937

[a] From Bagus [18]. The basis set includes 5 s-type and 12 p-type STOs (4 of each m quantum number). ζ's were varied as well as linear coefficients. The neutral ground state gives $E = -128.547$ a.u. (compare Table 11-2).

[14] The oscillator strength is a measure of the probability (i.e., intensity) of a transition. For a transition between states a and b in a $2n$-electron system it is commonly given by the formula

$$\frac{2}{3}(E_b - E_a)\left|\int \psi_a^* \left\{\sum_{i=1}^{2n} r_i\right\} \psi_b \, d\tau\right|^2$$

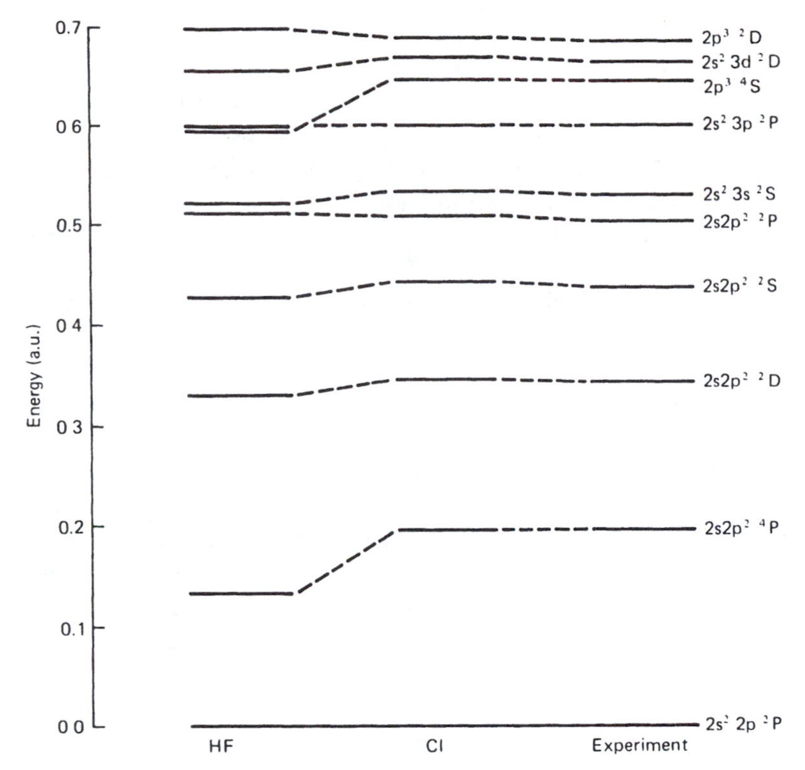

Figure 11-3 ▶ Transition energies in the C^+ ion as calculated by HF, CI, and as measured. (From Weiss [19].) Ionization from the ground state of C^+ occurs at 0.8958 a.u.

In brief, then, the evidence indicates that reasonably accurate atomic ionization energies can be obtained by high-quality SCF calculations on the neutral and ionized species (ΔSCF, not $-\epsilon$), but that transition energies and intensities require CI sufficient to account for much of the valence electron correlation.

Before we leave the subject of atoms, it should be pointed out that, for any atom, the expectation value $\bar{T}$ of the kinetic energy operator is equal to $-\bar{E}$ if the wavefunction has been optimized with respect to a scale factor in the coordinates r_1, r_2, etc. This relation, called the *virial relation*, is proved in Appendix 8. It is necessarily satisfied for any level of calculation that cannot be improved by replacing every r_i in ψ by ηr_i and allowing η to vary. Since the single- and double-ζ STO solutions have already been optimized with respect to such scale parameters, they satisfy the virial relation. Thus, for the beryllium atom, the single-ζ STO value for $\bar{E}$ is (Table 11-2) -14.556740 a.u., and so we know that $\bar{T} = +14.556740$ a.u. and $\bar{V} = -29.113480$ a.u. for this wavefunction (since $\bar{E} = \bar{T} + \bar{V}$). For the double-$\zeta$ wavefunction $\bar{T} = +14.572369$ a.u., etc. The Hartree–Fock wavefunction is, by definition, the lowest-energy solution achievable within a restricted (single determinantal[15]) wavefunction form. Use of a scale factor does not affect the wavefunction form. Hence, no further lowering of $\bar{E}$ below the HF level is possible in this way, and the HF energies $\bar{E}$, $\bar{T}$, and $\bar{V}$ must satisfy the virial relation also. Finally, the exact energies are the lowest achievable for

[15]For open-shell systems, more than one determinant may be needed to satisfy symmetry requirements. This is still considered a HF wavefunction.

any wavefunction. Again, scaling cannot lower the energy further, so these energies also satisfy the virial relation.

We turn next to *ab initio* calculations on molecules. First, let us compare HF and exact energies for molecules as we did for atoms and see how large the errors due to correlation are. The results are not too different from those for atoms having the same number of electrons, as shown in Table 11-4; that is, the correlation energies for molecules having ten electrons (CH_4, NH_3, H_2O, HF) are about the same as that for neon, whereas that for the 18-electron molecule H_2O_2 is more like the correlation energy for argon. But this is only a very rough rule of thumb. We have already indicated that the correlation energy in a molecule varies with bond length, a factor not present in atomic problems. In order to get a more meaningful idea of the capabilities of *ab initio* calculations on molecules, we must look more closely at specific examples.

A calculation on the OH radical, reported by Cade and Huo [21], provides a good example of the capabilities of the extended basis set LCAO-MO-SCF technique on a small molecule. Their final wavefunction for the ground state at an internuclear separation $R = 1.8342$ a.u. is presented in Table 11-5. A minimal basis set of STOs for OH would include 1s, 2s, $2p_x$, $2p_y$, and $2p_z$ STOs on oxygen and a single 1s AO on hydrogen. Cade and Huo chose a much more extensive basis. Oxygen is the site for two 1s, two 2s, one 3s, four 2p, one 4f, eight $2p_\pi$, two $3d_\pi$, and four $4f_\pi$ STOs. On hydrogen, there are two 1s, one 2s, one $2p_\sigma$, two $2p_\pi$, and two $3d_\pi$ STOs. (The π-type basis functions are indicated in Table 11-5 for only one of the two directions perpendicular to the O–H axis.) The orbital exponents for all of these STOs have been optimized, and the resulting wavefunction is of "near-Hartree–Fock" quality. The optimized ζ values appear in Table 11-5. The STO labeled $\sigma 2p'_o$ is located on oxygen and has the formula

$$\sigma 2p'_o = (2.13528)^{5/2}\pi^{-1/2} r \exp(-2.13528r)\cos\theta \tag{11-49}$$

TABLE 11-4 ► Estimated Hartree–Fock and Correlation Energies for Selected Molecules[a] and Atoms[b]

Molecule or atom	E (HF) (a.u.)	*E* (correlation) (a.u.)	Molecule or atom	E (HF) (a.u.)	*E* (correlation) (a.u.)
H_2	−1.132	−0.043	Ne	−128.547	−0.378
He	−2.862	−0.042	CO	−112.796	−0.520
BH_3	−26.403	−0.195	N_2	−108.994	−0.540
$O(^1D)$	−74.729	−0.262	Si (^1D)	−288.815	−0.505
CH_4	−40.219	−0.291	B_2H_6	−52.835	−0.429
NH_3	−56.225	−0.334	S (^1D)	−397.452	−0.606
H_2O	−76.067	−0.364	H_2O_2	−150.861	−0.688
HF	−100.074	−0.373	Ar	−526.817	−0.725

[a] From Ermler and Kern [20].
[b] See Table 11-2.

TABLE 11-5 ► Near Hartree–Fock Wavefunction for the OH Molecule in Its Ground-State Configuration ($1\sigma^2 2\sigma^2 3\sigma^2 1\pi^3$) at $R = 1.8342$ a.u.[a]

$\chi\sigma$	$C_{1\sigma}$	$C_{2\sigma}$	$C_{3\sigma}$	$\chi\pi$	$C_{1\pi}$
$\sigma 1s_O$ ($\zeta = 7.01681$)	0.94291	−0.25489	0.07625	$\pi 2p_O$ ($\zeta = 1.26589$)	0.37429
$\sigma 1s_O'$ (12.38502)	0.09313	0.00358	−0.00153	$\pi 2p_O'$ (2.11537)	0.46339
$\sigma 2s_O$ (1.71794)	−0.00162	0.46526	−0.20040	$\pi 2p_O''$ (3.75295)	0.23526
$\sigma 2s_O'$ (2.86331)	0.00418	0.55854	−0.18328	$\pi 2p_O'''$ (8.41140)	0.01023
$\sigma 3s_O$ (8.64649)	−0.03826	−0.02643	0.00550	$\pi 3d_O$ (1.91317)	0.02871
$\sigma 2p_O$ (1.28508)	−0.00055	0.05179	0.30153	$\pi 4f_O$ (2.19941)	0.00506
$\sigma 2p_O'$ (2.13528)	−0.00056	−0.07538	0.37791	$\pi 2p_H$ (1.76991)	0.02442
$\sigma 2p_O''$ (3.75959)	0.00115	0.01874	0.18390	$\pi 3d_H$ (3.32513)	0.00282
$\sigma 2p_O'''$ (8.22819)	0.00059	0.00229	0.00952		
$\sigma 3d_O$ (1.63646)	−0.00047	0.02437	0.04676		
$\sigma 3d_O'$ (2.82405)	0.00016	0.00845	0.01595		
$\sigma 4f_O$ (2.26641)	−0.00013	0.00882	0.01232		
$\sigma 1s_H$ (1.31368)	0.00150	−0.04651	0.21061		
$\sigma 1s_H'$ (2.43850)	−0.00034	0.09413	0.05113		
$\sigma 2s_H$ (2.30030)	0.00000	0.07654	0.04539		
$\sigma 2p_H$ (2.8052)	0.00018	0.01182	0.00999		

[a] From Cade and Huo [21].

There are three σ-type MOs and two π-type MOs to accommodate the nine electrons of this radical. One π-type MO is

$$\begin{aligned}\phi 1\pi_y = {} & 0.37429\pi 2p_{Oy} + 0.46339\pi 2p'_{Oy} + 0.23526\pi 2p''_{Oy} \\ & + 0.01023\pi 2p'''_{Oy} + 0.02871\pi 3d_{Oy} + 0.00506\pi 4f_{Oy} \\ & + 0.02442\pi 2p_{Hy} + 0.00282\pi 3d_{Hy}\end{aligned} \tag{11-50}$$

and the other occupied π-type MO would be the same except with x instead of y. (The z axis is coincident with the internuclear axis.) It is evident that writing out the complete wavefunction given in Table 11-5 would result in a very cumbersome expression. It is a nontrivial problem to relate an accurate but bulky wavefunction such as this to the kinds of simple conceptual schemes chemists like to use. One solution is to have a computer produce contour diagrams of the MOs. Such plots for the valence MOs 2σ, 3σ, and 1π of Table 11-5 are presented in Fig. 11-4.

Cade and Huo [21] carried out similar calculations for OH at 13 other internuclear distances and also for the united atom (fluorine) and the separated atoms in the states with which the Hartree–Fock wavefunction correlates. Some of their data are reproduced in Table 11-6. A plot of the electronic-plus-nuclear repulsion energies is given in Fig. 11-5 along with the experimentally derived curve. It is evident that the near HF curve climbs too steeply on the right, leading to too "tight" a potential well for nuclear motion and too small an equilibrium internuclear separation. This comes about because, as mentioned earlier, the HF solution dissociates to an incorrect mixture of states, some

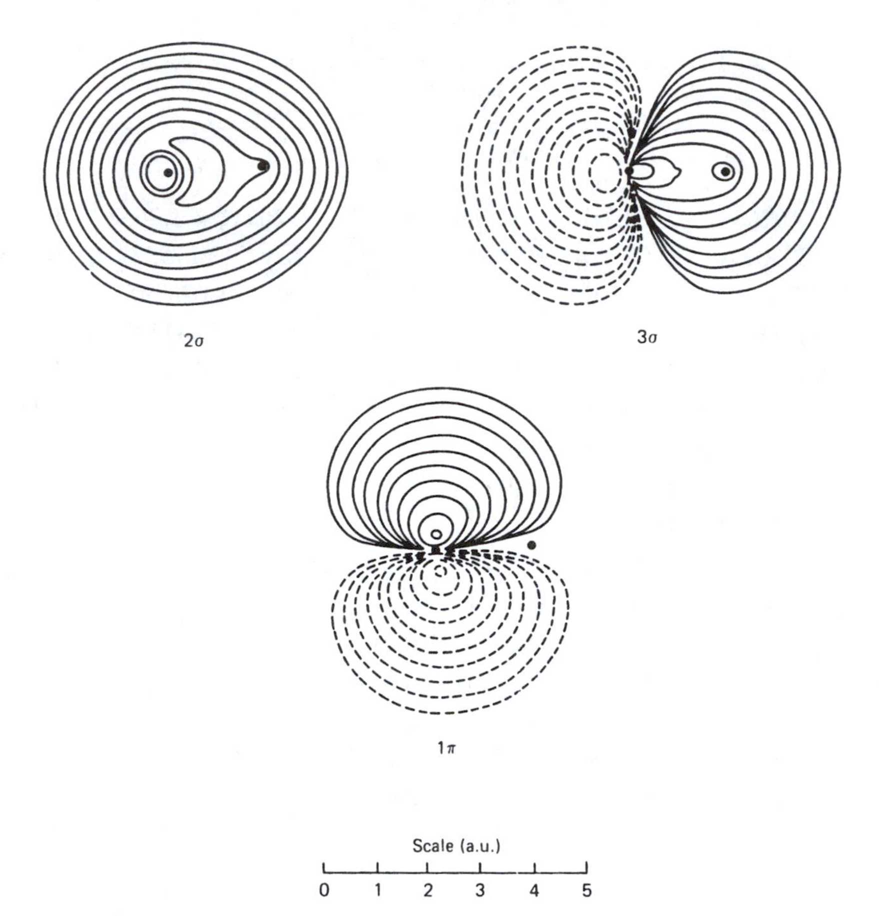

Figure 11-4 ▶ Contour plots of HF valence orbitals for OH as given in Table 11-5. (From Stevens et al. [22].)

TABLE 11-6 ▶ Spectroscopic Parameters and Dipole Moment for OH from Theoretical Curves and from Experiment[a]

Wavefunction	Dipole moment (debyes)	R_e (a.u.)	D_e (eV)	ω_e (cm^{-1})	$\omega_e x_e$ (cm^{-1})	α_e (cm^{-1})
SCF	1.780	1.795	8.831	4062.6	165.09	0.661
CI	1.655	1.838	4.702	3723.6	83.15	0.628
Experimental	1.66±.01	1.834	4.63	3735.2	82.81	0.714

[a]From Stevens et al. [22].

of which are ionic. It is possible to use the HF curve of Fig. 11-5 to derive theoretical values for molecular constants that can be compared to spectroscopic data. The results are displayed in Table 11-6, and they reflect the inaccuracy in the HF energy curve. Included there are the SCF and experimental values for the molecular dipole moment.

We turn now to the behavior of $\bar{V}/\bar{T}$ for the HF wavefunctions of Cade and Huo at various internuclear separations. The data appear in Table 11-7. Observe that the

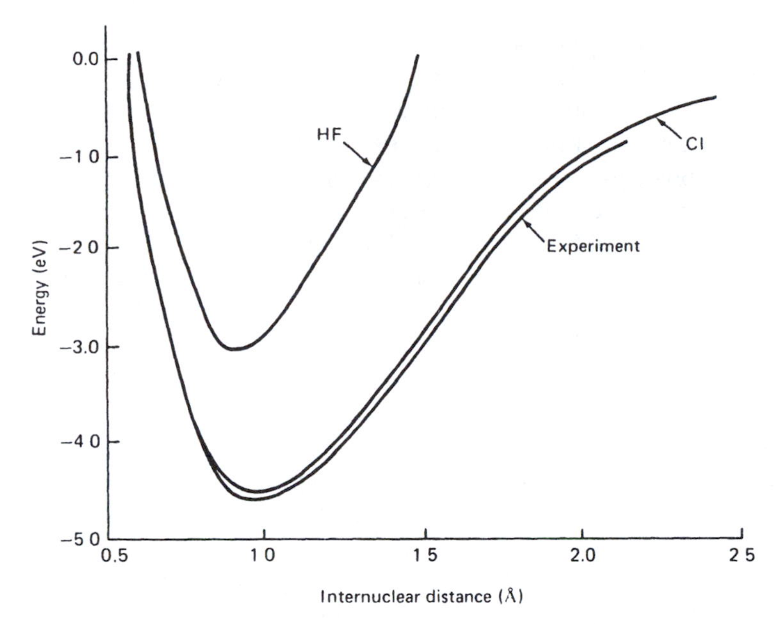

Figure 11-5 ▶ Theoretical and experimental energy curves for OH (from Stevens et al. [22].)

value of −2.00000 for $\bar{V}/\bar{T}$ occurs at three values of R: 0, ∞, and the point where $\bar{E}$ is a minimum. At $R=0$ and ∞, we are dealing with one or two atoms, for which we have already seen the HF solution should give $\bar{V}/\bar{T}=-2$. At intermediate R we have a diatomic molecule, for which the virial relation is (see Appendix 8)

$$2\bar{T}+\bar{V}+R\frac{\partial \bar{E}}{\partial R}=0 \tag{11-51}$$

There are three cases to consider. If $\partial\bar{E}/\partial R=0$, then $\bar{V}/\bar{T}=-2$. This will occur at the minimum of the potential energy curve (and also at any subsidiary maxima or minima). If $\partial\bar{E}/\partial R$ is negative, then, since $\bar{V}/\bar{T}=-2-(R/\bar{T})(\partial\bar{E}/\partial R)$ and $\bar{T}$ is

TABLE 11-7 ▶ HF Total Energies and $\bar{V}/\bar{T}$ for OH as a Function of Internuclear Distance[a]

R (a.u.)	E (a.u.)	$\bar{V}/\bar{T}$	R (a.u.)	E (a.u.)	$\bar{V}/\bar{T}$
0	−99.40933	−2.00000	1.90	−75.41837	−2.00129
1.40	−75.34382	−1.99076	2.00	−75.41140	−2.00225
1.50	−75.38378	−1.99398	2.10	−75.40163	−2.00300
1.60	−75.40696	−1.99651	2.25	−75.38372	−2.00380
1.70	−75.41829	−1.99850	2.40	−75.36367	−2.00433
1.75	−75.42065	−1.99933	2.60	−75.33582	−2.00474
1.795	−75.42127	−2.00000	2.80	−75.30822	−2.00492
1.8342	−75.42083	−2.00052	∞	−75.30939	−2.00000

[a] From Cade and Huo [21].

positive, $\bar{V}/\bar{T}$ will be algebraically higher than -2 (e.g.,-1.98). If $\partial \bar{E}/\partial R$ is positive, $\bar{V}/\bar{T}$ will be lower than -2. Thus, the values of $\bar{V}/\bar{T}$ in Table 11-7 reflect the slope of a line tangent to the potential energy curve at each R value.

As mentioned earlier, it is possible to at least partly include the effects of electron correlation by allowing determinants corresponding to other configurations to mix into the wavefunction. Such calculations have been performed for the OH radical by several groups, and the results of Stevens et al. [22] are included in Table 11-6 and Fig. 11-5. These data come from intermixing 14 configurations. It is evident that the inclusion of correlation through CI has markedly improved the agreement with experiment.

Many diatomic molecules have been treated at a comparable and higher level, and it is clear that *ab initio* calculations including electron correlation are capable of giving quite accurate molecular data. In cases in which the diatomic system is experimentally elusive, such calculations may be the best source of data available. A further example of this is provided in Table 11-8, in which are listed dipole moments for ground and some excited states of diatomic molecules. The dipole moments computed from near-HF wavefunctions contain substantial errors. It can be seen that CI greatly improves dipole moments. It has been observed that inclusion of singly excited configurations is very important in obtaining an accurate dipole moment.

As a general rule, CID correlates electron motion and therefore has a significant energy-lowering effect but has little effect on the one-electron distribution or related properties, like dipole moment. Inclusion of singly excited configurations (CISD) allows the one-electron distribution to shift in response to the change in calculated interelectronic repulsion. For example, the value of the ground-state dipole moment of CO (entry 4 of Table 11-8) is calculated at the CID level to be $-0.20D$ and at the CISD level to be $+0.12D$. Thus CID may be a suitable level of computational effort if the interest is in energy, but CISD is better if the interest is in one-electron properties.

TABLE 11-8 ► Calculated and Experimental Dipole Moments of Diatomic Molecules (in Debyes)

Molecule and polarity	State	HF at R_e[a]	CI at R_e[a]	Experiment	Reference
Li^+H^-	$X\,^1\Sigma^+$	6.002	5.86	5.83	[23]
C^+N^-	$X\,^2\Sigma^+$	2.301	1.465	1.45 ± 0.08	[24]
C^-N^+	$B\,^2\Sigma^+$	—	0.958	1.15 ± 0.08	[24]
C^-O^+	$X\,^1\Sigma^+$	-0.274	0.12	0.112 ± 0.005	[23]
C^+O^-	$A\,^3\Pi$	2.34	1.43	1.37	[25]
C^-S^+	$X\,^1\Sigma^+$	1.56	2.03	1.97	[23]
C^-S^+	$A\,^1\Pi$	-0.09	0.63	0.63 ± 0.04	[26]
C^-H^+	$X\,^2\Pi$	1.570	1.53	1.46 ± 0.06	[27]
O^-H^+	$X\,^2\Pi$	1.780	1.655	1.66 ± 0.01	[27]
F^-H^+	$X\,^1\Sigma$	1.942	1.805	1.797	[27]
N^-H^+	$X\,^3\Sigma^-$	1.627	1.537	Unknown	[27]

[a]Experimental R_e Value used.

Highly accurate properties can be obtained with more sophisticated electron correlation methods, such as CCSD(T) or MRCISD.

Other examples of the use of *ab initio* methods on small molecules are shown in Table 11-9, which displays some calculated properties for the electronic ground states of H_2 and N_2 as a function of method with the cc-pVTZ basis set. In the case of H_2, the SCF bond length and harmonic frequency are both slightly too large compared to experiment, but the dissociation energy is underestimated by more than 30 kcal/mole due to the lack of electron correlation. For this two-electron system, the CISD and CCSD methods are equivalent to a FCI and exhibit marked improvement compared to SCF. The remaining deviations from experiment at this level of theory can be attributed to the use of the finite cc-pVTZ basis set. The MP methods show systematic improvement with each order of perturbation theory, but even a fourth-order treatment of single and double excitations results in non-negligible errors compared to FCI for this simple system. The B3LYP hybrid density functional method is observed to yield very reliable properties in this case.

As might be expected due to its triple bond, the N_2 molecule is considerably more challenging for *ab initio* methods. With the cc-pVTZ basis set, the SCF dissociation energy is smaller than experiment by nearly a factor of 2. Appreciable differences are now observed between the CISD and CCSD results, with the latter being somewhat closer to experiment. In addition, triple excitations, as measured by the difference between CCSD and CCSD(T), are relatively important for N_2, raising the dissociation energy by nearly 9 kcal/mole. In contrast to the H_2 case, the results for N_2 using perturbation theory (MP2, MP3, MP4) display a disturbing oscillatory behavior. This type of result with MP methods has been the subject of several previous studies.[16]

TABLE 11-9 ▶ Calculated Equilibrium Bond Lengths, Harmonic Vibrational Frequencies, and Equilibrium Dissociation Energies for the Ground States of H_2 and N_2 with the cc-pVTZ Basis Set Compared to Experiment

	H_2			N_2		
	r_e(Å)	ω_e(cm^{-1})	D_e (kcal/mole)	r_e(Å)	ω_e(cm^{-1})	D_e (kcal/mole)
SCF	0.734	4587	83.7	1.067	2732	120.4
CISD	0.743	4409	108.4	1.089	2509	193.1
MP2	0.737	4526	103.6	1.114	2195	228.7
MP3	0.739	4476	107.1	1.090	2532	206.0
MP4	0.741	4441	108.0	1.113	2192	221.2
CCSD	0.743	4409	108.4	1.097	2424	207.7
CCSD(T)				1.104	2346	216.5
B3LYP	0.743	4419	110.3	1.092	2449	229.6
Expt. [28]	0.741	4403	109.5	1.098	2359	228.4

[16]See Dunning and Peterson [29] and references therein.

Lastly, it is again the case that the B3LYP method yields relatively accurate results for this molecule and appears to be comparable in quality to MP2.

The results shown in Table 11-10 explore the choice of basis set with the CCSD and CCSD(T) methods for the H_2 and N_2 molecules, respectively. A large dependence on basis set is observed in each case. The use of a minimal basis set, STO-3G, leads to large errors since it provides very few virtual orbitals for electron correlation. Just a double-ζ basis set, either 6-31G** or cc-pVDZ, is observed to be a great improvement. The systematic convergence of the correlation consistent basis sets is readily observed in these results. One should note that increasing the size of the basis set from cc-pVTZ to cc-pV5Z in N_2 results in an increase in D_e by nearly 9 kcal/mole. This implies that the highly accurate result for D_e shown in Table 11-9 for the MP2 level of theory with the cc-pVTZ basis set was clearly fortuitous. From these results it should be obvious that errors due to basis set incompleteness can often rival those due to inadequate electron correlation.

As shown above, the hybrid DFT method B3LYP can be competitive in accuracy to more computationally expensive methods, such as CCSD(T). In fact, recent benchmark calculations [30] have shown that for the calculation of thermochemical quantities like enthalpies of formation, B3LYP exhibits average errors of only 1–5 kcal/mol. While these are still more than a factor of two larger than the accuracy obtainable with coupled cluster methods, the much lower computational cost of B3LYP makes it a very attractive alternative. The accuracy of equilibrium bond lengths and harmonic vibrational frequencies calculated by B3LYP have also been shown to be very satisfactory. The accurate calculation of some molecular properties, however, is still a great challenge to hybrid DFT methods. In particular, reaction activation energies are often too small and van der Waals interactions can be qualitatively incorrect. Correcting these deficiencies is the goal of many second generation hybrid DFT functionals.[17]

We have seen that inclusion of electron correlation often improves the $\bar{E}$ versus R curve because it allows for variable ionic-covalent character in the wavefunction.

TABLE 11-10 ► Dependence on Basis Set Choice for the CCSD and CCSD(T) Properties of H_2 and N_2, respectively

	H_2/CCSD			N_2/CCSD(T)		
	r_e(Å)	ω_e(cm^{-1})	D_e (kcal/mole)	r_e(Å)	ω_e(cm^{-1})	D_e (kcal/mole)
STO-3G	0.735	5002	128.1	1.190	2145	147.8
6-31G**	0.738	4504	105.9	1.120	2342	201.6
cc-pVDZ	0.761	4383	103.6	1.119	2339	200.6
cc-pVTZ	0.743	4409	108.4	1.104	2346	216.5
cc-pVQZ	0.742	4403	109.1	1.100	2356	222.9
cc-pV5Z	0.742	4405	109.3	1.099	2360	225.1
Expt. [28]	0.741	4403	109.5	1.098	2359	228.4

[17] See, for instance, Zhao et al. [31] and Xu et al. [32].

However, there are some diatomic molecules that maintain a high degree of ionic character even when the nuclei are quite widely separated. NaCl is an example. For such systems, the Hartree–Fock energy curve is quite nearly parallel to the exact energy curve throughout the minimum energy region (i.e., the correlation energy is almost constant) and the theoretical values of spectroscopic constants agree quite well with experimental values. (*In vacuo*, an electron ultimately transfers from Cl^- to Na^+, and the experimental curve leads to neutral dissociation products, whereas the HF curve does not. This theoretical error affects the curve only at large R, however, and so has little effect on spectroscopic constants.) Schaefer [13] has reviewed this situation.

Of course, *ab initio* calculations have been performed on molecular systems much larger than the molecules referred to above. However, as one moves to molecules having four or more nuclei, one encounters a new difficulty: Integrals now appear that have the form

$$\langle ab \mid cd \rangle \equiv \langle \chi_a(1)\chi_b(2)\,|1/r_{12}|\,\chi_c(1)\chi_d(2)\rangle \tag{11-52}$$

where χ_a is a basis function located on nucleus a, etc. Such integrals have basis functions on four different nuclei and are referred to as four-center integrals. If the basis functions χ are STOs, such integrals are relatively slow to evaluate on a computer. If they are gaussian functions, the computation is much faster, and this is the main reason for using gaussian basis functions. But the number of such integrals becomes enormous for a reasonable basis set and a medium sized molecule. In fact, the number of such integrals grows as the fourth power of the number of basis functions. Thus, replacing each STO by, say, three gaussian functions, will lead to 3^4 times as many integrals to evaluate. Even though such integrals can be evaluated very rapidly, we eventually come to molecules of such a size that the sheer number of integrals makes for a substantial computing effort. The efficient calculation of molecular integrals continues to be an active research area, however, and new techniques have now diminished the importance of this bottleneck with reasonably sized gaussian basis sets on systems up to hundreds of atoms.

Modern quantum chemical programs have made high-quality calculations on reasonably large molecules tractable, but one is always balancing the level of accuracy against the computer time needed to achieve it. While a Hartree–Fock calculation on benzene with a cc-pVTZ basis set (264 contracted gaussian functions) might require just 4 minutes to complete on a given computer, inclusion of electron correlation at the MP2, CCSD, and CCSD(T) levels would require an additional 0.1, 4.3, and 11 times 4 minutes, respectively.

Numerous calculations have been reported for barriers to internal rotation in various molecules. The theoretical barriers agree best with experiment for molecules having threefold symmetry in the rotor. Self-consistent-field values are compared with experimental barrier values in Table 11-11. In every case, the theoretical energy curve predicts the correct stable conformation and even does reasonably well at predicting barrier height. The disagreement between different computed values of the barrier for the same molecule reflects differences in basis sets and, sometimes, differences in choices for bond length and angle made by different workers. The evidence to date suggests that *ab initio* calculations approaching the HF limit will ordinarily be within 20% of the experimental barrier. Even this level of accuracy is useful because experimental measurements of barriers in transient molecules or for excited molecules are often very rough, ambiguous, or nonexistent. Given the favorable cost and relative accuracy of

TABLE 11-11 ▶ Internal Rotation Barriers from Experiment and as Calculated by the LCAO-MO-SCF Method

	Barrier (kcal/mole)		
Molecule	SCF	Experiment	Reference
CH_3–CH_3	2.58	2.88	[33]
	2.88	—	[34]
CH_3–NH_2	1.12	1.98	[35]
	2.02	—	[34]
CH_3–OH	1.59	1.07	[34]
CH_3–CH_2F	2.59	3.33	[33]
CH_3–N=O	1.05	1.10	[36]
CH_3–CH=CH_2	1.25	1.99	[37]
cis-CH_3–CH=CFH	1.07	1.06	[37]
trans-CH_3–CH=CFH	1.34	2.20	[37]
CH_3–CH=O	1.09	1.16	[38]

DFT approaches compared to HF, even higher quality results might be expected with the use of methods such as B3LYP; hence, DFT is often now the method of choice for calculations on medium to large organic molecules.

A large number of *ab initio* calculations have been made on clusters of molecules. Many of these have sought to delineate the distance and angle dependence of the hydrogen bond strength between molecules like water or hydrogen fluoride. Xantheas et al. [39] have reported large basis set MP2 calculations on small water clusters, $(H_2O)_n$, where n ranged from 2–6. These calculations predict that there are four distinct isomers of the water hexamer ($n = 6$) whose relative energies lie within ~1 kcal/mole of each other. These kinds of results are of great usefulness in defining new effective interaction potentials involving water that can be used in large-scale molecular simulations of solvation phenomena. Re et al. [40] have calculated the structures and relative energies of sulfuric acid solvated by 1–5 water molecules using the B3LYP method to provide a fundamental understanding of acid ionization. In addition to investigating the interaction of water with both the cis and trans conformers of H_2SO_4, they found that just five water molecules were sufficient to make dissociation into HSO_4^- and H_3O^+ energetically favorable. The field of materials science is also benefitting from *ab initio* calculations, and studies of metal clusters and their absorbates are currently areas of high interest.

A great deal of attention has been given to the calculation by *ab initio* methods of energy surfaces for chemical reactions. For many years, such efforts were limited to reactions, such as $D + H_2 \rightarrow HD + H$, which involve only a small number of electrons and nuclei. Much more complicated systems are now being explored.

In setting out to perform such a calculation, one likes to have some idea of whether the correlation energy of the system will change significantly with nuclear configuration. If it does not, then a Hartree–Fock or MCSCF calculation will parallel the true energy surface. If the correlation energy does change, it is necessary to include some treatment of electron correlation in the calculation.

As a rough rule of thumb, one expects the correlation energy to change least when the reactants, the intermediate or transition state complex, and the products are all closed-shell systems, hence all approximately equally well described by a single-determinantal wavefunction. Some calculations on S_{N2} and radical reactions are summarized in Table 11-12. It can be seen that the S_{N2} reactions, which do involve closed-shell systems in the three stages mentioned above, are fairly insensitive to the inclusion of CI, whereas the radical reactions undergo extensive change of correlation energy.

The determination of the potential energy surface for the unimolecular rearrangement $HOCl \rightleftharpoons HClO$ by Peterson et al. [43] provides an example of a very accurate and exhaustive calculation on a fairly small molecule. Because there are only three nuclei, there are only three structural variables to explore, so the number of calculations needed to map out the surface is not too large. (Note that, with three geometric variables, the energy "surface" is really a four-dimensional hypersurface.) These authors were also interested in the reactions occurring on this surface, i.e., $Cl + OH \rightarrow HCl + O$ and $Cl + OH \rightarrow ClO + H$, which required a global representation of the surface that was constructed from over 1500 individual energies. Since the full energy surface involves bond breaking processes, MCSCF and MRCISD methods were utilized. Accurate relative energetics between HOCl, HClO, and the various dissociation asymptotes were obtained by carrying out calculations with a series of three correlation consistent basis sets at each geometry. This produced an approximate complete basis set (CBS) MRCISD energy surface. At the MRCISD CBS limit, HOCl was found to be more stable than HClO by 53.7 kcal/mole and the barrier for $HOCl \rightarrow HClO$ was predicted to be 73.5 kcal/mole above the HOCl minimum.

After determining an analytical representation of this surface from the individual energies, these authors carried out calculations of the full anharmonic vibrational spectrum of HOCl and HClO by solving the Schrödinger equation for nuclear motion. The HClO molecule has not yet been experimentally observed, but these calculations predict that the lowest three vibrational levels of this species lie below its dissociation threshold, so it should be detectable.

TABLE 11-12 ► Reaction Barrier Energies for Reactions as Calculated by *ab Initio* Methods

Reactant	Transition	Product	Reaction type	Reaction barrier (kcal/mole) SCF	CI (no.config.)	Reference
$H^- + CH_4$	$(CH_5)^-$	$CH_4 + H^-$	S_N2	59.3	55.2(6271)	[41]
$F^- + CH_3F$	$(FCH_3F)^-$	$CH_3F + F^-$	S_N2	5.9	5.9(26910)	[41]
$H^\bullet + CH_4$	$CH_5^\bullet$	$CH_3^\bullet + H_2$	Radical abstraction (axial)	35.2	18 (692)	[42]
$H^\bullet + CH_4$	$CH_5^\bullet$	$CH_4 + H^\bullet$	Radical exchange (inversion)	63.7	41.7(692)	[42]

The decisions regarding basis set and level of correlation can be daunting and in the past this sometimes discouraged nonspecialists from taking advantage of *ab initio* methods. However, there are now a wide range of programs that are available, which have made *ab initio* calculations amenable to theoreticians and experimentalists alike. The best known of these is undoubtedly the GAUSSIAN series of programs originally developed by the group of J.A. Pople. In this and other programs, one can conveniently choose from a large variety of available basis sets and methods to carry out energy evaluations or geometry optimizations and harmonic frequency calculations. These programs have brought about a revolution in the way that chemical research is done.

For small molecules ($\sim$1–5 nonhydrogen atoms) *ab initio* methods are sometimes more precise and reliable than experiment, especially for unstable systems. The saga of the energy difference between ground and excited CH_2 is one of the best known of these experimental–theoretical confrontations.[18]

In summary, *ab initio* calculations provide useful data on bond lengths and angles, molecular conformation and internal rotation barriers, for ground and excited states of molecules. They are also very useful for calculating accurate thermochemistry, ionization energies, oscillator strengths, dipole moments (as well as other one-electron properties) and excitation energies. If one has access to large blocks of computer time, *ab initio* calculations can reveal the nature of energy surfaces pertaining to chemical reactions or molecular associations, as in fluids. The accuracy of the calculation and the magnitude of the system are limited ultimately by computer speed and capacity.

11-17 Approximate SCF-MO Methods

At the beginning of this chapter it was stated that *ab initio* calculations require exact calculation of all integrals contributing to the elements of the Fock matrix, but we have seen that, as we encounter systems with more and more electrons and nuclei, the number of three- and four-center two-electron integrals becomes enormous, driving the cost of the calculation out of the reach of most researchers. This has led to efforts to find sensible and systematic simplifications to the LCAO-MO-SCF method—simplifications that remain within the general theoretical SCF framework but shorten computation of the Fock matrix.

Since many of the multicenter two-electron integrals in a typical molecule have very small values, the obvious solution to the difficulty is to ignore such integrals. But we wish to ignore them without having to calculate them to see which ones are small since, after all, the reason for ignoring them is to avoid having to calculate them. Furthermore, we want the selection process to be linked in a simple way to considerations of basis set. That is, when we neglect certain integrals, we are in effect omitting certain interactions between basis set functions, which is equivalent to omitting some of our basis functions part of the time. It is essential that we know exactly what is involved here, or we may obtain strange results such as, for example, different energies for the same molecule when oriented in different ways with respect to Cartesian coordinates.

A number of variants of a systematic approach meeting the above criteria have been developed by Pople and co-workers, and these are now widely used. The approximations

[18] See Goddard [44] and Schaefer [45].

are based on the idea of *neglect of differential overlap* between atomic orbitals in molecules.

Differential overlap dS between two AOs, χ_a and χ_b, is the product of these functions in the differential volume element dv:

$$dS = \chi_a(1)\chi_b(1)\,dv \qquad (11\text{-}53)$$

The only way for the differential overlap to be zero in dv is for χ_a or χ_b, or both, to be identically zero in dv. Zero differential overlap (ZDO) between χ_a and χ_b in *all* volume elements requires that χ_a and χ_b can never be finite in the same region, that is, the functions do not "touch." It is easy to see that, if there is ZDO between χ_a and χ_b (understood to apply in all dv), then the familiar overlap integral S must vanish too. The converse is not true, however. S is zero for any two *orthogonal* functions even if they touch. An example is provided by an s and a p function on the same center.

It is a much stronger statement to say that χ_a and χ_b have ZDO than it is to say they are orthogonal. Indeed, it is easy to think of examples of orthogonal AOs but impossible to think of any pair of AOs separated by a finite or zero distance and having ZDO. Because AOs decay exponentially, there is always some interpenetration.

The attractive feature of the ZDO approximation is that it causes all three- and four-center integrals to vanish. Thus, in a basis set of AOs χ having ZDO, the integral $\langle \chi_a(1)\chi_b(2)\,|1/r_{12}|\,\chi_c(1)\chi_d(2)\rangle$ will vanish unless $a \equiv c$ and $b \equiv d$. This arises from the fact that, if $a \neq c$, $\chi_a^*(1)\chi_c(1)$ is identically zero, and this forces the integrand to vanish everywhere, regardless of the value of $(1/r_{12})\,\chi_b^*(2)\chi_d(2)$.

It is not within the scope of this book to give a detailed description or critique of the numerous computational methods based on ZDO assumptions. An excellent monograph [46] on this subject including program listings is available. Some of the acronyms for these methods are listed in Table 11-13. In general, these methods have been popular because they are relatively cheap to use and because they predict certain properties (bond length, bond angle, energy surfaces, electron spin resonance hyperfine splittings, molecular charge distributions, dipole moments, heats of formation) reasonably well. However, they generally do make use of some parameters evaluated from experimental data, and some methods are biased toward good predictions of some properties, while other methods are better for other properties. For a given type of problem, one must exercise judgment in choosing a method.

As an example of the sort of chemical system that becomes accessible to study using such methods, we cite the valence-electron CNDO/2 calculations of Maggiora [56] on free base, magnesium, and aquomagnesium porphines. Such calculations enable us to examine the geometry of the complex (i.e., is the metal ion in or out of the molecular plane, and how is the water molecule oriented?), the effects of the metal ion on ionization energies, spectra, and orbital energy level spacings, and the detailed nature of charge distribution in the system.

Use of a combination of methods is often convenient. Novoa and Whangbo [57] studied theoretically the relative stabilities of di- and triamides in various hydrogen-bonded and nonhydrogen-bonded conformations, in both the absence and presence of solvent (CH_2Cl_2) molecules. There are many structural parameters to optimize in each of the conformations, and so high-level *ab initio* calculations for energy minimization of each class of structure would be prohibitively expensive. Instead, AM1 was used to determine the optimum geometry for each configuration, and then *ab initio* calculations

TABLE 11-13 ► Acronyms for Common Approximate SCF Methods

Acronym	Description
CNDO/1	Complete neglect of differential overlap. Parametrization Scheme no. 1 (Pople and Segal [47]).
CNDO/2	Parametrization scheme no. 2. Considered superior to CNDO/1 (Pople and Segal [48]).
CNDO/BW	Similar to above with parameters selected to give improved molecular structures and force constants. (See Pulfer and Whitehead [49] and references therein.)
INDO	Intermediate neglect of differential overlap. Differs from CNDO in that ZDO is not assumed between AOs on the same center in evaluating one-center integrals. This method is superior to CNDO methods for properties, such as hyperfine splitting, or singlet-triplet splittings, which are sensitive to electron exchange (Pople et al. [50]).
MINDO/3	Modified INDO, parameter scheme no. 3. Designed to give accurate heats of formation (Bingham et al. [51] and also Dewar [52]).
NDDO	Neglect of diatomic differential overlap. Assumes ZDO only between AOs on different atoms (Pople et al. [53]).
MNDO	Modified neglect of diatomic overlap. A semiempirically parametrized version of NDDO. Yields accurate heats of formation and many other molecular properties, but fails to successfully account for hydrogen bonding (Dewar and Thiel [54]).
AM1	"Austin Model 1." A more recent parametrization of NDDO that overcomes the weakness of MNDO in that it successfully treats hydrogen bonding. (Dewar et al. [55].)

(e.g., 6-31G** with MP2) were done for a few near-optimum geometries for each conformation to check the AM1 results.

Additional helpful information on standard programs available at *ab initio* and semiempirical levels—where to get them, how to use them, what they have been used for—is available in the very well-written reference handbook by Clark [58].

11-17.A Problems

11-1. Use the data in Table 11-3 to calculate the theoretical transition energies for Ne^+ when 1s and 2s electrons are excited into the 2p level. The experimental values are $2p \leftarrow 2s$, 0.989 a.u.; $2p \leftarrow 1s$, 31.19 a.u.

11-2. Use the data in Table 11-1 to estimate separately the errors in ionization energies for the three states due to

a) omission of electron correlation.
b) failure to allow electronic relaxation.

11-3. In Section 11-11, it is argued that neglect of electron correlation and electronic relaxation in setting $I_k^0 = -\epsilon_k$ causes errors of opposite sign that partly cancel.

Would this also occur when Koopmans' theorem is used to predict electron affinities? Why?

11-4. Demonstrate that, if

$$D_1 = \begin{vmatrix} a & c \\ b & d \end{vmatrix} \text{ and } D_2 = \begin{vmatrix} a & e \\ b & f \end{vmatrix} \text{ then } D_1 + \lambda D_2 = \begin{vmatrix} a & c+\lambda e \\ b & d+\lambda f \end{vmatrix}$$

11-5. A singly excited configuration ψ_1 differs by one orbital from the ground state ψ_0 and also by one orbital from certain doubly excited configurations ψ_2. Brillouin's theorem gives $\langle\psi_0|\hat{H}|\psi_1\rangle = 0$, but not $\langle\psi_1|\hat{H}|\psi_2\rangle = 0$. Where does the attempted proof to show that $\langle\psi_1|\hat{H}|\psi_2\rangle = 0$ break down?

11-6. Show that, if $\psi = c_0\psi_0 + c_1\psi_1 + c_2\psi_2 + \cdots + c_n\psi_n$, and if ψ is to be an eigenfunction of $\hat{A}$ with eigenvalue a_1, then it is necessary that all the $\psi_i (i = 0, \ldots, n)$ also be eigenfunctions of $\hat{A}$ with eigenvalues a_1.

11-7. How many distinct four-center coulomb and exchange integrals result when one has four nuclei, each being the site of five basis functions? Make no assumptions about symmetry or basis function equivalence or electron spin.

11-8. For a homonuclear diatomic molecule, which of the following singly excited configurations would be prevented for reasons of symmetry from contributing to a CI wavefunction for which the main "starting configuration" is $1\sigma_g^2 1\sigma_u^2 2\sigma_g^2 1\pi_u^4$?

a) $1\sigma_g^2 1\sigma_u^2 2\sigma_g^2 1\pi_u^3 1\pi_g$ (i.e., $1\pi_u \rightarrow 1\pi_g$)
b) $2\sigma_g \rightarrow 3\sigma_g$
c) $2\sigma_g \rightarrow 1\pi_g$
d) $1\sigma_g \rightarrow 3\sigma_g$

11-9. Write down the hamiltonian operator for electrons in the water molecule. Use summation signs with explicit index ranges. Use atomic units.

11-10. An SCF calculation on ground state H_2 at $R = 1.40$ a.u. using a minimal basis set gives a σ_g and a σ_u MO having energies

$$\epsilon_{\sigma_g} = -0.619 \text{ a.u.} \quad \epsilon_{\sigma_u} = +0.401 \text{ a.u.}$$

The nonvanishing two-electron integrals over these MOs are

$$\iint \sigma_g(1)\sigma_g(2)(1/r_{12})\sigma_g(1)\sigma_g(2)dv(1)dv(2) = 0.566 \text{ a.u.}$$

$$\iint \sigma_g(1)\sigma_u(2)(1/r_{12})\sigma_g(1)\sigma_u(2)dv(1)dv(2) = 0.558 \text{ a.u.}$$

$$\iint \sigma_g(1)\sigma_u(2)(1/r_{12})\sigma_g(2)\sigma_u(1)dv(1)dv(2) = 0.140 \text{ a.u.}$$

$$\iint \sigma_u(1)\sigma_u(2)(1/r_{12})\sigma_u(1)\sigma_u(2)dv(1)dv(2) = 0.582 \text{ a.u.}$$

a) Write down the Slater determinant for the ground state of H_2.
b) Calculate the SCF electronic energy for H_2 at $R = 1.40$ a.u.

c) Calculate the total (electronic plus nuclear repulsion) energy for H_2.
d) What is the bond energy for H_2 predicted by this calculation, assuming that the minimum total energy occurs at $R = 1.40$ a.u.?
e) Estimate the (vertical) ionization energy for H_2.
f) What is the value of the kinetic-plus-nuclear-attraction energy for one electron in ground-state H_2 according to this calculation?

References

[1] T. Koopmans, *Physica* **1**, 104 (1933).

[2] D. W. Smith and O. W. Day, *J. Chem. Phys.* **62**, 113 (1975).

[3] A. W. Potts and W. C. Price, *Proc. Roy. Soc.* (London) **A326**, 181 (1972).

[4] T. H. Dunning, Jr., R. M. Pitzer, and S. Aung, *J. Chem. Phys.* **57**, 5044 (1972).

[5] J. S. Binkley and J. A. Pople, *Int. J. Quantum Chem.* **9**, 229 (1975).

[6] C. Møller and M. S. Plesset, *Phys. Rev.* **46**, 618 (1934).

[7] P. Hohenberg and W. Kohn, *Phys. Rev.* **136**, B864 (1964).

[8] R. G. Parr and W. Yang, *Density Functional Theory of Atoms and Molecules.* Oxford University Press, Oxford, UK, 1989.

[9] W. Kohn and L. J. Sham, *Phys. Rev.* **140**, A1133 (1965).

[10] A. D. Becke, *Phys. Rev. A.* **38**, 3098 (1988).

[11] C. Lee, W. Yang and R. G. Parr, *Phys. Rev. B* **37**, 785 (1988); B. Miehlich, A. Savin, H. Stoll, and H. Preuss, *Chem. Phys. Lett.* **157**, 200 (1989).

[12] A. D. Becke, *J. Chem. Phys.* **98**, 5648 (1993); P. J. Stephens, F. J. Devlin, C. F. Chabalowski, and M. J. Frisch, *J. Phys. Chem.* **98**, 11623 (1994).

[13] H. F. Schaefer, III, *The Electronic Structure of Atoms and Molecules: A Survey of Rigorous Quantum-Mechanical Results.* Addison-Wesley, Reading, Massachusetts, 1972.

[14] W. J. Hehre, L. Radom, P. v.R. Schleyer and J. A. Pople, *Ab initio Molecular Orbital Theory.* Wiley, New York, 1986.

[15] K. Raghavachari, *Annu. Rev. Phys. Chem.* **42**, 615 (1991).

[16] C. Roetti and E. Clementi, *J. Chem. Phys.* **60**, 4725 (1974).

[17] A. Veillard and E. Clementi, *J. Chem. Phys.* **49**, 2415 (1968).

[18] P. S. Bagus, *Phys. Rev.* **139**, A619 (1965).

[19] A. W. Weiss, *Phys. Rev.* **162**, 71 (1967).

[20] W. C. Ermler and C. W. Kern, *J. Chem. Phys.* **61**, 3860 (1974).

[21] P. E. Cade and W. M. Huo, *J. Chem. Phys.* **47**, 614 (1967).

[22] W. J. Stevens, G. Das, and A. C. Wahl, *J. Chem. Phys.* **61**, 3686 (1974).

[23] S. Green, *J. Chem. Phys.* **54**, 827 (1971).

[24] S. Green, *J. Chem. Phys.* **57**, 4694 (1972).

[25] S. Green, *J. Chem. Phys.* **57**, 2830 (1972).

[26] S. Green, *J. Chem. Phys.* **56**, 739 (1972).

[27] W. J. Stevens, G. Das, A. C. Wahl, M. Krauss, and D. Neumann, *J. Chem. Phys.* **61**, 3686 (1974).

[28] K. P. Huber and G. Herzberg, *Molecular Spectra and Molecular Structure IV. Constants of Diatomic Molecules.* Van Nostrand, Princeton, 1979.

[29] T. H. Dunning, Jr. and K. A. Peterson, *J. Chem. Phys.* **108**, 4761 (1998).

[30] V. N. Staroverov and G. E. Scuseria, J. Tao, and J. P. Perdew, *J. Chem. Phys.* **119**, 12129 (2003); N. X. Wang and A. K. Wilson, *J. Chem. Phys.* **121**, 7632 (2004).

[31] Y. Zhao, B. J. Lynch, and D. G. Truhlar, *J. Phys. Chem. A* **108**, 2715 (2004).

[32] X. Xu, Q. Zhang, R. P. Muller, and W. A. Goddard III, *J. Chem. Phys.* **122**, 014105 (2005).

[33] L. C. Allen and H. Basch, *J. Amer. Chem. Soc.* **93**, 6373 (1971).

[34] L. Pedersen and K. Morokuma, *J. Chem. Phys.* **46**, 3741 (1967).

[35] W. H. Fink and L. C. Allen, *J. Chem. Phys.* **46**, 2261 (1967).

[36] P. A. Kollman and L. C. Allen, *Chem. Phys. Lett.* **5**, 75 (1970).

[37] E. Scarzafava and L. C. Allen, *J. Amer. Chem. Soc.* **93**, 311 (1971).

[38] R. B. Davidson and L. C. Allen, *J. Chem. Phys.* **54**, 2828 (1971).

[39] S. S. Xantheas, C. J. Burnham, and R. J. Harrison, *J. Chem. Phys.* **116**, 1493 (2002).

[40] S. Re, Y. Osamura, and K. Morokuma, *J. Phys. Chem. A* **103**, 3535 (1999).

[41] A. Dedieu, A. Veillard, and B. Roos, *Jerusalem Symp. Quantum Chem. Biochem.* **6**, 371 (1974).

[42] K. Morokuma and R. E. Davis, *J. Amer. Chem. Soc.* **94**, 1060 (1972).

[43] K. A. Peterson, S. Skokov, and J. M. Bowman, *J. Chem. Phys.* **111**, 7446 (1999).

[44] W. A. Goddard, III, *Science* **227**, 917 (1985).

[45] H. F. Schaefer, III, *Science* **231**, 1100 (1986).

[46] J. A. Pople and D. L. Beveridge, *Approximate Molecular Orbital Theory*. McGraw-Hill, New York, 1970.

[47] J. A. Pople and G. A. Segal, *J. Chem. Phys.* **43**, S136 (1965).

[48] J. A. Pople and G. A. Segal, *J. Chem. Phys.* **44**, 3289 (1966).

[49] J. D. Pulfer and M. A. Whitehead, *Can. J. Chem.* **51**, 2220 (1973).

[50] J. A. Pople, D. L. Beveridge, and P. A. Dobosh, *J. Chem. Phys.* **47**, 2026 (1967).

[51] R. C. Bingham, M. J. S. Dewar, and D. H. Lo, *J. Amer. Chem. Soc.* **97**, 1285 (1975).

[52] M. J. S. Dewar, *Science* **187**, 1037 (1975).

[53] J. A. Pople, D. P. Santry, and G. A. Segal, *J. Chem. Phys.* **43**, S129 (1965).

[54] M. J. S. Dewar and W. Thiel, *J. Amer. Chem. Soc.* **99**, 4899 (1977).

[55] M. J. S. Dewar, E. G. Zoebisch, E. F. Healy, and J. J. P. Stewart, *J. Amer. Chem. Soc.* **107**, 3902 (1985).

[56] G. M. Maggiora, *J. Amer. Chem. Soc.* **95**, 6555 (1973).

[57] J. J. Novoa and M.-H. Whangbo, *J. Amer. Chem. Soc.* **113**, 9017 (1991).

[58] T. Clark, *A Handbook of Computational Chemistry*. Wiley, New York, 1985.

Chapter 12

Time-Independent Rayleigh–Schrödinger Perturbation Theory

12-1 An Introductory Example

Imagine a city having one million resident wage earners. The city government plans to raise additional revenue by assessing each such resident a wage tax. This tax will not apply to wage earners residing in suburbs. The government estimates that a $10 assessment will bring in new revenues of $10 million. This estimate assumes that the city population before imposition of the tax will hold after the tax is imposed as well. But the tax will produce a slight change, or *perturbation*, in the economic climate of the city. It is true that the tax is small, and so few people are likely to move to the suburbs as a result of it. Therefore, it is probably fairly accurate to use the population of the city before the perturbation to calculate the change in revenue brought about by the perturbation. But, if the perturbation were large, say $1000 a head, we would expect a substantial migration of wage earners to the suburbs, and so the estimate produced by using the original unperturbed population would contain substantial error. Corrections should be made, therefore, to account for population changes produced by the perturbation.

This use of the *unperturbed* population to calculate the change in revenue is a crude example of a certain level of estimation (called "first order") in Rayleigh–Schrödinger perturbation theory.[1] We will now proceed to develop the theory more formally in the context of wavefunctions and energies. The above example has been presented to encourage the reader to anticipate that there is a lot of simple good sense in the results of perturbation theory even though the mathematical development is rather cumbersome and unintuitive.

12-2 Formal Development of the Theory for Nondegenerate States

Perturbation theory involves starting with a system with known hamiltonian, eigenvalues, and eigenfunctions, and calculating the changes in these eigenvalues and eigenfunctions that result from a small change, or *perturbation*, in the hamiltonian for the

[1] Other perturbation methods exist, but the Rayleigh–Schrödinger (R-S) theory is the oldest and the most widely used in quantum chemistry.

system. We restrict the discussion to stationary states of systems having hamiltonians that are not time dependent. Let the known, unperturbed system have H_0 as hamiltonian, and let the eigenfunctions ψ_i be orthonormal. Then

$$H_0\psi_i = E_i\psi_i \tag{12-1}$$

$$\int \psi_i^*\psi_j \, d\tau = \delta_{ij} \tag{12-2}$$

and the functions ψ form a complete set as discussed in Chapter 6. We are interested in the system with hamiltonian

$$H = H_0 + \lambda H' \tag{12-3}$$

where $\lambda H'$ is the perturbation.[2] (The parameter λ is a scalar quantity that will be convenient in the mathematical development of the theory. When the derivation is complete, we will set λ equal to unity so that it no longer appears explicitly in any formula, and then H' must account entirely for the perturbation.) The eigenvalues and eigenfunctions of the perturbed hamiltonian H are unknown. Let us symbolize them as W and ϕ, respectively. Then

$$H\phi_i = W_i\phi_i \tag{12-4}$$

It is clear that, when $\lambda = 0$, then $H = H_0$, $\phi_i = \psi_i$ and $W_i = E_i$. As λ is increased from zero, W_i and ϕ_i change in (we expect) a continuous way. In other words, W_i and ϕ_i are continuous functions of the variable parameter λ, and they are known at the particular value $\lambda = 0$. Therefore, we can expand them[3] as series in powers of λ about the point $\lambda = 0$. Thus, for a given state i,[4]

$$W_i = \lambda^0 W_i^{(0)} + \lambda^1 W_i^{(1)} + \lambda^2 W_i^{(2)} + \lambda^3 W_i^{(3)} + \cdots \tag{12-5}$$

Here we must remember that λ is a variable and $W_i^{(0)}$, $W_i^{(1)}$, etc. are constants. The superscript in parentheses is simply a label to tell us for which power of λ this constant is the coefficient. Since we know that $W_i = E_i$ when $\lambda = 0$, we see at once that $W_i^{(0)}$ in Eq. (12-5) is equal to E_i. Our problem is to evaluate $W_i^{(1)}$, $W_i^{(2)}$, etc. It is traditional to call $W_i^{(0)}$ (or E_i) the *unperturbed energy* or, sometimes, the *energy to zeroth order*. $\lambda W_i^{(1)}$ (which is just $W_i^{(1)}$ after λ is set equal to unity) is the *first-order correction to the energy*, and $W_i^{(0)} + \lambda W_i^{(1)}$ is the *energy to first order*. $\lambda^2 W_i^{(2)}$, $\lambda^3 W_i^{(3)}$, ..., etc. are the *second-*, *third-*, etc., *order corrections to the energy*. Normally, for expansion in a series to be useful, it is necessary for the series to converge at a reasonable rate. In most simple applications of perturbation theory, only a few orders of correction are made. Thus, one very commonly reads of energies calculated "to first order," or "to second order." Calculations to much higher orders are also made, but these are not as common.

---

[2] Some treatments expand H as $H_0 + \lambda H' + \lambda^2 H'' + \cdots$ and ultimately achieve working formulas that appear different from those we will achieve. In fact, they are equivalent. For an example of this alternative formulation, see Pauling and Wilson [1, Chapter 6].

[3] In effect, we assume ϕ_i and W to be analytic functions of λ in the range $0 \le \lambda \le 1$.

[4] It is important to recognize that henceforth the subscript i will refer to the state that we are studying as a function of λ.

In precisely the same manner, we can expand the unknown eigenfunction ϕ_i as a power series in λ:

$$\phi_i = \phi_i^{(0)} + \lambda\phi_i^{(1)} + \lambda^2\phi_i^{(2)} + \lambda^3\phi_i^{(3)} + \cdots \tag{12-6}$$

Since ϕ_i is a *function* of particle coordinates, $\phi_i^{(0)}$, $\phi_i^{(1)}$, etc. are also functions, but they are invariant to changes in λ. Again, it is clear that $\phi_i^{(0)} = \psi_i$. $\phi_i^{(0)}$ (or ψ_i) is the unperturbed or *zeroth-order wavefunction*, $\lambda\phi_i^{(1)}$ is the *first-order correction* to the wavefunction, etc.

Now we substitute Eqs. (12-3), (12-5), and (12-6) into (12-4) and obtain

$$\begin{aligned}(H_0 + \lambda H')&\left(\phi_i^{(0)} + \lambda\phi_i^{(1)} + \lambda^2\phi_i^{(2)} + \cdots\right)\\ &= \left(W_i^{(0)} + \lambda W_i^{(1)} + \lambda^2 W_i^{(2)} + \cdots\right)\left(\phi_i^{(0)} + \lambda\phi_i^{(1)} + \lambda^2\phi_i^{(2)} + \cdots\right)\end{aligned} \tag{12-7}$$

The variable in Eq. (12-7) is λ, and each power of λ is linearly independent of all other powers of λ. As indicated in Section 3-4D, this means that Eq. (12-7) can be satisfied for all values of λ only if it is satisfied for *each power* of λ separately. Collecting terms having the zeroth power of λ gives

$$H_0\phi_i^{(0)} = W_i^{(0)}\phi_i^{(0)} \tag{12-8}$$

However, we have already recognized that

$$\phi_i^{(0)} = \psi_i, \quad W_i^{(0)} = E_i \tag{12-9}$$

and Eq. (12-8) is simply a restatement of Eq. (12-1). Collecting terms from Eq. (12-7) containing λ to the first power we obtain

$$\lambda\left(H'\phi_i^{(0)} + H_0\phi_i^{(1)} - W_i^{(0)}\phi_i^{(1)} - W_i^{(1)}\phi_i^{(0)}\right) = 0 \tag{12-10}$$

This equality must hold for any value of λ, so the term in parentheses is zero. Hence, rearranging and making use of Eqs. (12-9), we have the first-order equation

$$\left(H' - W_i^{(1)}\right)\psi_i + (H_0 - E_i)\phi_i^{(1)} = 0 \tag{12-11}$$

Let us multiply this from the left by ψ_i^* and integrate:

$$\int \psi_i^* H'\psi_i\, d\tau - W_i^{(1)}\int \psi_i^*\psi_i\, d\tau + \int \psi_i^* H_0\phi_i^{(1)}\, d\tau - E_i\int \psi_i^*\phi_i^{(1)}\, d\tau = 0 \tag{12-12}$$

Using the hermitian property of H_0, it is easy to show that the third and fourth terms cancel, leaving

$$\boxed{W_i^{(1)} = \int \psi_i^* H'\psi_i\, d\tau} \tag{12-13a}$$

Thus we have arrived at an expression for the first-order correction to the energy in terms of known quantities. *It is the expectation value for the perturbation operator calculated using the wavefunction of the unperturbed system.* The analogy between this formula and the use of the population of the unperturbed city to calculate additional revenues from a new tax should be apparent. H' corresponds to the tax per wage earner, and $\int \psi_i^* \psi_i \, d\tau$ corresponds to the sum of wage earners in the city before the tax was imposed.

> **EXAMPLE 12-1** A one-dimensional box potential is perturbed so that it is raised by a constant amount, δ, in the left half of the box, and lowered by δ in the right half. What is the first-order change in energy for the lowest-energy state ψ_1? for ψ_2?

SOLUTION ▶ H' is antisymmetric for reflection through the box center, and ψ_1^2 is symmetric, so the integral giving $W_1^{(1)}$ equals zero by symmetry. The same argument applies to ψ_2. Another way to argue is to recognize that ψ_n^2 is symmetric in the box (for all n), which means that half of the distribution responds to the energy increase and half to the equal energy decrease, so the net energy change to first order is zero for all states. ◀

Equation (12-13a) can be written in bra-ket notation (see Appendix 9):

$$\boxed{W_i^{(1)} = \langle \psi_i | H' | \psi_i \rangle} \tag{12-13b}$$

or in the notation wherein an integral is indicated by affixing subscripts to the operator (i.e., as a matrix element)

$$\boxed{W_i^{(1)} = H'_{ii}} \tag{12-13c}$$

In most discussions of perturbation theory one of these alternative notations is used. We will continue our formal development from this point using bra-ket notation for integrals.

To find an expression for $\phi_i^{(1)}$, the first-order correction to the wavefunction for the ith state, we first recognize that we can expand $\phi_i^{(1)}$ in terms of the complete set of eigenfunctions ψ:

$$\phi_i^{(1)} = \sum_j c_{ji}^{(1)} \psi_j \tag{12-14}$$

The summation symbol suggests that ψ is a discrete set of functions. This need not be true. Contributions from functions whose eigenvalues are in a continuum would require integration rather than summation. However, we will use the sum symbol since most actual applications of perturbation theory in quantum chemistry invoke only discrete functions.

We now insert Eq. (12-14) into (12-11) to obtain

$$\left(H' - W_i^{(1)}\right) \psi_i + (H_0 - E_i) \sum_j c_{ji}^{(1)} \psi_j = 0 \tag{12-15}$$

Multiplying from the left by ψ_k^* and integrating yields

$$\langle\psi_k|H'|\psi_i\rangle - W_i^{(1)}\langle\psi_k|\psi_i\rangle + \sum_j c_{ji}^{(1)}\left(\langle\psi_k|H_0|\psi_j\rangle - E_i\langle\psi_k|\psi_j\rangle\right) = 0 \tag{12-16}$$

If $k = i$, this reduces to Eq. (12-13), as was shown above. If $k \neq i$, the terms in the sum all vanish except when $j = k$. Thus,

$$\langle\psi_k|H'|\psi_i\rangle + c_{ki}^{(1)}\left(\langle\psi_k|H_0|\psi_k\rangle - E_i\langle\psi_k|\psi_k\rangle\right) = 0, \quad (k \neq i) \tag{12-17}$$

or

$$\boxed{c_{ki}^{(1)} = \frac{\langle\psi_k|H'|\psi_i\rangle}{E_i - E_k}, \quad k \neq i} \tag{12-18}$$

Inserting this into Eq. (12-14) gives an expression for $\phi_i^{(1)}$:

$$\boxed{\phi_i^{(1)} = \sum_{j \neq i} \frac{\langle\psi_j|H'|\psi_i\rangle}{E_i - E_j}\psi_j} \tag{12-19}$$

This formula prescribes the way the first-order correction to the wavefunction is to be built up from eigenfunctions of the unperturbed system. We discuss this formula in detail later when considering an example.

Note that, if the perturbed state of interest (the ith) is degenerate with another state (the lth), Eqs. (12-18) and (12-19) blow up for k or j equal to l unless $\langle\psi_l|H'|\psi_i\rangle$ vanishes. Therefore, we restrict the theoretical discussion of this section to states of interest that are nondegenerate and discrete. (States of the system other than the ith may be degenerate or continuum states, however.)

If we extract the terms containing λ^2 from Eq. (12-7) and proceed, in the same way as above, to expand $\phi_i^{(2)}$ as a linear combination of unperturbed eigenfunctions,

$$\phi_i^{(2)} = \sum_j c_{ji}^{(2)}\psi_j \tag{12-20}$$

we arrive, after some manipulation, at the following formula for $W_i^{(2)}$:

$$\boxed{W_i^{(2)} = \sum_{j(\neq i)} \frac{\langle\psi_i|H'|\psi_j\rangle\langle\psi_j|H'|\psi_i\rangle}{E_i - E_j}} \tag{12-21}$$

Comparing this with Eq. (12-18) allows us to write

$$W_i^{(2)} = \sum_{j(\neq i)} \left|c_{ji}^{(1)}\right|^2 (E_i - E_j) \tag{12-22}$$

or comparing Eqs. (12-19) and (12-21) gives

$$W_i^{(2)} = \langle\psi_i|H'|\phi_i^{(1)}\rangle \tag{12-23}$$

The formula that emerges for $c_{ji}^{(2)}$ of Eq. (12-20) is

$$c_{ji}^{(2)} = \sum_{k(\neq i)} \frac{\langle \psi_j | H' | \psi_k \rangle \langle \psi_k | H' | \psi_i \rangle}{(E_i - E_k)(E_i - E_j)} - \frac{\langle \psi_i | H' | \psi_i \rangle \langle \psi_j | H' | \psi_i \rangle}{(E_i - E_j)^2}, \quad i \neq j \tag{12-24}$$

The boxed equations are "working equations" since they enable us to calculate the correction terms from the known eigenvalues and eigenfunctions of the unperturbed system. Equation (12-23) is interesting because it indicates that the second-order correction to the energy is calculable if we know the zeroth and first-order functions.[5] Comparing Eqs. (12-13) and (12-23) shows that, whereas the first-order correction for the energy is the average value for the perturbation operator with the unperturbed wavefunction, the second-order correction is an interaction element between two functions, and not an average value in the usual sense.

Higher-order correction terms may be found by proceeding in a similar way with λ^3, λ^4, etc. terms from Eq. (12-7). The equations become progressively more cumbersome and will be of no interest to us for applications to be considered in this book.

Having made use of the parameter λ to keep terms properly sorted, we can now dispense with it by setting it equal to unity. Then

$$H = H_0 + H' \tag{12-25}$$

$$W_i = E_i + W_i^{(1)} + W_i^{(2)} + \cdots \tag{12-26}$$

$$\phi_i = \psi_i + \phi_i^{(1)} + \phi_i^{(2)} + \cdots \tag{12-27}$$

Notice that ϕ_i is not normalized. The normalization coefficient needed will depend on the order to which ϕ_i has been calculated.[6]

12-3 A Uniform Electrostatic Perturbation of an Electron in a "Wire"

12-3.A Description of the System

Suppose that a small uniform electric field is applied to an electron somehow constrained to move on a line segment of length L. In the absence of this field, we assume the electron states to be described by the one-dimensional "box" wavefunctions discussed

[5] Löwdin [2] has shown that, if we know all the ϕ_i's up to $\phi_i^{(n)}$, we can calculate all the W_i's up to and including $W_i^{(2n+1)}$.

[6] One can guarantee normality up to second order in ϕ_i^2 by setting $c_{ii}^{(1)} = 0$, $c_{ii}^{(2)} = -\frac{1}{2}\Sigma_k \left|c_{ki}^{(1)}\right|^2$ (see Schiff [3]).

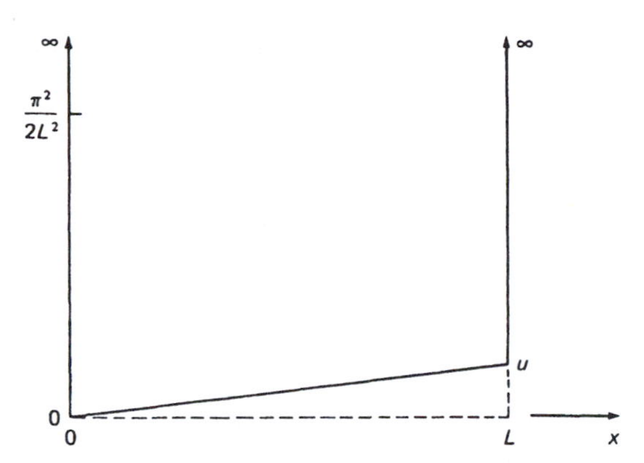

Figure 12-1 ▶ Potential of an electron in a line segment of length L in the presence of a uniform electric field. The perturbation is "small" if the potential change u across L is small compared to $\pi^2/2L^2$, the energy in atomic units of the lowest unperturbed state.

in Chapter 2. The electric field will be treated as a perturbation. The potential energy of the electron is sketched in Fig. 12-1. A uniform field produces a constant gradient (in potential energy) along the line segment. For purposes of discussion, we let the perturbation rise from zero at $x=0$ to u at $x=L$, but we shall see later that there is a degree of arbitrariness here. The perturbation, then, is given by

$$H' = ux/L, \quad 0 \le x \le L \tag{12-28}$$

where u, x, and L are measured in atomic units.

12-3.B The Energy to First Order

We now ask, what is the effect, to first order, of H' on the energy of the lowest-energy state of the electron? As described in the preceding section, this is obtained by calculating the average value of H' using the unperturbed wavefunction:

$$W_1^{(1)} = \int_0^L \left[\sqrt{2/L}\sin(\pi x/L)\right](ux/L)\left[\sqrt{2/L}\sin(\pi x/L)\right]dx \tag{12-29}$$

We now describe three ways to evaluate this integral. One way is to integrate explicitly, the other two ways involve simple inspection.

1. *Explicit Integration.* Factoring constants from Eq. (12-29) gives

$$W_1^{(1)} = 2u/L^2 \int_0^L x\sin^2(\pi x/L)\,dx \tag{12-30}$$

To achieve a common variable, we multiply x and dx each by π/L and outside by L^2/π^2, thereby keeping the value unchanged:

$$W_1^{(1)} = 2u/\pi^2 \int_{x=0}^{x=L} (\pi x/L)\sin^2(\pi x/L)d(\pi x/L) \tag{12-31}$$

Letting $\pi x/L = y$ and noting that $y=0,\ \pi$ when $x=0,\ L$, we have

$$W_1^{(1)} = 2u/\pi^2 \int_0^{\pi} y\sin^2 y\,dy \tag{12-32}$$

Standard tables (see also Appendix 1) lead to a value of $\pi^2/4$ for the integral, and

$$W_1^{(1)} = (2u/\pi^2)(\pi^2/4) = u/2 \tag{12-33}$$

To first order, then, the energy of the lowest-energy perturbed state is, in atomic units,

$$W = (\pi^2/2L^2) + u/2 \tag{12-34}$$

2. *Evaluation by Inspection: First Method.* Equation (12-34) is certainly a reasonable result since, as the potential is increased everywhere in the box (except at $x = 0$), we expect that the energy of the electron should also increase. The fact that the increase is such a simple quantity ($u/2$) suggests that there might be a simple way to understand this result, and this is indeed the case. Consider the distribution of the electron in the lowest unperturbed state, shown in Fig. 12-2. This distribution is symmetric about the midpoint of the line segment. Consequently, for each instant of time the unperturbed electron spends in element dx_1 of Fig. 12-2, it spends an equal instant in the symmetrically related element dx_2. In other words, for each instant the electron experiences a perturbation potential *less* than $u/2$, it experiences an instant of potential *greater* (by an equal amount) than $u/2$. Hence, the *average* potential must be precisely $u/2$. Since we know that ψ^2 is symmetric for *every state* in the unperturbed box, we can immediately extend our result and say that the first-order correction to the energy of every state is $u/2$.

   The ability to evaluate first-order energies by inspection is very useful. Even in cases where exact evaluation by this technique is not possible, it may still be useful in making an estimate or in checking the reasonableness of a computed result.

3. *Evaluation by Inspection: Second Method.* A variation of the above approach is sometimes useful. We begin by recognizing that, whereas H' is neither symmetric nor antisymmetric about the midpoint of the wire, we can make it antisymmetric by subtracting the constant $u/2$, as indicated in Fig. 12-3.

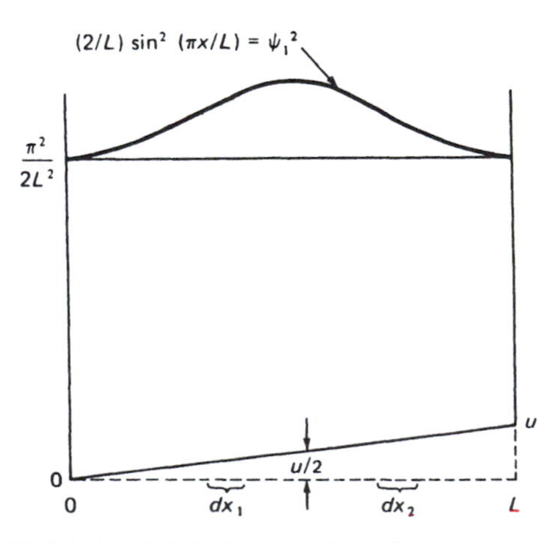

Figure 12-2 ▶ $\psi^2(n = 1)$ for a particle in the unperturbed box.

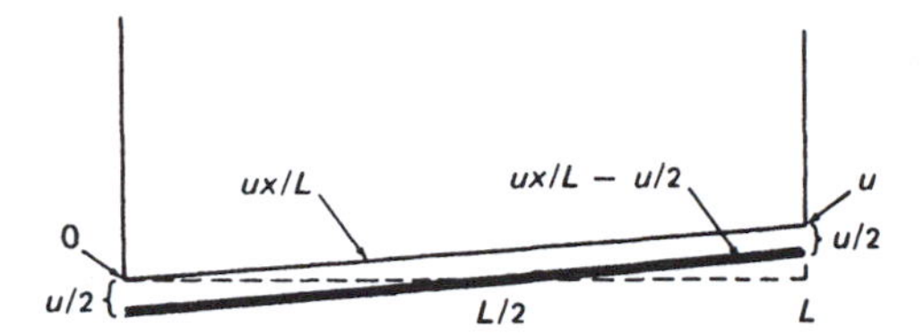

Figure 12-3 ▶ The function ux/L is unsymmetric about $L/2$, but $ux/L - u/2$ is antisymmetric about $L/2$.

By writing

$$H' = ux/L - u/2 + u/2 \equiv H'_{\text{anti}} + u/2 \tag{12-35}$$

we express H' as an antisymmetric function plus a constant. Our integral for $W_1^{(1)}$ becomes

$$W_1^{(1)} = \int_0^L \psi_{n=1} H'_{\text{anti}} \psi_{n=1}\, dx + \int_0^L \psi_{n=1}(u/2)\psi_{n=1}\, dx \tag{12-36}$$

The first integral vanishes because ψ^2 is symmetric. The second integral is just $u/2$ times unity since ψ is normalized. Hence, $W_1^{(1)} = u/2$.

EXAMPLE 12-2 A one-dimensional box potential is perturbed so that it is raised by a constant amount, δ, on the left side of the box but is unchanged on the right side. What is the first-order change in energy for ψ_1? for ψ_2?

SOLUTION ▶ One can argue, similarly to the reasoning in Example 12-1, that half of the distribution responds to the increase in potential, half to zero change, giving a net first-order increase of $\delta/2$. Or, one can turn the perturbation into an antisymmetric function plus a constant: $H' = H' - \delta/2 + \delta/2 = H'_{\text{anti}} + \delta/2$, leading to $\mathrm{W}_1^{(1)} = \delta/2$. The same result applies to ψ_2 and all higher states because ψ_n^2 is symmetric in the box for all n. ◀

12-3.C The First-Order Correction to ψ_1

How should we expect the lowest-energy wavefunction to change in response to the perturbation? Since we are dealing with the lowest-energy state, we might expect the electron to spend more time in the low-potential end of the box (the nonclassical result), and so the wavefunction should tend to become skewed, as shown in Fig. 12-4. In this figure it is demonstrated how the perturbed wavefunction can be resolved into an unperturbed wavefunction and a correction, or difference, function. Since this correction function must increase ψ on the low potential side and decrease ψ on the high potential side, it is clear that it must be close to antisymmetric in nature. According to Eq. (12-19), the first-order approximation $\phi_1^{(1)}$ to this correction function is formed by adding together small amounts of higher-energy wavefunctions. Comparing the correction function in Fig. 12-4 and these higher-energy wavefunctions (Fig. 12-5) leads us to expect that ψ_2 will be a heavy contributor, whereas ψ_3, being

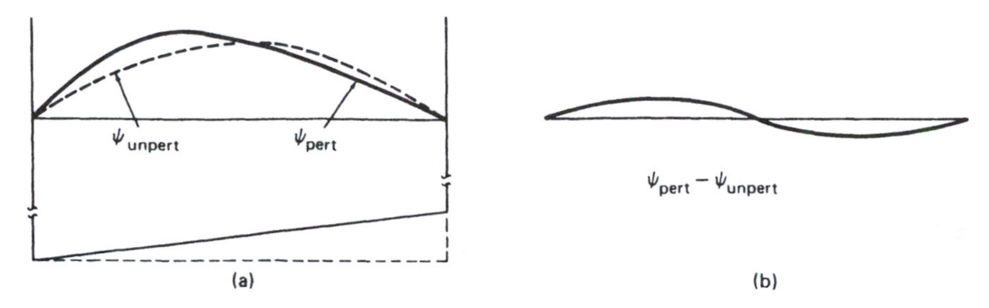

Figure 12-4 ▶ Sketches of the lowest-energy wavefunction (a) before and after perturbation and (b) the difference between them.

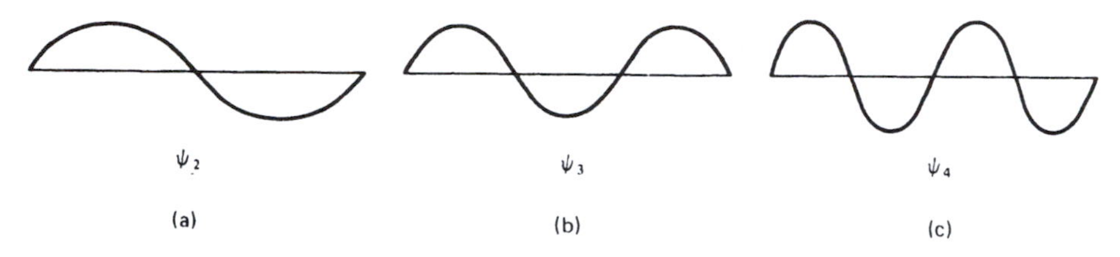

Figure 12-5 ▶ Sketches of the (a) second, (b) third, and (c) fourth ($n = 2, 3, 4$, respectively) wavefunctions ψ for the particle in the unperturbed box.

symmetric, will not contribute strongly. The mixing coefficient for ψ_2 is, according to Eq. (12-18),

$$c_{21}^{(1)} = \frac{\int_0^L \left(\sqrt{2/L}\sin 2\pi x/L\right)\left(ux/L\right)\left(\sqrt{2/L}\sin \pi x/L\right)dx}{-3\pi^2/2L^2} = \frac{32L^2u}{27\pi^4} \tag{12-37}$$

Thus, $c_{21}^{(1)}$ is positive, and ψ_2 contributes to $\phi_1^{(1)}$ in the manner expected. The mixing coefficient for ψ_3 is given by

$$c_{31}^{(1)} = \frac{\langle \psi_1 | H'_{\text{anti}} | \psi_3 \rangle}{E_1 - E_3} + \frac{\langle \psi_1 | u/2 | \psi_3 \rangle}{E_1 - E_3} = 0 + 0 \tag{12-38}$$

where we have used expression (12-35) for H'. The first integral vanishes because ψ_1 and ψ_3 are symmetric. The second integral vanishes because they are orthogonal. Clearly, *no* symmetric state will contribute to $\phi_1^{(1)}$.

Since ψ_4 is antisymmetric, it can contribute to $\phi_1^{(1)}$. On evaluation, we find that $c_{41}^{(1)}$ is about 2% of $c_{21}^{(1)}$ (Problem 12-5). $c_{41}^{(1)}$ is so much smaller than $c_{21}^{(1)}$ for two reasons. First, the integral $\langle \psi_1 | H' | \psi_4 \rangle$ in the numerator of $c_{41}^{(1)}$ is much smaller than $\langle \psi_1 | H' | \psi_2 \rangle$ in $c_{21}^{(1)}$. This means that the shifting of charge produced by adding ψ_4 to ψ_1 is much less helpful for lowering the energy than is that produced by adding ψ_2 to ψ_1. Examination of ψ_2 and ψ_4 (Fig. 12-5) reveals why this is so. ψ_2 causes removal of charge from the right-hand half of the box and accumulation of charge in the left-hand half. ψ_4 causes removal of charge from the second and fourth quarters (numbering from the left) and buildup of charge in the first and third quarters. On balance ψ_4 helps, but charge buildup in the third quarter is not desirable nor is removal of charge from the second quarter, and so ψ_4 is much less helpful than ψ_2. The second reason for $c_{41}^{(1)}$ being so small is that $E_1 - E_4$ in the denominator of $c_{41}^{(1)}$ is five times as big as $E_1 - E_2$ in $c_{21}^{(1)}$.

In general, mixing between states of widely different energies is discouraged by the formula for $\phi_i^{(1)}$.

The fact that a large contribution by ψ_j to $\phi_i^{(1)}$ is favored by large $\langle\psi_i|H'|\psi_j\rangle$ and small $|E_i - E_j|$ is strikingly similar to the situation found for variational calculations (Chapter 7). There, the mixing between two basis functions χ_i and χ_j is favored if $\langle\chi_i|H|\chi_j\rangle \equiv H_{ij}$ is large and if $|H_{ii} - H_{jj}|$ is small.

EXAMPLE 12-3 What sign should be expected for $c_{21}^{(1)}$ for the perturbation of Example 12-1? of 12-2?

SOLUTION ▶ Assuming that ψ_1 will behave anticlassically, and be skewed so as to decrease the probability distribution in the higher-energy region (left side), we expect $c_{21}^{(1)}$ to have the opposite sign of that in Eq. 12-37. Hence, it should be negative. This applies for both Examples 12-1 and 12-2. ◀

12-3.D The Role of an Additive Constant in H′

Review of the results of Sections 12-3B and C will show that addition of a constant to H' will change $W_i^{(1)}$ by the same constant for all states and have no effect on $\phi_i^{(1)}$. The change in energy of all states by a constant is equivalent to a relocation of the zero of energy and is normally of no interest. Our initial statement that the perturbation is produced by a uniform electric field left the additive constant in H' unspecified. We chose to set $H' = 0$ at $x = 0$, but we could have made any of an infinite number of choices for H' at $x = 0$. A more sensible choice would have been the antisymmetric function $H' = ux/L - u/2$ because this still leads to changes in wavefunctions due to the perturbation, but introduces no energy change to first order ($W_i^{(1)} = 0$ for all states using this H'). In other words, the antisymmetric function includes the relevant physics of the problem and excludes the trivial effects of a constant ($u/2$) potential change. Our choice of $H' = ux/L$ was made for pedagogical reasons.

12-3.E The Calculation of $W_1^{(2)}$

Since the constant first-order contribution to the energies of all the states is not physically interesting, let us examine the second-order contribution to the ground-state energy, $W_1^{(2)}$. Equation (12-23) shows that this is related to the first-order correction to the wavefunction, $\phi_1^{(1)}$, which, as we have already seen, causes the wavefunction to become skewed toward the low-potential end of the box. It is clear from Eq. (12-22) that, in calculating the coefficients for $\phi_1^{(1)}$, we have already done most of the work needed to find $W_1^{(2)}$. We saw earlier that $\phi_1^{(1)}$ is made primarily from ψ_2. For simplicity, we will neglect all higher-energy contributions, and so

$$\phi_1^{(1)} \cong c_2^{(1)}\psi_2 = (32L^2u/27\pi^4)\psi_2 \tag{12-39}$$

and, using Eq. (12-22),

$$W_1^{(2)} \cong (32L^2u/27\pi^4)^2(-3\pi^2/2L^2) \tag{12-40}$$

Thus, the effect of $W_1^{(2)}$ is to *lower* the energy.

We pause at this point to summarize our results. The perturbation raises the potential everywhere in the box. (This depends on our choice of an arbitrary constant.) The energy to first order is increased by the same constant amount for every state. This is easily seen by inspection, utilizing simple symmetry features of H' and the unperturbed wavefunctions. The *wavefunction* for the lowest-energy state is skewed, to first order, in a way that is energetically favorable as far as interaction with H' is concerned. The second-order contribution to the *energy* for this state is negative, reflecting this energetically favorable shift of charge. Thus far, everything behaves sensibly. We next examine the behavior of the second-lowest energy state, where some important new features occur.

EXAMPLE 12-4 For the ψ_1 case of Example 12-1, should the energy to second order be higher or lower than the unperturbed energy?

SOLUTION ▶ The energy to second order is $E_1 + W_1^{(1)} + W_1^{(2)}$. We have already seen that $W_1^{(1)} = 0$. Thus the question comes down to asking whether $W_1^{(2)}$ is positive or negative. The first-order correction to the wavefunction shifts charge from the higher-energy regions, hence produces a negative $W_1^{(2)}$. ◀

12-3.F The Effects of the Perturbation on ψ_2

We begin by examining $\phi_2^{(1)}$, the first-order correction to ψ_2. Inspection of Eq. (12-18) leads at once to the observation that $c_{21}^{(1)} = -c_{12}^{(1)}$. This means that, since ψ_2 contributes to $\phi_1^{(1)}$ with a positive coefficient, ψ_1 contributes to $\phi_2^{(1)}$ with a negative coefficient. A sketch of ψ_2 minus a small amount of ψ_1 will show that this has the effect of shifting charge from the left half to the right half of the box. This is just the reverse of what we found for the lowest-energy state.

The antisymmetry of Eq. (12-18) for interchange of i and k allows us to make the following general statement. Let a perturbation occur that raises or lowers the potential more in one region of space than in another. The first-order correction to a given wavefunction will contain higher-energy wavefunctions in a manner to cause charge to shift into regions of lowered (or less raised) potential and it will contain lower-energy wavefunctions in a manner to cause charge to shift into regions of raised (or less lowered) potential.

We have not yet completed our construction of $\phi_2^{(1)}$. We must calculate coefficients for contributions from the higher-energy functions ψ_3, ψ_4, etc. The state ψ_3 should contribute fairly strongly since it has the proper symmetry and is not too distant in energy from E_2:

$$c_{32}^{(1)} = \frac{\langle\psi_2|H'|\psi_3\rangle}{E_2 - E_3} = \frac{3}{5}\frac{32L^2u}{25\pi^4} \tag{12-41}$$

Therefore, ψ_3 contributes to $\phi_2^{(1)}$ with a coefficient about $\frac{3}{5}$ the magnitude of the contribution from ψ_1. The sign of $c_{32}^{(1)}$ is positive and a sketch of ψ_2 plus a small amount of ψ_3 will demonstrate that this causes charge shifting to the left in accord with our general statement above.

Since ψ_4 contributes nothing (by symmetry) and ψ_5 is fairly distant in energy, we will neglect all contributions to $\phi_2^{(2)}$ above ψ_3.

We have, then, two sizable contributions to $\phi_2^{(1)}$, each favoring charge shifts in opposite directions. Let us see how $W_2^{(2)}$ reflects this:

$$W_2^{(2)} \cong c_{12}^{(1)2}(E_2 - E_1) + c_{32}^{(1)2}(E_2 - E_3) \tag{12-42a}$$

$$\cong \left[c_{12}^{(1)2}(3) + (\frac{3}{5}c_{12}^{(1)})^2(-5)\right]\pi^2/2L^2 \tag{12-42b}$$

$$\cong 1.2c_{12}^{(1)2}\pi^2/2L^2 \tag{12-42c}$$

$W_2^{(2)}$ is the difference between energy contributions of opposite sign. The net result (energy increases) comes about in this case because ψ_1 contributes more heavily than ψ_3.

The fact that contributing wavefunctions from lower and higher energies affect $W_i^{(2)}$ oppositely is made evident by Eq. (12-22). Hence, we can extend our general statement above by adding that the second-order contribution to the energy from states below ψ_i in energy cause the energy of the jth state to go up, contributions from above cause it to go down.

Because the unperturbed energies of the particle in the box increase as n^2, any state (except the lowest) is closer in energy to states below than to states above. Hence, for at least some kinds of perturbation, we might expect these states to "feel" the effects of states at lower energies more strongly and to rise in energy (as far as $W^{(2)}$ is concerned). This is what happens in this example. There are no states below the lowest, and so $W_1^{(2)}$ cannot be positive, but $W_2^{(2)}$ is positive and it turns out that $W_i^{(2)}$ is positive for all higher i as well.

It is interesting to compare these results with those from classical physics. Classically, the particle moves most slowly at the top of the potential gradient and therefore spends most of its time there. The lowest-energy state has responded in the opposite manner, in a way we might call anticlassical. The second and all higher states have responded classically.

12-4 The Ground-State Energy to First-Order of Heliumlike Systems

The hamiltonian for a two-electron atom or ion with nuclear charge Z a.u. is (neglecting relativistic effects and assuming infinite nuclear mass)

$$H(1,2) = -\frac{1}{2}\left(\nabla_1^2 + \nabla_2^2\right) - Z/r_1 - Z/r_2 + 1/r_{12} \tag{12-43}$$

This may be written as a sum of one-electron operators and a two-electron operator:

$$H(1,2) = H(1) + H(2) + 1/r_{12} \tag{12-44}$$

where $H(i)$ is simply the hamiltonian for the hydrogenlike system with nuclear charge Z:

$$H(i) = -\frac{1}{2}\nabla_i^2 - Z/r_i \tag{12-45}$$

Since we know the eigenvalues and eigenfunctions for $H(i)$, we can let $H(1)+H(2)$ be the unperturbed hamiltonian with $1/r_{12}$ the perturbation. Such a perturbation is not very small, but it is of interest to see how well the method works in such a case. We have, therefore, for the lowest-energy state of the system:

$$H_0 = H(1) + H(2) \tag{12-46}$$

$$H' = 1/r_{12} \tag{12-47}$$

$$\psi_1 = (Z^3/\pi)\exp[-Z(r_1+r_2)] \tag{12-48}$$

$$E_1 = -Z^2/2 - Z^2/2 = -Z^2 \tag{12-49}$$

$$W_1^{(1)} = \langle\psi_1|1/r_{12}|\psi_1\rangle \tag{12-50}$$

All quantities are in atomic units. The unperturbed wavefunction [Eq. (12-48)] is simply the product of two one-electron 1s AOs. Because this is an eigenfunction of the system in the absence of interelectronic repulsion, it is too contracted about the nucleus.

The first-order correction to the energy is the repulsion between the two electrons in this overly contracted eigenfunction [Eq. (12-50)]. We have encountered this same repulsion integral in our earlier variational calculation on helium-like systems. The evaluation of this integral is described in Appendix 3. Its value is $5Z/8$. Hence, to first order,

$$W_1 = E_1 + W_1^{(1)} = -Z^2 + 5Z/8 \tag{12-51}$$

In Table 12-1, this result is compared with exact energies for the first ten members of this series. Several points should be noted:

1. The effect of the perturbation to first order is to increase the ground-state energy. This is expected since $1/r_{12}$ is always positive (i.e., repulsive).
2. The energy to first order is never below the exact ground-state energy. This is a general property of perturbation calculations as is easily proved (Problem 12-1).
3. The energy to first order is in error by a fairly constant amount throughout the series. For H^-, this gives a substantial *percentage* of error and fails to show H^- stable compared with an H atom and an unbound electron. For higher Z, this error becomes *relatively* smaller since $1/r_{12}$ becomes relatively less important compared with Z/r_i. The assumption that H' is a *small* perturbation is thus better fulfilled at large Z and results in $W_1^{(1)}$ being a much smaller correction relative to the total energy W_1. Because "large" and "small" are relative terms, they can be misleading. Since energies of chemical interest are often small differences between large numbers (see the discussion at the end of Chapter 7), errors that were originally relatively small can become relatively large after the subtraction. Therefore, even though perturbation terminology would suggest that the results at $Z=10$ are better than those at $Z=1$, this may not be the case for some practical applications.

TABLE 12-1 ▶ Comparison of Exact Energy (in atomic units) with Energy to First Order when $H' = r_{12}^{-1}$

Z	System	E_1	$W_1^{(1)}$	$E_1 + W_1^{(1)}$	E_{exact}^a	$E_{exact} - E_1 - W_1^{(1)}$	%Error
1	H^-	−1.0000	5/8 = 0.6250	−0.3750	−0.52759	−0.15259	28.92
2	He	−4.0000	5/4 = 1.2500	−2.7500	−2.90372	−0.15372	5.29
3	Li^+	−9.0000	15/8 = 1.8750	−7.1250	−7.27991	−0.15491	2.13
4	Be^{2+}	−16.0000	5/2 = 2.50000	−13.5000	−13.65556	−0.15556	1.14
5	B^{3+}	−25.0000	25/8 = 3.1250	−21.8750	−22.03097	−0.15597	0.71
6	C^{4+}	−36.0000	15/4 = 3.7500	−32.2500	−32.40624	−0.15624	0.48
7	N^{5+}	−49.0000	35/8 = 4.3750	−44.6250	−44.78144	−0.15644	0.35
8	O^{6+}	−64.0000	5.0000	−59.0000	−59.15659	−0.15659	0.26
9	F^{7+}	−81.0000	45/8 = 5.6250	−75.3750	−75.53171	−0.15671	0.21
10	Ne^{8+}	−100.0000	25/4 = 6.2500	−93.7500	−93.90680	−0.15680	0.17

[a] E_{exact} is the nonrelativistic energy to thirteenth order in Z^{-1}, truncated at the fifth decimal place. See Scherr and Knight [4].

12-5 Perturbation at an Atom in the Simple Hückel MO Method

Perturbation theory can be used to estimate the effect of a change in the value of the coulomb integral H_{kk} at carbon atom k. This is normally given the value of α, but it might be desirable to consider a different value due, for example, to replacement of an attached hydrogen by some other atom or group. We take the unperturbed Hückel MOs and orbital energies as our starting point and let the perturbed value of H_{kk} be $a + \delta\alpha$. Therefore,

$$H' = \delta\alpha \quad (\text{at center } k \text{ only}) \tag{12-52}$$

The first-order correction for the energy of ϕ_i, the ith MO, is

$$W_i^{(1)} = \langle\phi_i|H'|\phi_i\rangle \tag{12-53}$$

Now ϕ_i is a linear combination of the AOs χ:

$$\phi_i = \sum_j c_{ji}\chi_j \tag{12-54}$$

and so

$$W_1^{(1)} = \sum_j \sum_l c_{ji}^* c_{li} \langle\chi_j|H'|\chi_l\rangle \tag{12-55}$$

but H' is zero except at atom k, where it is $\delta\alpha$, and so the sum reduces to one term:

$$W_i^{(1)} = c_{ki}^* c_{ki}\,\delta\alpha \tag{12-56}$$

Summing over all the MOs times the number of electrons in each MO gives the first-order correction to the total energy:

$$W^{(1)} = \delta\alpha \sum_i n_i c_{ki}^* c_{ki} = \delta\alpha q_k \tag{12-57}$$

where q_k is the π-electron density at atom k. Thus the energy change to first order is equal to the change in H_{kk} times the unperturbed electron density at that atom. This can also be seen to be an immediate consequence of the alternative energy expression

$$E = \sum_l q_l \alpha_l + 2 \sum_{l<m} p_{lm}\beta_{lm} \tag{12-58}$$

which was derived in Chapter 8.

EXAMPLE 12-5 Consider the methylene cyclopropene molecule, C_4H_4, in the HMO approximation. (See Appendix 6 for data.) At which carbon will a perturbation involving α affect the total π energy the most? Calculate the energy to first order if the value of α at that atom increases to $\alpha + 0.1000\beta$. Calculate the energy change, to first order, of each of the four MOs.

SOLUTION ▶ The total energy will be most affected if the perturbation occurs at the atom having the greatest π-electron density, that is, at atom 4. To first order, the total π energy changes by $(0.1000\beta)(1.4881) = 0.1488\beta$, giving an energy to first order of $4\alpha + 4.9264\beta + 0.1488\beta = 4\alpha + 5.1112\beta$. Each MO energy drops by (using Eq. (12-56), from lowest to highest, $0.00794\beta, 0.06646\beta, 0.0000\beta, 0.02559\beta$. The sum of the first two of these, times 2 (because of having two electrons each), gives 0.1488β, which equals the first-order change in total π energy found above. ◀

Calculation of first-order corrections to the MOs proceeds in a straightforward manner using Eq. (12-19) (see Problem 12-17). One of the results of interest from such a calculation is the change in π-electron density at atom l due to a change in the coulomb integral at atom k. The differential expression is

$$\delta q_1 = (\partial q_l/\partial \alpha_k)\delta\alpha_k = \pi_{l,k}\delta\alpha_k \tag{12-59}$$

The quantity of interest $\pi_{l,k}$ is called the *atom-atom polarizability*. We will now derive an expression for $\pi_{l,k}$ in terms of MO coefficients and energies.

We assume that our MOs fall into a completely occupied set $\phi_1, \ldots, \phi_m$ and a completely empty set $\phi_{m+1}, \ldots, \phi_n$. We take $\delta\alpha$ to be positive, which means that center k is made less attractive for electrons. According to Eq. (12-19) the jth MO is, to first order,

$$\phi_j + \phi_j^{(1)} = \phi_j + \sum_{i=1, i\neq j}^{n} \frac{\langle\phi_j|H'|\phi_i\rangle}{E_j - E_i}\phi_i \tag{12-60}$$

where E_j is the orbital energy of ϕ_j. As indicated above, the integral vanishes except over AO k. Therefore,

$$\phi_j^{(1)} = \sum_{i=1, i\neq j}^{n} \frac{c_{kj}^* c_{ki}\,\delta\alpha_k}{E_j - E_i}\phi_i \tag{12-61}$$

Expanding ϕ_i in terms of AOs, we have

$$\phi_j^{(1)} = \delta\alpha_k \sum_{i=1, i\neq j}^{n} \frac{c_{kj}c_{ki}}{E_j - E_i} \sum_{l=1}^{n} c_{li}\chi_l \tag{12-62}$$

We can write this as

$$= \sum_{l=1}^{n} c_{lj}^{(1)}\chi_l \tag{12-63}$$

where

$$c_{lj}^{(1)} = \delta\alpha_k \sum_{i=1, i\neq j}^{n} \frac{c_{kj}c_{ki}c_{li}}{E_j - E_i} \tag{12-64}$$

The change in density at atom l equals the square of the perturbed wavefunction minus that of the unperturbed wavefunction and involves only the doubly occupied MOs $1 - m$:

$$\delta q_l = 2\sum_{j=1}^{m}\left[(c_{lj} + c_{lj}^{(1)})^2 - c_{lj}^2\right] \tag{12-65}$$

$$= 4\sum_{j=1}^{m} c_{lj}c_{lj}^{(1)} \tag{12-66}$$

where we have neglected $(c_{lj}^{(1)})^2$ because the perturbation is assumed small. Substituting Eq. (12-64) into (12-66) gives

$$\delta q_l = 4\delta\alpha_k \sum_{j=1}^{m} \sum_{i=l,i\neq j}^{n} \frac{c_{lj}c_{kj}c_{ki}c_{li}}{E_j - E_i} \tag{12-67}$$

This equation can be simplified further by recognizing that, for j *and* $i < m+1$, each term with denominator $E_j - E_i$ has a mate with denominator $E_i - E_j$. Hence, the terms differ only in sign and cancel. Therefore,

$$\sum_{j=1,i\neq j}^{m} \frac{c_{lj}c_{kj}c_{ki}c_{li}}{E_j - E_i} = 0 \tag{12-68}$$

and so

$$\delta q_l = 4\delta\alpha_k \sum_{j=1}^{m} \sum_{i=m+1}^{n} \frac{c_{lj}c_{kj}c_{ki}c_{li}}{E_j - E_i} \tag{12-69}$$

Comparing Eqs. (12-69) and (12-59) gives

$$\boxed{\pi_{lk} = 4\sum_{j=1}^{m} \sum_{i=m+1}^{n} \frac{c_{lj}c_{kj}c_{ki}c_{li}}{E_j - E_i}} \tag{12-70}$$

Related quantities, derivable in a similar manner, are the *bond-atom polarizability*, $\pi_{st,r}$, and the *bond-bond polarizability*, $\pi_{tu,rs}$:

$$\boxed{\pi_{st,r} = 2\sum_{j=1}^{m} \sum_{k=m+1}^{n} \frac{c_{rj}c_{rk}(c_{sj}c_{tk} + c_{tj}c_{sk})}{E_j - E_k}} \tag{12-71}$$

$$\boxed{\pi_{tu,rs} = 2\sum_{j=1}^{m} \sum_{k=m+1}^{n} \frac{(c_{rj}c_{sk} + c_{sj}c_{rk})(c_{tj}c_{uk} + c_{uj}c_{tk})}{E_j - E_k}} \tag{12-72}$$

These refer, respectively, to the change in bond order p_{st} induced by a perturbation $\delta\alpha_r$, and to the change in bond order p_{tu} induced by a change in β_{rs}.

The atom–atom polarizability formula (12-70) makes it evident that a large contribution to the change in electron density at atom l due to a change in the coulomb integral at atom k is favored if a filled and an empty MO exist that have large absolute coefficients on both atoms and are close in energy. However, another factor comes into play that usually confounds this expectation, as the following example shows:

EXAMPLE 12-6 Consider the naphthalene molecule, $C_{10}H_8$, in the HMO approximation. (See Appendix 6 for data.) If atom 1 were perturbed in a way to make it more attractive to π electrons, which of the atoms 2, 3, or 4, would you expect to experience the greatest population change?

SOLUTION ▶ The HOMO-LUMO contribution to the charge shift favors atom 4, since its coefficients in these MOs are larger than those for atoms 2 or 3. However, the actual data (not given in Appendix 6) are: $\pi_{1,2}=0.21342$, $\pi_{1,3}=-0.01771$, $\pi_{1,4}=0.13937$. So atom 2 experiences the greatest population change, atom 3 the least. Why is this? It results from the fact that we are summing products that can be positive or negative. Two atoms near each other are less likely to have nodal planes separating them, so their coefficients are more likely to be of the same sign giving fewer sign changes in their products. So the sum has less tendency to cancel itself. This is the way in which MO theory predicts a tendency for polarization effects to die off as distance increases—the well-known "inductive effect." This dying-off is not monotonic, as can be seen in the remaining atom-atom polarizabilities: $\pi_{1,5}=0.02325$, $\pi_{1,6}=-0.00643$, $\pi_{1,7}=0.03228$, $\pi_{1,8}=-0.02674$, $\pi_{1,9}=0.08889$, $\pi_{1,10}=-0.00356$. ◀

In general, π_{lk} can have either sign. This means that making atom k more attractive for electrons will, to first order, cause the π electron density to decrease at some carbons and increase at others. (This is necessary since the total π-charge density must be conserved.)

12-6 Perturbation Theory for a Degenerate State

Equations (12-19), (12-21), and (12-24) have denominators containing $E_i - E_j$. If ψ_i and ψ_j are degenerate, this leads to difficulty, alerting us to the fact that our earlier derivations do not take proper account of states of interest that are degenerate. The problem results from our initial expansion of ϕ_i as a power series in λ. The leading term in the expansion, $\phi_i^{(0)}$, is the wavefunction that ϕ_i becomes in the limit when $\lambda=0$. In nondegenerate systems there is no is ambiguity; $\phi_i^{(0)}$ has to be ψ_i. But if ψ_i is degenerate with ψ_j, any linear combination of them is also an eigenfunction. We need a method to determine how ψ_i and ψ_j should be mixed together to form the correct zeroth-order functions $\phi_i^{(0)}$ and $\phi_j^{(0)}$.

To find the conditions that $\phi_i^{(0)}$ and $\phi_j^{(0)}$ must satisfy, we return to the first-order perturbation equation (12-11) but with $\phi_i^{(0)}$ in place of ψ_i:

$$(H' - W_i^{(1)})\phi_i^{(0)} + (H_0 - E_i)\phi_i^{(1)} = 0 \tag{12-73}$$

Multiplying from the left by $\phi_j^{(0)*}$ and integrating gives (taking $\phi_i^{(0)}$ and $\phi_j^{(0)}$ to be orthogonal and H_0 hermitian)

$$\langle\phi_j^{(0)}|H'|\phi_i^{(0)}\rangle + (E_j - E_i)\langle\phi_j^{(0)}|\phi_i^{(1)}\rangle = 0 \tag{12-74}$$

If $E_i = E_j$, this gives

$$\langle\phi_j^{(0)}|H'|\phi_i^{(0)}\rangle = 0 \tag{12-75}$$

This means that the first-order perturbation equation is satisfied for degenerate states only when the wavefunctions "*diagonalize*" the perturbation operator H'. Therefore, our problem reduces to finding those linear combinations of ψ_i and ψ_j that will diagonalize H'. We have already seen two ways to accomplish this in earlier chapters. One way is to construct the matrix H', where $(\mathsf{H}')_{ij} = \langle\psi_i|H'|\psi_j\rangle$, and then find the unitary matrix C that diagonalizes H' in the similarity transformation $\mathsf{C}^\dagger\mathsf{H}'\mathsf{C}$. The elements in C are then the coefficients for the linear combinations of ψ_i and ψ_j, and the diagonal elements of the diagonalized matrix are equal to $\langle\phi_i^{(0)}|H'|\phi_i^{(0)}\rangle$ and $\langle\phi_j^{(0)}|H'|\phi_j^{(0)}\rangle$; they are the first-order corrections to the energy of $\phi_i^{(0)}$ and $\phi_j^{(0)}$. The other method, equivalent to the above, is to set up the determinantal equation

$$\begin{vmatrix} H'_{11} - E & H'_{12} \\ H'_{21} & H'_{22} - E \end{vmatrix} = 0 \tag{12-76}$$

and solve for the roots E. Substituting these back into the simultaneous equations corresponding to the determinant gives the coefficients with which ψ_i and ψ_j should be combined. The roots E are the first-order corrections to the energy and are identical to the diagonal matrix elements mentioned above.

In the event that the roots are equal, the perturbation has failed to split the degeneracy to first order and it is then necessary to proceed to higher order. We will not pursue this problem to higher orders, however.

If there are n degenerate states, the above procedure is simply carried out with an $n \times n$ matrix or determinant.

We will now give an example of perturbation theory for a degenerate state.

12-7 Polarizability of the Hydrogen Atom in the $n = 2$ States

In Chapter 7, the polarizability of the hydrogen atom in the 1s state was calculated by the variation method (see Section 7-4 and Problem 7-13). We now use perturbation theory for degenerate states to calculate to first order the polarizabilities of the $n = 2$ states.

The perturbation due to a z-directed uniform electric field is, in atomic units,

$$H' = -Fr\cos\theta \tag{12-77}$$

There are four degenerate states at the $n = 2$ level, giving us a 4×4 secular determinant. If we choose the real AOs as our basis set and let s, x, y, z represent these functions, the determinantal equation is

$$\begin{vmatrix} H'_{ss} - E & H'_{sz} & H'_{sy} & H'_{sx} \\ H'_{zs} & H'_{zz} - E & H'_{zy} & H'_{zx} \\ H'_{ys} & H'_{yz} & H'_{yy} - E & H'_{yx} \\ H'_{xs} & H'_{xz} & H'_{xy} & H'_{xx} - E \end{vmatrix} = 0 \tag{12-78}$$

where

$$H'_{yz} = \langle y|H'|z\rangle, \quad \text{etc.} \tag{12-79}$$

The operator H' is antisymmetric for reflection in the xy plane, but symmetric for reflection in the xz or yz planes. It follows from this that all integrals in Eq. (12-78) vanish except for H'_{sz} and H'_{zs}, which are equal. Assigning a value of x (not to be confused with the x coordinate) to these integrals, our equation becomes

$$\begin{vmatrix} -E & x & 0 & 0 \\ x & -E & 0 & 0 \\ 0 & 0 & -E & 0 \\ 0 & 0 & 0 & -E \end{vmatrix} = 0 \tag{12-80}$$

The block diagonal form indicates that this is a product of one 2×2 determinant and two 1×1's. Evidently, the proper zeroth-order wavefunctions must be a mixture of 2s and $2p_z$ AOs, with $2p_x$ and $2p_y$ being acceptable as they stand. The roots of the 1×1 determinants are zero, the roots of the 2×2 are $E_+ = +x$, $E_- = -x$. These lead to coefficients that produce the zeroth-order wave-functions

$$\phi_+^{(0)} = \left(1/\sqrt{2}\right)(2s + 2p_z) \tag{12-81}$$

$$\phi_-^{(0)} = \left(1/\sqrt{2}\right)(2s - 2p_z) \tag{12-82}$$

The roots $+x$ and $-x$ are the first-order corrections to the energy. We now proceed to calculate these quantities:

$$x = \langle 2s|H'|2p_z\rangle = \cdots = 3F \tag{12-83}$$

Our perturbation calculation indicates that the $n = 2$ level splits into three levels under the influence of an electric field. Since this splitting occurs to first order, it is sometimes called a *first-order Stark effect*. Since x is proportional to F, the splitting is linear in F, as indicated in Fig. 12-6.

It is instructive to compare these results with the behavior of the 1s state. In the first place, the effect of a uniform electric field on the 1s level is zero, to first order, because $\langle 1s|H'|1s\rangle$ vanishes for reasons of symmetry. Only when first-order corrections are made to the 1s wavefunction are energy effects seen, and these occur in the second-order energy terms. Therefore, the 1s state gives a *second-order Stark effect*, but no first-order effect. The 1s state gives no first-order effect because the spherically symmetric zeroth-order wavefunction has no electric dipole to interact with the field. But the proper zeroth-order wave functions for some of the $n = 2$ states, given by Eqs. (12-81) and (12-82), *do* provide electric dipoles in opposite directions that interact with the field to produce first-order energies of opposite signs.

Notice that the energy change for the second-order effect goes as F^2 [(Eq. (7-62)], whereas that for the first-order effect goes as F. These dependences are indicative of an *induced dipole* and a *permanent dipole*, respectively. The induced dipole for the 1s state depends on the field strength F, since, as F increases, the dipole moment increases (due to mixing in higher states). This induced dipole, which depends on F,

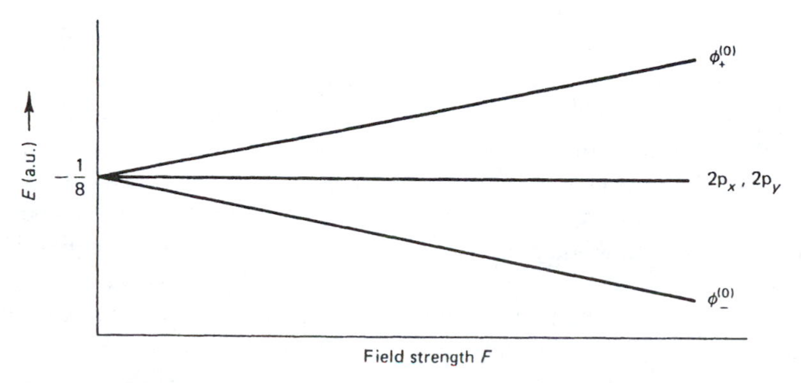

Figure 12-6 ▶ Energy to first order of $n = 2$ level of hydrogen as a function of uniform electric field strength (z-directed field).

then interacts with the field of strength F. Since both the size of the induced dipole and the energy of interaction with the field depend on F to the first power, the energy goes as F^2. For the $n=2$ wavefunctions $\phi_+^{(0)}$ and $\phi_-^{(0)}$, however, the dipole is permanent.[7] (The mixing to produce these states occurs even in the limit as F goes to zero.) This permanent dipole interacts with the field to give an energy depending on F instead of F^2.

The $2p_x$ and $2p_y$ AOs are not affected to first-order in the perturbation. At higher orders, these AOs (as well as $\phi_+^{(0)}$ and $\phi_-^{(0)}$) mix in higher-energy AOs of symmetries appropriate to produce induced dipoles. However, due to the equivalence of $2p_x$ and $2p_y$ with respect to H', the induced dipole is the same in both cases so that the degeneracy is still not lifted.

The *qualitatively* different behaviors of the $n=1$ and $n=2$ levels of hydrogen result from the fact that degenerate eigenfunctions can be mixed together at no energy expense, whereas the $n=1$ state can produce a dipole only by mixing in higher-energy states. This suggests that the polarizability of an atom or molecule should be strongly dependent on the availability of fairly low-energy unoccupied orbitals or wavefunctions, and this is indeed the case. As we proceed down the series H^-, He, Li^+, Be^{+2}, ..., etc., we find that the distance in energy between the occupied state and the higher-energy states increases. Also, we find that the systems become less and less polarizable. Again, if we compare helium and beryllium, both of which are ordinarily considered to have s^2 valence-state configurations, we find beryllium to be much more polarizable. This is due to the presence of empty 2p orbitals lying fairly close in energy to the 2s AO in beryllium. In fact, these two atoms are strongly reminiscent of the $n=1$, $n=2$ polarizabilities in hydrogen, the chief difference being that the 2s and 2p levels of beryllium are only close in energy—not actually degenerate.

12-8 Degenerate-Level Perturbation Theory by Inspection

Degenerate and nondegenerate perturbation theories share the feature that the first-order correction to the energy is the expectation value of the perturbation operator using the unperturbed wavefunction. Often we can evaluate or estimate this expectation value by inspection. The special problem for *degenerate* wavefunctions or orbitals is that there is an infinite number of ways to express the unperturbed wavefunctions, and only one of these ways is proper for finding $W^{(1)}$. The mathematical procedure for finding the proper set of functions tends to complicate and obscure the situation. *However, it is often easy to deduce the proper unperturbed wavefunctions by inspection.*

The clue for doing this comes from Eq. (12-76). It is apparent that this equation, which leads to the first-order corrections to the energy *and* to the proper unperturbed (or zeroth-order) wavefunctions, is exactly like the secular determinantal equation we would use for a variational calculation. Since the secular equation comes from a calculus-based determination of energy extrema, we know that the energies that come out of Eq. (12-76) must be the minimum and maximum possible values and that the proper zeroth-order functions must be those that give the maximum and minimum expectation values for H'. This leads immediately to a powerful insight: *For a given*

[7] In the language of hybridization (Chapter 13), $\phi_+^{(0)}$ and $\phi_-^{(0)}$ are sp hybrids.

perturbation, the proper zeroth-order degenerate wavefunctions to use for predicting the effect on energies are that set giving the greatest difference in response.

This rule can be seen operating in the hydrogen atom polarization example of the preceding section. The proper combination of 2s and $2p_z$ AOs is the pair of sp hybrid combinations giving the greatest energy increase and decrease. (The $2p_x$ and $2p_y$ AOs do not mix any 2s or $2p_z$ into themselves because to do so will have no energy-raising or -lowering effect.)

Another example is seen when changing the coulomb integral at an atom in the HMO-level treatment of a cyclic molecule. Such molecules always have some degenerate MOs, and so one must decide which linear combination is "proper" for evaluating the effect of this perturbation. The situation is analyzed most easily by recognizing that mixing such degenerate MOs together corresponds to rotating the positions of the nodes and antinodes in the molecule. So the question of finding the proper MOs becomes one of asking where the nodes and antinodes should be located. The greatest difference in response to changing the coulomb integral at an atom comes when one of the degenerate pair of MOs has a node (i.e., is zero) at that atom and the other MO has an antinode (i.e., has maximum absolute value) there. Suppose, for example, you are asked to evaluate the effect on a particular degenerate pair of MOs of such a perturbation at carbon number 3 in a cyclic molecule. When you look at those MOs in a tabulation, you may find that atom 3 is not at the node or antinode position for either MO. Instead, you might observe that the node of one MO and antinode of the other occur at atom number 4. You could set about mixing these MOs to move the node-antinode positions to atom 3, but it is much easier and just as correct to pretend that the perturbation is occurring at atom 4, since, in a cycle, all atoms are equivalent. This allows you to use the tabulated MOs to evaluate the perturbation (now at atom 4) without any modification. Once this is recognized, the problem can be completed by inspection, just as in nondegenerate cases: The MO energy for the MO that has its node at atom 4 is unaffected to first order by the perturbation, and the other is affected by an amount proportional to the square of the coefficient at atom 4. Thus, the MOs become nondegenerate and are described to zeroth order by these "proper" functions.

It is important to realize that the identity of the proper unperturbed wavefunctions depends on the nature of the perturbation. Thus, for the cyclic system discussed above (which we now assume has an even number of atoms), stretching the molecule so that two bonds on opposite sides of the cycle get longer is treated perturbatively with a different pair of proper MOs. We can quickly guess which MOs those are by seeking the pair that respond most differently to the bond stretching. Since any degenerate MO in this even–alternant cycle has at least one nodal plane perpendicular to the molecular plane, there is always a rotational position that places a nodal plane across the center of the two bonds being stretched. The mate to this MO then has an antinode in these bonds. That means that one MO is antibonding in these bonds and the other MO is bonding. This in turn means that one MO's energy will drop as the bond lengthens and the other MO's energy will rise. This, then, is the proper MO set to use for evaluating this perturbation to first order. Furthermore, these MOs are nondegenerate after the perturbation and are the zeroth-order approximation to the MOs of the perturbed system.

The two perturbations described above bring up a final important point. Suppose we have two different perturbations occurring simultaneously in a system involving degenerate MOs or state functions. The question then arises as to whether the proper zeroth-order functions are the same or different for the two perturbations. We have just seen that, if the perturbations involve substituting a new atom for a carbon and stretching a pair of bonds in a cyclic molecule, the proper zeroth-order MOs are different. This means that one of these perturbations favors a postperturbative pair of nondegenerate MOs that is different from the pair being favored by the other perturbation. In other words, these perturbations compete to produce different final MOs. The other possibility is that two perturbations have the same set of proper zeroth-order wavefunctions, which means that they cooperate to produce the same final MOs. The cooperative situation is the one that gives the largest energy effects, since a single set of proper zeroth-order functions is selected that produces the greatest difference in response to both perturbations simultaneously. Competing perturbations require a compromise among zeroth-order functions which involves diminution in the differences in energy response.

The reader may have realized that there is another way to select the proper zeroth-order wavefunctions, and that is to use symmetry. Recall that the proper wavefunctions $\psi_i^{(0)}$ and $\psi_j^{(0)}$ are those that "diagonalize the determinant," i.e., that cause $\langle\psi_i^{(0)}|H'|\psi_j^{(0)}\rangle$ to equal zero. In each of the examples cited above, this can be seen to be true. The H' operator corresponding to changing the coulomb integral at an atom, and nowhere else, is unaffected by reflection through a plane containing that atom. The MO placing a maximum at that atom is symmetric for such a reflection, and the MO placing a node there is antisymmetric. Hence the integrand is overall antisymmetric and the integral vanishes, in accord with the requirement that must be satisfied by proper zeroth-order wavefunctions. A similar argument applies for stretching or shrinking bonds, with the reflection plane now occurring at the bond midpoints. As a useful guide, then, *proper zeroth-order wavefunctions may be recognized as those having opposite symmetry for an operation by which the perturbation operator is unaffected.*

EXAMPLE 12-7 Consider the cyclopentadienyl radical in the HMO approximation. Assume it to be planar and pentagonal. (See Appendix 6 for data.) If α_3 is perturbed to become $\alpha + 0.1000\beta$, what are the MO energy changes, to first order?

SOLUTION ▶ The lowest-energy MO is nondegenerate, and all atoms have the same coefficient, so $\Delta E_1 = (0.4472)^2(0.1000\beta) = 0.0200\beta$. The second two MOs are degenerate, so we need to identify the proper zeroth-order combinations. One should be symmetric with respect to reflection through atom 3, the other antisymmetric. None of the tabulated MOs have these properties about atom 3, but they do have these properties for atom 1. Therefore, we proceed by pretending that atom 1 is the one being perturbed (it makes no difference). Then, $\Delta E_2 = (0.6325)^2(0.1000\beta) = 0.0400\beta$, $\Delta E_3 = 0.0000\beta$, $\Delta E_4 = \Delta E_2$, $\Delta E_5 = \Delta E_3$. ◀

12-9 Interaction Between Two Orbitals: An Important Chemical Model

Qualitative quantum-chemical discussion often relies on a simplified model wherein all interactions are neglected except for the primary one. For example, in considering how a Lewis base and a Lewis acid interact, one might consider only the highest occupied

MO (HOMO) of the base (electron donor) and the lowest unfilled MO (LUMO) of the acid (electron acceptor). Consideration of the ways two levels interact will provide some useful rules of thumb and also consolidate some of our earlier findings.[8]

We label the unperturbed orbitals and energies ψ_a, ψ_b, and E_a, E_b. We will work within the Hückel type of framework so that state energy differences may be written as orbital energy differences. The orbitals are allowed to interact with each other by virtue of close approach. We will assume that the perturbation felt by the orbitals is proportional to the overlap between them. (If the systems involved are ions, strong electrostatic interactions will exist as well. Therefore, we assume all systems to be neutral.)

Since the overlap of ψ_a on ψ_b is the same as that of ψ_b on ψ_a, the energy change to first order is the same for both levels and hence we ignore it (unless the levels are degenerate—an eventuality we discuss shortly).

The second-order contributions to the energies are:

$$W_a^{(2)} = \frac{|\langle \psi_a | H' | \psi_b \rangle|^2}{E_a - E_b} \tag{12-84}$$

$$W_b^{(2)} = \frac{|\langle \psi_a | H' | \psi_b \rangle|^2}{E_b - E_a} \tag{12-85}$$

Thus, the second-order energy $W_a^{(2)}$ is positive if $E_a > E_b$, negative if $E_a < E_b$. In other words, the higher-energy level is pushed up, or destabilized, through interaction with the level below, and the lower-energy level is stabilized by interaction with the level above. This behavior, which we noted also in connection with the perturbed particle in a wire, may be summarized by the statement: *interacting levels repel each other*. Equations (12-84) and (12-85) also make clear that *the levels repel each other more strongly the greater their interaction (numerator) and the smaller their energy separation (denominator)*.

If $E_a = E_b$, we must set up a 2×2 first-order perturbation determinant using ψ_a and ψ_b as basis, and find the two roots. Just as was true in the polarizability calculation of Section 12-7, we will find that one root lies above E_a, and the other lies an equal distance below. Thus, for *degenerate levels*, the repulsion between levels becomes a *first-order effect*.

The first-order mixing of ψ_a and ψ_b parallels the energy results in the expected way. If the second-order energy stabilizes the level, the orbitals are mixed in a bonding fashion. Since we have seen that it is the lower-energy orbital that is stabilized, we can state that, *if two orbitals interact, the lower-energy one of the two mixes into itself the higher-energy one in a bonding way, while the higher-energy orbital mixes into itself the lower one in an antibonding way* [5]. In short, "the upper combination takes the node" [5].

As an example of the utility of this model, consider the norbornadiene molecule (see Fig. 12-7a). According to our understanding of unsaturated systems, the double bonds in this molecule should behave like isolated ethylene double bonds and not like the conjugated bonds of butadiene. But these bonds are at an orientation and proximity allowing significant overlap between $2p_\pi$ AO lobes beneath the molecule. If we treat this system with our two-level model, the unperturbed MOs are the bonding

[8]The discussion in this section follows closely that of Hoffmann [5]. That article reviews a variety of illustrative applications of the two-orbital model.

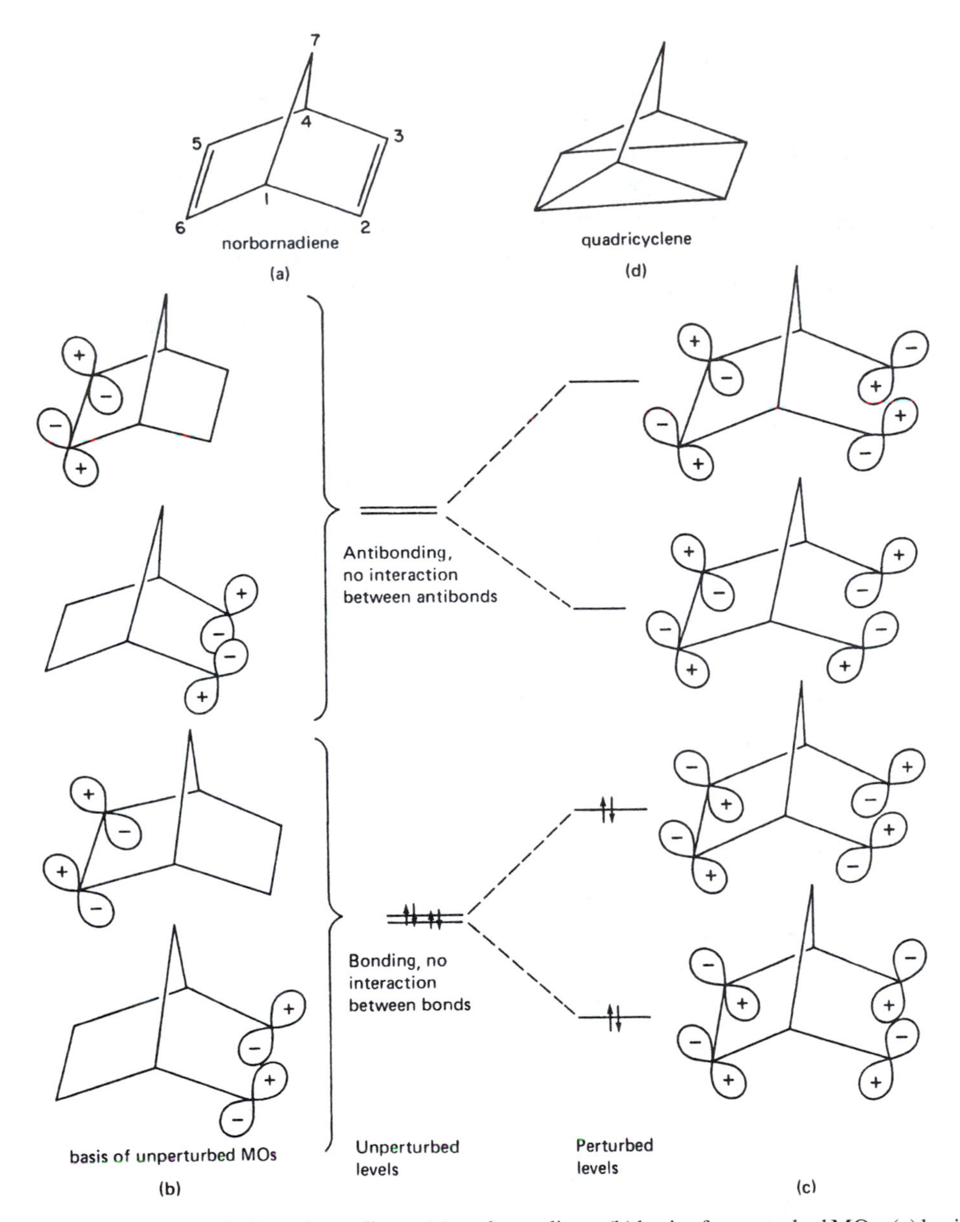

Figure 12-7 ▶ π MOs in norbornadiene: (a) norbornadiene; (b) basis of unperturbed MOs; (c) basis of perturbed MOs; (d) quadricyclene (after R. Hoffmann [5].)

and antibonding π MOs of two ethylene molecules (Fig. 12-7b). We expect the MO overlap across the molecule to split each of these levels as indicated in Fig. 12-7c, the bonding interaction producing stabilization in each case. (Combinations involving a bonding MO on one side and an antibonding MO on the other are ruled out because symmetry forces the interaction term to vanish. Even if this were not the case, the energy gap between these unperturbed MOs would make such contributions negligible compared to those from the degenerate MOs.)

A careful look at Fig. 12-7c reveals certain implications about norbornadiene. For instance, the ionization energy for norbornadiene should be smaller than that for norbornene, which has but one double bond. Experimental measurements support this contention. Also, since the highest occupied MO of norbornadiene is antibonding between carbons 3 and 5 and also 6 and 2, whereas the lowest empty MO is bonding between these carbons, excitation of an electron from the former MO to the latter should promote formation of quadricyclene (Fig. 12-7d). This compound is a common product in the photochemistry of norbornadienes.

12-10 Connection Between Time-Independent Perturbation Theory and Spectroscopic Selection Rules

Molecules may change to higher-or lower-energy states under the influence of incident light. Such processes are called, respectively, absorption and induced emission. Perturbation theory can be used to study such transitions induced by an external oscillating electromagnetic field. Here we briefly describe a rather simple connection between selection rules and the perturbation theory we have discussed in this chapter.

Let a molecule initially be in a state with wavefunction ψ_i. We are interested in the probability of a transition occurring to some final state with wavefunction ψ_f. A time-dependent perturbation treatment (not given here, but see Section 6-16) indicates that such a transition is probable only when the external field frequency ν satisfies the conservation of energy relation

$$\nu = |E_f - E_i|/h \tag{12-86}$$

Even when this condition is satisfied, however, we may find experimentally that the transition is so improbable as to be undetectable. Evidently some factor other than satisfaction of Eq. (12-86) is also involved.

Since the molecule is being subjected to light, it experiences fluctuating electric and magnetic fields. For ordinary (as opposed to magnetic) spectroscopy, the effects of the magnetic field are negligible compared to those of the electric field. Therefore, we ignore the former and imagine the molecule *at a particular instantaneous value* of an external electric field. As we have seen from earlier sections, this field causes polarization through the admixture of unperturbed wavefunctions ψ_1, ψ_2, etc., with ψ_i to form a perturbed wavefunction ϕ_i: For a normalized ϕ_i, we have, therefore,

$$\phi_i = c_i\psi_i + c_1\psi_1 + c_2\psi_2 + \cdots + c_f\psi_f + \cdots \tag{12-87}$$

At a later time, when the perturbation is over, the system returns to an unperturbed state. The probability of returning to the initial state is given by $c_i^*c_i$. The probability of going instead to the final state described by ψ_f is given by $c_f^*c_f$. If c_f is zero, there is no tendency for a transition to that final state to occur and the transition is said to be *forbidden*. If c_f is nonzero, the transition is *allowed*.

Since the coefficient c_f is given to first order by [see Eq. (12-18)],

$$c_f = \frac{\langle\psi_f|H'|\psi_i\rangle}{E_i - E_f} \tag{12-88}$$

where H' is the perturbation operator for the electric field component of the light, we can say at once that a transition between two states is forbidden (to first order) if $\langle\psi_f|H'|\psi_i\rangle$ vanishes.

One example of a forbidden transition is that between a singlet state and a triplet state of a system. We can see this at once since H' is an *electric* field and does not interact with spin *magnetic* moment. Therefore the orthogonality between ψ_i and ψ_f due to spin functions is uninfluenced by H' and the integral $\langle \psi_f | H' | \psi_\mathrm{i} \rangle$ must vanish. In effect, the perturbation H' causes the initial singlet state to become polarized by mixing in other singlet state functions, but gives no impetus for mixing in states of different multiplicity. Transitions between states of different multiplicity are said to be *spin forbidden*.

Another example of a forbidden transition is that between two different s-type states of a hydrogen atom. Such states have spherically symmetric wavefunctions, but H' (the electric field) is antisymmetric for reflection through a plane (to within an additive constant), and so $\langle \psi_f | H' | \psi_i \rangle$ must vanish for reasons of symmetry. It is easy to generalize this argument to other states of the hydrogen atom (p to p, d to d, etc.) and also to certain other atoms (e.g., the alkali metals) electronically similar to hydrogen.

Molecular electronic transitions can be understood from the same standpoint. The intense $\pi^* \leftarrow \pi$ transition in ethylene is a simple example.[9,10] Let us imagine that the molecule is oriented as shown in Fig. 12-8. Suppose we could somehow orient all our molecules this way (in a host matrix or in a crystal) and that we then subjected the sample to plane-polarized light. We will consider what should happen for light polarized in each of the directions x, y, z. First, let us consider the integral $\langle \pi | z | \pi^* \rangle$. The ethylene molecule has three reflection planes of symmetry, so we can examine the symmetry of the integrand with respect to each of these three reflections. Remember that, if the integrand is antisymmetric for any one of these reflections, the integral vanishes. The function π can be seen, from inspection of Fig. 12-8, to be symmetric for reflection in the xz and yz planes, antisymmetric for reflection in the xy plane. These observations are shown in Table 12-2, along with similar conclusions regarding the symmetries of the functions π^*, x, y, and z. The symbols "s" and "a" stand for "symmetric" and "antisymmetric." Our integral $\langle \pi | z | \pi^* \rangle$ can now be seen to have an integrand that is symmetric for reflection in the xz plane but antisymmetric for reflection in the xy

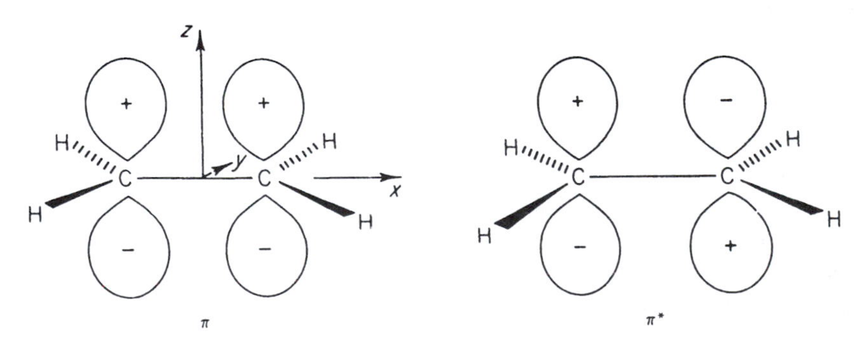

Figure 12-8 ▶ π and π^* MOs for ethylene. The C–C bond is coincident with the x axis and all nuclei lie in the xy plane.

---

[9]Here, π^* refers to the antibonding π_g MO, not to complex conjugation.

[10]In this discussion, we consider molecules of high symmetry, so that reflection planes of symmetry exist. For molecules of lower symmetry, one must resort to actual calculation of the c_f in Eq. (12-88).

TABLE 12-2 ► Symmetries of Functions under Reflection through Cartesian Coordinate Planes

	Symmetry operation		
Function	xy reflection	xz reflection	yz reflection
π	a	s	s
π^*	a	s	a
z	a	s	s
x	s	s	a
y	s	a	s
$\pi\ z\pi^*$	a a a = a	s s s = s	s s a = a
$\pi\ x\ \pi^*$	a s a = s	s s s = s	s a a = s
$\pi\ y\pi^*$	a s a = s	s a s = a	s s a = a

and yz planes. Hence, this integral vanishes, and this transition is forbidden insofar as light polarized perpendicular to the molecular plane is concerned. According to Table 12-2, symmetry also causes $\langle\pi|y|\pi^*\rangle$ to vanish. However, symmetry does not force $\langle\pi|x|\pi^*\rangle$ to vanish. Therefore, the $\pi^* \leftarrow \pi$ transition is allowed and is polarized along the C–C axis. Indeed, because of the significant spatial extension of the π and π^* MOs (compared to AOs), the integral $\langle\pi|x|\pi^*\rangle$ is relatively large, which means that the transition is not only allowed, but is *intense*.

It is physically reasonable that the $\pi^* \leftarrow \pi$ transition should be polarized along the C–C axis. If some π^* character is mixed with π, it is easy to see that this results in a shift of π charge from one carbon to the other—a shift along the x axis. Conversely, an electric field in the x direction will cause polarization along the x axis, hence mix π^* character into the π MO. When the perturbation is removed, a finite probability exists that the molecule will go to the state wherein π^* is occupied.

In the above example, we imagined all ethylene molecules to be identically oriented. Under those conditions, we would observe a maximum in $\pi^* \leftarrow \pi$ absorption when our incident light was polarized parallel to the molecular axis and zero absorption (ideally) when the polarization axis was perpendicular to the molecular axis. If the light is unpolarized, or if the ethylene is randomly oriented, as in liquid or gaseous states, the absorption is isotropic and allowed because there is always a certain degree of "overlap" between the molecular axes of most of the molecules and the direction of the electric field due to the light.

The example discussed above illustrates a simple and powerful rule for recognizing whether or not a given atomic or molecular orbital dipolar transition is allowed (assuming electron spin agreement): *A dipole transition is allowed between orbitals that have opposite symmetry for one and only one reflection plane.*[11] *The polarization of the transition will be perpendicular to that reflection plane.*

[11]This means that, for an allowed transition, one orbital has one and only one nodal plane where the other does not.

EXAMPLE 12-8 Selection rules for allowed dipolar transitions in a hydrogen atom include $\Delta l = \pm 1$, $\Delta m_l = 0, \pm 1$. Show that these rules are equivalent to the symmetry rule stated above.[a] Consider the following cases: $2s \leftarrow 1s$, $2p_x \leftarrow 1s$, $3d_{xy} \leftarrow 2s$, $3d_{xz} \leftarrow 2p_z$, $3d_{xy} \leftarrow 2p_z$.

[a]There is yet another way to rationalize these selection rules. Photons are bosons with spin quantum number of 1, hence they carry spin angular momentum. The total angular momentum of the atom or molecule that absorbs a photon must change to obey the conservation requirement consistent with the vector addition of orbital and photon angular momenta. The Δl rule handles that for one-electron atoms. Furthermore, for a unit change of l, the z-component of the absorber's orbital angular momentum can change by $+1$, 0, or -1 a.u. The Δm_l rule takes care of that. This rationalizes the atomic selection rules, but doesn't yield information on polarization or intensity as readily as the more physical approach described in the text, and it is less easily applied to molecules. Generally, however, the more ways one understands a thing, the better.

SOLUTION ▶ The $2s \leftarrow 1s$ transition is forbidden by the Δl rule (because $\Delta l = 0$) and also by the symmetry rule (because neither wavefunction has a nodal plane). This makes it easy to see that the Δl rule applies because it forces one of the orbitals to have one and only one more nodal plane than the other. The $2p_x \leftarrow 1s$ transition is allowed by the Δl rule. Since $2p_x$ is a linear combination of $2p_{+1}$ and $2p_{-1}$, the Δm_l rule is also satisfied, so the transition is allowed. When we examine symmetry, we note that there is only one plane of disagreement–the yz plane, so this approach also indicates that the transition is allowed, and that it is polarized in the x direction (perpendicular to the yz plane). The $3d_{xy} \leftarrow 2s$ transition is forbidden by the Δl rule since $\Delta l = 2$. The symmetry approach indicates that a transition is forbidden because the $3d_{xy}$ AO has two nodal planes, whereas $2s$ has none. The $3d_{xz} \leftarrow 2p_z$ transition has $\Delta l = 1$. $3d_{xz}$ is a combination of $3d_{+1}$ and $3d_{-1}$, and $2p_z$ is $2p_0$, so $\Delta m_l = \pm 1$. Therefore, this is an allowed transition. The symmetry approach shows $3d_{xz}$ with two nodal planes (xy and yz) and $2p_z$ with one (xy). Therefore, there is only one plane of symmetry disagreement (yz), so the transition is allowed and is x polarized. The $3d_{xy} \leftarrow 2p_z$ transition has $\Delta l = +1$. $3d_{xy}$ is a linear combination of $3d_{+2}$ and $3d_{-2}$, and $2p_z$ is $2p_0$, so $\Delta m_l = \pm 2$. Therefore, the transition is forbidden. The symmetry analysis shows $3d_{xy}$ having two nodal planes (xz and yz), and $2p_z$ having one (xy). Therefore, they disagree in symmetry in three planes, and the transition is forbidden. This last case makes it clear why the Δm_l rule is needed: When satisfied, it guarantees that all but one of the nodal planes in the two AOs coincide. (If $\Delta m_l = \pm 2$, then, taking the z axis to be vertical, one AO has two more vertical nodal planes than the other, so there must be at least two reflection planes of symmetry disagreement.) ◀

12-10.A Problems

12-1. Prove that the energy to first order for the lowest-energy state of a perturbed system is an upper bound for the exact energy of the lowest-energy state of the perturbed system, that is, that $E_0 + W_0^{(1)} \geq W_0$.

12-2. A one-dimensional box potential is perturbed as sketched in Fig. P12-2. From a consideration of the first three wavefunctions of the unperturbed system, ψ_1, ψ_2,

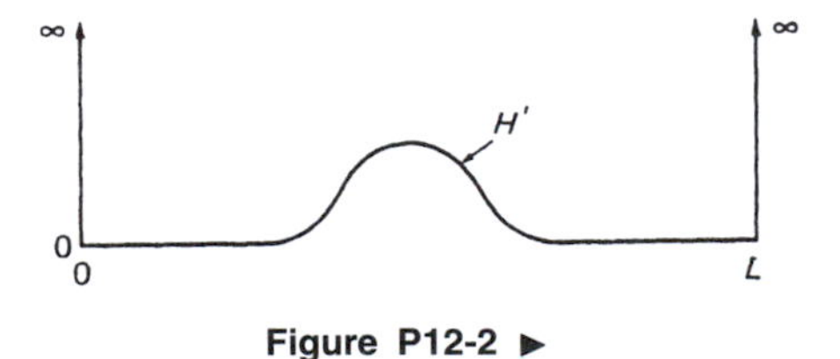

Figure P12-2 ▶

ψ_3, which will have its energy increased *most*, to first order, and which least? No explicit calculation is necessary to answer this question.

12-3. Evaluate by inspection the first-order contributions to the energies for the states shown in Fig. P12-3. In every case the unperturbed state is a particle in the *indicated* state in the one-dimensional, infinitely deep, square well.

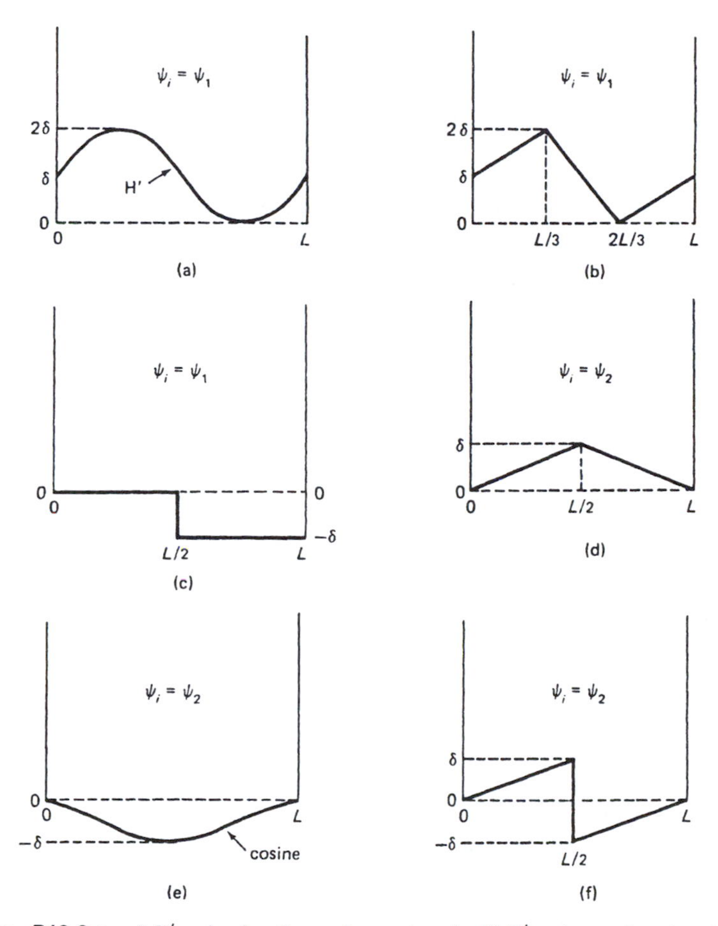

Figure P12-3 ▶ a) H' = sine function as shown, $\psi_i = \psi_1$. b) H' as shown, $\psi_i = \psi_1$. c) H' as shown; $\psi_i = \psi_1$. d) H' as shown; $\psi_i = \psi_2$. e) H' = cosine as shown, $\psi_i = \psi_2$. f) H' as shown; $\psi_i = \psi_2$.

12-4. a) Evaluate $W_2^{(1)}$ for the particle in a box with the perturbing potential shown in Fig. P12-4.

b) For the perturbation above:

1) What sign would you expect for $c_{21}^{(1)}$? Describe the effect of this on ϕ_1; on $W_1^{(2)}$.

2) What sign would you expect for $c_{12}^{(1)}$? Describe the effect of this on ϕ_2; on $W_2^{(2)}$.

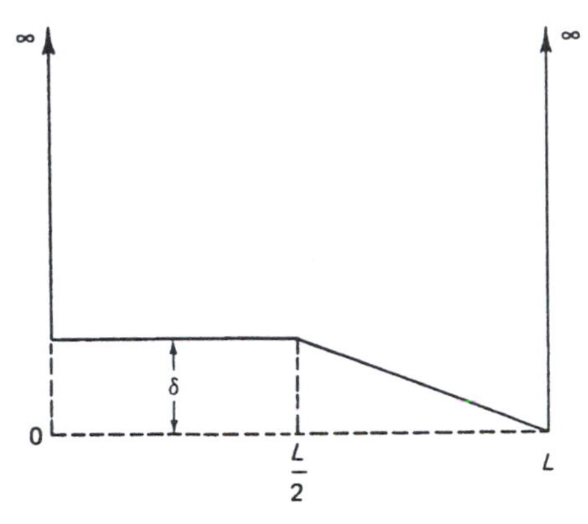

Figure P12-4 ▶

12-5. Calculate $c_{41}^{(1)}$ for the perturbed particle in a wire example discussed in Section 12-3. Show that $c_{41}^{(1)}$ is only about 2% as large as $c_{21}^{(1)}$. Use

$$\int_0^{\pi} y \sin my \sin ny \, dy = \left[(-1)^{m+n} - 1\right] 2mn/(m^2 - n^2)^2, \quad m, n = 1, 2, \ldots, m \neq n$$

12-6. An electron moves in a harmonic potential, $V = \frac{1}{2}kx^2$. What is the effect, to first order, on the energies of superimposing an electric field, $V' = Ex$? Explain your reasoning.

12-7. Calculate the energy to first order of He^+ in its lowest-energy state. Use the hydrogen atom in its ground state as your zeroth-order approximation. Use atomic units. Predict the signs (plus, zero, minus) of $c_{2s,1s}^{(1)}$ and $c_{2p_0,1s}^{(1)}$. Explain your reasoning.

12-8. The previous problem gives an energy to first order of $-3/2$ a.u. for He^+, using H as a starting point. Now reverse the process and try to get the ground-state energy of H to first order, using He^+ as starting point. Discuss the reasonableness of your answer in terms of a lower bound.

12-9. A hydrogen atom in its ground state is perturbed by imposition of a uniform z-directed electric field of strength F atomic units: Field $= -Fz = -Fr\cos\theta$.

a) What is the first-order change in energy experienced by the atom? Show your logic.
b) For the first-order correction to the ground-state *wavefunction*, we can consider $c_{2s,1s}^{(1)}$ and $c_{2p_z,1s}^{(1)}$ to be the coefficients for mixing in 2s and $2p_z$ character. Predict the signs of these coefficients and sketch their effects on the ground-state wavefunction.
c) Predict the sign of $W_{1s}^{(2)}$ and explain your reasoning.

12-10. What is the Hückel MO π-electron energy to first order of the molecule (**I**) if C_4 is perturbed by $H' = 0.1\beta$? (See Appendix 6 for Hückel data.)

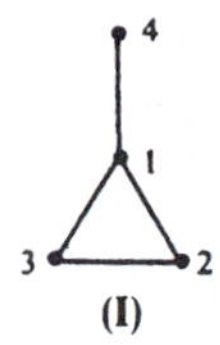

(I)

12-11. Consider the fulvene molecule (**II**) in the HMO approximation (see Appendix 6).

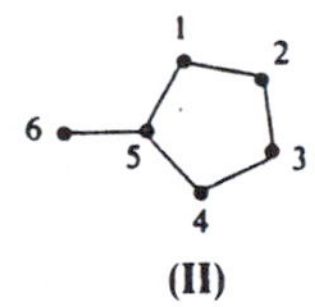

(II)

At which carbon will a perturbation involving α affect the total π-electron energy the least? Calculate to first order the energy of fulvene with $a_6 = a + 0.5\beta$. If a computer and Hückel program are available to you, calculate the energy for this perturbed molecule directly and compare with your first-order result.

12-12. Produce an expression for π_{kk}, the atom–atom *self*-polarizability, from Eq. (12-70). Can this, like π_{lk}, have either sign? What can you infer about the sign of δq_k when $\delta\alpha_k$ is positive? Discuss the physical sense of this.

12-13. Using data from Example 12-6, calculate the self-atom polarizability, $\pi_{1,1}$ for carbon 1 in naphthalene.

12-14. Consider the hexatriene system. (See Appendix 6 for data.) *One* of the two central atoms is perturbed so that α becomes replaced by $\alpha + h\beta$ on that one atom only.

a) What is the energy change, to first order, of the total π energy of hexatriene?
b) Which of the *occupied* MOs is perturbed the most (i.e., has the largest first-order contribution to the energy) by this perturbation?

12-15. Answer the following questions from examination of HMO data for benzcyclopentadienyl radical, C_9H_7, given in Appendix 6.

a) What is the *total π energy* to first order if atom 5 is replaced by a new atom X which contributes the same number of π electrons a carbon did and has $\alpha_x = \alpha + 0.5\beta$ and also has $\beta_{CX} = 0.9\beta$?
b) For the positive ion of this system, how do you think the self-atom polarizability should compare *qualitatively* at atoms 1 and 2? Make your prediction by inspection, rather than computation, and explain your thinking.

12-16. Use simple Hückel MOs for butadiene to calculate to first order the change in π energy that would result from closing *cis*-butadiene to cyclobutadiene (**III**). Repeat the approach for closing hexatriene to benzene. (**IV**). Which of these two systems benefits most from cyclic as opposed to linear topology? For each system, compare your energy to first order with the actual Hückel energy (see Appendix 6 for data).

12-17. The unperturbed Hückel MO energy levels of the allyl system are sketched in Fig. P12-17. The system is perturbed so that $H_{22} = \alpha + c\beta$ where c is positive.

a) Sketch the effects of this perturbation, to first order, on the energy levels.
b) Calculate to first order the perturbed MOs $\phi_1^{(1)}$, $\phi_2^{(1)}$, $\phi_3^{(1)}$ in terms of the unperturbed MOs ψ_1, ψ_2, ψ_3. Sketch the results in a manner that makes clear the nature of the change in each MO.

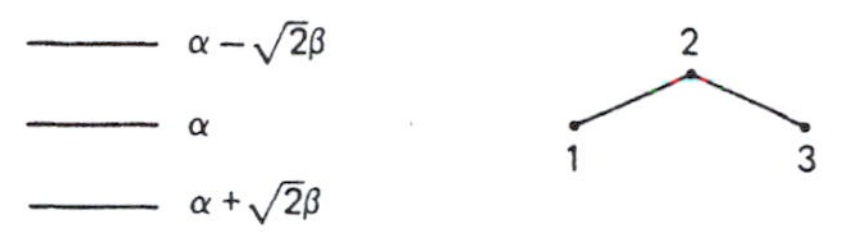

Figure P12-17 ▶

12-18. Calculate the atom–atom polarizabilities $\pi_{1,2}$ and $\pi_{1,3}$ for the allyl cation. What do your results indicate will happen to the π-electron densities at atoms 2 and 3 if atom 1 becomes more attractive?

12-19. The cyclopropenyl system is perturbed so that the Hückel matrix element $H_{22} = a + c\beta$, where c is positive.

a) Ascertain, by calculation or inspection, the appropriate zeroth-order degenerate wavefunctions for this situation.
b) Sketch the effects of the perturbation on the orbital energies to first order.

12-20. Given the HMO data in Appendix 6 for the cyclopentadienyl radical, calculate to first order all the orbital energies you would obtain if atom 2 were changed so that α_2 became $\alpha + 0.5\beta$.

12-21. Imagine two allyl radicals coming together to form benzene. Each allyl has three π MOs and benzene has six.

a) Sketch an energy level diagram showing how interaction between allyl orbitals can be related to the MO energies of benzene, at the simple Hückel level.
b) Show with sketches how the three lowest-energy zeroth-order allyl combination orbitals are related to the three lowest-energy benzene MOs.
c) Estimate to first order the energies of the benzene-like MOs coming from the *nonbonding* MOs of the allyl pair. Compare these to the relevant MO energies of benzene. (Remember that, when combining fragments, you must renormalize the functions)
d) What is the energy to first order of the lowest MO of the allyl combination, and how does it compare to the lowest benzene orbital energy?

12-22. Discuss the possible transitions $\pi_3 \leftarrow \pi_2$ and $\pi_4 \leftarrow \pi_2$ in butadiene (Here π_1 is the lowest-energy MO, etc.)

a) Are they dipole-allowed transitions?
b) If yes, what is the polarization?

TABLE P12-24[a]

Level	Energy	c_1	c_2	c_3	c_4	c_5	c_6	c_7
7	-2.01β	−0.238	0.500	−0.406	0.354	0.336	0.354	−0.406
6	-1.26β	−0.397	0.500	−0.116	−0.354	0.562	−0.354	−0.116
5	-1.00β	0	0	−0.500	0.500	0	−0.500	0.500
4 ↑	0.00	−0.756	0	0.378	0	−0.378	0	0.378
3 ↑↓	1.00β							
2 ↑↓	1.26β							
1 ↑↓	2.10β							

[a]Coefficients for $2p_z$ basis function forming the four highest MOs are listed. Overlap has been assumed to be negligible.

12-23. Calculate the dipole moment in the z direction for the states $\phi_{\pm}^{(0)}$ of Eq. (12-81). Now construct a variational function of the form

$$\phi = \cos(\alpha)2s + \sin(\alpha)2p_z$$

and maximize the z component of the dipole moment as a function of α. Are the states $\phi_{\pm}^{(0)}$ those of maximum dipole magnitude? Discuss why this is reasonable.

12-24. The benzyl radical, C_7H_7(V), has the Hückel energies and ground-state configuration given in Table P12-24.

The radical is trapped and oriented in an external reference system as shown. Light polarized in the x direction is beamed on the system, and the frequency varied until an absorption is observed. Assuming this to result from excitation of the unpaired electron, to which level has the electron been promoted?

(V)

12-25. The simple Hückel energies, occupation numbers, and coefficients for MOs in naphthalene (VI) are listed in Table P12-25. Assume that a single crystal of naphthalene is oriented so that each molecule is aligned with respect to an external coordinate system (VI).

a) Light polarized in the x direction is beamed on the crystal. Assuming that the electron is excited from the highest occupied MO, to which empty MOs could it be promoted by the x-polarized light?
b) Which transitions from the highest occupied MO would be allowed for y-polarized light?
c) Which ones would be allowed for z-polarized light?
d) Are any transitions from the highest occupied MO forbidden for nonpolarized light?

TABLE P12-25 ▶

Energies $-\beta$	Occupation no.	Coefficients									
		1	2	3	4	5	6	7	8	9	10
−2.303	2										
−1.618	2										
−1.303	2										
−1.000	2										
−0.618	2	−0.42	−0.26	0.26	0.42	0	−0.42	−0.26	0.26	0.42	0
+0.618	2	−0.42	0.26	0.26	0.42	0	−0.42	−0.26	0.26	0.42	0
+1.000	0	0	0.41	−0.41	0	0.41	0	−0.41	0.41	0	−0.41
+1.303	0	0.40	−0.17	−0.17	0.40	−0.35	0.40	−0.17	−0.17	0.40	−0.35
+1.618	0	−0.26	0.42	−0.42	0.26	0	−0.26	0.42	−0.42	0.26	0
+2.303	0	−0.30	0.23	−0.23	0.30	−0.46	0.30	−0.23	0.23	−0.30	0.46

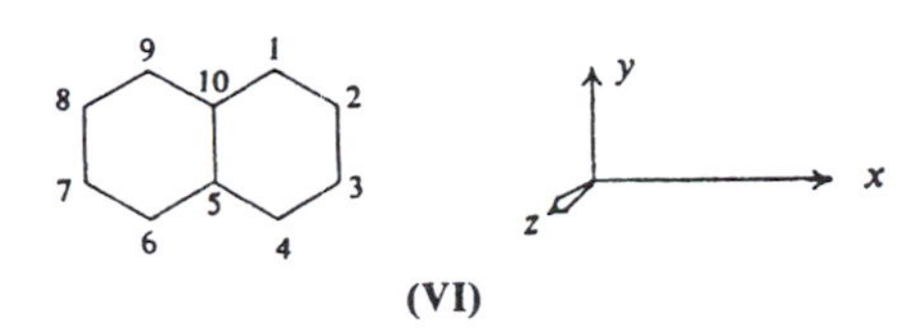

(VI)

12-26. A hydrogen atom in the 2s state is metastable, but, when passed through an electric field, its tendency to relax to the 1s state is greatly enhanced. Explain.

12-27. Using EHMO data for formaldehyde given in the Problems section of Chapter 10, decide which empty MOs can be reached by allowed dipolar transitions of an electron out of MO 7. For each allowed transition, state the polarization with respect to the molecular axis.

12-28. Square cyclobutadiene has a degenerate pair of nonbonding MOs. Calculate the energies of these MOs, to first order, when two carbons are replaced by atoms that contribute the same number of pi electrons as carbon did and are treated at the HMO level by replacing α with $\alpha + 0.5\beta$, and the substitutions occur at a) opposite and b) adjacent positions. Discuss your results in terms of cooperative vs. competing perturbations. Which compound should be more stable?

12-29 Calculate to first order the HMO energy changes for the occupied MOs of benzene that occur when the molecule is stretched so that an opposing pair of bonds have resonance integrals of 0.9β instead of 1.0β.

12-30 The intensity of a transition between initial (ψ_i) and final (ψ_f) states is dependent upon the transition dipole moment: $\vec{\mu} = e\int \psi_f^* \vec{r} \psi_i \, d\tau$. Calculate $\vec{\mu}$ in atomic units for $2p_0 \leftarrow 1s$ in a hydrogen-like ion having atomic number Z. Assume that the electron has the same spin in both states. (See Appendix **??** for the definition of 1 a.u. of electric dipole moment.)

Multiple Choice Questions

(Try to answer these without referring to the text.)

1. For the $n = 2$ states of a hydrogen atom in a uniform electric field of strength F,

a) the first-order correction to the energy equals zero for all four states.
b) there are four different values of the energy to first order.
c) the first-order correction to the energy is proportional to F for two states.
d) the first-order correction to the energy is proportional to F^2 for two states.
e) None of the above is a true statement.

2. Which one of the following statements is true for a hydrogen atom in its ground state that is placed in a uniform z-directed electric field?

a) The first-order correction to the energy equals zero.
b) The proper zeroth-order wavefunctions are $(1/\sqrt{2})(1s \pm 2p_z)$.
c) The second-order correction to the energy is positive.
d) The first-order correction to the wavefunction contains some $2s$ AO.
e) None of the above statements is true.

3. The hydrogen atom transition $3d_{xy} \leftarrow 2p_x$ is

a) dipole allowed and z-polarized.
b) dipole allowed and x-polarized.
c) dipole allowed and y-polarized.
d) dipole allowed and xy-polarized.
e) forbidden.

4. Which of the following integrals over all space vanish by symmetry for a hydrogen atom with nucleus at the coordinate origin? 1) $\int 1s\ y\ 2p_z dv$ 2) $\int 2p_z\ z\ 3d_{xy} dv$ 3) $\int 2p_z\ x\ 3d_{xz} dv$

a) 1 only.
b) 1 and 2 only.
c) 1 and 3 only.
d) 2 and 3 only.
e) 1, 2, and 3.

References

[1] L. Pauling and E. B. Wilson, *Introduction to Quantum Mechanics*. McGraw-Hill, New York, 1935.

[2] P. O. Löwdin, *J. Mol. Spectry*. **13**, 326 (1964).

[3] L. I. Schiff, *Quantum Mechanics*, 3rd ed. McGraw-Hill, New York, 1968.

[4] C. W. Scherr and R. E. Knight, *Rev. Mod. Phys*. **35**, 436 (1963).

[5] R. Hoffmann, *Accounts Chem. Res*. **4**, 1 (1971).

Chapter 13

Group Theory

13-1 Introduction

It should be evident by this point that symmetry requirements impose important constraints on MOs, and that one can often use symmetry arguments to tell what the MOs of a molecule must look like, in advance of calculation. We have also seen that one can often use symmetry arguments to tell whether or not an integral vanishes. Thus, we have made much use of symmetry already, but in an informal way. The formal use of symmetry, through group theory, will be described in this chapter. Knowledge of group theory augments one's power to use symmetry as a shortcut, and familiarity with formal symmetry notation is necessary to follow the literature of many areas of chemistry, particularly inorganic and quantum chemistry, and molecular spectroscopy.

13-2 An Elementary Example

Consider the following four operations, or commands:

1. Left face (L).
2. Right face (R).
3. About face (A).
4. Remain as you are (E).

These four operations constitute a group in the mathematical sense, and provide a convenient example with which to illustrate some definitions and terminology of group theory. The operations are called the *elements* of the group. Because there are four elements, this group is said to be of *order* four.

How do we know that these four operations constitute a group? Before answering this question, it is necessary to describe what is meant by a *product of* elements. Products of group elements are written in the usual algebraic manner. That is, the symbols are written side by side without an algebraic symbol. Just as xy is understood to be the product of x and y, LR is understood to be the product of "left face" and "right face." But elements of a group need not commute, and so the ordering of symbols in a product is important in group theory. The sequence of operations is understood to read from *right to left* in the product. Thus, LR means "right face" followed by "left face."

If we imagine a drill soldier carrying out the sequence of operations implied by LR, the result is a return of the soldier to his original position. That is, the product of

TABLE 13-1 ► Products of Elements in the Group of Four Commands

Second operation \ First operation	E	L	R	A
E	E	L	R	A
L	L	A	E	R
R	R	E	A	L
A	A	R	L	E

operations LR gives the same result as the single operation E. This is written $LR = E$. It is possible to write similar equations for all the product combinations in our group, and to arrange the results as in Table 13-1. An important fact to notice from this *group multiplication table* is that there is no *sequence* of operations in the group that cannot be accomplished by a *single* operation. All the possible positions open to the drill soldier through a sequence of two moves are also achievable through a single move. In other words, no product of elements takes us out of the group—the group exhibits *closure*.

The *reciprocal*, or *inverse* of an operation is that subsequent operation which returns the soldier to his original position. Hence, L is the reciprocal of R, which is expressed as $LR = E$, or $L = R^{-1}$. Examination of Table 13-1 shows that every column has E appearing once, which means that every element of our group has an inverse in the group.

The *associative* law is obeyed if, in general, the sequence of operations $A(BC) = (AB)C$, where the parentheses enclose the pair of elements the product of which is to be evaluated first. We can test whether our elements satisfy this law by trying out all the combinations. For example $L(RA) = LL = A$ and $(LR)A = EA = A$. One can quickly show that this group satisfies the associative law for all combinations of elements.

We are now in a position to define a group. A group is a set of elements that meets the following requirements:

1. The group contains the identity element (traditionally symbolized E) which corresponds to "make no change."

2. The group exhibits closure with respect to "multiplication." The product of any two elements in the group is a single element in the group.

3. There is a reciprocal in the group for every element of the group.

4. The elements of the group obey the associativity law $(AB)C = A(BC)$.

We mentioned that, in general, group elements need not commute. That is, it is possible that $AB \neq BA$. However, in this example, the elements do all commute. This results in Table 13-1 being symmetric about the main diagonal. Groups in which all the elements commute are called *abelian* groups.

13-3 Symmetry Point Groups

Our example of four operations of a drill soldier differs from the kinds of groups most commonly used in quantum chemistry because it is not a symmetry group. Assuming that the soldier begins by facing north, we are able to *distinguish* four distinct possible subsequent orientations for him (facing the four compass directions). After a *symmetry* operation, an object is understood to be *indistinguishable* from the object before the operation.

Whether or not an operation is a symmetry operation must be decided within the context of the object being operated on. For example, a square (devoid of identifying marks enabling us to distinguish one corner from another; see Fig. 13-1a) may be rotated by 90° about an axis perpendicular to the center (Fig. 13-1b), and be indistinguishable from its starting configuration. However, rotation by 120° does not lead to an indistinguishable configuration (Fig. 13-1c).

Therefore, rotation by 90° is a symmetry operation for a square, but rotation by 120° is not. However, for an equilateral triangle, rotation by 90° is not a symmetry operation, while rotation by 120° is. For a circle, both rotations are symmetry operations. The only operation that is a symmetry operation for *every* object is the identity operation E.

Another sort of symmetry operation pertains to infinite networks such as occur in idealized models of crystals. Here we can define operations that move an infinite line of cells or atoms to the left or right by a unit number of "steps," producing a configuration indistinguishable from the initial one (Fig. 13-2). Symmetry operations can be usefully categorized into those that leave at least one point in space unmoved (rotations, reflections, inversion) and those that do not (translations). The latter category is of importance in the fields of crystallography and solid-state physics. The former category is useful when we deal with systems that do not undergo unending periodic repetition in space, and it is with this category that we will be concerned in this chapter.

If we pick some object or shape of finite size and construct a group from symmetry operations for that shape, we have a *symmetry point group* for that object. We now illustrate how this is done for a ball-and-stick model of the ammonia molecule in its equilibrium nuclear configuration. It is not difficult to find operations that do no more than interchange identical hydrogen nuclei. There are a number of possible rotations about the z axis (Fig. 13-3). One could rotate by 120°, 240°, 360°, 480°, etc., either

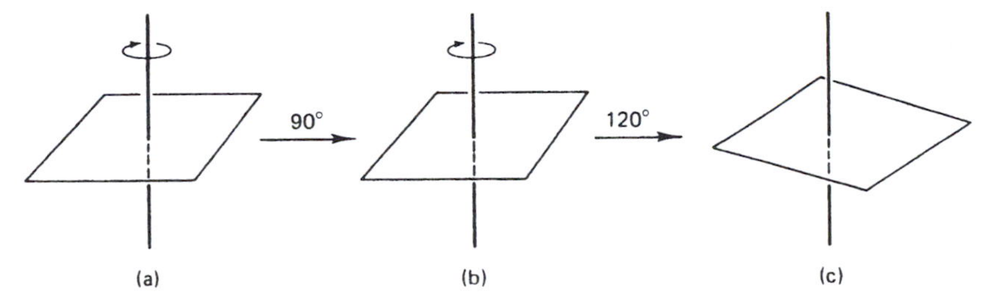

Figure 13-1 ▶ Effects of rotations on a square.

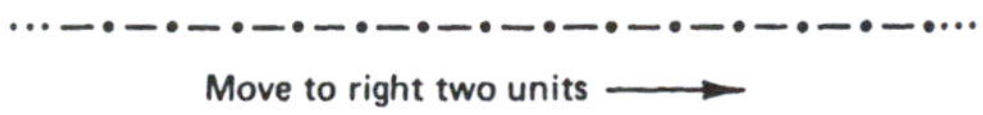

Move to right two units

Figure 13-2 ▶ Segment of an infinite repeating sequence of identical "cells".

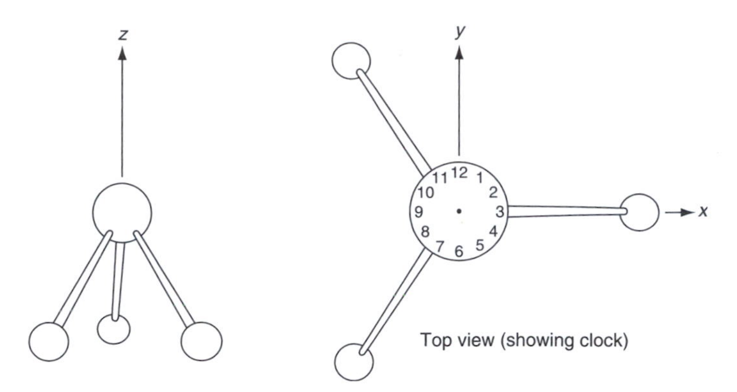

Figure 13-3 ▶ Orientation of a model of ammonia in Cartesian space.

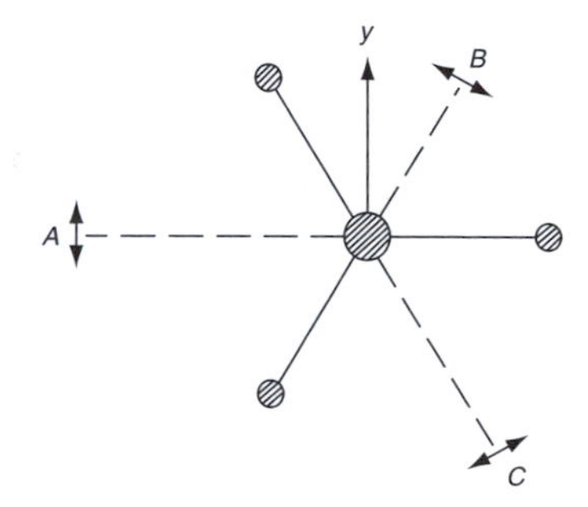

Figure 13-4 ▶ Three symmetry reflection planes for ammonia.

clockwise or counterclockwise. Then there are three reflection planes. One of these is the xz plane (labeled "A" in Fig. 13-4) and the other two are planes containing the z axis and one of the other N–H bonds. Also there is the identity operation E, which we know that we need to satisfy the general requirements for a group.

There is a problem with the rotations, and this is that we could produce an infinite number of them. How do we know when to stop? The answer is that we select only those that avoid "redundancies" among our indistinguishable configurations. Redundancies occur when two operations lead to "identical" indistinguishable configurations. For example, rotation by 360° is redundant with the identity operation because both operations leave the protons (which we imagine *for the moment* to be distinguishable) in the same locations. Similarly, rotation clockwise by 240° is redundant with rotation counterclockwise by 120° (Fig. 13-5). If we remove all such redundancies, we are left with only two rotations.

There is arbitrariness as to how we wish to describe these. If we let the first be clockwise rotation by 120°, the second could be described as either clockwise rotation

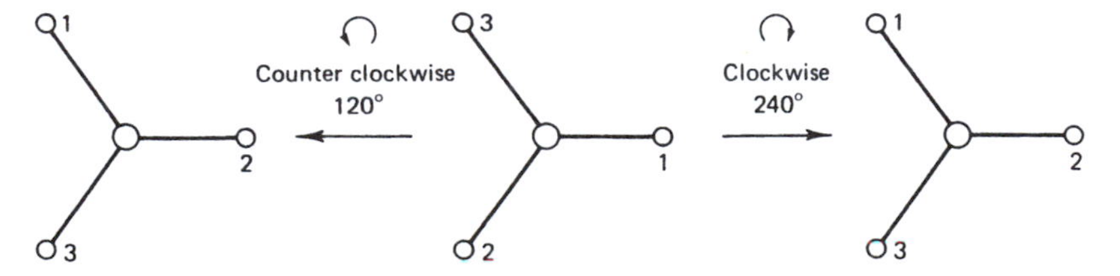

Figure 13-5 ▶ Identical configurations result from the two rotations shown.

by 240° or counterclockwise rotation by 120°. We will use the latter description. (Note that the clock is assumed to face $+\infty$ on the axis about which rotation occurs; see Fig. 13-3.)

Our list of symmetry operations is now as follows:

1. Identity (E).
2. Reflection through xz plane (A).
3. Reflection through plane B (B).
4. Reflection through plane C (C).
5. 120° clockwise rotation about z axis (D).
6. 120° counterclockwise rotation about z axis (F).

We now proceed to set up the multiplication table for these operations. It is suggested that the reader do this as an exercise, checking the result against Table 13-2. Construction of the multiplication table is more challenging in this case than in our earlier example.

Let us now use this table to see if the requirements for a group are satisfied by these symmetry elements. Since every product of two operations is equivalent to one of the operations of the set (i.e., since every spot in the table is occupied by one of our six symbols) the set exhibits closure. Also, the identity element is present in the set and occurs once in each column and row (Problem 13-2), and so every element has an inverse in the set. The associative law is satisfied, as one can establish by trying various examples, e.g., $D(CB) = DF = E$, $(DC)B = BB = E$, so that $D(CB) = (DC)B$. (In general, symmetry operations satisfy the associativity law.) Therefore, our set of six symmetry operations constitutes a symmetry point group of order 6.

Inspection of Table 13-2 reveals that it is not symmetric across its principal diagonal. For example, $CF = B$, and $FC = A$; this is not an abelian group. One might inquire whether this group is the smallest that we can set up for the ammonia model. Inspection of the multiplication table should convince the reader that the following subsets meet the requirements for a group: E; E, D, F; E, A; E, B; E, C. These *subgroups* can be

TABLE 13-2 ▶ Multiplication Table for Symmetry Operations of an Ammonia Molecule

Second operation \ First operation	E	A	B	C	D	F
E	E	A	B	C	D	F
A	A	E	D	F	B	C
B	B	F	E	D	C	A
C	C	D	F	E	A	B
D	D	C	A	B	F	E
F	F	B	C	A	E	D

distinguished from the full group by virtue of the fact that none of them exhausts all the possible physically achievable, nonredundant, indistinguishable configurations for the molecule.

EXAMPLE 13-1 BH_3, like ammonia, has a three-fold symmetry axis. However, BH_3 is planar. As a result, there is an extra symmetry operation—reflection through the molecular plane—that does not move any nuclei. Thus, both the identity operation and this reflection leave all nuclei unmoved. Is this an example of redundant operations?

SOLUTION ▶ These operations are not redundant. Even though all the nuclei stay fixed for both operations, reflection interchanges the spaces above and below the plane, whereas the identity operation changes nothing. Two operations are redundant only if they move all points in space in the same way. ◀

13-4 The Concept of Class

Imagine that we are subjecting our ammonia model to the various symmetry operations and that someone in a parallel universe is subjecting his ammonia model to symmetry operations too. Suppose that the parallel universe differs from ours in that everything is reflected through the xz plane. We shall refer to this as "the mirror A universe." We now pose the following problem. Suppose we initially have a particular configuration and our "mirror A" man has the corresponding configuration (achieved by reflecting through plane A) as shown in Fig. 13-6. If the mirror A man now performs some symmetry

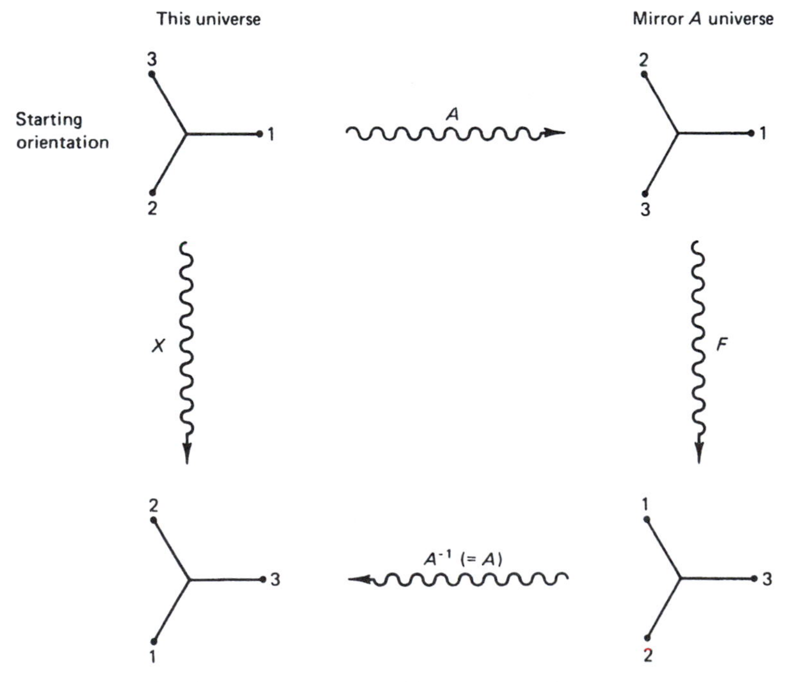

Figure 13-6 ▶ Operation X in this universe parallels operation F in the mirror A universe.

operation on his model, say, counterclockwise rotation by 120°, F, he ends up with the final configuration shown at the lower right of the figure. (We continue to define "clockwise" from clocks in our universe.) For us to arrive at the *corresponding* final orientation, what operation must we perform? We symbolize this unknown operation X. We can find out what our final configuration must be by taking the *reciprocal* of reflection A on the mirror man's final configuration. (A takes us from here to there; A^{-1} takes us from there to here.) But $A^{-1} = A$, so we obtain the result at the lower left of the figure. It is evident that X must be D, clockwise rotation by 120°. This is not surprising. It is related to the fact that, when you turn counterclockwise before a mirror, your image appears to turn clockwise.

We can repeat the solution to this problem in a more formal way by requiring that the operation X *followed* by A (into the mirror A universe) bring us to the same configuration as operation A *followed by* F. That is, we seek X such that $AX = FA$. Multiplying both sides from the left by A^{-1} gives $X = A^{-1}FA$. This is merely a mathematical statement of our above discussion. It says, "What operation is equivalent to the process of reflection through plane A followed by counterclockwise rotation by 120° followed by the inverse of reflection through plane A?" Using our multiplication table, we see that $A^{-1} = A$, and so $X = AFA = AB = D$. Therefore, $D = A^{-1}FA$.

We can find the mirror A universe equivalents to all the symmetry operations in the group in a similar way. Thus,

$$A^{-1}EA = E, \quad A^{-1}AA = A, \quad A^{-1}BA = C,$$
$$A^{-1}CA = B, \quad A^{-1}DA = F, \quad A^{-1}FA = D$$

There is no need to stop here. We can imagine other parallel universes corresponding to reflections B and C and rotations D and F. (The E universe is the one we inhabit.) We can use the same sort of technique to accumulate corresponding operations for all these universes. The results are given in Table 13-3.

If we examine this table, we note certain patterns. The operation E in our universe corresponds to E in all the universes. The reflections (A, B, and C) always correspond to reflections, and the rotations (D and F) always correspond to rotations. This makes physical sense. If we "do nothing" (E) in our universe, we expect the people in the other universes to do nothing also. If we rotate by 120°, we expect the people to perform

TABLE 13-3 ▶ $X = Y^{-1}ZY$ as a Function of Y and Z

Y \ Z	E	A	B	C	D	F
E	E	A	B	C	D	F
A	E	A	C	B	F	D
B	E	C	B	A	F	D
C	E	B	A	C	F	D
D	E	C	A	B	D	F
F	E	B	C	A	D	F

rotations by 120° too, although not necessarily always in the same direction. The argument for reflection is the same. These are examples of three *classes* of operation. The formal definition of class is as follows: If P and Q in a group have the property that $X^{-1}PX = P$ or Q and $X^{-1}QX = P$ or Q for all members X in the group, then P and Q belong to the same class. Physically, operations in the same class are of the "same kind"—all reflections, all rotations, etc.

EXAMPLE 13-2 BH_3, like ammonia, has three planes of reflection symmetry that contain the three-fold rotational axis. However, BH_3, being planar, also has a reflection operation through the molecular plane. Is this reflection operation in the same class as the other three?

SOLUTION ▶ No. In both molecules, the three planes each interchange two hydrogens and keep one fixed in space. The extra reflection plane in BH_3 keeps all three hydrogens fixed. If we reflect so as to keep all hydrogens fixed, there is no way a person in a symmetry-related universe could make a corresponding reflection by using a reflection that interchanges two hydrogens and keeps one fixed. ◀

We shall see that classes are important subdivisions of groups. Note, however, that a class need not be a subgroup. For instance, D and F constitute a class, but, as the class does not include E, it is not a subgroup.

In discussing the concept of class, it is unnecessary to postulate parallel universes, and the reader should not be disturbed by this pedagogical device. The people in the "other universes" are merely working with ammonia models that have been reflected or rotated with respect to the model orientation that we chose in Fig. 13-3. Operations in the same class are simply operations that become interchanged if our coordinate system is subjected to one of the symmetry operations of the group.

13-5 Symmetry Elements and Their Notation

There are five kinds of symmetry operations that one can utilize to move an object through a maximum number of indistinguishable configurations. One is the trivial identity operation E. Each of the other kinds of symmetry operation has an associated *symmetry element*[1] in the object. For example, our ammonia model has three reflection operations, each of which has an associated *reflection plane* as its symmetry element. It also has two rotation operations and these are associated with a common *rotation axis* as symmetry element. The axis is said to be *three-fold* in this case because the associated rotations are each one-third of a complete cycle. In general, rotation by $2\pi/n$ radians is said to occur about an n-fold axis. Another kind of operation—one we have encountered before is inversion, and it has a *point of inversion* as its symmetry element. Finally, there is an operation known as *improper rotation*. In this operation, we first rotate the object by some fraction of a cycle about an axis, and then reflect it through a plane perpendicular to the rotation axis. The axis is the symmetry element and is called an *improper* axis.

The following two examples should help clarify the nature of these symmetry elements. Consider first the ethane molecule in its eclipsed conformation (**I**).

[1]This term is not to be confused with a *group element*.

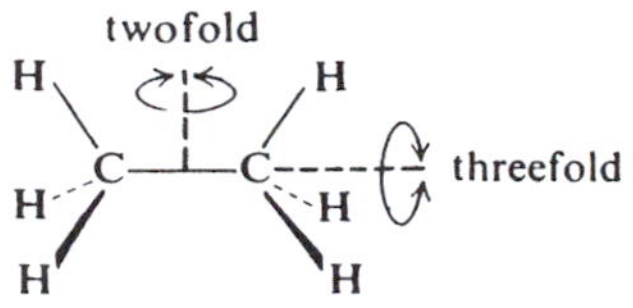

(I)

The following symmetry *elements* can be identified. (It is important that you satisfy yourself that you see these elements in the sketch.)

1. One three-fold axis coincident with the C–C bond.
2. Three two-fold axes perpendicular to the C–C bond and intersecting its midpoint.
3. Three reflection planes, each containing the C–C bond and a pair of C–H bonds.
4. One reflection plane perpendicular to the C–C bond and bisecting it.
5. No point of inversion.
6. One three-fold *improper* axis coincident with the C–C bond.

[Note that these operations are to be applied to the *rigid* molecule. Movement of only one methyl group and not the other (i.e., internal rotation) is not allowed.]

Now let us consider staggered ethane (**II**) (i.e., one methyl group rotated 60° from its eclipsed position). Its symmetry elements are

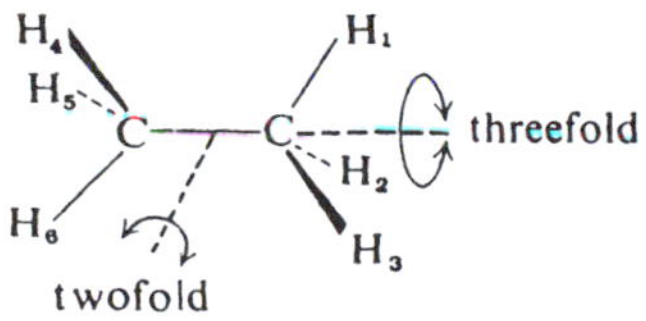

(II)

1. One three-fold axis coincident with the C–C bond.
2. Three two-fold axes perpendicular to the C–C bond and intersecting its midpoint.
3. Three reflection planes, each containing the C–C bond and a pair of C–H bonds.
4. No reflection plane perpendicular to the C–C bond.
5. One point of inversion at the midpoint of the C–C bond.
6. One six-fold *improper* axis coincident with the C–C bond.

In going from the eclipsed to the staggered conformation, we have lost a reflection plane perpendicular to the C–C bond, gained a point of inversion, and changed the order of the improper axis.

Usually it is not difficult to "see" most elements of symmetry in molecules, the exception being improper axes, which tend to be a little tricky. The six-fold improper rotation in staggered ethane is not too hard to envision when it is applied once. If we rotate clockwise by 60° and reflect, H_1 replaces H_5, H_5 replaces H_2, etc. It is a little more difficult when we try to imagine applying this operation twice in succession. The reader should recognize that *two* 60° rotations and *two* reflections result in H_1 replacing H_2, H_2 replacing H_3, H_4 replacing H_5, etc.

EXAMPLE 13-3 BH_3 has a three-fold axis and a reflection plane in the molecular plane. Does BH_3 have a three-fold improper rotation axis as a necessary member of its group?

SOLUTION ▶ A three-fold improper axis coincident with the "regular" three-fold rotational axis appears to have the correct symmetry characteristics: It rotates each hydrogen to the next position and then interchanges the spaces above and below the plane. Do we need it in the group? Only if there is no other **single** operation that accomplishes the same thing. There is no such operation, so this improper axis is indeed a nonredundant symmetry element. ◀

A notation for these symmetry elements (and for the operations related to them) has come to be generally accepted in chemistry (exclusive of crystallography, which uses a different notation). This notation is summarized in Table 13-4. The symmetry elements of eclipsed ethane would be, by these conventions, C_3, $3C_2$, σ (perpendicular to C–C), 3σ (containing C–C), S_3. It often occurs that several classes of reflection planes and axes for proper or improper rotations are present in a system. In eclipsed ethane, we can imagine that the three reflection planes containing the C–C bond might all be in the same class, that is, might be interchanged by a symmetry operation. In fact, it is easy to see that they are interchanged by rotations about the threefold axis. However, none of these could ever be equivalent to the reflection perpendicular to the C–C bond because this reflection interchanges the carbons, whereas the others do not. For this reason, we keep separate tally of the different classes of reflection in eclipsed ethane, rather than simply writing "4σ."

There is a geometric convention that aids discussion of molecules having axes of rotation. One looks for a unique axis C_n (usually the axis of highest order) and imagines this axis to be vertical (coincident with the z axis). Then a reflection plane perpendicular to this axis is *horizontal* and is labeled σ_h. Planes containing this axis are necessarily vertical and are subdivided into *dihedral* and *vertical* planes. *Dihedral* planes must contain the unique reference axis, C_n, and must also bisect the angles between two-fold

TABLE 13-4 ▶ Symbols for Symmetry Elements and Operations

Symmetry operation	Symmetry element	Symbol
"Do nothing" (identity)	None	E
Rotation by $2\pi/n$ radians	An n-fold (proper) axis	C_n
Reflection through plane	A plane	σ
Inversion through a point	A point	i
Rotation through $2\pi/n$ radians followed by reflection through a plane perpendicular to the rotation axis	An n-fold (improper) axis	S_n

axes perpendicular to C_n. Such planes are labeled σ_d. Vertical planes that do not bisect two-fold axes[2] are labeled σ_v. In eclipsed ethane, the C_3 axis is the principal axis, so the C–C bond is oriented vertically. Therefore, our symmetry *elements* are labeled $C_3, 3C_2, \sigma_h, 3\sigma_v, S_3$. (The vertical planes are labeled σ_v because they *contain*, rather than bisect, the two-fold axes that are perpendicular to the principal axis.) For staggered ethane, we have $C_3, 3C_2, i, 3\sigma_d, S_6$. (Here the vertical planes do *not* contain the two-fold axes, but do bisect them. Hence, we label them σ_d. Making a simple sketch will aid in clarifying this distinction.)

Once one has recognized the set of symmetry elements associated with a given object, it is a straightforward matter to list the symmetry *operations* associated with the set. Simplest are the operations associated with elements σ and i, because each such element gives rise to only one operation. Proper and improper axes are somewhat more complicated. Let us return to our ammonia molecule for illustration of this. There we had a threefold axis C_3 and we noted that we could rotate by $2\pi/3\,(C_3^+)$ to get one configuration, and $4\pi/3\,[C_3^+C_3^+ = (C_3^+)^2]$ to get another. Alternatively we could choose to rotate by $2\pi/3$ and $-2\pi/3\,(C_3^-)$, the latter easily being shown to be equivalent to $(C_3^+)^2$. But if we rotate by $(C_3^+)^3$, we return to our original configuration. That is, $(C_3^+)^3 = E$. Therefore, a C_3 axis produces two unique operations. The reader can easily generalize this to the statement that a C_n axis yields $n-1$ unique symmetry operations $C_n, C_{n^2}, C_{n^3}, \ldots, C_n^{n-1}$, where we assume rotation to be in the clockwise direction in all cases. Benzene has a C_6 axis with five associated nonredundant symmetry operations $(C_6, C_6^2, C_6^3, C_6^4, C_6^5)$. Now C_6^3 is equivalent to a rotation by π radians, an operation we normally write as C_2. Similarly, $C_6^2 = C_3$. The result of such reductions is the set of operations $C_6, C_3, C_2, C_3^2, C_6^5$ for the C_6 axis of benzene. There are several classes of rotation here. If we use a compressed notation, we can write this set as $2C_6, 2C_3, C_2$. This indicates that the presence of a sixfold axis implies the existence of coincident twofold and threefold axes. However, these implied elements are not listed for such a system since their operations are all contained in the set of operations of the C_6 axis.

Improper axes can also be associated with several symmetry operations. We noted earlier that S_6, applied twice in succession, results in a simple $2\pi/3$ rotation about the S_6 axis. In other words, we can write the set $S_6, S_6^2, S_6^3, S_6^4, S_6^5$ as S_6, C_3, S_2, C_3^2. We stop at S_6^5 because $S_6^6 = E$ due to the combination of C_6^6 and an *even* number of reflections. S_6^3 is equivalent to S_2 because it contains three rotations by $2\pi/6$ and an odd number of reflections, and S_2 means one rotation by π and one reflection. The operation S_2, however, is easily shown to be equivalent to an inversion, and so we have, using a compressed notation, $2S_6, 2C_3, i$ associated with the S_6 axis. Since we have already explicitly listed the elements C_3 and i in our set of elements for staggered ethane (or any other system containing an S_6 axis) only the $2S_6$ operations are unique to the S_6 axis. The generalization of this case is that an S_{2n} axis with *odd n* implies that elements C_n and i are also present. Of the $2n-1$ operations associated with S_{2n}, $n-1$ are preempted by the C_n axis and 1 by the element i leaving $n-1$ operations to be attributed to the S_{2n} axis. [If $n=1$, we have S_2, C_1, and i as elements. But $C_1 = E$, so we ignore it. There is only one operation here $(2\cdot 1 - 1 = 1)$ and it is preempted by i. Therefore, S_2 has no unique operations and it is not listed as a symmetry element.] For S_{2n} with *n even*, i is not implied and there are n unique operations.

[2] In certain cases there will be two geometrically nonequivalent sets of vertical planes that cannot be distinguished on the basis of bisecting two-fold axes (e.g., in square-planar XeF_4). In such cases the convention is to take the σ_v planes to be those containing the larger number of atoms.

EXAMPLE 13-4 How many operations must be associated with S_4 in the square-planar molecule $XeCl_4$?

SOLUTION ▶ This molecule possesses a C_2 and a C_4 axis coincident with S_4. C_2 usurps S_4^2 since the latter has two reflections that cancel, making it just like C_2. (C_2 also usurps C_4^2.) $S_4^4 = E$. Only two operations, S_4^1 and S_4^3, survive as unique symmetry operations ◀

In eclipsed ethane, we have the element S_3, which generates operations S_3, S_3^2, S_3^3, S_3^4, S_3^5, S_3^6, ... In deciding where to stop here, we note that $S_3^3 \neq E$ because we have here an *odd* number of reflections. Therefore, $S_3^3 = \sigma$, where we understand this to be reflection through the plane perpendicular to the S_3 axis. For eclipsed ethane, this is σ_h. The element S_3^6 has an even number of reflections and is equal to E. How about S_3^4? This has an even number of reflections, so it is identical to C_3^4 which is the same as $C_3^3 C_3 = EC_3 = C_3$. Thus, our S_3 element has operations that are consistent with the presence of C_3 and σ_h elements. We note that eclipsed ethane was indeed found to have these elements. Once again, we allow these elements to preempt their operations from S_3. This leaves only two unique operations, S_3 and S_3^5, corresponding to clockwise and counterclockwise improper rotations by $2\pi/3$ radians. In general, S_{2n+1} will occur only when C_{2n+1} and σ (perpendicular to C_{2n+1}) are also present. These preempt a total of $2n+1$ operations from the total of $2(2n+1) - 1 = 4n+1$ we can achieve with S_{2n+1} alone; and we are left with $2n$ operations for S_{2n+1}. The above conclusions are summarized in Table 13-5.

The set of symmetry *operations* is contrasted to the set of symmetry *elements* for eclipsed and staggered ethane in Table 13-6. Note that there are 12 symmetry operations in each case. By setting up the multiplication table for either of these sets of 12 operations, we can show that the mathematical requirements for a group are satisfied. Thus, each of these sets of symmetry operations constitutes a separate group of order 12.

The multiplication table for the ammonia model is given in Table 13-7 in terms of the symmetry symbols just described. (A, B, C are taken to be σ_1, σ_2, σ_3, respectively.)

It would be possible, using what has been described up to this point, to construct a list of symmetry operations for any object we please. Then we could test the set to see if it satisfied the various requirements for a group. In practice, this is not done. It turns out that (1) only a rather limited number of distinct kinds of symmetry are possible, and (2) in each case, the symmetry operations for the object do form a group. The practical question, then, is, given an object, to which of the known groups does it

TABLE 13-5 ▶ Number of Operations Associated with Symmetry Elements

Element	(Other elements present)	No. operations
i	—	1
σ	—	1
C_n	—	$n-1$
S_{2n}	$n = \text{odd}: C_n$ and i present	$n-1$
	$n = \text{even}: C_n$ present	n
S_{2n+1}	C_{2n+1}, σ	$2n$

TABLE 13-6 ► Symmetry Operations and Elements in Ethane

Molecule	Elements	Operations
Eclipsed (C_3, S_3; C_2 (1 of 3); σ_h; σ_v (1 of 3))		E
	C_3	$2C_3$
	$3C_2$	$3C_2$
	σ_h	σ_h
	$3\sigma_v$	$3\sigma_v$
	S_3	$2S_3$
		Group order: 12
Staggered (C_3, S_6; σ_d (1 of 3); σ_i (for S_6); i; C_2 (1 of 3))		E
	C_3	$2C_3$
	$3C_2$	$3C_2$
	i	i
	$3\sigma_d$	$3\sigma_d$
	S_6	$2S_6$
		Group order: 12

TABLE 13-7 ► Multiplication Table for the Ammonia Molecule

Second operation \ First operation	E	σ_1	σ_2	σ_3	C_3^+	C_3^-
E	E	σ_1	σ_2	σ_3	C_3^+	C_3^-
σ_1	σ_1	E	C_3^+	C_3^-	σ_2	σ_3
σ_2	σ_2	C_3^-	E	C_3^+	σ_3	σ_1
σ_3	σ_3	C_3^+	C_3^-	E	σ_1	σ_2
C_3^+	C_3^+	σ_3	σ_1	σ_2	C_3^-	E
C_3^-	C_3^-	σ_2	σ_3	σ_1	E	C_3^+

belong? Once this is settled, we merely look up the tabulated properties of that group and save ourselves the effort of working through all the details.

13-6 Identifying the Point Group of a Molecule

One can decide to which point group a molecule belongs by systematically looking for certain symmetry elements. Each symmetry point group has a unique group symbol, so basically one tries to figure out which group symbol is appropriate for a given molecule. A flowchart that serves this purpose is displayed in Fig. 13-7. As an illustration, we will work out the group symbol for staggered ethane. The first few questions in the flowchart

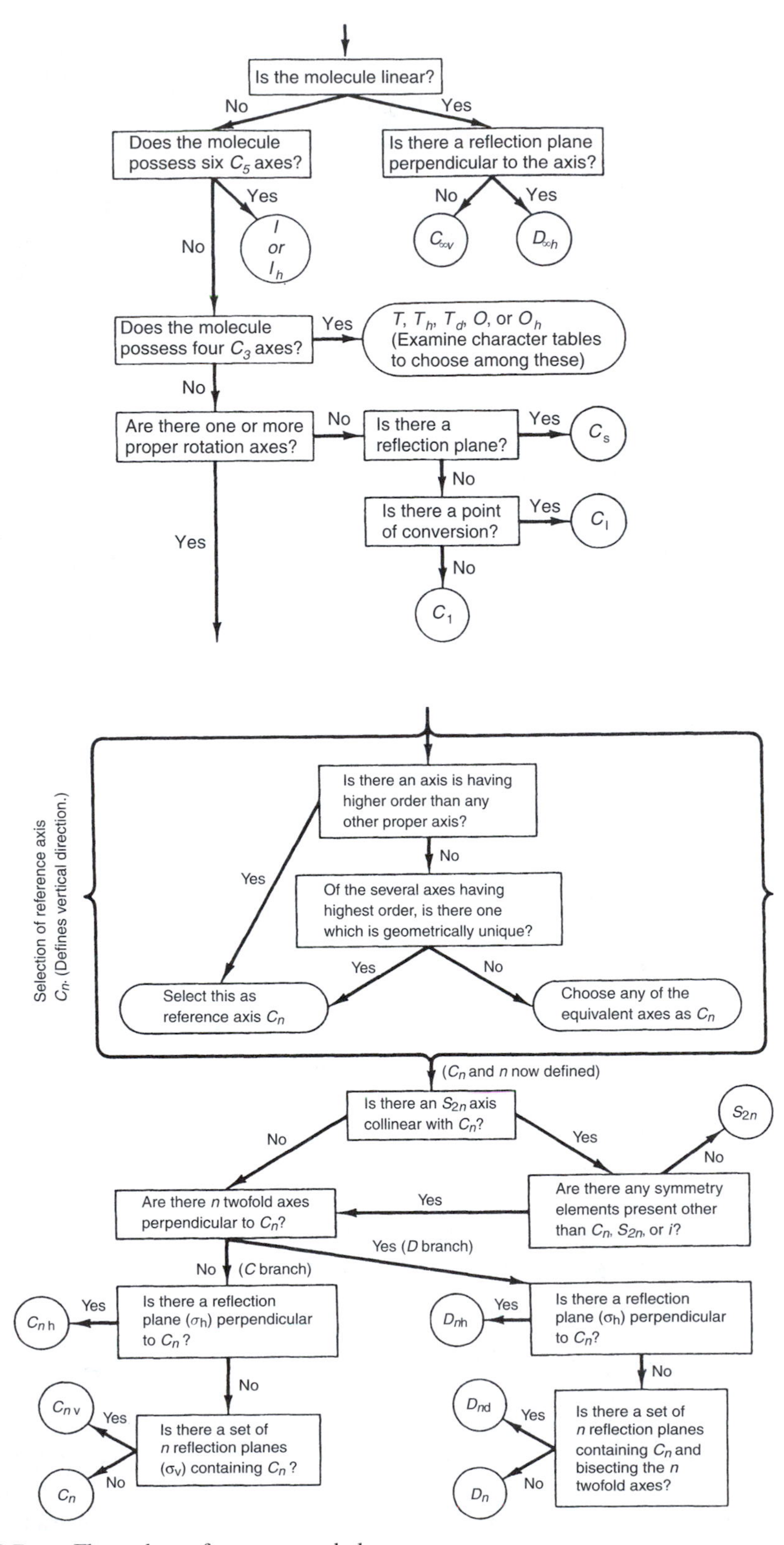

Figure 13-7 ► Flow scheme for group symbols.

check to see whether the molecule belongs to one of several special groups of very high symmetry. Since staggered ethane is not linear, and since it does not possess the symmetry of any of the regular polyhedra, we move to the question concerning proper rotation axes. We have noted earlier that there are several proper rotation axes in this molecule. We next consider which one should be selected as the reference axis. The axis of highest order is threefold and there is only one of these; therefore, we choose it as our reference axis. This defines n as 3 for the remainder of the flowchart, and determines the vertical direction. Next we look to see if there is an S_6 axis coincident with the C_3 axis, and we find that there is. Checking to see if symmetry elements other than C_3, S_6, and i exist, we find that they do indeed, and that among them are three twofold axes perpendicular to the reference axis C_3. This leads us along the "D branch" of the diagram, which means that, whatever else we find, the major symbol for our group will be D. Next we look for a σ_h reflection plane. Finding none, we next seek a set of reflection planes containing C_3 and also bisecting the various C_2 axes. Such dihedral planes are present, so our group symbol is D_{3d} for staggered ethane. The reader may verify that eclipsed ethane follows the same route out to the D branch, but there we *do* find a σ_h (horizontal) plane, and so eclipsed ethane has D_{3h} symmetry. A simple exercise, suggested at this point, is to ascertain the symmetry symbol for ammonia. (The answer is given in the next section.)

EXAMPLE 13-5 What is the group symbol for (planar) BH_3?

SOLUTION ▸ Starting at the top of Fig. 13-7, the answers "no, no, no, yes" bring us to the box for selection of the reference axis. There is one three-fold (proper) rotational axis, and no other of higher order, so this defines n (= 3) and "vertical." Coming out of the box we get answers "no, yes, yes," giving D_{3h} as the group symbol. Notice that we arrive at this conclusion without having had to use all the symmetry elements of the molecule (S_3 or vertical reflection planes). ◂

13-7 Representations for Groups

We have seen that the group of symmetry operations for the ammonia molecule leads to a particular group table of product operations (Table 13-7). Let us now see if we can assign a *number or matrix* to each symmetry operation such that the products of numbers satisfy the same group multiplication table relationships as do the products of symmetry operations. If we can find such a set of numbers or matrices, we say we have a *representation* for the group.

It is always possible to produce a trivial representation by simply assigning the number $+1$ to each operation. Then any operator relation, such as $\sigma_1\sigma_2 = C_3^+$ is necessarily satisfied by the numbers since $1 \cdot 1 = 1$. Nontrivial representations exist too (except for the C_1 group). For example, the group for molecules (like ammonia) having C_{3v} symmetry can be represented by the set of numbers $+1, -1, -1, -1, +1, +1$, respectively, for the operations $E, \sigma_1, \sigma_2, \sigma_3, C_3^+, C_3^-$ in Table 13-7. Then the relation $\sigma_2 C_3^+ = \sigma_3$ is paralleled by the representation since $(-1)(+1) = -1$, etc. It is also possible to find a representation for the C_{3v} group wherein each operation corresponds to a 2×2 matrix. This is shown in Table 13-8, along with the two other representations already mentioned. The reader is encouraged to test that, for instance, $\sigma_2 C_3^+ = \sigma_3$ in this matrix representation.

TABLE 13-8 ▶ Representations for the C_{3v} Group

	E	σ_1	σ_2	σ_3	C_3^+	C_3^-
Γ_1	1	1	1	1	1	1
Γ_2	1	−1	−1	−1	1	1
Γ_3	$\begin{pmatrix}1 & 0\\0 & 1\end{pmatrix}$	$\begin{pmatrix}1 & 0\\0 & -1\end{pmatrix}$	$\begin{pmatrix}-1/2 & \sqrt{3}/2\\\sqrt{3}/2 & +1/2\end{pmatrix}$	$\begin{pmatrix}-1/2 & -\sqrt{3}/2\\-\sqrt{3}/2 & 1/2\end{pmatrix}$	$\begin{pmatrix}-1/2 & \sqrt{3}/2\\-\sqrt{3}/2 & -1/2\end{pmatrix}$	$\begin{pmatrix}-1/2 & -\sqrt{3}/2\\\sqrt{3}/2 & -1/2\end{pmatrix}$

TABLE 13-9 ▶ A Two-Dimensional Unit Matrix Representation for the C_{3v} Group

	E	σ_1	σ_2	σ_3	C_3^+	C_3^-
Γ'	$\begin{pmatrix}1 & 0\\0 & 1\end{pmatrix}$	$\begin{pmatrix}1 & 0\\0 & 1\end{pmatrix}$	$\begin{pmatrix}1 & 0\\0 & 1\end{pmatrix}$	$\begin{pmatrix}1 & 0\\0 & 1\end{pmatrix}$	$\begin{pmatrix}1 & 0\\0 & 1\end{pmatrix}$	$\begin{pmatrix}1 & 0\\0 & 1\end{pmatrix}$

The symbol Γ_i is often used to stand for the ith representation of a group. $\Gamma_i(R)$ refers to the representation of the particular operation R in the ith representation. Thus, $\Gamma_2(\sigma_3) = -1$. In our ammonia example, Γ_1 and Γ_2 are *one-dimensional* representations and Γ_3 is *two-dimensional*. The representation Γ_1 is really the simplest member of a set of equally trivial representations, namely all representations made from unit matrices. Thus, the representation Γ' (Table 13-9) is just as successful a representation as is Γ_1. We could obviously use unit matrices of arbitrary dimension in constructing a representation. We can think of such representations as being "built up" from Γ_1:

$$\Gamma'(R) = \begin{pmatrix}\Gamma_1(R) & 0\\0 & \Gamma_1(R)\end{pmatrix} \quad \text{etc.}$$

It is possible to extend this approach by building up representations from mixtures of $\Gamma_1, \Gamma_2, \Gamma_3$. Thus,

$$\Gamma''(R) = \begin{pmatrix}\Gamma_2(R) & 0 & 0 & 0\\0 & \Gamma_1(R) & 0 & 0\\0 & 0 & \multicolumn{2}{c}{\Gamma_3(R)}\\0 & 0 & & \end{pmatrix}$$

would produce the representation given in Table 13-10. It is easy to show that these matrices multiply together in a way that parallels the group table for symmetry oper-

TABLE 13-10 ▶ A C_{3v} Group Representation Built from Other Representations

	E	σ_1	σ_2
Γ'' (R)	$\begin{pmatrix}1 & 0 & 0 & 0\\0 & 1 & 0 & 0\\0 & 0 & 1 & 0\\0 & 0 & 0 & 1\end{pmatrix}$	$\begin{pmatrix}-1 & 0 & 0 & 0\\0 & 1 & 0 & 0\\0 & 0 & 1 & 0\\0 & 0 & 0 & -1\end{pmatrix}$	$\begin{pmatrix}-1 & 0 & 0 & 0\\0 & 1 & 0 & 0\\0 & 0 & -\frac{1}{2} & \sqrt{3}/2\\0 & 0 & \sqrt{3}/2 & \frac{1}{2}\end{pmatrix}$

ations, since block diagonal matrices always multiply in a block-for-block manner. Hence, the upper left 1×1 blocks (Γ_2) multiply among themselves, the 1×1 blocks in the 2,2 positions (Γ_1) multiply among themselves, and the 2×2 blocks at the lower right (Γ_3) multiply among themselves. Since each of these sets conforms to the group table, it follows that their composite Γ'' will also. It is evident that there is no limit to the number of representations we could build up in this manner.

Representations like Γ' and Γ'' are called *reducible* representations because they are composites of one or more smaller-dimensional representations and hence can be decomposed into those smaller representations. Any representation that cannot be reduced, or decomposed, into smaller representations is said to be *irreducible*.

How can we tell whether a given representation is reducible or not? If it is already one dimensional, it is obviously irreducible. If it is multidimensional and, like Γ' or Γ'', *uniformly* block diagonalized throughout the whole set of symmetry operations, it is plainly reducible. The trouble comes when it is multidimensional but not block diagonalized, for example, Γ_3. It is sometimes the case that such representations are reducible. To illustrate this point, let us go back to Γ'', which we know is reducible, and imagine transforming this representation by multiplying every matrix from the right by β and from the left β^{-1}, where β is some 4×4 unitary matrix (see Chapter 9). This will give us a new set of six 4×4 matrices that no longer will necessarily be block diagonalized. Yet it is easy to show that these new matrices, Γ^{β}, are still a representation for our group. We prove this by showing that, if three matrices from Γ'' (call them A'', B'', C'') are related by $A''B'' = C''$, then the corresponding transformed matrices A^{β}, B^{β}, C^{β}, satisfy $A^{\beta}B^{\beta} = C^{\beta}$. If this is true, then the ability of Γ'' to correspond to the group table relationships is retained by Γ^{β}. The proof is as follows:

Let

$$A''B'' = C''$$

Then

$$A^{\beta}B^{\beta} = \beta^{-1}A''\beta\beta^{-1}B''\beta = \beta^{-1}A''B''\beta = \beta^{-1}C''\beta \equiv C^{\beta}$$

The point we are trying to make here is that a reducible representation like Γ'' can be put forth in many guises, each corresponding to a different choice of β, and many of these will not be block diagonal in form. (Representations that differ only by a unitary transformation are said to be *equivalent*. Thus, Γ^{α} and Γ^{β} are equivalent representations.) To show that such a representation as Γ^{β} is reducible and also to reveal its component representations, we could back transform it by finding the matrix β and calculating

$$\beta\Gamma^{\beta}\beta^{-1} = \beta\beta^{-1}\Gamma''\beta\beta^{-1} = \Gamma''$$

σ_3	C_3^+	C_3^-
$\begin{pmatrix} -1 & 0 & 0 & 0 \\ 0 & 1 & 0 & 0 \\ 0 & 0 & -\frac{1}{2} & -\sqrt{3}/2 \\ 0 & 0 & -\sqrt{3}/2 & \frac{1}{2} \end{pmatrix}$	$\begin{pmatrix} 1 & 0 & 0 & 0 \\ 0 & 1 & 0 & 0 \\ 0 & 0 & -\frac{1}{2} & \sqrt{3}/2 \\ 0 & 0 & -\sqrt{3}/2 & -\frac{1}{2} \end{pmatrix}$	$\begin{pmatrix} 1 & 0 & 0 & 0 \\ 0 & 1 & 0 & 0 \\ 0 & 0 & -\frac{1}{2} & -\sqrt{3}/2 \\ 0 & 0 & \sqrt{3}/2 & -\frac{1}{2} \end{pmatrix}$

Therefore, deciding whether or not a multidimensional representation is reducible is the same as deciding whether it can be uniformly block-diagonalized through a common unitary transformation. It turns out that there is no single unitary transformation capable of diagonalizing all the 2×2 matrices of Γ_3, and so Γ_3 is an irreducible two-dimensional representation. (We shall show later that this can be established without "trying out" an infinite number of transformations.) For any group, a goal is to find all the inequivalent, irreducible representations possible. Anything beyond this is superfluous information. For our C_{3v} (ammonia) group, Γ_1, Γ_2, and Γ_3 exhaust the possibilities and constitute a complete set of inequivalent irreducible representations.

13-8 Generating Representations from Basis Functions

The reader may wonder how one goes about discovering nontrivial representations like Γ_3.[3] A convenient way to do this will now be described, and we will show at this stage the connection between quantum mechanics and the group theory of symmetry operations.

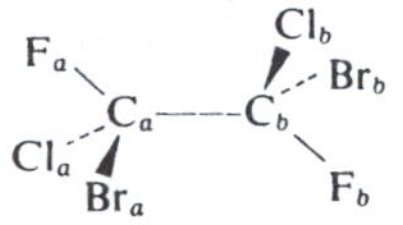

(III)

Consider the molecule shown in (**III**). This molecule has but one nontrivial symmetry element—a point of inversion. According to our flowchart, this places it in the C_i point group. The only symmetry operations here are E and i. Now consider two functions, f_1 and f_2. Let f_1 be located on one end of the molecule. For instance, let f_1 be $1s_{F_a}$, a 1s AO centered on the fluorine atom on the left side of the molecule. Let f_2 be a similar function on the other side of the molecule, $1s_{F_b}$. Now let us see what happens to these functions when they are acted upon by our symmetry operations E and i:

$$Ef_1 = f_1, \quad Ef_2 = f_2, \quad if_1 = f_2, \quad if_2 = f_1$$

We see that f_1 and f_2 are interchanged by inversion. Let us try to find numbers to *represent* these results. Clearly, replacing E by $+1$ will give the correct result. However, to obtain the effect of operation by i we need to *interchange* f_1 and f_2. We cannot achieve this by multiplying by a number, since f_1 and f_2 are linearly independent. If, however, we rewrite the effect of i as

$$i\begin{pmatrix} f_1 \\ f_2 \end{pmatrix} = \begin{pmatrix} f_2 \\ f_1 \end{pmatrix}$$

it becomes clear that i can be represented by the matrix $\binom{01}{10}$. Thus, use of the functions $1s_{F_a}$ and $1s_{F_b}$ has generated a two-dimensional representation shown in Table 13-11. (The representation for E has been put into a 2×2 matrix form to be in dimensional agreement with the representation for i.) $1s_{F_a}$ and $1s_{F_b}$ are called a *basis* for Γ. Clearly,

[3] Indeed, the reader may wonder why we even worry about all this. Patience.

TABLE 13-11 ▶ A Representation for the C_i Group

C_i	E	i	Basis
Γ	$\begin{pmatrix} 1 & 0 \\ 0 & 1 \end{pmatrix}$	$\begin{pmatrix} 0 & 1 \\ 1 & 0 \end{pmatrix}$	$(1s_{F_a}, 1s_{F_b})$

any pair of identical functions symmetrically placed with respect to the point of inversion would generate the same two-dimensional representation, and so there is nothing very special about $1s_{F_a}$ and $1s_{F_b}$ as a basis. Is Γ reducible? It is easily argued that it must be. Any unitary 2×2 transformation will have no effect on the matrix for E since $\beta^{-1}1\beta = \beta^{-1}\beta = 1$. We are thus at liberty to look for any unitary transformation β that diagonalizes the second matrix. Since this is a nonsingular matrix (its determinant is unequal to zero), it should be diagonalizable, and it is if we take

$$\begin{pmatrix} 1/\sqrt{2} & 1/\sqrt{2} \\ -1/\sqrt{2} & 1/\sqrt{2} \end{pmatrix} \begin{pmatrix} 0 & 1 \\ 1 & 0 \end{pmatrix} \begin{pmatrix} 1/\sqrt{2} & -1/\sqrt{2} \\ 1/\sqrt{2} & 1/\sqrt{2} \end{pmatrix} = \begin{pmatrix} 1 & 0 \\ 0 & -1 \end{pmatrix}$$

Now that our 2×2 representation has been diagonalized to two 1×1 "blocks," we can rewrite our representations as shown in Table 13-12.

Our first choice of basis generated a reducible representation. What bases would we need to generate the 1×1 representations Γ_1 and Γ_2? To generate Γ_1, we need a basis function that turns into itself when it is inverted. One possibility is $f_1 = 1s_{F_a} + 1s_{F_b}$. If we sketch this function (Fig. 13-8a), it is evident that it is regenerated unchanged by inversion. The mathematical demonstration is

$$if_1 = i(1s_{F_a} + 1s_{F_b}) = i\,1s_{F_a} + i\,1s_{F_b} = 1s_{F_b} + 1s_{F_a} = f_1$$

TABLE 13-12 ▶ Reduced Representations for the C_i Group

C_i	E	i	Basis
Γ_1	1	1	?
Γ_2	1	-1	?

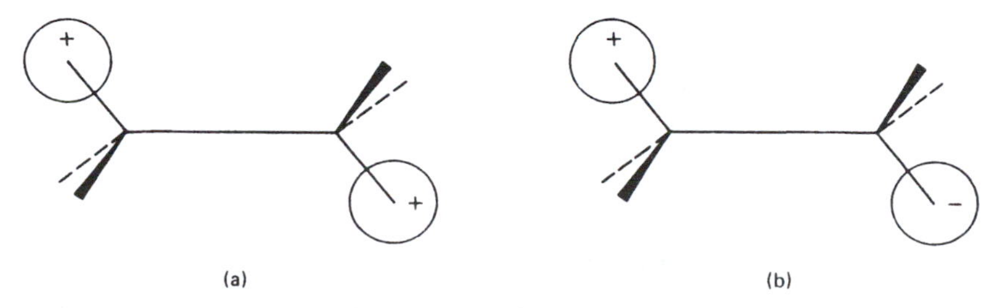

Figure 13-8 ▶ Basis functions for irreducible representations of the C_i group.

f_1 is *symmetric* for operations E and i. For Γ_2, we need some function f_2 that turns into *minus* itself upon inversion. $f_2 = 1s_{F_a} - 1s_{F_b}$ would serve, as Fig. 13-8b shows, and as is demonstrated mathematically by

$$if_2 = i(1s_{F_a} - 1s_{F_b}) = i\,1s_{F_a} - i\,1s_{F_b} = 1s_{F_b} - 1s_{F_a} = -f_2$$

f_2 is symmetric for E and antisymmetric for i. (Notice that the way in which $1s_{F_a}$ and $1s_{F_b}$ needed to be mixed to produce a diagonal representation is indicated by the coefficients in the *columns* of the matrix we used to diagonalize the original 2×2 representation. The matrix β not only diagonalizes our Γ, it also tells us how to mix the original bases in order to arrive at bases for the *irreducible* representations.)

This example demonstrates an important fact: *In order to generate a one-dimensional representation for a group, we need a basis function that is either symmetric or antisymmetric for every symmetry operation in the group.* This provides the point of connection with quantum mechanics. We showed much earlier (Chapter 2) that any *nondegenerate* wavefunction (or MO) must be symmetric or antisymmetric for every operation that leaves the hamiltonian unchanged. Since symmetry operations leave the hamiltonian unchanged (they merely interchange identical nuclei), it follows that nondegenerate wavefunctions or orbitals are symmetric or antisymmetric for every operation in the symmetry point group of the molecule. This means that *every nondegenerate wavefunction or orbital is a basis for a one-dimensional representation for its molecular point group.* Indeed, it can be proved that *every wavefunction or orbital, even if degenerate, is a member of a basis for an irreducible representation* for the point group of the molecule. The irreducible representation produced by an n-fold degenerate set of orbitals will be n-dimensional. (We indicate how this comes about in a later section.) Therefore, knowing wavefunctions enables one to generate representations. More importantly, knowing representations allows us to say something about wavefunctions. For example, the representation table (Table 13-12) for the C_i group (which is now complete—this group has only two irreducible inequivalent representations) tells us the following: (1) The molecule FClBrC–CBrClF in the conformation pictured has no degeneracies due to symmetry in its electronic states or in MO energies (if we do a calculation at the MO level). We can tell this because only one-dimensional representations exist for this group. (2) Every wavefunction for an electronic state or orbital must be either symmetric or antisymmetric for inversion. Therefore, the following AO combinations are feasible for MOs insofar as symmetry is concerned:

$$c(1s_{F_a} - 1s_{F_b}),$$
$$c_1(1s_{F_a} + 1s_{F_b}) + c_2(1s_{Cl_a} + 1s_{Cl_b}) + c_3(1s_{Br_a} + 1s_{Br_b})$$
$$+ c_4(1s_{C_a} + 1s_{C_b}) + c_5(2s_{F_a} + 2s_{F_b}) + \cdots$$

The following are disallowed:

$$c\,1s_{F_a},$$
$$c_1(2s_{F_a} + 2s_{F_b}) + c_2(2s_{Cl_a} - 2s_{Cl_b})$$

Our halogenated ethane molecule is a good starting example because it belongs to such a simple group. But let us now return to the C_{3v} group of the ammonia molecule and continue developing the relations between group theory and MO theory. For convenience, the molecule is again sketched (Fig. 13-9) to show its orientation with respect to Cartesian axes. (The z coordinate is the principal axis.)

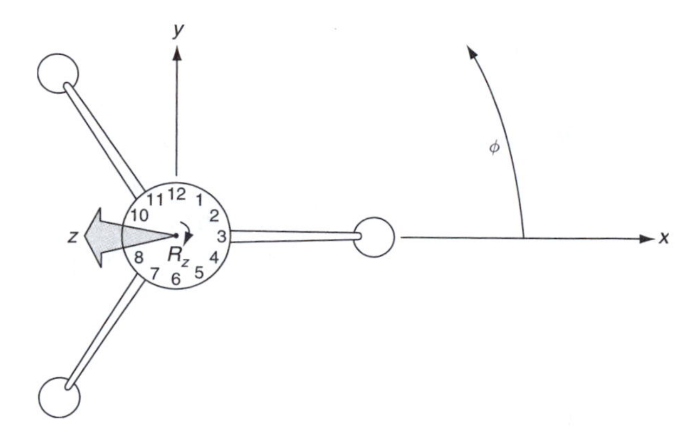

Figure 13-9 ▶ Orientation of a model of ammonia in Cartesian space.

The normal practice in group theory is to use Cartesian coordinates or linear combinations of such coordinates as basis functions[4] for generating many of the representations of a group. Therefore, we begin by examining the z coordinate to see what becomes of it under the various symmetry operations in the C_{3v} group. The results are easily seen to be

$$Ez = +1z, \quad \sigma_1 z = +1z, \quad \sigma_2 z = +1z,$$
$$\sigma_3 z = +1z, \quad C_3^+ z = +1z, \quad C_3^- z = +1z$$

Thus, z is a basis for the totally symmetric representation Γ_1 of Table 13-8. The coordinates x and y can also be used as bases. Here things get more complicated. Clockwise rotation of x by $2\pi/3$ radians (to give x') causes it to end up in a position where it must be expressed as a resultant of both x and y (Fig. 13-10). Simple trigonometry requires that, for a general rotation through the angle ϕ, the unit vector x' has an x coordinate of $\cos\phi$ and a y coordinate of $\sin\phi$. Similarly, a rotation of y to a new position designated y' must yield a new x coordinate of $-\sin\phi$ and a new y coordinate of $\cos\phi$. Thus, for a rotation through the angle ϕ, we have

$$C_\phi \begin{pmatrix} x \\ y \end{pmatrix} = \begin{pmatrix} x' \\ y' \end{pmatrix} = \begin{pmatrix} x\cos\phi + y\sin\phi \\ -x\sin\phi + y\cos\phi \end{pmatrix}$$

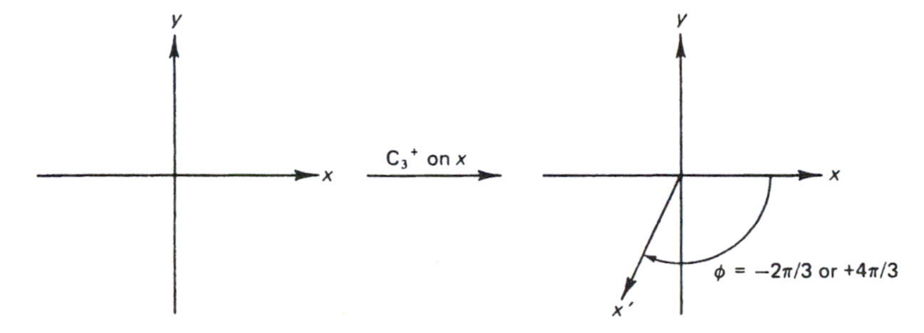

Figure 13-10 ▶ Result of clockwise rotation of x by $2\pi/3$ to produce the new vector x'.

[4]The function corresponding to a coordinate is not exactly the same thing as the coordinate itself. The z coordinate is a ray running perpendicular to the xy plane through the coordinate origin. The *function* z is the altitude above (or below) the xy plane at *every* point, regardless of whether it is on the z axis; z^2 is the square of the altitude at every point, etc. The behavior of these functions upon rotation, reflection, etc. is the same as that of the coordinate or product of coordinates.

The rotation operator C_ϕ is thus represented as

$$C_\phi = \begin{pmatrix} \cos\phi & \sin\phi \\ -\sin\phi & \cos\phi \end{pmatrix} \tag{13-1}$$

In the case at hand, *clockwise* rotation by $2\pi/3$ radians is a *decrease* of $2\pi/3$ in ϕ. Substituting $\phi = -2\pi/3$ for $C_3{}^+$ and $\phi = +2\pi/3$ for $C_3{}^-$ gives us the following 2×2 representations:

$$C_3^+ : \begin{pmatrix} -\frac{1}{2} & +\frac{\sqrt{3}}{2} \\ -\frac{\sqrt{3}}{2} & -\frac{1}{2} \end{pmatrix} \quad C_3^- : \begin{pmatrix} -\frac{1}{2} & -\frac{\sqrt{3}}{2} \\ +\frac{\sqrt{3}}{2} & -\frac{1}{2} \end{pmatrix}$$

For reflection σ_1, it is easy to see that x is unmoved and y goes into minus itself. Maintaining our dimensionality of two, this gives

$$\sigma_1 : \begin{pmatrix} 1 & 0 \\ 0 & -1 \end{pmatrix}$$

For σ_2, x, and y again move into positions x' and y', expressible as resultants of the original x and y vectors (Fig. 13-11). Thus, $x' = -\frac{1}{2}x + \frac{\sqrt{3}}{2}y$, $y' = \frac{\sqrt{3}}{2}x + \frac{1}{2}y$, and σ_2 has the representation

$$\sigma_2 : \begin{pmatrix} -\frac{1}{2} & \frac{\sqrt{3}}{2} \\ \frac{\sqrt{3}}{2} & \frac{1}{2} \end{pmatrix}$$

Similarly, σ_3 is easily shown to have the representation

$$\sigma_3 : \begin{pmatrix} -\frac{1}{2} & -\frac{\sqrt{3}}{2} \\ -\frac{\sqrt{3}}{2} & \frac{1}{2} \end{pmatrix}$$

The operation E does not move either x or y and is represented by the two-dimensional unit matrix. This completes our use of x and y, and we see that together they generate the two-dimensional representation Γ_3. What should we use next? We have not yet generated Γ_2, and so we know that we need to look for another basis. A basis that is

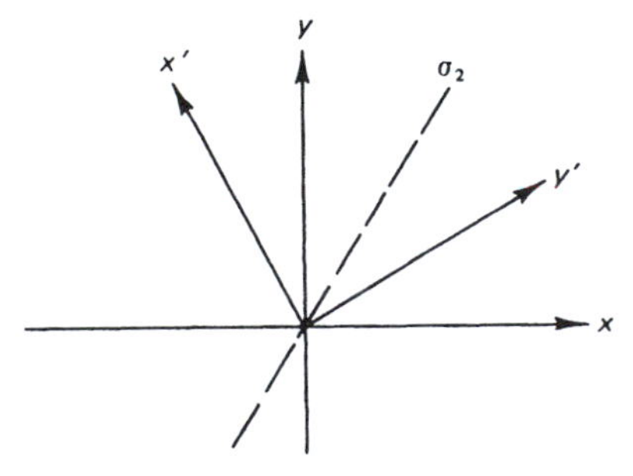

Figure 13-11 ▶ Result of reflection through the σ_2 plane of x and y to produce new vectors x' and y'.

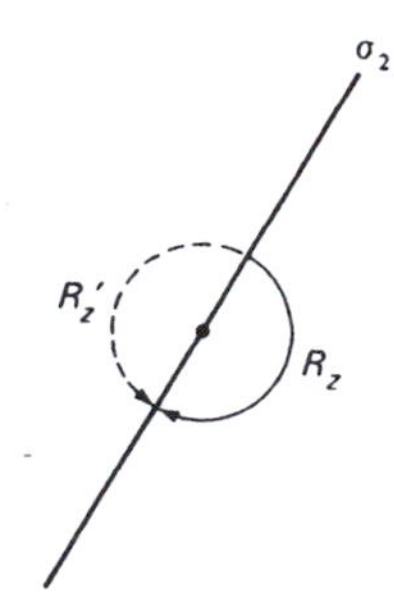

Figure 13-12 ▶ Effect of a reflection on the direction of ϕ.

frequently employed in group theory is a *direction of rotation* about some symmetry axis. In our case, the symmetry axis is the z axis, and we let R_z stand for a *direction of rotation* about this axis. (*Which* direction we choose, clockwise or counterclockwise, is arbitrary.) Now we consider how R_z is affected by the symmetry operations. C_3^+ and C_3^- have no effect since they merely shift the origin of the ϕ coordinate but do not affect the direction in which ϕ increases or decreases. Put another way, the direction of motion of the hands of a clock is not affected by rotating the clock about an axis perpendicular to its face. However, reflections σ_1, σ_2, and σ_3 will cause the direction to be reversed (Fig. 13-12). Therefore,

$$ER_z = +1R_z, \quad \sigma_1 R_z = -1R_z, \quad \sigma_2 R_z = -1R_z$$
$$\sigma_3 R_z = -1R_z \quad C_3^+ R_z = +1R_z \quad C_3^- R_z = +1R_z$$

and we see that R_z is a basis for Γ_2.

13-9 Labels for Representations

Thus far, we have labeled our representations Γ_1, Γ_2, etc. We will now describe rules for a more meaningful symbolism—one that has become standard. The rules are as follows:

1. A one-dimensional representation is given the main symbol A or B. A is used if the representation is symmetric for rotation by $2\pi/n$ about the n-fold principal axis, B if it's antisymmetric. (For C_1, C_s, and C_i groups, which have no principal axis, the symbol is always A.) A two-dimensional representation has the main symbol E. Three-dimensional representations are symbolized T (or sometimes F), and four-dimensional representations are symbolized G.

2. Subscripts may be applied as follows. If there are C_2 axes perpendicular to the principal axis, then a subscript 1 (2) means that the basis for the representation is symmetric (antisymmetric) for such a rotation. In the absence of such C_2 axes, vertical reflection planes are used instead, if present. If there is an inversion center, subscripts g (*gerade*) and u (*ungerade*) refer to the basis for the representation being respectively symmetric or antisymmetric for inversion.

3. Superscripts may be applied as follows: If there is a σ_h plane, a single prime means the basis for the representation is symmetric for that reflection; a double prime means it is antisymmetric.

TABLE 13-13 ► Representation Table for the C_{3v} Group

C_{3v}	E	σ_1	σ_2	σ_3
A_1	1	1	1	1
A_2	1	-1	-1	-1
E	$\begin{pmatrix} 1 & 0 \\ 0 & 1 \end{pmatrix}$	$\begin{pmatrix} 1 & 0 \\ 0 & -1 \end{pmatrix}$	$\begin{pmatrix} -\frac{1}{2} & \sqrt{3}/2 \\ \sqrt{3}/2 & \frac{1}{2} \end{pmatrix}$	$\begin{pmatrix} -\frac{1}{2} & -\sqrt{3}/2 \\ -\sqrt{3}/2 & \frac{1}{2} \end{pmatrix}$

Use of these conventions enables us to write our C_{3v} representation as shown in Table 13-13. (The symbol E in the left-most column of the table should not be confused with the E in the top row. The former labels a two-dimensional representation, the latter refers to the identity operation.) At the right-hand side of the table are listed, for each representation, the bases described above plus a few others. It is convenient for chemical applications to list all Cartesian combinations up to the second power, as has been done here. One can go to powers as high as one pleases (z^{99} is a basis for A_1), but first and second powers are of most common practical use in chemistry. The use of parentheses and commas for bases for the E representation [e.g., (xz, yz)] indicate which *pairs* of functions may be selected as bases for this two-dimensional representation.

13-10 Some Connections Between the Representation Table and Molecular Orbitals

It is possible, by inspecting Table 13-13, to predict certain properties of MOs or wavefunctions for a molecule having C_{3v} symmetry. Suppose that we did an MO calculation on ammonia. What does this table tell us to expect? In the first place, it tells us that there are only three MO symmetry types possible. We might find nondegenerate MOs having A_1 symmetry (totally symmetric), or A_2 symmetry (antisymmetric for reflections). It is also possible for MOs to exist that form bases for E representations. Since such bases are intermixed by some operations, however, these MOs cannot be either symmetric or antisymmetric for every operation. *Therefore, they must be degenerate.* And, since the intermixing occurs only within pairs of such bases, the MOs must be *doubly* degenerate. Thus, inspection of the representation table tells us at once that nondegenerate and doubly degenerate MOs are possible for ammonia. Since the symmetry requirements apply to wavefunctions as well as to one-electron MOs, the table tells us also that ammonia can have *states* with wavefunctions whose *spatial* symmetries are A_1, A_2, or E. It is conventional to label MOs with lowercase symmetry symbols, a_1, a_2, e, and state functions with uppercase symbols.

The representation table also tells us which basis AOs can appear in the various MOs. For instance, if we used a set of valence AOs for NH_3, we would have a $2p_z$ AO on nitrogen, oriented along the C_3 axis of the molecule, and also $2p_x$ and $2p_y$ AOs perpendicular to C_3. Now $2p_x$, $2p_y$, and $2p_z$ transform like x, y, and z, respectively (if they are centered at a common point on the C_3 axis); just as z is symmetric for all symmetry operations of the group, so is $2p_z$. This means that the $2p_z$ AO can be expected to appear only in MOs with a_1 symmetry. $2p_x$ and $2p_y$ must appear together

C_3^+	C_3^-	
1	1	$z, x^2 + y^2, z^2$
1	1	R_2
$\begin{pmatrix} -\frac{1}{2} & \sqrt{3}/2 \\ -\sqrt{3}/2 & -\frac{1}{2} \end{pmatrix}$	$\begin{pmatrix} -\frac{1}{2} & -\sqrt{3}/2 \\ \sqrt{3}/2 & -\frac{1}{2} \end{pmatrix}$	$(x, y)(R_x, R_y)(x^2 - y^2, xy)(xz, yz)$

in e-type MOs. What about 2s on nitrogen? Intuitively, we know that it, like $2p_z$, is unaffected by all the operations, and so it should appear only in a_1 MOs. The table indicates this by listing z^2 and $x^2 + y^2$ as bases for the A_1 representation. Since these have the same symmetry, their sum retains the symmetry, and so $x^2 + y^2 + z^2 = r^2$ is also a basis for A_1. Since r^2 is spherically symmetric, this indicates that any spherically symmetric function on nitrogen is a basis for the A_1 representation.

If we were to use an expanded basis set of AOs, including 3d AOs on nitrogen, the group table tells us that $3d_{z^2}$ would appear in the nondegenerate a_1 MOs and that $3d_{x^2-y^2}$, $3d_{xy}$, $3d_{xz}$, and $3d_{yz}$ would appear in degenerate e-type MOs. (Recall that $3d_{z^2}$ is really $3d_{z^2-r^2}$ or $3d_{2z^2-x^2-y^2}$.)

What about the 1s AOs on the H's? Can the representation table tell us in which MOs these will appear? It turns out that it can, and that they go into both a_1- and e-type MOs, but we will defer showing how this can be told from the table until later in the chapter.

Readers may begin to appreciate the usefulness of group theory in quantum chemistry when they consider that, simply by assigning ammonia to the C_{3v} group and looking up the representation table, we are able to say that a minimum-valence basis set MO calculation will produce nondegenerate, totally symmetric (a_1) MOs containing N_{2s}, N_{2p_z}, and H_{1s} AOs, and doubly degenerate (e) MOs containing N_{2p_x}, N_{2p_y}, and H_{1s} AOs. (No a_2 MOs will appear because none of our AOs are a basis for that representation.)

13-11 Representations for Cyclic and Related Groups

Cyclic groups are the groups $C_2, C_3, C_4, \ldots, C_n$ containing only the $n - 1$ rotation operations and the identity operation E. We devote a separate section to these because there are some special problems connected with finding and labeling representations for these groups. (This section is off the mainstream of development of this chapter and may be skipped if desired.)

Let us consider the operations associated with the n-fold proper axis oriented along the z axis and ask what will become of the function $f = \exp(i\phi)$ as it is rotated clockwise about this axis by $2\pi/n$ radians. (We entertain the idea that $\exp(i\phi)$ might be a convenient basis for a representation since such functions were found to be eigenfunctions for the particle-in-a-ring problem in Chapter 2.) Since the clockwise direction is opposite to the normal direction of the ϕ coordinate, the effect of the rotation is to put $f(\phi)$ where $f(\phi - 2\pi/n)$ used to be. To see how the function after rotation compares to that before rotation, we must compare $\exp(i\phi)$ with $\exp[i(\phi - 2\pi/n)]$. That is, the representation R_f, such that $C_n^+ \exp(i\phi) = R_f \exp(i\phi)$, is given by $\exp(i\phi)/\exp[i(\phi - 2\pi/n)]$, or

TABLE 13-14a ▶ Partial Representation for the C_3 Group

C_3	E	C_3	C_3^2	
A	1	1	1	z, R_z
	1	$\exp(2\pi i/3)$	$\exp(4\pi i/3)$	$\exp(i\phi)$

TABLE 13-14b ▶ Partial Representation for the C_3 Group

C_3	E	C_3	C_3^2	$\epsilon = \exp(2\pi i/3)$
A	1	1	1	z, R_z
	1	ϵ	ϵ^*	$\exp(i\phi)$

$\exp(2\pi i/n)$, which equals $\cos(2\pi/n) + i\sin(2\pi/n)$. For various fractions of a cycle (i.e., various n), R_f takes on different values:

n:	1	2	3	4	5	6	7	8
R_f:	1	-1	$\exp(2\pi i/3)$	i	$\exp(2\pi i/5)$	$\exp(\pi i/3)$	$\exp(2\pi i/7)$	$\exp(\pi i/4)$

The important point here is that $f = \exp(i\phi)$ is a basis for a one-dimensional representation for C_n since a rotation turns f into a constant times f and not into some other function. For the C_3 group, then, we could write a partial representation table as shown in Table 13-14a. Now $\exp(4\pi i/3)$ is equal to $\exp(-2\pi i/3)$, and $\exp(-2\pi i/3)$ is the complex conjugate of $\exp(2\pi i/3)$. If we let $\epsilon \equiv \exp(2\pi i/3)$, we can write the table as shown in Table 13-14b. If $f = \exp(i\phi)$ is a satisfactory basis, $f^* = \exp(-i\phi)$ is also acceptable, since it is linearly independent of f. If one calculates the representations for $n = 1, 2, 3, \ldots$ as before, one finds that the numbers R_{f^*} are the complex conjugates of R_f. Therefore, we can immediately expand our table as shown in Table 13-14c. Since the existence of $\exp(i\phi)$ as a basis for a one-dimensional representation always implies that $\exp(-i\phi)$ exists as a basis, these sorts of one-dimensional representations always occur in pairs. *It is conventional* to combine these with braces and refer to them with the symbol E, which we claimed earlier is reserved for two-dimensional representations. Note, however, that *a pair of one-dimensional representations is not the same as a two-dimensional representation*, and we must broaden our definition of the symbol E to include both types of situation.

Our representation table for the C_3 group now looks almost the way one would find it in a standard tabulation. However, instead of listing $\exp(i\phi)$ and $\exp(-i\phi)$ as bases,

TABLE 13-14c ▶ Representation for the C_3 Group

C_3	E	C_3	$C_3{}^2$	$\epsilon = \exp(2\pi i/3)$
A	1	1	1	z, R_z
E	$\{1$	ϵ	$\epsilon^*\}$	$\exp(i\phi)$
	$\{1$	ϵ^*	$\epsilon\}$	$\exp(-i\phi)$

TABLE 13-15 ► Representation for the C_3 Group

C_3	E	C_3	C_3^2	
A	1	1	1	z, R_z
E	$\begin{pmatrix} 1 & 0 \\ 0 & 1 \end{pmatrix}$	$\begin{pmatrix} -\frac{1}{2} & \frac{\sqrt{3}}{2} \\ -\frac{\sqrt{3}}{2} & \frac{1}{2} \end{pmatrix}$	$\begin{pmatrix} -\frac{1}{2} & -\frac{\sqrt{3}}{2} \\ \frac{\sqrt{3}}{2} & -\frac{1}{2} \end{pmatrix}$	(x, y)

the convention is to list x and y. In relating Cartesian to spherical polar coordinates, $x = r\sin\theta\cos\phi$ and $y = r\sin\theta\sin\phi$. When we are concerned only with changes in ϕ, x goes as $\cos\phi$, y as $\sin\phi$, and, since $\sin\phi$ and $\cos\phi$ are expressible as linear combinations of $\exp(i\phi)$ and $\exp(-i\phi)$, it follows that x and y are *equivalent* to $\exp(\pm i\phi)$ as bases. If we had started out with x and y as bases, we would have found that, for rotations, these are intermixed, leading to a truly two-dimensional representation. In fact, we worked out the effects of C_3 and C_3^2 on x and y earlier for the C_{3v} group. There we found that C_3 and $C_3^2 (= C_3^-)$ were represented by the two-dimensional matrices shown in Table 13-15. But this E representation is reducible to the two one-dimensional representations through the unitary matrix

$$\mathsf{U} = \frac{1}{\sqrt{2}}\begin{pmatrix} 1 & i \\ i & 1 \end{pmatrix}$$

The resulting block-diagonalized representation is

$$\begin{array}{ccc} E & C_3 & C_3^2 \\ \begin{pmatrix} 1 & 0 \\ 0 & 1 \end{pmatrix} & \begin{pmatrix} -\frac{1}{2} - \frac{i\sqrt{3}}{2} & 0 \\ 0 & -\frac{1}{2} + \frac{i\sqrt{3}}{2} \end{pmatrix} & \begin{pmatrix} -\frac{1}{2} + \frac{i\sqrt{3}}{2} & 0 \\ 0 & -\frac{1}{2} - \frac{i\sqrt{3}}{2} \end{pmatrix} \end{array}$$

which can be separated into two one-dimensional representations:

$$\begin{array}{ccc} E & C_3 & C_3^2 \\ 1 & -\frac{1}{2} - \frac{i\sqrt{3}}{2} & -\frac{1}{2} + \frac{i\sqrt{3}}{2} \\ 1 & -\frac{1}{2} + \frac{i\sqrt{3}}{2} & -\frac{1}{2} - \frac{i\sqrt{3}}{2} \end{array}$$

Since $\epsilon = \exp(2\pi i/3) = \cos(2\pi/3) + i\sin(2\pi/3) = -\frac{1}{2} + \frac{i\sqrt{3}}{2}$, we recognize that this pair of one-dimensional representations is the same as the pair we found earlier. Furthermore, we note that U is precisely the transformation that turns x (i.e., $\cos\phi$) and y (i.e., $\sin\phi$) back into $\exp(\pm i\phi)$ (to within a constant multiplier):

$$\mathsf{U}\begin{pmatrix} x \\ y \end{pmatrix} = \frac{1}{\sqrt{2}}\begin{pmatrix} 1 & i \\ i & 1 \end{pmatrix}\begin{pmatrix} \cos\phi \\ \sin\phi \end{pmatrix} = \frac{1}{\sqrt{2}}\begin{pmatrix} \cos\phi + i\sin\phi \\ i\cos\phi + \sin\phi \end{pmatrix} = \begin{pmatrix} (1/\sqrt{2})\exp(i\phi) \\ (-i/\sqrt{2})\exp(-i\phi) \end{pmatrix}$$

Thus, (x, y) produce a *reducible*, two-dimensional representation *equivalent* to the *irreducible* representation given by $\exp(\pm i\phi)$. The reason for listing (x, y) as bases

TABLE 13-16 ▶ Irreducible Representations for the C_3 Group

C_3	E	C_3	C_3^2		$\epsilon = \exp(2\pi i/3)$
A	1	1	1	z, R_z	$z^2, x^2+y^2, x^2-y^2, xy$
E	$\left\{\begin{matrix}1 \\ 1\end{matrix}\right.$	$\begin{matrix}\epsilon \\ \epsilon^*\end{matrix}$	$\left.\begin{matrix}\epsilon^* \\ \epsilon\end{matrix}\right\}$	(x, y) (R_x, R_y)	(yz, xz)

is simply that most applications of the table are made to real functions (e.g., chemists usually prefer to work with $2p_x$ and $2p_y$ AOs rather than with $2p_{+1}$ and $2p_{-1}$). Our final form for the table, then, is that shown in Table 13-16, where additional bases have been listed.

The preceding discussion has been within the context of the C_n groups. One might ask what sort of symmetry operation is needed to make it impossible for $\exp(i\phi)$ to be a basis for a one-dimensional representation. The answer is, any operation that causes a reversal in the direction of the coordinate ϕ. For then, $\exp(i\phi) \to \exp(-i\phi)$, and our basis function has turned into another independent function rather than into a constant times itself. Therefore, the presence of any symmetry operation that reverses the direction of motion of the hands of a clock will suffice to prevent representations of the ϵ, ϵ^* sort. Operations that reverse clock direction are σ_d, σ_v, and C_2' (perpendicular to the principal axis). Clock direction is unaffected by σ_h, i, and S_n. Therefore, we can expect ϵ, ϵ^* types of representations to occur in groups of types C_n, C_{nh}, S_n, all of which have a C_n axis but no σ_d, σ_v, or C_2' elements.

13-12 Orthogonality in Irreducible Inequivalent Representations

We come now to a very important point regarding representations. We will illustrate our arguments with the representation table (Table 13-13) for the C_{3v} group. Notice the following features of that table:

1. If we choose the A_1 representation, square all the numbers, and sum over all six symmetry operations, we get 6 as a result.

2. If we do the same thing with the A_2 representation, we get the same result.

3. If we do the same thing for the upper left-hand elements (the 1, 1 elements) of the E representation, we get 3 as a result.

4. If we do the same thing for each of the other positions in the E representation, the result is 3 each time.

In general, the result of this procedure for any irreducible representation in any group will be the order of the group divided by the dimension of the representation. That is, if h is the order of the group, l_i is the dimension of representation Γ_i, and $\Gamma_i^{(j,k)}(R)$

is the number in the (j, k) position of the matrix representing symmetry operation R, then the mathematical formula that corresponds to our general statement is

$$\sum_R \left| \Gamma_i^{(j,k)}(R) \right|^2 = \frac{h}{l_i} \tag{13-2}$$

Note that the *absolute* square is used to accommodate the ϵ, ϵ^* type representations of cyclic groups.

We can conceive of the set of six numbers for A_1 as being a vector of six elements. The six numbers of A_2 constitute a second vector, and E provides four more six-dimensional vectors. If each such vector is multiplied by $\sqrt{l_i/h}$, then each vector is normalized.

Now let us examine these vectors regarding their orthogonality. If we take the scalar product of the unnormalized vectors A_1 and A_2, the result is zero:

$$\underbrace{\begin{pmatrix} 1 & 1 & 1 & 1 & 1 & 1 \end{pmatrix}}_{A_1} \underbrace{\begin{Bmatrix} 1 \\ -1 \\ -1 \\ -1 \\ 1 \\ 1 \end{Bmatrix}}_{A_2} = 0$$

The reader may quickly verify that the scalar product of any two *different* vectors from among the set of six in the C_{3v} representation table is zero. Thus these six vectors are orthogonal. Once again, this result always holds between "representation vectors" in irreducible inequivalent representations for any group. Combining this orthogonality property with the normality property mentioned earlier, we have

$$\sum_R \left[\sqrt{l_i/h}\,\Gamma_i^{(k,l)}(R) \right]^* \left[\sqrt{l_i/h}\,\Gamma_j^{(m,n)}(R) \right] = \delta_{i,j}\delta_{k,m}\delta_{l,n} \tag{13-3}$$

where Γ_i and Γ_j are understood to be irreducible and, if $i \neq j$, inequivalent. This relation, sometimes called "the great orthogonality theorem," is of central importance in group theory. Its essence is captured by the statement that "irreducible inequivalent representations are comprised of orthogonal vectors." We do not prove the theorem in this book,[5] but we do make considerable use of Eq. (13-3).

One immediate result of the relation is that it enables us to tell when we have completed the task of finding all the inequivalent irreducible representations of a group. If we consider the C_{3v} group, for example, we note that it is of order six, since there are six symmetry operations. This means that each "representation vector" will have six elements, i.e., is a vector in six-dimensional space. The maximum number of orthogonal vectors we can have in six-dimensional space is six. Therefore, the number of representation vectors cannot exceed the order of the group. Furthermore, since the number of vectors provided by an n-dimensional representation is n^2 (e.g., E is two-dimensional and gives four vectors), we can state that *the sum of the squares of the*

[5] See Bishop [1] or Eyring et al. [2, Appendix VI].

dimensions of the inequivalent irreducible representations of a group cannot exceed the order of the group.

In fact, it can be proved[5] that this sum of squares of dimensions must *equal* the order of the group when all such representations are included. That is,

$$\sum_{i}^{\substack{\text{all inequivalent}\\ \text{irreproducible representations}}} l_i^2 = h \tag{13-4}$$

Thus, the fact that the squares of the dimensions of the A_1, A_2, and E representations for the $C_{3\text{v}}$ group add up to six, which is the order of the group, indicates that no more irreducible representations exist (except those that are equivalent to those we already have).

EXAMPLE 13-6 Count up all the symmetry operations you can think of for (planar) BH_3. Can you guess from this how many irreducible representations exist for this molecule and what their dimensions are?

SOLUTION ▸ The operations are: E, C_3^+, C_3^-, σ_1, σ_2, σ_3, σ_h, S_3^+, S_3^-, C_2, C_2', C_2'' or 12 operations, so the group order is 12. There must be at least one irreducible representation of order one, so we cannot have three two-dimensional representations. We could have: 12 one-dimensional representations; eight one-dimensional and one two-dimensional representations; four one-dimensional and two two-dimensional representations. From what has been presented so far, all three of these are possible. ◂

13-13 Characters and Character Tables

Thus far, we have defined representations and shown how they may be generated from basis functions. We have distinguished between reducible and irreducible representations and have indicated that there is an unlimited number of *equivalent* representations corresponding to any given two- or higher-dimensional representation. An example of a pair of equivalent, reducible, two-dimensional representations, derived in Section 13-11, is given in Table 13-17. Equivalent representations are related through unitary transformations, which are a special kind of similarity transformation (see Chapter 9), and two matrices that differ only by a similarity transformation have the same

TABLE 13-17 ▸ Equivalent Representations for the C_3 Group

C_3	E	C_3	C_3^2
$\Gamma_{x,y}$	$\begin{pmatrix} 1 & 0 \\ 0 & 1 \end{pmatrix}$	$\begin{pmatrix} -\frac{1}{2} & -\sqrt{3}/2 \\ \sqrt{3}/2 & -\frac{1}{2} \end{pmatrix}$	$\begin{pmatrix} -\frac{1}{2} & \sqrt{3}/2 \\ -\sqrt{3}/2 & -\frac{1}{2} \end{pmatrix}$
$\Gamma_{\exp(\pm\phi)}$	$\begin{pmatrix} 1 & 0 \\ 0 & 1 \end{pmatrix}$	$\begin{pmatrix} -\frac{1}{2} - i\sqrt{3}/2 & 0 \\ 0 & -\frac{1}{2} + i\sqrt{3}/2 \end{pmatrix}$	$\begin{pmatrix} -\frac{1}{2} + i\sqrt{3}/2 & 0 \\ 0 & -\frac{1}{2} - i\sqrt{3}/2 \end{pmatrix}$

TABLE 13-18 ► Characters for the C_{3v} Group

C_{3v}	E	σ_1	σ_2	σ_3	$C_3{}^+$	$C_3{}^-$	
A_1	1	1	1	1	1	1	z, x^2+y^2, z^2
A_2	1	−1	−1	−1	1	1	R_z
E	2	0	0	0	−1	−1	$(x, y)(R_x, R_y)$
							$(x^2-y^2, xy)(xz, yz)$

trace, or *character* (Problem 9-7), which is defined as the sum of the diagonal elements of a matrix. The matrices in Table 13-17 exemplify this fact, their characters being respectively 2, −1, −1 for E, C_3, and C_3^2 in both representations. The generally accepted symbol for the character of operation R in representation Γ_i is $\chi_i(R)$, and the mathematical definition is

$$\chi_i(R) = \sum_j \Gamma_i^{j,j}(R) \tag{13-5}$$

In practice, it is the *characters* of irreducible representations that are used in most chemical applications of group theory. This means that one needs only the *character table* for a group, rather than the whole representation table. For the C_{3v} group, the character table is displayed in Table 13-18. Comparison with Table 13-13 will make clear that the character is merely the sum of diagonal elements. (For one-dimensional representations, the character and the representation are identical.) Since the representation for the *identity operation* is always a unit matrix, the character for this operation is always the same as the dimension of the representation. Hence, the first character in a row tells us the dimension of the corresponding representation.

An immediate consequence of the orthogonality theorem for representations is that the vectors resulting from *characters* are orthogonal too. This is trivially obvious for characters of one-dimensional representations. For multidimensional representations, the character vector is simply the sum of the representation vectors in diagonal positions. If a given vector (say, the A_1 vector) is orthogonal to each of these (say, E_{11} and E_{22}), then it is orthogonal to their sum; that is, in terms of the vector notation of Chapter 9, if $\tilde{\mathsf{a}}_1\mathsf{e}_{11} = 0$ and $\tilde{\mathsf{a}}_1\mathsf{e}_{22} = 0$, then $\tilde{\mathsf{a}}_1(\mathsf{e}_{11} + \mathsf{e}_{22}) = 0$.

Inspection of Table 13-18 reveals a curious thing. For any given row of characters, *all operations in the same class have the same character*. There is a fairly simple reason for this. We have indicated already that operations in the same class are operations that can be interchanged merely by group reflections, rotations, etc., of the symmetry elements in the group, but we have seen that such changes are mathematically effected through *similarity transformations*. This means that representations for operations in the same class are interchangeable *via* similarity transformations. That is, the matrix representing, say, σ_1 (in the E representation of C_{3v}) can be made equal to the matrix representing σ_2 through a similarity transformation:

$$\mathsf{T}^{-1}\sigma_1\mathsf{T} = \sigma_2 \tag{13-6}$$

(From our group Table 13-7, we can ascertain that T must be the matrix representing σ_3.) Now, the two sides of Eq. (13-6) must have the same character since they are identical 2 × 2 matrices, but the left-hand side must have the same character as σ_1

TABLE 13-19 ► The Standard Short-Form Character Table for the C_{3v} Group

C_{3v}	E	3σ	$2C_3$	
A_1	1	1	1	z, x^2+y^2, z^2
A_2	1	-1	1	R_z
E	2	0	-1	$(x, y)(R_x, R_y)(x^2-y^2, xy)(xz, yz)$

since character is unchanged by a similarity transformation. Therefore, σ_1 and σ_2 have the same character.

We can take advantage of the above rule to write our character table in abbreviated form, illustrated for the C_{3v} group in Table 13-19. This is the standard form for character tables. A collection of such tables appears in Appendix 11.

That characters must be equal in the same class is a restriction on our character vectors. In Table 13-19 it is made evident that, in the C_{3v} group, our vectors really only have three independent variables each, one for each class. These are properly thought of, then, as vectors in three-dimensional space (with weighting factors 1, 3, and 2 for E, σ, and C_3, respectively). There can be no more than three such vectors that are orthogonal, and so we are left with the result that the number of inequivalent irreducible representations in a group cannot exceed (and is in fact equal to[6]) the number of classes in the group. This result, together with the fact that the sum of squares of dimensions of inequivalent irreducible representations must equal the order of the group, often suffices to tell us in advance how many representations there are and what their dimensions are. For our C_{3v} group, the order is six and there are three classes. Hence, we know that there are three representations and that the squares of their dimensions sum to six. The problem reduces to: "What three positive integers squared, sum to six?" There is only one answer: 1, 1, and 2. The fact that there will *always* be a totally symmetric one-dimensional representation also helps pin down the possibilities. For example, can one have a group of order eight and only two classes? There is no way this can happen. Two classes would mean two representations. If both were E type, their dimensions squared would indeed sum to eight. But one of them must be one-dimensional, and there is no way the other can square to seven.

EXAMPLE 13-7 How many classes of symmetry operation can you count for (planar) BH_3? Can you use this to select one of the possibilities from Example 13-12?

SOLUTION ► The classes are E, σ_v, C_3, C_2, σ_h, S_3. That's six classes, so there are only six irreducible representations. There are four one-dimensional and two two-dimensional representations. ◄

Our collected list of conditions that the characters in a completed table must satisfy is as follows:

1. There must be a one-dimensional representation having all characters equal to $+1$.
2. The leading character in each row (i.e., the character for operation E) must equal the dimension of the representation.

[6]See Bishop [1].

3. The sum of the squares of the leading characters must equal the order of the group.

4. The number of rows in the character table must equal the number of classes in the group.

5. The absolute squares of the characters in a given row (times the weighting factor for each class if the abbreviated form is used) equals the order of the group. (Character vectors are normalized) This results directly from Eq. (13-2).

6. The character vectors are orthogonal (again, using weighting factors, if appropriate).

This is a fairly large number of restrictions, and may suffice to allow one to produce the character table for a group without ever actually producing representations. For example, consider the C_{4v} group, which has the operations $E, 2C_4, C_2, 2\sigma_v$, and $2\sigma_d$. The group thus has order eight and five classes. There must be five representations, and the only way their dimensions can square to eight is if four of them are one-dimensional and one is two-dimensional. This already enables us to write Table 13-20. It is not

TABLE 13-20 ► Partial C_{4v} Character Table

C_{4v}	E	$2C_4$	C_2	$2\sigma_v$	$2\sigma_d$
A_1	1	1	1	1	1
Γ_1	1				
Γ_2	1				
Γ_3	1				
E	2				

difficult to find a way to make Γ_1 orthogonal to A_1. We simply place -1 in some places to produce four products of -1 and four of $+1$. Three possibilities are shown in Table 13-21. These are orthogonal not only to A_1, but to each other as well, and so we

TABLE 13-21 ► Partial C_{4v} Character Table

E	$2C_4$	C_2	$2\sigma_v$	$2\sigma_d$
1	1	1	-1	-1
1	-1	1	1	-1
1	-1	1	-1	1

have found the characters for Γ_1, Γ_2, and Γ_3. The characters for the E representation must have squares that sum to eight and also be orthogonal to all four one-dimensional representation vectors. One possibility is fairly obvious. Since the characters for operations E and C_2 are $+1$ in all the one-dimensional representations, we could take the characters for the E representation to be

$$\begin{array}{ccccc} E & 2C_4 & C_2 & 2\sigma_v & 2\sigma_d \\ 2 & 0 & -2 & 0 & 0 \end{array}$$

Other cases that meet the normality condition are

$$\begin{array}{ccccc} 2 & \pm 1 & 0 & \pm 1 & 0 \\ 2 & 0 & 0 & \pm 1 & \pm 1 \\ 2 & \pm 1 & 0 & 0 & \pm 1 \end{array}$$

But none of these is orthogonal to all the one-dimensional sets. Therefore, the complete character table (except for the basis functions) for the C_{4v} group is shown in Table 13-22, where the symbols A_2, B_1, B_2 are consistent with symmetry or antisymmetry for C_4 and σ_v as described in Section 13-9.

TABLE 13-22 ► Completed C_{4v} Character Table

C_{4v}	E	$2C_4$	C_2	$2\sigma_v$	$2\sigma_d$
A_1	1	1	1	1	1
A_2	1	1	1	−1	−1
B_1	1	−1	1	1	−1
B_2	1	−1	1	−1	1
E	2	0	−2	0	0

13-14 Using Characters to Resolve Reducible Representations

It was pointed out earlier that several irreducible representations can be combined into a larger-dimensional reducible representation. Our example was

$$\begin{pmatrix} \Gamma_2 & 0 & 0 \\ 0 & \Gamma_1 & 0 \\ 0 & 0 & \Gamma_3 \end{pmatrix} \equiv \begin{pmatrix} A_2 & 0 & 0 \\ 0 & A_1 & 0 \\ 0 & 0 & E \end{pmatrix} = \Gamma'$$

which is a four-dimensional representation (since Γ_3 is two-dimensional). A matrix built up in this way is symbolized $A_2 \oplus A_1 \oplus E$. It is easy to see that the characters of the reducible representation Γ' are simply the *sums of characters* for the individual irreducible component representations (since the diagonal elements of A_2, A_1, and E all lie on the diagonal of Γ'). Thus, the characters of Γ' are

$$\begin{array}{rccc} & E & 3\sigma & 2C_3 \\ \Gamma': & 4 & 0 & 1 \end{array}$$

Furthermore, no matter how Γ' is disguised by a similarity transformation, its character vector is unchanged. Now, suppose you were given the representation Γ', disguised through some similarity transformation so as to be nonblock diagonal and asked to tell which irreducible representations were present. How could you do it? One way would be to find the similarity transformation that would return the representation to block diagonal form. But there is a much simpler way. We can test to see if the character vector of Γ' is orthogonal to the character vectors of each of our irreducible representations. For instance, if Γ' contained only A_2 and E, its character vector would be orthogonal to that of A_1 because the character vectors of A_2 and E are orthogonal to that of A_1. If A_1 is present in Γ', then Γ' and A_1 character vectors are not orthogonal. Indeed, the *amount of* A_1, A_2, or E present can be found by making use of the character vector normality relation. This is illustrated for Γ' by the following equations:

$$A_1: \frac{1}{6}(1\cdot 4+3\cdot 1\cdot 0+2\cdot 1\cdot 1)=1$$

$$A_2: \frac{1}{6}(1\cdot 4+3\cdot -1\cdot 0+2\cdot 1\cdot 1)=1$$

$$E: \frac{1}{6}(2\cdot 4+3\cdot 0\cdot 0+2\cdot -1\cdot 1)=1$$

In general, for $\Gamma' = c_1\Gamma_1 \oplus c_2\Gamma_2 \oplus \cdots \oplus c_i\Gamma_i \oplus + \cdots$

$$c_i = (1/h)\sum_R \chi_i(R)\chi'(R) \tag{13-7}$$

where h is the order of the group. This technique for resolving a reducible representation into its component irreducible representations is very useful in quantum chemistry, as we shall see shortly.

13-15 Identifying Molecular Orbital Symmetries

We stated earlier that any MO must be a basis for an irreducible representation. Given a set of computed MOs, how does one decide which representation each MO is a basis for? One does this by comparing the MOs with the characters in the character table. Some examples will make this clear. In Table 13-23 are extended Hückel data for NH_3, oriented as shown in Fig. 13-13. NH_3 belongs to the C_{3v} group, having the character table shown in Tables 13-18 and 13-19. The minimal valence basis set of seven AOs leads to seven MOs. Notice that MOs 5 and 6 and also 2 and 3 are degenerate. Therefore, these MOs must be bases for two-dimensional representations, and are assigned the symbol e (lower case for MOs). The other MOs are all given the main symbol a. There are two possibilities for these MOs—a_1 or a_2. These differ in their characters for reflection, a_1 being symmetric, and a_2 antisymmetric. If we look at the eigenvectors for MOs 1, 4, and 7, we see that they contain the 1s AOs on each hydrogen with equal sign and magnitudes. Since reflection always interchanges two hydrogens, these MOs are clearly all symmetric for reflection, and so we label them all a_1. Our result, then, is

MO:	1	2	3	4	5	6	7
Symmetry:	a_1	e	e	a_1	e	e	a_1

Next, consider staggered ethane, which we have earlier assigned to the D_{3d} point group. Orbital energy levels and sketches of the MOs appear in Fig. 13-14. The character table for the D_{3d} group is given in Table 13-24.

As before, we observe that certain of the orbitals have the same energies, so we assign such doubly-degenerate MOs the main symbol e. These MOs are either symmetric or

TABLE 13-23 ▶ Extended Hückel Data for NH_3

MO no.	MO energy (a.u.)	MO occupancy
1	0.7494	0
2	0.1279	0
3	0.1279	0
4	−0.4964	2
5	−0.5955	2
6	−0.5955	2
7	−1.0178	2

	Eigenvectors						
AO\MO	1	2	3	4	5	6	7
N(2s)	1.2946	0.0000	0.0000	−0.1715	0.0000	0.0000	0.7387
N($2p_z$)	−0.4369	0.0000	0.0000	−0.9628	0.0000	0.0000	0.0214
N($2p_x$)	0.0000	1.0275	0.0000	0.0000	0.6498	0.0000	0.0000
N($2p_y$)	0.0000	0.0000	−1.0275	0.0000	0.0000	0.6498	0.0000
H_1(1s)	−0.7166	−1.0399	0.0000	0.0656	0.4422	0.0000	0.1561
H_2(1s)	−0.7166	0.5200	0.9006	0.0656	−0.2211	0.3829	0.1561
H_3(1s)	−0.7166	0.5200	−0.9006	0.0656	−0.2211	−0.3829	0.1561

antisymmetric for inversion and are accordingly subscripted g or u, respectively. All of the nondegenerate MOs must have the main symbol a, since b does not appear in the D_{3d} table. The character table indicates that a_1 and a_2 differ in that they are respectively symmetric and antisymmetric for two-fold rotations that switch the molecule end for end. The u, g subscripts again refer to inversion. Inspection of the figures enables us to decide which symbols are appropriate in each case. The resulting symmetry designations are included in Fig. 13-14.

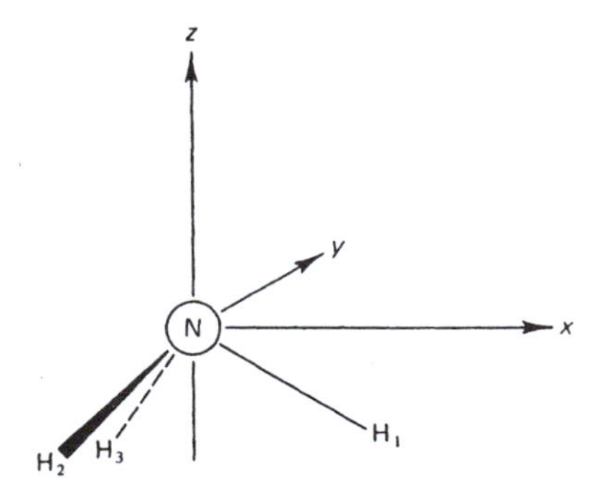

Figure 13-13 ▶ Orientation of ammonia with respect to Cartesian axes.

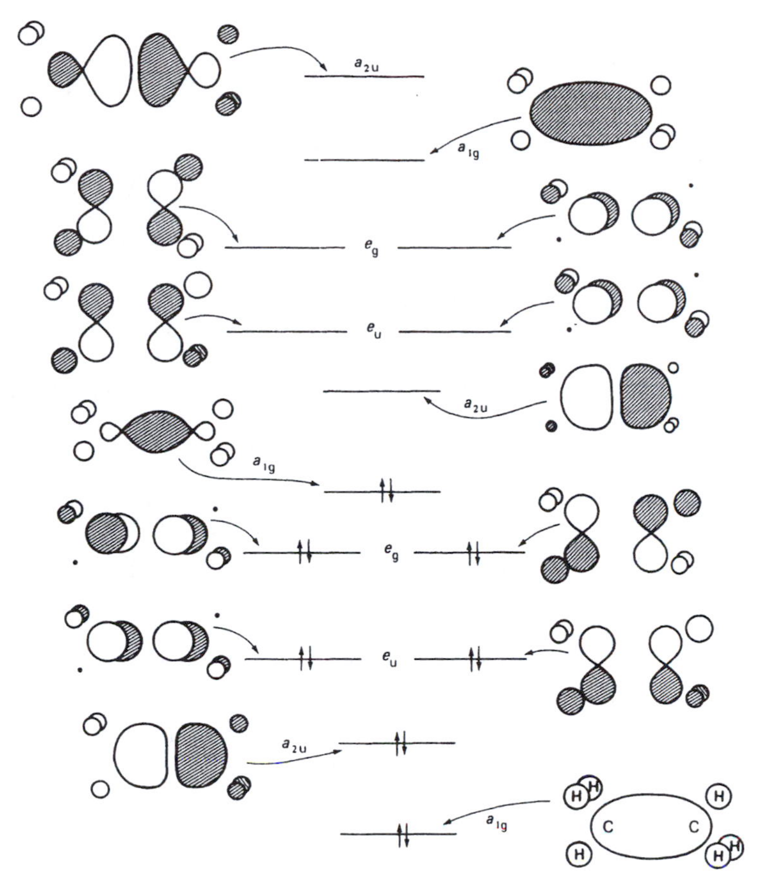

Figure 13-14 ▶ Valence MOs for staggered ethane. The energy level spacings have been altered for convenience. The *hatched* areas have a negative sign.

TABLE 13-24 ▶ Characters for the D_{3d} Point Group

D_{3d}	E	$2C_3$	$3C_2$	i	$2S_6$	$3\sigma_d$
A_{1g}	1	1	1	1	1	1
A_{2g}	1	1	−1	1	1	−1
E_g	2	−1	0	2	−1	0
A_{1u}	1	1	1	−1	−1	−1
A_{2u}	1	1	−1	−1	−1	1
E_u	2	−1	0	−2	1	0

13-16 Determining in Which Molecular Orbital an Atomic Orbital Will Appear

Earlier, we noted that the $2p_z$ AO in ammonia will contribute to a_1 MOs because $2p_z$ transforms like z, and z is listed as a basis for the a_1 representation. If the basis functions were not listed, we could have reached the same conclusion simply by

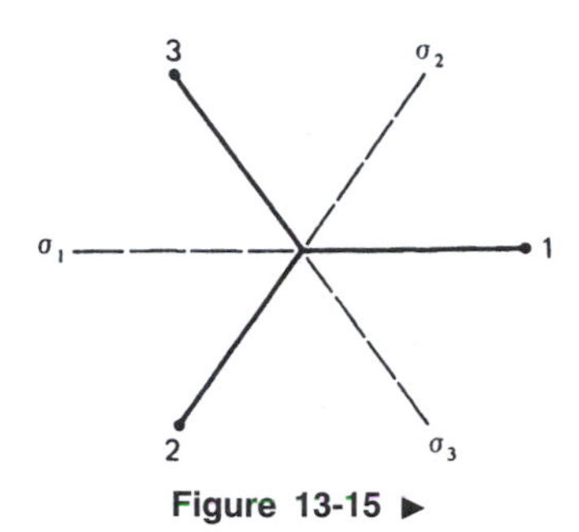

Figure 13-15 ▶

taking $2p_z$ and putting it through all the symmetry operations to produce a representation:

$$E2p_z = +1 \cdot 2p_z$$
$$\sigma_i 2p_z = +1 \cdot 2p_z, \quad i = 1, 2, 3$$
$$C_3^{\pm} 2p_z = +1 \cdot 2p_z$$

The representation contains only $+1$, so $2p_z$ obviously "has" a_1 symmetry.

When we come to the 1s AOs on hydrogens in NH_3, we cannot use the list of basis functions. It includes only coordinates originating on the principal axis, and the hydrogens are not on that axis. Here we *must* generate a representation. Let us try to do this by putting one of the hydrogen atoms, H_1, through the various symmetry operations (also see Fig. 13-15):

$$EH_1 = +1H_1$$
$$\sigma_1 H_1 = +1H_1$$
$$\sigma_2 H_1 = H_3$$
$$\sigma_3 H_1 = H_2$$
$$C_3^+ H_1 = H_2$$
$$C_3^- H_1 = H_3$$

Since some of these operations interchange H_1 with H_2 or H_3, we are not achieving a one-dimensional representation. We must take all three functions together and work out a three-dimensional representation. Thus,

$$E\begin{pmatrix} H_1 \\ H_2 \\ H_3 \end{pmatrix} = \begin{pmatrix} H_1 \\ H_2 \\ H_3 \end{pmatrix}, \quad E: \begin{pmatrix} 1\,0\,0 \\ 0\,1\,0 \\ 0\,0\,1 \end{pmatrix}, \quad \chi(E) = 3$$

$$\sigma_1\begin{pmatrix} H_1 \\ H_2 \\ H_3 \end{pmatrix} = \begin{pmatrix} H_1 \\ H_3 \\ H_2 \end{pmatrix}, \quad \sigma_1: \begin{pmatrix} 1\,0\,0 \\ 0\,0\,1 \\ 0\,1\,0 \end{pmatrix}, \quad \chi(\sigma_1) = 1$$

$$C_3^+\begin{pmatrix} H_1 \\ H_2 \\ H_3 \end{pmatrix} = \begin{pmatrix} H_2 \\ H_3 \\ H_1 \end{pmatrix}, \quad C_3^+: \begin{pmatrix} 0\,1\,0 \\ 0\,0\,1 \\ 1\,0\,0 \end{pmatrix}, \quad \chi(C_3^+) = 0$$

and similarly for σ_2, σ_3, and C_3^-. But we are only going to use the characters χ, and we know that σ_2 and σ_3 must have the same character as σ_1 (i.e., 1) and C_3^- must have the same character as C_3^+ (i.e., 0). Our character table for the representation resulting from the basis of three hydrogen 1s AOs then, is

C_{3v}	E	3σ	$2C_3$	
Γ_{3H}	3	1	0	(H_1, H_2, H_3)

Notice that a "one" on the diagonal of a 3×3 representation matrix has the effect of keeping a hydrogen 1s AO in place. Thus, the E operation keeps all three H's unmoved, has three "ones" on the diagonal, and has a character of 3. The σ operations each leave but one hydrogen unmoved, have a single "one" on the diagonal, and have a character of 1. C_3 leaves no hydrogen unmoved and has a character of zero. *In general, the characters of such representations are the numbers of functions not moved by the various operations.* Thus, we could have written down the above characters for Γ_{3H} without figuring out the representation matrices.

It is evident that Γ_{3H} must be reducible, since it is of higher dimension than anything in the C_{3v} character table. To resolve Γ_{3H} into its irreducible components, we use the relation (13-7):

$$A_1: \frac{1}{6}(1 \cdot 3 \cdot 1 + 3 \cdot 1 \cdot 1 + 2 \cdot 0 \cdot 1) = 1$$

$$A_2: \frac{1}{6}(1 \cdot 3 \cdot 1 + 3 \cdot 1 \cdot -1 + 2 \cdot 0 \cdot 1) = 0$$

$$E: \frac{1}{6}(1 \cdot 3 \cdot 2 + 3 \cdot 1 \cdot 0 + 2 \cdot 0 \cdot -1) = 1$$

Therefore, $\Gamma_{3H} = A_1 \oplus E$. We conclude from this that hydrogen 1s AOs can appear in MOs having a_1 or e symmetry. Table 13-23 indicates that this is correct.

EXAMPLE 13-8 Given only the MO sketches of Fig. 13-14 (and no energies), is there an easy way to select the MOs that are degenerate?

SOLUTION ▸ Nondegenerate MOs must be either symmetric or antisymmetric for every symmetry operation of the molecule. Failure to obey this criterion indicates a degenerate MO. For instance, the two highest energy MOs in the figure are symmetric or antisymmetric for every reflection, rotation, improper rotation, or inversion, hence are nondegenerate. The next two MOs, however, are unsymmetrical for some of the operations. Most easily seen is their unsymmetrical behavior for rotation by 120° about the C–C axis. These MOs must be degenerate. ◂

13-17 Generating Symmetry Orbitals

Consider the lowest-energy extended Hückel MO for NH_3. It is

$$\phi_7 = 0.7387 N_{2s} + 0.0214 N_{2p_z} + 0.1561\ 1s_1 + 0.1561\ 1s_2 + 0.1561\ 1s_3$$

We see that the hydrogen AOs all have the same sign and magnitude in this MO. In fact, we noted above that this *must* happen in all a_1 MOs of NH_3 for reasons of symmetry. If an MO is to be symmetric for all the reflection and rotation operations of the C_{3v} group, there is no other combination that will be adequate. When we have several equivalent atoms, and symmetry forces their AOs to appear in MOs in certain combinations, we refer to those combinations as *symmetry orbitals*. Thus, $\phi_{a_1}^H = N(1s_1 + 1s_2 + 1s_3)$ is the a_1 symmetry orbital for the hydrogens in ammonia (N is a normalizing constant).

One can use the character table to generate symmetry orbitals. This is done by picking any one of the AOs, say $1s_1$, and operating on it with each symmetry operation *times the character for each operation* in the representation of interest. The sum of all these operations is an unnormalized symmetry orbital. Thus, for the a_1 representation in NH_3,

$$\begin{aligned}\phi_{a_1}^H &= E\cdot 1\cdot 1s_1 + \sigma_1\cdot 1\cdot 1s_1 + \sigma_2\cdot 1\cdot 1s_1 + \sigma_3\cdot 1\cdot 1s_1 + C_3^+\cdot 1\cdot 1s_1 + C_3^-\cdot 1\cdot 1s_1\\ &= 1s_1 + 1s_1 + 1s_3 + 1s_2 + 1s_2 + 1s_3\\ &= 2(1s_1 + 1s_2 + 1s_3)\end{aligned}$$

If we normalize (ignoring overlap between 1s AOs on different centers), we obtain

$$\phi_{a_1}^H = \frac{1}{\sqrt{3}}(1s_1 + 1s_2 + 1s_3)$$

To generate symmetry orbitals of e symmetry is a little more involved. First we pick a 1s AO and do just as before, using now the characters for e rather than a_1:

$$\begin{aligned}\phi_e{}^H &= E\cdot 2\cdot 1s_1 \quad + 0\cdot \text{all reflections} \quad + (-1)\cdot C_3^+\cdot 1s_1 \quad + (-1)\cdot C_3^-\cdot 1s_1\\ &= 2 1s_1 \qquad\qquad\qquad\qquad\quad - 1s_2 \qquad\qquad\qquad - 1s_3\end{aligned}$$

Normalization yields

$$\phi_e{}^{H'} = \frac{1}{\sqrt{6}}(2\ 1s_1 - 1s_2 - 1s_3)$$

Because e symmetry is manifested by doubly-degenerate MOs, we need to find a mate for ϕ_e^H. We can try to do this by repeating the above procedure except operating on $1s_2$ instead of $1s_1$. This yields

$$\psi_e^H = \frac{1}{\sqrt{6}}(2\ 1s_2 - 1s_1 - 1s_3)$$

But this function is not orthogonal to ϕ_e^H. (The overlap is $-\frac{1}{2}$.) To achieve orthogonality, we resort to Schmidt orthogonalization (Chapter 6):

$$\psi_e^{H'} = \psi_e^H - S\phi_e^H$$

Upon expansion and renormalization, this gives

$$\psi_e^{H'} = \frac{1}{\sqrt{2}}(1s_2 - 1s_3)$$

The data in Table 13-23 show that the e-type MOs do contain the hydrogen 1s AOs in just the manner prescribed by symmetry. For example, MO 5 has hydrogen 1s coefficients 0.4422, −0.2211, −0.2211, a combination similar to that in ϕ_e^H. The degenerate mate, MO 6, has hydrogen coefficients of 0.0000, 0.3829, −0.3829, similar to $\psi_e^{H'}$.

Chemists who have had some experience in these matters are likely to prefer thinking in terms of symmetry orbitals. Thus, in thinking of ammonia, they are likely to take as a minimal valence basis set the functions: N_{2s}, N_{2p_x}, N_{2p_y}, N_{2p_z}, $\frac{1}{\sqrt{3}}(1s_1 + 1s_2 + 1s_3)$, $\frac{1}{\sqrt{6}}(21s_1 - 1s_2 - 1s_3)$, $\frac{1}{\sqrt{2}}(1s_2 - 1s_3)$. (We continue to ignore overlap between 1s AOs.) Since they know that it is not possible for bases of different symmetries to mix (it would produce an MO of mixed symmetry), they know at once that they can have a_1 MOs from mixtures of N_{2s}, N_{2p_z} and $(1/\sqrt{3})(1s_1 + 1s_2 + 1s_3)$ and e-type MOs from the remaining functions. In effect, they have used symmetry to partition their functions into two subsets that do not interact with each other. This means that, in the MO calculation, the hamiltonian matrix will have no mixing elements between members of different subsets. This is indicated schematically in Fig. 13-16. They also know in advance that there will be *three* MOs of a_1 symmetry (since only three basis functions have that symmetry) and *four* of e symmetry (two degenerate pairs). It is interesting to see how strongly symmetry controls the nature of NH_3 MOs.

Because symmetry orbitals depend on symmetry and not on finer details of molecular structure, they recur again and again in molecules of similar symmetry. For instance, the a_1 and e combinations of hydrogen coefficients discussed above for ammonia will be found for hydrogen AOs in staggered or eclipsed ethane (see the drawings in Fig. 13-14, for instance) and for carbon AOs in MOs for cyclopropenyl (Chapter 8). Because symmetry requirements transcend the differences between various approximate methods

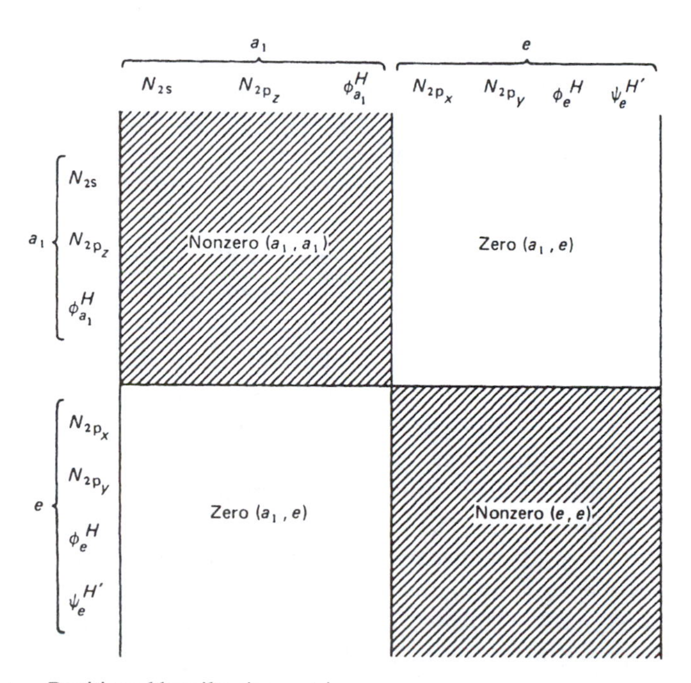

Figure 13-16 ▶ Partitioned hamiltonian matrix.

for solving the Schrödinger equation, we expect these symmetry patterns to appear in MOs for, say, ammonia, at extended Hückel, CNDO, INDO, MINDO, or *ab initio* levels of computation.

13-18 Hybrid Orbitals and Localized Orbitals

The concept of a *hybridized orbital* is often encountered in the literature, especially in introductory discussions of bonding theory. While this concept is not essential to MO theory, it is used enough to justify a brief discussion.

In the previous section, we showed how one could transform from a basis set of STOs to a basis set of symmetry orbitals. Since these two sets are related through a unitary transformation, they are *equivalent* and must lead to the same MOs when we do a linear variation calculation. However, there are an infinite number of unitary transformations available, and so the set of symmetry orbitals is only one of an infinite number of possible equivalent bases. Of course, this set has the unique advantage of being a set of bases for representations of the symmetry group, which makes it easy to work with. Another set of equivalent basis functions are the *hybrid orbitals*. These have the distinction of being the functions that are concentrated along the directions of bonds in the system. Consider, for example, methane, which was discussed in detail in Chapter 10. The minimal basis set of valence STOs on carbon can be transformed to form four tetrahedrally directed hybrids:

$$\phi_1 = \frac{1}{2}(\mathrm{s} + \mathrm{p}_x - \mathrm{p}_y + \mathrm{p}_z), \quad \phi_2 = \frac{1}{2}(\mathrm{s} - \mathrm{p}_x + \mathrm{p}_y + \mathrm{p}_z)$$

$$\phi_3 = \frac{1}{2}(\mathrm{s} - \mathrm{p}_x - \mathrm{p}_y - \mathrm{p}_z), \quad \phi_4 = \frac{1}{2}(\mathrm{s} + \mathrm{p}_x + \mathrm{p}_y - \mathrm{p}_z)$$

One of these hybrids, ϕ_4, is shown in Fig. 13-17 and can be seen to point toward one of the hydrogen atoms. Because the square of each hybrid consists of one part s AO to three parts p AO, these are called sp^3 hybrids (pronounced s-p-three). The reason for focusing on sp^3 hybrids in this case is that they have physical appeal since they point along the C–H bonds and therefore seem to have a more natural relation to the electron-pair bond approach of G. N. Lewis. This is deceptive, however, because the set of four carbon sp^3 orbitals is completely equivalent to the set of four carbon STOs. The sum of squares of the hybrids is spherically symmetric just as is the sum of squares of STOs. Thus, even though each hybrid is directed toward a hydrogen, the electron density due to all four occupied hybrids is spherically symmetric. Furthermore, after the linear variation is performed, the MOs that are produced contain *mixtures* of hybrids to give us the exact same delocalized MOs produced from STOs. No single MO consists of just one hybrid and one hydrogen 1s AO, and therefore no single MO can be identified with one C–H bond. We conclude then that hybrid orbitals are one of an infinite number of choices of basis, that they have an appealing appearance because of their concentration in bond regions, but that no concentration of charge in the molecule results as a consequence of using hybrids rather than STOs. (Some concentration of charge in the bonds does result from overlap between basis functions on carbon and those on hydrogens, but this occurs to exactly the same degree for the various equivalent basis sets.)

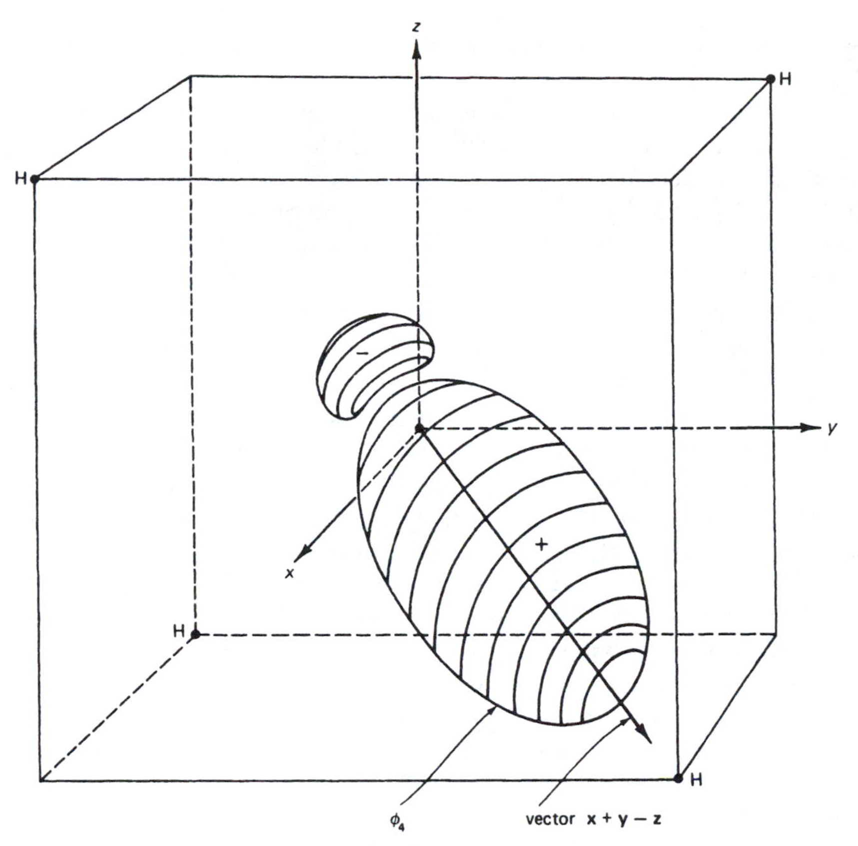

Figure 13-17 ▶ A sketch of the hybrid $\phi_4 = \frac{1}{2}(s + p_x + p_y - p_z)$. The direction of the hybrid is coincident with that of a vector having the same x, y, z dependence as ϕ_4. The other three hybrids are identical except that they point toward the other three H atoms. (The coordinate system here is rotated with respect to that used in Chapter 10.)

Another example is planar CH_3^+ (D_{3h}). As before, we can mix our minimal valence STOs on carbon to produce hybrids pointing toward the hydrogens. If one hydrogen is on the $+y$ axis, the hybrids are

$$\psi_1 = \frac{1}{\sqrt{3}}s + \frac{\sqrt{2}}{\sqrt{3}}p_y,$$
$$\psi_2 = \frac{1}{\sqrt{3}}s + \frac{1}{\sqrt{2}}p_x - \frac{1}{\sqrt{6}}p_y,$$
$$\psi_3 = \frac{1}{\sqrt{3}}s - \frac{1}{\sqrt{2}}p_x - \frac{1}{\sqrt{6}}p_y$$

Each of these hybrids, when squared, is one part s to two parts p and is called an sp^2 hybrid. The coefficients for the p AOs are determined from simple vector considerations: One merely calculates how x and y vectors must be combined to produce resultant vectors pointing toward the corners of an equilateral triangle. The resulting hybridized basis set for CH_3^+ is ψ_1, ψ_2, ψ_3, plus the $2p_z$ STO on carbon and a 1s STO on each hydrogen. As before, the sum of the squares of ψ_1, ψ_2, ψ_3, and $2p_z$ is spherically symmetric.

We turn now to *localized orbitals*. We have been emphasizing that one can subject *basis sets* to unitary transformations without making any physical difference. A similar rule applies for *filled molecular orbitals* in a determinantal wavefunction. These too can be subjected to unitary transformations without affecting the *total* energy or *total* electronic distribution for the system. (We have encountered this fact before. See, for instance, Appendix 7.) Thus, we have the capability of altering the appearance of the individual orbitals in the wavefunction without affecting the wavefunction itself. Chemists tend to think of the electrons in molecules as being paired in bond regions, lone pairs, and inner shells, but MOs are delocalized and do not reflect this viewpoint. But by carrying out unitary transformations, we can attempt to produce orbitals that are more localized without sacrificing any of the properties of the overall wavefunction. For methane, we could mix our four delocalized occupied MOs together to try to produce four new orbitals, each one concentrated in a different C–H bond region. One can do this, but it is important to realize that these localized orbitals are not eigenfunctions of an energy operator, for they have been produced by mixing eigenfunctions having different energies. Furthermore, the localization is never complete in any system of physical interest. Each localized orbital always contributes at least slightly to charge buildup in regions outside that of its primary localization. For instance, a localized C–H_1 orbital in methane will have small "residual" components at H_2, H_3, and H_4.

What we have been discussing in this section are various kinds of *equivalent* orbitals. At the level of *basis sets*, we have indicated that a minimal valence basis set of STOs is equivalent to a set of symmetry orbitals and also to a set of hybrid orbitals (as well as an infinite number of other possibilities). At the level of *molecular orbitals* we have indicated that the set of occupied delocalized MOs is equivalent to an infinity of transformed sets, some of which will tend to be localized in regions chemists associate with bonds, lone pairs, or inner shells. One's choice among the possibilities for *basis* is a matter of taste. However, at the MO level, the *delocalized* MOs have two features that are sometimes advantageous. The first is that their energies are eigenvalues for some energy operator (the Fock operator in SCF theory). These are related in a simple way to ionization energies and electron affinities *via* Koopmans' theorem. Hence, delocalized MOs are more appropriate when considering photoelectron spectra, etc. The second advantage of delocalized MOs is that they display in a clear way the symmetry requirements on the system because they are bases for representations. Hence, these MOs are the most appropriate to use when one is using MO phase relations to infer the nature of certain intra- or intermolecular interactions. (See Chapter 14 for examples.) When delocalized MOs are mixed to form localized orbitals, these energy and symmetry features become partially disguised.

13-19 Symmetry and Integration

Throughout this book, the usefulness of symmetry to determine whether an integral vanishes has been emphasized. It should come as no surprise, therefore, that the formal mathematics of symmetry—group theory—is also useful for this purpose.

The basic idea we have used all along is that, if an integrand is antisymmetric for any symmetry operation, it must have equal positive and negative regions, which cancel on integration. If there is *no* symmetry operation for which the integrand is antisymmetric, then the integral need not vanish. (It still might vanish, but not because of symmetry.)

The group theoretical equivalent of this is as follows. Suppose that we have an integral over the integrand f:

$$\int f\,dv = ?$$

The function f is identified as being related to some symmetry point group. (Examples are given shortly.) We want to know what representation f is a basis for. If f produces a representation containing A_1, then f has some totally symmetric character and the integral need not vanish. But if f is devoid of A_1 character, the integral vanishes by symmetry since all other representations are antisymmetric for at least one operation. Our problem, therefore, is to decide which irreducible representations are present in the representation that is produced by the integrand f.

In quantum chemistry, the integrand of interest is often a product of wavefunctions (or orbitals) and operators. For example, the hamiltonian matrix $\hat{H}$ contains integrals of the form

$$H_{ij} = \int \psi_i^* \hat{H} \psi_j \, d\tau$$

We know that $\hat{H}$ is invariant for any symmetry operation of the group, and so $\hat{H}$ has A_1 symmetry. ψ_i and ψ_j are assigned symmetries by comparing their behaviors under various operations with the group character table, as illustrated earlier. Thus, it is fairly easy to ascertain the symmetries of the various *parts* of the integrand. The problem is to determine the symmetry of the *product* $\psi_i^* \hat{H} \psi_j$.

To develop a rule for products, let us consider the simplest case—the one-dimensional representations. Suppose that ψ_1 and ψ_2 are bases for one-dimensional representations Γ_1 and Γ_2. Then, for some symmetry operation R

$$R\psi_1 = \chi_1(R)\psi_1, \quad R\psi_2 = \chi_2(R)\psi_2$$

where χ is a character $(1, -1, \epsilon, \text{or } \epsilon^*)$. If we operate on the *product* $\psi_1\psi_2$ with R, we obtain[7]

$$R\psi_1\psi_2 = (R\psi_1)(R\psi_2) = \chi_1(R)\psi_1\chi_2(R)\psi_2 = \chi_1(R)\chi_2(R)\psi_1\psi_2$$

That is, *the characters for the product $\psi_1\psi_2$ are equal to the products of the characters for ψ_1 and ψ_2.* We have demonstrated the rule for one-dimensional representations, but it can be proved for higher-dimensional cases as well. In group theory, the product of two functions, like $\psi_1\psi_2$, is referred to as a *direct product* to distinguish it from a product of symmetry operations, like $\sigma_3 C_3^+$. The symbol for a direct product is $\otimes$.

For the C_{3v} group, the characters for some direct products of bases for irreducible representations are shown in Table 13-25. The direct product x^2 has as characters the product of characters of E times itself. These characters $(4, 0, 1)$ do not agree with any of the irreducible representation character sets, and so $E \otimes E$ is reducible. We can tell, in fact, that $E \otimes E$ is four-dimensional from the leading character. To resolve $E \otimes E$, we employ the formula (13-7), which gives $E \otimes E = A_1 \otimes A_2 \otimes E$, and fits the observation that $E \otimes E$ is four-dimensional. The other direct products listed

[7]That $R\psi_1\psi_2 = (R\psi_1)(R\psi_2)$ is not always obvious to the student, but it should be evident that operating on (say reflecting) the function $\psi_1\psi_2$ gives the same result as reflecting ψ_1 and ψ_2 separately and then taking the product.

TABLE 13-25 ▶ Characters for Direct Products of Bases for Irreducible Representations of C_{3v}

C_{3v}	E	3σ	$2C_3$		
A_1	1	1	1	z	x^2+y^2, z^2
A_2	1	-1	1	R_z	
E	2	0	-1	$(x, y), (R_x, R_y)$	$(x^2-y^2, xy)(xz, yz)$
$E \otimes E$	4	0	1	$x^2 \quad y^2 \quad x^2z^2 \quad xy$	
$A_2 \otimes E$	2	0	-1	$R_zx \quad R_zy$	
$A_2 \otimes A_2$	1	1	1	R_z^2	
$A_1 \otimes A_2 \otimes E$	2	0	-1	zR_zx	

in Table 13-26 (R_zx, R_z^2, zR_zx, etc.) all give character sets indicative of irreducible representations. We see that $A_2 \otimes E = E$, $A_2 \otimes A_2 = A_1$, $A_2 \otimes A_1 \otimes E = E$.

There is a general rule that is illustrated by these examples: *A direct product of bases for two irreducible representations contains A_1 character if and only if the two irreducible representations are the same*. That is, if f_i is a basis for Γ_i and f_j is a basis for Γ_j and $f_i f_j$ is a basis for $\Gamma_{i,j}$, where

$$\Gamma_{i,j} \equiv \Gamma_i \otimes \Gamma_j = c_1 A_1 \oplus c_2\Gamma_2 \oplus \cdots \oplus c_i\Gamma_i \oplus c_j\Gamma_j \oplus \cdots$$

then $c_1 \neq 0$ if and only if $\Gamma_i = \Gamma_j$.

Another rule is that, if Γ_i is A_1, then $\Gamma_{i,j} = \Gamma_j$; that is, multiplying a function f_2, by a totally symmetric function f_1 gives a product with the symmetry of f_2.

Now we are in a position to decide whether the integral of a product of functions and operators will vanish. For our examples, we will continue to use orbitals, operators, and coordinates from the ammonia molecule. Some of these quantities, segregated according to symmetry, are given in Table 13-26.

TABLE 13-26 ▶ Operators and Orbitals for Ammonia Classified by Symmetry

a_1	e
N_{2s}	N_{2px}
$N2_{pz}$	N_{2py}
$(1/\sqrt{3})(1s_1 + 1s_2 + 1s_3)$	$(1/\sqrt{6})(2 \cdot 1s_1 - 1s_2 - 1s_3)$
ϕ_1, ϕ_4, ϕ_7 } MOs(see Table 13-23)	$(1/\sqrt{2})(1s_2 - 1s_3)$
$\hat{H}$	ϕ_2, ϕ_3, ϕ_5, ϕ_6 } MOs
z	x
	y

EXAMPLE 13-9 Indicate whether each of the following integrals must vanish due to symmetry.

1. $\int N_{2p_z} \hat{H} N_{2p_x} dv$

2. $\int N_{2p_x} \hat{H} \frac{1}{\sqrt{2}}(1s_2 - 1s_3) dv$

SOLUTION ▶

1. $\int N_{2p_z} \hat{H} N_{2p_x} dv$: The symmetries of the three functions in the integrand are respectively A_1, A_1, E. The direct product has symmetry E. There is no A_1. The integral vanishes.

2. $\int N_{2p_x} \hat{H} \frac{1}{\sqrt{2}}(1s_2 - 1s_3) dv$: The symmetries are E, A_1, E. The direct product is therefore $E \otimes E \otimes A_1 = A_1 \oplus A_2 \oplus E$. Since A_1 is present, the integral need not vanish.

◀

These two examples are related to the block diagonalization of the matrix H, discussed in a previous section. The zero blocks in that matrix correspond to integrals between functions of different symmetry. Since $\hat{H}$ is of A_1 symmetry, it has no influence on the symmetry of the integrand. If ψ_i and ψ_j have different symmetries, their direct product cannot contain A_1 symmetry and the integral over $\psi_i \hat{H} \psi_j$ must vanish. The reader can now understand how the computational procedure guarantees MOs of "pure" symmetry (i.e., bases of irreducible representations). If two basis functions ψ_i and ψ_j differ in symmetry, there will be a zero value for H_{ij}. A zero H_{ij} means that mixing ψ_i and ψ_j will produce no energy lowering. Hence, the variation procedure will not mix these functions together in the same MO, so the MO will not be of mixed symmetry.

EXAMPLE 13-10 Indicate whether each of the following integrals must vanish due to symmetry.

1. $\int \phi_1 x \phi_4 dv$

2. $\int \phi_3 y \phi_5 dv$

SOLUTION ▶

1. $\int \phi_1 x \phi_4 dv$: Integrals of this sort are involved in calculating spectral intensities. The symmetries are $A_1 \otimes E \otimes A_1 = E$ and the integral vanishes. This means a transition between states corresponding to an electron going from ϕ_1 to ϕ_4 (or ϕ_4 to ϕ_1) is forbidden for x-polarized light and an oriented molecule (see Section 12-9).

2. $\int \phi_3 y \phi_5 dv$: The symmetry here is $E \otimes E \otimes E$, which gives characters $8, 0, -1$. This resolves into $A_1 \oplus A_2 \oplus 3E$ and the integral need not vanish. Corresponding transitions are "y allowed."

◀

EXAMPLE 13-11 In Chapter 12 , we indicated that the $\pi^* \leftarrow \pi$ transition in ethene is dipole-allowed and polarized parallel to the molecular axis. Verify this from the character table for ethene, after determining the group symbol.

SOLUTION ▶ Referring to Fig. 13-7, ethene yields the answers "no, no, no, yes, {no," so choose any one. We arbitrarily take C_2 along the C–C axis} "no, yes," (go to D branch) "yes", so D_{2h}. Our choice of principal axis puts the C–C bond vertical. Using that we can identify the π MO as B_{2u} and the π^* MO as B_{3g}. The product of characters for $B_{2u}B_{3g} = 1\ 1 - 1 - 1 - 1 - 1\ 1\ 1$, which can be seen to have B_{1u} symmetry. So $\int \pi^*(x, y, \text{or } z)\pi\, dv \neq 0$ only if x, y, or z also has B_{1u} symmetry. z does, so the transition is allowed and is z-polarized (i.e., it is a parallel transition.) ◀

In this chapter we have seen how formal group theory can be used to characterize MO symmetries, construct symmetry orbitals, and indicate whether integrals vanish by symmetry. It is true that one can perform MO calculations and get correct results without explicitly considering symmetry or group theory, since the computational procedures satisfy symmetry considerations automatically. But group theory allows a much deeper understanding of the constraints that symmetry places on a problem and often leads to significant shortcuts in computation.

A notable feature of group theory is its hierarchy of concepts. At the lowest level are the symmetry operations and the basis functions they operate on. At the intermediate level are the representations for the group, produced from the basis functions. At the highest level are the characters, produced from the representations. The characters provide the "handles" that we actually work with, but our interest is often focused on the basis functions to which they are related. This tends to lend an air of unreality to the use of group theory. An aim of this chapter has been to avoid this feeling of unreality by dispensing with formal proofs, and instead illustrating relationships through investigation of examples. Further insight should come from solving the problems at the end of this chapter.

13-19.A Problems

13-1. Do the following operations constitute a group? "come 90° to port" (P) "come 90° to starboard" (S) "steady as she goes" (E).

13-2. The text indicates that every element in the group has an inverse if E appears in each column of the multiplication table. But E also appears once in each row. What does this mean?

13-3. Consider the group of four operations of the drill soldier (Section 13-2).

a) To which symmetry point group is this set of four operations isomorphic (i.e., which group has the same product relationship)?
b) Based on the *mathematical* definition of class and Table 13-1, how many classes are there in this group?
c) Based on your physical intuition about kinds of operation, how many classes would you have anticipated for this group? If there is a discrepancy between (b) and (c), try to explain it.

13-4. The C_{3v} (ammonia) group is of order six. This is the same as the number of ways one can place three hydrogens at the three corners of an equilateral triangle $(3 \cdot 2 \cdot 1)$. When we consider the C_{4v} group, we have 24 ways we can place four hydrogens at the corners of a square $(4 \cdot 3 \cdot 2 \cdot 1)$. But the C_{4v} group only has order eight. Explain.

13-5. For each of the molecules (**IV**)-(**VII**),

CH_2Cl_2 (**IV**) B_2H_6 (**V**) $AuCl_4$ (**VI**) CH_4 (**VII**)

a) list the symmetry elements,
b) calculate the number of symmetry operations for each element,
c) obtain the order of the group,
d) determine the group symmetry symbol,
e) check your results for (a-c) against the appropriate character table in Appendix 11.

13-6. a) Demonstrate that U is a unitary matrix.
b) Demonstrate that $U^{\dagger}AU$ is diagonal.

$$U = \begin{pmatrix} \frac{1}{\sqrt{2}} & -\frac{1}{\sqrt{2}} \\ \frac{1}{\sqrt{2}} & \frac{1}{\sqrt{2}} \end{pmatrix}, \quad A = \begin{pmatrix} 0 & 1 \\ 1 & 0 \end{pmatrix}$$

13-7. Assign the following molecules to point groups, look up their character tables, and indicate in each case whether one could expect doubly degenerate MOs.

a) C_6H_6
b) CH_2Cl_2
c) B_2H_6 (see Problem 13-5)
d) Staggered C_2H_6
e) Staggered CH_3CCl_3

13-8. Consider the planar molecule CO_3^{2-} (**VIII**). The oxygen atoms are at the corners of an equilateral triangle.

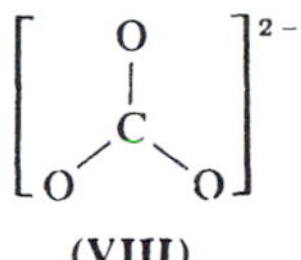

(**VIII**)

a) What is the point group of this molecule?
b) Using the appropriate character table, assign a symmetry symbol to each of MOs (**IX**)–(**XII**).

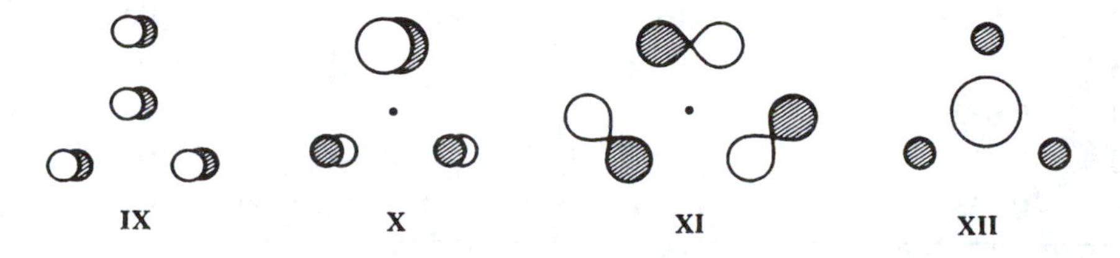

13-9. Table P13-9 gives the eigenvalues and eigenvectors resulting from an extended Hückel calculation of the allene molecule (**XIII**). The molecule is aligned as shown with respect to Cartesian axes. Ascertain the point group for this molecule. Using the character table for this group, assign a symmetry symbol to each MO. Is a transition from the highest occupied MO level to the lowest unoccupied level allowed by symmetry for any direction of polarization?

H_7, H_6, $C_5=C_1=C_2$, H_3, H_4; axes x, y, z

(XIII)

13-10. Consider the water molecule, oriented as shown in Fig. P13-10 with the y axis perpendicular to the molecular plane and the z axis bisecting the H–O–H angle.

a) Figure out as many nonredundant symmetry operations for this molecule as you can, and set up their multiplication table. Ascertain that you have a *group* of operations by checking closure, etc.
b) Use the functions z, R_z, x, and y as bases to set up a character table for this group.

TABLE P13-9 ▶ Extended Hückel Molecular Orbitals for Allene

		C_1				C_2		
MO no.	Energy (a.u.)	2s	$2p_z$	$2p_x$	$2p_y$	2s	$2p_z$	$2p_x$
1	1.7868	0	0	1.58	0	−1.23	0	0.58
2	1.4916	1.52	0	0	0	−1.01	0	0.57
3	0.4927	−0.54	0	0	0	−0.51	0	−0.75
4	0.3681	0	0	0.67	0	0.36	0	0.83
5	0.3231	0	0	0	−0.34	0	0	0
6	0.3231	0	0.34	0	0	0	−0.11	0
7	−0.2619	0	0	0	0.85	0	0	0
8	−0.2619	0	0.85	0	0	0	−0.78	0
9	−0.4326	0	0.56	0	0	0	0.67	0
10	−0.4326	0	0	0	−0.56	0	0	0
11	−0.4881	0	0	−0.49	0	−0.08	0	0.41
12	−0.5558	0	−0.16	0	0	0	−0.04	0
13	−0.5558	0	0	0	0.16	0	0	0
14	−0.6221	0.43	0	0	0	−0.14	0	−0.32
15	−0.8102	0	0	−0.13	0	−0.44	0	−0.02
16	−0.9363	0.47	0	0	0	0.36	0	−0.02

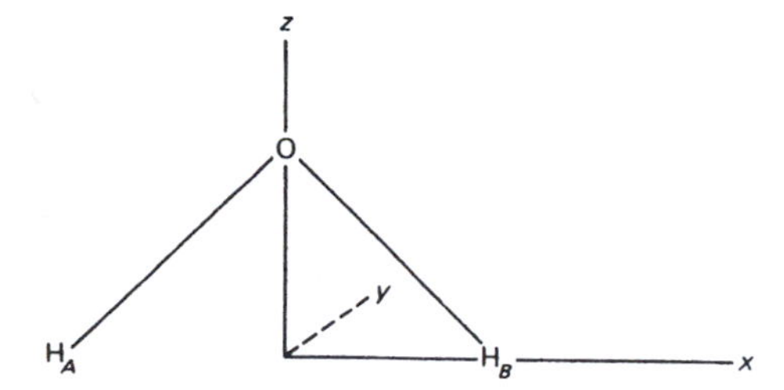

Figure P13-10 ▶ Relation of the water molecule to cartesian axes.

c) Use the resulting table to find out the symmetries of all MOs that can contain 1s AOs on the hydrogens.
d) Produce the symmetry combinations of these AOs that can appear in the MOs of water.

13-11. Consider the *trans*-chlorobromotetramine cobalt (III) ion (**XIV**).

a) Find the appropriate symmetry elements for this molecule and set up their group multiplication table. (Ignore the hydrogens on the ammonias.) Check the multiplication table to be sure all requirements for a mathematical group are satisfied.

Atomic Orbital Coefficients					C_5			
$2p_y$	h_3	h_4	2s	$2p_z$	$2p_x$	$2p_y$	h_6	h_7
0	0.30	0.30	1.23	0	0.58	0	−0.30	−0.30
0	0.24	0.24	−1.01	0	−0.57	0	0.24	0.24
0	0.61	0.61	−0.51	0	0.75	0	0.61	0.61
0	−0.58	−0.58	−0.36	0	0.83	0	0.58	0.58
1.27	−0.88	0.88	0	0	0	0.11	0	0
0	0	0	0	−1.27	0	0	0.88	−0.88
0.01	−0.21	0.21	0	0	0	−0.78	0	0
0	0	0	0	0.01	0	0	−0.21	0.21
0	0	0	0	−0.10	0	0	−0.16	0.16
0.10	0.16	−0.16	0	0	0	−0.68	0	0
0	0.17	0.17	0.08	0	0.41	0	−0.17	−0.17
0	0	0	0	−0.48	0	0	−0.44	0.44
−0.48	−0.44	0.44	0	0	0	−0.04	0	0
0	−0.24	−0.24	−0.14	0	0.32	0	−0.24	−0.24
0	−0.18	−0.18	0.44	0	−0.02	0	0.18	0.18
0	0.07	0.07	0.36	0	0.02	0	0.07	0.07

z
Cl
y
H_3N — NH_3
Co
H_3N — NH_3
Br
x

(XIV)

b) Use the following as bases for representations: $z, R_z, x, y, x^2 - y^2, xy$. Make sure you get all the inequivalent irreducible representations allowed in the group by checking $\Sigma_i l_i^2 = h$.

c) Set up the character table. Now ascertain which symmetry orbitals contain 2s orbitals of nitrogen. Give the symmetry combinations of those AOs that appear in these symmetry orbitals.

13-12. Use the relationships that must exist among characters to complete the following tables. Include proper symbols for the representations, but do not include bases.

(a)

	E	C_2	σ_v	σ_v'

(b)

	E	$2C_3$	$3C_2$	σ_h	$2S_3$	$3\sigma_v$

13-13. The D_5 group has four classes of operation and has order 10. How many inequivalent irreducible representations are there and what are their dimensions?

13-14. Consider the structure shown in Fig. P13-14.

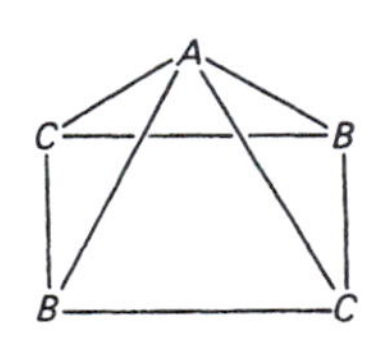

Figure P13-14 ▶ Square pyramid.

a) Figure out the symmetry elements and operations for this molecule.
b) What is the group order and number of classes?
c) How many inequivalent irreducible representations are there and what are their dimensions?
d) Ascertain the group symbol and compare your answers with the character table in Appendix 11.

13-15. A group has the following representations: $A_1, A_2, B_1, B_2, E_1, E_2$. What is the group order and how many classes are there?

13-16. Find the matrices that transform s and p STOs into sp^3 and sp^2 hybrids (Section 13-18). Demonstrate that these are unitary matrices.

13-17. It has been argued (Section 13-18) that sp^2 hybrid orbitals are appropriate basis functions for CH_3^+ (D_{3d}). Could one use a basis set of sp^3 hybrid orbitals for this system?

13-18. In each of the following cases, resolve the given character set. If these were characters of integrands, would the integral vanish by symmetry? For example,

$$C_{3v}: 4\ 0\ 1 \quad A_1 \oplus A_2 \oplus E \quad \text{No.}$$

a) C_{4v}: 5 −1 1 1 −3
b) D_{2h}: 3 −1 −1 3 −1 3 3 −1
c) D_{3d}: 8 2 0 0 0 0

13-19. Referring to the data in Problem 13-9, which integrals below must vanish by symmetry?

a) $\int \phi_6 x \phi_{10}\, dv$
b) $\int \phi_6 y \phi_{10}\, dv$
c) $\int \phi_1 x \phi_2\, dv$
d) $\int \phi_3 x \phi_{14}\, dv$
e) $\int \phi_2 z \phi_4\, dv$

13-20. Consider the possible electronic excitations of staggered ethane from its occupied $1e_g$ MO to the various empty MOs of symmetry a_{1g}, a_{2u}, e_g, and e_u. Which of these are symmetry allowed and how are they polarized?

13-21. Referring to the data in Problem 12-24, which transitions from MO 4 can be induced by y-polarized light? (*Note*: the molecular y axis is coincident with the symmetry z axis.)

13-22. The order of a group for a given object is equal to the number of equivalent locations for any nonspecial point in space about the object. Without referring to tables, predict the order of the group for

a) a square (both faces identical).
b) a cube (all faces identical).

Multiple Choice Questions

(Answer the following questions without referring to text or tables.)

1. Generate the symmetry operations for water, and choose the true statement from the set below.

a) H_2O has two C_2 axes.
b) H_2O belongs to a group of order three.
c) H_2O belongs to a group having exactly three classes.
d) H_2O can have only nondegenerate MOs.
e) None of the above statements is true.

2. Which one of the following statements is true?

a) All reflections in a group must belong to the same class.
b) All linear molecules must have an inversion center.
c) An S_6 axis has exactly two nonredundant operations.

d) Use of a basis set of hybrid orbitals built from a minimal basis set of STOs leads to a different variationally minimized energy than does use of the minimal basis set of STOs without hybridization.
e) None of the above is a true statement.

3. Which one of the following is *not* a symmetry element for eclipsed ethane?

a) i
b) S_3
c) C_2
d) σ_v
e) σ_h

4. Which of the following *nondegenerate* MOs is (are) possible for the CO_2 molecule?

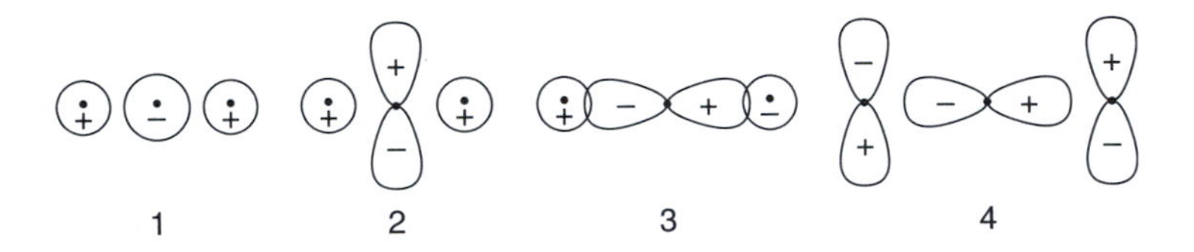

a) 1 only.
b) All four are possible.
c) 1, 2, and 3 only.
d) 2 and 3 only.
e) 1 and 3 only.

5. The presence of an S_4 axis guarantees the presence of

a) a point of inversion.
b) a σ_h reflection plane.
c) a C_4 axis.
d) a C_2 axis.
e) None of the above is correct.

6. A molecule having only the E symmetry operation can have

a) only nondegenerate MOs.
b) MOs of any degeneracy.
c) an infinitely degenerate level.
d) only doubly-degenerate MOs.
e) None of the above is correct.

7. An MO that is symmetric or antisymmetric for every symmetry operation of a molecule

a) must be degenerate.
b) cannot be variationally altered.
c) must be the lowest- or highest-energy MO.
d) must be nondegenerate.
e) None of the above is correct.

8. The character for a reducible representation of C_4 on a square planar molecule equals

a) 0
b) 1
c) 2
d) −2
e) 4

9. Consider the planar molecule-ion $NOCl_2^+$.

$$\left[\begin{array}{l} \mathrm{Cl} \diagdown \\ \quad \mathrm{N{=}O} \\ \mathrm{Cl} \diagup \end{array} \right]^+$$

The number of reflection planes, σ, three-fold rotation axes, C_3, two-fold rotation axes, C_2, and inversion centers, i, that are symmetry elements for this molecule-ion axis

	σ	C_3	C_2	i
a.	1	0	1	0
a)	2	0	1	0
b)	2	0	1	1
c)	2	1	2	0

d) None of the above is correct.

References

[1] D. M. Bishop, *Group Theory in Chemistry*. Oxford University Press, London and New York, 1973.

[2] H. Eyring, J. Walter, and G. E. Kimball, *Quantum Chemistry*. Wiley, New York, 1944.

Chapter 14

Qualitative Molecular Orbital Theory

14-1 The Need for a Qualitative Theory

Ab initio and semiempirical computational methods have proved extremely useful. But also needed is a simple conceptual scheme that enables one to predict the broad outlines of a calculation in advance, or else to rationalize a computed result in a fairly simple way. Chemistry requires conceptual schemes, simple enough to carry around in one's head, with which new information can be evaluated and related to other information. Such a theory has developed alongside the mathematical methods described in earlier chapters. We shall refer to it as qualitative molecular orbital theory (QMOT). In this chapter we describe selected aspects of this many-faceted subject and illustrate QMOT applications to questions of molecular shape and conformation, and reaction stereochemistry.

14-2 Hierarchy in Molecular Structure and in Molecular Orbitals

We seek a simple qualitative approach to the question, "How does the total energy of a system change as the nuclei move with respect to each other?" This question is very broad, encompassing the phenomena of molecular structure and chemical reactivities.

It is useful to distinguish three kinds of process that can occur as nuclei move. One of these is the process in which two nuclei move closer together or farther apart, with their separation being somewhere around a bond length, either at the outset or the conclusion of the motion (or both). This process includes the breaking or forming of bonds and also the stretching or compressing of bonds. It also includes the forcing together of two species that will not bond (e.g., He with He). Let us refer to this as a *nearest-neighbor* interaction, even though the two nuclei need not be bonded in the usual chemical sense. The second process is the changing of the bond angle between two nuclei bonded to a third. The changing of the H–O–H angle in water is an example. A necessary consequence of such a change in angle is a change in distance between the two moving nuclei (here H–H). In geometries normally of interest, however, this distance is somewhat greater than a typical bond length throughout the entire process. We refer to this process as *bond-angle change*. The third process is the rotation of one part of a system with respect to the other about some axis (usually a single bond in the system). An example is rotation of one methyl group in ethane with respect to the other.

We refer to this process as a *torsional angle change*, or an *internal rotation*. Such a change will produce changes in distances between nuclei located on opposite ends of the torsional axis, but these distances typically remain several times as great as a bond length throughout the entire process.

Chemists have long recognized that the energies associated with these three kinds of change fall into a loose hierarchy, with nearest-neighbor interactions having the greatest effect on energy, bond-angle changes having a smaller effect, and torsional angle changes the least. Indeed, for this reason spectral transitions corresponding to stretching, bending, and torsional modes are found in different regions of the electromagnetic spectrum.

Most nearest-neighbor interactions are a consequence of the *connectedness*, or *topology*, of a molecule. This aspect of molecular structure, sometimes called the *first-order*, or *primary* structure, is the first aspect one considers when establishing structure, and it is the first aspect that chemists became aware of, historically. The *second-order* aspect of structure concerns the bond angles. Once those are at least roughly known, one can go on to consider the *third-order*, or *tertiary* structure resulting from torsional energetics. The last aspect is usually referred to as the *conformation* of the system.

It is advantageous to discuss MOs from a similar viewpoint. If we wish to guess the nature of the MOs of a system (i.e., where they have their nodes) and the MO energy order, we first consider the topology of the system. (Recall that this is the *only* thing that the simple Hückel method considers.) Elementary arguments lead to a fair approximation of the appearance and relative energies of the MOs. Next, we can consider bending the system, bending the MOs along with it. By judging whether MO energies will rise or fall in this process, we shall show that one can often predict whether the molecule will be more stable in the linear or bent form. Finally, when we know the second-order structure, we can imagine the various conformational possibilities, allowing the MOs to be carried along with the nuclei. Again, by judging how the MO energies respond, it is possible to make predictions as to which conformation is most stable.

In order to formulate rules for QMOT, we will return to the H_2^+ molecule ion and the H_2 molecule. Then, using insights gained there, we shall consider more complicated systems.

14-3 H_2^+ Revisited

In Chapter 7 we used the linear variation method to solve the minimal basis H_2^+ problem. However, symmetry conditions alone suffice to force the solutions to be

$$\psi_{\sigma_g} = 1/\sqrt{2(1+S)}(1s_a + 1s_b), \quad E_g(\text{el}) = (H_{aa} + H_{ab})/(1+S) \qquad (14\text{-}1)$$

$$\psi_{\sigma_u} = 1/\sqrt{2(1-S)}(1s_a - 1s_b), \quad E_u(\text{el}) = (H_{aa} - H_{ab})/(1-S) \qquad (14\text{-}2)$$

where H_{aa} and H_{ab} are negative energies, and S is the (positive) overlap integral between 1s AOs on nuclei a and b. The energies $E_g(\text{el})$ and $E_u(\text{el})$ are upper bounds for the *electronic* energy of the σ_g and σ_u states. The total energies are obtained by adding the internuclear repulsion energy $1/R$ a.u., where R is the internuclear distance in atomic units.

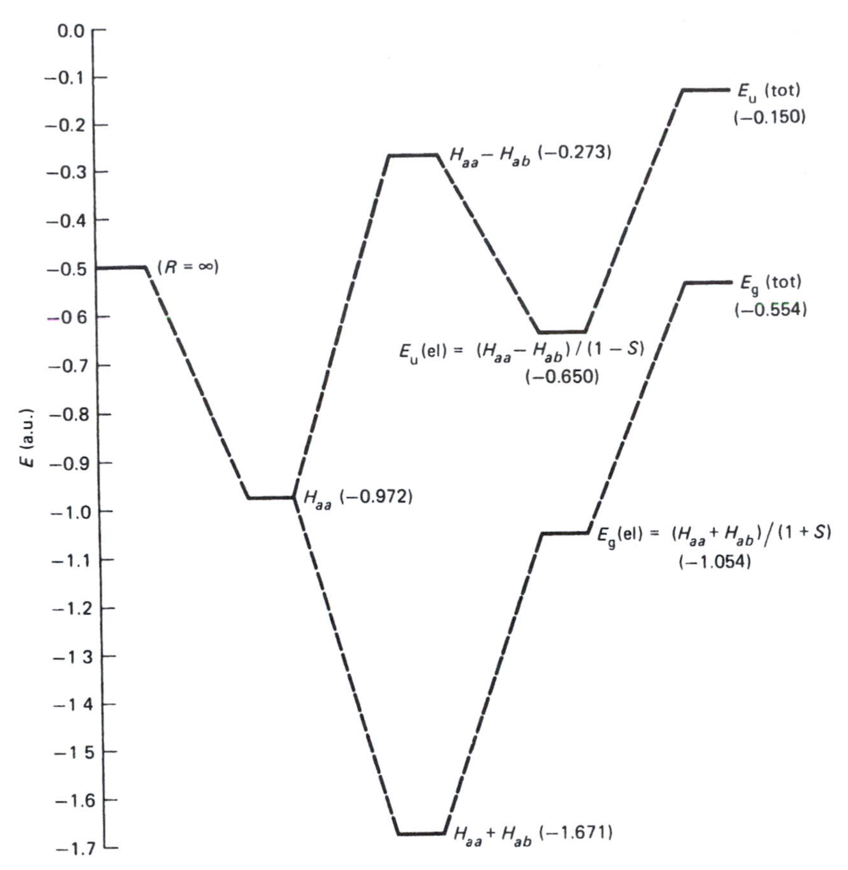

Figure 14-1 ▶ Steps in compiling total energies for minimal basis LCAO-MO calculation on H_2^+ at $R = 2$ a.u.

It is instructive to consider the total energies, E_g(tot) and E_u(tot) as the result of an artificial step-by-step procedure as indicated in Fig. 14-1. We begin with a 1s AO on nucleus a at $E = -\frac{1}{2}$ a.u. If we now bring proton b up to a distance of 2 a.u., but do not allow the 1s AO to respond, we get an energy lowering due to the increased nuclear attraction now felt by the electron in $1s_a$. At $R = 2$ a.u., this additional attraction lowers the energy by 0.472 a.u., giving $H_{aa} = -0.972$ a.u. Let us refer to this as the *frozen AO in molecule* energy. (It may be thought of as the energy to first order due to a "perturbation" caused by the approach of proton b, although this is hardly a small perturbation.) As our next step, we include H_{ab}. This term equals -0.699 a.u., and it splits the energy evenly around H_{aa}, as shown in the figure. The resulting energies are the expectation values for the electronic energies of the *unnormalized* (due to omission of overlap) functions $2^{-1/2}(1s_a \pm 1s_b)$. Two things have happened in this step: The electronic charge has become delocalized over both centers, and *it has changed in amount so that we no longer have one electron in each MO.* Because the overlap term S (equal to 0.586 at $R = 2$ a.u.) has not yet come into the calculation, we have here the energies due to $1 + S$, or 1.586 electron in the σ_g MO, and $1 - S$, or 0.414 electron in the σ_u MO.

Before proceeding to the next step, let us examine the two energies at this point to see how they compare for equal amounts of charge. The 0.414 electronic charge in σ_u is associated with an energy of -0.273 a.u., whereas the 1.586 charge in σ_g corresponds to -1.671 a.u. It is clear that the σ_g MO is *inherently* of lower energy (relative to $E=0$) per unit charge, due to the detailed nature of kinetic and nuclear electronic energies. (We will not concern ourselves with these detailed aspects, however.)

Our next step is to normalize the charges by dividing by $1 \pm S$. This essentially adds 0.586 electron to σ_u and subtracts 0.586 from σ_g. From the figure, we see that this lowers the energy of σ_u and raises that of σ_g. This is reasonable: Adding more charge to an MO of negative energy should make its energy contribution more negative, removal of charge should make it less negative. A very important fact, though, is that σ_g rises more than σ_u goes down. This simply reflects our observation in the preceding paragraph that σ_g charge "has lower energy per unit charge," so removal of it "costs more" than is gained by putting it into σ_u. A useful summary of the effect of renormalization is: Renormalization tends to cancel the energy level changes due to nuclear motion. The higher an energy level is, the less effective is this cancellation. Therefore, net energy changes (due to nuclear motion) tend to be greatest for the antibonding member of a bonding-antibonding pair of levels, and, in general, greater for a higher-energy MO (bonding or antibonding) than for a lower-energy MO.

The final step is to add the internuclear repulsion energy of 1/2 a.u. to each level. It is important to notice that the final *total* energy levels are related to the initial H atom energies in almost the identical way that $E_g(\text{el})$ and $E_u(\text{el})$ are related to the "atom-in-molecule" energy H_{aa} (see Fig. 14-2) because the energy lowering due to the original $1s_a$ electron being attracted by proton b (-0.472 a.u. at $R=2$ a.u.) is fairly close in magnitude to the repulsion between the protons ($+0.5$ a.u.). We expect this near cancellation to hold as long as R is large enough that proton b is "outside" the charge cloud due to the $1s_a$ electron. In other words, the *total* energies should lie above or below the *separated-*

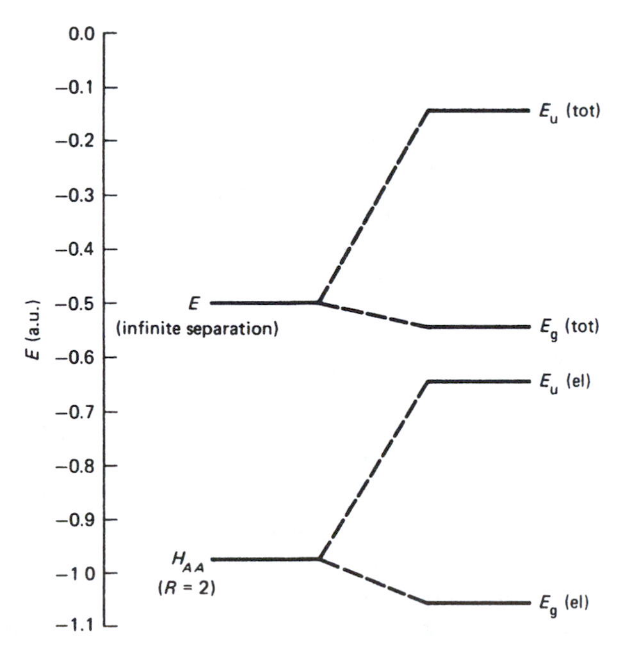

Figure 14-2 ▶ Total and electronic energies for bonding and antibonding states of H_2^+ compared to appropriate reference energies.

atom energy in much the same way that the *electronic energies* lie above or below the *"atom-in-molecule" energy, if the internuclear separation is not too small.* [We know that, as R approaches zero, the internuclear repulsion approaches infinity, whereas the exact values of E_g(el) and E_u(el) respectively, approach the united atom He^+ energies of −2 a.u. (1s) and −0.5 a.u. ($2p_\sigma$), and the cancellation ultimately breaks down.]

From this discussion of H_2^+, the following ideas emerge:

1. The energy of an MO is lowered by bonding interactions, raised by antibonding interactions, relative to the appropriate reference energy.
2. Antibonding interactions are inherently more destabilizing than bonding interactions are stabilizing.
3. A parallel exists between electronic energy change and total energy change if we are careful about reference energies and internuclear distances.

But H_2^+ is atypical of the systems we will be interested in treating by QMOT. We will usually be considering neutral, many-electron systems, and H_2^+ is a charged, one-electron system. Therefore, we turn next to the molecule H_2 to consider how well the above ideas carry over to a more typical system.

EXAMPLE 14-1 The simplest MO treatment of H_2 would follow an HMO-like procedure, and assume that two electrons in the $1\sigma_g$ MO should give a total energy for H_2 of $2E_g(\text{tot}, H_2^+)$, or $2(-0.554\,\text{a.u.}) = -1.108\,\text{a.u.}$ The correct total energy for H_2 is -1.1745 a.u. (Table 7-4), so the agreement is better than one might expect. Is this because our procedure is physically correct?

SOLUTION ▶ There are several errors in the procedure. Most obvious is that E(tot) for H_2^+ includes the repulsion between two protons, so our value for H_2 contains double that nuclear repulsion. But both molecules have a pair of protons. So how come the result agrees so well with the exact value? For one thing, doubling E(tot) for H_2^+ fails to include interelectronic repulsion, so we fail to include any interelectronic repulsion but we include too much internuclear repulsion. In addition, R_e for H_2 is about 3/4 that for H_2^+, so V_{nn} in H_2 is really about 4/3 V_{nn} in H_2^+. Also, V_{ne} in these two species differs, both because the protons are not separated by the same distance and because the electron cloud is swollen by the presence of interelectronic repulsion. Because the electronic potential energy changes, the kinetic energy changes too. We are observing here the result of partial cancellation of several errors. ◀

14-4 H_2: Comparisons with H_2^+

We first consider the ramifications of the neutrality and nonpolarity of H_2. Imagine that we have a ground-state hydrogen atom H_a and we allow another similar atom, H_b, to approach it. At values of R in excess of, say, 2 a.u., we expect the perturbation felt by H_a to be much smaller than was the case in our H_2^+ discussion. This is because, where before we had approach by a *charged* particle, here we have approach by a neutral atom. The attraction between the electron on H_a and the proton on H_b is counterbalanced by repulsion between the electron on H_a and that on H_b. This means that the frozen atom-in-molecule energy for H_2 is fairly close to the energy of the isolated atom. This simplifies our qualitative treatment for neutral molecules since it means we can meaningfully compare molecular electronic energies directly with unperturbed atomic electronic energies.

We now come to a consideration of the orbital energies of H_2. Here we will find that there are important differences between a one-electron system like H_2^+ and multielectronic systems. Consider the lowest ($1\sigma_g$) MO of H_2. According to the description of *ab initio* theory in Chapter 11, the "energy of" this MO at some internuclear separation R [call it $E_{1\sigma_g}(R)$] is equal to the energy of an electron *in* this MO (i.e., the orbital energy is equal to the one-electron energy). This energy results from kinetic and nuclear attraction components and from repulsion for the other electron. *But where is this other electron*? If we are considering the ground state of H_2, it is in the $1\sigma_g$ MO also. That gives us an energy we might call $E_{1\sigma_g}(R, 1\sigma_g^2)$. However, we might instead be considering an excited state of H_2, perhaps with the configuration $1\sigma_g 1\sigma_u$. Depending on whether we choose the symmetric or antisymmetric combination for the spatial part of the wavefunction, we shall be considering an excited singlet or triplet state. In each case, repulsion for the $1\sigma_g$ electron will be different. Thus, we already have three orbital energies for this lowest MO, and we could continue getting more by considering other excited states of H_2. Clearly, the orbital energies in a multielectronic molecule are dependent on the state being considered.

Another important feature of one-electron energies in multielectronic systems is that they do not add up to the total electronic energy. That is, $2E_{1\sigma_g}(R, 1\sigma_g^2)$ is not equal to the electronic energy of H_2 at R. As pointed out in Chapter 11, this is because $E_{1\sigma_g}(E, 1\sigma_g^2)$ includes the interelectronic repulsion, so $2E_{1\sigma_g}(R, 1\sigma_g^2)$ counts the interelectronic repulsion twice.

One might imagine that, since MO energies in multielectronic systems are state-dependent and cannot be simply summed to give the electronic energy, it is hopeless to use them as a basis for explanation or prediction. However, this would be too pessimistic. It turns out that the *qualitative* features of MO energies are not all that sensitive to change of state (see Fig. 14-3) and, besides, we are usually concerned with the ground state or with excited states wherein most of the electrons remain in orbitals occupied in the ground state. Also, the extra measure of interelectronic repulsion that is

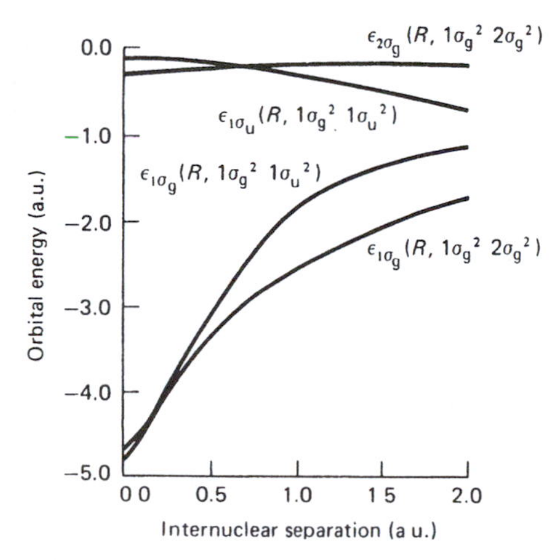

Figure 14-3 ▸ Orbital energies as function of internuclear separation for He_2. The two curves for $\epsilon_{1\sigma_g}$ are qualitatively similar, although far from identical. (From Yarkony and Schaefer [1].)

included in the sum of one-electron energies can sometimes be expected to compensate roughly for the as yet unincluded internuclear repulsion energy. This rough equality requires that our system be neutral (so that the total number of repelling negative charges is equal to the total amount of positive charge) and nonpolar (so that the loci of charge are similar). We expect the equality to be quite good at large internuclear separation, where the electrons of each atom repel those of others as though they were centered on the nuclei, and to become progressively poorer as the atoms get closer together and undergo interpenetration of charge clouds. If we accept this rough equality, then we are permitted to approximate the change in *total* energy of the system (when the nuclei move), as equal to the change in the sum of one-electron energies.[1]

14-5 Rules for Qualitative Molecular Orbital Theory

The considerations discussed in the two preceding sections provide a rather loose justification for the following QMOT rules related to *total energy changes when nuclei are moved.*

1. An increase in overlap population between two AOs tends to lower the energy of an MO; a decrease tends to raise it.
2. The effects of a given amount of orbital overlap population change on orbital energy are much more pronounced for higher-energy MOs. A corollary is: The destabilizing effect of an antibonding relation between two AOs tends to be greater than the stabilizing effect of a bonding relation, other factors (AO identities, distances) being equal.
3. The sum of changes of one-electron energies should approximately equal the *total* energy change if we are treating a neutral, nonpolar system and if the nuclei that are moving with respect to each other are separated by a distance of a normal bond length or more.

Rule (2) and its corollary are due to the different effect of renormalization on orbitals of different energy. Note that rule three refers only to energy *changes*. We do *not* expect the sum of one-electron energies to equal the total energy. This becomes obvious in the limit of infinite separation of atoms (e.g., in CO_2) in which there is no internuclear repulsion, but interelectronic repulsion *within* each atom is still being counted twice.

These rules are normally applied only to MOs made from valence-shell AOs. Inner-shell electrons are only very weakly perturbed in normal nuclear motions. The resulting small changes, while detectable and useful for analytical purposes, are inconsequential compared with valence electron effects.

14-6 Application of QMOT Rules to Homonuclear Diatomic Molecules

If we apply our rules from the preceding section to the motion of a pair of identical nuclei, each carrying a 1s AO, we obtain the orbital energy versus R curve shown in Fig. 14-4. The σ_g MO energy is predicted to drop as overlap population

[1]For comments on this point, see Ruedenberg [2].

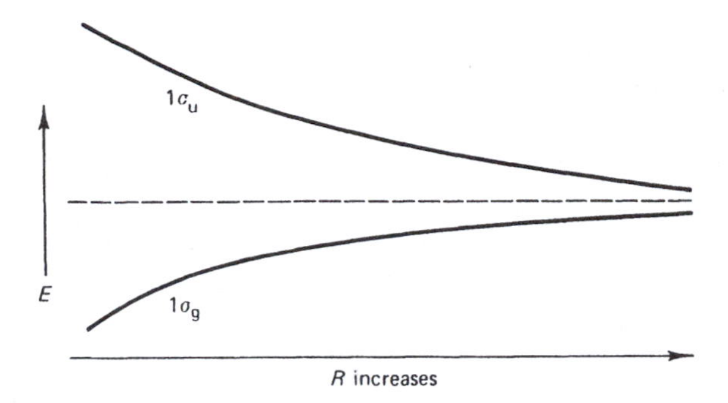

Figure 14-4 ▶ Energy *versus* internuclear separation R for two nuclei with 1s AOs, derived from QMOT rules.

increases, the σ_u to rise as out-of-phase overlap increases, and σ_u rises faster than σ_g drops. In QMOT, we argue from this figure that one-, two-, or three-electron homonuclear diatomic molecules should be stable ($H_2^+, H_2, H_2^-, He_2^+$) and that four-electron molecules (He_2) should be unstable. Since a similar figure holds for molecules in which the valence orbital is 2s, 3s, etc., we might also expect systems like $Li_2^+, Li_2, Li_2^-, Be_2^+, Na_2^+, Na_2, Na_2^-, Mg_2^+$ to exist, but not Be_2 or Mg_2. This set of predictions is a reasonable starting point, and in many cases agrees with observation (see e.g., Table 7-2). By the time we get to the Na–Mg series, however, the energy difference between 3s and 3p AOs is so small that it is unrealistic to treat these as "pure" 3s cases. Consequently, we must put less faith in simple predictions for those molecules. The comparison between our QMOT guess about these molecules and experimental or *ab initio* observation is shown in Table 14-1.

The reader may have noticed that we have violated one of our QMOT conditions by considering nonneutral systems. One is presumably less safe in including such systems, although several of them have been studied and found to fit the QMOT prediction.

TABLE 14-1 ▶ Stability of Some Homonuclear Diatomics as Predicted by QMOT and as Observed from Experiment or *ab initio* Calculation

Molecule	QMOT	Experimental or *ab initio*	Molecule	QMOT	Experimental or *ab initio*
H_2^+	Stable	Stable	Be_2^+	Stable	Stable
H_2	Stable	Stable	Be_2	Not Stable	Not Stable
H_2^-	Stable	Stable[a]	Na_2^+	Stable	Stable
He_2^+	Stable	Stable	Na_2	Stable	Stable
He_2	Not Stable	Not Stable	Na_2^-	Stable	?
Li_2^+	Stable	Stable	Mg_2^+	Stable	?
Li_2	Stable	Stable	Mg_2	Not Stable	Stable (?)
Li_2^-	Stable	Stable			

[a]This molecule–ion is unstable with respect to losing an electron and forming this neutral molecule at its lowest energy, but it is stable with respect to dissociation into a neutral atom and a negative ion in their ground states.

Apparently, if one is concerned only with the question of the presence or absence of a valley in the total energy curve (and not with relative depths of valleys or relative R_e values), deviation from neutrality by *one* electron is not too damaging (as we saw for H_2^+). However, for doubly positive diatomics, like He_2^{2+}, the internuclear repulsion dominates so that there is no stable species.

One can add p-type AOs to the two nuclei and expand the orbital energy plot, as illustrated in Fig. 14-5. Filling in the orbital levels with electrons leads to the prediction that B_2 is singly bonded, C_2 is doubly bonded, N_2 triply bonded, O_2 doubly bonded, F_2 singly bonded, and Ne_2 not bonded. These systems, with their singly charged relatives, have been described in Chapter 7, and we will say no more about them here except that the agreement between QMOT rules and observations is quite respectable (see Fig. 14-6).

An important distinction exists between the energy versus R curves drawn in Fig. 14-5 and those in Fig. 7-17. The latter set was constructed by sketching orbital energies for the united-atom and separated-atom limits, and linking these energies together using symmetry agreement and the noncrossing rule. The former set was constructed by sketching orbital energies for the separated atoms, and then using QMOT rules to decide which curves go up, and which go down in energy as R decreases.

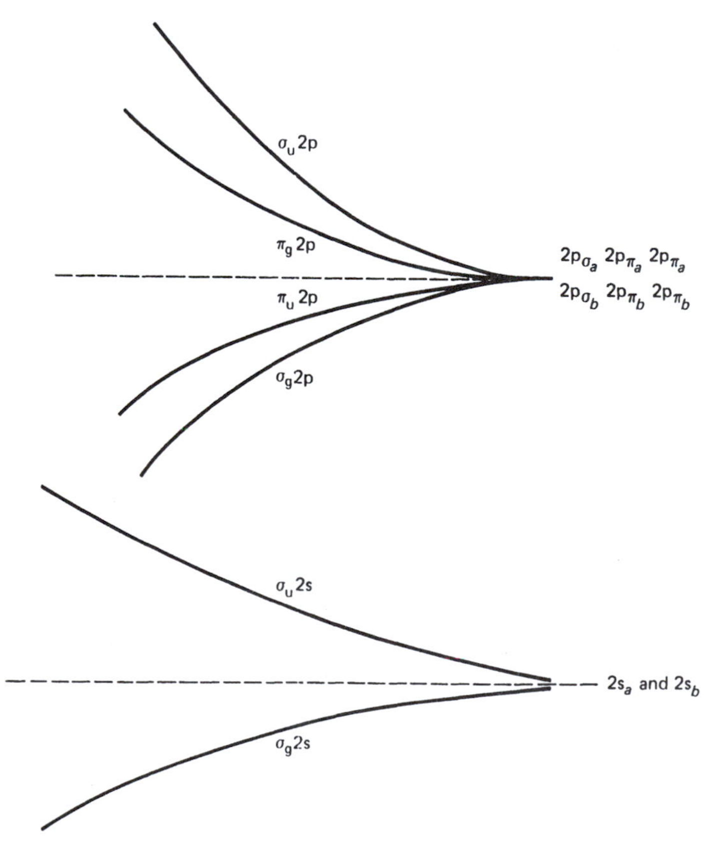

Figure 14-5 ▶ Qualitative sketches of homonuclear diatomic MO energies as a function of R based on QMOT rules.

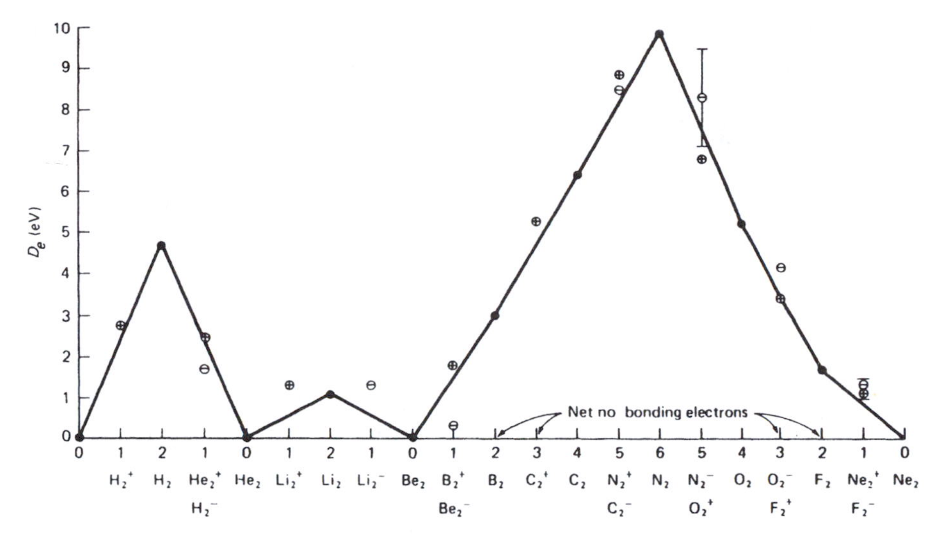

Figure 14-6 ▶ D_e versus *aufbau* sequence for homonuclear diatomic molecules and ions. The tendency for systems having more net bonding electrons to have a greater D_e is adhered to fairly well (see Table 7-2 for data).

When we wish to emphasize this distinction, we will refer to these as *two-sided* and *one-sided* correlation diagrams, respectively.

According to the simple QMOT viewpoint, we should expect a bonding MO to be lower in energy than the AOs that comprise it. Likewise, an anti-bonding MO should be higher in energy. We can test this by comparing first ionization energies (IEs) for molecules against those for the constituent atoms. (Recall that Koopmans' theorem equates orbital energy to ionization energy.) If the highest occupied MO (HOMO) of the molecule is bonding, the IE for the molecule should be greater than that for the atom. If it is antibonding, the IE of the molecule should be smaller. Strictly speaking, the HOMO of the molecule may consist of mixtures of AOs on each atom. In that case, we should compute a "valence state ionization energy" for the atom. Also, for heteronuclear molecules AB, we need to know what percentage of the HOMO to identify with each atom in order to obtain a suitable average atomic IE to compare with the IE of the molecule. Even if we ignore these corrections, however, and simply compare experimental IEs, the anticipated behavior is shown nicely by diatomic molecules, as indicated in Table 14-2.

There is much similarity between the QMOT rules and the assumptions inherent in the extended Hückel method described in Chapter 10. There, also, a bonding interaction is equated with energy lowering and an antibonding interaction with an energy rise. Furthermore, the sum of orbital energies is assumed to change in parallel with the *total* energy change, even though the internuclear repulsion is not included. We noted that this assumption limits the range of validity for EHMO calculations to relative nuclear motions at distances that are on the order of a bond length or greater. The EHMO method is, in essence, the numerical equivalent to the qualitative MO approach, and such calculations can serve as a guide for developing qualitative explanations in complicated situations or for producing numbers that enable one to compare QMOT with experimental or *ab initio* results.

TABLE 14-2 ▶ Comparison of Molecular and Atomic First Ionization Energies[a]

Molecule	HOMO bonding (b) or anitbonding (a)	IE (eV)	IE of molecule expected $>$ or $<$ than atomic average	IE atoms	Does experiment agree with theory?
H_2	b	15.427	$>$	13.598	Yes
He_2	a	$\sim$22.0	$<$	24.46	Yes
Li_2	b	5.12	$>$	5.363	No
B_2	b	$\sim$9.5	$>$	8.257	Yes
C_2	b	12.0 ± 0.6	$>$	11.267	Yes
N_2	b	15.576	$>$	14.549	Yes
O_2	a	12.06	$<$	13.618	Yes
F_2	a	15.7	$<$	17.426	Yes
Ne_2	a	20.1	$<$	21.47	Yes
Si_2	b	7.4 ± 0.3	$>$	8.15	No
Cl_2	a	11.48 ± 0.1	$<$	13.02	Yes
Br_2	a	10.53	$<$	11.85	Yes
I_2	a	9.3	$<$	10.457	Yes
CN	b	14.5 ± 0.5	$>$	11.27 (C), 14.55 (N)	Yes
NO	a	9.25	$<$	13.62 (O), 14.55 (N)	Yes
CO	b	14.013	$>$	13.62 (O), 11.27 (C)	Yes
CS	b	11.8, 119	$>$	11.27 (C), 10.36 (S)	Yes
ICl	a	10.3	$<$	10.46 (I), 13.02 (Cl)	Yes
IBr	a	9.98	$<$	10.46 (I), 11.85 (Br)	Yes

[a] Data are from [3] or from Table 7-2.

14-7 Shapes of Polyatomic Molecules: Walsh Diagrams

In this section we will describe how the rules and concepts of QMOT enable one to rationalize and predict molecular shapes. The earliest systematic treatment of this problem was given by Walsh,[2] whose approach has been extended by others, particularly Gimarc [6].[3]

We begin by considering the symmetric triatomic class of molecules HAH, where A is any atom. Such molecules can be linear or bent. Walsh's approach predicts which are linear, which are bent, and sometimes which of two bent molecules is more bent.

We approach the problem in the following way. First we sketch the valence MOs for the generalized linear molecule HAH, deciding which is lowest, second lowest, etc. in energy. Then we imagine bending the molecule and argue whether each MO should go up or down in energy on the basis of our QMOT rules. This produces a chart of orbital energies versus bond angle—a one-sided correlation diagram. Finally, we use this diagram to argue that HAH will be linear or bent, depending on how many valence electrons HAH has and on how they are distributed among the MOs.

The first problem is to sketch the MOs for linear HAH and decide their energy order. Probably the simplest way to do this is through use of symmetry and perturbation arguments. We know that the linear molecule belongs to the $D_{\infty h}$ point group, possessing a center of inversion and a reflection plane through the central atom and perpendicular to the HAH axis. This means that the two hydrogen 1s AOs ($1s_1$ and $1s_2$) will appear in MOs in the symmetry combinations $\phi_g = 1s_1 + 1s_2$, $\phi_u = 1s_1 - 1s_2$, where g and u stand for *gerade* and *ungerade*, respectively (see Chapters 7 and 13 for a background discussion). Recognizing this, we can next consider which MOs will result from interactions between the valence AOs on atom A and the symmetry orbitals ϕ_g and ϕ_u. A perturbation-type diagram for this appears in Fig. 14-7. We have assumed that only the valence s and p AOs on A are involved in bonding. Extension to include d AOs is possible.

On the right side of Fig. 14-7, the symmetry orbitals ϕ_g and ϕ_u are shown to be slightly split. This reflects the greater stability of the in-phase, or bonding, combination. However, the splitting is slight because the hydrogen atoms are quite far apart (so that atom A can fit between them). These two levels sandwich the separated-atom limit of $-\frac{1}{2}$ a.u. or -13.6 eV.

The AOs of atom A are sketched on the left. Their energies are arranged so that the 2p energies are about the same as the ϕ_g, ϕ_u symmetry orbital energies on the right. (For example, AO energies used for nitrogen in EHMO calculations are: $2s \sim -25$ eV, $2p \sim -13$ eV.) For the linear molecule, we can label the s and p AOs on atom A as σ or π and g or u.

To generate the MO energy-level pattern from the interactions between these AOs on A and ϕ_g, ϕ_u, we use the following rules from perturbation theory (see Chapter 12):

1. Interactions occur only between orbitals of identical symmetry.
2. Interactions lead to larger splittings if the interacting orbitals are closer in energy (overlap considerations being equal).

The resulting energy levels appear in the central column of Fig. 14-7.

We need sketches of the MOs whose energy level pattern we have just approximated. We can guess the qualitative appearance of these by recalling that, when two orbitals

[2] See Walsh [4] and the papers immediately following. See also Mulliken [5].

[3] For a critical review of the theoretical validity of Walsh's method, see Buenker and Peyerimhoff [7].

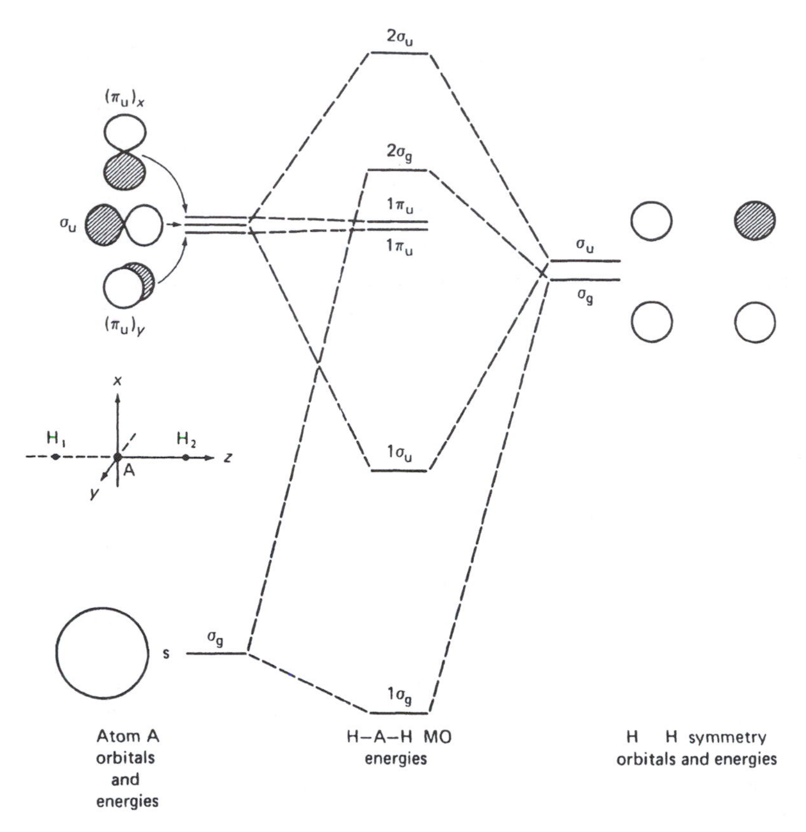

Figure 14-7 ▶ Orbital energies of MOs formed by interaction of σ_u and σ_g symmetry orbitals on $H_1 \ldots H_2$ with valence s and p AOs on central atom A. The π AOs and MOs are degenerate but are sketched as slightly split.

interact to give splitting, the lower energy corresponds to a bonding interaction, the higher energy to antibonding. Thus, for example, Fig. 14-7 indicates that the $1\sigma_g$ MO is a bonding combination of the 2s AO on A and the σ_g symmetry orbital, whereas $2\sigma_g$ is the antibonding combination. The π_u MOs are simply the p_π AOs on A, since there is nothing of the same symmetry for them to interact with. The six MOs for the HAH molecule are sketched in Fig. 14-8a.

Before proceeding, notice that the lowest two MOs are A–H bonding, the next two are nonbonding, the highest two are antibonding. This is an example of the way in which topological, or nearest-neighbor, interactions govern the gross features of energy ordering. In fact, we could have generated this same set of MOs and energy order by simply sketching all of the MOs we could think of that were symmetric or antisymmetric for relevant symmetry operations and then putting the bonding ones lowest (with s lower than p), nonbonding next, and antibonding highest (again recognizing that greater s character should yield lower energy). (Note that the nonbonding MOs are not symmetric or antisymmetric for arbitrary rotations about the C_∞ axis. Hence, they must form a basis for a representation of dimension greater than one. Hence, they are degenerate. See Chapter 13 for detailed discussion.) Electrons in the two lowest MOs produce A–H bonding. Electrons in the next two MOs have little effect on A–H bonding and, in fact, constitute what a chemist normally thinks of as lone pairs. Electrons in the two highest MOs tend to weaken the A–H bonds. We normally do not worry about questions of

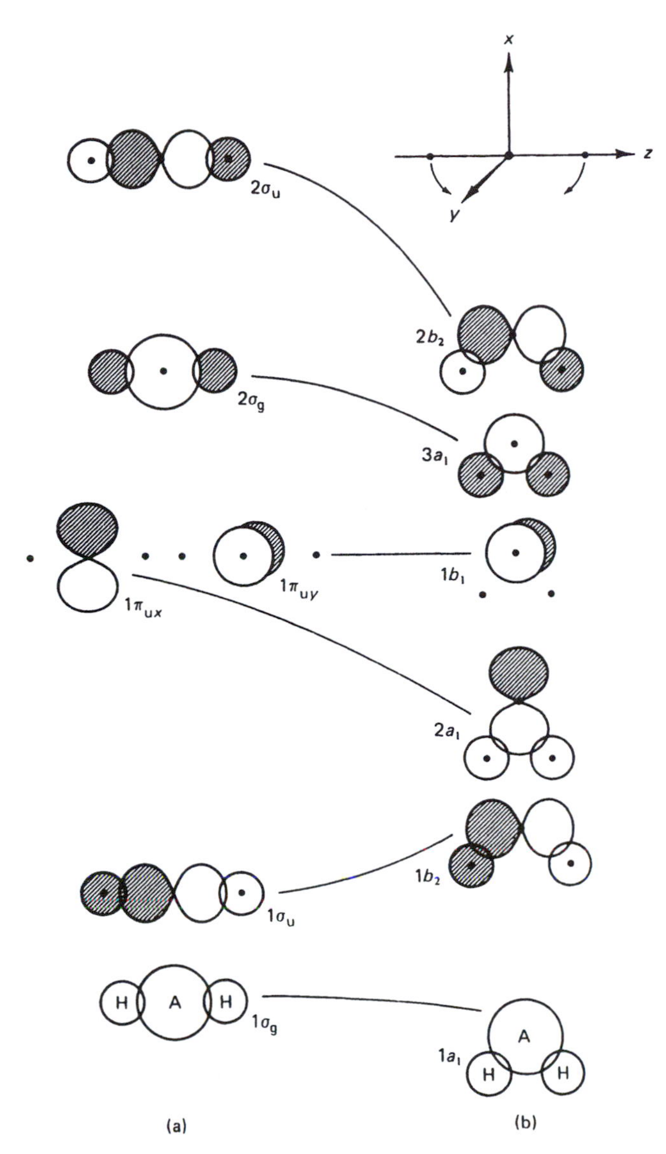

Figure 14-8 ▶ The Walsh-type correlation diagram for HAH: (a) linear $D_{\infty h}$; (b) bent C_{2v}. The cross-hatched parts of MOs have opposite sign from open parts.

shape for systems where these highest two MOs are filled because such an HAH system is not even bonded (i.e., we do not worry about second-order structure if there is no stable first-order structure). Consequently, these two highest MOs will be omitted in many energy-versus-angle diagrams, although there are certain cases where they can be useful (e.g., in singly excited configurations) (see Problem 14-5).

EXAMPLE 14-2 Referring to Fig. 14.8, give the QMOT rationale for the MOs in the lower-half of the figure having energies below those in the upper half. What is the rationale for the lowest (a_{1g}) MO being at lower energy that the second-lowest (a_{2u}) MO, and why are these both below the next (e_u) MOs?

SOLUTION ► The lowest-energy half of the set are all C–H bonding MOs, the other half all being C–H antibonding. Because there are so many C–H bonds, their nature determines this separation. Within the C–H bonding set, the lowest-energy pair differ from the others in having 2s AOs on carbon, which are sufficiently lower in energy than 2p AOs to give them lower energy. The a_{1g} MO is C–C bonding, hence has lower energy than the a_{2u} MO, which is C–C antibonding. ◄

We now consider how the MO energies change upon bending the molecule. As $1\sigma_g$ is bent, the two hydrogen AOs move closer together. Because they have the same phase, this leads to an overlap increase, but it occurs over a fairly long distance, being a second-nearest neighbor interaction. Therefore, there is an energy lowering, but it is not very large. Bending $1\sigma_u$ leads to two changes. First, the 1s AOs move away from the axis of maximum concentration of the 2p AO. This causes a substantial loss of overlap and a substantial increase in energy. Second, the 1s AOs move closer to each other. They disagree in phase, and this also tends to increase the energy, although it is a relatively small effect because it occurs between second-nearest neighbors. In the linear molecule, $1\pi_{ux}$ and $1\pi_{uy}$ MOs contain no contribution from hydrogen 1s because the hydrogens are in a nodal plane in each case. If we imagine that, in bending the molecule, we keep the hydrogens in the xz plane, then we see that the hydrogens are remaining in the nodal plane for $1\pi_{uy}$, but have moved away from the nodal plane of $1\pi_{ux}$. Once this happens, the 1s AOs are no longer forbidden by symmetry from contributing to the π_x MO, and a "growing in" of 1s AOs occurs, leading to an MO like the $2a_1$ MO drawn in Fig. 14-8. This behavior is more complicated than we observed for the lower-energy MOs because it involves more than a mere distortion of an existing MO. With a little experience, this additional complication is easily predicted (or, one can do an EHMO calculation and "peek" at the answer by sketching out the MOs contained in the output; see Problem 14-6). The effect on the energy of this $2a_1$ orbital is quite large because several things happen, all of which are energy lowering:

1. For a given amount of 1s AO, the bonding overlap with 2p increases with bending.
2. For a given amount of 1s AO, the bonding overlap between the two hydrogens increases with bending.
3. Since the amount of 1s AO present is not constant, but increases with bending (from zero in the linear configuration) the rate of energy lowering due to (1) and (2) is further augmented.

Also, this is a fairly high-energy MO, and QMOT rule (3) tells us to expect such MOs to respond more dramatically to overlap changes. Finally, the overlap of a 1s AO with a 2p AO on another nucleus varies as $\cos\theta$, where θ is zero when the 2p AO points directly at the 1s AO. This means that the *rate of change of overlap with angle* is much greater around $\theta = 90°$ than at $\theta = 0°$ (see Problem 14-8). This is another reason for thinking that $1\pi_{ux} \rightarrow 2a_1$ will drop in energy much faster than $1\sigma_u \rightarrow 1b_2$ will rise.

The other π MO, $1\pi_{uy}$, undergoes no changes in overlap since the hydrogen atoms remain in the nodal plane throughout the bending process. Therefore, QMOT arguments predict no energy change for this MO.

The highest two MOs change in energy in ways that should be obvious to the reader, based on the above examples. Since these are the highest-energy MOs, they should show further enhanced sensitivity to overlap changes.

The MOs for the bent form are labeled in accordance with the symmetry notation for the C_{2v} point group, with a and b meaning symmetric and antisymmetric, respectively, for rotation about the two-fold axis and 1 and 2 being analogous symbols for reflection in the plane containing the C_2 axis and perpendicular to the molecular plane. The lowest valence MO of each symmetry type is numbered "1" despite the fact that lower-energy inner-shell orbitals exist.

Because of the very qualitative nature of the arguments leading to Fig. 14-8, no effort is made to attach a numerical scale, either for energy or angle.

We are now in a position to see how predictions based on our Walsh-type correlation diagram compare with experimental data. For molecules having only one or two valence electrons, we expect the preferred shape in the ground state to be bent. Examples are H_3^+ and LiH_2^+, both of which have been shown by experiment and/or accurate calculation to be bent. Molecules with three or four valence electrons should be linear, since the $1\sigma_u - 1b_2$ energy rise is much greater than the energy change for the lower MO. Examples are BeH_2^+, BeH_2, and BH_2^+, which are indeed linear. Addition of one or two more electrons now brings the $1\pi_u - 2a_1$ MO into play, and we have already argued that the energy change of this MO should be considerably greater than that of $1\sigma_u - 1b_2$. Basically we have here a competition between a filled MO that favors the linear form and a higher, partially or completely filled MO favoring the bent form. We therefore might reasonably expect molecules in which $2a_1$ is *singly* occupied to be bent, and molecules in which it is *doubly* occupied to be *more* bent. Occupancy of the $1\pi_u - 1b_1$ MO should have no effect on angle. Thus, that a molecule like $BH_2(1a_1)^2(1b_2)^2(2a_1)$ has an equilibrium bond angle of 131°, whereas $SiH_2(1a_1)^2(1b_2)^2(2a_1)^2$ has an angle of 97°, $NH_2(1a_1)^2(1b_2)^2(2a_1)^2(1b_1)$ has 103° and $H_2O(1a_1)^2(1b_2)^2(2a_1)^2(1b_1)^2$ has 105° is in pleasing accord with these simple ideas.

Changes of angle upon electronic excitation also agree well with the correlation diagram. The triplet state of SiH_2 resulting from the $2a_1 \rightarrow 1b_1$ excitation has a wider angle (124°) than does the ground state (97°). The excited singlet corresponding to the same excitation has a comparable angle (126°). NH_2, when excited from $\ldots(2a_1)^2(1b_1)$ to $(2a_1)(1b_1)^2$ opens from 103° to 144°. The isoelectronic PH_2, under similar excitation, opens from 92° to 123°. There are other examples to support the validity of the HAH diagram, but these suffice to illustrate that this qualitative approach has considerable generality and utility. Note again that singly charged cations appear to fit QMOT predictions despite the fact that there is less theoretical basis for success here.

EXAMPLE 14-3 SiH_2 and H_2O have rather similar H-A-H angles of 97° and 105°, respectively. How would you expect these angles to compare for SiH_2^+ and H_2O^+?

SOLUTION ▶ SiH_2 loses an electron from the $2a_1$ MO, and should open up to an angle in the 130° range. H_2O loses an electron from the $1b_1$ MO, so its angle should not be greatly affected. (The value for H_2O^+ is 110.5°.) ◀

Walsh-type correlation diagrams have been constructed and discussed for many systems, among them AH_3, HAB, HAAH, BAAB, H_2AAH_2, B_2AAB_2, H_3AAH_3.[4] It is not appropriate that these all be described here. We will briefly discuss two more cases that bring in some additional features.

Molecules with HAB configuration lack the high symmetry of HAH, and this means that the MOs are not as highly symmetry-determined as in HAH. Probably the simplest

[4] See Gimarc [6].

way to arrive at sketches for linear HAB MOs is to start with AB MOs (similar to A_2 MOs) and add the 1s AO of H in bonding and antibonding modes to form linear MOs. The results for the seven lowest-energy MOs (all A–H bonding or nonbonding) are seen at the left side of Fig. 14-9. (Diatomic MOs were discussed in Chapter 7.)

We now imagine the hydrogen atom to move away from the AB axis, as shown, and use our QMOT rules to decide whether the energies should rise or fall. As before, we expect overlap changes between H and A (nearest neighbors) to have a greater effect on energy than those between H and B.

The bent molecule has only one symmetry element, namely a reflection plane containing the nuclei. An a' MO is symmetric under this reflection, a'' is antisymmetric.

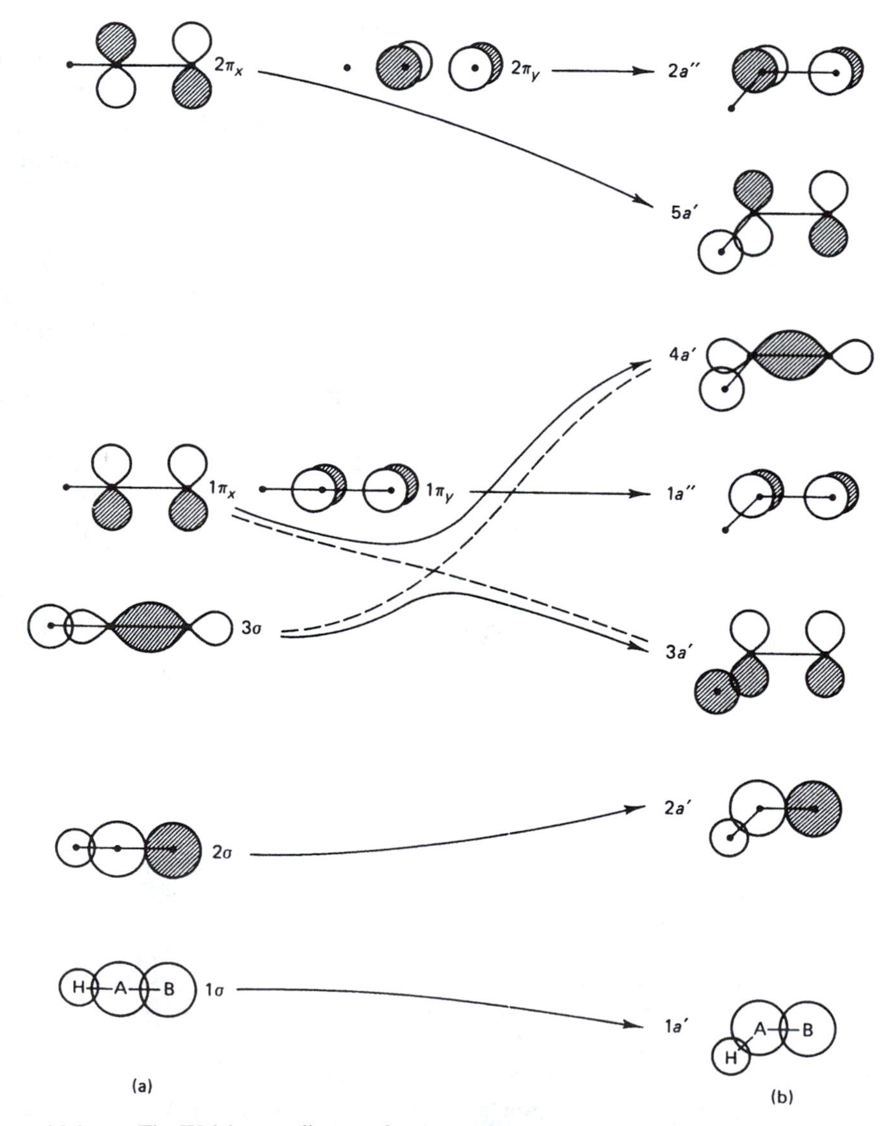

Figure 14-9 ▶ The Walsh-type diagram for the HAB system: (a) linear $C_{\infty v}$; (b) bent C_s. (After Gimarc [6].) Reprinted with permission from *Accounts Chem. Res.* **7**, 384 (1974). Copyright by the American Chemical Society.

This paucity of symmetry types means that many of the correlation lines in the diagram refer to the same symmetry. Hence, it is not too surprising that some of our correlations run into conflict with the noncrossing rule. In Fig. 14-9, dashed lines are drawn from 3σ to $4a'$ and from $1\pi_x$ to $3a'$. These dashed lines connect the MO drawings in the manner expected if we ignore the noncrossing rule and simply bend the MOs along with the molecule. They are, as it were, *intended* correlations. But these lines are associated with the same symmetry a' and hence cannot cross. Instead we have an *avoided* crossing, as 3σ switches course and connects with $3a'$ and $1\pi_x$ goes to $4a'$ (solid lines in Fig. 14-9). Avoided crossings are not uncommon in quantum chemistry, and they occur in curves referring to *state energy* as well as orbital energy. A generalized sketch exemplifying the idea is shown in Fig. 14-10. The actual energy change (as a function of bond length, angle change, or whatever process is occurring) may show an intermediate maximum or minimum as a result of the avoided crossing. (The dashed lines are energies we predict by "forgetting" to allow the two functions of the same symmetry to be mixed in the variational procedure. The error involved in this is small if the two functions are of dissimilar energy. As they grow closer in energy, the error grows worse, and the deviation between solid and dashed lines gets bigger as the dashed lines converge.)

A classic example of such an intermediate maximum is seen in an excited state of H_2, illustrated in Fig. 14-11. Such maxima are important in understanding high-energy processes because they provide a means for some molecules to exist in bound vibrational states even while unstable with respect to dissociation products. Such states are called *metastable* states.

The HAB Walsh diagram of Fig. 14-9 rationalizes the fact that the ten-valence electron HCN is linear in the ground state (due to $1\pi_x - 4a'$) but bent in the first excited state. Such molecules as HNO, HNF, and HOCl, with 12–14 electrons, are bent in both ground and excited states because there is always at least one electron in the $5a'$ level.

The final system we shall consider is the H_3AAH_3 system. We will show how QMOT can be used to understand why diborane (B_2H_6) has a bridged structure whereas ethane (C_2H_6) does not, why ethane prefers to be staggered (D_{3d}) rather than eclipsed (D_{3h}), and what geometry changes we might expect if ethane is forced into the eclipsed conformation.

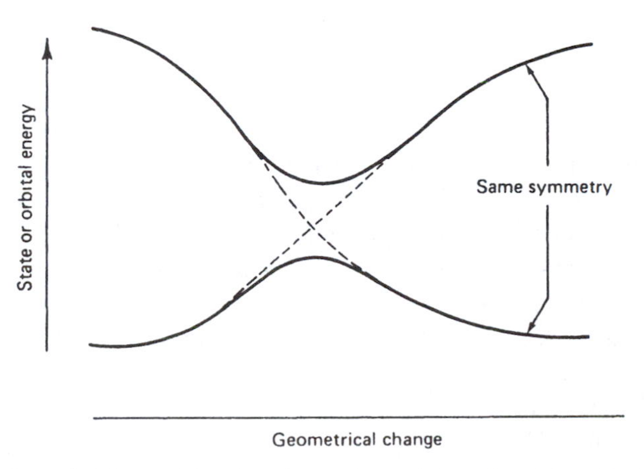

Figure 14-10 ▶ Intended (dashed lines) and actual correlations between orbitals or states having the same symmetry.

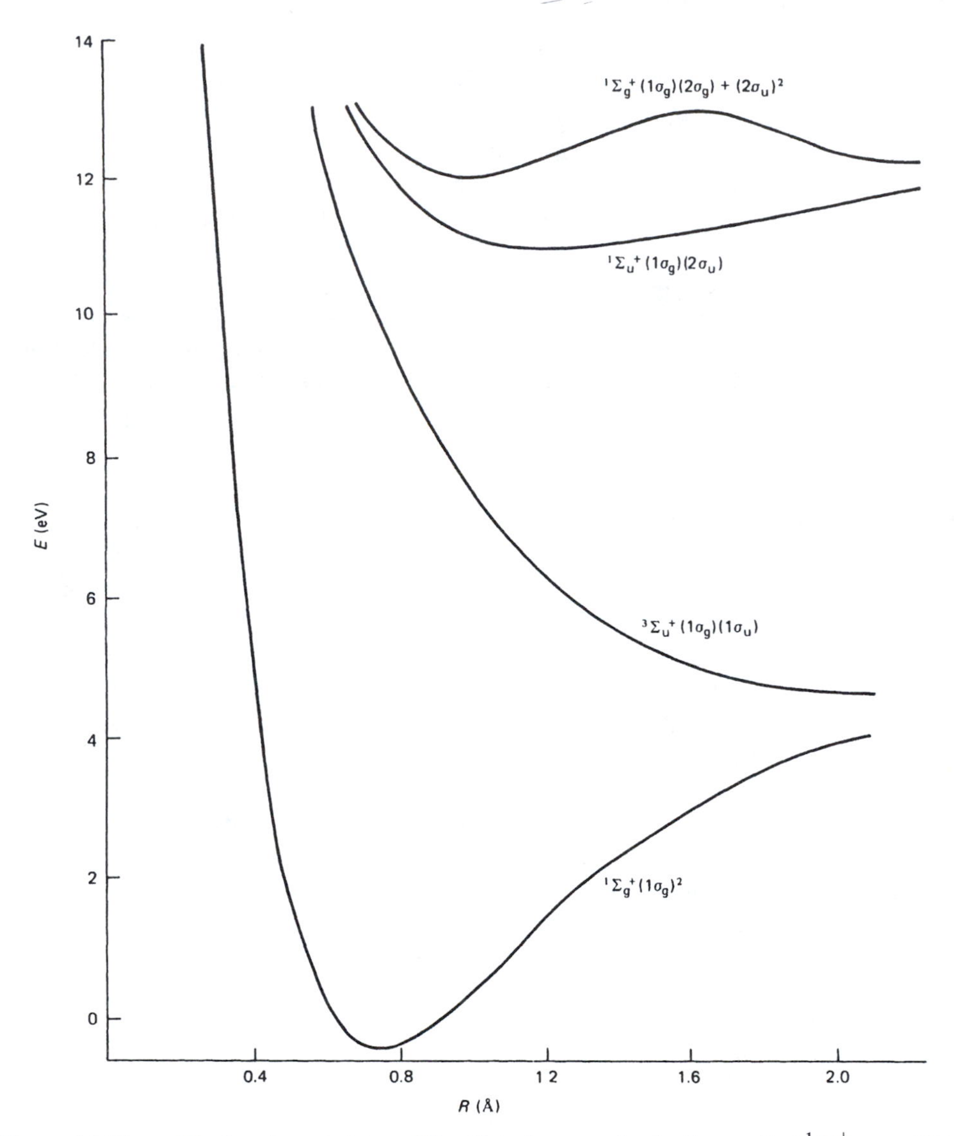

Figure 14-11 ▶ The four lowest states of H_2. Note the curve for the first excited $^1\Sigma_g^+$ state. This shape results from an avoided crossing. (From Sharp [8].)

Gimarc's diagram relating MOs for A_2H_6 in D_{3d} staggered and D_{2h} bridged shapes is shown in Fig. 14-12. The MOs on the left are all H–A bonding and are arranged pretty much in the order one would expect on the basis of the A–A bonding. The lowest two are composed mainly of valence s AOs on atoms A. Next come the π bonding combinations, then the p_σ bond, followed finally by the π antibonds. The $1e_u\pi$ bonding levels lie below the $2a_{1g}p_\sigma$ bonding level because the former have greater ability to overlap with the hydrogens.

There are only two MOs that show much energy change the molecule goes from D_{3d} to D_{2h} geometry. These are the $2a_{1g} \rightarrow 2a_g$ and the $1e_g \rightarrow 1b_{2g}$. In the former case, the energy drops because two hydrogens have moved from positions off axis of one p lobe into positions off axis of two p lobes, thereby increasing the total amount

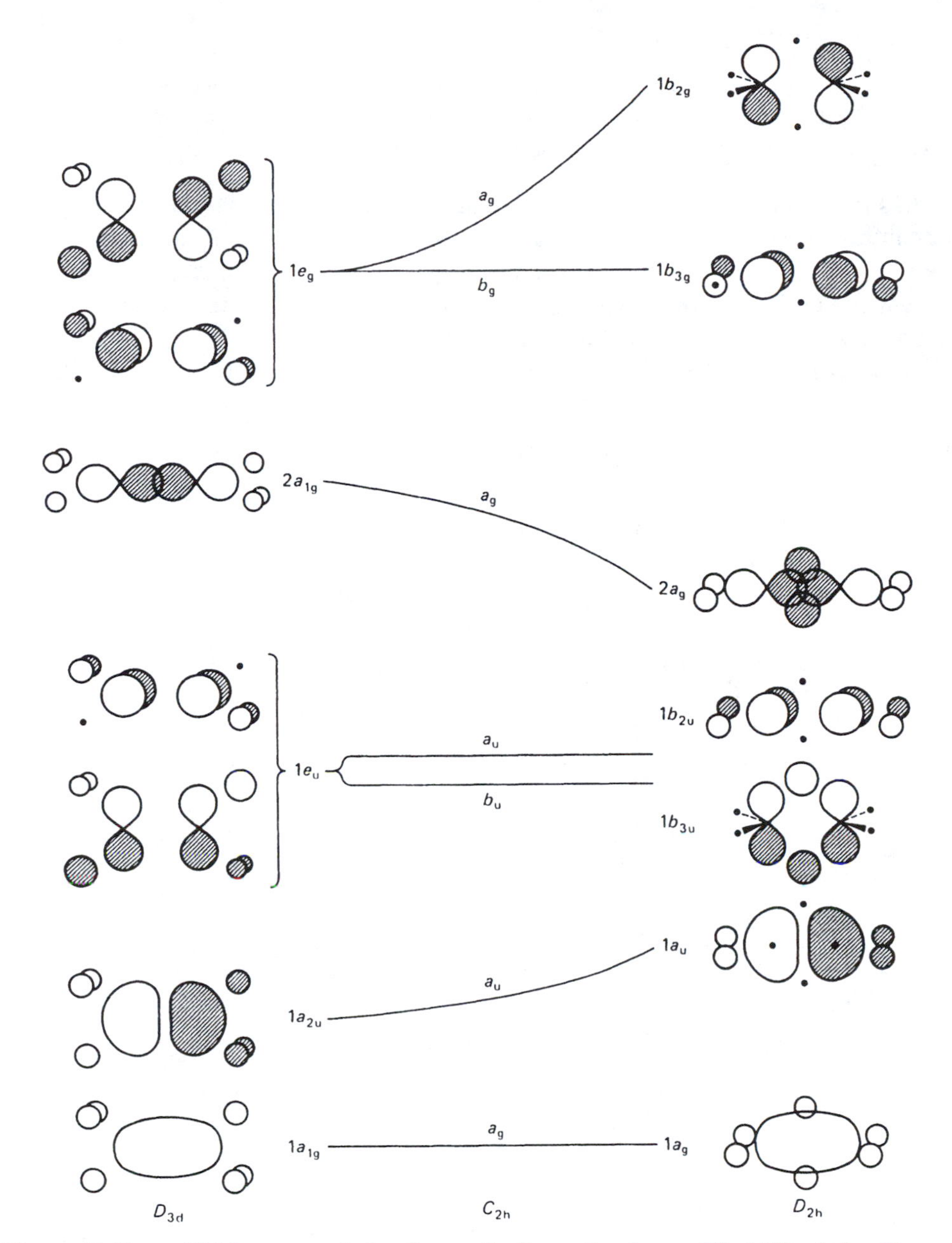

Figure 14-12 ▶ Walsh-type correlation diagram for $D_{3d} - D_{2h}$ shapes of H_3AAH_3. (After Gimarc [6].) Reprinted with permission from *Accounts Chem. Res.* **7**, 384 (1974). Copyright by the American Chemical Society.

of overlap. The energy in the latter case rises very markedly because the change in geometry places all six hydrogens into nodal planes, greatly reducing the overlap.

The prediction is that a 10- or 12-valence electron A_2H_6 system should favor a bridged D_{2h} geometry over D_{3d} but a 14-valence electron system should prefer D_{3d} over D_{2h}. Diborane (12 valence electrons) and ethane (14) have structures consistent with this.

EXAMPLE 14-4 What structure should be expected for $B_2H_6^-$?

SOLUTION ▶ Assuming that the extra electron occupies the $1b_{2g}$ MO, one might expect that $B_2H_6^-$ would have the D_{3d} structure of ethane. It depends on whether or not one electron in the $1b_{2g}$ MO suffices to overcome the effects of the doubly occupied MOs. (Experiments and *ab initio* calculations indicate that $B_2H_6^-$ has the same shape as ethane.) ◀

In structural problems such as this, one must be careful that, when comparing two possible molecular shapes, one is not overlooking other possibilities that might be even more stable. For instance, even though the diagram in Fig. 14-12 indicates that ethane should prefer D_{3d} geometry to D_{2h}, it says nothing about D_{3d} relative to D_{3h}, the eclipsed form. For this we must construct another diagram, shown in Fig. 14-13. Here the principal energy changes occur in the doubly degenerate *e*- type MOs. Recall that

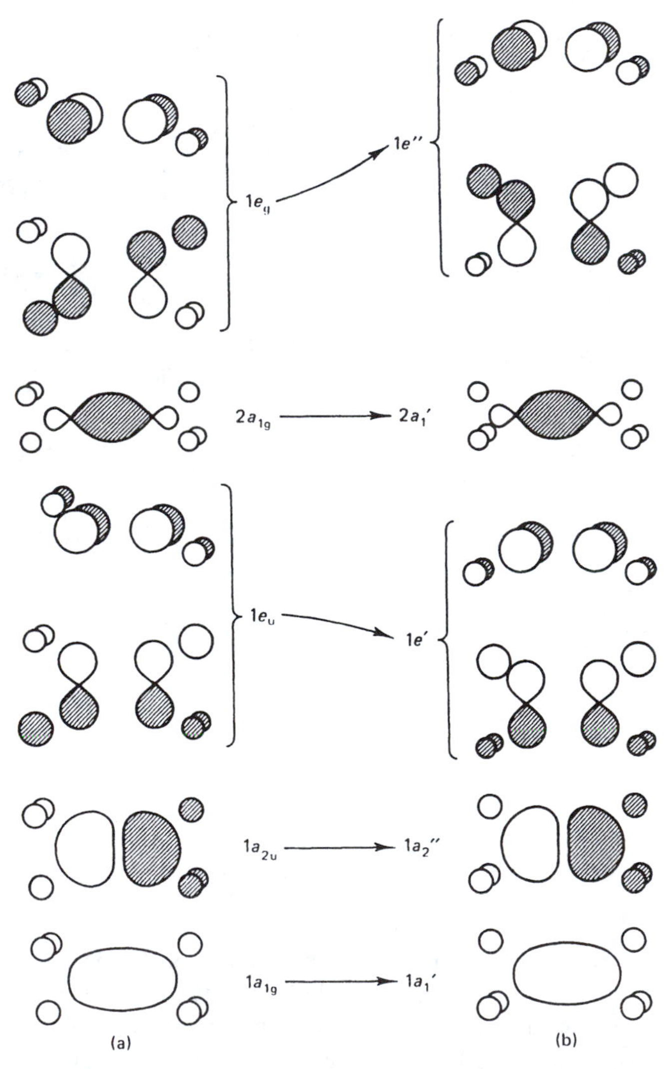

Figure 14-13 ▶ Walsh-type diagram for D_{3d}–D_{3h} A_2H_6: (a) D_{3d} (staggered); (b) D_{3h} (eclipsed). Note that the *e*-type MOs for the two forms do not turn into each other by rotating about the C–C bond. This is really a two-sided correlation diagram, the high symmetry of each form determining the MOs. Then QMOT rules are used to decide how similar MOs on the two sides should relate in energy. (See Lowe [9].)

degenerate MOs need not be symmetric or antisymmetric for all symmetry operations of the molecular point group. This results, in this case, in the *e*-type MOs being much more unbalanced, or lopsided, than the nondegenerate *a*-type MOs. As a result, the changes in overlap population upon rotation are much bigger for *e*-than for *a*-type MOs. The higher-energy $e_g - e''$ set of MOs dominates the lower energy $e_u - e'$ set for two reasons; it is at higher energy and hence more sensitive to overlap change, and the coefficients on the hydrogens are bigger due to the central nodal plane, which reduces the size of the MO in the A–A bond, forcing it to be larger elsewhere. In sum, a 14-valence electron A_2H_6 molecule prefers the staggered (D_{3d}) form because the long-range H $\cdots$ H antibonding in $1e''$ dominates the long range bonding in $1e'$. We can describe this as "nonbonded repulsion" between hydrogens at opposite ends of the molecule.

The energy changes in Fig. 14-12 are much larger than those in Fig. 14-13. In the former, we are charting energy changes associated with changes in bond angle and even molecular topology. Overlap changes are large and occur between nearest neighbors. In Fig. 14-13 we are charting energy changes associated with internal rotation. Here the overlap changes occur between third-nearest neighbors and are very small. The order in which the possibilities have been examined—D_{3d} versus D_{2h} followed by D_{3d} versus D_{3h}—is thus sensible in that we are considering the grosser energy changes first.

One can go even further and guess the qualitative changes in C–C, C–H distances, and C–C–H angle if ethane is forced into the eclipsed conformation. We argue that the $1e_g \rightarrow 1e''$ pair of MOs suffer the greatest overlap change, losing population between the vicinal hydrogens. We must renormalize the MO to compensate for this loss, just as we had to in H_2^+, discussed earlier. To renormalize, the MO $1e''$ is multiplied by a factor slightly greater than unity. This magnifies the π antibond between the carbons and the bonding between carbon and hydrogens. Thus, we expect eclipsed ethane to have a slightly lengthened C–C bond, slightly shorter C–H bonds, and a larger C–C–H angle (the latter presumably mainly due to the increased vicinal repulsion brought about by overlap changes in the original rotation). *Ab initio* calculations[5] support these predictions.

14-8 Frontier Orbitals

We have indicated that higher-energy MOs tend to undergo more pronounced energy changes upon overlap change due to distortion of the nuclear frame. This fact has led to a shortcut method for guessing the results of full orbital correlation diagrams of the sort we have already discussed. One merely considers what the energy behavior will be for the highest occupied MO (HOMO) and bases the prediction entirely on that MO, ignoring all the others. Fukui[6] was the first to draw attention to the special importance of the HOMO. He also noted that, in certain reactions in which the molecule in question acted as an electron acceptor, the lowest unfilled MO (LUMO) of the molecule (before it has accepted the electrons) is the important one. These two MOs are called the *frontier* MOs. It sometimes happens that the second-highest-energy occupied MO undergoes a

---

[5] Stevens [10] finds that the C–C distance increases by 0.01 Å, the C–H distance decreases by 0.001 Å, and the C–C–H angle opens by 0.3°.

[6] See Fujimoto and Fukui [11].

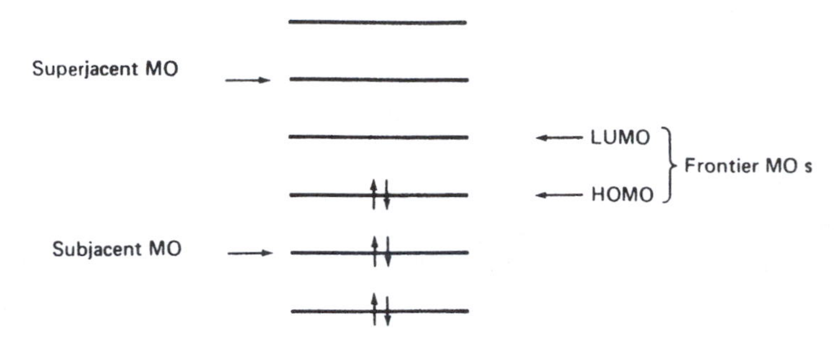

Figure 14-14 ▶ HOMO and LUMO frontier MOs and subjacent and superjacent MOs.

much bigger overlap change than the HOMO does, and therefore dominates the process. Recognition of this fact has led to a special name for the MOs just below the HOMO and just above the LUMO. They are called *subjacent* and *superjacent* MOs (see Fig. 14-14).

An example of the frontier MO approach is provided by reconsidering the staggered versus eclipsed conformation for ethane. We expect the highest-energy, lopsided MO to be the π antibonding $e_g - e''$ degenerate pair. Qualitative molecular orbital theory rules lead us to expect this pair to have higher energy in the eclipsed form. Therefore, we expect ethane to be staggered.

The same approach can be used to predict the conformation of dimethylacetylene, $H_3C–C\equiv C–CH_3$. The HOMO is again a degenerate pair of π-type MOs. (Generally speaking, one assumes that occupied π MOs are higher in energy than occupied σ MOs, and this is often true. Even when it is not, however, the π-type MOs often tend to be more lopsided and hence to dominate because their overlap changes are greater.) The two degenerate HOMOs are delocalized over the entire molecule and can be expected to have the following characteristics:

1. They will be orthogonal to each other.
2. They will be bonding in the central C≡C region, helping to establish multiple bond character there.
3. They will be antibonding in the C–C single bond regions, thereby cancelling out double-bond character from a lower set of π-type MOs.
4. They will be C–H bonding.

[Use of rules (2)–(4) often suffices to establish the qualitative nature of HOMOs of hydrocarbons.] The results of all these conditions are the MOs sketched in Fig. 14-15. Observe that the end-to-end hydrogen overlap is most positive in the eclipsed conformation. This leads to the prediction that this molecule is more stable in the eclipsed conformation. Since the hydrogens are so far apart, the overlap change is expected to be very small. *Ab initio* calculations indicate that dimethylacetylene is more stable in the eclipsed form and that it has a barrier of less than 0.02 kcal/mole.

As another example of frontier orbital usage, consider the methyl rotation barrier in propene, $H_3C–CH{=}CH_2$. Here the HOMO should be π-bonding in the double bond, antibonding in the single bond, and C–H bonding in the methyl group. This MO is sketched in Fig. 14-16 for the two possible conformations. The end-to-end antibonding in this MO is greatest for case (b), and so conformation (a) is favored. Indeed, it has been observed that, *in general*, a threefold rotor attached to a double bond

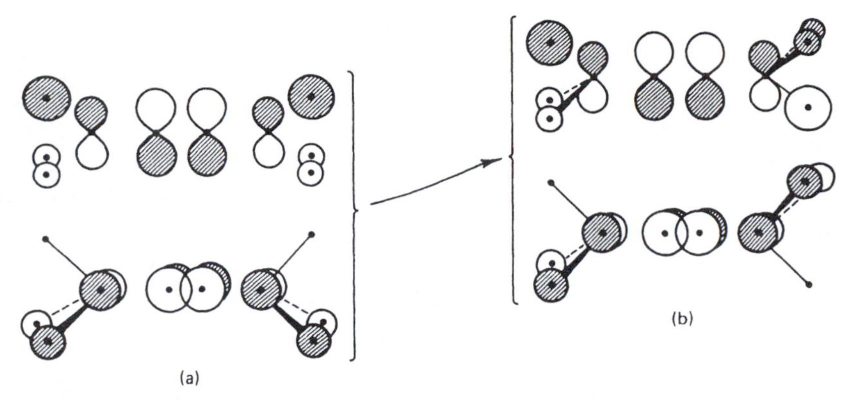

Figure 14-15 ▶ Degenerate HOMOs for dimethylacetylene, (a) eclipsed, (b) staggered, lead to the prediction that this molecule should prefer the eclipsed conformation.

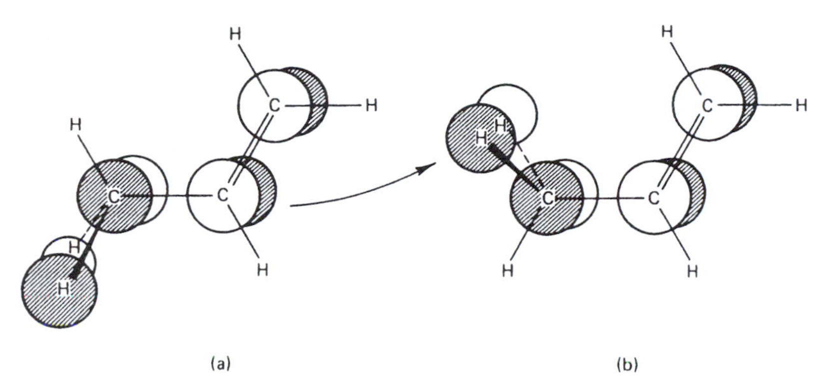

Figure 14-16 ▶ HOMO of propene in two conformations. Methyl group (a) eclipses and (b) staggers the double bond. The energy is lower in (a).

prefers to eclipse the double bond. A few examples are acetaldehyde ($H_3C–CH=O$), *N*-methylformaldimine ($H_3C–N=CH_2$), nitrosomethane ($H_3C–N=O$), and vinyl silane ($H_3Si–CH=CH_2$).

Notice that the HOMO of Fig. 14-16 is qualitatively similar to one of the $1e_g - 1e''$ HOMOs of ethane. The QMOT frontier orbital argument for the stability of staggered ethane is basically the same as that for the stability of form (a) in propene. Observe that this MO also resembles the HOMO of 1,3-butadiene, and would lead to the prediction that the trans form of this molecule is more stable than the cis. This is, in fact, observed to be the case.

It is important to bear in mind that the frontier orbital approach is an approximation to an approximation. It is not always easy to know when one is on safe ground. Of the examples mentioned here, ethane and dimethylacetylene are safest because the overlap changes are small. Hence, the perturbation is slight, and our assumption that the MOs of the two forms are essentially identical, except for AO overlap changes, is quite accurate. Also, symmetry is high, and so we know that σ MO overlap changes will be smaller than lopsided π MO overlap changes. Propene and butadiene are risky. Here the whole molecular framework is lopsided, so overlap changes are large in σ as well as in π MOs. Indeed, if one performs EHMO calculations on these molecules, one finds that σ-type MO energies change much more than does the π HOMO energy. This comes about because of the fairly close approach by some of the hydrogens in these molecules. Much of this σ energy change cancels out among the several σ MOs. In propene, the cancellation is so complete that the π HOMO energy change is almost the same as the total EHMO energy change. In butadiene, however, the π HOMO accounts for only about one-third of the total energy change. Even though the frontier orbital method has an astonishing range of qualitative usefulness (we shall see more applications shortly), it is clear that caution is needed.

EXAMPLE 14-5 How should the bond lengths and angles of propene change if the methyl group rotates from the stable conformation (Fig. 14-16a) to the unstable one (Fig. 14-16b)?

SOLUTION ▶ The HOMO for form b has a bit more negative overlap, which requires a slightly larger normalizing coefficient. This increases the influence of the HOMO, so we expect a slight shortening of the two out-of-plane C–H bonds and the C=C bond, a lengthening of the C–C single bond, and a slight opening of the H–C–H angle for the out-of-plane C–H bonds. No changes are predicted for the in-plane C–H bond lengths or angles as a result of HOMO renormalization. (However, other MOs are also being renormalized, so other changes will result. However, the HOMO-induced changes should dominate.) ◀

14-9 Qualitative Molecular Orbital Theory of Reactions

It has been found possible to extend and amplify QMOT procedures so that they apply to chemical reactions. One of the most striking examples of this was application to unimolecular cyclization of an open conjugated molecule (e.g., *cis*-1,3-butadiene, closing to cyclobutene). This type of reaction is called an *electrocyclic* reaction. The details of the electrocyclic closure of *cis*-1,3-butadiene are indicated in Fig. 14-17.

If we imagine that we can keep track of the terminal hydrogens in butadiene (perhaps by deuterium substitution as indicated in the figure) then we can distinguish between two products. One of them is produced if the two terminal methylene groups have rotated in the same sense, either both clockwise or both counterclockwise, to put the two inside atoms of the reactant (here D atoms) on opposite sides of the plane of the four carbon atoms in the product. This is called a *conrotatory* (cŏn′ · rō · tā′ · tory) closure. The other mode rotates the methylenes in opposite directions (*disrotatory*) to give a product wherein the inside atoms appear on the same side of the C_4 plane.

A priori, we do not know whether the reaction follows either of these two paths. Figure 14-17 depicts processes where both methylene groups rotate by equal amounts

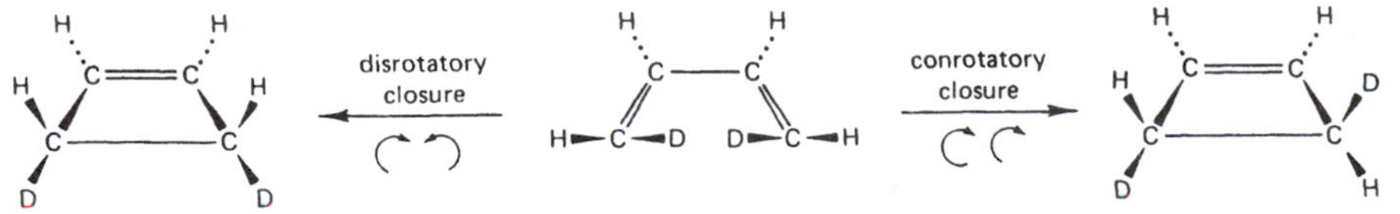

Figure 14-17 ▶ Two idealized modes of electrocyclic closure of *cis*-1,3-butadiene.

as the reaction proceeds. This is an extreme case of what is known as a *concerted* process. The two processes occur together, or in concert. The opposite extreme is a *nonconcerted*, or *stepwise* process, wherein one methylene group would rotate all the way (90°) and only after this was completed would the other group begin to rotate. This process would lead to an intermediate having a plane of symmetry (ignoring the difference between D and H), which means that the second methylene group would be equally likely to rotate either way, giving a 50–50 mixture of the two products pictured in Fig. 14-17.

One can make a case for the reaction having some *substantial degree of concertedness* (by which we mean that the second methylene should be partly rotated before the first one is finished rotating). The reaction involves destruction of a four-center conjugated π system and formation of an isolated π bond and a new C–C σ bond. Energy is lost in the dissolution of the old bonds, and gained in formation of the new ones. Therefore, we expect the lowest-energy path between reactants and products to correspond to a reaction coordinate wherein the new bonds start to form before the old ones are completely broken. But the new σ bond cannot form to any significant extent until *both* methylene groups have undergone some rotation. Thus, concertedness in breaking old bonds and forming new ones is aided by some concertedness in methylene group rotations. (Note that concertedness does not necessarily imply absence of an intermediate. If the reaction surface had a local minimum at a point at which both methylenes were rotated by 45°, it would not affect the argument at all.)

Because they knew that many electrocyclic reactions are observed to be *stereospecific* (i.e., give ~100% of one product or the other in a reaction like that in Fig. 14-17), Woodward and Hoffmann [12] sought an explanation of a qualitative MO nature. They used frontier orbitals and argued how their energies would change with a con- or disrotatory motion, due to changes in overlap. For butadiene in its ground state, the HOMO is the familiar π MO shown in the center of Fig. 14-18. The figure indicates that the interaction between p–π AOs on terminal carbons is favorable for bonding in the region of the incipient σ bond only in the conrotatory case. Therefore, the prediction is that, for concerted electrocyclic closure, butadiene in the ground state should prefer to go by a conrotatory path. When the reaction is carried out by heating butadiene (thermal reaction), which means that the reactant is virtually all in the ground electronic state, the product is indeed purely that expected from conrotatory closure.

One can also carry out electrocyclic reactions photochemically. The excited butadiene now has an electron in a π MO that was empty in the ground state. This MO was the lowest unoccupied MO (LUMO) of ground-state butadiene, pictured in Fig. 14-19. One can see that the step to the next-higher MO of butadiene has just introduced one

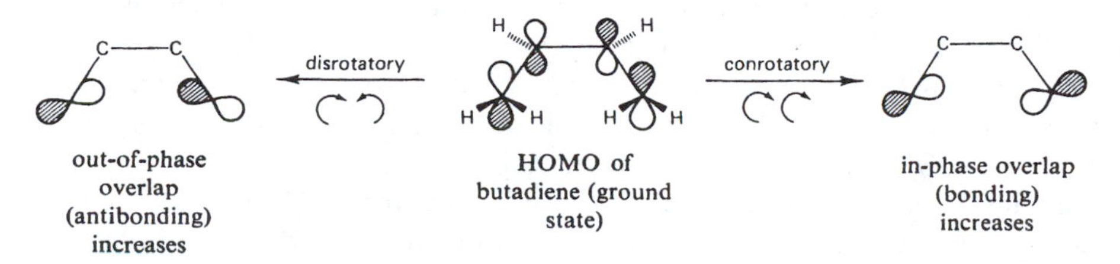

Figure 14-18 ▶ The HOMO of ground state *cis*-1,3-butadiene as it undergoes concerted closure by either mode.

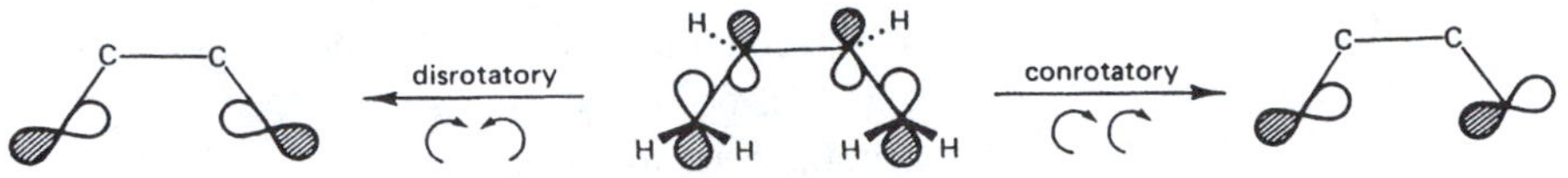

in-phase overlap (bonding) increases

disrotatory

LUMO of butadiene (ground state) or HOMO of butadiene (first excited state)

conrotatory

out-of-phase overlap (anti-bonding) increases

Figure 14-19 ▶ The HOMO of the first excited state of *cis*-1,3-butadiene as it undergoes closure by either mode.

more node, reversing the phase relation between terminal π AOs, and reversing the predicted path from con-to disrotatory. Experimentally., the photochemical reaction is observed to give purely the product corresponding to disrotatory closure. (It is not always obvious which empty MO becomes occupied in a given photochemical experiment. One assumes that the LUMO of the ground state is the one to use, but there is some risk here.)

EXAMPLE 14-6 Is the LUMO $\leftarrow$ HOMO transition dipole-allowed for *cis*-1,3-butadiene?

SOLUTION ▶ The HOMO is antisymmetric for reflection through the symmetry plane that bisects the molecule, and the LUMO is symmetric for this reflection. This is the only symmetry reflection plane where the MOs have opposite symmetry, so the transition is dipole-allowed (and is polarized from one side of the molecule towards the other). The group theory approach for this C_{2v} molecule is that the HOMO has a_2 symmetry, the LUMO has b_1 symmetry, their product has b_1 symmetry, and, since x also has b_1 symmetry, the transition is allowed and is x-polarized (where x is colinear with the central C–C bond). ◀

One might worry about the fact that we are looking at only a part of one MO, thereby ignoring a great deal of change in other MOs and other parts of the molecule. However, much of this other change, while large, is expected to be about the same for either of the two paths being compared. The large overlap changes between p–π AOs on terminal and inner carbon atoms, for instance, are about the same for either mode of rotation. This approach, then, is focused first on the frontier orbitals, which are guessed as being most likely to dominate the energy change, and second on those changes in the frontier orbitals that will differ in the two paths.

This method is trivially extendable to longer systems. Hexatriene closes to cyclohexadiene in just the manner predicted by the frontier orbitals. The only significant change in going from butadiene to hexatriene is that we go from four to six π electrons. This means that the HOMO for hexatriene has one more node than that for butadiene (or, the HOMO for a $2n$ π-electron system is like the LUMO for a $2n-2$ π-electron system insofar as end-to-end phase relations are concerned). The net effect is that the predictions for hexatriene are just the reverse of those for butadiene. That is, hexatriene closes thermally by the disrotatory mode and photochemically by the conrotatory mode. The general rule, called a *Woodward–Hoffmann rule*, is this: the thermal electrocyclic reactions of a k π-electron system will be disrotatory for $k = 4q + 2$, conrotatory for $k = 4q$ ($q = 0, 1, 2, \ldots$); in the first excited state these relationships are reversed.[7]

It is possible to treat electrocyclic reactions in another way, namely, via a two-sided correlation diagram approach. This was first worked out by Longuet-Higgins and Abrahamson [14]. Only orbitals (occupied *and* unoccupied) that are involved in bonds being made or broken during the course of the reaction are included in the diagram. For butadiene, these are the four π MOs already familiar from simple Hückel theory. For cyclobutene, they are the two π MOs associated with the isolated 2-center π bond and the two σ MOs associated with the new C–C σ bond. These orbitals and their energies are shown in Fig. 14-20.

In cyclobutene, the σ and σ^* MOs are assumed to be more widely split than the π and π^* because the p_σ AOs overlap more strongly. Also, the σ MO is assumed lower than π_1 of butadiene. However, these details are not essential. All we have to be certain of is that we have correctly divided the occupied from the unoccupied MOs on the two sides. The dashed line in Fig. 14-20 separates these sets.

Next we must decide which symmetry elements are preserved throughout the idealized reactions we wish to treat. Let us consider first the reactants and products. These have C_{2v} symmetry, that is, a two-fold rotational axis, C_2, and two reflection planes σ_1

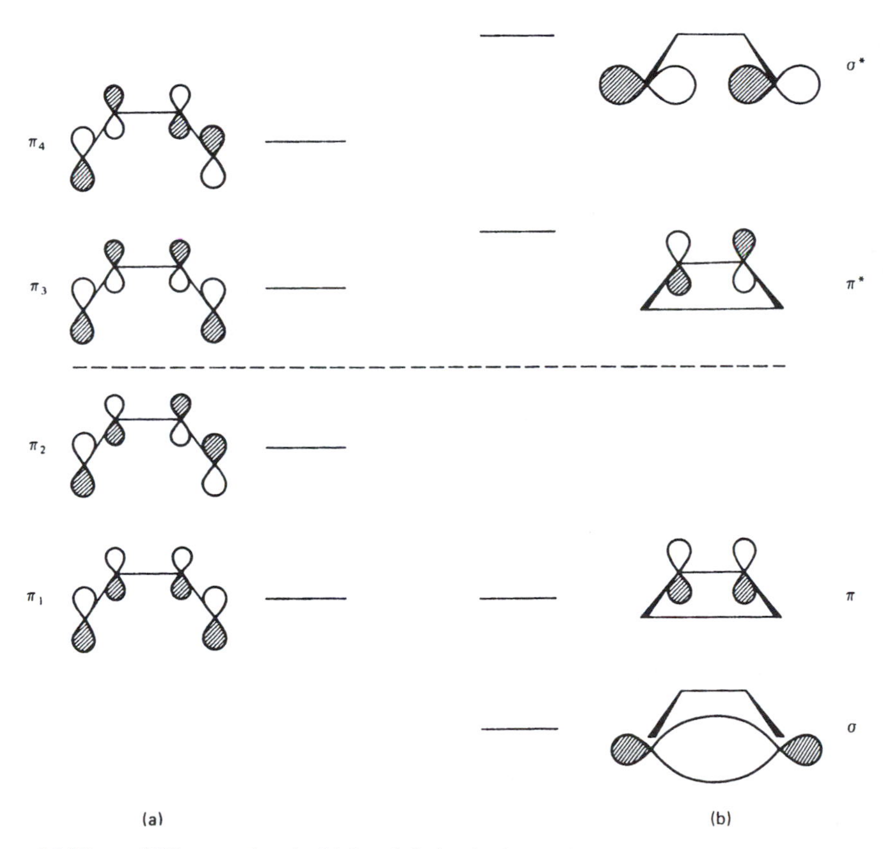

Figure 14-20 ▶ MOs associated with bonds being broken or formed in the electrocyclic closure of (a) *cis*-1,3-butadiene to (b) cyclobutene.

[7] See Woodward and Hoffmann [13, p. 45].

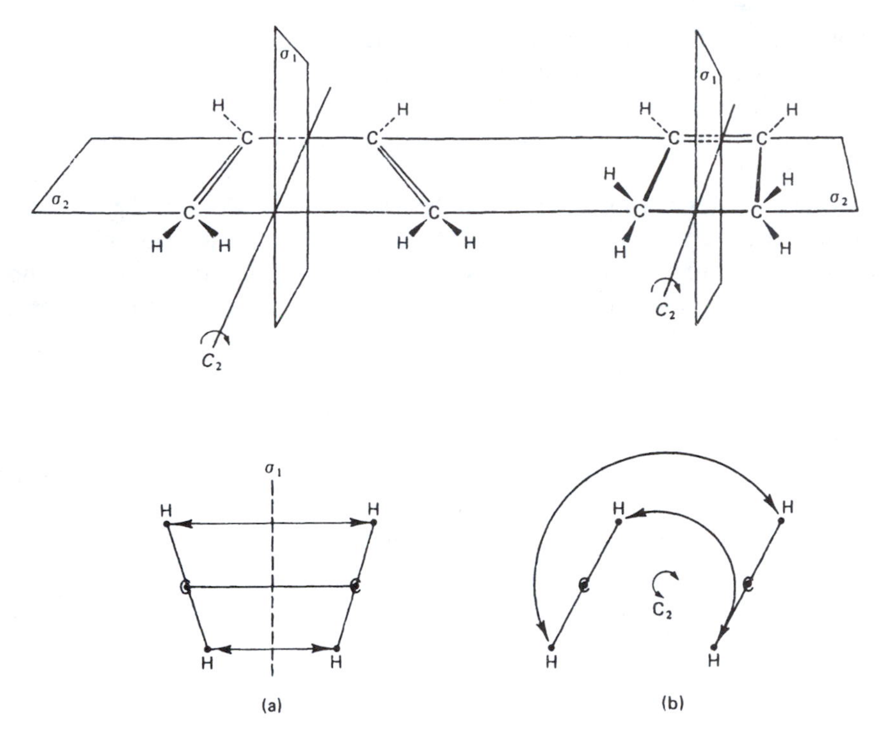

Figure 14-21 ▶ Sketches illustrating that the conrotatory mode (b) preserves the C_2 axis while the disrotatory mode (a) preserves the reflection plane σ_1.

and σ_2 containing the C_2 axis (see Fig. 14-21). A conrotatory twist preserves C_2, but, during the intermediate stages between reactant and product, σ_1 and σ_2 are lost as symmetry operations. A disrotatory twist preserves σ_1 but destroys C_2 and σ_2. Therefore, when we connect energy levels together for the disrotatory mode, we must connect levels of the same symmetry for σ_1, but for the conrotatory mode, they must agree in symmetry for C_2. The σ_2 plane applies to neither mode and is therefore ignored. The symmetries for each MO are easily determined from examination of the sketches in Fig. 14-20, and are given in Table 14-3. These assignments lead to two different correlation diagrams, one for each mode. It is conventional to arrange these as shown in Fig. 14-22.

TABLE 14-3 ▶ Symmetries for C_{2v} MOs

MO	σ_1	C_2	MO	σ_1	C_2
Butadiene			Cyclobutene		
π_1	S[a]	A	σ	S	S
π_2	A	S	π	S	A
π_3	S	A	π^*	A	S
π_4	A	S	σ^*	A	A

[a] S is symmetric; A antisymmetric.

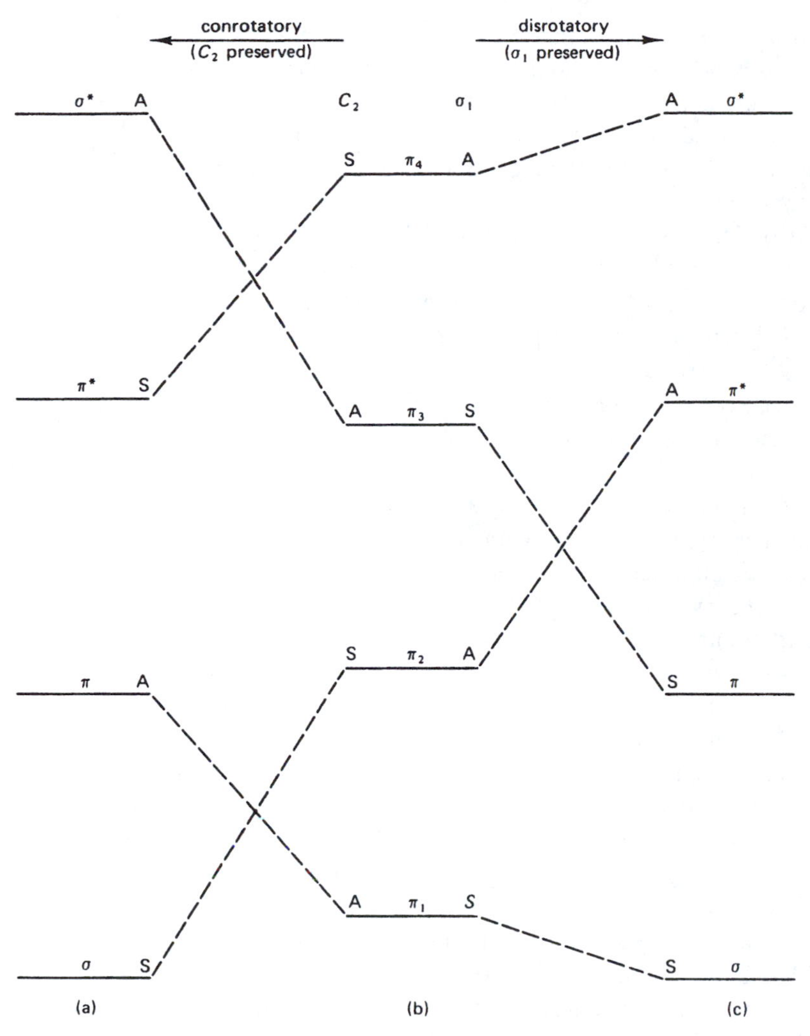

Figure 14-22 ▶ A pair of two-sided correlation diagrams (one for each mode) for the electrocyclic reactions of *cis*-1,3-butadiene: (a) cyclobutene; (b) butadiene; (c) cyclobutene.

There is curve crossing in these diagrams, but it is always lines of different symmetry that cross, and so no violation of the noncrossing rule occurs.

If we are considering a thermal reaction, the lowest two π MOs of butadiene are occupied. These correlate with the lowest two MOs of cyclobutene if the conrotatory mode is followed, and the thermal conversion of *cis*-butadiene to cyclobutene by a conrotatory closure is said to be *symmetry allowed.* The other mode correlates π_2 with an empty cyclobutene MO (π^*). Taking this route moves the reactant toward doubly excited cyclobutene. (Even though we might anticipate de-excitation somewhere along the way, the energy required in early stages would still be much higher than would be needed for the symmetry-allowed mode.) This is said to be a *symmetry-forbidden* reaction.

If we now imagine photo-excitation of *cis*-butadiene to have generated a state associated with the configuration $\pi_1^2\pi_2\pi_3$, and trace the fate of this species for the two modes of reaction, we note that the disrotatory route leads to cyclobutene in the configuration $\sigma^2\pi\pi^*$ while the conrotatory mode gives $\sigma\pi^2\sigma^*$. Both of these are excited, but the former corresponds to the lowest excited configuration ($\pi \rightarrow \pi^*$) while the second corresponds to a very high-energy excitation ($\sigma \rightarrow \sigma^*$). Therefore, the former is "allowed" (since it goes from lowest excited reactant to lowest excited product) and the latter is "forbidden."

The two-sided correlation diagrams of Fig. 14-22 thus lead to the same predictions as the frontier orbital maximization of overlap approach. The difference between these approaches is as follows: The frontier-orbital approach requires sketching the HOMO and then judging overlap changes upon nuclear motion using QMOT reasoning. The two-sided correlation diagram approach requires sketching all the MOs (occupied and unoccupied) of both reactant and product involved in bonds breaking or forming, ordering the corresponding energy levels, and finding symmetry elements preserved throughout the reaction. Once all this is done, the levels are connected by correlation lines *without* reliance on QMOT reasoning. Some qualitative reasoning enters in the ordering of energy levels (levels with more nodes have higher energy), but the two-sided correlation diagram technique is the more rigorous method of the two and tends to be preferred whenever the problem has enough symmetry to make it feasible. Reliance on frontier orbitals is more common for processes of lower symmetry.

Concern is sometimes expressed about the apparent restrictions resulting from use of symmetry in correlation diagram arguments. One can imagine the butadiene cyclization occurring with less-than-perfect concertedness, the two methylene groups rotating by different amounts as the reaction proceeds. But that would destroy all symmetry elements. Will our symmetry-based arguments still pertain to such an imperfectly concerted reaction coordinate? Again, if we label certain sites by substituting deuteriums for hydrogens as shown in Fig. 14-17, the symmetry will be destroyed. Do our predictions still apply? One can answer these questions affirmatively by reasoning in the following way. If we had a collection of nuclei and electrons, and we could move the nuclei about in arbitrary ways and study the ground-state energy changes, experience tells us that the energy would be found to change in a smooth and continuous way. We can think of the energy as a hypersurface, with hyperdimensional "hills," "valleys," and "passes." Now, in a few very special nuclear configurations, identical nuclei would be interrelated by symmetry operations, and we would be able to make deductions on group-theoretical grounds. Such deductions would only strictly apply to those symmetric configurations, but they would serve as indicators of what the energy is like in nearby regions of configuration space. Thus, the correlation diagram indicates that a *perfectly* concerted thermal electrocyclic reaction of butadiene will require much less energy to go conrotatory as opposed to disrotatory. The inference that a less-perfectly concerted reaction will have a similar preference is merely an assumption that it is easier to pass through the mountains *in the vicinity* of a low pass than a high one. Experience also leads us to expect that substituting for H a D (or even a CH_3) will have little effect on the MOs, even though, strictly speaking, symmetry is lost. In essence, we work with an ideal model and use chemical sense to extend the results to less ideal situations, just as we do when, in applying the ideal gas equation of state to real gases, we avoid the high-pressure, low-temperature conditions under which we know the oversimplifications in the ideal gas model will lead to *significant* error.

It is possible to combine information on *orbital* symmetries and energies to arrive at *state* symmetries and energies. Then one can construct a correlation diagram for states.[8] We now demonstrate this for the dis- and conrotatory reactions just considered.

Each orbital occupation scheme is associated with a net symmetry for any given symmetry operation. Character tables could be used to assign these symmetries, but this is not necessary. All we need to use is the fact that, in multiplying functions together, symmetries follow the rules: S × S = S, A × A = S, S × A = A. Thus, any *doubly* occupied MO in a configuration will contribute symmetrically to the final result. To ascertain the net symmetry, then, we focus on the partly filled MOs. The symmetries for C_2 and σ_1 of ground and some excited configurations of butadiene and cyclobutene are listed in Table 14-4. Included are the cyclobutene configurations that result from intended correlations of various butadiene configurations. (For instance, $\pi_1^2\pi_2^2$ butadiene has an

TABLE 14-4 ▶ Symmetries and Intended Correlations of Some Configurations of Butadiene and Cyclobutene

	Symmetry for		Cyclobutene "intended" configuration	
Configuration	σ_1	C_2	Con	Dis
Butadiene				
$\pi_1{}^2\pi_2{}^2$	S[a]	S	$\pi^2\sigma^2$	$\sigma^2\pi^{*2}$
$\pi_1{}^2\pi_2\pi_3$	A	A	$\pi^2\sigma\sigma^*$	$\sigma^2\pi^*\pi$
$\pi_1{}^2\pi_2\pi_4$	S	S	$\pi^2\sigma\pi^*$	$\sigma^2\pi^*\sigma^*$
$\pi_1\pi_2{}^2\pi_3$	S	S	$\pi\sigma^2\sigma^*$	$\sigma\pi^{*2}\pi$
$\pi_1\pi_2{}^2\pi_4$	A	A	$\pi\sigma^2\pi^*$	$\sigma\pi^*\sigma^{*2}$
$\pi_1{}^2\pi_3^2$	S	S	$\sigma^2\sigma^{*2}$	$\sigma^2\pi^2$
⋮				
Cyclobutene				
$\sigma^2\pi^2$	S	S		
$\sigma^2\pi\pi^*$	A	A		
$\sigma\pi^2\pi^*$	A	S		
$\sigma\pi^2\pi^*$	A	S		
$\sigma\pi^2\sigma^*$	A	A		
$\sigma^2\pi^{*2}$	S	S		
$\sigma^2\pi^*\sigma^*$	S	A		
⋮				

[a] S is symmetric; A antisymmetric.

[8] Actually, we shall be looking at simple products of MOs, or *configurations*. Each configuration is associated with one or more states and gives the proper symmetry for these states as well as an approximate average energy of all the associated states. Hence, the treatment described here gives a sort of *average* state correlation diagram. It might be more accurately called a *configuration* correlation diagram.

intended correlation with $\sigma^2\pi^{*2}$ cyclobutene if the disrotatory mode is followed. This is inferred from the *orbital* correlation diagram, Fig. 14-22.)

Assuming that the energies of states associated with these configurations fall into groups roughly given by sums of orbital energies, we obtain the two-sided diagram shown in Fig. 14-23. Only a few of the configurations are interconnected, to keep the diagram simple. Note that the ground-state configuration of butadiene correlates directly with the ground state of cyclobutene for conrotatory closure, but has an intended correlation with a doubly excited configuration in the disrotatory mode. This intended correlation would violate the noncrossing rule by crossing another line of S symmetry, so that the actual curve turns around and joins onto the ground state level for cyclobutene. The effect of the intended correlation with a high-energy state is to produce a significant barrier to reaction. The figure shows that, for the first excited configuration, the high-energy barrier occurs for the opposite mode of reaction. State correlation diagrams

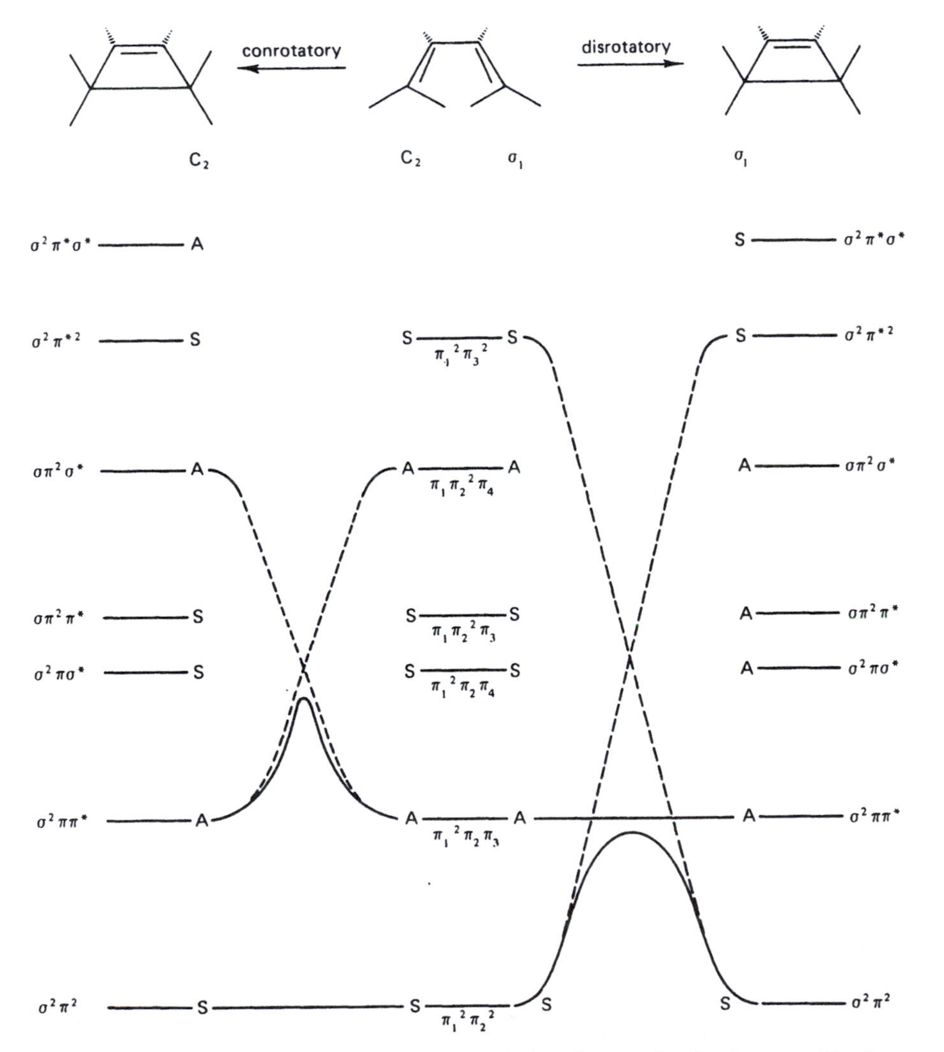

Figure 14-23 ▶ A state or configuration correlation diagram for the electrocyclic closure of *cis*-1,3-butadiene.

thus convert a "symmetry-forbidden" orbital correlation diagram into a high-activation-energy barrier: the conclusions are the same using either diagram.

Another kind of reaction that is formally closely related to the electrocyclic reaction is the *cycloaddition* reaction, exemplified by the Diels–Alder reaction between ethylene and butadiene to give cyclohexene (**I**). Such reactions are classified in terms of the

(I)

number of centers between the points of connection. Thus, the Diels–Alder reaction is a [4 + 2] cycloaddition reaction. One can conceive of several distinct geometrical possibilities for a concerted mechanism for such a reaction. The two new σ bonds can be envisioned as being formed on the same face (suprafacial) (**II**) or opposite faces (antarafacial) (**III**) of each of the two reactants. The various possibilities are illustrated in Fig. 14-24. Qualitative MO theory is used to judge which process is energetically most favorable. One has a choice between the two-sided correlation diagram and the

(II) **(III)**

frontier orbital approach. We demonstrate the latter[9] since it is simpler. Both methods lead to the same conclusion. In the course of this reaction, electrons become shared between the π systems of butadiene and ethylene. This is accomplished, to a rough approximation, by interaction between the HOMO of butadiene and the LUMO of ethylene and also between the LUMO of butadiene and the HOMO of ethylene. Let us consider the former interaction. The MOs are shown in Fig. 14-25 and the overlapping regions are indicated for the four geometric possibilities. Inspection of the sketches indicates that the two MOs have positive overlap in the regions of *both* incipient σ bonds only for the [4s + 2s] and [4a + 2a] modes. Therefore, the prediction is that these modes proceed with less activation energy and are favored. Now let us turn to the other pair of MOs, namely the LUMO of butadiene (**IV**) and the HOMO of ethylene (**V**).

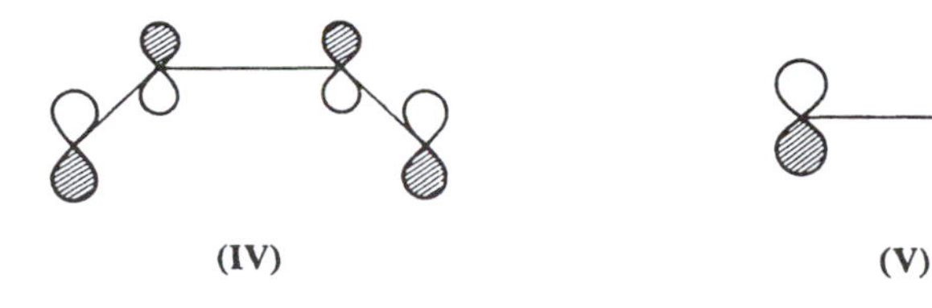

(IV) **(V)**

[9] See Hoffmann and Woodward [15].

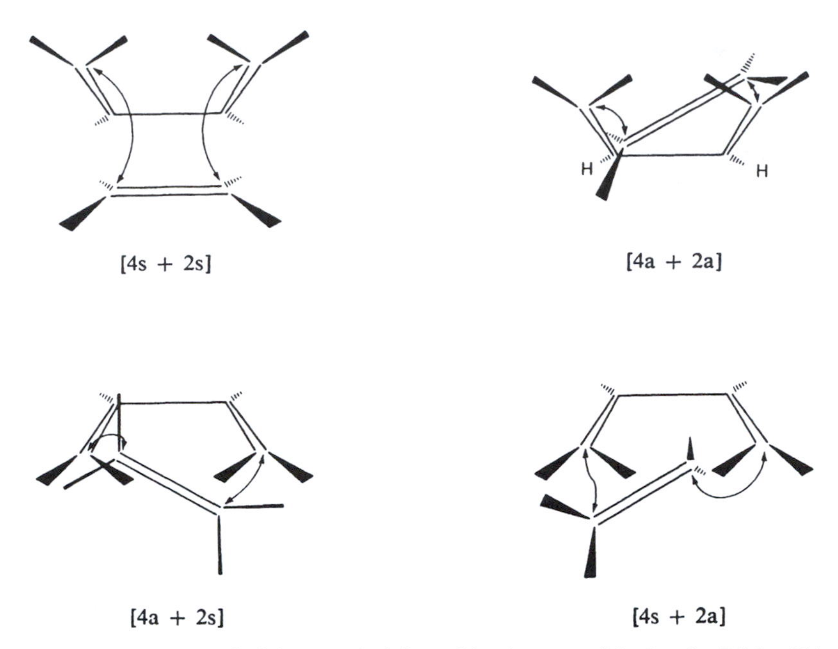

Figure 14-24 ▶ Four suprafacial–antarafacial combinations possible for the Diels–Alder 2 + 4 cycloaddition reaction.

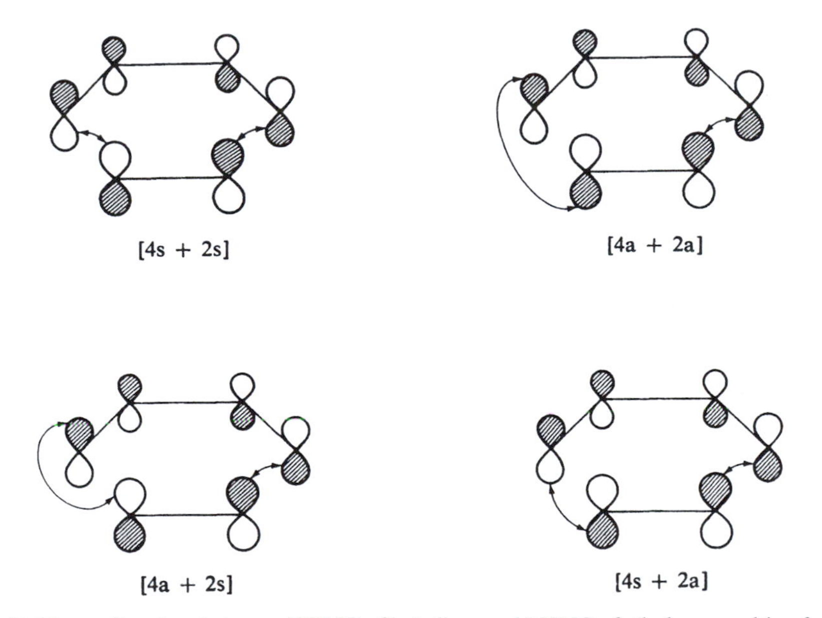

Figure 14-25 ▶ Overlaps between HOMO of butadiene and LUMO of ethylene resulting from four interactive modes pictured in Fig. 14-24. The geometries for the four modes would all differ. These drawings are highly stylized.

Note that, for each MO, the end-to-end phase relationship is reversed from what it was before. Two symmetry reversals leave us with no net change in the intermolecular phase relations. It is easy to see, therefore, that these MOs also favor the [s, s] and [a, a] modes. In cycloaddition reactions of this sort, one need analyze only one HOMO-LUMO pair in order to arrive at a prediction. Extension to longer molecules or to photochemical cycloadditions proceeds by the same kinds of arguments presented for the electrocyclic reactions.

Another type of reaction to which qualitative MO theory has been applied is the *sigmatropic shift* reaction, where a hydrogen migrates from one carbon to another and simultaneously a shift in the double bond system occurs. An example is given in Fig. 14-26.

The usual treatment of this reaction[10] involves examining the HOMO for the system at some intermediate stage in the reaction where the hydrogen has lost much of its bonding to its original site and is trying to bond onto its new site. At this stage, the HOMO of the molecule becomes like that of the nonbonding MO of an odd alternant hydrocarbon (Fig. 14-27) with a slightly bound hydrogen on one end. The suprafacial mode is favored in this particular case because the hydrogen can maintain positive overlap simultaneously with its old and new sites—the new bond can form as the old

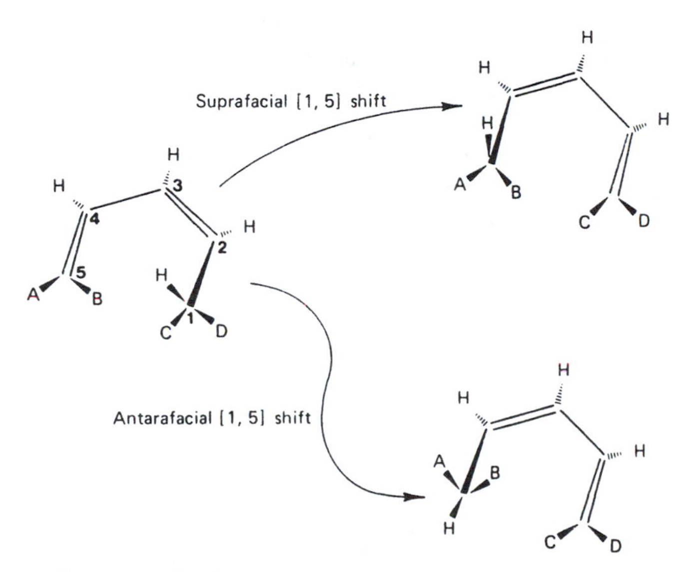

Figure 14-26 ▶ The two possible distinct products resulting from a shift of a hydrogen from position 1 to position 5 in a substituted 1,3-pentadiene. Groups A, B, C, D are deuterium atoms, methyl groups, etc., enabling us to distinguish the products.

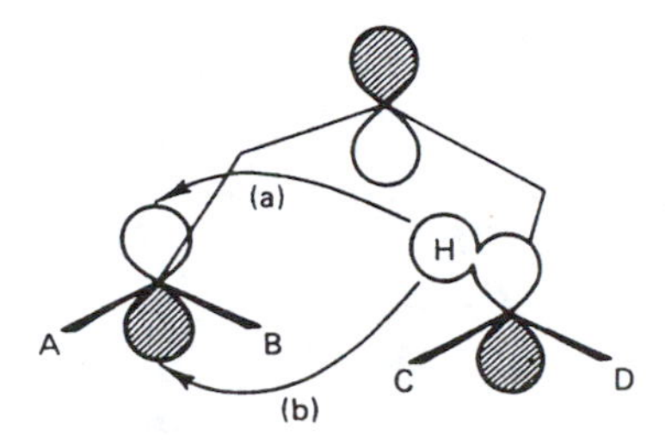

Figure 14-27 ▶ Phase relations in the HOMO for (a) suprafacial and (b) antarafacial [1, 5] sigmatropic shifts.

[10] See Woodward and Hoffmann [16].

bond breaks. This is not possible for the antarafacial [1, 5] shift. However, the [1, 7] shift prefers the antarafacial mode.

Many other types of chemical reaction have been rationalized using qualitative MO theory. The association of S_N2 reactions with Walden inversion (i.e., the adding group attacks the opposite side of an atom from the leaving group) is rationalized by arguing that an approaching nucleophile will donate electrons into the LUMO of the substrate. The LUMO for CH_3Cl is shown in Fig. 14-28. A successful encounter between CH_3Cl and a base results in a bond between the base and the carbon atom, so the HOMO of the base needs to overlap the p AO of carbon in the LUMO of Fig. 14-28a. Attack at the position marked "1" in the figure is unfavorable because any base MO would be near a nodal surface, yielding poor overlap with the LUMO. Therefore, attack at site 2 is favored. As the previously empty LUMO of CH_3Cl becomes partially occupied, we expect a loss of bonding between C and Cl. Also, negative overlap between the forming C-base bond and the three "backside" hydrogens should encourage the latter to migrate away from the attacked side, as indicated in Fig. 14-28b.

The tendency of a high-energy, occupied MO of the base to couple strongly with the LUMO of a molecule like CH_3Cl depends partly on the energy agreement between these MOs. If they are nearly isoenergetic, they mix much more easily and give a bonded combination of much lower energy. Molecules where the HOMO is high tend to be polarizable bases. A high-energy HOMO means that the electrons are not very well bound and will easily shift about to take advantage of perturbations. Such bases react readily with molecules having a low-energy LUMO (Fig. 14-29a). This corresponds to a "soft-base-soft-acid" interaction in the approach of Pearson [17]. When the HOMO and LUMO are in substantial energy disagreement, orbital overlap becomes less important as a controlling mechanism, and simple electrostatic interactions may dominate. This

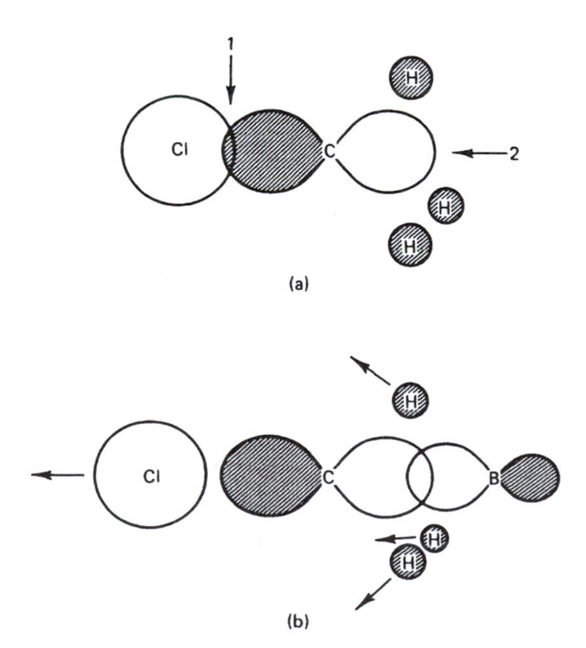

Figure 14-28 ► (a) The LUMO of CH_3Cl. (b) Positive overlap between HOMO of base B and LUMO of CH_3Cl increases antibonding between C and Cl and also repels H atoms from their original positions.

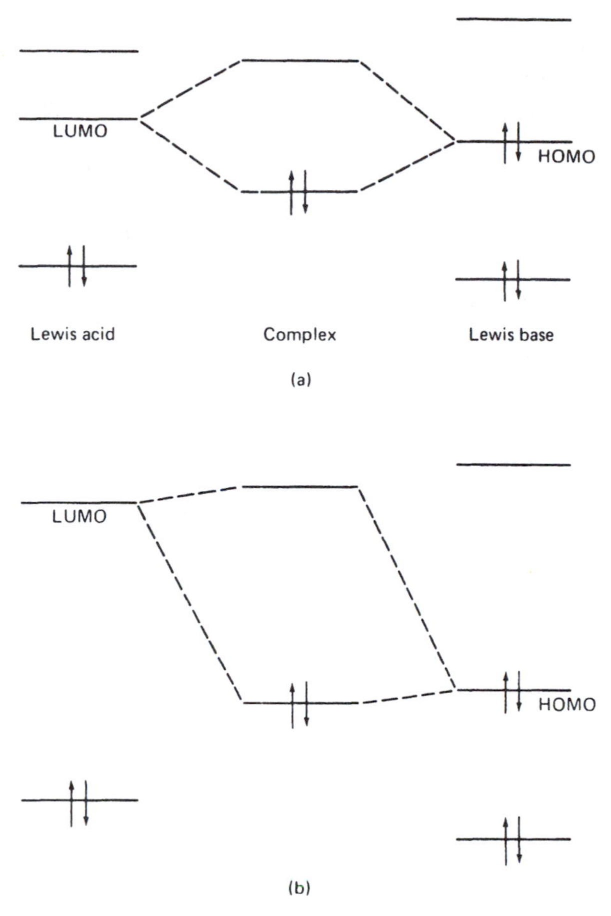

Figure 14-29 ▶ LUMO–HOMO interactions and splitting for (a) nearly degenerate levels, (b) well-separated levels.

is a "hard-acid-hard-base" situation. Thus, we expect QMOT rules to apply to soft-soft, rather than hard-hard interactions.

The reader may, by this time, begin to appreciate the very wide scope of QMOT and the large number of variations of a common theme that have been used. In this chapter we have given only a few representative examples. We have not described all the variations or all types of application. For fuller treatment, the reader should consult specialized books on this subject, some of which have been referred to in this chapter [18].

14-9.A Problems

14-1. Calculate to first order the electronic energy of a hydrogen atom in its 1s state and in the presence of an additional proton at a distance of 2 a.u. What is the *total* energy to first order? Repeat for distances of 1 and 3 a.u. (See Appendix 3.)

14-2. Evaluate and graph the effects of dividing $H_{aa} \pm H_{ab}$ by $1 \pm S$ for each of the following cases: $H_{aa} = 0, -5, -10, +10, -20$. In each case, let $H_{ab} = -5$, $S = 0.5$. Does the QMOT rule that antibonding interactions are more destabilizing than bonding interactions are stabilizing apply in all cases? Is the bonding level always the lower of the two? Does the QMOT expectation appear to be better followed by very low-energy levels, or by higher-energy levels?

14-3. Table P14-3 is a list of electron affinities (in electron volts) of certain molecules and atoms. Can you rationalize the molecular values relative to the atomic values using QMOT ideas?

TABLE P14-3 ▶ Atomic and Molecular Electron Affinities (in electron volts)

H (0.75)	Cl (3.61)	C_2 (3.4)	O_2 (0.45)	Cl_2 (2.38)
C (1.26)	Br (3.36)	CN (3.82)	F_2 (3.08)	Br_2 (2.6)
N (0.0 ± 0.2)	I (3.06)	N_2 (-16)	S_2 (1.67)	I_2 (2.55)
O (1.46)	S (2.08)	CO (< -1.8)	SO (1.13)	ICl (1.43)
F (3.40)	H_2 (~ -2)	CS (0.21)	FCl (1.5)	IBr (2.6)

14-4. For some time, it was uncertain whether the ground states of CH_2 and NH_2^+ are singlets $(2a_1)^2$ or triplets $(2a_1)(1b_1)$. The ground-state geometries of these systems have HAH angles of 134°(CH_2) and 140–150°(NH_2^+). Based on other data described in the text for HAH systems, would you say these angles are more consistent with a singlet or a triplet ground state? Assuming that the first excited state is the other multiplicity, should the first excited state be more or less bent than the ground state?

14-5. Based on Fig. 14-8, what should happen to the geometry of H_2O upon $3a_1 \leftarrow 1b_1$ excitation?

14-6. Carry out EHMO calculations for CH_2 at HCH angles of 180°, 150°, 120°, and 90°. (Use a constant C–H bond distance of about 1 Å in all cases.) From an examination of the MO coefficients, sketch and assign symmetry symbols to each MO. Plot the energies versus angle. Critically discuss your computed results compared to Fig. 14-8. If there are differences, try to rationalize them. (Remember that any EHMO orbital energy change can be analyzed in terms of Mulliken population changes, as discussed in Chapter 10.)

14-7. Using the EHMO energy formula [Eq. (10-25)], analyze the contributions to the orbital energy change between 180° and 120° that you calculated for the $1b_2$ MO in Problem 14-6. What percentage of the energy change comes from loss of overlap between 1s and 2p AOs? From antibonding between hydrogens? Compare this with the discussion in the text.

14-8. If overlap between an s AO and a p AO goes as $\cos\theta$ (for constant and finite R) (Fig. P14-8), what is the mathematical expression for the rate of change of overlap with angle? Calculate the effect on overlap of a 30° shift, starting from $\theta = 0$. Calculate the effect of a 30° shift from $\theta = 90°$.

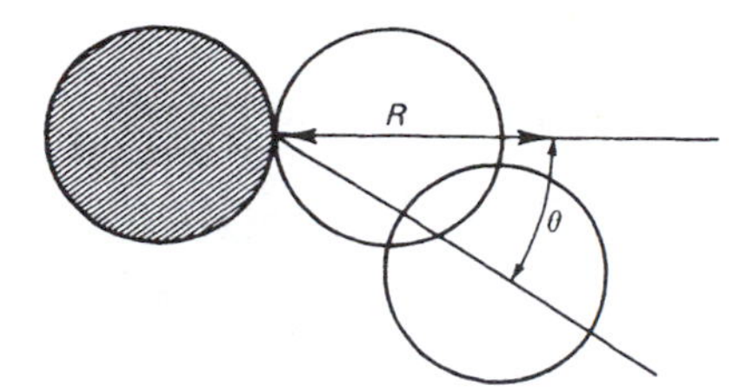

Figure P14-8 ▶

14-9. Use symmetry to help establish an energy level pattern and MO sketches for planar AH_3 (equilateral triangular). Use QMOT rules to produce the correlation diagram for planar versus pyramidal AH_3. Based on your reasoning, which of the following should be planar? $BH_3, CH_3^+, BeH_3^-, NH_3, PH_3, H_3O^+, CH_3^-$. Can you think of any other shapes that might be examined as possibilities for AH_3 systems?

14-10. Use an EHMO program to generate MOs for CO_2 at 180°, 150°, 120°, and 90°. Sketch the MOs, characterize their symmetries, and construct an orbital energy correlation diagram for this molecule. Indicate what causes each MO energy to rise or fall. Based on your figure, would you expect the following molecules to be linear or bent? $BeCl_2, C_3, CO_2, N_3^-, NO_2, O_3, F_2O$.

14-11. Should a $\pi_u \leftarrow \pi_g$ electronic transition for ozone cause the O–O–O angle to increase or decrease according to Walsh-type arguments? Sketch the MOs and indicate your reasoning. [The notation π_g and π_u refers to the MOs in the linear molecule. For the equilibrium bent structure, these MOs are: $\pi_g \rightarrow a_2, b_2; \pi_u \rightarrow a_1, b_1$.]

14-12. Consider the electrocyclic reaction wherein the allyl *anion* closes to form a cyclopropenyl π anion (**VIII**). (The negative charge in the cyclopropyl anion may be thought of as resulting from double occupancy of a p–π AO on the singly protonated carbon.) Sketch the orbitals being formed or destroyed in

(VIII)

this process. Determine the orbital symmetries for the symmetry operations conserved in conrotatory and disrotatory modes of closure. Set up an orbital correlation diagram and decide which mode is more likely for thermal and photochemical reactions.

14-13. Construct an orbital correlation diagram for the "broadside" 2 + 2 cycloaddition of two acetylenes to form cyclobutadiene (**IX**). Is the reaction likely to proceed

(IX)

through an intermediate of square planar geometry? Assuming this geometry, would reaction be easier thermally or photochemically?

14-14. For a system having no symmetry elements (except E), what is the result of the noncrossing rule?

14-15. a) Construct an orbital correlation diagram for the $2 + 4$ cycloaddition (Diels–Alder) reaction discussed in the text.
b) Construct a *state* correlation diagram for this reaction.

14-16. Based on the orbital relations discussed in the text and extensions of these relations to other cases, formulate generalized verbal rules (Woodward–Hoffmann rules) for cycloadditions and sigmatropic shift reactions.

References

[1] D. R. Yarkony and H. F. Schaefer, III, *J. Chem. Phys.* **61**, 4921 (1974).

[2] K. Ruedenberg, *J. Chem. Phys. 66*, 375 (1977).

[3] Ionization potentials, appearance potentials, and heats of formation of gaseous positive ions, NSRDS-NBS 26, Natl. Bur. Stand. (1969).

[4] A. D. Walsh, *J. Chem. Soc.* p. 2260 (1953).

[5] R. S. Mulliken, *Rev. Mod. Phys.* **14**, 204 (1942).

[6] B. M. Gimarc, *Accounts Chem. Res.* **7**, 384 (1974).

[7] R. J. Buenker and S. D. Peyerimhoff, *Chem. Rev.* **74**, 127 (1974).

[8] T. E. Sharp, *Atomic Data* **2**, 119 (1971).

[9] J. P. Lowe, *J. Amer. Chem. Soc.* **92**, 3799 (1970).

[10] R. M. Stevens, *J. Chem. Phys.* **52**, 1397 (1970).

[11] H. Fujimoto and K. Fukui, in *Chemical Reactivity and Reaction Paths* (G. Klopman, ed.). Wiley (Interscience), New York, 1974.

[12] R. B. Woodward and R. Hoffmann, *J. Amer. Chem. Soc.* **87**, 395 (1965). See also R. Hoffmann, *Angew. Chem. Int. Ed.* **43**, 2, 2004, for historical insights.

[13] R. B. Woodward and R. Hoffmann, *The Conservation of Orbital Symmetry*. Academic Press, New York, 1970.

[14] H. C. Longuet-Higgins and E. W. Abrahamson, *J. Amer. Chem. Soc.* **87**, 2045 (1965).

[15] R. Hoffmann and R. B. Woodward, *J. Amer. Chem. Soc.* **87**, 2046 (1965).

[16] R. B. Woodward and R. Hoffmann, *J. Amer. Chem. Soc.* **87**, 2511 (1965).

[17] R. G. Pearson, ed., *Hard and Soft Acids and Bases*. Dowden, Hutchison, and Ross, Stroudsburg, PA, 1973.

[18] B. M. Gimarc, *Molecular Structure and Bonding*. Academic Press, New York 1979.

Chapter 15

Molecular Orbital Theory of Periodic Systems

15-1 Introduction

A structure is periodic in space, and hence has a *periodic potential*, if a subunit of the structure can be found that generates the entire structure when it is repeated over and over while traversing one or more spatial coordinates. Thus, a regular polymer can be "generated" mentally by translating a unit cell along an axis, sometimes with accompanying rotations. A crystalline solid results from such translations along three coordinates. In an analogous sense, some *molecules* are periodic. Benzene is an example since it can be generated by rotating a C–H unit in 60° increments about a point which ultimately becomes the molecular center.

We have already seen (Chapters 8 and 13) that the orbital energies, degeneracies, and coefficients in benzene are to a great extent determined by symmetry. But these symmetry constraints can also be viewed as resulting from the cyclic "periodicity" of benzene, and this finite cyclic periodicity is, in essential ways, like the extended infinite periodicity of, say, regular polyacetylene or of graphite. It is reasonable, therefore, that some of the aspects of the energies and orbitals in these *extended* periodic structures are closely related to those in *cyclic* periodic systems.

The concepts commonly used to describe periodic polymers, surfaces, or solids are closely related to those we have already developed for molecules, but their terminology and depiction are not familiar to most chemists. Our goal in this chapter is to develop an understanding of these concepts by progressing from more familiar cyclic systems, like benzene, to polymers, and to show how qualitative MO concepts apply to periodic systems in general.

15-2 The Free Particle in One Dimension

We will first examine the particle moving parallel to the x coordinate with no variation in potential energy. (Let $V = 0$.) This system is periodic because every segment of x is the same as every other, and it provides a convenient starting point for discussion of periodic systems in general.

The quantum-mechanical solutions for this system are discussed in Chapter 2. The relevant points are:

1. All non-negative energies are possible because there are no boundary conditions (beyond the conditions applying to well-behaved functions).

2. Except for $E = 0$, all energy levels are doubly degenerate.

3. The pair of linearly independent wavefunctions associated with each doubly degenerate energy E can be combined in an infinite number of ways to produce different resultant wavefunction pairs. Two of these pairs are especially convenient. These are

$$\psi_+ = \exp\left[\frac{\sqrt{2mE}}{\hbar}ix\right] \tag{15-1a}$$

$$\psi_- = \exp\left[\frac{-\sqrt{2mE}}{\hbar}ix\right] \tag{15-1b}$$

and

$$\psi_{\sin} = \sin\left[\frac{\sqrt{2mE}}{\hbar}x\right] \tag{15-2a}$$

$$\psi_{\cos} = \cos\left[\frac{\sqrt{2mE}}{\hbar}x\right] \tag{15-2b}$$

4. The exponential forms [Eqs. (15-1ba, b)] are also eigenfunctions for the momentum operator, $\hat{p}_x = (\hbar/i)d/dx$. Thus ψ_+ corresponds to a particle moving parallel to the x axis with momentum $+\sqrt{2mE}$ (i.e., toward $x = +\infty$), and ψ_- corresponds to motion toward $x = -\infty$. We can associate the double degeneracy of the energies with the fact that there is no difference in the potential felt by the particle, regardless of whether it moves from left to right or right to left: The system has two equivalent directions.

5. The real forms [Eqs. (15-2ba, b)] give oscillating particle distributions (except when $E = 0$). In a given pair, the sine puts maximum particle density where the cosine puts its minimum, and vice versa. As E increases, the frequency of nodes increases. These wavefunctions are not "pure" momentum states because they are not eigenfunctions of $\hat{p}_x$ (except when $E = 0$). They are formed by mixing of the exponential forms. However, these functions are sometimes more convenient to work with, especially pictorially, and they are just as good as the exponential forms as long as we are concerned with *energy* rather than momentum.

Equations (15-1) and (15-2) are sometimes written with $\sqrt{2mE}/\hbar$ replaced by the symbol k, with k having any value from zero to infinity. Evidently, k *in this system* is proportional to particle momentum, hence to the square root of the kinetic energy, T. (Since $V = 0$, the kinetic energy is identical to the total energy, E.) For a free particle, de Broglie's relation holds, so p is proportional to $1/\lambda$, where λ is the de Broglie wavelength. This means that k is proportional to the number of de Broglie wavelengths

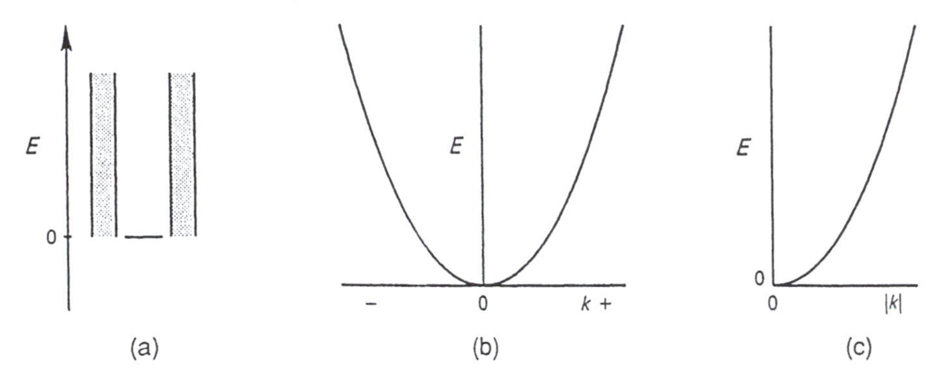

Figure 15-1 ▶ Three ways of picturing the energies of the free particle. (b) and (c) show the dependence of E on k. In (c), the energies are understood to be degenerate (except at $E=0$).

per unit of distance. Consequently, k is often referred to as the *wavenumber*. It is the de Broglie-wave-equivalent to the wavenumber of light or of any other classical harmonic wave.

Summarizing, *for the free particle*,

$$k \propto p_x \propto \sqrt{E} = \sqrt{T} \propto 1/\lambda \tag{15-3}$$

There are several ways in which the solutions of the Schrödinger equation for periodic structures are displayed graphically. One way is the familiar one of drawing all the energy levels against a vertical energy scale, as in Fig. 15-1a. (Since all energies are allowed for the free particle, this gives a single energy at $E=0$ and a degenerate continuum, or "band", from $E>0$ to ∞, rather than a series of discrete lines.) Alternatively, we can plot E versus k to obtain the parabolic plot of Fig. 15-1b. This can be simplified, though, by plotting only one arm of the parabola (Fig. 15-1c), making implicit the fact that, for each solution with wavenumber k, there is a degenerate solution with wavenumber $-k$. Graphs that relate energy to k for periodic polymers, surfaces, or solids are called *band diagrams*. Figure 15-1c is the band diagram for the one-dimensional free particle.

Another quantity of great physical importance is the *number* of states near a particular energy value. This is called the "density of states" (DOS). If we consider Fig. 15-1c, we can see that the states associated with $0<|k|<1$ all lie within a certain range which we call $E_0 - E_1$ in Fig. 15-2a. If we assume that all values of k are equally likely (we show this to be true in Section 15-5), then there is an equal (infinite) number of states associated with $1<|k|<2$, and these lie in the larger range $E_1 - E_2$. Hence, the states in the range $E_1 - E_2$ are less densely packed than those in the range $E_0 - E_1$. It is not hard to show that the density of states for the free particle drops off as $1/\sqrt{E}$, as plotted in Fig. 15-2b.

Ultimately we will consider one-dimensional periodic structures (polymers) with *varying* potentials caused by the presence of nuclei and other electrons, and we will find that these systems retain some of the above features. In particular, degeneracy due to directional equivalence and choice of real or complex forms for orbitals persist. Also, the quantity k retains its meaning as a wavenumber (in a restricted sense). However, k loses its simple relation to momentum and energy (because the electrons possess potential energy as well as kinetic energy), and, typically, certain energies become disallowed. This changes the infinite band to band segments separated by gaps in energy.

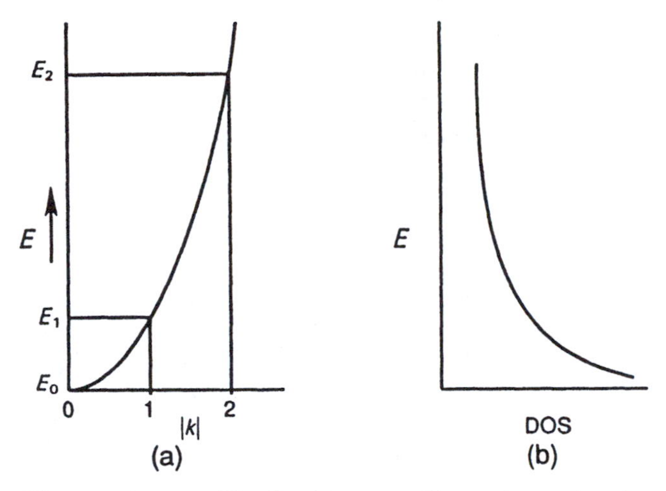

Figure 15-2 ▶ The states between $|k| = 0$ and 1 are equal in number to those between $|k| = 1$ and 2, but the former set is packed into a smaller energy range, producing a greater density of states.

15-3 The Particle in a Ring

A particle of mass m constrained to move in the angular coordinate ϕ about a ring of radius r with $V(\phi) = 0$ is the *cyclic* analog of the free particle. This system, also described in Chapter 2, is similar in many ways to the one just discussed. Except for the case of $E = 0$, all solutions are doubly degenerate and describable with real (trigonometric) or complex (exponential) functions. The quantum number j (we will use j in cyclic systems, k in linear systems) is proportional to the angular momentum, to the root of the energy, to the reciprocal wavelength, and to the number of de Broglie waves in one circuit of the ring. However, since the number of waves in the ring must be integral, we have a periodic boundary condition that restricts j to integer values and leads to a discrete energy spectrum as opposed to the continuous spectrum of the free particle.

The diagrams summarizing relations between energies, j, and number of states are collected in Fig. 15-3.

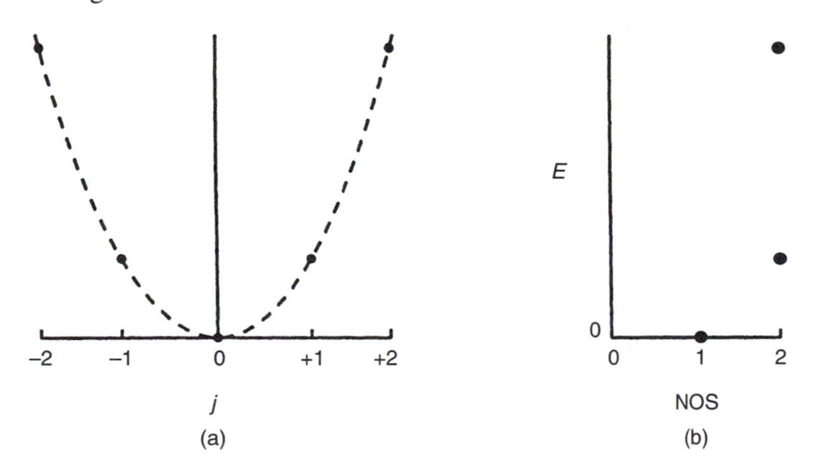

Figure 15-3 ▶ (a) The energy for a particle in a ring has parabolic dependence on j, but exists only when j is an integer. (b) The number of states versus energy. This is the discrete-state analog of the density-of-states plot for a continuum of energies.

Cyclic structures with *varying* potentials exist (e.g., benzene), and such systems retain the degeneracy and real or complex orbital features. Also, j continues to have meaning (restricted) with respect to wavelength and number of nodes. However, angular momentum and energy are no longer simply related to j.

15-4 Benzene

We next examine benzene—a cyclic system having a nonuniform potential. We consider only the π electrons as described by the simple Hückel method, since the features we wish to point out are already present at that elementary level. A formula is presented (without derivation) in Chapter 8 [Eq. (8-52)] for the MOs of molecules like benzene. This gives, for the jth MO of benzene,

$$\phi_j = \sum_{n=1}^{6} \{6^{-1/2} \exp[2\pi i j(n-1)/6]\} 2p_\pi(n), \qquad j = 0, 1, \ldots, 5 \tag{15-4}$$

where $i = \sqrt{-1}$ and $2p_\pi(n)$ is a $2p_\pi$ atomic orbital centered on the nth carbon atom. The energy for MO ϕ_j is equal to [Eq. (8-51)]

$$\epsilon_j = \alpha + 2\beta \cos(2\pi j/6). \tag{15-5}$$

If $j = 0$, all the exponentials equal one and Eq. (15-5) gives $\epsilon_0 = \alpha + 2\beta$, which can be recognized as the lowest-energy π MO, and Eq. (15-4) gives $\phi_0 = (1/\sqrt{6})[2p_\pi(1) + 2p_\pi(2) + 2p_\pi(3) + 2p_\pi(4) + 2p_\pi(5) + 2p_\pi(6)]$. This gives all the $2p_\pi$ AOs the same phase, which is what we expect for the totally bonding, lowest-energy π MO.

If $j = 1$, $\epsilon_1 = \alpha + 2\beta\cos(\pi/3) = \alpha + 2\beta(1/2) = \alpha + \beta$. This is an energy in the second-lowest level, which is a degenerate level. The corresponding MO is $\phi_1 = (1/\sqrt{6})\{[\exp(0)]2p_\pi(1) + [\exp(\pi i/3)]2p_\pi(2) + [\exp(2\pi i/3)]2p_\pi(3) + [\exp(\pi i)] \times 2p_\pi(4) + [\exp(4\pi i/3)]2p_\pi(5) + [\exp(5\pi i/3)]2p_\pi(6)\}$. This somewhat formidable-looking MO is complex. Because ϕ_1 is one of a degenerate *pair* of MOs (the other one is ϕ_5), we can mix the complex MOs to obtain real ones that are still eigenfunctions. The real forms of the benzene MOs are described in Section 8-10D. The MO formulas produced by Eq. (15-4) are real for nondegenerate cases but are usually complex for degenerate MOs. However, it is always possible to mix any pair of degenerate complex MOs to produce a pair of degenerate real MOs. Notice that the complex MOs place the same electron density at each carbon whereas the real forms do not. The real MOs show nodes at various points in the ring (see Fig. 8-13) just as the real wavefunctions for the free particle show nodes at various points in x.

In these formulas, j is restricted to the integer values ranging from 0 to 5, giving six MOs. If one tries other integer values of j in formulas (15-4) and (15-5), one simply reproduces members of the above set. Indeed, *any* six sequential integer values for j produces the same set of solutions as does the sequence 0–5. In particular, the set $-2, -1, 0, 1, 2, 3$ is perfectly acceptable and provides a match with the conventions of solid-state physics and chemistry.

The energies for the six unique MOs of benzene are reproduced over and over again if we allow the index j to run beyond the specified range. This is depicted in Fig. 15-4a, and it is easy to see that the same six energies result for any six contiguous j values.

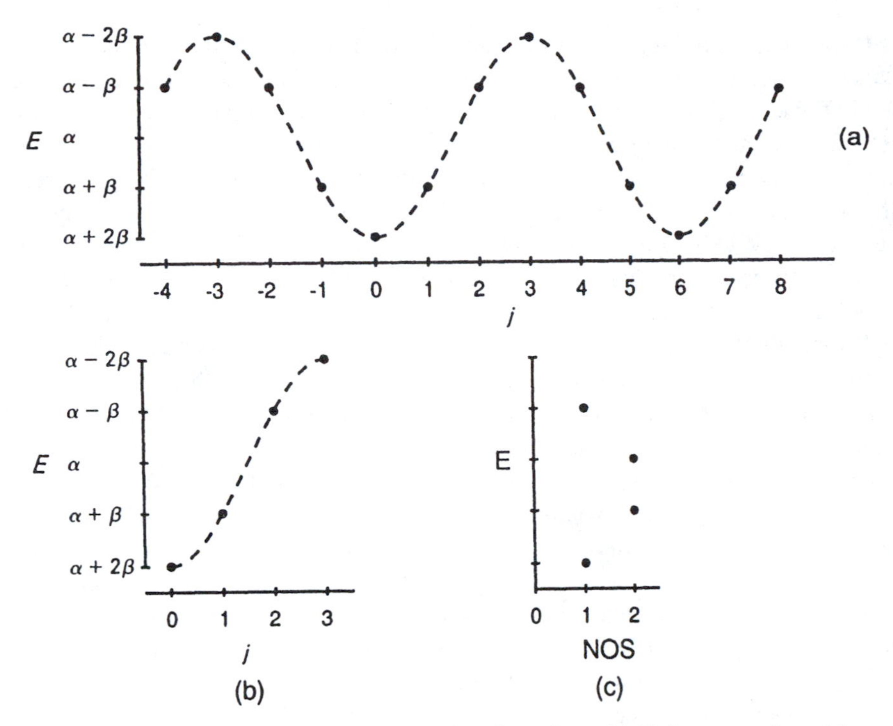

Figure 15-4 ▶ (a) Energies for benzene as a function of j. (b) Unique energies of benzene. (c) Number of states at each energy. All data refer to π energies at the simple Hückel level.

The set from -2 to 3, alluded to above, can be replotted in condensed form over the range 0–3 (Fig. 15-4b) if we keep in mind the fact that the energies are degenerate except for the first and last. The "number of states" diagram for benzene is sketched in Fig. 15-4c.

If we consider a monocycle with thousands of carbon atoms, Eq. (15-5) gives (condensed) results as sketched in Fig. 15-5a. The very large number of energies still lie in the range $\alpha + 2\beta$ to $\alpha - 2\beta$ with the cosine wave now stretched over a much larger range of integers. The well-separated points of Fig. 15-4b coalesce into the line of

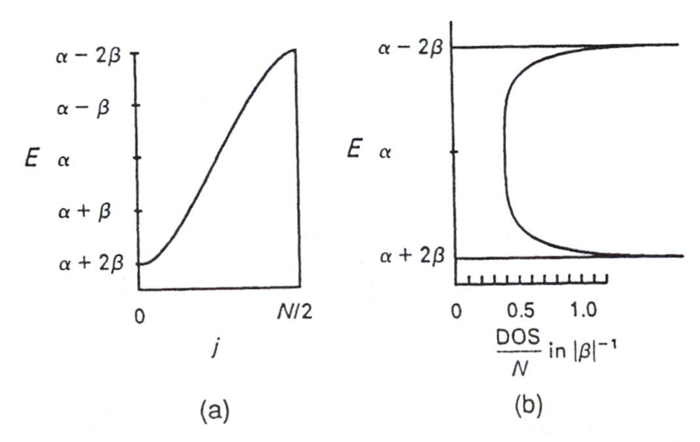

Figure 15-5 ▶ (a) Hückel energies and (b) density of states for a very large number of carbons in a cycle.

Fig. 15-5a, which now appears almost continuous. The shape of the curve in Fig. 15-5a makes it evident that the states around $j = 0$ and $j = N/2$ are closer together in energy than are those around $j = N/4$, giving the density of states curve shown in Fig. 15-5b.

There is a qualitative difference between what we found for the particle in a ring and benzene. In the former case, k can increase without limit, and energy keeps increasing as the square of k. In benzene, increasing j beyond the prescribed range simply causes the energy to cycle back and forth between $\alpha + 2\beta$ and $\alpha - 2\beta$. Why do these systems behave so differently in this regard? The energy of the free particle increases as we fit more and more waves into a circle of fixed radius, obtaining, therefore, shorter de Broglie wavelengths. The Hückel energy of a benzene π MO depends on the extent of bonding and antibonding character between adjacent π AOs. When j is zero, all the interactions are bonding and $\epsilon = \alpha + 2\beta$. As j increases, bonding interactions disappear, ultimately to be replaced by antibonding interactions. At $j = 3$, all interactions are antibonding and $\epsilon = \alpha - 2\beta$. It certainly makes sense that this is the highest energy an MO can have because this is the most antibonding arrangement imaginable. But what happens *mathematically* to make this work out? Let us examine the exponential functions in Eq. (15-4) for two values of j, say $j = 3$ and $j = 9$, to see if we can resolve this question. An immediate problem confronts us: The function is complex, and sketching a complex function is not convenient. However, because the complex MOs can always be mixed to form real MOs, we can let the mixing occur within Eq. (15-4) itself to give sine and cosine equivalents. For example,

$$\phi_{j\cos} = \sum_{n=1}^{6} \{6^{-1/2} \cos[2\pi j(n-1)/6]\} 2p_\pi(n). \tag{15-6}$$

(For some values of j, this expression will yield MOs that are not normalized.) Equation (15-6) tells us that MO ϕ_j has six terms, each term being a $2p_\pi$ AO times a coefficient. The values of the coefficients are given by the term in curly brackets. This term produces a discrete set of values, since n is a discrete set of integers, but we can sketch a *continuous* function [by replacing $(n-1)/6$ with the continuous variable ϕ] and then locate the places on this "coefficient wave" that correspond to the discrete points of interest. Sketches for this "coefficient wave" when $j = 3$ and $j = 9$ are shown in Fig. 15-6. The special points of interest, where actual coefficient values are given

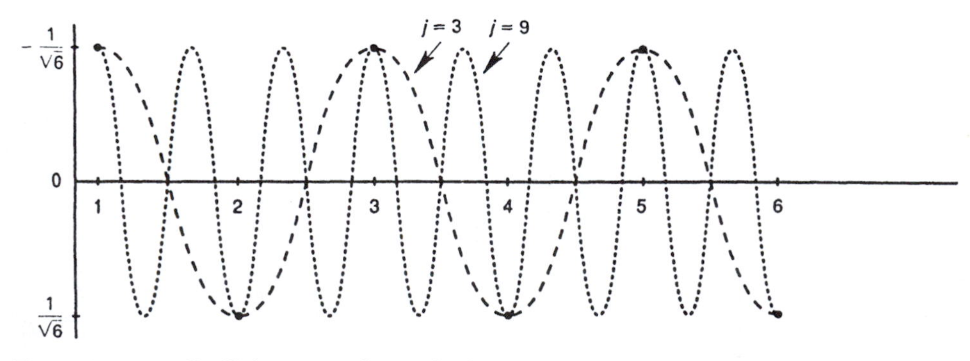

Figure 15-6 ▶ Coefficient waves for $j = 3$ (*dashed curve*) and $j = 9$ (*dotted curve*). Both curves intercept the same coefficient values ($\pm 1/\sqrt{6}$) at the positions related to carbon atoms 1–6, so the curves produce the same MO.

for benzene MOs, are also shown (at the positions labeled 1–6), and it is obvious that the same set of coefficients is produced by each function. Increasing j has caused the coefficient wave to oscillate with a shorter wavelength, but the extra oscillations do not register because they occur *between* the special points of interest where we are sampling the function (i.e., at the points where the carbon nuclei reside). In other words, the extra oscillations in our benzene coefficient functions are just "empty wiggling" and have no effect on the MO, so we cut off the j range where the meaningful wiggling starts to become empty wiggling. The reason that the particle in a ring does not show this behavior is that the exponential (or sine, cosine) functions in that system actually *are* the wavefunctions, while in benzene these functions are sampled only at discrete points where AOs are located, and the resulting coefficients are then used to produce MOs.

Notice that the MOs for benzene as given by Eq. (15-4) are produced from an equation having the following form: There is an exponential term that, for each MO, supplies the coefficients for various carbons in the molecule, and there is a basis set of functions located on the various carbons. This basis set has the same "periodicity" as does the molecule. (In the case of benzene, we have so far taken this to be six identical $2p_\pi$ AOs.) *Wavefunctions for all periodic systems have this same form—an exponential (or equivalent trigonometric) expression times a periodic basis.* This is the content of *Bloch's theorem*, which we prove in the next section.

EXAMPLE 15-1 What coefficients are generated for a benzene MO by a coefficient wave of the type shown in Fig. 15-6 with $j=-3$? $j=0$?

SOLUTION ▶ The arguments of the cosine function of Eq. (15-6) will be the negative of those of the $j=3$ case. Because the cosine is symmetric about an argument of zero, we obtain the same curve for $j=-3$ as for $j=+3$, hence the same MO coefficients. This demonstrates that the MO generated by $j-6$ is the same as that generated by j. For $j=0$, the cosine arguments all vanish, the cosines all equal $+1$, and the curve of Fig 15-6 becomes a straight line at a coefficient value of $+1/\sqrt{6}$. This produces the lowest-energy MO of benzene. ◀

15-5 General Form of One-Electron Orbitals in Periodic Potentials—Bloch's Theorem

We will establish in this section the general mathematical nature of eigenfunctions for periodic hamiltonians. This is the content of *Bloch's theorem*. The actual proof will be carried through for *cyclic* "periodic" systems, and those results will then be extended to noncyclic periodic systems. The rationale for this is that a nearly infinite linear periodic structure can be treated as cyclic without introducing error. That is, a sufficiently long extended chain of atoms can be assumed to have the same wavefunctions and energies as the same chain joined end to end to form a cycle. For short chains and rings, this is not the case; the lowest-energy MO for hexatriene is not as low in energy as that for benzene, nor is it uniform over the whole chain (as it is in benzene). But for long enough chains, the difference becomes negligible.

Bloch's theorem, as it applies to periodic cycles, states that eigenfunctions have the form

$$\Psi_j(\phi)=\exp(ij\phi)U_j(\phi) \tag{15-7}$$

where $i=\sqrt{-1}$, j is an integer and $U_j(\phi)$ has the periodicity of the cycle. If the cycle belongs to the C_n point group, then $U(\phi)=U(\phi+2\pi/n)$. Functions of the form of Eq. (15-7) are called *Bloch functions*. It was stated in Section 13-8 that every wavefunction or orbital is a member of a basis for an irreducible representation for the point group of the system, so it follows that the wavefunctions we seek to characterize have to be bases for irreducible representations of the C_n point group. These groups are discussed in Section 13-11, where it is shown that the functions $\epsilon=\exp(i\phi)$ and $\epsilon^*=\exp(-i\phi)$ are bases for irreducible representations for these groups, which is to say that they are eigenfunctions for the rotation operator, no matter what the value of the integer n in C_n. We consider the more general functions $\exp(\pm ij\phi)$ with j an integer (so that the function obeys the cyclic continuity condition) and ask what happens when such functions are subjected to a rotation. Let the rotation operator be symbolized $\mathbf{C}_n^q$ for a rotation of q times $2\pi/n$. The n-fold cyclic system is invariant to this rotation if q is an integer. It is not difficult to show that (see Section 13-11)

$$\mathbf{C}_n^q f(\phi)=f(\phi-2\pi q/n) \tag{15-8}$$

(For example, clockwise rotation by 60° brings to each point in the circle the value that used to be 60° in the counterclockwise direction.) Then

$$\begin{aligned}\mathbf{C}_n^q \exp(ij\phi) &= \exp[ij(\phi-2\pi q/n)]\\ &= \exp(-2\pi iqj/n)\exp(ij\phi)\end{aligned} \tag{15-9}$$

Equation (15-9) shows that $\exp(ij\phi)$ is an eigenfunction for the rotation operator, so $\exp(ij\phi)$ is a basis for an irreducible representation, with j any integer. We can build additonal flexibility into this eigenfunction if we write an exponential function in a more complicated way: $\exp[i(j+nN)\phi]$ with N also an integer. We know that this must still be a basis function because $(j+nN)$ must still be an integer. If we operate with $\mathbf{C}_n^q$, we find that (Problem 15-6)

$$\mathbf{C}_n^q \exp[i(j+nN)\phi]=\exp(-2\pi iqj/n)\exp[i(j+nN)\phi] \tag{15-10}$$

This shows that our newest, most general, exponential has the *same eigenvalue* for $\mathbf{C}_n^q$ *no matter what integer value we choose for* N. [Observe that the eigenvalue in Eq. (15-10) depends on q, j, and n, but not N.] Therefore, since we have an unlimited number of choices for the integer N, we have an unlimited number of *degenerate* eigenfunctions for $\mathbf{C}_n^q$ for each choice of j. We can mix together such degenerate eigenfunctions in any way we please and still have an eigenfunction. Let us take the linear combination

$$\sum_{N=-\infty}^{\infty} A_N \exp[i(j+nN)\phi] \tag{15-11}$$

This can be factored into the form

$$\exp(ij\phi)\sum_{N=-\infty}^{\infty} A_N \exp(inN\phi) \tag{15-12}$$

The sum is a function of ϕ which we can call $U(\phi)$. It has the same periodicity in ϕ as the cycle because n is equal to the number of identical cells in the cycle and N is an integer. *It is actually the general Fourier series expansion of a periodic function.* For $N = 0$, the contribution is constant (A_0) at all ϕ. For $N = \pm 1$, the functions repeat n times around the cycle and clearly have the periodicity of the system. For $N = \pm 2$ the functions repeat $2n$ times, so they obviously still have the periodicity of the system. Since all the individual members of the sum have the periodicity of the system, so does the sum itself. The specific nature of the resultant function $U(\phi)$ depends on the choice of coefficients A_N.

We have shown that

$$\exp(ij\phi)U(\phi) \tag{15-13}$$

with $U(\phi)$ n-fold periodic is a general mathematical form for bases for irreducible representations for the C_n group. But wavefunctions, $\psi(\phi)$, for a periodic potential must also be bases for such irreducible representations. Therefore, wavefunctions must be expressible in this form:

$$\psi_j(\phi) = \exp(ij\phi)U(\phi), \qquad j = 0, \pm 1, \pm 2, \ldots \tag{15-14}$$

We are *not* claiming that *all* functions of the form Eq. (15-13) are wavefunctions for the n-fold periodic potential. Rather, we are claiming that all such wavefunctions belong to the class of functions represented by Eq. (15-13). Only certain choices of $U(\phi)$ will serve to make Eq. (15-13) become a wavefunction for the system, and the choice of an appropriate function $U(\phi)$ is not necessarily the same at different values of j. To make this dependence on j explicit, we include it as a subscript on U:

$$\psi_j(\phi) = \exp(ij\phi)U_j(\phi), \qquad j = 0, \pm 1, \pm 2, \ldots \tag{15-15}$$

with $U_j(\phi)$ n-fold periodic. This completes the proof of Bloch's theorem for periodic cycles.

Before extending this result to noncyclic systems, a number of comments and clarifications should be made. First, the function $U_j(\phi)$ may still seem to be something of a mystery. Think of it this way. For benzene π MOs we know we need to create a basis set having two properties: It should be like a $2p_\pi$ AO near any carbon atom, and it should be the same at each carbon atom. Bloch's theorem does not comment on whether U should look like a $2p_\pi$ AO at a carbon atom, but it does require that it be the *same* at each carbon atom, that it be the same at the midpoints between adjacent carbons, that it be the same at a point one bohr above each carbon atom, etc. In short, U must be symmetric for the six-fold rotation operation. It is up to *us* to figure out what that six-fold symmetric basis set should look like in detail, and to construct U appropriately.

Second, U depends on j. What does this mean? Suppose we compare the lowest- and highest-energy π MOs for benzene. The lowest is totally bonding and the highest is totally antibonding. Ordinarily, we use the identical set of $2p_\pi$ AOs as basis functions for both MOs, but this is not required. Perhaps the p_π AOs for the bonding MO would be more appropriately chosen to be slightly larger, so they overlapped better, and maybe those for the antibonding MO should be slightly smaller, to reduce antibonding interactions. These objectives could be achieved, for example, by making U a mixture of $2p_\pi$ and $3p_\pi$ AOs on each center and letting the nature of the mixture depend on j. This makes for a more involved variational calculation, so it is normally not included in cases like this one where such effects are expected to be small. ($3p_\pi$ character should not mix in to a significant extent because the $3p_\pi$ AO is considerably higher in energy than the $2p_\pi$ AO.) However, there are many cases where a unit cell contains several AOs in the same symmetry class and of similar energy (e.g., $2s, 2p_\sigma$), and it is then quite important to allow the mix of these to vary with j. A given band, for example, could start out being mainly $2s$ at low energies and end up being mainly $2p_\sigma$ at the high-energy end. This means that the appropriate description of the function U must be redetermined for each choice of wavenumber j.

Third, careful comparison of Eqs. (15-15) and (15-4) shows that they are not exactly the same. Equation (15-15) instructs us to find a periodic function $U_j(\phi)$ and multiply it by $\exp(ij\phi)$ at *every point in* ϕ. Think of the sine or cosine related to the exponential and imagine what this means as we multiply it times a $2p_\pi$ on some carbon. Say the cosine is *increasing* in value as it sweeps clockwise past the carbon nucleus at 2:00 on a clock face. This produces a product of cosine and $2p_\pi$ that is unbalanced—smaller toward 1:00 than toward 3:00, because the cosine wave modulates $U_j(\phi)$ *everywhere*. But Eq. (15-4) is different. It instructs us to take the value of the cosine at 2:00 and simply multiply the $2p_\pi$ AO on that atom by that number. The $2p_\pi$ AO is not caused to become unbalanced. Only its *size* in the MO is determined by the cosine. Equation (15-4) is called a *Bloch sum*. Such sums are approximations to Bloch functions, but any errors inherent in this form are likely to be quite small if the basis functions and unit cell are sensibly chosen. (Using Bloch sums is similar in spirit to the familiar procedure of approximating a molecular wavefunction as a linear combination of basis functions.)

Extending Bloch's theorem to linear periodic structures requires identifying an appropriate substitute for the coordinate ϕ and reconsidering the appropriate values for j (which we will call k in noncyclic systems). We require that these substitutions yield eigenfunctions and eigenvalues for the linear system that are the same as what results when it is treated as a cycle. Let us suppose that we have some very large number, n, of unit cells in the cycle, with a repeat distance of a. This means that the cycle has a C_n axis and also a circumference of na, so the linear coordinate (call it x) ranges from 0 to na as the angular coordinate ϕ ranges from 0 to 2π. When we move s steps around the cyclic polymer, the exponential's argument changes by a factor of $2\pi ijs/n$, with j an integer. The same number of steps should give us the same effect on the exponential for the linear polymer. If we choose our exponential's argument to be of the form ikx, then s steps creates the factor $iksa$, so we require that $iksa = 2\pi ijs/n$. This gives $k = 2\pi j/na$, with j an integer. We see that, for extremely large values of n, k becomes an effectively continuous variable (points separated by $2\pi/na$). Also, we see that k is *uniformly* distributed ($j = 0, \pm 1, \pm 2$, etc.), which affects our DOS calculations, as was mentioned in Section 15-2.

We can also ask about the *range* of k, which should correspond to the 0–2π range of ϕ. We expect "empty wiggling" to occur when the sin or cos equivalent of $\exp(ikx)$ has a wavelength shorter than $2a$. The largest nonredundant value of k should come when $\cos(kx)$ goes from $\cos(0)$ to $\cos(2\pi)$ as x goes from 0 to $2a$. Therefore $kx = k(2a) = 2\pi$, so k has a range of $2\pi/a$. Since the convention is to center k about zero, k runs from $-\pi/a$ to π/a. (Note that this range becomes infinite for the free particle in a constant potential, for which a is zero.) The range of k over which all unique wavefunctions for the periodic system are produced once and only once ($-\pi/a < k \leq \pi/a$) is called the *first Brillouin zone* (FBZ).

For noncyclic, one-dimensional systems, then, eigenfunctions have the form

$$\psi_k = \exp(ikx)U_k(x), \qquad -\pi/a < k \leq \pi/a \tag{15-16}$$

Just as was true for cyclic cases, we can approximate this Bloch functional form with a Bloch sum:

$$\psi_k = \sum_{s=0}^{n-1} \exp(iksa)U_k(sa), \qquad -\pi/a < k \leq \pi/a \tag{15-17}$$

In chemistry, it is Bloch *sums* that are normally used.

Two- or three-dimensional periodic systems can be treated similarly, with separate k vectors for each of the independent translational directions.

15-6 A Retrospective Pause

We have reviewed some familiar systems and presented Bloch functions and sums for one-dimensional cyclic and linear periodic systems. The practical lesson to this point is that one can generate useful one-electron wavefunctions for some such systems simply by choosing a basis set of AOs that is identical from one unit cell to the next and then modulating that set with exponential (or sine and cosine) functions. This is most easily done by using the exponential or trigonometric functions as "coefficient waves" from which coefficient values are plucked at points where a unit cell basis set is centered (i.e., by taking Bloch sums). In a case like the π electrons of benzene, each unit cell is a carbon atom and the basis set is a $2p_\pi$ AO at each atom. The coefficients determined by $\exp(ij\phi)$ with $j = 0, \pm 1, \pm 2, 3$ serve to define the wavefunctions completely. Once these are known, the energy values follow. In cases where several AOs are present in each unit cell the situation is complicated because the mix of these AOs may change as j (or k) changes. This added complexity is easily handled by computer programs working within the paradigm of the variational method. What is important for us to understand as chemists is what the computer is doing in such cases, why it is doing it, and what it means for the chemical and physical properties of the system.

15-7 An Example: Polyacetylene with Uniform Bond Lengths

Consider the molecule produced by addition polymerization of acetylene. The standard chemical representation for the all *trans* version of the molecule is shown in Fig. 15-7a. The resonance diagram makes it plausible that this polymer, like benzene, will have equal C–C bond lengths, intermediate in value between the lengths of single and double bonds. For the moment let us assume this to be true and examine the MOs for the π orbitals of this system. To keep matters simple, we begin by treating the system at the simple Hückel level. This means that the hydrogen atoms can be ignored. Also, since only nearest-neighbor π interactions are accounted for, we can pretend that the carbon framework is linear (Fig. 15-7b).

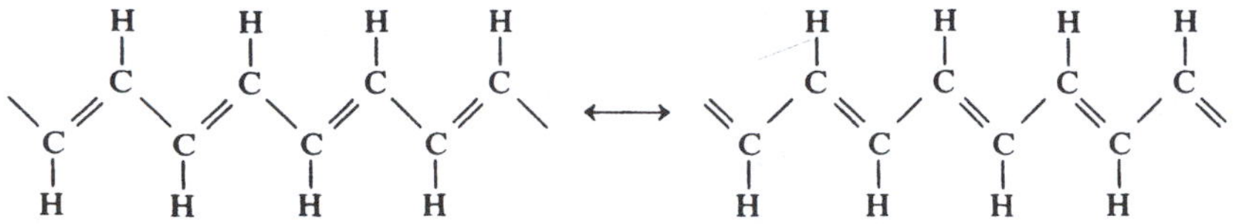

(a)

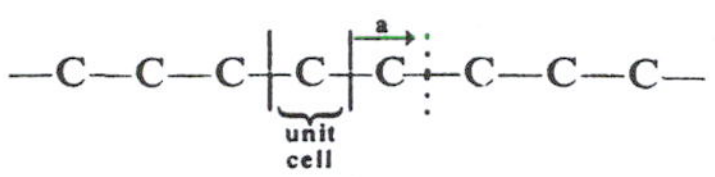

(b)

Figure 15-7 ▶ (a) All-*trans* polyacetylene in its two "resonance structures." (b) The linear model appropriate for a nearest-neighbor, π-only calculation.

First we must choose our unit cell. The simplest choice is to make it a single carbon atom with the characteristic translational distance being a C–C bond length, shown as a in Fig. 15-7b.

Next we must choose the periodic basis set $U_k(x)$ for the π MOs. The normal choice for this (and also the choice dictated by the simple Hückel method) is a $2p_\pi$ AO on each carbon. Furthermore, it is normal to assume that this basis does not change as we go from one MO to another, so we can dispense with the subscript k on U. (Hence, no variational calculation will be necessary in finding the wavefunctions for this system.)

Equation (15-17) tells us how to produce wavefunctions. We simply take $\exp(ikx)$ times our basis set, with k taking on all values between $-\pi/a$ and π/a, and pick off values at discrete points corresponding to carbon atom positions. Since it is more convenient to work with the real forms of solutions, we in effect choose a *pair* of k values (e.g., $-\pi/4a$ and $\pi/4a$) so that we can generate a pair of trigonometric coefficient waves $[\cos(\pi x/4a)$ and $\sin(\pi x/4a)]$. The π MOs for regular polyacetylene produced from Bloch sums are shown in Fig. 15-8 for selected values of k. These can be used to illustrate some important points:

1. The $k = 0$ situation merely reproduces the basis set.

2. As $|k|$ increases, the coefficient wave goes to shorter wavelength and more nodes.

3. Only one MO results from each of the $\pm\pi/a$ extremes of the k range—these are nondegenerate solutions.

4. The pair of MOs drawn for $\pm\pi/4a$ are degenerate, as are those for $\pm\pi/2a$. Therefore, they can be mixed, either to generate complex functions or else simply to shift the phase. For example, the pair at $k = \pm\pi/2a$ can be added or subtracted to produce the equally valid alternative pair shown in Fig. 15-9. In this way a phase shift of any degree could be created. There is no requirement that the MOs be "lined up" in a simple way with the atoms, as we have done in Figs. 15-8 and 15-9. (It is, however, usually more convenient.)

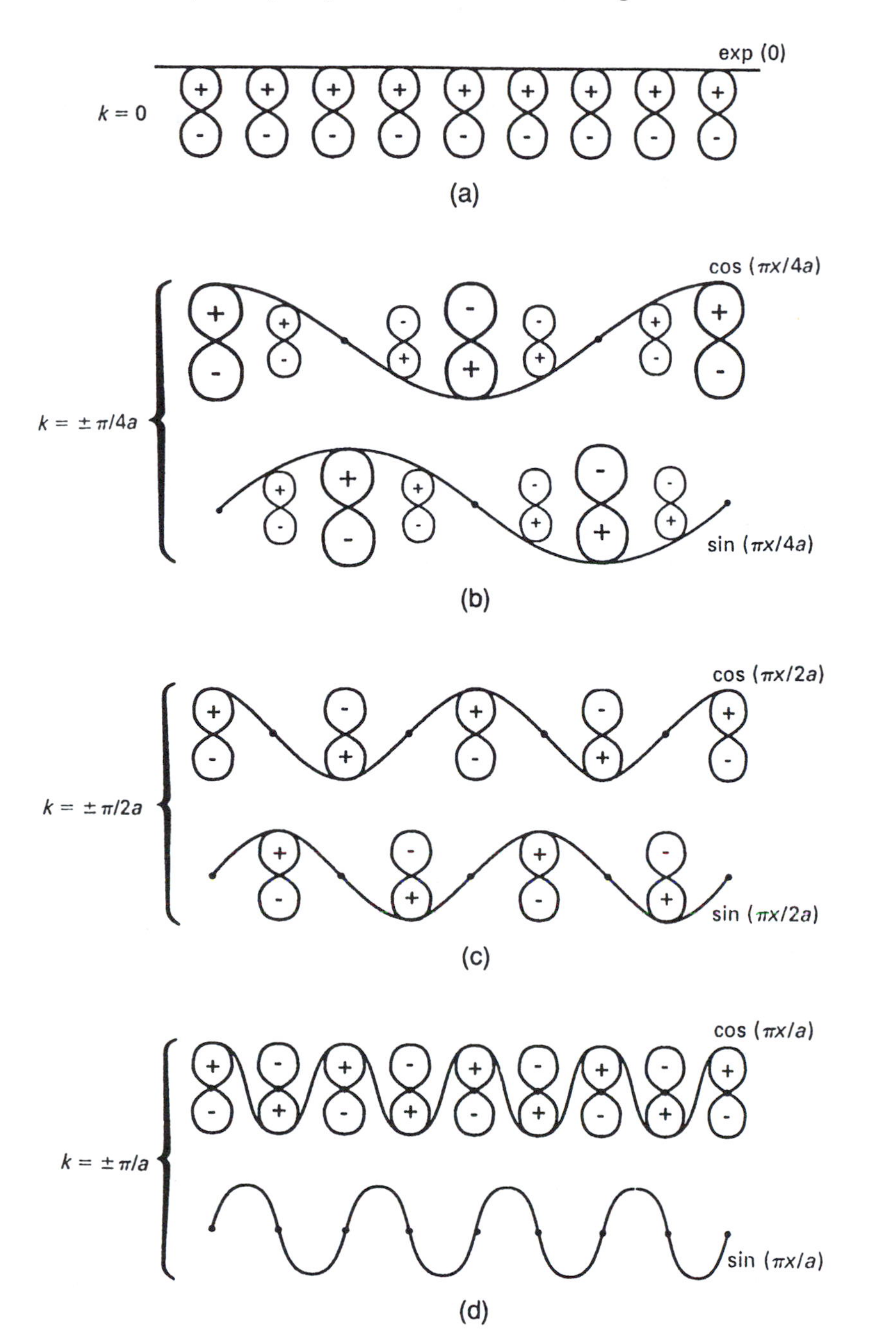

Figure 15-8 ▶ Coefficient waves times a periodic basis set of one $2p_\pi$ AO on each carbon. (a) $k = 0$, so the coefficient wave is a constant and the periodic basis is unmodulated. (b)–(d) Plus and minus k-value exponentials are mixed to give trigonometric coefficient waves. In (d), one of these waves has nodes at every carbon so no Bloch sum function exists for this case.

Once we have the MOs, also sometimes called "crystal orbitals" (COs) for infinite systems, we can consider their energies. In general, these would come from computations involving nuclear-electronic interactions, kinetic energies, etc. But the simple Hückel method takes the much easier approach of relating energy to bond order.

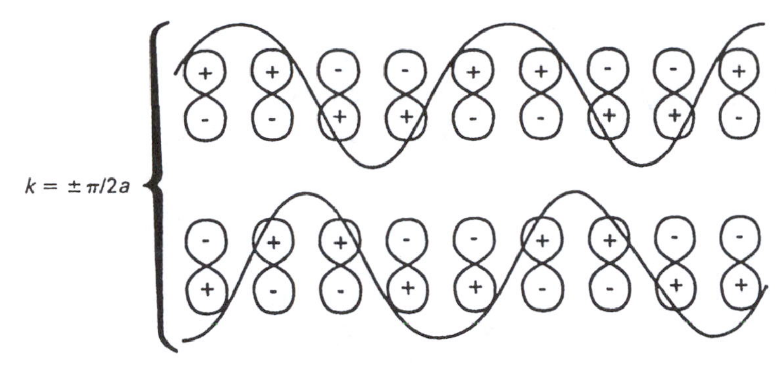

Figure 15-9 ▶ An alternative to the Bloch sums shown in Fig. 15-8c. These have the same coefficient wavelength but the wave is displaced by $a/2$.

(See Section 8-9.) We can see at once that the MO at $k = 0$ is totally bonding. By analogy with the totally bonding MO of benzene, its energy is $\alpha + 2\beta$. The pair of MOs due to $k = \pm\pi/2a$ are nonbonding. This is obvious from Fig. 15-8, but is also not hard to see in Fig. 15-9, which shows the MOs as having equal numbers of bonding and antibonding interactions. These MOs, then, have $E = \alpha$. Finally, the totally antibonding MO at $k = \pi/a$ has an energy of $\alpha - 2\beta$. (Note that the pairing theorem holds. Compare the coefficients of this MO with those at $k = 0$.)

EXAMPLE 15-2 What is the energy of an MO in Fig. 15-8 having $k = \pm\pi/4a$?

SOLUTION ▶ The figure shows that these are triads of atoms having bonding interactions within themselves. Interactions between triads are nonbonding, since only nearest-neighbor interactions are taken into account in the HMO method. So E lies between α and $\alpha + 2\beta$. Within a triad, the coefficient ratios (if we choose the cosine set) go as $\cos(-\pi/4)$, $\cos(0)$, $\cos(\pi/4)$ or $\sqrt{2}/2$, 1, $\sqrt{2}/2$. Normalizing for a triad gives $1/2$, $1/\sqrt{2}$, $1/2$. Eq. (8-58) gives $E = \alpha + 2\beta(1/2\sqrt{2} + 1/2\sqrt{2}) = \alpha + 1.414\beta$. This is the same as the result we obtained for the allyl radical in Section 8-6, which makes sense since, because they are nonbonding to their neighbors in this CO, they behave like isolated allyl systems. ◀

The Hückel π MO energies for regular polyacetylene can be plotted versus k. The result appears in Fig. 15-10 along with a plot of the density of states (DOS). Each carbon atom brings one π electron to the polymer, and these fill the lower-energy half of the MOs, so the highest occupied MO (HOMO) for the polymer is at $E = \alpha$, $|k| = \pi/2a$. Because the CO energies are identical for $|k|$ and $-|k|$, we can observe all the unique *energies* by plotting over the range $0 \leq k \leq \pi/a$. The k-range that produces all unique energies (rather than all the independent wavefunctions) is called the *reduced first Brillouin zone* (RFBZ).

The term *Fermi energy* is often used in reference to electrons having the highest energy in the ground state of the system. Here, if we ignore the effects of thermal energy, the Fermi energy is the HOMO energy. When there is a gap between the HOMO and LUMO energies, there is some disagreement in use of this term. Physicists, for good theoretical reasons, place the Fermi level midway between HOMO and LUMO energies; chemists often continue to equate it to the HOMO energy.

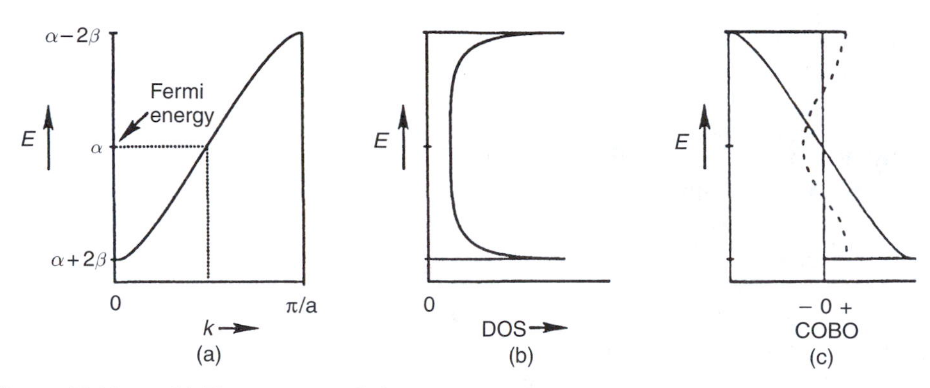

Figure 15-10 ▶ (a) Energy versus k for a chain of π AOs as calculated by the simple Hückel method with Bloch sums. (b) Sketch of the variation of density of states versus energy. (c) Sketch of the crystal orbital bond order (COBO) function. *Solid line* is bond orders between neighbors. *Dashed line* is bond orders between second-nearest neighbors (reduced by an arbitrary factor of five to reflect smaller extent of overlap between π AOs on second-nearest-neighbor atoms). (Adapted from Hoffmann et al. [1].)

It is apparent that the energy span of the π band depends on the magnitude of β and that this in turn should depend on a, the distance between the carbon atom unit cells. The simple Hückel method in fact *assumes* β to be roughly proportional to AO overlap. (The extended Hückel method makes the assumption explicit.) This means that the π band of polyacetylene should become "wider" as the distance between carbons decreases. Standard terminology is to refer to the span between lowest and highest energies in a band as the *band width*, even though normal graphical representations show this as a vertical distance.)

There is yet another graphical representation of the relationships within these MOs that is useful to chemists. This is a plot of the amount of Mulliken overlap population between AOs as a function of E. This quantity, called *crystal orbital overlap population* (COOP), allows one to see at a glance how the bonding interactions change in a band. The simple Hückel analog of COOP is the *crystal orbital bond order* (COBO). Figure 15-10c shows the COBO for the linear polyacetylene π system as a function of k. The lower-energy COs have a high degree of net bond order, but this falls off to become zero at $E = \alpha$ and then negative at higher energies. This is just what we know must be happening. The Hückel MO energies are, after all, directly proportional to bond order, so low energy *must* go with large positive net bond order, etc. Also, the pairing theorem holds, so the nearest-neighbor COBO curve must behave antisymmetrically through $E = \alpha$. Although this particular COBO curve is not especially subtle, it is a good first example because we understand so well what it is telling us.

A useful feature of COOP or COBO curves is that one has complete freedom as to *which* AOs (or groups of AOs) one can look at in this way. For example, the dashed line in Fig. 15-10c shows the overlap populations between π AOs on *second-nearest* neighbors in linear polyacetylene. (Even though all overlaps are formally assumed to be zero in the Hückel method, once we *have* the MOs we are at liberty to calculate the overlaps that actually exist between the AOs in these MOs.) We see that the 1,3 interactions start out with positive overlap at low E, decrease to a negative value at $E = \alpha$, and then rise to a positive value at high energy. This behavior is less obvious than the 1,2 COOP curve, though examination of the MOs at $k = 0, \pi/2a$, and π/a

in Fig. 15-8 makes sense of it. (COOP curves are often shown without an explicit numerical overlap scale because they are usually used for qualitative comparisons.)

EXAMPLE 15-3 Compare Fig. 15-8(c) to Fig. 15-9 with regard to second-nearest-neighbor COBO values.

SOLUTION ▸ The values should be equal since these are equivalent orbitals. It is easy to see that the value should be negative in Fig. 15-8(c) because half of the next-nearest-neighbor coefficient products are negative, and the other half are zero. In Fig. 15-9, they are all negative. Because the normalization constant in 15-8(c) is larger by a factor of $\sqrt{2}$, the sum is the same in each case. These are nonbonding orbitals, so their energies equal α. Fig. 15-8(c) shows that the corresponding second-nearest-neighbor COBO is indeed negative. ◂

The π system of polyacetylene is a convenient first example. It provides elementary examples of the unfamiliar (to chemists) band, DOS, and COOP plots. Also, the MOs are trivial to generate because they are *identical* to the Bloch sums of the basis functions. We have already seen similar behavior in the context of diatomic molecules, where the analog to a Bloch sum is a symmetry orbital (SO). (See Chapter 7.) In the case of H_2, we saw that use of a minimal basis set ($1s_A$ and $1s_B$) gives two symmetry orbitals ($1s_A \pm 1s_B$) and that these are *identical* to the MOs for H_2. However, when we go to larger basis sets (1s, 2s, $2p_x$, $2p_y$, $2p_z$ on each center) we find that symmetry orbitals may no longer be the same as MOs—some MOs are *mixtures* of symmetry orbitals of like symmetry (e.g., the $2s_{\sigma g}$ SO mixes with the $2p_{\sigma g}$ SO). We will now explore a case where the analogous mixing of Bloch sums occurs for a polymer.

Consider what happens to our treatment of the linear chain of carbon atoms when we use an all-valence basis set (2s, $2p_{x,y,z}$) on each carbon. Since our chain lies along the x-coordinate, the symmetries of the 2s and $2p_x$ AOs is σ, while that for $2p_y$ and $2p_z$ is π. We can use the extended Hückel (EH) method to deal with this basis. At very large internuclear distances, interactions between AOs on different atoms vanish, so our energy diagram becomes the simple two-level picture for isolated atoms (Fig. 15-11a). As the atoms move closer together, bands of finite width begin to develop (Fig. 15-11b). At first, while the bands are still fairly narrow compared to the distance between E_{2s} and E_{2p}, we can describe them as almost pure s and pure p in character. That is, the MOs are essentially identical to Bloch sums of the basis set at this point. The 2s band rises in energy from a totally bonding set of 2s AOs to a totally antibonding set of 2s AOs. The $2p_y$ and $2p_z$ bands (degenerate) behave similarly. The $2p_x$ band looks peculiar in that it *drops* in energy as k increases. Sketching out the Bloch sums for $2p_x$ at $k=0$ and π/a (Fig. 15-12) shows why this happens. The $2p_x$ AOs in their original basis set arrangement are antibonding. The $k=0$ Bloch sum keeps them that way. At the other extreme, $k=\pi/a$ and the Bloch sum reverses the sign of every second coefficient, making all of the interactions bonding. This is just the opposite of what happens for the other bands. We say that the 2s, $2p_x$, and $2p_y$ bands "run up" as $|k|$ increases and that the $2p_x$ band "runs down." The simple rule is that bands run up when unit cell basis functions are *symmetric* across the unit cell (so they have phase agreement on the two ends) and run down for antisymmetric basis functions. Note that the $2p_x$ band is wider than the $2p_y$, $2p_z$ band because the $2p_x$ orbitals overlap more strongly due to their orientation. Note also how the DOS plot (Fig. 15-11c) reflects the degeneracy of

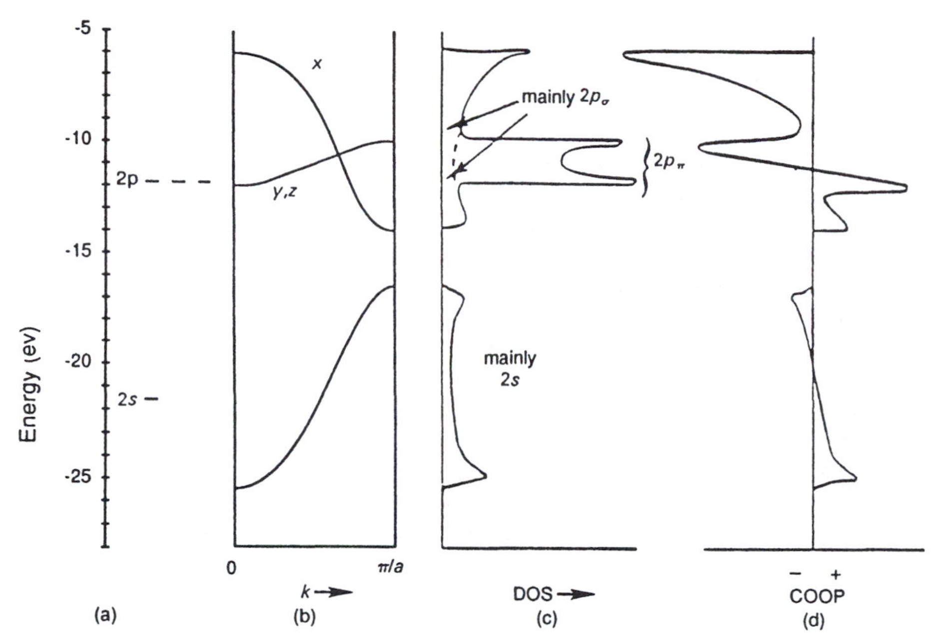

Figure 15-11 ▶ (a) Valence AO energies for an EHMO calculation of a linear chain of carbon atoms. (b) EHMO band diagram calculated at an interatomic distance of 220 pm. (c) DOS diagram and (d) COOP curve between adjacent carbons, for the same calculation. (Adapted from Hoffmann et al. [1].)

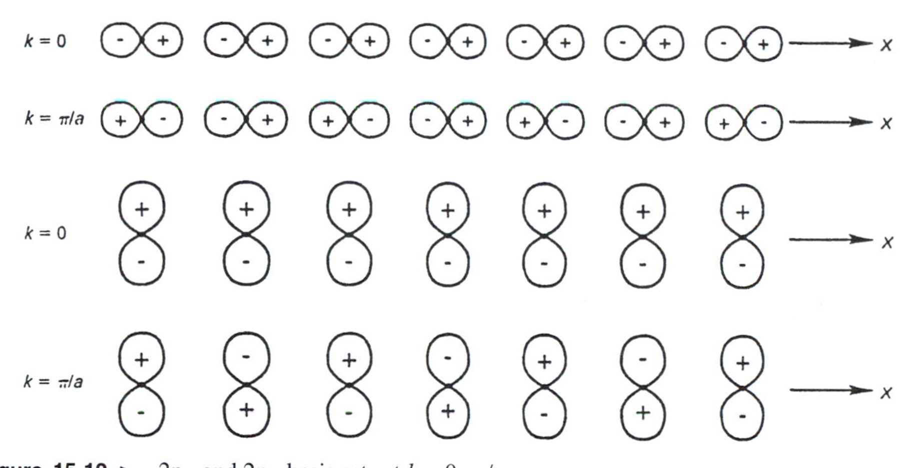

Figure 15-12 ▶ $2p_x$ and $2p_y$ basis sets at $k = 0, \pi/a$.

the $2p_y$, $2p_z$ (or $2p_\pi$) band as a large area. Because the data leading to Fig. 15-11 are produced by the *extended* Hückel (EH) method, we see the characteristic magnification of antibonding effects (Chapter 10), leading to asymmetries in DOS and COOP curves.

The variational method underlying EH calculations on this system evaluates interactions between various Bloch sums at each k value. Because the Bloch sums (BSs) for 2s and $2p_x$ AOs have the same symmetry (σ), a nonzero hamiltonian matrix element may exist between them, permitting mixing. However, the higher energy of the $2p_x$ BS compared to the 2s BS discourages much mixing as long as the overlap between them is small. At large distances, the overlap is indeed small, so we see the relatively "pure" bands of Fig. 15-11. Since the $2p_\pi$ band disagrees in symmetry with the σ BSs, it remains unmixed with them regardless of C–C distance.

EXAMPLE 15-4 How should the areas compare in the DOS of 2s and 2p regions in Fig. 15-11?

SOLUTION ▶ To the extent that these regions remain "pure," we should have three times more area in the p regions because, with three times as many basis functions, there are three times as many states. (The integral over DOS gives the number of states.) ◀

Recalculating the band diagram for all-valence linear carbon at a smaller C–C distance gives the band diagram in Fig. 15-13a. Several things have happened: The lowest band has dropped significantly in energy at $k = 0$ and risen slightly at $k = \pi/a$, so it is wider. The π band has gotten slightly wider too. The highest band has shot up to very high energies and has undergone a noticeable change of shape: It has a large hump at intermediate k values that did not show at large C–C separation. What is responsible for these changes?

Even without BS mixing we should expect band widths to increase as a result of increased overlap between AOs on adjacent carbons. Hence, at first glance the lowest-energy band could be simply a more spread out 2s band. This cannot be right, though, because the EH method on a pure 2s band would give an antibonding-caused energy rise at $k = \pi/a$ that is *larger* than the bonding-caused energy drop at $k = 0$. Instead, we are seeing a *smaller* rise. Something must be happening to either enhance the drop at $k = 0$ or counteract some of the rise at $k = \pi/a$. The DOS plots with 2s and $2p_\sigma$ portions shaded (Fig. 15-13b, c) indicate what is happening. We see that the lowest

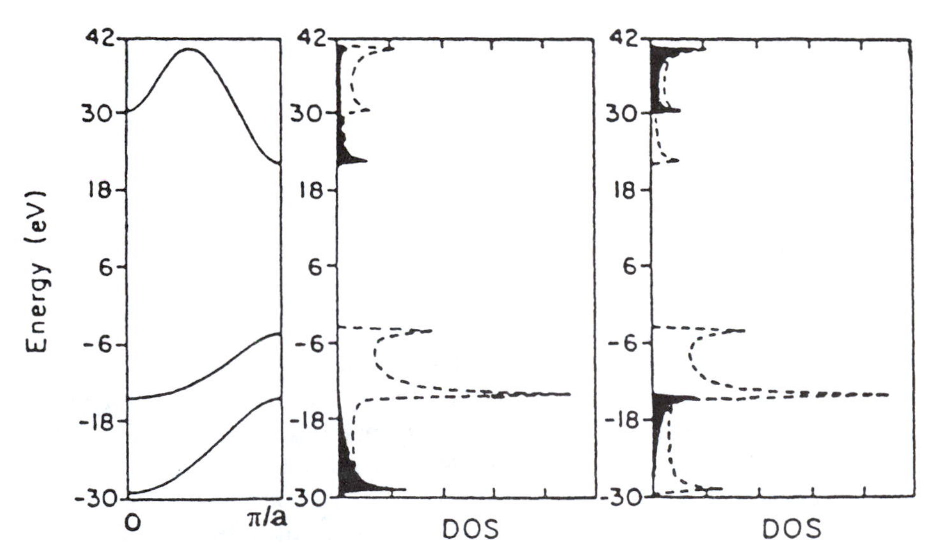

Figure 15-13 ▶ (a) EHMO band diagram for a linear chain of carbon atoms separated by 140 pm. (b) DOS plot, with 2s AO contributions shaded. (c) DOS plot with $2p_\sigma$ AO contributions shaded. (From Hoffmann et al. [1].)

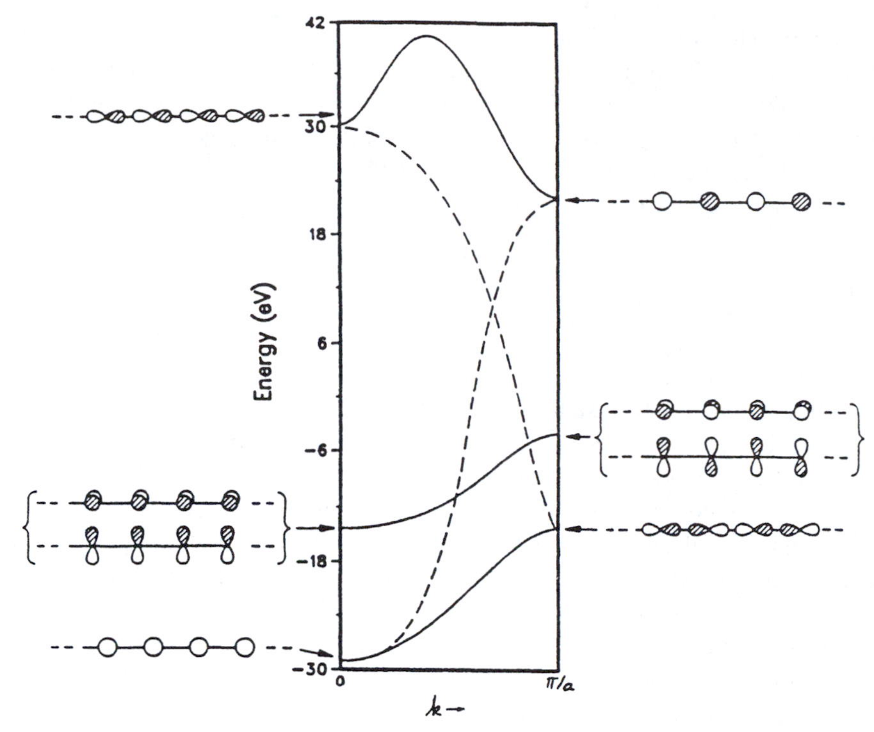

Figure 15-14 ▶ Band diagram for a linear chain of carbon atoms showing some COs and intended correlations. Positive and negative phases are indicated by shading or lack of shading. (From Hoffmann et al. [1].)

band is still primarily 2s for small k, but becomes mainly $2p_\sigma$ at larger k. The highest band shows opposite behavior. It is mostly $2p_\sigma$ at smaller k and 2s at larger k. Mixing is occurring between Bloch sums.

Another way to see these features is to sketch the MOs that the computer program reports for various k values. These are shown for $k = 0, \pi/a$ in Fig. 15-14. It is obvious that the lowest band is all 2s at $k = 0$ and all $2p_\sigma$ at $k = \pi/a$, whereas the highest band is exactly the reverse. The bands have switched character. The pure 2s band we were discussing, that should have gone to very high energy, is represented by a dashed line in Fig. 15-14, and the pure $2p_\sigma$ band by another, and these dashed lines do just what we argued they should.

The variational program is figuring out, at each value of k, how to get the minimum-energy σ band, minimum-energy π band, etc. and is not seeking to maintain 2s or $2p_\sigma$ purity in the bands. Evidently, for the lowest band, it finds pure 2s to be lowest in energy at $k = 0$ and pure $2p_\sigma$ to be lowest at $k = \pi/a$. At intermediate k values it finds a mixture to be best.

The dashed lines in Fig. 15-14 are the bands for Bloch sums of *pure* 2s and $2p_\sigma$ AOs. They differ from the bands for variational COs, where the Bloch sums are mixed. Even though they are not real bands, these dashed lines are nevertheless useful for guessing in advance of calculation what a band diagram will look like. The dashed lines are sometimes referred to as *intended correlations* because they show where the band that starts out as 2s, for instance, "intends" to be at $k = \pi/a$. These two dashed lines are forced to cross at some value of k, but we have seen (Section 7-6) that such crossings are not allowed for wavefunctions of like symmetry (i.e., Bloch sums that have nonzero overlap). Because the two dashed lines belong to Bloch sums of the same symmetry, the noncrossing rule prevents their crossing, forcing the alternative correlation scheme shown by the solid lines. We will see later that an easy way to construct some qualitative band diagrams is to draw them first for simple Bloch sums and then to reroute some of the lines to remove forbidden crossings.[1]

15-8 Electrical Conductivity

The all-valence band diagram of Fig. 15-14 still leads to the prediction that the π band is half filled. Many texts make the point that metallic conductivity results when there are empty MOs immediately above the Fermi energy in a bulk solid, so we should consider whether or not polyacetylene is a metallic conductor. This leads us into the topic of electrical conductivity in one-dimensional periodic systems.

The essence of electrical conductivity is electron flux, and this requires net electron momentum. Consider the "band" for the one-dimensional free particle, shown in Fig. 15-1. If we had a band of this kind partially filled with electrons, the levels would be occupied from the lowest up to the Fermi level. For each occupied state $\exp(ikx)$ there would be an occupied mate $\exp(-ikx)$. In short, each electron having momentum $k\hbar$ would be balanced by one having momentum $-k\hbar$. Net momentum and electrical current would be zero. But if we apply a voltage, states $\exp(ikx)$ and $\exp(-ikx)$ lose their degeneracy. The state corresponding to motion in the direction of lower potential energy for the electron will now have lower energy than that for equal momentum in the opposite direction. As a result, some states that were originally above the Fermi level will move below it because they correspond to motion in the "right" direction, while others that were originally below the Fermi level will move above it. The electrons readjust their occupancies to fit the new scheme, and we now have more states occupied that correspond to momentum in the right direction and fewer corresponding to motion in the wrong direction. Electric current flows. Notice that this mechanism will work for very small applied voltages (metallic conductivity) only if there are empty MOs *just* above the Fermi energy.

If the first empty levels are separated from the Fermi level by a modest energy gap (on the order of $k_B T$, where k_B is Boltzmann's constant, 1.38×10^{-23} J K^{-1}), then a small population of electrons will exist in the "empty" orbitals at thermal equilibrium. This allows electrical conductivity (though typically much lower than that normal for metals) which increases with temperature—the situation in *intrinsic semiconductors*. If the gap is large, no conductivity occurs except at extreme voltages and the material is an *insulator*.

The above discussion is based on a free-particle wavefunction. In real materials the potential along a coordinate is not constant, so the band structure becomes more complicated than the parabola of Fig. 15-1. Also, charge-density adjustments occur that tend to screen out the applied field. However the basic requirement for metallic conductivity continues to be the absence of a gap between the highest filled and lowest empty MOs. Therefore, if the π band of polyacetylene really is partly filled, pure

[1]Apparent symmetry disagreement of bands for reflection through a plane perpendicular to the polymer axis and bisecting an atom or a bond does not lead to allowed crossing. This is explained in Section 15-13.

polyacetylene should be a good (metallic) conductor. However, we will show next that there are other factors operating that always open up a band gap just above the Fermi level in one-dimensional systems, thereby preventing metallic conductivity.

15-9 Polyacetylene with Alternating Bond Lengths—Peierls' Distortion

We now reconsider our assumption of uniform C–C bond lengths. Since the molecules leading up to polyacetylene (butadiene, hexatriene, etc.) have alternating long and short C–C bonds, it is reasonable to ask whether this variation disappears completely in the limit of the infinite polymer.

We return to the π-only system at the simple Hückel level. This allows us to continue treating the polymer as though it were a straight chain of carbon atoms (Fig. 15-15a). Since we are considering the possibility that the bond lengths alternate, we can no longer take a single carbon as our unit cell. (That would require two different translation distances.) Instead we take two atoms, say bonded through the shorter distance, and let the translation distance, a, equal the sum of the two bond lengths (Fig. 15-15b).

The band diagram undergoes a marked change in appearance when we change from a one-atom to a two-atom unit cell, quite aside from the fact that we have two bond distances. To show this, we will first work out the two-atom unit cell band diagram for equal bond lengths.

The most obvious change comes in the range of k. Recall that $-\pi/a < k \leq \pi/a$. But our new choice of unit cell gives a new translation distance, a, equal to twice its previous value, so the numerical range of k is cut in half.

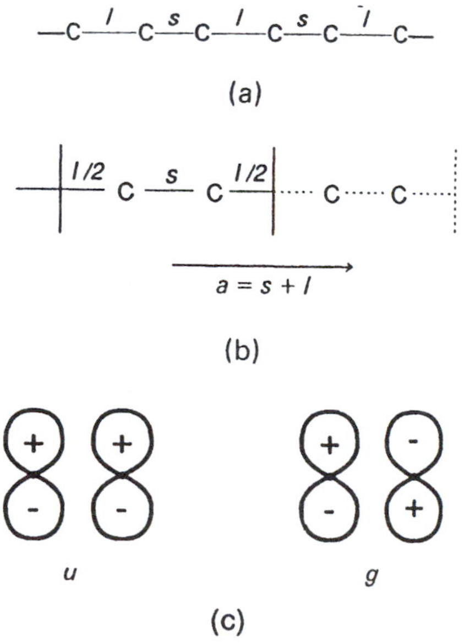

—C—l—C—s—C—l—C—s—C—l—C—

(a)

$l/2$ C s C $l/2$ ····· C ····· C ·····

$a = s + l$

(b)

u g

(c)

Figure 15-15 ▶ (a) A section of a linear carbon polymer with alternating bond lengths. (b) Unit cell and translation distance. (c) *Ungerade* (u) and *gerade* (g) basis sets for a unit cell.

The other big change is in the basis set U. Before it was one $2p_\pi$ AO on each carbon. Now, however, we must have two AOs for each unit cell, and so we need to consider the most useful combinations to choose as basis. Generally we seek choices that will minimize subsequent mixing of Bloch sums, because evaluation of such mixing requires variation calculations and a computer. Bloch sums will not mix if they disagree in symmetry (for operations for which the polymer hamiltonian is invariant). Clearly, inversion through the center of a C–C unit cell is a symmetry operation for the whole polymer, so let us choose the g and u combinations of $2p_\pi$ AOs shown in Fig. 15-15c. Since these basis functions cannot mix without producing a polymer MO of mixed symmetry (which is not allowed unless the MOs are degenerate), we can take the Bloch sums of these bases as identical to the *nondegenerate* COs at least.

We can now construct the band diagram. At $k=0$, our two basis functions give two Bloch sums, shown in Fig. 15-16a. The *ungerade* combination at $k=0$ is obviously totally bonding over the whole molecule, has energy $\alpha+2\beta$, and is identical to the MO we obtained at $k=0$ when we chose one carbon atom per unit cell. (Compare with Fig. 15-8.) The *gerade* basis function at $k=0$ is totally antibonding, has energy $\alpha-2\beta$, and is the same as what we previously had at $k=\pi/a$ (with a equal to *one* C–C bond length). These, then, are the COs we previously indicated to be nondegenerate. When we shift to a equal to two C–C bond lengths, we find at $k=\pi/a$ that the g and u bases produce the polymer MOs shown in Fig. 15-16b. These are easily seen to be nonbonding MOs ($E=\alpha$) exactly like those in Fig. 15-9. When we start the band diagram at $k=0$, we find two energies ($E=\alpha\pm 2\beta$). The u basis has the same phase on the ends of the unit cell, so it runs up (from $\alpha+2\beta$) as k increases, ending up at $E=\alpha$ when $k=\pi/a$. The g basis has opposite phases on the ends of the unit cell, runs down from $\alpha-2\beta$ as k increases, and finishes at $E=\alpha$ too, as shown in Fig. 15-17. The Fermi level is still at $E=\alpha$, but this point now occurs on the right edge of the band diagram instead of at the midpoint.

In going from a uniform chain of π AOs with one carbon per unit cell to an identical chain with two carbons per unit cell, we have not changed anything physically. We have merely changed our choice of representation. Therefore we must obtain the same MOs and energies either way, and it appears that we do, at least at the edges of the diagram. However, the appearance of the band diagram is affected strongly. The relation between

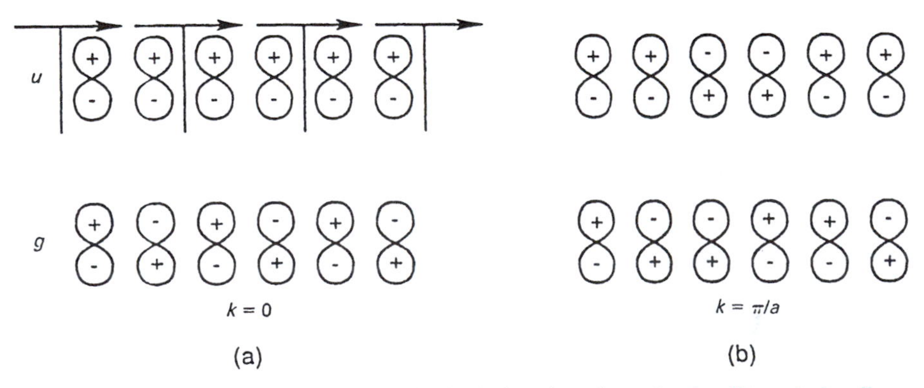

Figure 15-16 ▶ Results of translating u and g basis functions through unit cell translation distances a, modulated by $\cos(kx)$ with $k=0$ and π/a.

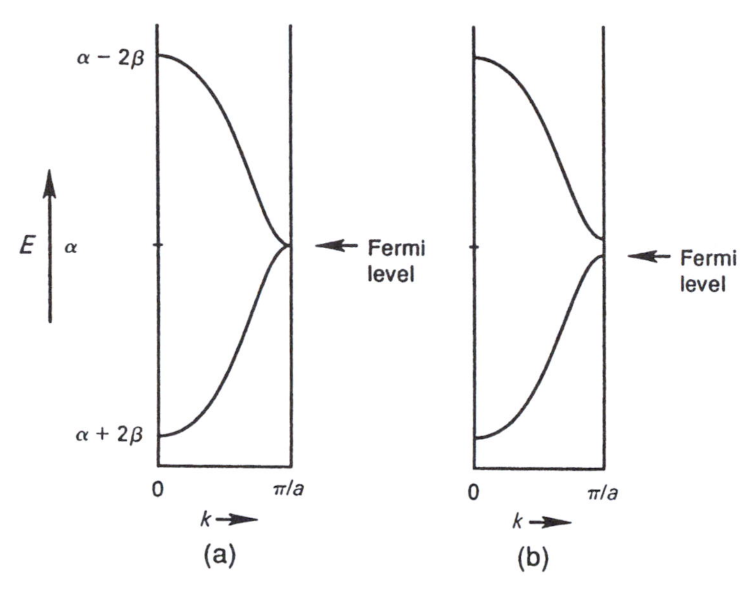

Figure 15-17 ▶ π-band diagram from using two-atom unit cell of Fig. 15-15 with (a) $s = l$, (b) $s \neq l$.

the two band diagrams turns out to be very simple, though. Comparison of Figs. 15-17a and 15-10a shows that the band diagram for the two-atom unit cell is generated from that for the one-atom unit cell by simply folding the latter through a vertical line halfway across the diagram. There is always a smallest possible entity that can be chosen as the unit cell, and it is always possible to select any multiple of this for the unit cell and still generate the periodic structure. The band diagrams based on smaller unit cells can always be converted to those based on larger cells by the process of folding.

Note that what was *one* π band in the earlier band diagram becomes *two* π bands when we double the unit cell size. We shall refer to the lower of these two as the π band and the other as the π^* band, the asterisk indicating net antibonding character. What was previously a degenerate level in the middle of the $|k|$ range is now at the right edge of the diagram, with each member belonging to a different band (though still degenerate).

There is a subtlety that needs to be addressed at this point. All that has been said up to now suggests that no energy changes occur as a result of our doubling the size of the unit cell. This is misleading. The Bloch sums resulting at intermediate k values are *not* identical for corresponding states in these two representations. This reflects the fact that, as k varies for a single-carbon unit cell, the coefficient of the p_π AO can change from one carbon to the next. For the two-carbon unit cell, the AO coefficients are locked together in pairs. We cannot get a CO that looks like the one in Fig. 15-8b, for instance, using a Bloch sum based on the π_u basis set in a two-carbon unit cell. Doubling the unit cell forces the Bloch sums to be a more "coarse-grained" approximation. What saves the day is that the *variational* procedure mixes together the π and π^* Bloch sums at intermediate k values in precisely the manner needed to undo the result of this limitation. The lesson we take from this is that, if we seek to estimate band wavefunctions and energies from Bloch sums *without variational modification*, we make the smallest error if we use the smallest possible unit cell, since this gives the finest-grained first approximation.

Now that we understand the band diagram for a two-atom unit cell, we can consider the effects of bond-length alternation. This is most easily treated as a perturbation on the uniform bond-length system, with half of the bonds getting longer and the intervening half getting shorter. It will suffice to consider the effects of this perturbation to first order and to consider only the four MOs shown in Fig. 15-16. Looking first at the $k=0$ MOs, we see that making half of the bonds longer will raise the energy of the bonding MO, but making the other half of the bonds shorter will lower the energy. Hence we expect little net energy change. The totally antibonding MO will likewise respond oppositely to each type of change, so the effects will again cancel. To first order, then, we expect the perturbation to have little effect on the left side of the π band diagram.

On the right side, the situation is much more interesting. The MOs here (Fig. 15-16b) are degenerate (both nonbonding, $E=\alpha$), so we must be wary. There is only one pair (among the infinite number of possible mixtures) that is "proper" for evaluating this particular perturbation. From Chapter 12 we know that the proper pair is that which *responds most differently* to the perturbation. Consider first the MOs shown in Fig. 15-8c. It is easy to see that neither of these would undergo any energy change as bond lengths change because there is no π AO overlap at all between nearest neighbors. On the other hand, the versions of these MOs shown in Fig. 15-9 or 15-16b respond strongly to the perturbation. One of them is bonding across the bonds being shortened and antibonding across the bonds being lengthened, and its energy will be lowered by both factors. The other is antibonding across the shortening bond, bonding across the lengthening bond, so its energy will rise. Clearly, this choice of MOs leads to the most different responses to this perturbation, so this is the prediction we use—that the endpoints of the curves at the right side of the band diagram will split apart, one going up and one going down (Fig. 15-17b). States at slightly smaller $|k|$ are slightly less affected, etc., so the curves split apart over a range, not just at their termini.

Will the polymer elect to make this bond-length distortion? Yes, because the *occupied* MOs are affected only by energy lowering. All of the MOs of increasing energy are empty. This situation is analogous to what we saw in Chapter 8 for molecules having partly occupied degenerate MOs. In that context, the molecule undergoes what is called Jahn–Teller distortion. For infinite periodic materials, the phenomenon is called a *Peierls distortion*.

We should consider whether the perturbation affects the other bands (σ bands) in a manner to counteract the effect on the π bands. Although some splitting does occur for those bands too, the effect on overall energy is much smaller because these bands are either completely occupied, so that the effects of energy rise and lowering are both registered, or else are empty, so that neither effect registers. Thus, it is the partly filled band that determines the overall energy change due to the perturbation.

We have been dealing with a half-filled band, but it is not difficult to show that a suitable distortion can be found to produce a gap at the Fermi energy for any fractional degree of filling.

We arrive, then, at the conclusion that a one-dimensional periodic structure having a partly filled band is unstable with respect to a distortion in bond lengths that will produce a gap at the Fermi level. This means that polyacetylene cannot be a metallic conductor, though it might be a semiconductor if the gap is not too large.

Finally, we must address the question of how *any* metal can be a conductor. How can sodium, say, have a partly filled band without a band gap? Why doesn't the metal relax into a lower-energy structure having a gap at the Fermi energy? The answer is that, while one-dimensional systems can always find a distortion that has opposite effects on the energies of the MOs above and below the Fermi level, multidimensional systems cannot. A distortion that would work to split the MOs around the Fermi level according to their behavior in one coordinate does not necessarily work when we examine what is happening in the other coordinates. A kind of coincidence is needed that is not usually found except in crystals called quasi–one-dimensional. As a result, two- and three-dimensional crystals with partly filled bands can exist without undergoing spontaneous structural reorganization and losing metallic conductivity.

15-10 Electronic Structure of All-*Trans* Polyacetylene

We finish our discussion of this polymer by examining the band structure for the *non*linear, all *trans* structure with alternating bond lengths (Fig. 15-18a) as calculated by the extended Hückel method.

The EHMO band structure appears in Fig. 15-18b. There are 10 valence AO basis functions per unit cell (two carbons and two hydrogens) so we get 10 lines in the band diagram. Note that, since the chain is no longer linear, the $2p_y$ and $2p_z$ bands are no longer degenerate. One of these is still perpendicular to the molecular plane, hence is still of π symmetry, but the other now lies in the plane and has σ symmetry.

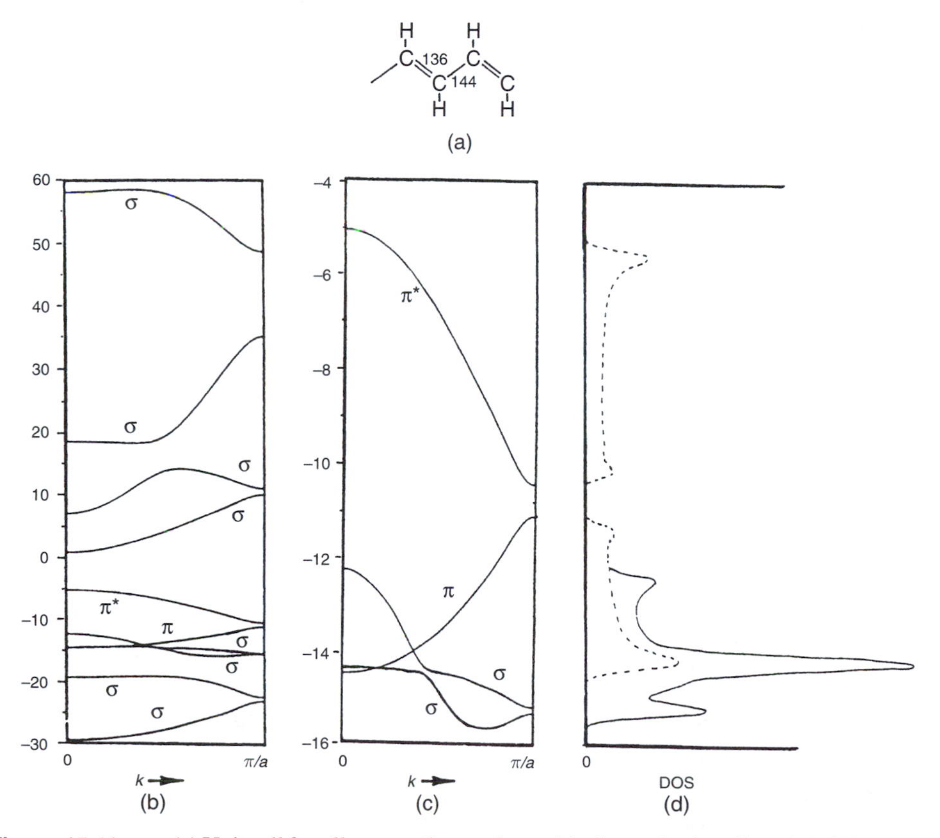

Figure 15-18 ▶ (a) Unit cell for *all-trans* polyacetylene with alternating bond lengths of 136 pm and 144 pm. (b) Valence-band diagram calculated by the EHMO method. All bands have gaps at $k = \pi/a$. (c) Magnified band diagram showing region near the Fermi level. (d) Density of states in region of Fermi level. *Dashed line* shows π-band portion. (Parts a–c adapted from Hoffmann et al. [1].)

A gap resulting from bond-length alternation occurs at the right edge for each of the bands, though it is quite small in bands associated with MOs that have small C–C overlap. A density-of-states plot for the occupied valence σ bands and the π bands shows the gap to be just above the highest-energy π-band level.

With two exceptions, the EHMO band structure shown in Fig. 15-18a, b for *alternating* bond lengths is almost indistinguishable from that for uniform bond lengths (not shown). The only significant differences are that (1) the avoided crossing between σ levels shown in 15-18b becomes a true crossing in the more symmetric, uniform bond-length case, and (2) all the gaps at $k = \pi/a$ disappear for uniform C–C bond lengths.

15-11 Comparison of EHMO and SCF Results on Polyacetylene

So far our entire discussion of band theory has been in the context of Hückel-type models. It is possible to do band calculations using a self-consistent-field (SCF) approach wherein the contributions to one-electron energies resulting from kinetic energy, nuclear-electron attraction energy, and electron–electron repulsion and exchange energies are each explicitly calculated as described in Chapter 11. As pointed out there, the total energy is not the same as the sum of one-electron energies in such calculations, since this would double-count the interelectronic interactions. Also, the energies for virtual orbitals are not given the same physical interpretation as for occupied orbitals. For a calculation on a neutral n-electron system, the occupied MO energies are appropriate for an electron in the neutral system interacting with the $n - 1$ other electrons. The virtual MO energies are appropriate for an *additional* electron (electron $n + 1$) interacting with the n other electrons (with the proviso that the original n electrons have not reacted to the presence of electron $n + 1$). Thus, the virtual MOs refer to a different system (the unrelaxed negative ion) than the occupied MOs. We will see that this affects the nature of band diagrams.

In contrast to SCF calculations, Hückel methods evaluate orbital energies entirely in terms of nodal behavior. Simple Hückel theory calculates bond orders and converts these to energies. As more and more nodes go into the MO, the bond orders drop from net positive through zero (nonbonding) to net negative, and the MO energies rise accordingly. Since overlap is neglected, the MO coefficients remain similar in absolute value over this range, so the high-energy antibonding MOs are elevated above zero energy as much as the bonding MOs are depressed below it, as the pairing theorem requires. The extended Hückel method is similar in spirit but evaluates the overlaps between AOs explicitly in order to arrive at interaction elements that are tailored to explicit AO types, the particular kinds of atoms involved, and the explicit distance between them. Importantly, this method *does* include AO overlap in MO normalization, which leads to very much larger coefficients in the highly noded MOs and a resulting "inflation" on the high-energy end of the energy spectrum. However, because these high-energy MOs are usually not occupied with electrons, they do not affect predictions and can generally be ignored. In contrast to the SCF situation, there is no qualitative difference between the interpretation appropriate for an empty and a filled simple or extended Hückel MO. The lowest empty MO is simply an MO that is a little more antibonding than the highest occupied MO. If two Hückel MOs are equally antibonding,

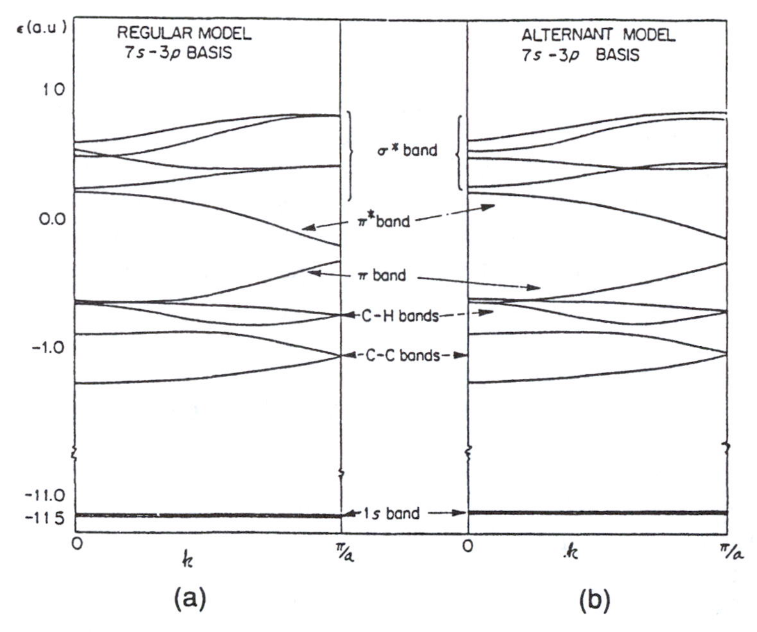

Figure 15-19 ▶ SCF band diagrams computed for all-*trans*-polyacetylene. (a) All C–C bonds 1.39 Å. (b) Alternating bond lengths of 1.3636 Å and 1.4292 Å. (From André and Leroy [2].)

they have equal energy, even if one is occupied and the other is not, whereas this is not the case for two *ab initio* SCF MOs.

We can see the results of these factors when we compare the band diagram for polyacetylene calculated by an SCF procedure with that from an EHMO program. Figure 15-19 shows the SCF bands resulting from *all-trans* polyacetylene with uniform and with alternating C–C bond lengths. Comparing these with the EHMO results (Fig. 15-18a and discussion in previous section), we see that the occupied valence bands agree quite well in the two approaches (the five lowest curves in Fig. 15-18a and the third through seventh lowest curves in Fig. 15-19b). This is important if these methods are to agree in predictions of relative energies of various polymer structures. The highest-energy band in the EHMO diagram is much higher than in the SCF diagram as a result of the inflationary factor described above. This is not a problem as long as we do not try to use this band for predictions. The most important difference is that the SCF band diagram for *uniform* C–C distances shows a sizable π–π^* band gap, while the EHMO method would show no gap there. The SCF gap gets larger when bond-length alternation is introduced, while the EHMO gap becomes finite, so the methods agree that bond-length alternation increases gap size.

The apparent disagreement about the existence of a gap in the uniform polymer is really not a disagreement in light of what we have seen about the different ways these methods define orbital energies. The EHMO method indicates that the highest π and lowest π^* COs have the same amount of overlap-induced bonding energy (the MOs are the same, merely differing in phase). The SCF method indicates that the *least* stable electron in the neutral polymer (at the top of the π band) is at lower energy than an electron can achieve in the best CO it can find in the (unrelaxed) negative ion (at the bottom of the π^* band). The EHMO method is suggesting that the uniform polymer should be a conductor (if we could make it stay uniform). The SCF method is telling us (via Koopmans' theorem) that the energy needed to remove an electron from neutral polyacetylene is larger than the energy released when neutral polyacetylene gains an electron (without relaxing). It would be incorrect to use the EHMO band diagram for a Koopmans' theorem analysis and say that the ionization energy and electron affinity of polyacetylene have the same absolute value. It would also be improper to argue that the large gap seen in this SCF calculation on uniform polyacetylene means such a system would not be a conductor. Both types of calculation have advantages and disadvantages, and these include their appropriateness for revealing various properties. We will continue to use Hückel methods in this chapter.

15-12 Effects of Chemical Substitution on the π Bands[2]

We have seen that Peierls' distortion produces a band gap at the Fermi level in polyacetylene. Another type of change we can consider is replacement of one kind of atom in the unit cell with another. While this is not a modification that the polymer can make spontaneously, it is a way for us to predict what sort of changes in band structure would result if substitutional relatives of polyacetylene were synthesized.

Consider what would happen if one C–H group in the two-carbon unit cell were replaced by N. In making this substitution, we lose a hydrogen atom and its 1s AO, so the all-valence band diagram would have 9 lines rather than 10. Also, we replace one carbon with a more electronegative atom having greater attraction for its electrons. At the simple Hückel level, this would lead to use of a coulomb integral for nitrogen that is lower in energy than that for carbon: $\alpha_{N\cdot} = \alpha_C + h_{N\cdot}\beta$. According to Table 8-3, $h_{N\cdot}$ is 0.5. At the extended Hückel level, valence state ionization energies for nitrogen would replace those for carbon. *Ab initio* methods would account for the change by increasing the nuclear charge from 6 a.u. to 7. Because the results of all these methods agree qualitatively, we continue the discussion in terms of the simple Hückel method since all relevant calculations can be done mentally.

We again start with uniform bond lengths and consider only the π electrons. It is not difficult to predict the changes, to first order, that will occur at $k = 0$ when N replaces C–H. Referring once more to Fig. 15-16, we note that both the π and π^* bands place equal amounts of π-electron density on every atom. Therefore, each energy should drop at $k = 0$ by an amount reflecting the fact that each electron now spends half of its time in a nitrogen 2p AO. At the simple Hückel level this is easily seen to be an energy drop of 0.25β. At $k = \pi/a$, we once again confront the question as to which of the infinite number of linear combinations of degenerate COs is the correct set to use in evaluating the substitutional perturbation. Both members of the pair we used to evaluate bond-length alternation (shown in Fig. 15-16b) place equal density on every carbon, so both COs would drop from α to $\alpha + h\beta/2$ and remain degenerate. Maximum difference in response comes from using the pair shown in Fig. 15-8c because one of these COs places all of its density on the set of atoms that become nitrogen while the other places all of its density on the atoms that remain carbon. The first-order change predicted by the Hückel method, based on the principle of "maximum difference in response" from degenerate-level perturbation theory, is that one level drops to $\alpha + h\beta$ while the other remains at α. The results of these considerations are sketched in Fig. 15-20a.

[2]This section follows closely the discussion of Lowe and Kafafi [3].

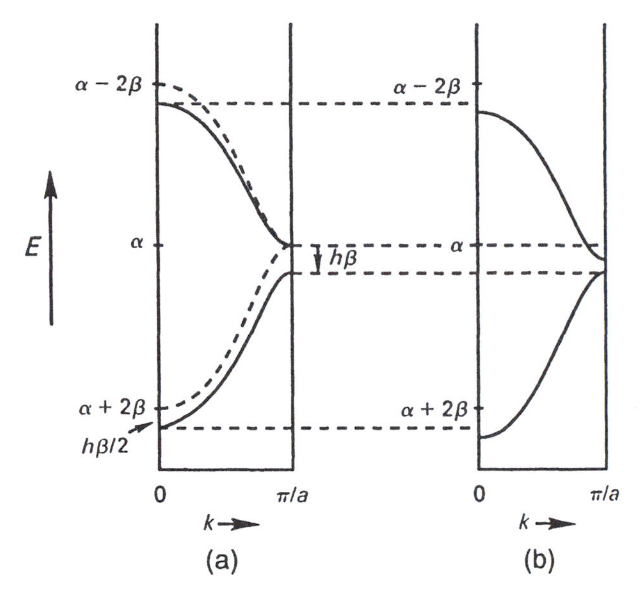

Figure 15-20 ▶ (a) π band of polyacetylene (simple Hückel method) with alternating C–H groups replaced by nitrogen. (b) Same as (a) except that the remaining C–H groups have been replaced by an atom for which the correction factor h is half as large as that for nitrogen.

We can go further and consider the effects of replacing the other C–H groups with something else. If we continue the perturbational treatment using the original polyacetylene functions, it is easy to see that now it will be the *other* band that will be affected at $k = \pi/a$ (Fig. 15-20b).

Since substitution removes the degeneracy at $k = \pi/a$, we now find the situation changed if we go on to consider alternating bond lengths. Whereas before we found that the uniform structure would distort to open a gap, we now find a gap already present due to substitution. Furthermore, the COs now describing the nondegenerate states at the upper and lower edges of the gap are the ones that we saw earlier are *not* affected by bond-length changes. This means that the substitutional change strongly suppresses the tendency of the polymer to have alternating bond lengths. (To first order for degenerate states, perturbations will counteract each other unless they share the same set of proper zeroth-order wavefunctions. See Section 12-8.)

There is a lot going on that we are ignoring. The C–N length is different from the C–C length, the COs change when we go to higher levels of perturbation theory, and Hückel methods are very approximate and become less reliable in systems with polar bonds. It should nevertheless be clear that, if one were looking for ways to predict the effects of chemical substitution on band gap, this sort of approach would provide the conceptual framework.

15-13 Poly-Paraphenylene—A Ring Polymer

We turn now to the π COs of poly-paraphenylene (PPP). This system serves as an excellent illustrator of principles introduced in earlier sections.[3]

[3]This section follows closely the discussion of Lowe et al. [4].

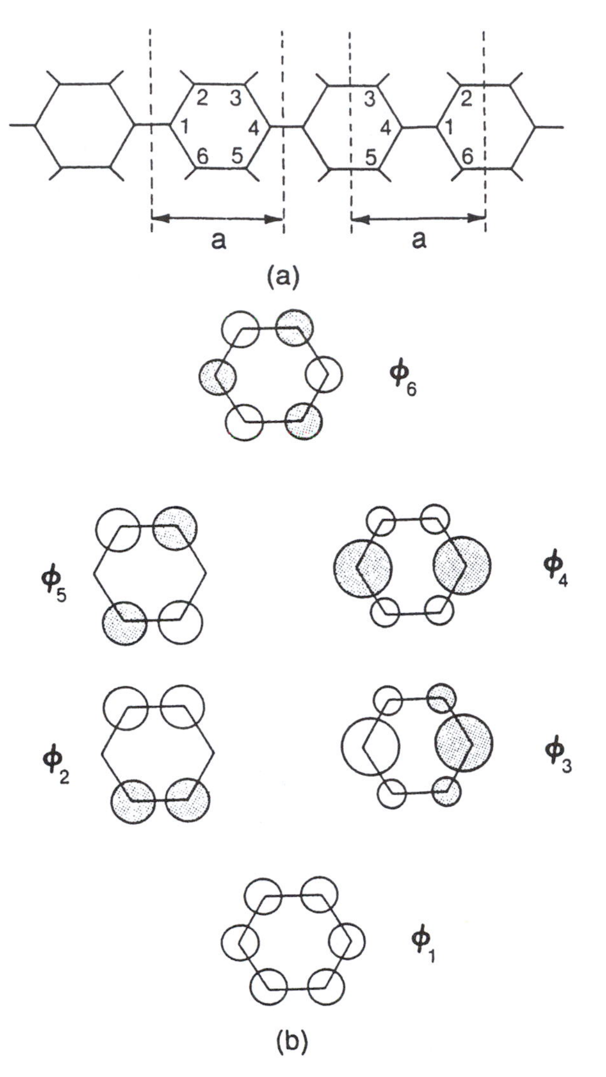

Figure 15-21 ► (a) Two of the choices for the unit cell of PPP. (b) The π MOs of benzene. Shading indicates negative phase. (From Lowe et al. [4].)

As its name indicates, PPP results when benzene molecules are attached in a planar chain, with each benzene being linked to two neighbors at *para* (opposite) positions (Fig. 15-21a). There is an infinity of ways we could select the repeating segment, or unit cell, for this polymer. Two of the most obvious are indicated in Fig. 15-21a, but any portion of length a (or na) would serve. Since selection of the unit cell determines the basis set of MOs we will use to construct the COs, we choose the unit cell on the left and start with the familiar π MOs of benzene, sketched in Fig. 15-21b.

Two of the benzene MOs, ϕ_1 and ϕ_6, are nondegenerate. There are two degenerate pairs: ϕ_2, ϕ_3 and ϕ_4, ϕ_5. Our first task is to decide how these MOs will respond to the perturbation resulting from linking them to a pair of neighbors in an infinite chain. We will assess the interaction energies at the simple Hückel level, since the computations are so trivial that the reasoning is less obscured. Before the perturbation, then, these MOs are at energies $E_{\phi_1} = \alpha + 2\beta$; $E_{\phi_2,\phi_3} = \alpha + \beta$; $E_{\phi_4}, E_{\phi_5} = \alpha - \beta$; $E_{\phi_6} = \alpha - 2\beta$, where β is a negative quantity.

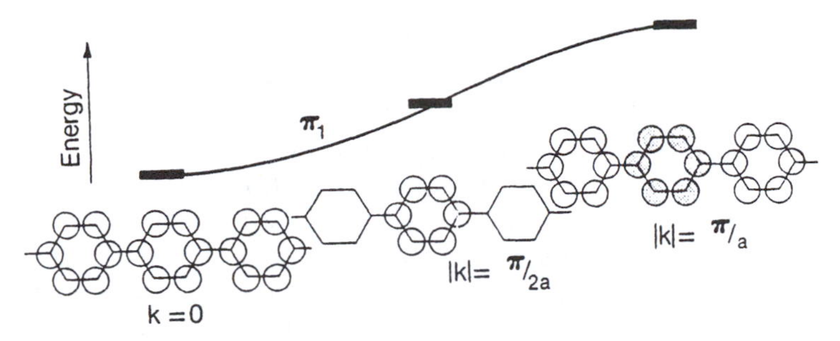

Figure 15-22 ▶ Crystal orbitals for PPP constructed as Bloch sums from ϕ_1 at $|k| = 0, \pi/2a, \pi/a$. (From Lowe et al. [4].)

We first consider the lowest-energy MO, ϕ_1. We want to know the energy that results when ϕ_1 is a component of COs in which $|k|$ ranges from 0 to π/a. We can estimate this to first order by imagining that the MO does not undergo any change when it is incorporated into the CO. That is, it remains six $2p\pi$ AOs, all with the same coefficient $(1/\sqrt{6})$, just as in the molecule. When $k = 0$, the MO repeats itself [times $\exp(ika) = \exp(0) = 1$] for each translation through the distance a, giving the pattern shown at left in Fig. 15-22. An energy of $\alpha + 2\beta$ results from the unchanged bond order *within* each benzene, but there is now some additional bond order *between* molecules. Recalling that the bond order between AOs j and k in MO i equals $c_{j,i}c_{k,i}$ this is $(1/\sqrt{6})(1/\sqrt{6}) = 1/6$. Since MO energy involves 2β times bond order, such new bonds are worth $|\beta|/3$. Since there is one such new bonding interaction per ring, the energy of the CO at $k = 0$ is lowered by $|\beta|/3$ to $\alpha + 2.33\beta$. Qualitatively, the MOs have been linked in a bonding manner so the first-order energy is lower than that for an isolated MO. At $k = \pi/a$, the MO ϕ_1 is multiplied by $\exp(ika) = \exp(i\pi) = \cos(\pi) + i\,\sin(\pi) = -1$, for each translation through the distance a, resulting in the situation sketched at the right of Fig. 15-22. When we analyze this to first order, all is as before except that the additional interaction is antibonding, so the energy *rises* by $|\beta|/3$ to $\alpha + 1.67\beta$. At the halfway point, when $k = \pi/2a$, ϕ_1 oscillates half as fast, going through zero on every second unit cell, giving the middle sketch of Fig. 15-22. Obviously, ϕ_1 is not interacting at all with a companion on either side, so its energy remains unperturbed at $\alpha + 2\beta$. The resulting CO energy curve, labeled π_1, runs up (by $2\beta/3$ to first order) because the unit cell MO has phase agreement at the two points of attachment to the polymer chain.

The analysis for ϕ_6 is identical except that there is phase *disagreement* at the points of attachment, no matter which opposite set of carbons we choose. This means that the band π_6 runs down. Since the coefficients all have the same absolute value as in ϕ_1, the first-order interaction magnitudes are the same, so π_6 runs from $\alpha - 2.33\beta$ at $k = 0$ through $\alpha - 2\beta$ at $k = \pi/2a$ to $\alpha - 1.67\beta$ at $k = \pi/a$.

Analysis of the degenerate MOs requires that we work with the proper zeroth-order versions. We need to ask, for each level, which orthogonal pair responds most differently to the perturbation resulting from linking onto para carbons. Alternatively, we can select a reflection plane that does not move the linkage sites and then seek the orthogonal pair having opposite symmetries for reflection through that plane. Such a plane is the one perpendicular to the benzene plane and intersecting both of the linking sites. Either of these approaches leads to the conclusion that the proper zeroth-order MOs are pairs such that one MO places a node at the linking carbons and the other places an antinode there. Looking at Fig. 15-21b, we see that we can use these MOs as they are sketched provided we assume that the linking occurs at the carbons located at the 3 o'clock and 9 o'clock positions (i.e., atoms 1 and 4 in Fig. 15-21a). Once we ascertain this starting point for analysis, the rest is easy. COs π_2 and π_5, resulting from ϕ_2 and ϕ_5, are unaffected by the polymerization perturbation because they are zero at the linking sites. π_3 runs down because the AO phases disagree at the linking sites, but π_4 runs up by an equal amount because phases agree. π_3 and π_4 are wider bands than π_1 and π_6 because the coefficients at the linking sites are larger. In fact, they are $\sqrt{2}$ times larger (see Appendix 6 for coefficients), so the bands are twice as wide. All of these considerations are brought together in Fig. 15-23.

The first-order simple HMO band diagram of Fig. 15-23 must be considered tentative because it shows two curve crossings and we have not yet considered whether these are allowed by symmetry. The noncrossing rule tells us that such crossings are forbidden if the COs agree in symmetry for all operations, so it is necessary to compare the symmetries of π_2 and π_3 and also of π_4 and π_5. Examination of the COs sketched in Fig. 15-23 makes it clear that, in each pair, there is symmetry disagreement for reflection through a plane perpendicular to the polymer plane and containing the polymer axis. (Call this σ_1.) Therefore, the crossings are allowed.

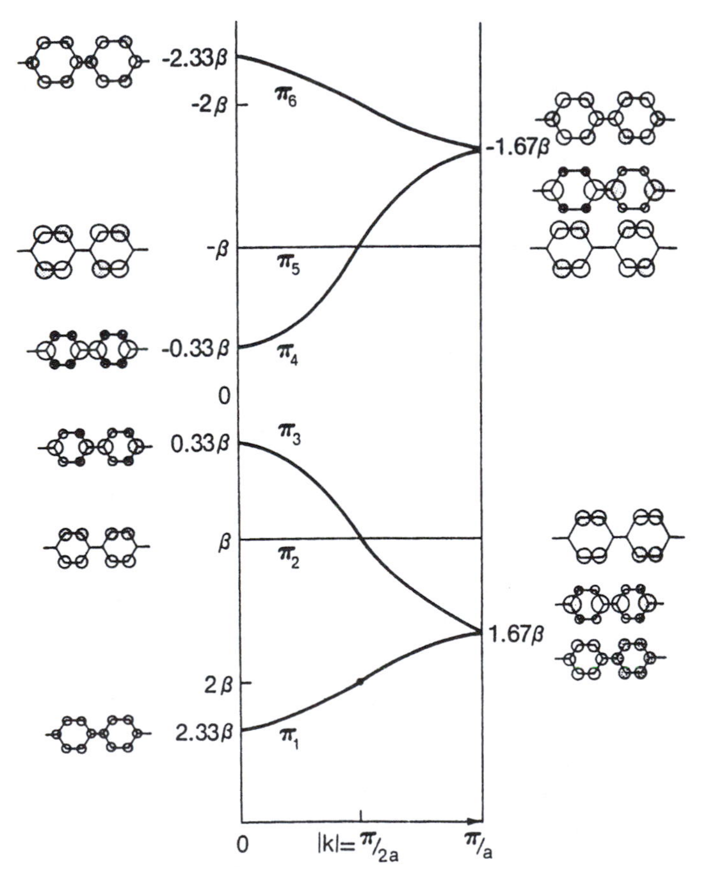

Figure 15-23 ▶ First-order simple Hückel band diagram for PPP showing fragments of COs at $|k| = 0, \pi/a$. The curves are sketched between points computed only at $|k| = 0, \pi/2a$, and π/a. (From Lowe et al. [4].)

One might select a different plane, say the one that is perpendicular to the polymer axis and bisects a benzene ring. (Call it σ_2.) This also appears to be a plane of symmetry disagreement between π_4 and π_5, but, surprisingly, this disagreement alone would *not* allow crossing. This subtlety arises from the existence of two degenerate COs for each value of $|k|$ (except at 0 and π/a). For example, π_4 at $|k| = \pi/2a$ is multiplied by either $+i$ or $-i$ for each translation by a. For convenience in picturing COs, we take appropriate linear combinations to achieve the real forms that correspond to the $+, 0, -, 0, +, 0, -$ pattern of a cosine wave and the corresponding $0, +, 0, -, 0, +, 0$ sine wave pattern, just as we did for Figs. 15-8b and c. These real pairs for π_4 and π_5 are shown in Fig. 15-24. Evidently, if one member of a pair is symmetric for the reflection σ_2, the other is antisymmetric. This is simply a manifestation of a general property of sine and cosine functions and obviously holds for all the bands at intermediate $|k|$ values. Now we can see that the apparent symmetry disagreement we found between π_4 and π_5 for σ_2 reflection is only part of the story and that, if one of the *real* π_4 COs disagrees with a real π_5 CO in symmetry, the *other* real π_4 must agree with it. This in turn means that, when we convert back to exponential versions, which are sine-cosine mixtures, there will always be some symmetry agreement between π_4 and π_5, so the bands still cannot cross (unless there is some other symmetry disagreement). *Therefore, symmetry with respect to a reflection perpendicular to the k axis is not useful in band-crossing analyses.* (A more general approach to such analyses is to assign symmetry

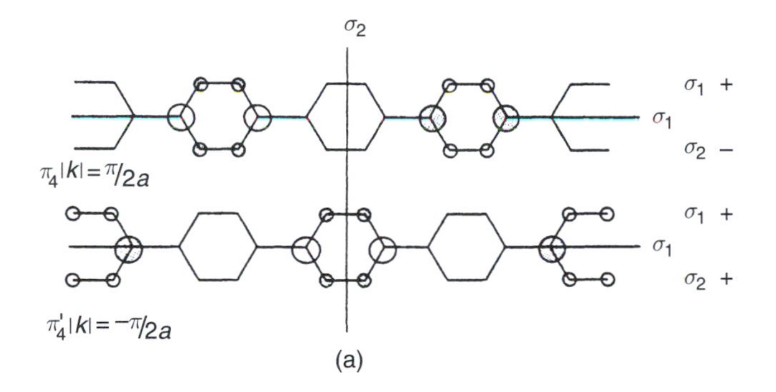

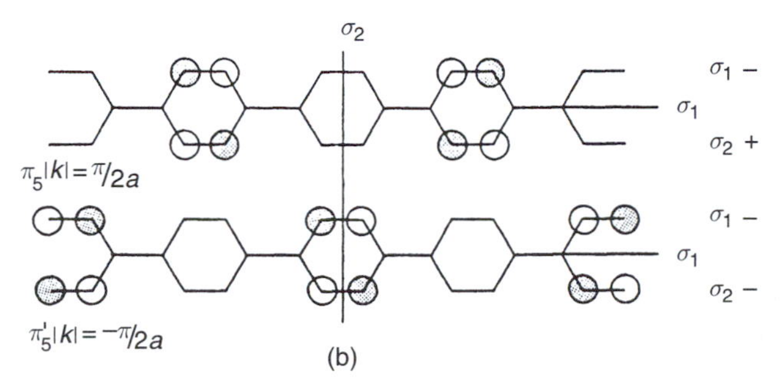

Figure 15-24 ▶ Sine and cosine versions of two crystal orbitals of PPP. (a) π_4 at $|k| = \pi/2a$. (b) π_5 at $|k| = \pi/2a$. σ_1 and σ_2 indicate where reflection planes intersect the molecular plane. (From Lowe et al. [4].)

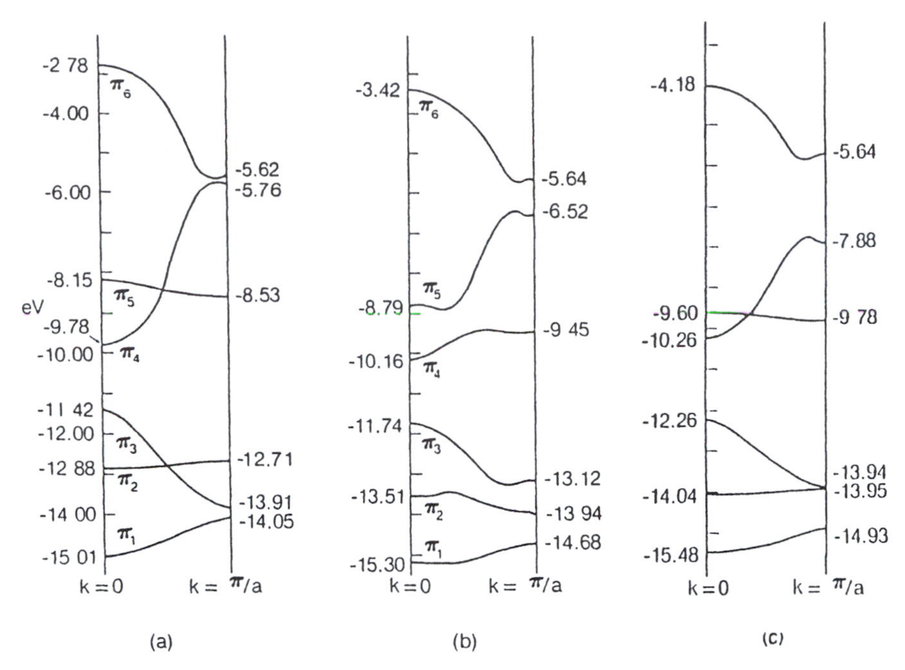

Figure 15-25 ▶ EHMO π-energy bands in eV for (a) PPP, (b) PPP-N2. (c) PPP-N2N6. All C–C and C–N bond lengths are set at 1.40 Å, all angles at 120°. (From Lowe et al. [4].)

labels to the COs, which effectively reveals symmetry agreement or disagreement for all symmetry operations of the polymer.)

A variational EHMO band calculation results in the diagram of Fig. 15-25a. There is much similarity with the simple Hückel first-order band diagram of Fig. 15-23, but there are also some differences. The π_2 and π_5 lines are not exactly horizontal, so these bands now have finite width. This results from the ability of AOs separated by two or more bonds to interact weakly in EHMO calculations. The π_4 and π_6 bands are much wider than π_1 and π_3, a result of the inflation at high energies characteristic of EHMO calculations. The CO coefficients indicate that some mixing of Bloch sums has occurred. For instance, in π_1 at $k=0$ not all carbon coefficients are equal: The ones on the linking *para* carbons are larger. Variational mixing has managed this because it produces more overlap population and therefore a lower minimum energy root. Finally, the degeneracies at $k=\pi/a$ of π_1 and π_3 and also of π_4 with π_6 are lost. The gap between π_4 and π_6 results from an avoided crossing. We can tell that this is the case because the EHMO-computed wavefunction for π_6 at $k=\pi/a$, when sketched (not shown), looks like the second-highest CO sketched at the right of Fig. 15-23. That is, it is made up of ϕ_4-type monomer MOs, so it is the "intended correlation" point for π_4. But π_4 and π_6 are both symmetric for σ_1, hence cannot cross. On the other hand, examination of the CO coefficients at $k=\pi/a$ for π_1 and π_3, shows that this gap does *not* result from an avoided crossing. These bands merely fail to meet as they run up or down to their intended correlation points.

The effects of chemical substitution on the band structure of this system can be analyzed using perturbation arguments, just as was done earlier. Consider the effect, to first order, of replacing the C–H group at position 2 (see Fig. 15-21a) with a nitrogen atom

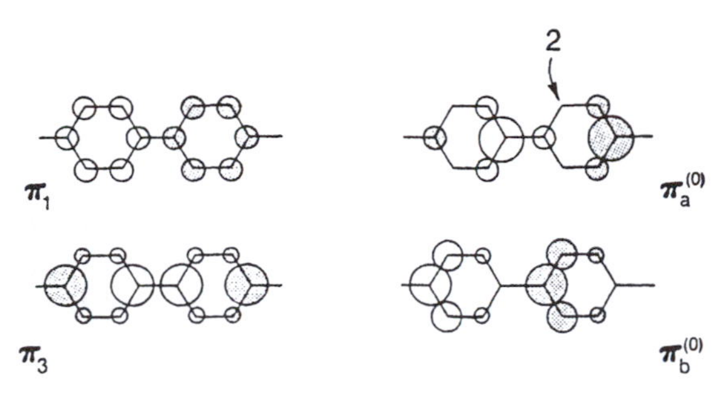

Figure 15-26 ▶ π_1 and π_3 at $|k| = \pi/a$, shown at left as Bloch sums from the benzene basis, are degenerate and mix under the influence of a perturbation at position 2 to produce proper zeroth-order functions $\pi_a^{(0)}$ and $\pi_b^{(0)}$. (From Lowe et al. [4].)

to form what we will code as PPP-N2. Because of nitrogen's greater electronegativity (more negative valence-state ionization energy) we expect any band's energy to be lowered at each k value by an amount proportional to the amount of electronic charge the band places at position 2. All of the COs place some charge at position 2 when $k = 0$ (see Fig. 15-23), so all of these band energies should drop on the left side of the band diagram. It appears that all the COs place charge at position 2 for $k = \pi/a$ also, but now we have degeneracies (or near-degeneracies) to consider. We need to discover the linear combinations of degenerate COs that lead to zeroth-order COs having the greatest *difference* in their responses to the perturbation. Clearly it should be possible to mix π_3 and π_4 to achieve cancellation at position 2. The orthogonal mate for this new CO is correspondingly *larger* at position 2 because the charge placed at any site by this *pair* of COs must not vary with CO mixing. Figure 15-26 shows the results of mixing π_1 and π_3 to produce $\pi_a^{(0)}$ and $\pi_b^{(0)}$. Note that $\pi_a^{(0)}$ is zero not only at position 2 but also at position 6, due to the symmetries of π_1 and π_3 with respect to a σ_1 reflection. We expect, then, that $\pi_a^{(0)}$ will be unaffected by substitution at position 2, whereas $\pi_b^{(0)}$ will be strongly lowered in energy. A similar analysis results for COs π_4 and π_6. Looking at the results of an EHMO *variational* calculation on PPP-N2 (Fig. 15-25b), we find that these expectations are borne out: All band energies at $k = 0$ are lowered, as are all but two at $k = \pi/a$. These two exceptions (π_2 and π_6) are essentially unaffected by nitrogen substitution at position 2.

Notice that the π_2–π_3 and π_4–π_5 crossings in PPP become avoided crossings in PPP-N2. This results from the loss of σ_1 reflection as a symmetry operation for the system when nitrogen is substituted at position 2. Whereas these COs can have different symmetries for σ_1 reflection in PPP, this is not possible for PPP-N2. (There is only one point-symmetry operation for PPP-N2, namely a reflection through the molecular plane. All π COs are antisymmetric for this. Hence no crossings are possible among π bands.)

When we replace yet another C–H (at position 6) with nitrogen, giving PPP-N2N6, we find (Fig. 15-25c) that the band energies all move down some more in energy except for the two that did not move before, because, as we saw, the corresponding COs for these two are also zero at position 6. This substitution restores σ_1 reflection as a symmetry operation, so the π_4–π_5 crossing recurs. Accidental degeneracy now occurs for π_2 and π_3 at $k = \pi/a$.

15-14 Energy Calculations

Suppose we wish to compare the calculated energy of regular polyacetylene to that of a bond-alternating version, or that of PPP with PPP-N2. How do we get energies from the band calculations for these two cases? In a Hückel-type calculation of a *molecule*, we simply add up the one-electron energies. In a band calculation we are faced with a very large set of one-electron energies—one for each k value. How do we deal with this?

Consider again the simple Hückel band diagram for regular polyacetylene (Fig. 15-17a). This band diagram indicates that electrons in the π CO at $k=0$ have an energy of $\alpha+2\beta$, those at $k=\pi/a$ have $E=\alpha$, and those at $k=\pi/2a$ have an energy somewhat below $a+\beta$. It is fairly obvious that the CO average energy is somewhat below $\alpha+\beta$. Each CO has two electrons delocalized over the polymer, but the net number of electrons *per unit cell* is two. Therefore, with an average π CO energy of below $a+\beta$ and two π electrons per unit cell, we have a π energy per unit cell below $2a+2\beta$. To be more precise would require a more precise average energy for the CO. [Even without such a calculation, however, we can see that bond alternation causes the average energy of the occupied π band to drop (Fig. 15-17b), so the π-electron energy per unit cell is lower for the alternating structure.]

The problem, then, is to calculate an accurate average energy for each of the filled bands, the average being over the first Brillouin zone, and then multiply each such average energy by the number of electrons in that band, per unit cell. The sum of these is the Hückel total energy per unit cell. (This procedure assumes that the bands are not partially filled.)

The practical difficulty in doing this is that each variational calculation is carried out at a single point in k space. Thus, for the variational bands sketched in Fig. 15-23, an EHMO calculation was made at $k=0$, another was made at $k=\pi/2a$, and a third was made at $k=\pi/a$. Then lines were drawn to connect the energy points found at these k values, with consideration being given to the noncrossing rule. Obviously, this produces a rather qualitative diagram. To refine it would require making additional variational calculations at intermediate points in k space. Thus, we must be concerned with the trade-off between accuracy of final average energy and effort needed to achieve well-characterized band energies from which to calculate that average.

Achieving accurate average CO energies from values at a few k points is possible if those points are sensibly chosen and if appropriate weighting factors are employed. The problem of choosing the correct few points and their weight factors has been worked out by Chadi and Cohen [5] for multidimensional systems of various symmetries. Although this is a matter of real practical interest, we will not explore it further here.

15-15 Two-Dimensional Periodicity and Vectors in Reciprocal Space

In one-dimensional problems there is a single translation direction, a single step size a, and a single accompanying variable k. It is a simple matter in such problems to treat a and k as scalars and to plot band energies versus k in the range of the first Brillouin zone, $-\pi/a<k\leq\pi/a$. Step size a has dimensions of length and k of reciprocal length to give an argument for $\exp(ikna)$ that is dimensionless.

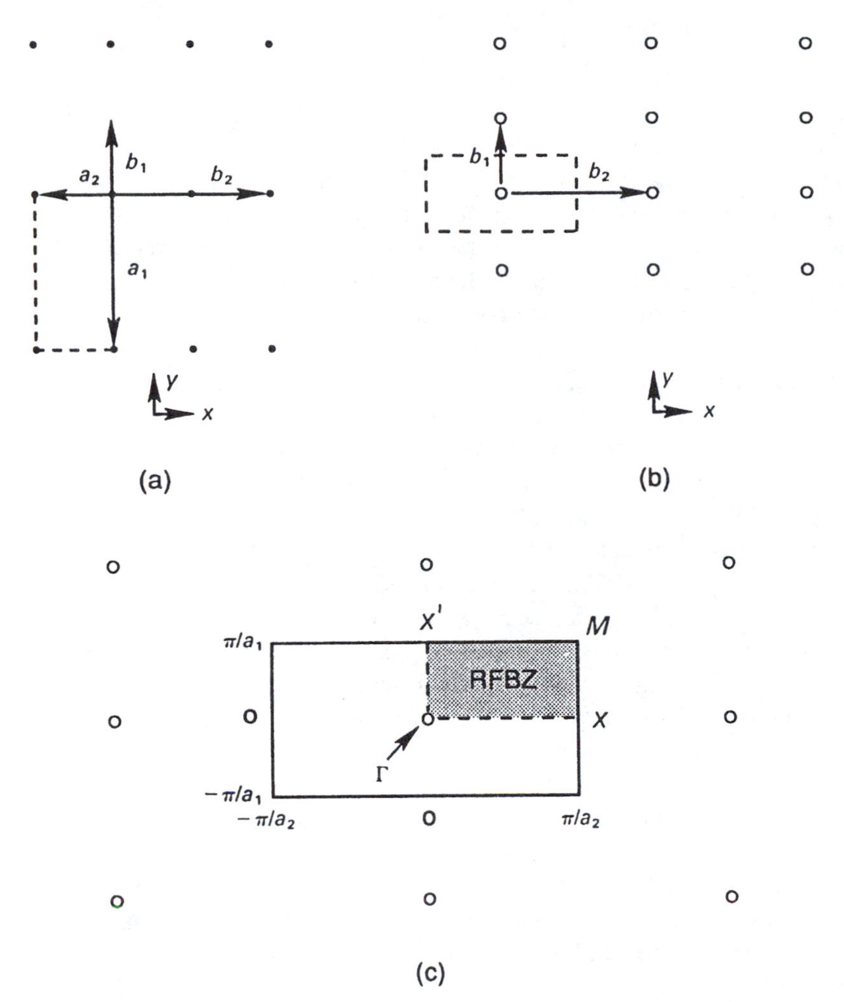

Figure 15-27 ▶ (a) A two-dimensional rectangular lattice (*solid points*) showing translation vectors $\mathbf{a}_1$ and $\mathbf{a}_2$ as well as reciprocal vectors $\mathbf{b}_1$ and $\mathbf{b}_2$. (b) The reciprocal lattice (*hollow points*) with the first Brillouin zone outlined by dashed lines. (c) The first Brillouin zone with the reduced first Brillouin zone (RFBZ) shaded.

In multidimensional systems we must keep track of k values for waves oriented along different directions, and we need to calculate each CO energy at a point corresponding to k values for each of these directions. This makes treatment of such problems considerably more complicated. It is convenient and conventional to handle all this in terms of vectors. We will outline the vector treatment here, using the two-dimensional rectangular lattice shown in Fig. 15-27a as an example.

For a three-dimensional crystal, we label the translation vectors for a unit cell in real space $\mathbf{a}_1$, $\mathbf{a}_2$, and $\mathbf{a}_3$. Any point $\mathbf{r}$ in the unit cell can then be expressed in terms of vectors $\mathbf{a}$. For the vectors in *reciprocal space*, we use the symbols $\mathbf{b}_1$, $\mathbf{b}_2$, and $\mathbf{b}_3$. The position $\mathbf{k}$ in reciprocal space can then be expressed in terms of vectors $\mathbf{b}$. Vectors $\mathbf{a}$ have dimensions of length, and $\mathbf{b}$ of reciprocal length.

The vectors $\mathbf{b}$ in reciprocal space are defined according to the formula

$$\mathbf{b}_i = 2\pi(\mathbf{a}_j \times \mathbf{a}_k)/(\mathbf{a}_i \cdot \mathbf{a}_j \times \mathbf{a}_k) \tag{15-18}$$

This makes each vector $\mathbf{b}_i$ orthogonal to the plane of the "other" two $\mathbf{a}$ vectors, $\mathbf{a}_j$ and $\mathbf{a}_k$. In the case of a two-dimensional crystal, the missing vector $\mathbf{a}_k$ is assumed to be a unit vector perpendicular to the plane of $\mathbf{a}_i$ and $\mathbf{a}_j$. The denominator of Eq. (15-18) is a scalar with the same value in all three cases, so the length of $\mathbf{b}_i$ is greater if the vector product of $\mathbf{a}_j$ and $\mathbf{a}_k$ is greater, i.e., if $|\mathbf{a}_j||\mathbf{a}_k|\sin\phi$ is greater, where ϕ is the angle between $\mathbf{a}_j$ and $\mathbf{a}_k$. In the example of Fig. 15-17a, all angles are the same, so the fact that $\mathbf{a}_1$ is twice as long as $\mathbf{a}_2$ makes $\mathbf{b}_1$ half as "long" (in reciprocal distance) as $\mathbf{b}_2$. The "length" of $\mathbf{b}_i$ is $2\pi/|\mathbf{a}_i|$, which means that these vectors are of appropriate *dimension* to define edges for the first Brillouin zone. However, we will see that they need some adjustment in *position*.

As is indicated in Fig. 15-27a, moving a lattice point through translations **a** generates the real, or *direct lattice*. Moving a point in reciprocal space through translations **b** generates the *reciprocal lattice*. Figure 15-27a and b indicate that a real lattice with orthogonal vectors **a** results in a reciprocal lattice with orthogonal vectors **b** but that the reciprocal lattice is rotated with respect to the real one. When the vectors **a** are not orthogonal, the reciprocal lattice may look quite different from the real one.

According to Fig. 15-27b, the zone defined by vectors **b** runs from 0 to $2\pi/|\mathbf{a}|$. We seek an FBZ that runs from $-\pi/|\mathbf{a}|$ to $\pi/|\mathbf{a}|$ in each direction. We can arrange this by drawing lines connecting one reciprocal lattice point to all its nearest neighbors and then bisecting all these lines with perpendicular bisectors. This gives the rectangle bounded by dashed lines in Fig. 15-27b. In essence, we are finding a cell in reciprocal space that fills up that space when translated by steps **b** and that also associates every point in reciprocal space with the reciprocal lattice point to which it is closest. This is sometimes called a *Wigner–Seitz unit cell*. It is also the FBZ for the crystal. In the example at hand, the FBZ looks like the zone defined by the **b** vectors, simply shifted, but in more complex systems, these may look quite different. Generally, the FBZ continues to have the symmetry of the real crystal. Thus, both the direct lattice and the FBZ in Fig. 15-27 are symmetric for reflections in the xz and yz planes.

We have seen that one-dimensional systems have degenerate COs for equal values of $|k|$ and so, if we wish to portray only the unique energies of the system, we need consider only the range from 0 to π/a. The analogous situation in two or three dimensions is that symmetrically equivalent positions **k** give degenerate COs. Hence we need consider only a symmetrically unique portion of the FBZ—*the reduced first Brillouin zone* (RFBZ). Since the FBZ is symmetric for the reflections mentioned above, the RFBZ is the quadrant shown in Fig. 15-27c.

It is standard to use the label Γ for the central point of the FBZ. Certain other unique points are often labeled with symbols such as K, L, M, and X. These points often generate waves having the same periodicity as the crystal.

If we choose a basis of one s-type AO on each *real-space* lattice point, then the COs resulting from the special *reciprocal-space* lattice points labeled in Fig. 15-27c appear as shown in Fig. 15-28. The point Γ has coordinate intercepts (or k_1, k_2 values) of (0,0), so $\exp(i\mathbf{k}\cdot n\mathbf{a}) = 1$ and the s AO is translated without change from point to point, giving the totally bonding CO of Fig. 15-28a. Point X corresponds to $(0, \pi/a_2)$, meaning that no phase change occurs as we move through the direct lattice by steps of $\mathbf{a}_1$ but that the phase reverses for each step by $\mathbf{a}_2$ (Fig. 15-28b). Point M is for $(\pi/a_1, \pi/a_2)$, so phase reversal occurs for a step by either $\mathbf{a}_1$ or $\mathbf{a}_2$, resulting in the totally antibonding

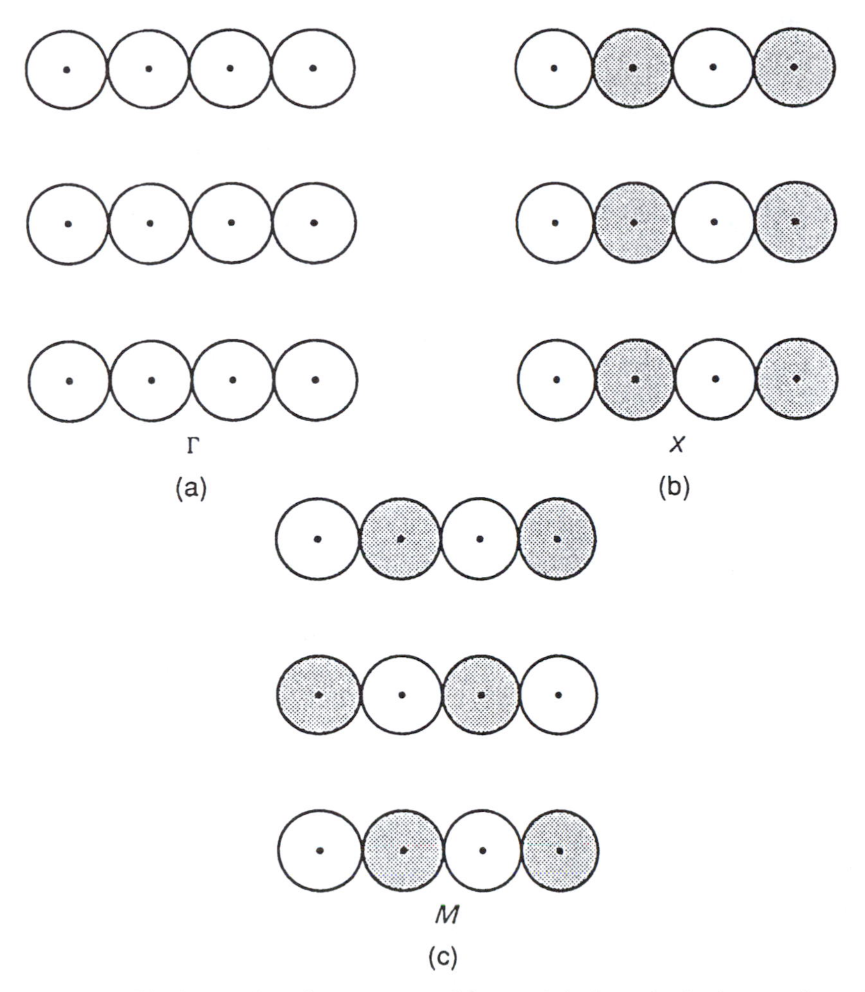

Figure 15-28 ▶ Bloch sums based on an s-type AO at each lattice point in the two-dimensional rectangular lattice. $(k_1, k_2) =$ (a) (0,0), (b) $(0, \pi/a_2)$, (c) $(\pi/a_1, \pi/a_2)$.

situation in Fig. 15-28c. The waves for points X and M can be seen to indeed have the periodicity of the lattice.

Comparison of these three COs leads us to expect lowest energy at Γ, higher energy at X, and highest energy at M. A complete map of the CO energies would require a surface lying over the RFBZ. Instead of showing this surface, it is standard to plot the energy as a function of various straight-line paths connecting special points—e.g., from Γ to X, from X to M, from M to X', from Γ to M. Such line plots are portrayed side by side on a common k axis, as in Fig. 15-29, even though the identity of k (k_1, k_2, or some combination) changes from panel to panel.

15-16 Periodicity in Three Dimensions—Graphite

Graphite is an important commercial material. Some of its uses derive from the fact that it is a very good conductor of electricity—nearly as good as metals. The most stable crystalline form of graphite (Bernal graphite) is depicted in Fig. 15-30a. It comprises

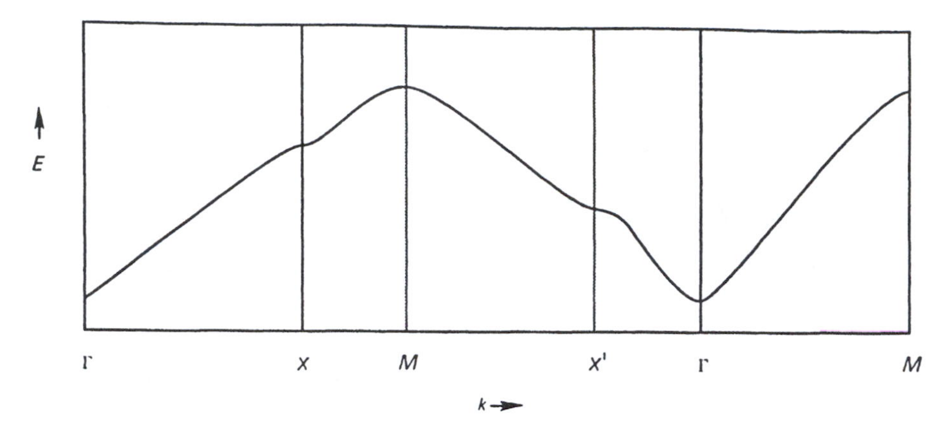

Figure 15-29 ▶ Energy versus k along various paths in the RFBZ for the system of s-type AOs in a rectangular two-dimensional lattice.

carbon atoms in two-dimensional sheets having hexagonal symmetry. These are stacked so that half of the carbons in a sheet lie directly between carbons in adjacent sheets while the other half lie between empty hexagon centers. The atoms within a sheet are strongly covalently bonded to their neighbors, with nearest-neighbor bond lengths of 1.42 Å. The sheets are separated by 3.5 Å, indicative that the forces between them are very weak. (This explains why they fracture into sheetlike fragments that can slide over each other like a pile of playing cards, making ground graphite a good lubricant.) A sensible approach to understanding the electronic structure of graphite, then, is to first analyze the two-dimensional sheet and then consider the perturbation involved in forming layers of sheets.

A section of a two-dimensional sheet, the translation vectors $\mathbf{a}_1$ and $\mathbf{a}_2$, and the primitive unit cell are sketched in Fig. 15-30b. The primitive unit cell contains two carbon atoms. The angle between $\mathbf{a}_1$ and $\mathbf{a}_2$ is 120°. Notice that the lattice points in real space, at the vector origin and at the termini of integral numbers of steps from the origin, fall at the centers of hexagons, where there are no atoms. This illustrates that there is no particular identification of direct lattice points with atoms or other structural features of the molecules that constitute a crystal. In the present instance the direct lattice, shown in Fig. 15-30c, looks like a graphite crystal that has been rotated by 30° with an extra point in the center of each hexagon. In general, even though the direct lattice need not look identical to the crystal, it will have the same symmetry as the crystal, in this case hexagonal. In Fig. 15-30d is shown the relation between the vectors **a** and the reciprocal space vectors **b**. Since $\mathbf{b}_1$ must be perpendicular to $\mathbf{a}_2$, etc., it follows that the angle between **b** vectors is 60°. The reciprocal lattice generated by the **b** vectors appears in Fig. 15-30e and is again a centered-hexagon pattern, but rotated with respect to the direct lattice. The FBZ is constructed by drawing lines from one reciprocal lattice point to all its nearest neighbors and then cutting them with perpendicular bisectors. This yields a hexagon, as shown in Fig. 15-30f. Also shown is the RFBZ. The entire FBZ can be filled by putting the RFBZ through the symmetry operations of the hexagonal sheet, so it includes all the points not equivalent by symmetry.

Points of special interest in the RFBZ are labeled Γ, M, and K. The point Γ corresponds to the (k_1, k_2) values (0,0). This is a unique point in the FBZ, so it produces one nondegenerate Bloch sum for each basis function in the unit cell. The point M

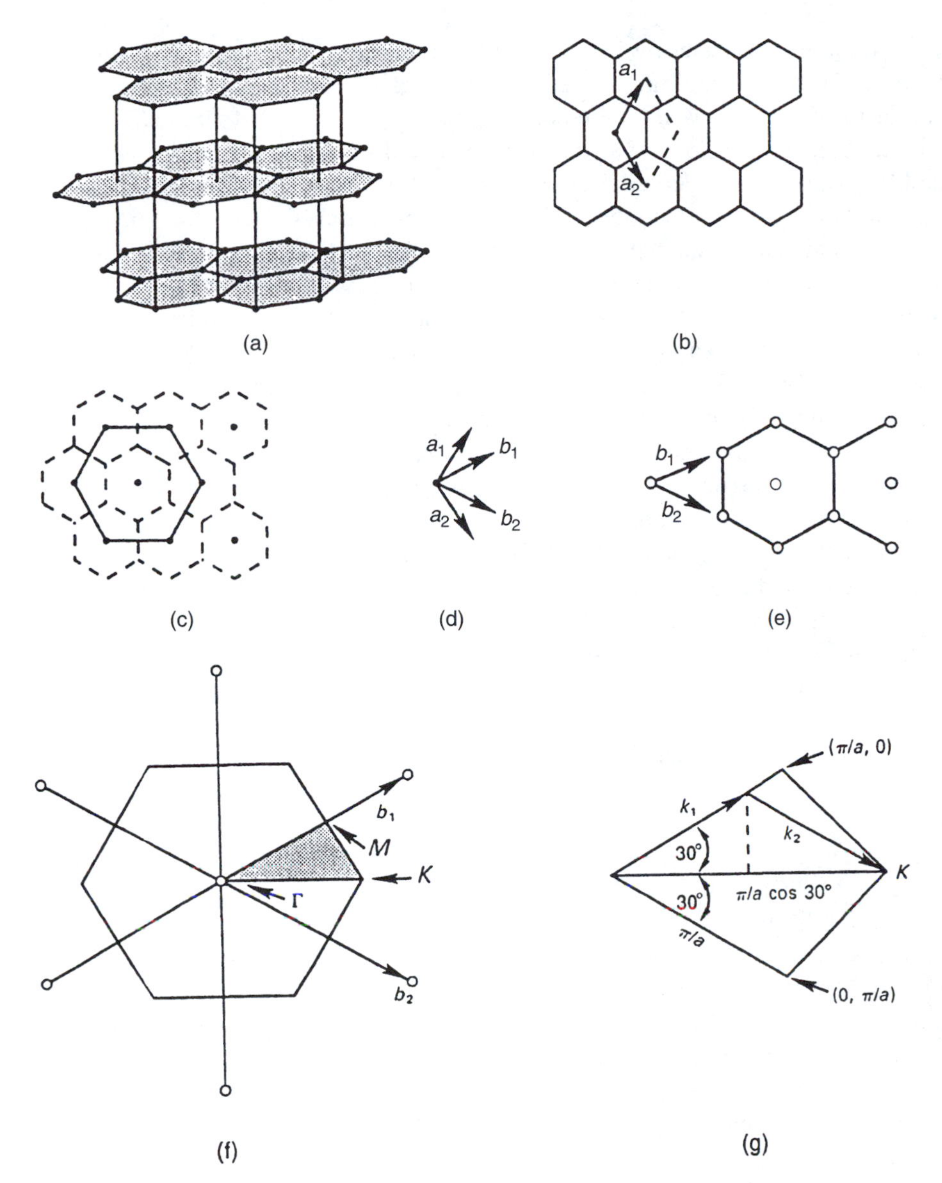

Figure 15-30 ▶ (a) The crystal structure of Bernal graphite. Each point represents a carbon atom. (b) Top view of a hexagonal sheet of graphite showing the unit cell and translation vectors for the two-dimensional structure. The "origin point" of the vectors does not correspond to an atom location. (c) The direct lattice produced by translations of the "origin point" of the **a** vectors. The full lines emphasize that this is a lattice of centered hexagons. The graphite structure is indicated with dashed lines. (d) Reciprocal lattice vectors **b** are shown in relation to vectors **a**. (e) The reciprocal lattice, produced by translating the origin point of the **b** vectors. (f) Construction of the FBZ by bisecting lines connecting one reciprocal lattice point to all of its nearest neighbors. The RFBZ is shaded. (g) Geometry of the relation between point K and the vectors connecting it to the origin.

corresponds to $(\pi/a, 0)$. We might expect this to be a six-fold degenerate point because it is one of six symmetrically equivalent points in the FBZ. However, the point diametrically opposite this in the FBZ, at $(-\pi/a, 0)$, does not produce a separate independent Bloch sum because, if we combine these complex functions to form a real (cosine) wave, the other real (sine) wave puts a node at every real lattice point, meaning that the basis function is multiplied everywhere by zero, which is not acceptable. This is similar to what we saw in the one-dimensional case, which led us to exclude $-\pi/a$ from the range of k. So here again we recognize that two points on opposite sides of the FBZ where perpendicular bisectors intersect lines between reciprocal lattice points really only account for one independent state. This means that the point M represents three independent Bloch sums rather than six. We shall have more to say about this later. The point labeled K does not lie on a **b** vector, so must be expressible as a sum of two vectors. Figure 15-30g shows the relevant construction. Since the lowest ray has length π/a, the central ray has length $\pi/(a\cos 30°)$. Then the length of k_1 or k_2 must be $\pi/a(2\cos^2 30°)$, or $2\pi/3a$. So $(k_1, k_2) = (2\pi/3a, 2\pi/3a)$ at K. As will be explained later, K does not have reduced degeneracy as M does.

Since there are two carbon atoms per unit cell and four valence AOs per carbon, we have eight basis functions per unit cell for a minimal valence basis set calculation. Therefore, we expect eight bands. Since the system is planar we can distinguish two π AOs and six σ AOs. We know that these two symmetry types will not interact for any **k**, so we can predict or interpret the σ and π band behaviors independently.

Let us begin by making some predictions about the two π bands. As was the case for polyacetylene, the two basis sets that are simplest to work with are the bonding (π_u) and antibonding (π_g) combinations in a unit cell. We shall label the band resulting from the bonding function π and the other band π^*. At the point Γ, the unit cell basis functions are simply translated along $\mathbf{a}_1$ and $\mathbf{a}_2$ with no change in phase, yielding the COs sketched in Fig. 15-31a. The π CO is bonding in every bond, π^* is antibonding, so these will be widely separated in energy. Indeed, there is no way they could become more widely separated by intermixing, so we can assume that, if we were to do a variational calculation, these Bloch sums would not be modified.

At point M, translation by each step along $\mathbf{a}_1$ is accompanied by phase reversal because $k_1 = \pi/a_1$, whereas translation along $\mathbf{a}_2$ involves no change because $k_2 = 0$ (Fig. 15-31b). Now we find each carbon to be bonding to two neighbors and antibonding to the third in the π CO and the reverse in the π^*CO. Once again, it is not hard to see that any mixing of these two would decrease their energy difference (Problem 15-24), so these Bloch sums should be unchanged in a variational calculation. Therefore, the π and π^* bands should be separated in energy at M, but only by about one third as much as at Γ. As we mentioned above, the COs sketched in Fig. 15-31b are each degenerate with two other COs corresponding to symmetrically equivalent points in the FBZ, for example, the points $(0, \pi/a)$ and $(\pi/a, -\pi/a)$. Sketching these (Problem 15-25) shows that they are like the CO already sketched except rotated by $\pm 60°$.

Point K is more complicated to analyze because it has k values that do not produce COs having the periodicity of the crystal. This means that the decreased degeneracy we find for high-symmetry points like X in Section 15-15, caused because one of the real coefficient waves has a node at every unit cell, does not occur at K. (If the wave lacks the periodicity of the crystal, it cannot put a node at every unit cell, by definition.) That means we will have two π and two π^* solutions at K. Also, the fact that these waves lack the periodicity of the crystal makes them more complicated to sketch. Nevertheless, undaunted, we plunge ahead. Since K corresponds to $(2\pi/3a, 2\pi/3a)$, we expect steps along either $\mathbf{a}_1$ or $\mathbf{a}_2$ in real COs to be modulated by either $\cos(2n\pi/3)$ or $\sin(2n\pi/3)$. For $n = 0, 1, 2, 3$, this gives the following repeating series for the cosine

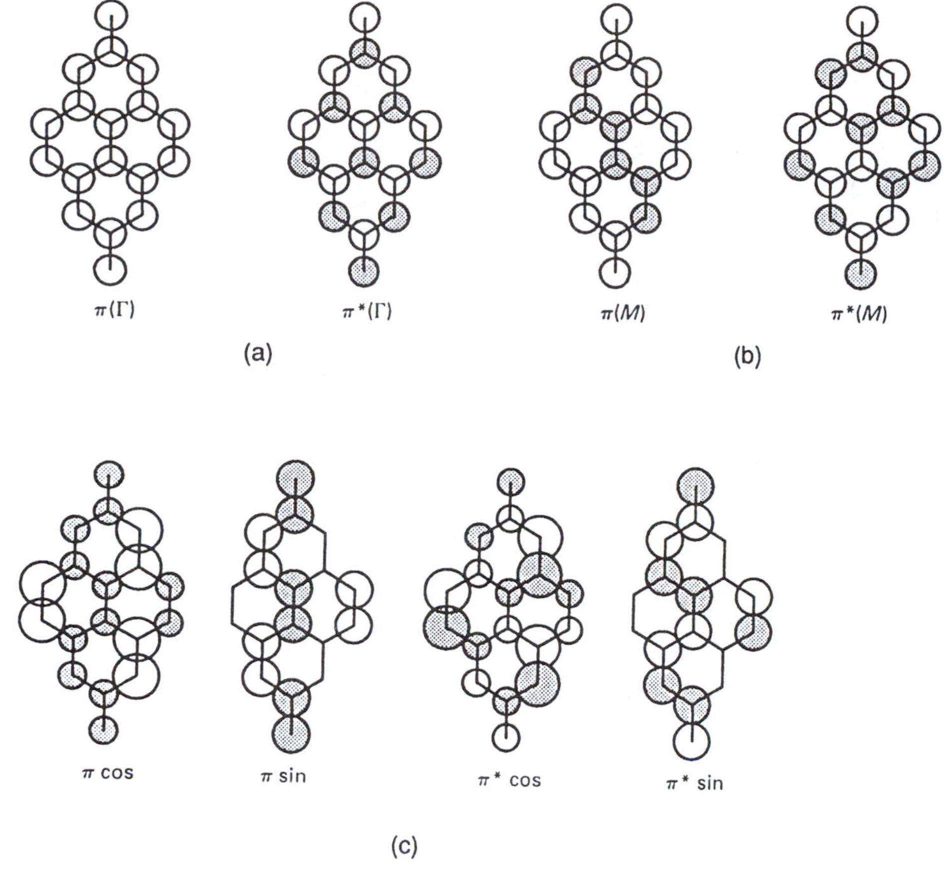

Figure 15-31 ▶ π and π^* COs at the point (a) Γ, (b) M, (c) K.

and sine cases, respectively: $1, -0.5, -0.5, 1$ and $0, 0.866, -0.866, 0$. The results of using these factors to modulate translations of the π_u and π_g basis functions are sketched in Fig. 15-31c. Inspection of these sketches shows that the extent of bonding and antibonding interactions around any carbon exactly cancels in every case. This means that the π and π^* bands are degenerate at K. Each of the pairs of COs sketched in Fig. 15-31c is degenerate with two other pairs that are rotated by $\pm 60°$. Therefore, the degeneracy of this energy level is 12.

The band diagram for two-dimensional graphite is sketched in Fig. 15-32 for a circuit around the RFBZ. The π and π^* bands behave essentially as we anticipated, being widely separated at Γ, much less separated at M, and degenerate at K. We note that there are three σ bands at lower energy and three more at higher energy, corresponding to predominantly bonding and antibonding situations, respectively. However, these bands are less easily analyzed because the 2s, $2p_x$, and $2p_y$ AOs undergo varying extents of mixing with each other as k varies. That is, these are bands where the variational method causes extensive mixing of Bloch sums. Nevertheless, we can see patterns that would not be difficult to investigate. For example, we can see that the π band and one of the σ bands run up from Γ to M, whereas the other two σ bands run down. This is consistent with the fact that the 2s and $2p_\pi$ AOs relate to corresponding AOs in adjacent unit cells in one way, while $2p_\sigma$ AOs relate in the opposite way (Problem 15-28).

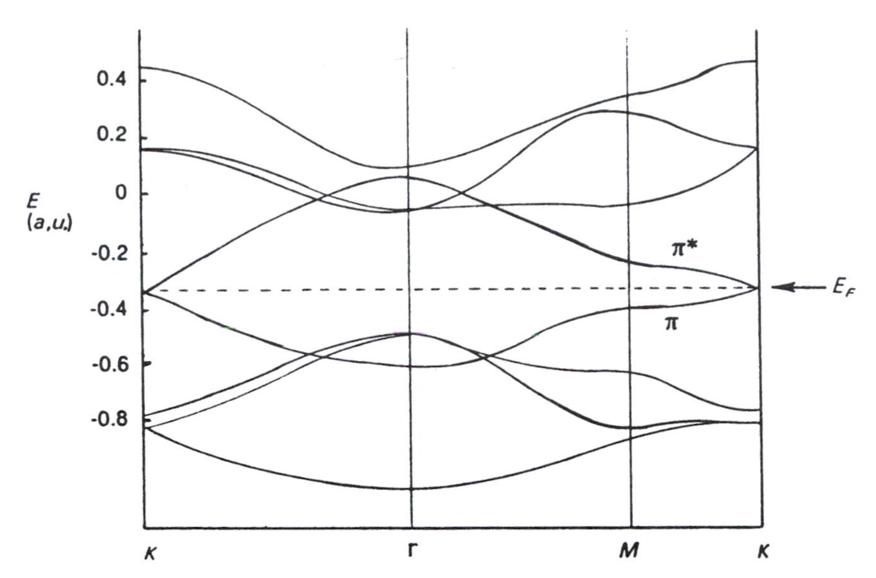

Figure 15-32 ▶ Valence-band diagram for two-dimensional graphite. (Adapted from Painter and Ellis [6].)

There are eight valence electrons per unit cell in neutral graphite. For the band diagram of Fig. 15-32 it is simple to figure out the band-population scheme. Eight electrons require four bands, and we see that the four lowest-energy bands lie below the other four bands for *all* values of k. This means that the four lowest-energy bands are filled and the others are empty. Therefore, the Fermi level comes at the top of the π band, which occurs at K. We see that there is an empty band (π^*) at the same energy, so there is no gap at the Fermi level. Thus the two-dimensional lattice is predicted to be a good electrical conductor. We could examine the possibility that lattice deformation would split the levels to produce a gap (a Peierls distortion), but, since we have structural information indicating that graphite does in fact have hexagonal symmetry, we will not pursue that point.

If there were crossing of one of the σ^* COs with the π CO over part of the range of k, then some fraction of the electronic population would occupy the low-energy portion of the σ^* band at the expense of the high-energy portion of the π band. Analysis of this would require more effort, but the result would still predict good conductivity since, instead of a full band "touching" the bottom of an empty band, we would have two partially filled bands.

Next we consider the effects of stacking two-dimensional sheets to form the three-dimensional crystal.[4] As Fig. 15-30a indicates, the stacking pattern repeats the orientation of a sheet after one intervening layer in an ABABAB stacking pattern. This means that the three-dimensional unit cell must contain carbon atoms from two layers. The unit cell now has three associated translation vectors. The two "intrasheet" translations $\mathbf{a}_1$ and $\mathbf{a}_2$ are as before, and the "intersheet" translation $\mathbf{a}_3$ is perpendicular to the planes of the sheets and 7.0 Å long. Because $\mathbf{a}_3$ is the longest vector in real space, $\mathbf{b}_3$ is the shortest vector in reciprocal space, leading to a reciprocal lattice where sheets

[4]A discussion of effects due to other stacking arrangements can be found in LaFemina and Lowe [7].

of centered hexagons are layered with short intersheet distances. The resulting FBZ, produced by perpendicular bisecting planes (since we are now working in three dimensions), appears in Fig. 15-33a. The RFBZ is similar to that in Fig. 15-30f except that it now has a third dimension, over which k_3 ranges from 0 to π/a_3 (Problem 15-26).

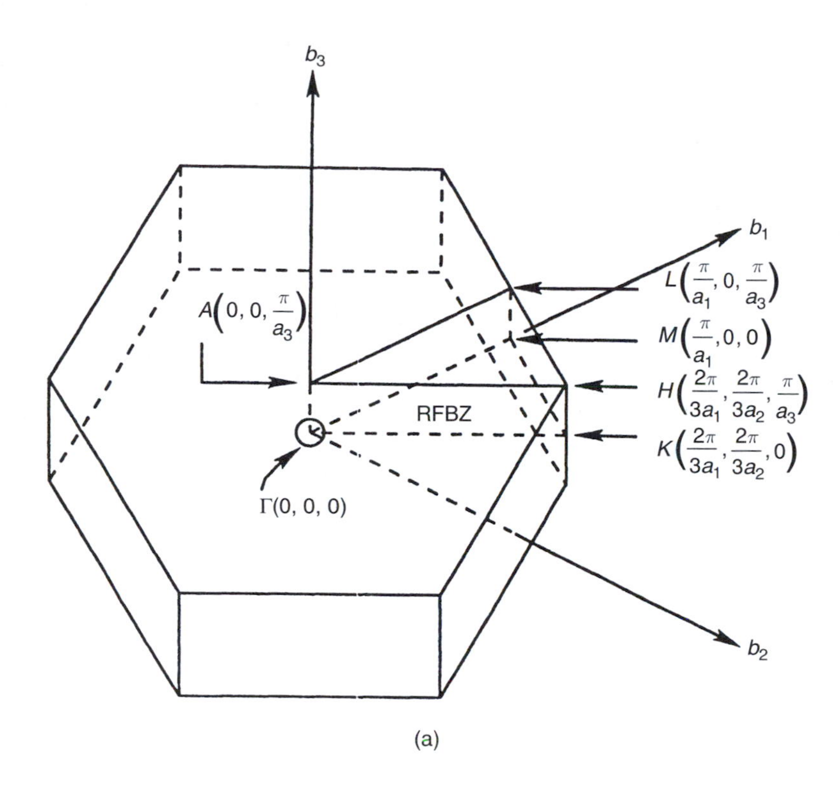

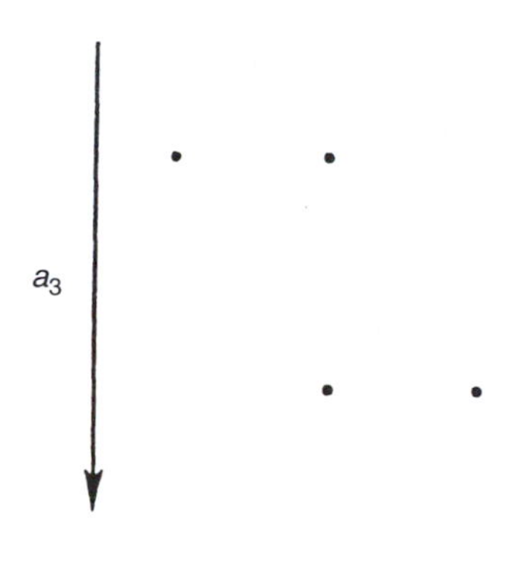

Figure 15-33 ▶ (a) The FBZ and RFBZ for three-dimensional Bernal graphite. (b) View of the unit cell for Bernal graphite, as seen along an axis parallel to the hexagonal sheets. (Compare to Fig. 15-30a.)

Let us see if we can predict the modifications in energies of COs that should occur at some of the special points in the RFBZ as a result of intersheet interactions. Because three-dimensional Bernal graphite is symmetric for reflection through the plane of a hexagonal sheet, we continue to have a valid distinction between σ and π COs. We expect intersheet interactions to be small, so we can take a perturbational approach. The interactions involving π COs should be largest since p_π AOs project farthest from the plane, so we will restrict our attention to the π bands.

A side view of the unit cell appears in Fig. 15-33b. Obviously, some of the intersheet interactions are already included in the unit cell, and the others will occur between unit cells. We will first consider the energy changes due to new interactions within the cell. These do not depend on k values. Then we will consider energy changes caused by stacking the unit cells. These depend on k_3. We will ignore all interactions except those between carbons directly above or below each other in adjacent sheets.

We begin with the situation where the π and π^* bands are nondegenerate, e.g., at Γ or M. We know from our two-dimensional treatment that each covalently bonded pair of carbons in the unit cell will have a π_u MO when we are dealing with the π band and a π_g MO for the π^* band. (See Fig. 15-31a, b.) But we have a choice of arranging these in the unit cell to give long-range bonding or antibonding between layers (Fig. 15-34). These two arrangements cause the energies to rise or fall slightly (indicated by $E\uparrow$ or $E\downarrow$), splitting both the π and π^* bands.

Next we examine what happens when we stack these unit cells. Consider first the point Γ. Here $k_3 = 0$, so we stack the unit cells with no change. If we start with π_{anti} of Fig. 15-34a, we obtain Fig. 15-35a, which shows that the unit cells interact in an antibonding manner. Here, then, is a Bloch sum that is antibonding within and also between unit cells. The symbol $E\uparrow\uparrow$ indicates two destabilizing interactions. Each arrow represents the same amount of energy because the interlayer distance within a unit cell is the same as that between unit cells and also because there are equal numbers of inter- and intralayer interactions in the crystal. Similar sketches easily demonstrate how the other three unit-cell function sets of Fig. 15-34 behave at Γ: π_{bond}, $E\downarrow\downarrow$; π^*_{anti}, $E\uparrow\uparrow$; π^*_{bond}, $E\downarrow\downarrow$. Therefore, at Γ, the π and π^* bands are both split.

Now let us consider point A. Here we stack the unit cells together with a reversal in sign. The situation for π_{anti} is shown in Fig. 15-35b. The intercell interaction has reversed from the $k_3 = 0$ case. This means that the $E\uparrow\uparrow$ and $E\downarrow\downarrow$ cases from Γ now are $E\uparrow\downarrow$ and $E\downarrow\uparrow$. The levels are *not* split by interlayer effects at A.

When we consider points M and L, we again are dealing with the same π_u and π_g functions in the unit cell (see Fig. 15-31b), so we obtain the same results with $k_3 = 0$, or π/a_3. The π and π^* bands are split at M by the same amount as at Γ, and at L they are not split at all.

At point K, the π and π^* bands are degenerate. This means we must ask which are the proper zeroth-order two-dimensional COs with which to analyze the perturbation. We know that these will be four COs produced by mixing the four COs shown in Fig. 15-31c. This mixing is determined by solving a 4×4 determinantal equation, which, as was pointed out in Chapter 12, is the same equation we would use to discover the COs having minimum and maximum energy changes as a result of the perturbation. As a shortcut, then, we can simply look for the COs that respond most differently. Our analysis at Γ and M tells us that the π_u and π_g bases will give a splitting of all the

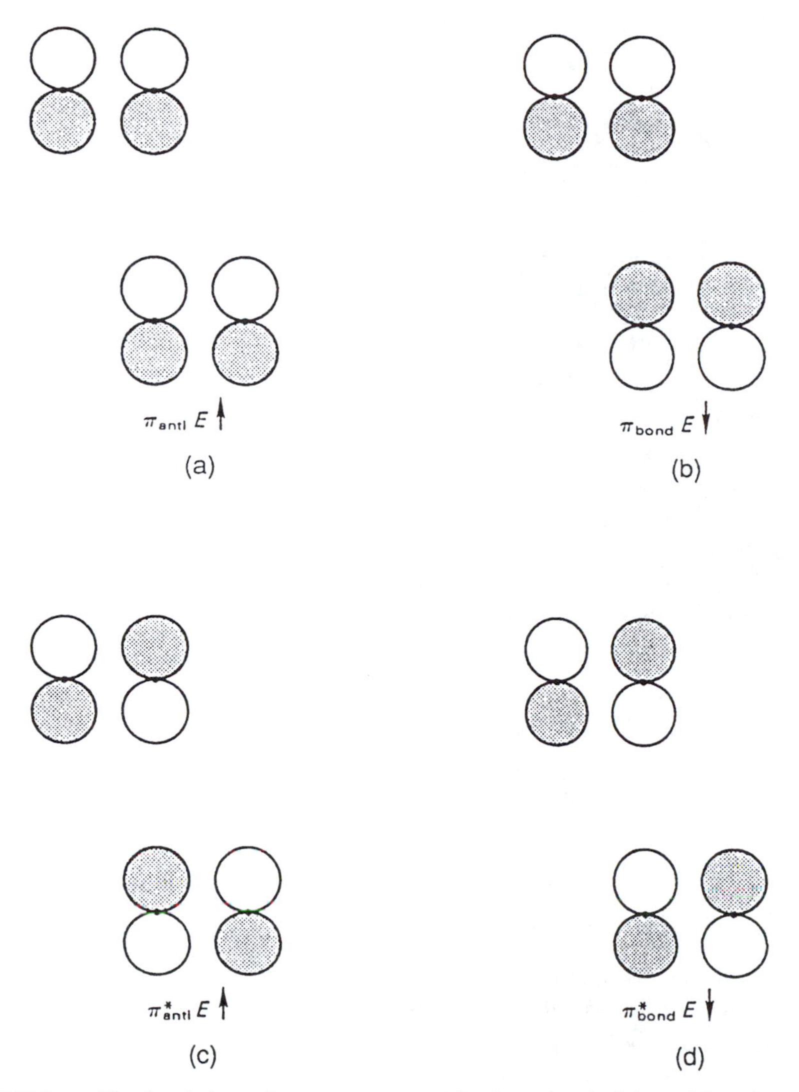

Figure 15-34 ▶ The four independent arrangements of π (or π_u) and π^* (or π_g) functions in a unit cell. Labels "anti" and "bond" refer to the interactions between AOs directly above/below each other. $E\uparrow$ and $E\downarrow$ indicate whether these interactions raise or lower the energy.

COs. Not bad, but we can do even better by mixing the COs pictured in Fig. 15-31c. Adding and subtracting the $\cos\pi^*$ and $\cos\pi$ COs produces two new functions whose unit cells (in two dimensions) have a $2p_\pi$ AO on one carbon and nothing on the other, and vice versa. (In effect, we are taking the sum and difference of the π_u and π_g basis functions that we constructed in the first place.) These new basis functions give us four new unit cell bases for the three-dimensional case, shown in Fig. 15-36. Two of these place all the $2p_\pi$ AOs on atoms that lie directly above and below each other, so these rise or fall in energy due to the intracell interactions. The other two place all the $2p_\pi$ AOs on carbons that do *not* have neighbors directly above or below them, so these do not undergo energy change. The rise and fall here are larger than for the cases pictured in Fig. 15-34 because the AO coefficients are larger when the basis set exists on

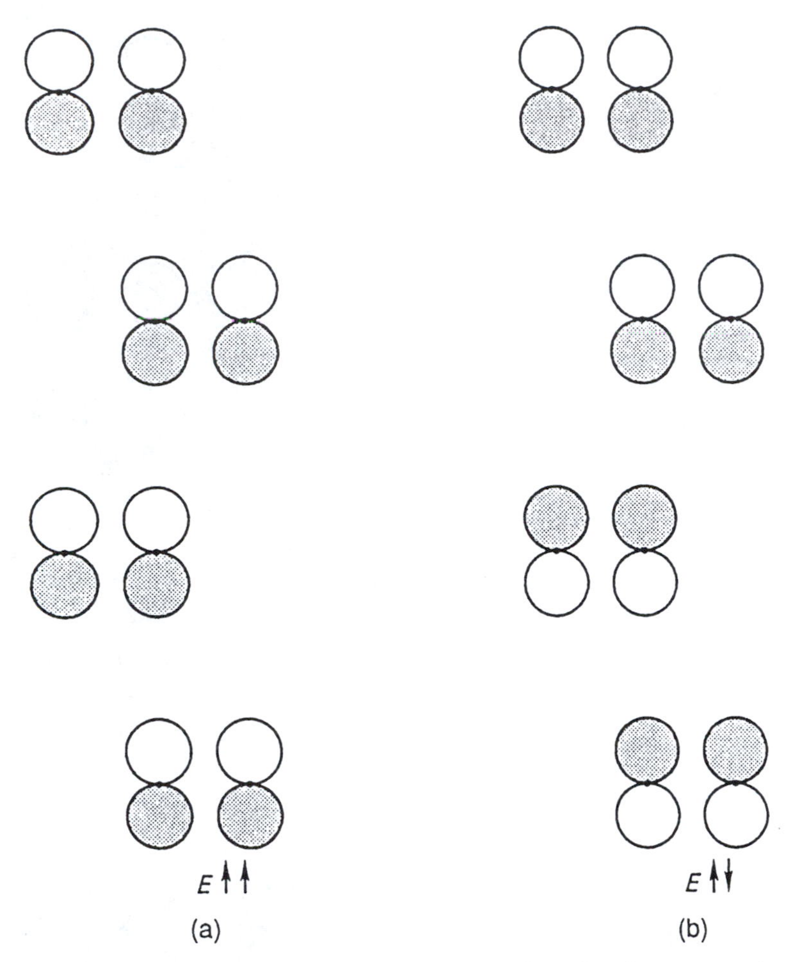

Figure 15-35 ▶ (a) The result of translating a π_{anti} unit-cell basis function one step along a_3 with no change of sign or value ($k_3 = 0$). (b) Same as (a) except now sign reverses ($k_3 = \pi/a$).

fewer atoms. These, then, are the proper zeroth-order functions with which to analyze the stacking perturbation. (The argument here is equivalent to saying that a calculus min-max problem will select roots $-2, 0, 0, +2$ in preference to $-1, -1, +1, +1$.)

When we put these unit-cell bases together at K, where $k_3 = 0$, we find that the functions that do not interact within the cell do not interact between cells either (Fig. 15-37a). These two cases remain unsplit in energy. The cases that interact within the cell interact again and in the same way between cells, so these two cases are split even more (Fig. 15-37b). At L, the intercell interactions reverse but the intracell interactions are unchanged. The $E^{\cdot}$ cases are unaffected, but $E\uparrow$ and $E\downarrow$ become $E\uparrow\downarrow$ and $E\downarrow\uparrow$ (Fig. 15-37c), just as at A and H, so the splitting disappears at L.

A sketch of the π and π^* bands for a route around the edges of the RFBZ appears in Fig. 15-38. In general, no splitting of these bands occurs for points on the top of the RFBZ. For points on the bottom, splitting is larger at K than at Γ or M.

Computing the electronic energy per unit cell for graphite requires knowing the average energy of the occupied bands, not only over the paths shown but also over the interior points of the RFBZ. The method of Chadi and Cohen [5] can be used to select

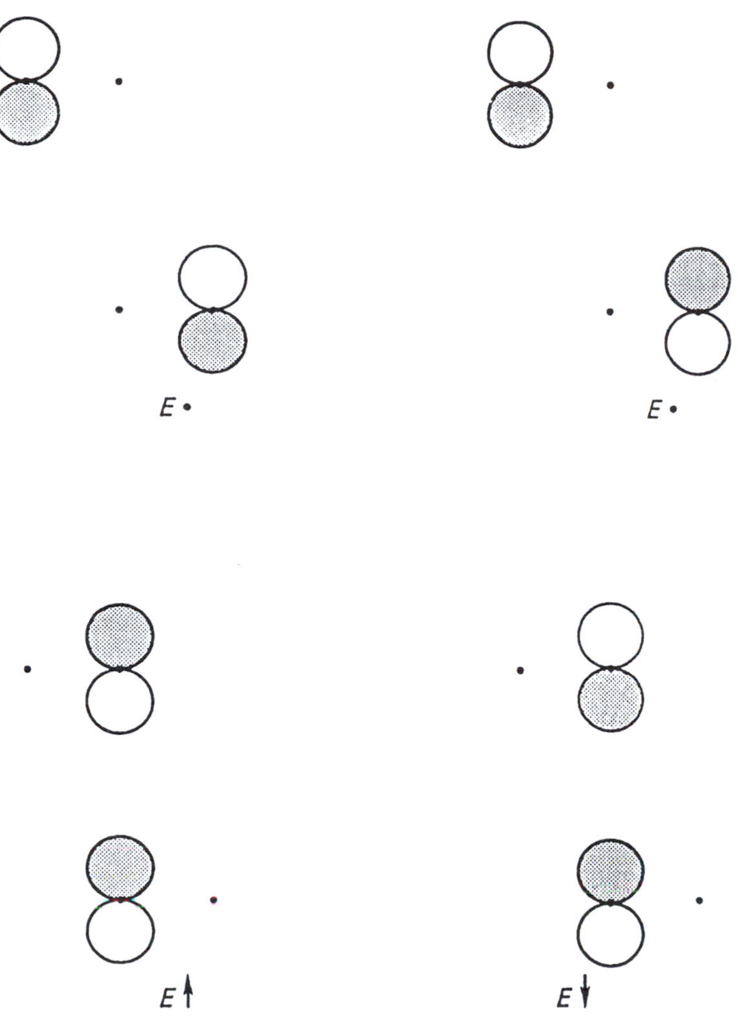

Figure 15-36 ▶ The four independent unit-cell functions that result from requiring maximum difference in response to the interlayer interaction. $E^{\cdot}$ indicates no energy change.

an optimized set of weighted points in the RFBZ for purposes of calculating accurate average energies.

One might wonder whether the splittings depicted in Fig. 15-38 have any detectable consequences. One change that occurs as a result of this splitting is that the Fermi level, E_F, now refers to π electrons in COs that are not split along K–H by the perturbation, i.e., in COs like the one pictured in Fig. 15-37a. This means that the electrons having the highest energy are predicted to be in $2p_\pi$ AOs at carbons that do not have carbon atoms directly above or below them in the crystal. This comes about because the electrons that do have such neighbors are lowered slightly in energy due to weak bonding between layers. This is relevant because there exists an experimental technique called scanning tunneling microscopy (STM) that detects the locations of surface electronic charge having energy near the Fermi level. When STM maps are made of electron distribution at a clean graphite surface, a trigonal pattern is seen, rather than the expected hexagonal one. Evidently, not all the atoms are "seen" equally well. The trigonal pattern is

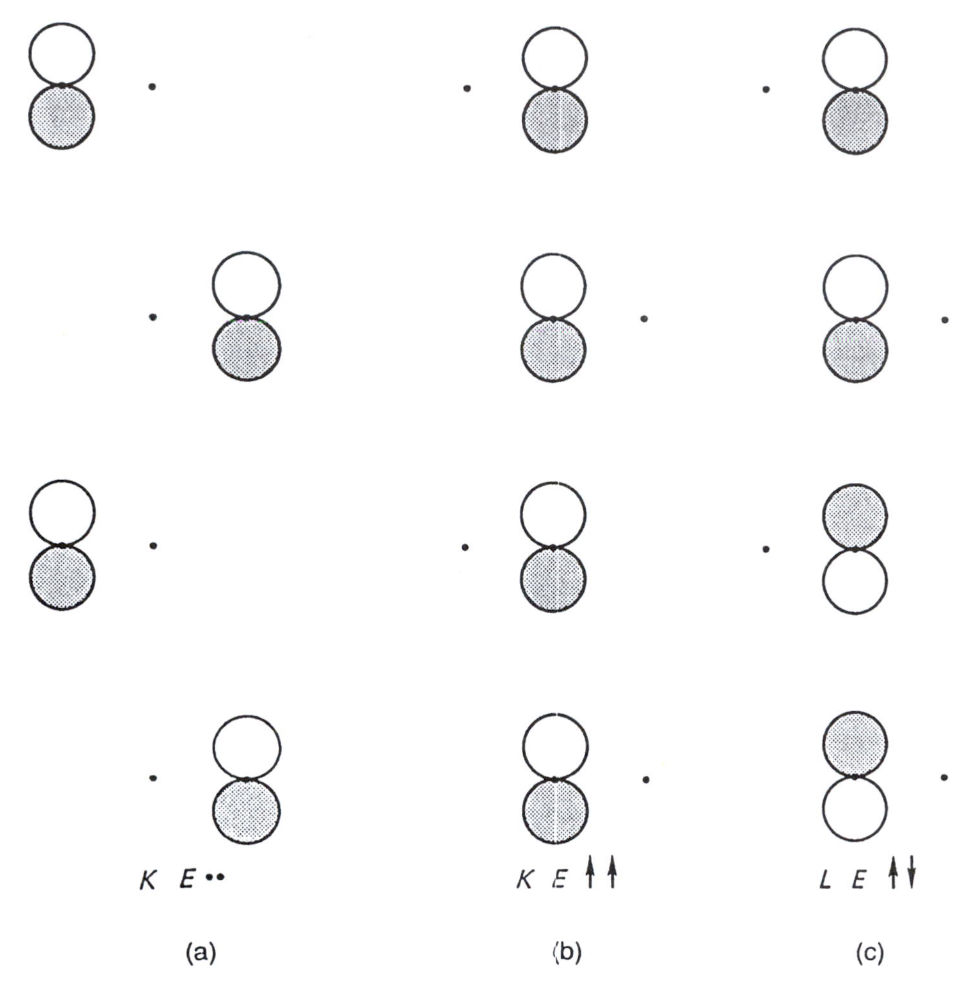

Figure 15-37 ▶ (a and b) Result of translating two of the functions from Fig. 15-36 along a_3 at point K ($k_3 = 0$, so no change in function). (c) Same as (b) except now at point L ($k_3 = \pi/a$, so function reverses sign).

consistent with more charge being detected over every-other atom. If the device is tuned to include electronic charge from deeper energies, the familiar hexagonal pattern of graphite emerges. This suggests that the former experiment detects charge in the unsplit, horizontal band shown between K and H, corresponding to COs on atoms without neighbors in the adjacent layer, and that the latter measurement detects charge in both that band and the one just below it. (Of course, the situation at the surface is not identical to that in the bulk, but surface carbons having carbons directly below them will still have their $2p_\pi$ energies lowered somewhat by the weak bonding interaction.)

15-17 Summary

Understanding or predicting the nature of the electronic band structure for a periodic material requires that we identify a unit cell and accompanying basis set in real space and then move this basis set along translation vectors $\mathbf{a}$, possibly with accompanying

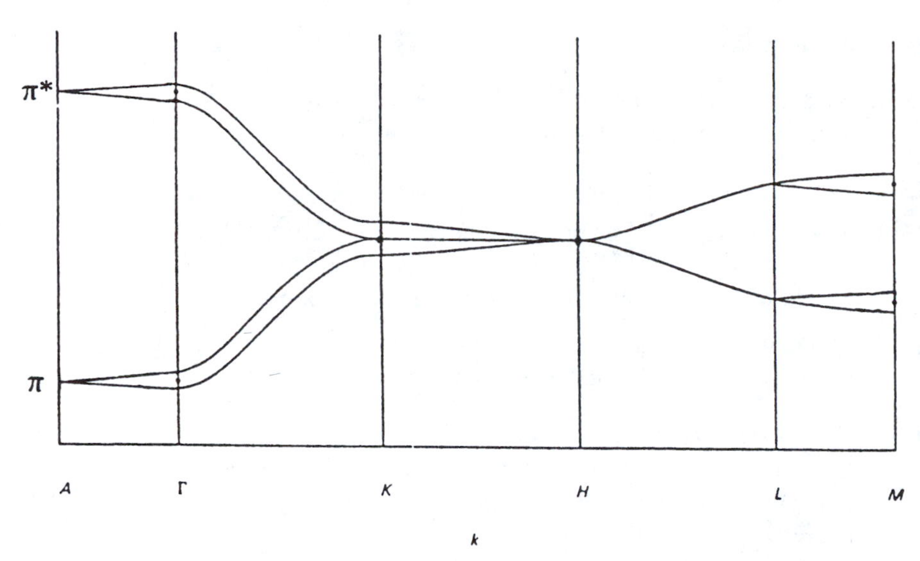

Figure 15-38 ▶ Sketch of π and π^* band energies for a circuit around the RFBZ of three-dimensional graphite. Heavy dots at Γ, K, H, and M represent energies for the isolated sheet. Splittings at Γ and M should be about half that at K.

rotations. Each time the basis set is moved to a different cell, it is multiplied by a coefficient that can be taken from a formula of the type $\cos(kn\pi)$, where n is the number of steps taken through **a** and k is the wavenumber for the wave that determines the coefficient. This process generates Bloch sums. Sometimes Bloch sums are the same as the COs, sometimes they are not. The reasons they might be different are that (1) there may be intended correlations that violate the noncrossing rule and (2) the basis set mixture in a unit cell may change with k in order to take better advantage of the changing interaction between unit cells, even when there is no problem with band crossings.

Degeneracies at the Fermi level result in Peierls distortions in one-dimensional systems, opening a band gap. Degenerate levels, whether at a gap or not, also allow one to predict which energies at the edges of the band diagram will be unaffected by certain chemical substitutions. Both of these phenomena are easily analyzed using the principle of "maximum difference in response" from degenerate-level perturbation theory.

The values of k giving independent COs fall into a zone of reciprocal space called the first Brillouin zone. A subspace within it, called the reduced first Brillouin zone (RFBZ), contains all the k values that are needed to calculate every energy in the band system. Certain points at corners of the RFBZ are of special interest because they are easy to use for constructing Bloch sums. They often give waves having the same periodicity as the real lattice, which means that the cosine Bloch sum will give alternating positive and negative coefficients at each unit cell and the sine Bloch sum will give zero at each unit cell, hence not exist. This results in reduced degeneracy.

Points in the FBZ that are related by symmetry to a point in the RFBZ correspond to COs that differ by a symmetry operation (e.g., a rotation) from the CO produced by the point in the RFBZ.

Readers interested in further treatment of either the chemical or physical aspects of periodic systems should consult references at the end of this chapter [1-11].

15-17.A Problems

15-1. Work out the relations in Eq. (15-3) (for a particle of mass m) so that it can be written as equalities rather than proportionalities.

15-2. Derive the functional form for Fig. 15-2b.

15-3. a) Obtain MOs and energies for benzene using Eqs. (15-4) and (15-5) with $j = 0 - 5$.
b) Show that the results for $j = 6$ are identical to those for $j = 0$.

15-4. a) Derive the mathematical formula for the curve of Fig. 15-5b.
b) What would this function be if DOS/N were constant over the energy range?

15-5. For benzene, $j = 1$ and $j = -1$ give different complex degenerate MOs. Produce these and find linear combinations that are real and orthogonal. Sketch these real MOs and compare them with those tabulated in Appendix 6 for the same energy.

15-6. Demonstrate that Eq. (15-10) is correct if N is an integer.

15-7. Using sketches, show that the Bloch sum for $(C_2H_2)_n$ at $|k| = \pi/3a$ is different from that for $(CH)_{2n}$ at $|k| = \pi/6a$. Which do you think has the lower energy? What will happen in a variation calculation on each of these two cases?

15-8. Suppose an infinite straight chain of atoms existed with a band of s orbitals that was one-third filled. Indicate what sort of bond-length pattern should result from a Peierls distortion.

15-9. Figure 15-14 indicates that the $2s$ and $2p_\sigma$ Bloch functions avoid crossing. Use symmetry to argue why this is so.

15-10. In connection with Fig. 15-18b, the text indicates that the avoided crossing between σ levels for alternating bond lengths becomes an allowed crossing when uniform bond lengths are used. This means that there is a symmetry disagreement for uniform bond lengths that does not exist for alternating bond lengths. The new symmetry element that comes into existence with uniform bond lengths is a two-fold screw axis. (The operation corresponding to this element is to rotate the polymer by 180° about the axis and also to translate it parallel to the axis.) Show, using sketches, that these σ bands disagree in symmetry for such a symmetry operation.

15-11. Crystalline H_2(H–H $\cdots$ H–H $\cdots$ H–H·) is an insulator. It is thought that it should become a metallic conductor at extremely high pressures. Explain why this is so.

15-12. A heteronuclear diatomic molecule A–B crystallizes end to end to form a linear chain: ...A–B $\cdots$ A–B $\cdots$ A–B. The internuclear distance between molecules is significantly longer than that within molecules. The monomer has a bonding valence σ MO and a higher energy σ^* MO:

Sketch a band structure diagram for the crystal showing qualitatively the energies of the crystal orbitals as a function of $|k|$ in the first Brillouin zone. Sketch the appearances of the COs for the σ and σ^* bands at $|k| = 0$ and π/a.

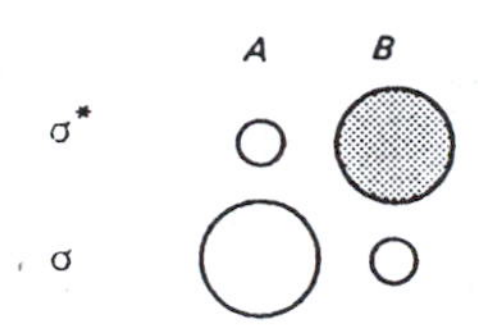

15-13. André and Leroy [2] report an ionization energy of 8.19 eV and an electron affinity of −4.09 eV (i.e., energy is released in forming the anion) from applying Koopmans' theorem to their *ab initio* calculations on *trans*-polyacetylene with regular bond lengths. a) What is the energy of the HOMO? b) Of the LUMO? c) What is the value of the energy gap? d) What should happen to the values of these three numbers when the bond lengths are allowed to alternate?

15-14. It is stated in Section 15-11 that it is improper to interpret a Hückel band diagram having zero gap as meaning that the ionization energy (from the HOMO energy) equals the electron affinity (from the equal LUMO energy). Yet we do use HOMO and LUMO Hückel energies to estimate the IE and EA in Chapter 8. What is the explanation for this apparent disagreement?

15-15. In a variational EHMO calculation on PPP, the coefficients for π_1 at $k = 0$ are 0.37 at linking carbons (1 and 4) and 0.28 at the others. This results from mixing the Bloch function π_1 at $k = 0$ (with all coefficients the same) with another Bloch function at $k = 0$. Which Bloch function, π_x, can mix with π_1 at $k = 0$ to accomplish this modification? (See Fig. 15-23 for sketches.) What happens to the energy of the π_x band at $k = 0$ as a result of mixing with π_1? What happens to the coefficients in π_x?

15-16. When PPP-N2 is substituted at position 6, the band energy values of −13.94 eV and −5.64 eV at $k = \pi/a$ are not affected. Would you expect the same thing to happen if substitution occurred at position 5 instead of 6? Why or why not?

15-17. The PPP diagram of Fig. 15-23 is constructed from Bloch sums built from benzene MOs. We could have used the MOs for the other choice of unit cell shown in Fig. 15-21a. Obtain the appropriate unit cell simple Hückel MOs and energies from Appendix 6 and construct the appropriate PPP band diagram. Discuss the differences with Fig. 15-23.

15-18. Predict the first-order Hückel-level energy band diagram for

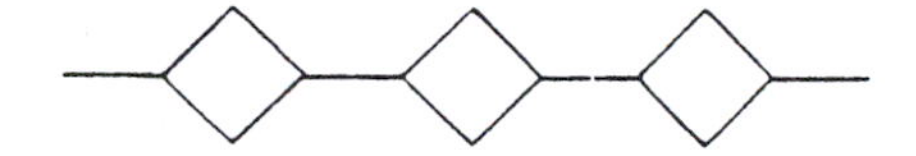

Show the actual band-edge energies in terms of α and β. Show how the edge energies at $k = 0$ and π/a would shift if substitution occurred in all the nonlinking sites so that $\alpha' = \alpha + 0.25\beta$, with no change in β. What if substitution occurred in only one nonlinking site in each monomer?

15-19. Sketch the CO for point X' in Fig. 15-27c for s-type unit-cell basis functions. Comparing with Fig. 15-28, how would you expect the energy of this CO to compare to the energies of COs at Γ, X, and M?

15-20. Figure out the (k_1, k_2) coordinates and sketch the COs for s-type basis functions for the following points in the RFBZ of Fig. 15-27c. a) Halfway between X and M. b) Halfway between X' and M. () Halfway between Γ and X. d) Halfway between Γ and M. Rank these in order of energy.

15-21. Sketch COs based on a $2p_x$ AO at each lattice point for the RFBZ of Fig. 15-27c, at points Γ, X, and X'. Rank according to predicted energy.

15-22. What degeneracy would you anticipate for energies associated with points Γ, X, M, and X' of Fig. 15-27c?.

15-23. The rectangular lattice of Fig. 15-27a leads to a rectangular RFBZ. What shape RFBZ would result for a square lattice?

15-24. It is stated in the text that mixing $\pi(M)$ and $\pi^*(M)$ of Fig. 15-31b will decrease the difference in their energies. Sketch the functions $\pi(M) \pm \pi^*(M)$ and comment on their energies.

15-25. For the FBZ of Fig. 15-30f, locate the points $(0, \pi/a)$ and $(\pi/a, -\pi/a)$. Sketch the π CO corresponding to the first of these and compare it with the π CO for $(\pi/a, 0)$, sketched in Fig. 15-31b. How are these COs related?

15-26. What physical situation allows us to use the half-range of $0 \le k_3 \le \pi/a_3$ for the RFBZ pictured in Fig. 15-33a?

15-27. Give a general argument for there being no interlayer-induced splitting of π bands for any point at the top of the RFBZ of Fig. 15-33a. What does your argument predict for σ bands?

15-28. For two-dimensional graphite, sketch the Bloch sums at Γ and M resulting from a bonding pair of $2p_x$ AOs in each unit cell. Let x be the bisector of $\mathbf{a}_1$ and $\mathbf{a}_2$. Which of these COs would you expect to have higher energy?

References

[1] R. Hoffmann, C. Janiak, and C. Kollmar, *Macromolecules* **24**, 3725 (1991).

[2] J.-M. André and G. Leroy, *Int. J. Quantum Chem.* **5**, 557 (1971).

[3] J. P. Lowe and S. A. Kafafi, *J. Amer. Chem. Soc.* **106**, 5837 (1984).

[4] J. P. Lowe, S. A. Kafafi, and J. P. LaFemina, *J. Phys. Chem.* **90**, 6602 (1986).

[5] D. J. Chadi and M. L. Cohen, *Phys. Rev. B* **8**, 5747 (1973).

[6] G. S. Painter and D. E. Ellis, *Phys. Rev. B* **1**, 4747 (1970).

[7] J. P. LaFemina and J. P. Lowe, *Int. J. Quantum Chem.* **30**, 769 (1986).

[8] J. C. Slater, *Quantum Theory of Molecules and Solids*, Vol. 2. McGraw-Hill, New York, 1965.

[9] R. Hoffmann, *Solids and Surfaces: A Chemist's View of Bonding in Extended Structures*. VCH Publishers, New York, 1988.

[10] N. W. Ashcroft and N. D. Mermin, *Solid State Physics*. Holt, Rinehart and Winston, New York, 1976.

[11] C. Kittel, *Introduction to Solid State Physics*, 7th ed. Wiley, New York, 1996.

Appendix 1

Useful Integrals

$$\int x^n e^{ax}\,dx = (x^n e^{ax}/a) - (n/a)\int x^{n-1}e^{ax}\,dx$$

$$\int_0^\infty x^n e^{-ax}\,dx = (n!/a^{n.+1}) = \Gamma_{n+1}(a), \qquad n > -1, a > 0$$

$$\int_0^\infty e^{-ax^2}\,dx = \frac{1}{2}\sqrt{\pi/a}$$

$$\int_0^\infty x e^{-ax^2}\,dx = 1/2a$$

$$\int_0^\infty x^2 e^{-ax^2}\,dx = \frac{1}{4}\sqrt{\pi/a^3}$$

$$\int_0^\infty x^3 e^{-ax^2}\,dx = 1/2a^2$$

$$\int_0^\infty x^{2n} e^{-ax^2}\,dx = \frac{1\cdot 3\cdot\cdots\cdot(2n-1)}{2^{n+1}}\sqrt{\frac{\pi}{a^{2n+1}}}$$

$$\int_0^\infty x^{2n+1} e^{-ax^2}\,dx = \frac{n!}{2a^{n+1}}$$

$$\int_1^\infty e^{-ax}\,dx = \frac{e^{-a}}{a}$$

$$\int_0^1 e^{-ax}\,dx = (1/a)(1-e^{-a})$$

$$\int_1^\infty x e^{-ax}\,dx = (e^{-a}/a^2)(1+a)$$

$$\int_0^1 x e^{-ax}\,dx = (1/a^2)[1-e^{-a}(1+a)]$$

$$\int_1^\infty x^2 e^{-ax}\,dx = (2e^{-a}/a^3)(1+a+a^2/2)$$

$$\int_0^1 x^2 e^{-ax}\,dx = (2/a^3)[1-e^{-a}(1+a+a^2/2)]$$

$$\int_1^\infty x^n e^{-ax}\,dx = (n!e^{-a}/a^{n+1})\sum_{k=0}^{n} a^k/k! \equiv A_n(a)$$

$$\int_y^\infty x^n e^{-ax}\,dx = (n!e^{-ay}/a^{n+1})\sum_{k=0}^{n}(ay)^k/k!$$

$$\int_{-1}^{+1} e^{-ax}\,dx = (1/a)(e^a - e^{-a})$$

$$\int_{-1}^{+1} xe^{-ax}\,dx = (1/a^2)[e^a - e^{-a} - a(e^a + e^{-a})]$$

$$\int_{-1}^{+1} x^n e^{-ax}\,dx = (-1)^{n+1}A_n(-a) - A_n(a)$$

$$\int_{-1}^{+1} x^n\,dx = \begin{cases} 0, & n = 1, 3, 5, \ldots \\ 2/(n+1), & n = 0, 2, 4, \ldots \end{cases}$$

$$\int \sin x\,dx = -\cos x$$

$$\int \cos x\,dx = \sin x$$

$$\int \sin^2 x\,dx = \frac{x}{2} - \frac{\sin 2x}{4}$$

$$\int \cos^2 x\,dx = \frac{x}{2} + \frac{\sin 2x}{4}$$

$$\int x\sin x\,dx = \sin x - x\cos x$$

$$\int x\cos x\,dx = \cos x + x\sin x$$

$$\int x\sin^2 x\,dx = \frac{x^2}{4} - \frac{x\sin 2x}{4} - \frac{\cos 2x}{8}$$

$$\int x\cos^2 x\,dx = \frac{x^2}{4} + \frac{x\sin 2x}{4} + \frac{\cos 2x}{8}$$

Appendix 2

Determinants

A determinant is a scalar calculated from an ordered set of elements according to a specific evaluation recipe. The elements are ordered in a square array of rows and columns, bounded at left and right by straight vertical lines. For instance, (A2-1) is a 2×2 determinant:

$$\begin{vmatrix} x & -i \\ 2 & y^2 \end{vmatrix} \tag{A2-1}$$

The recipe for evaluating a 2×2 determinant is: From the product of the elements on the principal diagonal (upper left to lower right) subtract the product of the other two elements. Thus, (A2-1) has the value $xy^2 + 2i$.

Larger determinants are evaluated by a process that reduces them step by step to a linear combination of smaller determinants until, finally, they are all 2×2's, which are then evaluated as above. The process of reduction involves the concept of a *cofactor*. As our example, we use the 4×4 determinant (A2-2), symbolized $|M|$, where M is the array of elements within the vertical bars:

$$|M| = \begin{vmatrix} a_{11} & a_{12} & a_{13} & a_{14} \\ a_{21} & a_{22} & a_{23} & a_{24} \\ a_{31} & a_{32} & a_{33} & a_{34} \\ a_{41} & a_{42} & a_{43} & a_{44} \end{vmatrix} \tag{A2-2}$$

The elements are numbered so that the first index tells which row, and the second index which column, the element is in. The cofactor of element a_{11} is defined as the determinant obtained by removing the row and column containing a_{11}. We see in (A2-2) that striking out row 1 and column 1 gives us a 3×3 determinant (dashed outline) as cofactor of a_{11}. Symbolize this cofactor as $|A_{11}|$.

To evaluate the determinant $|M|$, we expand in terms of cofactors. We begin by choosing any row or column of M. (We will choose row 1.) Then we write a linear combination containing every element in this row or column times its cofactor:

$$|M| = a_{11}|A_{11}| - a_{12}|A_{12}| + a_{13}|A_{13}| - a_{14}|A_{14}| \tag{A2-3}$$

The sign of each term in the linear combination is determined as follows. If the sum of row and column indices is even, the sign is plus. If the sum is odd, the sign is minus. Since the indices of a_{12} and a_{14} sum to odd numbers, they are minus in (A2-3).

The method of expanding in cofactors is successively applied until a large determinant is reduced to 3×3's or 2×2's that can be evaluated directly (see Problem A2-1). Thus, a 5×5 is first expanded to five 4×4's and each 4×4 is expanded to four 3×3's giving a total of 20 3×3's. This method becomes extremely clumsy for large determinants.

Some useful properties of determinants, symbolized $|M|$, are stated below without proof. The reader should verify that these are true using 2×2 or 3×3 examples, or by examining Eq. (A2-3).

1. Multiplying every element in *one* row or *one* column of M by the constant c multiplies the value of $|M|$ by c.
2. If every element in a row or column of M is zero, then $|M| = 0$.
3. Interchanging two rows or columns of M to produce M' results in $|M'| = -|M|$; i.e., it reverses the sign of $|M|$.
4. Adding to any row (column) of M the quantity c times any other row (column) of M does not affect the value of the determinant.
5. If two rows or columns of M differ only by a constant multiplier, then $|M| = 0$.

A2-1 Use of Determinants in Linear Homogeneous Equations

Suppose that we seek a nontrivial solution for the following set of linear homogeneous equations:

$$a_1x + b_1y + c_1z = 0 \tag{A2-4}$$

$$a_2x + b_2y + c_2z = 0 \tag{A2-5}$$

$$a_3x + b_3y + c_3z = 0 \tag{A2-6}$$

Here, x, y, and z are unknown and the coefficients a_i, b_i, c_i are given. Let us collect the coefficients into a determinant $|M|$,

$$|M| = \begin{vmatrix} a_1 & b_1 & c_1 \\ a_2 & b_2 & c_2 \\ a_3 & b_3 & c_3 \end{vmatrix} \tag{A2-7}$$

As before, let $|A_1|$ be the cofacter of a_1, etc. Now, multiply Eq. (A2-4) by $|A_1|$ (since $|A_1|$ is a determinant, it is just a number, and so this is a scalar multiplication), Eq. (A2-5) by $-|A_2|$, and Eq. (A2-6) by $|A_3|$ and add the results to get

$$x(a_1|A_1| - a_2|A_2| + a_3|A_3|) + y(b_1|A_1| - b_2|A_2| + b_3|A_3|) \\ + z(c_1|A_1| - c_2|A_2| + c_3|A_3|) = 0 \tag{A2-8}$$

The coefficient of x is just $|M|$. The coefficients of y and z correspond to determinants having two identical rows and hence are zero. Therefore,

$$|M|x = 0 \tag{A2-9}$$

In order for x to be nonzero (i.e., nontrivial) it is necessary that $|M|=0$. This is a result that is very useful. *The condition that must be met by the coefficients of a set of linear homogeneous equations in order that nontrivial solutions exist is that their determinant vanish.*

A2-2 Problems

A2-1. Expand the 3×3 determinant (Fig. PA2-1), by cofactors and show that this result is equivalent to the direct evaluation of the 3×3 by summing the three products parallel to the main diagonal (*solid arrows*) and subtracting the three products parallel to the other diagonal (*dashed arrows*).

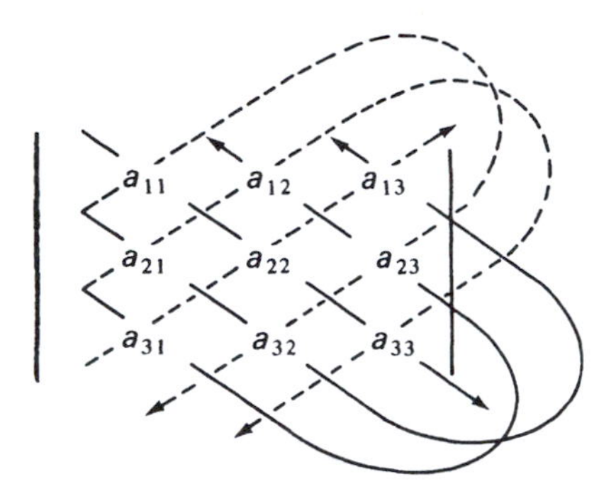

Figure PA2-1 ▶

A2-2. Evaluate (a) $\begin{vmatrix} 0 & 1 \\ 2 & 1 \end{vmatrix}$ (b) $\begin{vmatrix} 1 & 1 & 0 \\ 0 & 1 & 1 \\ 1 & 1 & 1 \end{vmatrix}$ (c) $\begin{vmatrix} 1 & 2 & 0 & 1 \\ 3 & 0 & 1 & 4 \\ 1 & 1 & 0 & 1 \\ 0 & 2 & 1 & 1 \end{vmatrix}$ (d) $\begin{vmatrix} x & 2 \\ 1 & x \end{vmatrix} = 0$ for x

A2-3. Verify that the coefficient of y in Eq. (A2-8) is zero.

A2-4. Consider the following set of linear homogeneous equations:

$$4x + 2y - z = 0, \quad 3x - y - 2z = 0, \quad 2y + z = 0$$

Do nontrivial roots exist?

A2-5. Find a value for c that allows nontrivial solutions for the equations

$$cx - 2y + z = 0, \quad 4x + cy - 2z = 0, \quad -8x + 5y - cz = 0$$

A2-6. Five properties of determinants have been listed in this appendix. (a) Prove statement (5) is true assuming statements (1)–(4) are true. (b) Demonstrate statements (1)–(4) using simple examples.

Appendix 3

Evaluation of the Coulomb Repulsion Integral Over 1s AOs

Evaluation of

$$\iint 1s(1)1s(2)(1/r_{12})1s(1)1s(2)\,dv(1)\,dv(2) \tag{A3-1}$$

where

$$1s(1) = \sqrt{\zeta^3/\pi}\exp(-\zeta r) \tag{A3-2}$$

may be carried out in two closely related ways.[1] Each method is instructive and sheds light on the other, and so we will give both of them here.

The first method works from a physical model and requires knowledge of two features of situations governed by the $1/r^2$ force law (e.g., electrostatics, gravitation). Suppose that there exists a spherical shell in which charge or mass is distributed uniformly, like the soap solution in a soap bubble. The first feature is that a point charge or mass *outside* the sphere has a potential due to attraction (or repulsion) by the sphere that is identical to the potential produced if the sphere collapsed to a point at its center (conserving mass or charge in the process). Thus, the electrostatic interaction between two *separated* spherical charge distributions may be calculated as though all the charge were concentrated at their centers. The second feature is that the potential is identical for all points *inside* the spherical shell; that is, if the core of the earth were hollow, a person would be weightless there. There would be no tendency for that person to drift toward a wall or toward the center.

Armed with these facts, we can evaluate the integral. First, we remark that all the functions in the integrand commute. This enables us to write Eq. (A3-1) as

$$\iint 1s^2(1)(1/r_{12})1s^2(2)dv(1)dv(2) \tag{A3-3}$$

The functions $1s^2(1)$ and $1s^2(2)$ are just charge clouds for electrons 1 and 2, and the integral is evidently just the energy of repulsion between the clouds. Suppose (see Fig. 3-1) that, at some instant, electron 1 is at a distance r_1 from the nucleus. What is its energy of repulsion with the charge cloud of electron 2? The charge cloud of electron 2 can be divided into two parts: the charge inside a sphere of radius r_1 and the charge outside that sphere. From what we just said, electron 1 experiences a repulsion due to the cloud *inside* the sphere that is the same as the repulsion it would feel if that

[1] Other methods, not discussed here, also exist. See, for example, Margenau and Murphy [1, pp. 382–383].

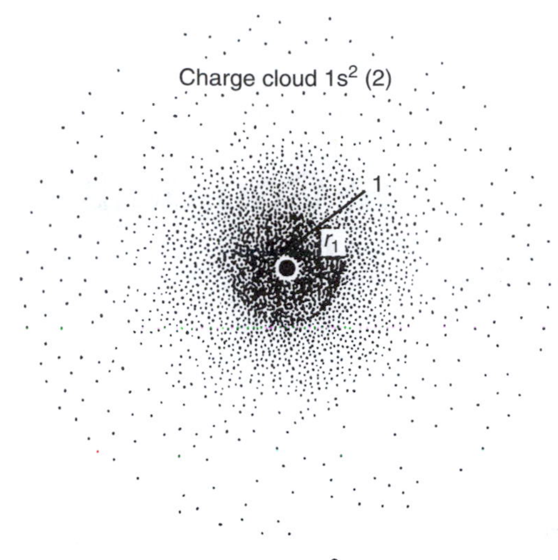

Figure A3-1 ► Sketch of spherical charge cloud $1s^2(2)$ with electron 1 at a distance r_1 from the nucleus.

part of the cloud were collapsed to the center. We can calculate the energy due to this repulsion (call it the "inner repulsion energy") by dividing the product of charges by the distance between them:

$$\begin{aligned}\text{inner repulsion energy} &= (\text{fraction of charge cloud 2 inside } r_1) \times 1/r_1 \\ &= 4\pi \int_0^{r_1} 1s^2(2) r_2^2 dr_2 \times 1/r_1 \qquad \text{(A3-4)}\end{aligned}$$

where the factor 4π comes from integrating over θ_2, ϕ_2. Electron 1 also experiences repulsion from charge cloud 2 *outside* the sphere of radius r_1. But, from what we said above, electron 1 would experience this same "outer repulsion" no matter where it was inside the inner sphere. Therefore we will calculate the energy due to this repulsion as though electron 1 were at the center, since this preserves spherical symmetry and simplifies the calculation. It follows that all the charge in a thin shell of radius r_2 repels electron 1 through an effective distance of r_2. Integrating over all such shells gives

$$\text{outer repulsion energy} = 4\pi \int_{r_1}^{\infty} (1/r_2) 1s^2(2) r_2^2 dr_2 \qquad \text{(A3-5)}$$

The total energy of repulsion between charge cloud 2 and electron 1 at r_1 is the sum of inner and outer repulsive energies. But electron 1 is not always at r_1. Therefore, we must finally integrate over all positions of electron 1, weighted by the frequency of their occurrence:

$$\begin{aligned}\text{repulsive energy} = 16\pi^2 \int_0^{\infty} 1s^2(1) &\left\{ (1/r_1) \int_0^{r_1} 1s^2(2) r_2^2 dr_2 \right. \\ &\left. + \int_{r_1}^{\infty} 1s^2(2) r_2 dr_2 \right\} r_1^2 dr_1 \qquad \text{(A3-6)}\end{aligned}$$

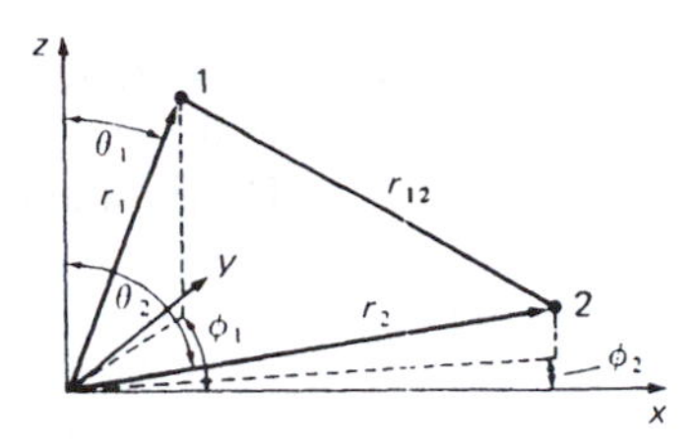

Figure A3-2 ▶

The inner integrals contain but one variable, r_2, and can be evaluated with the help of Appendix 1. After they are performed, the integrand depends only on r_1 and this is also easily integrated yielding $5\zeta/8$ as the result. A positive value is necessary since a net repulsion exists between two clouds of like charge.

The second method of evaluation is more mathematical and more general. The function $1/r_{12}$ is expressible[2] as a series of terms involving associated Legendre functions:

$$\frac{1}{r_{12}} = \sum_{l=0}^{\infty} \sum_{m=-l}^{+l} \frac{(l-|m|)!}{(l+|m|)!} \frac{r_<^l}{r_>^{l+1}} P_l^{|m|}(\cos\theta_1) P_l^{|m|}(\cos\theta_2) \exp[im(\phi_1 - \phi_2)] \quad \text{(A3-7)}$$

This infinite series will give the distance between particles 1 and 2 located at positions r_1, θ_1, ϕ_1 and r_2, θ_2, ϕ_2 (Fig. A3-2). All we need to do is pick the larger of r_1 and r_2 and call that $r_>$, the other being $r_<$, and substitute those numbers into the formula. We also need all the Legendre functions for $\cos\theta_1$ and $\cos\theta_2$, and values of $\exp[im(\phi_1 - \phi_2)]$ for all integral values of m. Putting all this together as indicated by Eq. (A3-7) would give a series of numbers whose sum would converge to the value of $1/r_{12}$. While this is an exceedingly cumbersome way to calculate the distance between two points, it turns out that use of the formal expression (A3-7) enables us to integrate Eq. (A3-1). This comes about because the Legendre functions satisfy the relation

$$\int_0^{\pi} P_l^{|m|}(\cos\theta) P_{l'}^{|m|}(\cos\theta) \sin\theta \, d\theta = \frac{2}{2l+1} \frac{(l+|m|)!}{(l-|m|)!} \delta_{ll'}. \quad \text{(A3-8)}$$

The first Legendre polynomial P_0 is equal to unity. Therefore, $1s(i)$, which has no θ dependence, can be written

$$1s(i) = (\zeta^3/\pi)^{1/2} \exp(-\zeta r_i) P_0(\cos\theta_i) \quad \text{(A3-9)}$$

Thus, in the integral (A3-1) there will be an integration over θ_1 of the form

$$\int_0^{\pi} P_0(\cos\theta_1) P_l^{|m|}(\cos\theta_1) P_0(\cos\theta_1) \sin\theta_1 \, d\theta_1 \quad \text{(A3-10)}$$

for each term in the sum (A3-7) (and a similar integral over θ_2). However since $1^2 = 1$, $[P_0(\cos\theta_1)]^2 = P_0(\cos\theta_1)$, and integral (A3-10) becomes [by Eq. (A3-8)]

$$\int_0^{\pi} P_0(\cos\theta_1) P_l^{|m|}(\cos\theta_1) \sin\theta_1 \, d\theta_1 = 2\delta_{0l} \quad \text{(A3-11)}$$

[2] See Eyring et al. [2, Appendix 5].

[m must equal zero here, otherwise the integral over ϕ will vanish.] In other words, all terms of the sum over l and m vanish except the first term, for which $l = m = 0$. This gives that the $1/r_{12}$ operator is equal to $1/r_>$. Hence, the repulsion integral is

$$16\pi^2 \int_0^\infty \int_0^\infty \left(\frac{1s^2(1)1s^2(2)r_1^2 r_2^2}{r_>} \right) dr_2 dr_1 \tag{A3-12}$$

where $r_>$ is the greater of r_1, r_2. Suppose that we integrate over r_2 first. As r_2 changes value, it is sometimes smaller, sometimes larger than a particular value of r_1. When it is larger, $r_>$ is r_2. When it is smaller, $r_>$ is r_1. Putting this argument into mathematical form gives

$$16\pi^2 \int_0^\infty \left\{ \int_0^{r_1} \frac{1s^2(1)1s^2(2)r_1^2 r_2^2 dr_2}{r_1} + \int_{r_1}^\infty \frac{1s^2(1)1s^2(2)r_1^2 r_2^2 dr_2}{r_2} \right\} dr_1 \tag{A3-13}$$

Since the variable of integration in the two inner integrals is r_2, the quantities $1s^2(1)$, r_1^2, and r_1 may be brought outside these inner integrals, giving us the same equation (A3-6) that we obtained by the first method. This second method is more generally useful because it can be used when repulsions involving p, d, etc. charge clouds are calculated. In these cases, terms involving $l = 1, 2$, etc., become nonvanishing, but the series generally truncates after a few terms.

References

[1] H. Margenau and D. M. Murphy, *The Mathematics of Physics and Chemistry*. Van Nostrand-Reinhold, Princeton, New Jersey, 1956.

[2] H. Eyring, J. Walter, and G. E. Kimball, *Quantum Chemistry*. Wiley, New York, 1944.

Appendix 4

Angular Momentum Rules

A4-1 Introduction

In Chapters 4, 5, and 7, we deal with the angular momentum of electrons due to their orbital motions and spins and also with angular momentum due to molecular rotation. The conclusions we arrive at are based partly on physical arguments related to experience with macroscopic bodies. While this is the most natural way to introduce such concepts, it is not the most rigorous. In this appendix we shall demonstrate how use of the postulates discussed in Chapter 6 leads to all of the relationships we have introduced earlier. The approach is based entirely on the mathematical properties of the relevant operators, making no appeal to physical models.

A4-2 The Classical Expressions for Angular Momentum

According to postulate II (Section 6-3), we must first find the classical physical expressions for the quantity of interest in terms of x, y, z, p_x, p_y, p_z, and t. We know that the angular momentum for an object of mass m moving with velocity v in a circular orbit of radius r is

$$\mathbf{L} = m\mathbf{r} \times \mathbf{v} = \mathbf{r} \times \mathbf{p} \tag{A4-1}$$

where the $\times$ symbol means we are taking a cross-product of vectors. To examine this in detail, we must resolve $\mathbf{r}$ and $\mathbf{p}$ into x, y, and z components:

$$\mathbf{r} = \mathbf{i}x + \mathbf{j}y + \mathbf{k}z \tag{A4-2}$$

$$\mathbf{p} = m\mathbf{v} = m\mathbf{i}\frac{dx}{dt} + m\mathbf{j}\frac{dy}{dt} + m\mathbf{k}\frac{dz}{dt} = \mathbf{i}p_x + \mathbf{j}p_y + \mathbf{k}p_z \tag{A4-3}$$

Here $\mathbf{i}, \mathbf{j}, \mathbf{k}$ are unit vectors pointing, respectively, along the x, y, z Cartesian axes. A cross product is taken by expanding the following determinant:

$$\mathbf{L} = \mathbf{r} \times \mathbf{p} = \begin{vmatrix} \mathbf{i} & \mathbf{j} & \mathbf{k} \\ x & y & z \\ p_x & p_y & p_z \end{vmatrix} \tag{A4-4}$$

to give

$$\mathbf{L} = \mathbf{i}(yp_z - zp_y) + \mathbf{j}(zp_x - xp_z) + \mathbf{k}(xp_y - yp_x) \tag{A4-5}$$

The coefficient for **i** is the magnitude of the x-component of angular momentum, etc. That is,

$$\mathbf{L} = L_x\mathbf{i} + L_y\mathbf{j} + L_z\mathbf{k} \tag{A4-6}$$

where we now see that

$$L_x = yp_z - zp_y \tag{A4-7}$$
$$L_y = zp_x - xp_z \tag{A4-8}$$
$$L_z = xp_y - yp_x \tag{A4-9}$$

(The simple cyclic x, y, z relationships among these formulas can be used as a memory aid.) We will also be interested in the square of the angular momentum, L^2:

$$L^2 = \mathbf{L}\cdot\mathbf{L} = (L_x\mathbf{i} + L_y\mathbf{j} + L_z\mathbf{k})(L_x\mathbf{i} + L_y\mathbf{j} + L_z\mathbf{k}) = L_x^2 + L_y^2 + L_z^2 \tag{A4-10}$$

where we have used the orthonormality of the unit vectors ($\mathbf{i}\cdot\mathbf{i} = 1, \mathbf{i}\cdot\mathbf{j} = 0$, etc). Note that L^2 is a *magnitude, not* a vector.

This gives the classical expressions we need for the x, y, and z components of angular momentum and for the square of its magnitude.

A4-3 The Quantum-Mechanical Operators

Postulate II tells us what to do next: Replace p_x with $(\hbar/i)\partial/\partial x$, and similarly for p_y, p_z. The resulting operators are:

$$\hat{L}_x = -i\hbar\left(y\frac{\partial}{\partial z} - z\frac{\partial}{\partial y}\right) \tag{A4-11}$$
$$\hat{L}_y = -i\hbar\left(z\frac{\partial}{\partial x} - x\frac{\partial}{\partial z}\right) \tag{A4-12}$$
$$\hat{L}_z = -i\hbar\left(x\frac{\partial}{\partial y} - y\frac{\partial}{\partial x}\right) \tag{A4-13}$$

and

$$\hat{L}^2 = \hat{L}_x^2 + \hat{L}_y^2 + \hat{L}_z^2 \tag{A4-14}$$

All of these operators are hermitian.

Now that we have the quantum-mechanical operators, we are free to transform them to other coordinate systems. In spherical coordinates, they are[1]

$$\hat{L}_x = i\hbar\left(\sin\phi\frac{\partial}{\partial\theta} + \cot\theta\cos\phi\frac{\partial}{\partial\phi}\right) \tag{A4-15}$$
$$\hat{L}_y = -i\hbar\left(\cos\phi\frac{\partial}{\partial\theta} - \cot\theta\sin\phi\frac{\partial}{\partial\phi}\right) \tag{A4-16}$$

[1] See Eyting et al. [1, p. 40.]

$$\hat{L}_z = -i\hbar \frac{\partial}{\partial \phi} \tag{A4-17}$$

$$\hat{L}^2 = -\hbar^2 \left(\frac{\partial^2}{\partial \theta^2} + \cot\theta \frac{\partial}{\partial \theta} + \frac{1}{\sin^2\theta} \frac{\partial^2}{\partial \phi^2} \right)$$
$$= -\hbar^2 \left(\frac{1}{\sin\theta} \frac{\partial}{\partial \theta} \sin\theta \frac{\partial}{\partial \theta} + \frac{1}{\sin^2\theta} \frac{\partial^2}{\partial \phi^2} \right) \tag{A4-18}$$

In atomic units, the quantity $\hbar$ becomes unity and does not appear. We will use atomic units henceforth in this appendix.

A4-4 Commutation of Angular Momentum Operators with Hamiltonian Operators and with Each Other

We have indicated (Chapter 4) that a rotating classical system experiencing torque maintains E, L_z, and $|\mathbf{L}|$ (or, equivalently, L^2) as constants of motion, but not L_x or L_y. We might anticipate a similar situation in quantum mechanics. This would mean that a state function ψ would be an eigenfunction for $\hat{H}$, $\hat{L}^2$, and $\hat{L}_z$ but not for $\hat{L}_x$ or $\hat{L}_y$. This in turn requires that $\hat{H}$, $\hat{L}^2$, and $\hat{L}_z$ commute with each other, but that $\hat{L}_x$ and $\hat{L}_y$ do not commute with all of them.

We first consider $\hat{H}$ and $\hat{L}^2$. Note that $\hat{L}^2$ appears in ∇^2 [compare Eq. (A4-18) with (4-7)]:

$$\nabla^2 = \frac{1}{r^2} \frac{\partial}{\partial r} r^2 \frac{\partial}{\partial r} - \frac{1}{r^2} \hat{L}^2 \tag{A4-19}$$

Also, since L^2 does not contain the variable r, $\hat{L}^2$ commutes with any function depending only on r. Since $\hat{L}^2$ must also commute with itself, it follows that $\hat{L}^2$ and ∇^2 commute, i.e., that

$$\left[\hat{L}^2, \nabla^2\right] = 0 \tag{A4-20}$$

If V in a hamiltonian operator is a function of r only, then

$$\left[\hat{L}^2, \hat{H}\right] = 0 \tag{A4-21}$$

This proves that $\hat{H}$ for a hydrogenlike ion commutes with $\hat{L}^2$.

From the expressions for $\hat{L}^2$ and $\hat{L}_z$ in spherical polar coordinates, it is obvious that

$$\left[\hat{L}_z, \hat{L}^2\right] = 0 \tag{A4-22}$$

$\hat{L}_z$ does not operate on functions of r, and so $\hat{L}_z$ commutes with ∇^2. For a spherically symmetric system, $V = V(r)$ and we have that

$$\left[\hat{L}_z, \hat{H}\right] = 0 \tag{A4-23}$$

Thus, we have shown that $\hat{H}$, $\hat{L}^2$, $\hat{L}_z$ all commute in a system having a spherically symmetric potential.

Now let us test $\hat{L}_x$ and $\hat{L}_y$ with each other and also with $\hat{L}^2$.

$$\left[\hat{L}_x, \hat{L}_y\right] = \hat{L}_x\hat{L}_y - \hat{L}_y\hat{L}_x \tag{A4-24}$$

We will examine this using the Cartesian system. We write out each operator product and subtract:

$$\begin{aligned}\hat{L}_x\hat{L}_y &= -\left(y\frac{\partial}{\partial z} - z\frac{\partial}{\partial y}\right)\left(z\frac{\partial}{\partial x} - x\frac{\partial}{\partial z}\right)\\ &= -\left(y\frac{\partial}{\partial x} + yz\frac{\partial^2}{\partial z\partial x} - z^2\frac{\partial^2}{\partial y\partial x} - yx\frac{\partial^2}{\partial z^2} + zx\frac{\partial^2}{\partial y\partial z}\right)\end{aligned} \tag{A4-25}$$

$$\hat{L}_y\hat{L}_x = -\left(zy\frac{\partial^2}{\partial x\partial z} - xy\frac{\partial^2}{\partial z^2} - z^2\frac{\partial^2}{\partial x\partial y} + x\frac{\partial}{\partial y} + xz\frac{\partial^2}{\partial z\partial y}\right) \tag{A4-26}$$

Subtracting (A4-26) from (A4-25):

$$\left[\hat{L}_x, \hat{L}_y\right] = -\left(y\frac{\partial}{\partial x} - x\frac{\partial}{\partial y}\right) = \left(x\frac{\partial}{\partial y} - y\frac{\partial}{\partial x}\right) = i\hat{L}_z \tag{A4-27}$$

A similar approach to other operator combinations gives

$$\left[\hat{L}_y, \hat{L}_z\right] = i\hat{L}_x \tag{A4-28}$$

$$\left[\hat{L}_z, \hat{L}_x\right] = i\hat{L}_y \tag{A4-29}$$

(Notice the x, y, z cyclic relation.) These commutation relations do not depend on choice of coordinate system. Use of the r, θ, ϕ coordinate system would give the same results. Evidently, these operators do not commute with each other since their commutators are unequal to zero.

From this point we will dispense with the carat symbol, since the context of the discussion makes it obvious that we are referring to operators.

We still have not checked L^2 with L_x and L_y. We will now show that

$$[L_x, L^2] = \left[L_x, (L_x^2 + L_y^2 + L_z^2)\right] = 0 \tag{A4-30}$$

We proceed by finding the commutator of L_x with L_x^2, L_y^2, and L_z^2 individually. It is obvious that the first of these, L_x, L_x^2 equals zero. The second can be evaluated as follows.

$$L_xL_y - L_yL_x = iL_z \text{ (from A4-27)} \tag{A4-31}$$

We multiply (A4-31) from the left by L_y:

$$L_yL_xL_y - L_y^2L_x = iL_yL_z \tag{A4-32}$$

We multiply (A4-31) from the right by L_y:

$$L_xL_y^2 - L_yL_xL_y = iL_zL_y \tag{A4-33}$$

We sum (A4-32) and (A4-33):

$$L_xL_y^2 - L_y^2L_x = \left[L_x, L_y^2\right] = i(L_yL_z + L_zL_y) \quad \text{(A4-34)}$$

A similar strategy, starting with (A4-29), multiplying from left and right by L_z, and summing, yields

$$L_z^2L_x - L_xL_z^2 = -\left[L_x, L_z^2\right] = i(L_zL_y + L_yL_z) \quad \text{(A4-35)}$$

Equation A4-34 minus A4-35 plus zero (from L_x, L_x^2) is equal to (A4-30) and is easily seen to equal zero.

A similar proof gives

$$\left[L_y, L^2\right] = 0 \quad \text{(A4-36)}$$

It follows at once that, since L_x and L_y operate only on θ and ϕ and since ∇^2 contains all θ and ϕ terms in the form of L^2:

$$\left[L_x, \nabla^2\right] = \left[L_y, \nabla^2\right] = 0 \quad \text{(A4-37)}$$

If the potential V for a system is independent of θ and ϕ, then

$$\left[L_x, H\right] = \left[L_y, H\right] = 0 \quad \text{(A4-38)}$$

We have shown that H for an atom and L^2 commute with each other and also with L_x, L_y, and L_z, but that the latter three operators do not commute with each other. Therefore, we can say that there exists a set of simultaneous eigenfunctions for H, L^2, and one of L_x, L_y, L_z, but not the other two. We choose L_z to be the privileged operator. This means that an atom can exist in states having "sharp" values of energy, *magnitude* of angular momentum (hence square of L), and z component of angular momentum, but not x or y components.

H and L^2 will commute if V in H is independent of θ and ϕ (central field potential). H and L_z will commute if V is independent of ϕ, even if it is dependent on θ. This is the case for any *linear* system. Therefore, M_L continues to be a sharp quantity for linear molecules like H_2 or C_2H_2, but total angular momentum value does not, because L^2 does not commute with H, so L is not a good quantum number. This is why the main term symbol for a linear molecule is based on M_L, whereas the main term symbol for an atom is based on L.

A4-5 Determining Eigenvalues for L^2 and L_z

We will now make use of our operator commutation relations to determine the nature of the eigenvalues for L^2 and L_z. We begin by defining two new operators:

$$L_+ = L_x + iL_y \quad \text{(A4-39)}$$

$$L_- = L_x - iL_y \quad \text{(A4-40)}$$

These are called "step-up" and "step-down" operators, respectively, or "raising" and "lowering" operators. They correspond to no observable property and are not hermitian.

They have been devised solely because they are useful in formal analysis of the sort we are doing here. The reason for their names will become apparent soon.

We now prove the following theorem for the step-up operator L_+:

$$L_zL_+ = L_+(L_z + 1) \tag{A4-41}$$

Expanding L_+ gives

$$L_zL_+ = L_z(L_x + iL_y) = L_zL_x + iL_zL_y \tag{A4-42}$$

We now add zero in the form [see Eq. A4-29]

$$-(L_zL_x - L_xL_z - iL_y) = 0 \tag{A4-43}$$

to obtain

$$L_zL_+ = L_xL_z + iL_y + iL_zL_y \tag{A4-44}$$

We again add zero, this time in the form [see Eq. A4-28]

$$iL_yL_z - iL_zL_y + L_x = 0 \tag{A4-45}$$

This results in

$$L_zL_+ = L_xL_z + iL_y + iL_yL_z + L_x \tag{A4-46}$$

Rearranging,

$$L_zL_+ = (L_x + iL_y)L_z + L_x + iL_y = (L_x + iL_y)(L_z + 1) = L_+(L_z + 1) \tag{A4-47}$$

This proves Eq. A4-41. An analogous procedure proves the analogous relation for the step-down operator L_-:

$$L_zL_- = L_-(L_z - 1) \tag{A4-48}$$

Our ultimate goal is to discover the nature of the eigenvalues of L^2 and L_z. Since these two operators commute, there must exist a common set of eigenfunctions for them. Let us symbolize these simultaneous eigenfunctions of L^2 and L_z with the symbol Y. A given eigenfunction Y will be associated with an eigenvalue for L^2 and a (possibly) different eigenvalue for L_z. To keep track of these, we use subscript labels l and m, defined in the following manner:

$$L^2Y_{l,m} = k_lY_{l,m} \tag{A4-49}$$

$$L_zY_{l,m} = k_mY_{l,m} \tag{A4-50}$$

Now we make use of our "ladder" operators. It is not difficult to show (Problem A4-1) that

$$L_zL_+Y_{l,m} = (k_m + 1)L_+Y_{l,m} \tag{A4-51}$$

This shows that operating on $Y_{l,m}$ with the step-up operator L_+ produces a new function and that this new function is also an eigenfunction of L_z. Furthermore, the new eigenfunction has an eigenvalue that is greater by one than the eigenvalue (k_m) of the original eigenfunction. This is the reason L_+ is called a step-up operator. The analogous relation for L_- is

$$L_z L_- Y_{l,m} = (k_m - 1) L_- Y_{l,m} \tag{A4-52}$$

The import of Eqs. (A4-51) and (52) is that a set of eigenfunctions for L_z exists with eigenvalues separated by unity.

We now use the ladder operators in another pair of useful operator relations (Problem A4-2):

$$L^2 = L_+ L_- + L_z^2 - L_z \tag{A4-53}$$

$$L^2 = L_- L_+ + L_z^2 + L_z \tag{A4-54}$$

It is also possible to show (Problem A4-3) that

$$L_+ Y_{l,m} = C_+ Y_{l,m+1} \tag{A4-55}$$

$$L_- Y_{l,m} = C_- Y_{l,m-1} \tag{A4-56}$$

where C_+ and C_- are constants. Equations A4-55 and A4-56 tell us that the ladder operators change the functions Y in a manner such that their eigenvalues for L^2 do *not* change.

We will now show that

$$k_l \geq k_m^2 \tag{A4-57}$$

We begin with the obvious relation

$$(L^2 - L_z^2) Y_{l,m} = (k_l - k_m^2) Y_{l,m} \tag{A4-58}$$

But, from Eq. A4-14,

$$L^2 - L_z^2 = L_x^2 + L_y^2 \tag{A4-59}$$

and so

$$(L_x^2 + L_y^2) Y_{l,m} = (k_l - k_m^2) Y_{l,m} \tag{A4-60}$$

Equation A4-60 is remarkable because it indicates that the functions $Y_{l,m}$ (which are *not* eigenfunctions of L_x or L_y, since these do not commute with L_z) are nevertheless eigenfunctions for the *combination* $L_x^2 + L_y^2$.

We will now show that the eigenvalues of Eq. A4-60 must be positive definite, which suffices to prove that $k_l \geq k_m^2$. We proceed by recognizing that the operators L_x and L_y *do* possess eigenfunctions, though they are different from the functions $Y_{l,m}$. Let us symbolize them f and g, respectively. Then

$$L_x f_i = x_i f_i \tag{A4-61}$$

$$L_y g_i = y_i g_i \tag{A4-62}$$

Because L_x and L_y are hermitian operators, x and y are real numbers and the function sets $\{f\}$ and $\{g\}$ are complete and can be assumed to be orthonormal. We can therefore expand the functions Y in terms of either set:

$$Y_{l,m} = \sum_i c_i^{l,m} f_i = \sum_j d_j^{l,m} g_j \tag{A4-63}$$

Substituting Eqs. A4-63 into A4-60 and operating:

$$
\begin{aligned}
(k_l - k_m^2)Y_{l,m} &= (L_x^2 + L_y^2)Y_{l,m} = L_x^2 Y_{l,m} + L_y^2 Y_{l,m} \\
&= L_x^2 \sum_i c_i^{l,m} f_i + L_{y^2} \sum_j d_j^{l,m} g_j \\
&= \sum_i c_i^{l,m} x_i^2 f_i + \sum_j d_j^{l,m} y_j^2 g_j
\end{aligned} \tag{A4-64}
$$

We can isolate $k_l - k_m^2$ by multiplying from the left by $(Y_{l,m})^*$ and integrating. On the left:

$$
\int (Y_{l,m})^* (k_l - k_m^2) Y_{l,m} dv = (k_l - k_m^2) \int (Y_{l,m})^* Y_{l,m} dv = (k_l - k_m^2) \tag{A4-65}
$$

On the right:

$$
\begin{aligned}
&\int \sum_p (c_p^{l,m} f_p)^* \sum_i c_i^{l,m} x_i^2 f_i \, dv + \int \sum_q (d_q^{l,m} g_q)^* \sum_j d_j^{l,m} y_j^2 g_j dv \\
&= \sum_p \sum_i (c_p^{l,m})^* c_i^{l,m} x_i^2 \int (f_p)^* f_i \, dv + \sum_q \sum_j (d_q^{l,m})^* d_j^{l,m} y_j^2 \int (g_q)^* g_j \, dv
\end{aligned} \tag{A4-66}
$$

But $\{f\}$ and $\{g\}$ are orthonormal sets, and so Eq. (A4-65-66) becomes

$$
(k_l - k_m^2) = \sum_i |c_i^{l,m}|^2 x_i^2 + \sum_j |d_j^{l,m}|^2 y_i^2 \tag{A4-67}
$$

Since x and y are real, neither sum has negative terms, so $(k_l - k_m^2) \geq 0$. (If we resort to a physical argument, we see the reasonableness of this since $L_x^2 + L_y^2$ is the square of the angular momentum projection in the x, y plane.)

Knowing that $k_l \geq k_m^2$, we now consider the implications of applying the step-up operator many times to $Y_{l,m}$ and then operating with L_z:

$$
\begin{aligned}
L_z L_+ L_+ L_+ \cdots L_+ Y_{l,m} &= (k_m + 1 + 1 + 1 \cdots + 1) Y_{l,m+1+1+1\cdots+1} \\
&= k_{m'} Y_{l,m'}
\end{aligned} \tag{A4-68}
$$

Eventually we will get a $k_{m'}$ that is too large to satisfy $k_{m'}^2 \leq k_l$. At that point the series must terminate, which means that $L_+ Y_{l,m'} = 0$. Let us call the maximum-value k_m reached in this way k_{max}. Use of L_- likewise gives us a lowest possible value of k_m which we call k_{min}. The corresponding eigenfunctions are labeled $Y_{l,\text{max}}$ and $Y_{l,\text{min}}$. Since

$$
L_+ Y_{l,\text{max}} = 0 \tag{A4-69}
$$

it follows that

$$
L_- L_+ Y_{l,\text{max}} = 0 \tag{A4-70}
$$

But, substituting for $L_+ L_-$ from Eq. A4-54, this can be written

$$
(L^2 - L_z^2 - L_z) Y_{l,\text{max}} = 0 \tag{A4-71}
$$

so

$$k_l^2 - k_{\max}^2 - k_{\max} = 0 \tag{A4-72}$$

or

$$k_l^2 = k_{\max}^2 + k_{\max} \tag{A4-73}$$

A similar treatment on $L_- L_+ Y_{l,\min}$ yields

$$k_l^2 = k_{\min}^2 - k_{\min} \tag{A4-74}$$

From these two expressions for k_l, we have

$$k_{\min}^2 - k_{\max}^2 = k_{\min} + k_{\max} \tag{A4-75}$$

This relation can be satisfied only if

$$k_{\max} = -k_{\min} \tag{A4-76}$$

What has been shown is that, for functions $Y_{l,?}$, the eigenvalue of L^2, k_l, determines the maximum value of k_m through $k_l = k_{\max}^2 + k_{\max}$, that the minimum value of k_m is $-k_{\max}$, and that intermediate values of k_m are given by $k_{\max} - 1, k_{\max} - 2$, etc. If we let the value of $k_{\max}$ be symbolized by l, then we have that $k_l = l^2 + l = l(l+1)$, so that

$$L^2 Y_{l,m} = l(l+1) Y_{l,m} \tag{A4-77}$$

If we let the value of k_m be symbolized by m, then

$$L_z Y_{l,m} = m Y_{l,m} \tag{A4-78}$$

For a given value of l, m can take on the values

$$m = l, l-1, l-2, \ldots, -l+1, -l \tag{A4-79}$$

There are only two possible scenarios for such a set of numbers m. One set is the integers; e.g., if $l = 3$, $m = 3, 2, 1, 0, -1, -2, -3$. The other set is the half-integers; e.g., if $l = 3/2$, $m = 3/2, 1/2, -1/2, -3/2$. Either way, there are $2l + 1$ allowed values of m.

All of the above relations have been derived from the commutation relations for the angular momentum operators. They hold for:

Electron orbital angular momentum	L^2, L_z
Electron spin angular momentum	S^2, S_z
Resultant of orbital and spin momenta	J^2, J_z
Molecular rotational angular momentum	J^2, J_z
Nuclear spin angular momentum	I^2, I_z

If we seek analytical expressions for the functions $Y_{l,m}$, we can start with the assumption that they are separable into products of two kinds of functions, one type depending only on θ, the other only on ϕ:

$$Y_{l,m} = \Theta(\theta)\Phi(\phi) \tag{A4-80}$$

Since $L_z = -i\partial/\partial\varphi$, since $L_z Y_{l,m} = m Y_{l,m}$, and furthermore since $Y_{l,m}$ must be single valued, it follows that $\Phi(\phi) = (1/\sqrt{2\pi})\exp(im\phi)$, with $m = 0, \pm 1, \pm 2, \ldots$. Note that m must be an integer. We have the curious result that, if $Y_{l,m}$ is separable into θ and ϕ parts, the eigenvalues for L_z cannot belong to the half-integer set mentioned above.

The separable, analytic functions $Y_{l,m}$ with l an integer are the spherical harmonics described in Chapter 4 and long known to classical physics. Indeed, the symbol $Y_{l,m}$ has come to stand for those functions. For systems involving half-integer l and m values, no analytical functions of the usual sort can be written. Instead, matrices and vectors are used that manifest the correct relationships. Thus, for the spin of a single electron, Pauli used the following representation:

$$\alpha = \begin{pmatrix} 1 \\ 0 \end{pmatrix}, \quad \beta = \begin{pmatrix} 0 \\ 1 \end{pmatrix}, \quad \langle\alpha|\alpha\rangle = \begin{pmatrix} 1 & 0 \end{pmatrix}\begin{pmatrix} 1 \\ 0 \end{pmatrix} = 1$$

$$S_z = \frac{1}{2}\begin{pmatrix} 1 & 0 \\ 0 & -1 \end{pmatrix}, \quad S_x = \frac{1}{2}\begin{pmatrix} 0 & 1 \\ 1 & 0 \end{pmatrix}, \quad S_y = \frac{1}{2}\begin{pmatrix} 0 & -i \\ i & 0 \end{pmatrix}$$

A4-5.1 Problems

A4-1. Prove Equation A4-51.

A4-2. Prove Equation A4-53.

A4-3. Prove Equation A4-55. What does this equation imply about L^2 and L_+?

A4-4. Evaluate $\langle Y_{l,m}|L_x|Y_{l,m}\rangle$ using only operator relations from this appendix.

A4-5. Demonstrate that the Pauli spin matrices and vectors satisfy the following relations:

a) $\langle\alpha|\beta\rangle = 0$
b) $S_+\beta = \alpha$
c) $S_+\alpha = 0$
d) $[S_x, S_y] = iS_z$

A4-6. Use the Pauli spin matrices to evaluate $S^2\alpha$.

Reference

[1] H. Eyring, J. Walter, and G. E. Kimball, *Quantum Chemistry*. Wiley, New York, 1944.

Appendix 5

The Pairing Theorem[1]

A5-1 Pairing of Roots and Relation Between Coefficients for Alternant Systems

The HMO assumptions are

$$H_{ij} = \begin{cases} \alpha & \text{if } i = j \\ \beta & \text{if } i \neq j \text{ are bonded together} \\ 0 & \text{if } i \neq j \text{ are not bonded together} \end{cases} \tag{A5-1}$$

$$S_{ij} = \delta_{ij} \tag{A5-2}$$

$$\phi_k = \sum_i c_{ik}\chi_i \tag{A5-3}$$

where χ_i is a $2p_\pi$ AO on carbon i.

An alternant hydrocarbon can be labeled with asterisks to demonstrate the existence of two sets of carbon centers in the molecule such that no two atoms in the same set are nearest neighbors (see Section 8-9). The following discussion pertains to alternant systems.

The simultaneous equations leading to Hückel energies and coefficients are of the form

$$c_i x + c_j + c_k + c_l + \cdots = 0 \tag{A5-4}$$

where $x = (\alpha - E)/\beta$. Atoms j, k, l must be bonded to atom i if c_j, c_k, c_l are to be unequal to zero. Hence, atoms j, k, and l belong to one set of atoms, and atom i belongs to the other set.

If we have already found a value of x and a set of coefficients satisfying the simultaneous equations (A5-4), it is easy to show that these equations will also be satisfied if we insert $-x$ and also reverse the signs of the coefficients for one set of centers *or* the other. If we reverse the coefficient signs for the set j, k, l, we obtain, on the left-hand side

$$c_i(-x) - c_j - c_k - c_l - \cdots \tag{A5-5}$$

which is the negative of Eq. A5-4 and hence still equals zero. If we reverse the sign of c_i, we have

$$-c_i(-x) + c_j + c_k + c_l + \cdots \tag{A5-6}$$

which is identical to Eq. A5-4.

[1] See Coulson and Rushbrooke [1].

This proves that each root of an alternant hydrocarbon at $x(\neq 0)$ has a mate at $-x$ and that their associated coefficients differ only in sign between one or the other sets of atoms. Note that, if $x=0$, the c_i term vanishes, leaving coefficients for only one set of centers. Reversing all signs in this case corresponds to multiplying the *entire* MO by -1, which does not generate a new (linearly independent) function. Thus, it is possible for an alternant system to have a single, unpaired root at $x=0$. It is *necessary* for odd alternants to have such a root. An even alternant may have a root at $x=0$, but, if it has one such root, it must have another since, in the end, there must be an even number of roots.

A5-2 Demonstration That Electron Densities Are Unity in Ground States of Neutral Alternant Hydrocarbons

From n AOs result n MOs. The AOs as well as the MOs can be assumed normalized and (in the HMO method) orthogonal. If each AO contains one electron, the electron density at each AO is unity. If these AOs are combined to form MOs, and *each MO* has one electron, each AO electron density would still be unity, since the set of all singly occupied MOs is just a unitary transformation of the set of all singly occupied AOs. (The matrix equivalent of these statements is

$$\mathsf{C}^{\dagger}\mathsf{C}=1=\mathsf{C}\mathsf{C}^{\dagger} \tag{A5-7}$$

The left equality means that the sum of squares (absolute) of coefficients over *all atoms* in *one* MO is unity, so the MO is normalized. The right equality means that the sum of squares of coefficients over *one* atom in all MOs is unity.)

For an alternant hydrocarbon, however, the *squares* of coefficients in an MO at $E=\alpha+k\beta$ are identical to those in the MO at $E=\alpha-k\beta$. Therefore, no change in electron density will result if each electron in the upper half of our MO energy spectrum is shifted to its lower-energy mate. Thus, the resulting state, which is the neutral ground state, still has unit electron density at each AO.

A5-3 A Simple Method for Generating Nonbonding MOs

An immediate consequence of Eq. A5-4 is that the coefficients *for any nonbonding MO*, for which $x=0$ by definition, satisfy the following simple rule: *The coefficients on all the atoms attached to any common atom sum to zero.* Consider the nonbonding MOs below:

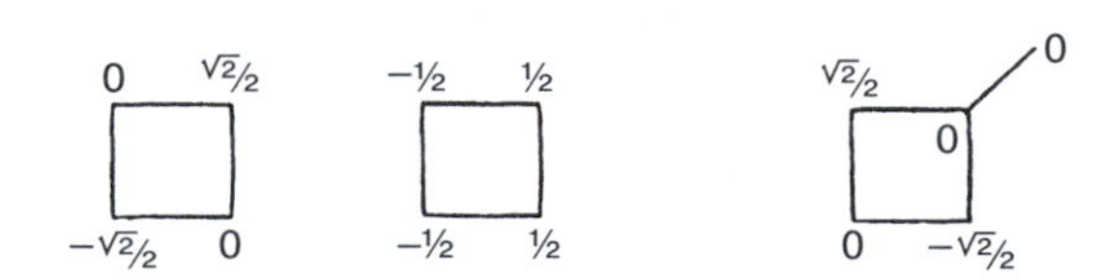

It is clear that, no matter which atom one chooses as reference, the sum of coefficients of attached atoms is zero.

It is possible to use this observation to generate nonbonding MOs without the aid of a computer or tabulation. For odd-alternant systems such an MO is guaranteed to exist, and the recipe is especially easy, so we start with these. The procedure is:

1. Divide the centers into asterisked and unasterisked sets, as described in Chapter 8. One set will have fewer centers. Set all of the coefficients in this set equal to zero.
2. Choose one of the nonzero sites and set its coefficient to be x. Then work around the molecule, setting other coefficients to values needed to satisfy the "sum to zero" rule.
3. When finished, evaluate x by requiring that the sum of squares of coefficients equal unity.

For example, consider the naphthyl system:

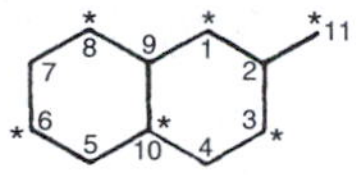

The asterisked centers are more numerous, so we set the others to zero.

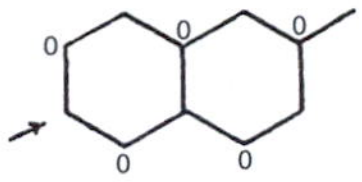

Now we set one of the nonzero sites equal to x. Let us use the one marked with an arrow. Then we move around the molecule, setting the other values.

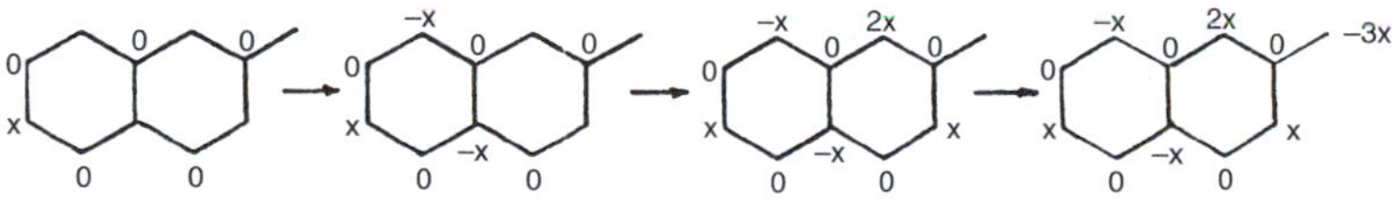

Now we set the sum of squares to unity: $17x^2 = 1; x = 0.242$. The final nonbonding MO is

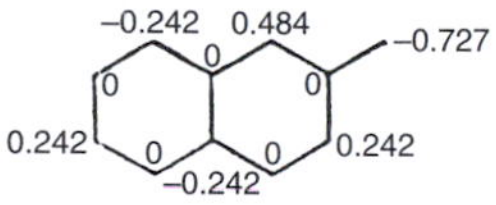

If an even alternant has a nonbonding MO, then it must have a pair of them. Each one corresponds to setting a different subset of coefficients equal to zero.

This simple recipe would be little more than a parlor trick were it not for the fact that the nonbonding MO is often very important in determining a molecule's chemical or physical properties. For example, the neutral molecule used above as an example is alternant, hence has π-electron densities of one at every carbon. However the *spin* density is controlled by the nonbonding MO, since that is where the unpaired electron is, so the ESR splitting pattern should correspond to a series of coupling constants proportional to the squares of these nonbonding MO coefficients (if we ignore negative spin density).

If the neutral radical is ionized to the cation, the electron is lost from the nonbonding MO, and so the deficiency of electronic density (i.e., the positive charge) appears on carbons with nonzero coefficients, in proportion to the squares of the coefficients. In the case of our previous example, the cation charge distribution is predicted to be

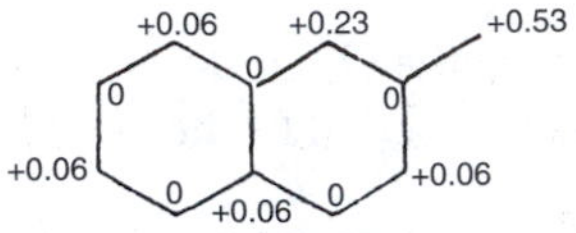

If we were to modify the molecule chemically by attaching a methyl group (which donates π electron charge *via* hyperconjugation) or by substituting for carbon a more electronegative nitrogen atom, we could hope to influence the ease of ionization. But it is apparent that such modifications will have greatest effect if they occur at sites where the largest changes in electron density occur upon ionization, i.e., at positions 11 and 1. Modifying the molecule at a position where a zero coefficient exists in the nonbonding MO (i.e., positions 4, 5, or 7) should have little effect on the ease of carbocation formation.

Reference

[1] C. A. Coulson and G. S. Rushbrooke, *Proc. Cambridge Phil. Soc.* **36**, 193 (1940).

Appendix 6

Hückel Molecular Orbital Energies, Coefficients, Electron Densities, and Bond Orders for Some Simple Molecules

Each molecule is labeled as alternant or nonalternant. For alternants, only the occupied MO data are tabulated since the remainder may be generated by use of the pairing theorem (see Appendix 5). Bond orders are tabulated for only one bond from each symmetry-equivalent set in a molecule.

x	HMO root $= (\alpha - E)/\beta$
n	number of electrons in MO when molecule is in neutral ground state
c_i	LCAO-MO coefficient of AO at atom i
q_i	π-electron density on atom i
p_{ij}	π-bond order between atoms i and j
E_π	total π energy of the molecule $=$ sum of π-electron energies

Molecules in this tabulation are grouped according to the number of centers in the conjugated system. (In all cases, it is assumed that the system is planar and undistorted which, in some cases, is not correct.)

Index of System Tabulated

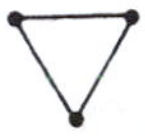

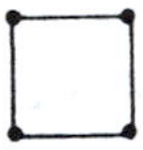

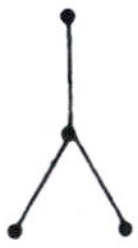

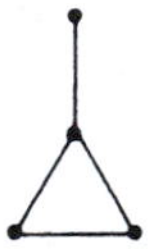

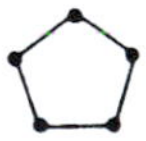

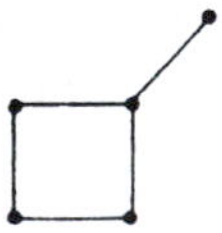

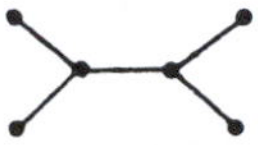

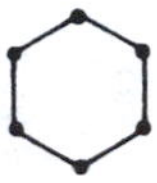

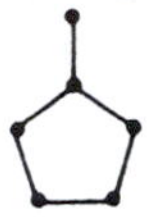

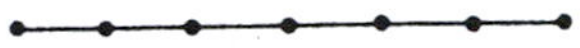

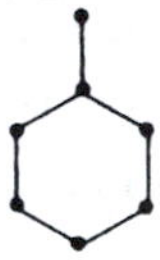

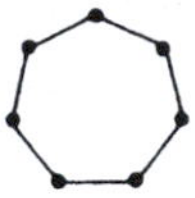

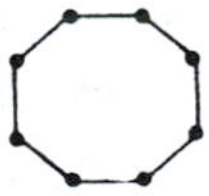

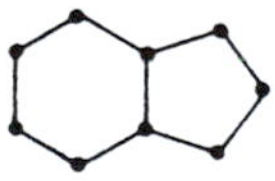

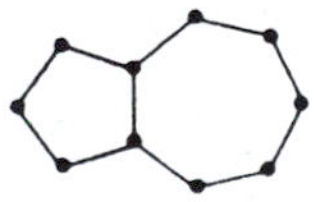

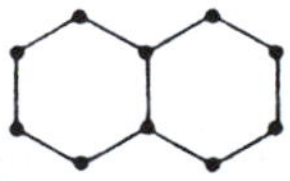

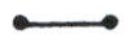

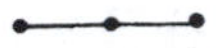

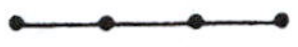

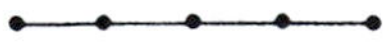

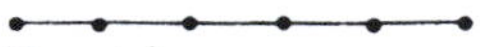

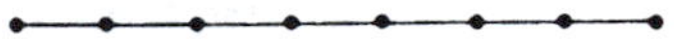

Two Centers

Ethylene (alternant) C_2H_2

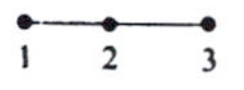

all $q = 1.0$, $p_{12} = 1.0$,
$E_\pi = 2\alpha + 2\beta$

n	x	c_1	c_2
2	−1.000	0.7071	0.7071

Three Centers

Allyl radical (alternant) C_3H_5

all $q = 1.0$, $p_{12} = 0.707$,
$E_\pi = 3\alpha + 2.8284\beta$

n	x	c_1	c_2	c_3
2	−1.4142	0.5000	0.7071	0.5000
1	0.0000	0.7071	0.0000	−0.7071

Cyclopropenyl radical (nonalternant) C_3H_3

all $q = 1.0$, $p_{12} = 0.5$,
$E_\pi = 3\alpha + 3.0000\beta$

n	x	c_1	c_2	c_3
2	−2.0000	0.5774	0.5774	0.5774
$\frac{1}{2}$	1.0000	−0.8165	0.4082	0.4082
$\frac{1}{2}$	1.0000	0.0000	0.7071	−0.7071

Four Centers

Butadiene (alternant) C_4H_6

all $q = 1.0$, $p_{12} = 0.8944$, $p_{23} = 0.4472$, $E_\pi = 4\alpha + 4.4721\beta$

n	x	c_1	c_2	c_3	c_4
2	−1.6180	0.3718	0.6015	0.6015	0.3718
2	−0.6180	0.6015	0.3718	−0.3718	−0.6015

Cyclobutadiene (alternant) C_4H_4

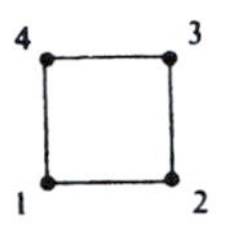

all $q = 1.0$, $p_{12} = 0.5$, $E_\pi = 4\alpha + 4.000\beta$

n	x	c_1	c_2	c_3	c_4
2	−2.0000	0.5000	0.5000	0.5000	0.5000
1	0.0000	0.5000	0.5000	−0.5000	−0.5000
1	0.0000	0.5000	−0.5000	−0.5000	0.5000

2-Allylmethyl (alternant) C_4H_6

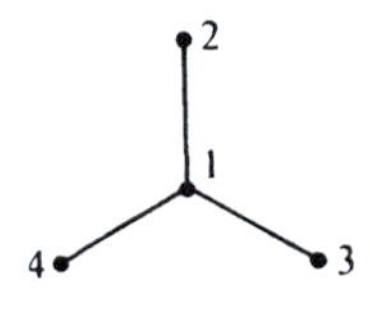

all $q = 1.0$, $p_{12} = 0.5774$, $E_\pi = 4\alpha + 3.4641\beta$

n	x	c_1	c_2	c_3	c_4
2	−1.7320	0.7071	0.4082	0.4082	0.4082
1	0.0	0.0000	0.7071	−0.7071	0.0000
1	0.0	0.0000	0.4082	0.4082	−0.8165

Methylene cyclopropene C_4H_4 (nonalternant)

$E_\pi = 4\alpha + 4.9624\beta$, $p_{12} = 0.4527$, $p_{23} = 0.8176$, $p_{14} = 0.7583$

n	x	c_1	c_2	c_3	c_4
2	−2.1701	0.6116	0.5227	0.5227	0.2818
2	−0.3111	0.2536	−0.3682	−0.3682	0.8152
0	1.0000	0.0000	0.7071	−0.7071	0.0000
0	1.4812	0.7494	−0.3020	−0.3020	−0.5059
	$q_i =$	0.8768	0.8176	0.8176	1.4881

Five Centers

Pentadienyl radical (alternant) C_5H_7

all $q = 1.0$, $p_{12} = 0.7887$, $p_{23} = 0.5774$,
$E_\pi = 5\alpha + 5.4641\beta$

n	x	c_1	c_2	c_3	c_4	c_5
2	−1.7320	0.2887	0.5000	0.5774	0.5000	0.2887
2	−1.0000	0.5000	0.5000	0.0000	−0.5000	−0.5000
1	0.0000	0.5774	0.0000	−0.5774	0.0000	0.5774

Cyclopentadienyl radical (nonalternant) C_5H_5

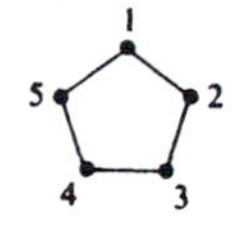

all $q = 1.0$, $p_{12} = 0.5854$, $E_\pi = 5\alpha + 5.8541\beta$

n	x	c_1	c_2	c_3	c_4	c_5
2	−2.0000	0.4472	0.4472	0.4472	0.4472	0.4472
3/2	−0.6180	0.6325	0.1954	−0.5117	−0.5117	0.1954
3/2	−0.6180	0.0000	−0.6015	−0.3718	0.3718	0.6015
0	1.6180	0.6325	−0.5117	0.1954	0.1954	−0.5117
0	1.6180	0.0000	0.3718	−0.6015	0.6015	−0.3718

Cyclobutadienylmethyl radical (alternant) C_5H_5

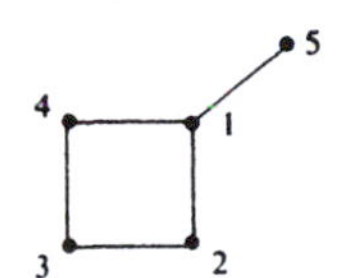

all $q = 1.0$, $p_{12} = 0.3574$, $p_{23} = 0.6101$
$p_{15} = 0.8628$, $E_\pi = 5\alpha + 5.5959\beta$

n	x	c_1	c_2	c_3	c_4	c_5
2	−2.1358	0.5573	0.4647	0.4351	0.4647	0.2610
2	−0.6622	−0.4351	0.1845	0.5573	0.1845	−0.6572
1	0.0000	0.0000	−0.7071	0.0000	0.7071	0.0000

Six Centers

Hexatriene (alternant) C_6H_8

all $q = 1.0$, $p_{12} = 0.8711$, $p_{23} = 0.4834$, $p_{34} = 0.7848$, $E_\pi = 6\alpha + 6.9879\beta$

n	x	c_1	c_2	c_3	c_4	c_5	c_6
2	−1.8019	0.2319	0.4179	0.5211	0.5211	0.4179	0.2319
2	−1.2470	0.4179	0.5211	0.2319	−0.2319	−0.5211	−0.4179
2	−0.4450	0.5211	0.2319	−0.4179	−0.4179	0.2319	0.5211

Butadiene-2,3-bimethyl (alternant) C_6H_8

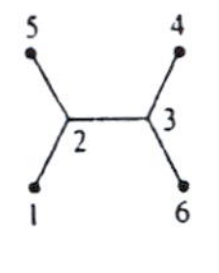

all $q = 1.0$, $p_{12} = 0.6667$, $p_{23} = 0.3333$, $E_\pi = 6\alpha + 6\beta$

n	x	c_1	c_2	c_3	c_4	c_5	c_6
2	−2.0000	0.2887	0.5774	0.5774	0.2887	0.2887	0.2887
2	−1.0000	0.4082	0.4082	−0.4082	−0.4082	0.4082	−0.4082
2	0.0000	−0.5000	0.0000	0.0000	−0.5000	0.5000	0.5000

Benzene (alternant) C_6H_6

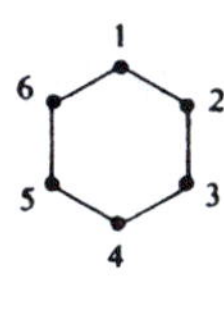

all $q = 1.0$, $p_{12} = 0.6667$, $E_\pi = 6\alpha + 8\beta$

n	x	c_1	c_2	c_3	c_4	c_5	c_6
2	−2.0000	0.4082	0.4082	0.4082	0.4082	0.4082	0.4082
2	−1.0000	0.0000	0.5000	0.5000	0.0000	−0.5000	−0.5000
2	−1.0000	0.5774	0.2887	−0.2887	−0.5774	−0.2887	0.2887

Fulvene (nonalternant) C_6H_6

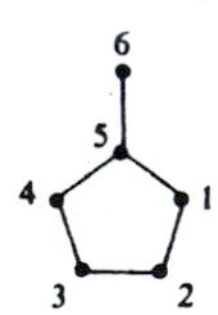

$E_\pi = 6\alpha + 7.4659\beta$, $p_{12} = 0.7779$, $p_{23} = 0.5202$, $p_{45} = 0.4491$, $p_{56} = 0.7586$

n	x	c_1	c_2	c_3	c_4	c_5	c_6
2	−2.1149	0.4294	0.3851	0.3851	0.4294	0.5230	0.2473
2	−1.0000	0.0000	0.5000	0.5000	0.0000	−0.5000	−0.5000
2	−0.6180	0.6015	0.3718	−0.3718	−0.6015	0.0000	0.0000
0	0.2541	−0.3505	0.2795	0.2795	−0.3505	−0.1904	0.7495
0	1.6180	−0.3718	0.6015	−0.6015	0.3718	0.0000	0.0000
0	1.8608	−0.4390	0.1535	0.1535	−0.4390	0.6635	−0.3566
	$q_i =$	1.0923	1.0730	1.0730	1.0923	1.0470	0.6223

Seven Centers

Heptatrienyl radical (alternant) C_7H_9

all $q = 1.0$, $E_\pi = 7\alpha + 8.0547\beta$
$p_{12} = 0.8155$, $p_{23} = 0.5449$, $p_{34} = 0.6533$

n	x	c_1	c_2	c_3	c_4	c_5	c_6	c_7
2	−1.8478	0.1913	0.3536	0.4619	0.5000	0.4619	0.3536	0.1913
2	−1.4142	0.3536	0.5000	0.3536	0.0000	−0.3536	−0.5000	−0.3536
2	−0.7654	0.4619	0.3536	−0.1913	−0.5000	−0.1913	0.3536	0.4619
1	0.0000	−0.5000	0.0000	0.5000	0.0000	−0.5000	0.0000	0.5000

Benzyl radical (alternant) C_7H_7

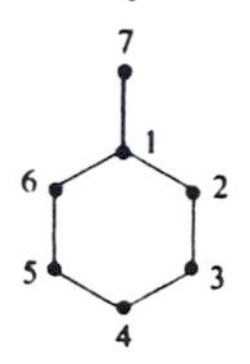

all $q = 1.0$, $p_{12} = 0.5226$, $p_{23} = 0.7050$,
$p_{34} = 0.6350$, $p_{17} = 0.6350$, $E_\pi = 7\alpha + 8.7206\beta$

n	x	c_1	c_2	c_3	c_4	c_5	c_6	c_7
2	−2.1010	0.5000	0.4063	0.3536	0.3366	0.3536	0.4063	0.2380
2	−1.2593	−0.5000	−0.1163	0.3536	0.5615	0.3536	−0.1163	−0.3970
2	−1.0000	0.0000	0.5000	0.5000	0.0000	−0.5000	−0.5000	0.0000
1	0.0000	0.0000	−0.3780	0.0000	0.3780	0.0000	−0.3780	0.7560

Cycloheptatrienyl radical (nonalternant) C_7H_7

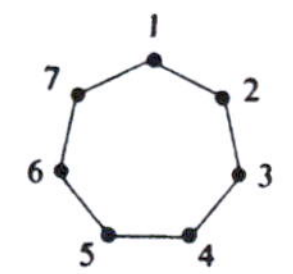

all $q = 1.0$, $p_{12} = 0.6102$, $E_\pi = 7\alpha + 8.5429\beta$

n	x	c_1	c_2	c_3	c_4	c_5	c_6	c_7
2	−2.0000	0.3780	0.3780	0.3780	0.3780	0.3780	0.3780	0.3780
2	−1.2470	−0.5345	−0.3333	0.1189	0.4816	0.4816	0.1189	−0.3333
2	−1.2470	0.0000	−0.4179	−0.5211	−0.2319	0.2319	0.5211	0.4179
$\frac{1}{2}$	0.4450	0.5345	−0.1189	−0.4816	0.3333	0.3333	−0.4816	−0.1189
$\frac{1}{2}$	0.4450	0.0000	0.5211	−0.2319	−0.4180	0.4180	0.2319	−0.5211
0	1.8019	−0.5345	0.4816	−0.3333	0.1189	0.1189	−0.3333	0.4816
0	1.8019	0.0000	0.2319	−0.4179	0.5211	−0.5211	0.4179	−0.2319

Eight Centers

Octatetraene (alternant) C_8H_{10}

all $q = 1.0$, $p_{12} = 0.8621$, $p_{23} = 0.4948$, $p_{34} = 0.7581$, $p_{45} = 0.5288$, $E_\pi = 8\alpha + 9.5175\beta$

n	x	c_1	c_2	c_3	c_4	c_5	c_6	c_7	c_8
2	−1.8794	0.1612	0.3030	0.4082	0.4642	0.4642	0.4082	0.3030	0.1612
2	−1.5321	0.3030	0.4642	0.4082	0.1612	−0.1612	−0.4082	−0.4642	−0.3030
2	−1.0000	−0.4082	−0.4082	0.0000	0.4082	0.4082	0.0000	−0.4082	−0.4082
2	−0.3473	0.4642	0.1612	−0.4082	−0.3030	0.3030	0.4082	−0.1612	−0.4642

Cyclooctatetraene (alternant) C_8H_8

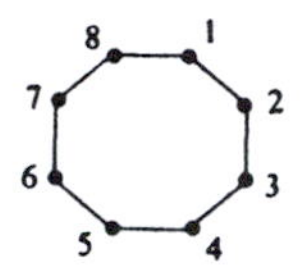

all $q = 1.0$, $p_{12} = 0.6035$, $E_\pi = 8\alpha + 9.6568\beta$

n	x	c_1	c_2	c_3	c_4	c_5	c_6	c_7	c_8
2	−2.0000	0.3536	0.3536	0.3536	0.3536	0.3536	0.3536	0.3536	0.3536
2	−1.4142	0.3536	0.0000	−0.3536	−0.5000	−0.3536	0.0000	0.3536	0.5000
2	−1.4142	0.3536	0.5000	0.3536	0.0000	−0.3536	−0.5000	−0.3536	0.0000
1	0.0000	0.3536	0.3536	−0.3536	−0.3536	0.3536	0.3536	−0.3536	−0.3536
1	0.0000	0.3536	−0.3536	−0.3536	0.3536	0.3536	−0.3536	−0.3536	0.3536

Nine Centers

Benzcyclopentadienyl radical (nonalternant) C_9H_7

$p_{12} = 0.6592$, $p_{49} = 0.6071$, $p_{56} = 0.6630$, $p_{18} = 0.4790$, $p_{45} = 0.6363$, $p_{89} = 0.5118$, $E_\pi = 9\alpha + 11.8757\beta$

n	x	c_1	c_2	c_3	c_4	c_5	c_6	c_7	c_8	c_9
2	−2.3226	0.3203	0.2758	0.3203	0.2988	0.2259	0.2259	0.2988	0.4681	0.4681
2	−1.5450	−0.3114	−0.4031	−0.3114	0.2689	0.4934	0.4934	0.2689	−0.0780	−0.0780
2	−1.1935	0.2992	0.0000	−0.2992	−0.4841	−0.2207	0.2207	0.4841	0.3571	−0.3571
2	−0.7293	−0.2054	−0.5634	−0.2054	0.0935	−0.3454	−0.3454	0.0935	0.4136	0.4136
1	−0.2950	0.5428	0.0000	−0.5428	0.3355	0.2591	−0.2591	−0.3355	0.1601	−0.1601
0	0.9016	0.1548	−0.3434	0.1548	−0.5424	0.2852	0.2852	−0.5424	0.2038	0.2038
0	1.2950	−0.2591	0.0000	0.2591	−0.1601	0.5428	−0.5428	0.1601	0.3355	−0.3355
0	1.6952	0.4840	−0.5711	0.4840	0.1884	−0.0699	−0.0699	0.1884	−0.2495	−0.2495
0	2.1935	−0.2207	0.0000	0.2207	0.3571	−0.2992	0.2992	−0.3571	0.4841	−0.4841
	$q_i =$	0.9571	1.1119	0.9571	0.9218	0.9920	0.9920	0.9218	1.0730	1.0730

Ten Centers

Azulene (nonalternant) $C_{10}H_8$

$p_{12} = 0.6560$, $p_{4,10} = 0.5858$, $p_{56} = 0.6389$,
$p_{19} = 0.5956$, $p_{45} = 0.6640$, $p_{9,10} = 0.4009$,
$E_\pi = 10\alpha + 13.3635\beta$

n	x	c_1	c_2	c_3	c_4	c_5	c_6	c_7	c_8	c_9	c_{10}
2	−2.3103	0.3233	0.2799	0.3233	0.2886	0.1998	0.1730	0.1998	0.2886	0.4670	0.4670
2	−1.6516	−0.2678	−0.3243	−0.2678	0.1909	0.4333	0.5247	0.4333	0.1909	−0.1180	−0.1180
2	−1.3557	0.2207	0.0000	−0.2207	−0.4841	−0.3571	0.0000	0.3571	0.4841	0.2992	−0.2992
2	−0.8870	−0.2585	−0.5829	−0.2585	0.2186	−0.1598	−0.3603	−0.1598	0.2186	0.3536	0.3536
2	−0.4773	0.5428	0.0000	−0.5428	0.1601	0.3355	0.0000	−0.3355	−0.1601	0.2591	−0.2591
0	0.4004	−0.0632	0.3158	−0.0632	0.4699	0.1023	−0.5109	0.1023	0.4699	−0.2904	−0.2904
0	0.7376	−0.2992	0.0000	0.2992	−0.3571	0.4841	0.0000	−0.4841	0.3571	0.2207	−0.2207
0	1.5792	0.4364	−0.5527	0.4364	−0.0844	0.2697	−0.3416	0.2697	−0.0844	−0.1365	−0.1365
0	1.8692	−0.2500	0.2675	−0.2500	−0.3233	0.4045	−0.4328	0.4045	−0.3233	0.1998	0.1998
0	2.0953	−0.2591	0.0000	0.2591	0.3355	−0.1601	0.0000	0.1601	−0.3355	0.5428	−0.5428
	$q_i =$	1.1729	1.0466	1.1729	0.8550	0.9864	0.8700	0.9864	0.8550	1.0274	1.0274

Naphthalene (alternant) $C_{10}H_8$

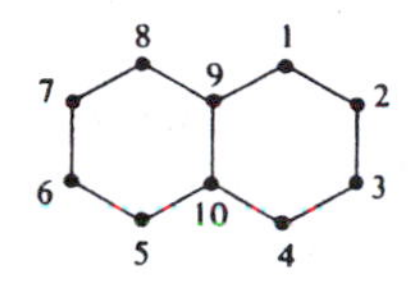

all $q = 1.0$, $E_\pi = 10\alpha + 13.6832\beta$
$p_{12} = 0.7246$, $p_{23} = 0.6032$, $p_{19} = 0.5547$, $p_{9,10} = 0.5182$

n	x	c_1	c_2	c_3	c_4	c_5	c_6	c_7	c_8	c_9	c_{10}
2	−2.3028	0.3006	0.2307	0.2307	0.3006	0.3006	0.2307	0.2307	0.3006	0.4614	0.4614
2	−1.6180	0.2629	0.4253	0.4253	0.2629	−0.2629	−0.4253	−0.4253	−0.2629	0.0000	0.0000
2	−1.3028	0.3996	0.1735	−0.1735	−0.3996	−0.3996	−0.1735	0.1735	0.3996	0.3470	−0.3470
2	−1.0000	0.0000	−0.4082	−0.4082	0.0000	0.0000	−0.4082	−0.4082	0.0000	0.4082	0.4082
2	−0.6180	0.4253	0.2629	−0.2629	−0.4253	0.4253	0.2629	−0.2629	−0.4253	0.0000	0.0000

Appendix 7

Derivation of the Hartree–Fock Equation

This appendix is divided into two parts. In the first section we develop the formula for the expectation value $\bar{E} = \langle\psi|H|\psi\rangle$ for the case in which ψ is a single determinantal wavefunction over MOs. In the second section we derive the Hartree–Fock equation by requiring $\bar{E}$ to be stationary with respect to variations in ψ.

A7-1 The Expansion of $\bar{E}$ in Terms of Integrals over MOs

We limit discussion to the case in which ψ is a single, closed-shell determinant. We will develop our arguments by referring to a four-electron example:

$$\psi_4 = (4!)^{-1/2}\left|\phi_1(1)\bar{\phi}_1(2)\phi_2(3)\bar{\phi}_2(4)\right| \tag{A7-1}$$

Recall that this is the shorthand formula for a Slater determinant. Each ϕ is a normalized MO, the MOs are assumed to be orthogonal, and a bar signifies that an electron possesses β spin. As we develop our arguments within the context of ψ, we will generalize them to apply to the general $2n$ electron closed-shell wavefunction

$$\psi_{2n} = [(2n!)]^{-1/2}|\phi_1(1)\bar{\phi}_1(2)\phi_2(3)\bar{\phi}_2(4)\cdots\phi_n(2n-1)\bar{\phi}_n(2n)| \tag{A7-2}$$

When ψ_4 is expanded according to the rule for determinants (Appendix 2), we obtain 4! products. We note the following features of the expanded form.

1. There is one product, occurring with coefficient $+1$, which is identical to the product appearing in the shorthand form of Eq. (A7-1). We refer to this as the "leading term."

2. An equivalent way of expressing a Slater determinant is via the expression (for ψ_4)

$$\psi_4 = (4!)^{-1/2}\sum_P(-1)^p P\left(\phi_1(1)\bar{\phi}_1(2)\phi_2(3)\bar{\phi}_2(4)\right) \tag{A7-3}$$

   Here P stands for all the sequences of permutations of electron labels that lead to different products (i.e., P is a permutation *operator*), and p is the number of pairwise permutations in a given sequence. For ψ_4 there are 4! sequences P, the simplest being "no permutations" (hence, $p=0$) which produces the leading term. Then there are single permutations, such as $P_{1,2}$ (with $p=1$), which produces the term $-\phi_1(2)\bar{\phi}_1(1)\phi_2(3)\bar{\phi}_2(4)$. There are also double permutations, etc. According to Eq. (A7-3), any term differing from the leading term by an odd number of permutations will appear with coefficient -1. We will be particularly concerned with products that differ from the leading term by a single permutation.

3. A single permutation may be made to occur between electrons in MO's with the same spins or different spins. In the latter case, two electrons in the singly permuted product will disagree in spin with their counterparts in the leading term. In the former case, no such spin disagreement will exist.

4. Terms also appear in ψ_4 corresponding to more than a single permutation of electron indices.

It is useful to pick a representative example of each type of product mentioned above. For ψ_4, we have the following:

Leading term:	$\phi_1(1)\bar{\phi}_1(2)\phi_2(3)\bar{\phi}_2(4)$	
Singly permuted term:	$\phi_1(1)\bar{\phi}_1(4)\phi_2(3)\bar{\phi}_2(2)$;	Spin agreement: $(P_{2,4})$
Singly permuted term:	$\phi_1(2)\bar{\phi}_1(1)\phi_2(3)\bar{\phi}_2(4)$;	Spin disagreement: $(P_{1,2})$
Doubly permuted term:	$\phi_1(2)\bar{\phi}_1(4)\phi_2(3)\bar{\phi}_2(1)$;	$(P_{1,2}P_{1,4})$

Upon expanding $\langle\psi_4|\hat{H}|\psi_4\rangle$, we obtain a set of 4! products on both the left and right-hand sides of $\hat{H}$:

$$\bar{E} = (4!)^{-1}\int\{\phi_1^*(1)\bar{\phi}_1^*(2)\phi_2^*(3)\bar{\phi}_2^*(4) - \phi_1^*(1)\bar{\phi}_1^*(4)\phi_2^*(3)\bar{\phi}_2^*(2) - \cdots\}$$
$$\times\hat{H}(1,2,3,4)\{\phi_1(1)\bar{\phi}_1(2)\phi_2(3)\bar{\phi}_2(4) - \phi_1(1)\bar{\phi}_1(4)\phi_2(3)\bar{\phi}_2(2) - \cdots\}d\tau \quad \text{(A7-4)}$$

This can be expanded into a set of integrals, one for each term on the left:

$$\bar{E} = (4!)^{-1}\Big\{\int \phi_1^*(1)\bar{\phi}_1^*(2)\phi_2^*(3)\bar{\phi}_2^*(4)\hat{H}(1,2,3,4)\big[\phi_1(1)\bar{\phi}_1(2)\phi_2(3)\bar{\phi}_2(4)$$
$$- \phi_1(1)\bar{\phi}_1(4)\phi_2(3)\bar{\phi}_2(2) - \phi_1(2)\bar{\phi}_1(1)\phi_2(3)\bar{\phi}_2(4) - \cdots\big]d\tau$$
$$- \int \phi_1^*(1)\bar{\phi}_1^*(4)\phi_2^*(3)\bar{\phi}_2^*(2)\hat{H}(1,2,3,4)\big[\phi_1(1)\bar{\phi}_1(2)\phi_2(3)\bar{\phi}_2(4)$$
$$- \phi_1(1)\bar{\phi}_1(4)\phi_2(3)\bar{\phi}_2(2) - \phi_1(2)\bar{\phi}_1(1)\phi_2(3)\bar{\phi}_2(4) - \cdots\big]d_\tau\Big\} \quad \text{(A7-5)}$$

etc. In Eq. (A7-5), $\bar{E}$ is a sum of 4! integrals, each containing one term from the set on the left of $\hat{H}$ and all 4! from the set on the right.

At first, it might seem that we must evaluate all of the 4! integrals in Eq. (A7-5). But it can be shown that these integrals, times their $+1$ or -1 coefficients, are all equal to each other, enabling us to write $\bar{E}$ as 4! times the first integral:

$$\bar{E} = \int \phi_1^*(1)\bar{\phi}_1^*(2)\phi_2^*(3)\bar{\phi}_2^*(4)\hat{H}(1,2,3,4)$$
$$\sum_P(-1)^P P(\phi_1(1)\bar{\phi}_1(2)\phi_2(3)\bar{\phi}_2(4))\,d_\tau \quad \text{(A7-6)}$$

The demonstration that the various integrals in Eq. (A7-5), times their coefficients, are equal to each other is as follows. Consider the second integral in Eq. (A7-5). Note that, if we permute electrons 2 and 4 in that integral, we restore the term on the left of $\hat{H}$ to its original "leading term" order, thereby making that product identical to its counterpart in the first integral. Furthermore, if we carry out this permutation throughout the whole of the integrand of integral number 2 (i.e., in $\hat{H}$, in all 4! products to the right of $\hat{H}$, and in $d\tau$), we will not affect the value of the integral. (Recall that, for example,

$$\int_0^1 x^2 dx \int_1^2 y^3 dy \equiv \int_0^1 y^2 dy \int_1^2 x^3 dx$$

The result of this permutation $P_{2,4}$ on the second integral in Eq. (A7-5) is

$$\int \phi_1^*(1)\bar{\phi}_1^*(2)\phi_2^*(3)\bar{\phi}_2^*(4)\hat{H}(1,4,3,2)\left[\phi_1(1)\bar{\phi}_1(4)\phi_2(3)\bar{\phi}_2(2) - \phi_1(1)\bar{\phi}_1(2)\phi_2(3)\bar{\phi}_2(4) - \cdots\right]d\tau \quad \text{(A7-7)}$$

Now $\hat{H}$ is invariant under exchange of electron indices, and the set of products to the right of $\hat{H}$ in Eq. (A7-7) is the same set we had before, but their order is changed, and the whole product evidently differs by a factor of -1 from the set in the first integral. Therefore, we can say that the first and second integrals of Eq. (A7-5) have the same absolute value but different signs. However, the fact that these integrals contribute to $\bar{E}$ with opposite signs cancels the sign disagreement. In this way, every integral in Eq. (A7-5) can be compared to the leading integral and Eq. (A7-6) verified. This much simplified expression for $\bar{E}$ is, for the $2n$-electron case

$$\bar{E} = \int \phi_1^*(1)\bar{\phi}_1^*(2)\cdots\phi_n^*(2n-1)\bar{\phi}_n^*(2n)\hat{H}(1,2,\ldots,2n) \left[\sum_P (-1)^p P(\phi_1(1)\bar{\phi}_1(2)\cdots\phi_n(2n-1)\bar{\phi}_n(2n))\right] d\tau \quad \text{(A7-8)}$$

At this point we write out $\hat{H}$ more explicitly. It is, in atomic units,

$$\hat{H}(1,2,\ldots,2n) = \sum_{i=1}^{2n}\left(-\frac{1}{2}\nabla_i^2 - \sum_{\mu}^{\text{nuclei}} Z_\mu/r_{\mu i}\right) + \sum_{i=1}^{2n-1}\sum_{j=i+1}^{2n} 1/r_{ij} \quad \text{(A7-9)}$$

$$= \sum_{i=1}^{2n} H_{(i)}^{\text{core}} + \sum{}' 1/r_{ij} \quad \text{(A7-10)}$$

Here, $\sum'$ is a shorthand symbol for the double sum in Eq. (A7-9). We see that $\hat{H}$ is composed of one-electron operators, $H_{(i)}^{\text{core}}$, which deal with the kinetic and nuclear-electron attraction energies for electron i, and two-electron operators for interelectronic repulsion. The internuclear repulsion is omitted since, for a given nuclear configuration, it is simply a constant that can be added to the electronic energy. Note for future reference that $\hat{H}$ has no dependence on electron spin coordinates.

If we insert the expression (A7-10) for $\hat{H}$ into Eq. (A7-6) for $\bar{E}$, it is easy to to see that we can expand the result into separate sets of integrals over one-and two-electron operators. We consider first the one-electron integrals. For ψ_4, these are

$$\int (\phi_1^*(1)\bar{\phi}_1^*(2)\phi_2^*(3)\bar{\phi}_2^*(4)) \times \sum_{i=1}^{4} H_{(i)}^{\text{core}} [\phi_1(1)\bar{\phi}_1(2)\phi_2(3)\bar{\phi}_2(4) - \phi_1(1)\bar{\phi}_1(4)\phi_2(3)\phi_2(2) - \cdots] \, d\tau \qquad \text{(A7-11)}$$

Upon expanding this, the first set of integrals we obtain is

$$\int \phi_1^*(1)\bar{\phi}_1^*(2)\phi_2^*(3)\bar{\phi}_2^*(4) \left[H_{(1)}^{\text{core}} + H_{(2)}^{\text{core}} + H_{(3)}^{\text{core}} + H_{(4)}^{\text{core}}\right] \phi_1(1)\bar{\phi}_1(2)\phi_2(3)\bar{\phi}_2(4) d\tau \qquad \text{(A7-12)}$$

This can be expanded again. The first integral contains the operator $H_{(1)}^{\text{core}}$, which does not act on electrons 2-4. This allows a separation into a product of integrals as follows:

$$\int \phi_1^*(1)\bar{\phi}_1^*(2)\phi_2^*(3)\bar{\phi}_2^*(4) H_{(1)}^{\text{core}} \phi_1(1)\bar{\phi}_1(2)\phi_2(3)\bar{\phi}_2(4) \, d\tau \qquad \text{(A7-13)}$$

$$= \int \phi_1^*(1) H_{(1)}^{\text{core}} \phi_1(1) \, d\tau_1 \int \bar{\phi}_1^*(2)\bar{\phi}_1(2) \, d\tau_2 \int \phi_2^*(3)\phi_2(3) \, d\tau_3 \int \bar{\phi}_2^*(4)\bar{\phi}_2(4) \, d\tau_4 \qquad \text{(A7-14)}$$

The last three integrals are overlap integrals and are all unity by virtue of normality of the MOs. The first integral, a "core integral," is normally symbolized H_{11}. Here the subscripts refer to the MO index, not the electron index:

$$H_{ii} = \int \phi_i^*(1) H_{(1)}^{\text{core}} \phi_i(1) \, d\tau_1 \qquad \text{(A7-15)}$$

Therefore, Eq. (A7-13) equals H_{11}. By continued expansion, Eq. (A7-12) can be shown to be equal to $H_{11} + H_{11} + H_{22} + H_{22} = 2(H_{11} + H_{22})$. In this case, we have been dealing with identical MO products on the two sides of the operator. As we continue evaluating the expansion of Eq. (A7-12), we next encounter an integral in which the products differ by a permutation, namely,

$$-\int \phi_1^*(1)\bar{\phi}_1^*(2)\phi_2^*(3)\bar{\phi}_2^*(4) \left[H_{(1)}^{\text{core}} + H_{(2)}^{\text{core}} + H_{(3)}^{\text{core}} + H_{(4)}^{\text{core}}\right] \phi_1(1)\bar{\phi}_1(4)\phi_2(3)\bar{\phi}_2(2) \, d\tau \qquad \text{(A7-16)}$$

Again, for $H^{\text{core}}(1)$, this may be written

$$-\int \phi_1^*(1) H_{(1)}^{\text{core}} \phi_1(1) \, d\tau_1 \int \bar{\phi}_1^*(2)\bar{\phi}_2(2) \, d\tau_2 \int \phi_2^*(3)\phi_2(3) \, d\tau_3 \int \bar{\phi}_2^*(4)\bar{\phi}_1(4) \, d\tau_4 \qquad \text{(A7-17)}$$

Orbital orthogonality will cause the second and fourth integrals to vanish. In general, if the two products differ by one or more permutations, they will have two or more sites of disagreement. Upon expansion, at least one disagreement will occur in an overlap integral, causing the integral to vanish. Thus, except for the integral involving identical products [Eq. (A7-12)], all the integrals obtained by expansion of Eq. (A7-11) vanish. Our result, generalized to the $2n$-electron case, is (see Appendix 11 for bra-ket notation)

$$\left\langle \psi_{2n} \left| \sum_{i=1}^{2n} H_{(1)}^{\text{core}} \right| \psi_{2n} \right\rangle = \sum_{i=1}^{n} 2H_{ii} \tag{A7-18}$$

We now turn to integrals containing two-electron operators. Consider, for example, an integral that has products differing in two places,

$$-\left\langle \phi_1(1)\bar{\phi}_1(2)\phi_2(3)\bar{\phi}_2(4) \left| 1/r_{13} \right| \phi_1(3)\bar{\phi}_1(2)\phi_2(1)\bar{\phi}_2(4) \right\rangle \tag{A7-19}$$

This can be partly separated into a product of integrals over different electron coordinates.

$$\text{(A7-19)} = -\langle \phi_1(1)\phi_2(3)|1/r_{13}|\phi_1(3)\phi_2(1)\rangle \langle \bar{\phi}_1(2)|\bar{\phi}_1(2)\rangle \langle \bar{\phi}_2(4)|\bar{\phi}_2(4)\rangle \tag{A7-20}$$

Observe that the two disagreements are inside the two-electron integral, and that the overlap terms both show complete internal agreement and are therefore equal to unity. It is clear that, if our two products differed in *more* than two places, at least one such disagreement would appear in an overlap integral, causing the whole integral to vanish. Therefore, two-electron integrals need be considered only if they involve products differing by zero or one permutations. Let us consider these two possibilities separately. If there are no disagreements, we have for ψ_4,

$$\begin{aligned}
&\left\langle \phi_1(1)\bar{\phi}_1(2)\phi_2(3)\bar{\phi}_2(4) \left| \sum{}' 1/r_{ij} \right| \phi_1(1)\bar{\phi}_1(2)\phi_2(3)\bar{\phi}_2(4) \right\rangle \\
&= \langle \phi_1(1)\bar{\phi}_1(2)|1/r_{12}|\phi_1(1)\bar{\phi}_1(2)\rangle + \langle \phi_1(1)\phi_2(3)|1/r_{13}|\phi_1(1)\phi_2(3)\rangle \\
&+ \langle \phi_1(1)\bar{\phi}_2(4)|1/r_{14}|\phi_1(1)\bar{\phi}_2(4)\rangle + \langle \bar{\phi}_1(2)\phi_2(3)|1/r_{23}|\bar{\phi}_1(2)\phi_2(3)\rangle \\
&+ \langle \bar{\phi}_1(2)\bar{\phi}_2(4)|1/r_{24}|\bar{\phi}_1(2)\bar{\phi}_2(4)\rangle + \langle \phi_2(3)\bar{\phi}_2(4)|1/r_{34}|\phi_2(3)\bar{\phi}_2(4)\rangle
\end{aligned} \tag{A7-21}$$

These integrals give the coulombic repulsion between electrons in MOs. They are symbolized J_{ij}, where

$$J_{ij} = \left\langle \phi_i(1)\phi_j(2) \left| 1/r_{12} \right| \phi_i(1)\phi_j(2) \right\rangle \equiv \langle ij|ij\rangle \tag{A7-22}$$

Here ϕ_i and ϕ_j may be associated with either spin. Because the operator and MOs commute, the integrand can be rearranged to give

$$J_{ij} = \int \phi_i^*(1)\phi_i(1)(1/r_{12})\phi_j^*(2)\phi_j(2)\, d\tau_1 d\tau_2 \equiv (ii|jj) \tag{A7-23}$$

Some people prefer this form because it places the two mutually repelling charge clouds on the two sides of the operator. It is important to realize that the parenthetical expression $(ii|jj)$ and the bra-ket shorthand $\langle ij|ij\rangle$ correspond to *different conventions* for electron index order, and are really the same integral. Returning to Eq. (A7-21), we see that it is equal to

$$J_{11} + J_{12} + J_{12} + J_{12} + J_{12} + J_{22} = \sum_{i=1}^{2} \left(J_{ii} + \sum_{j \neq i} 2J_{ij} \right) \tag{A7-24}$$

We must now consider the case where the products differ by a single permutation, hence in two places. An example has been provided in Eq. (A7-19). We noted that, when the operator $1/r_{ij}$ corresponds to electrons i and j in the positions of disagreement, the overlap integrals are all unity. Otherwise, at least one overlap integral vanishes. An integral like that in Eq. (A7-20) is called an exchange integral.

Exchange integrals can occur only when the product on the right of the operator differs from the leading term by a single permutation. Hence, exchange integrals always enter with a coefficient of -1.

We noted earlier that two classes of singly permuted products exist. One class involves permutations between electrons of like spin. In such a case, ϕ_i and ϕ_j appear throughout the integral K_{ij} with spin agreement. For cases where electrons of different spin have been permuted, spin disagreement forces the exchange integral to vanish. (Since $1/r_{ij}$ is not a spin operator, the integration over spin coordinates factors out and produces a vanishing integral if spins disagree.)

The result of all this is that each singly permuted product can give $-K_{ij}$ if the permutation is between electrons of like spin in MOs ϕ_i and ϕ_j, and zero otherwise. For ψ_4, the acceptable permutations can be seen to be electron 1 with 3 and electron 2 with 4, both of these occurring between ϕ_1 and ϕ_2 space MOs. Hence, the contribution to E is $-2K_{12}$. Combining this with J terms gives

$$\left\langle \psi_4 \left| \sum{}' 1/r_{ij} \right| \psi_4 \right\rangle = J_{11} + 4J_{12} - 2K_{12} + J_{22} \tag{A7-25}$$

From the definitions of J and K, it is apparent that

$$J_{ij} = J_{ji}, \quad K_{ij} = K_{ji}, \quad K_{ii} = J_{ii}$$

This allows us to rewrite Eq. (A7-25) as

$$\begin{aligned} &2J_{11} - K_{11} + 2J_{12} - K_{12} + 2J_{21} - K_{21} + 2J_{22} - K_{22} \\ &= \sum_{i=1}^{2} \sum_{j=1}^{2} (2J_{ij} - K_{ij}) \end{aligned} \tag{A7-26}$$

Generalizing to the $2n$-electron, closed-shell case and adding in our one-electron contribution,

$$\bar{E} = \langle \psi_{2n} | \hat{H} | \psi_{2n} \rangle = 2 \sum_{i=1}^{n} H_{ii} + \sum_{i=1}^{n} \sum_{j=1}^{n} (2J_{ij} - K_{ij}) \tag{A7-27}$$

This is the desired expression for $\bar{E}$ in terms of integrals over MOs ϕ_i for a single-determinantal, closed-shell wavefunction.

A7-2 Derivation of the Hartree–Fock Equations

To find the "best" MOs, we seek those that minimize $\bar{E}$, that is, those MOs ϕ for which $\bar{E}$ is stationary to small variations $\delta\phi$. But there is a restriction in the variations $\delta\phi$. The MOs can only be varied in ways that do not destroy their orthonormality since this property was assumed in deriving Eq. (A7-27). This means that, for proper variations

$\delta\phi$ at the minimum $\bar{E}$, both $\bar{E}$ and all the MO overlap integrals $S_{ij} \equiv \langle\phi_i|\phi_j\rangle$ must remain constant. (S_{ij} must equal unity when $i = j$, zero otherwise.) If $\bar{E}$ and S_{ij} are constant, any linear combination of them is constant too. Thus, we may write that, at the minimum $\bar{E}$,

$$c_0\bar{E} + \sum_i \sum_j c_{ij} S_{ij} = \text{constant} \tag{A7-28}$$

for our restricted type of $\delta\phi$. This equation will hold for *any* set of coefficients c as long as $\delta\phi$ is of the proper *restricted* nature. However, it is possible to show that, for a particular set of coefficients, Eq. (A7-28) is satisfied at minimum $\bar{E}$ for *any* small variations $\delta\phi$. The particular coefficients are called *Lagrangian multipliers*. They are of undetermined value thus far, but their values will become known in the course of solving the problem. The technique, known as "Lagrange's method of undetermined multipliers" is from the calculus of variations.[1] The Lagrangian multipliers will ultimately turn out to be essentially the MO energies. For future convenience we write Eq. (A7-28) in the form

$$\bar{E} - 2\sum_i \sum_j \epsilon_{ij} S_{ij} = \text{constant for } \delta\phi \tag{A7-29}$$

where we now understand $\delta\phi$ to be unrestricted and ϵ_{ij} to be some unknown special set of constants. The stability of the quantity on the left-hand side of Eq. (A7-29) may be expressed as follows:

$$\delta E - 2\delta \sum_i \sum_j \epsilon_{ij} S_{ij} = 0 \tag{A7-30}$$

or, expanding $\bar{E}$,

$$2\sum_{i=1}^{n} \delta H_{ii} + \sum_{i=1}^{n} \sum_{j=1}^{n} (2\delta J_{ij} - \delta K_{ij}) - 2\sum_{i=1}^{n} \sum_{j=1}^{n} \epsilon_{ij}\delta S_{ij} = 0 \tag{A7-31}$$

The variations occur in the MOs ϕ, and so

$$\delta S_{ij} = \int \delta\phi_i^*(1)\phi_j(1)\,d\tau_1 + \int \phi_i^*(1)\,\delta\phi_j(1)\,d\tau_1 \tag{A7-32}$$

$$\delta H_{ii} = \int \delta\phi_i^*(1) H_{(1)}^{\text{core}} \phi_i(1)\,d\tau_1 + \int \phi_i^*(1) H_{(1)}^{\text{core}}\,\delta\phi_i(1)\,d\tau_1 \tag{A7-33}$$

$$\begin{aligned}\delta J_{ij} = &\int \delta\phi_i^*(1)\phi_j^*(2)(1/r_{12})\phi_i(1)\phi_j(2)\,d\tau_1 d\tau_2 \\ &+ \int \phi_i^*(1)\delta\phi_j^*(2)(1/r_{12})\phi_i(1)\phi_j(2)\,d\tau_1 d\tau_2 + \text{complex conjugates}\end{aligned} \tag{A7-34}$$

It is convenient to define a coulomb operator $\hat{J}_i(1)$ as

$$\hat{J}_i(1) = \int \phi_i^*(2)(1/r_{12})\phi_i(2)\,d\tau_2 \tag{A7-35}$$

[1] For an introduction to this topic, see Margenau and Murphy [1].

Using this definition we can rewrite Eq. (A7-34) as

$$\delta J_{ij} = \int \delta\phi_i^*(1)\hat{J}_j(1)\phi_i(1)\,d\tau_1 + \int \delta\phi_j^*(1)\hat{J}_i(1)\phi_j(1)\,d\tau_1 + \text{complex conjugates} \tag{A7-36}$$

In the same spirit, we define an exchange operator $\hat{K}_i$, which, because it involves an orbital exchange, must be written in the context of an orbital being operated on:

$$\hat{K}_i(1)\phi_j(1) = \int \phi_i^*(2)(1/r_{12})\phi_j(2)\,d\tau_2\phi_i(1) \tag{A7-37}$$

This enables us to write δK_{ij}

$$\delta K_{ij} = \int \delta\phi_i^*(1)\hat{K}_j(1)\phi_i(1)\,d\tau_1 + \int \delta\phi_j^*(1)\hat{K}_i(1)\phi_j(1)\,d\tau_1 + \text{complex conjugates}$$

Employing the operators $\hat{J}$ and $\hat{K}$, Eq. (A7-31) can be written as follows:

$$2\sum_i \int \delta\phi_i^*(1)\left[H_{(1)}^{\text{core}}\phi_i(1) + \sum_j (2\hat{J}_j(1) - \hat{K}_j(1))\phi_i(1) - \sum_j \epsilon_{ij}\phi_j(1)\right]d\tau_1 + 2\sum_i \int \delta\phi_i(1)\left[H_{(1)}^{\text{core}*}\phi_i^*(1) + \sum_j (2\hat{J}_j^*(1) - \hat{K}_j^*(1))\phi_i^*(1) - \sum_j \epsilon_{ij}^*\phi_j^*(1)\right]d\tau_1 = 0 \tag{A7-38}$$

Here we have made use of the hermitian properties of H^{core}, $\hat{J}$, and $\hat{K}$, and also the relation $\epsilon_{ji}\int \delta\phi_j(1)\phi_i^*(1)\,d\tau_1 = \epsilon_{ij}\int \delta\phi_i(1)\phi_j^*(1)\,d\tau_1$, which is merely an index interchange.

Since the variations $\delta\phi_i^*$ and $\delta\phi_i$ are independent, each half of Eq. (A7-38) must independently equal zero. Hence, we can select either half for further development. We will select the first half. This equation indicates that the sum of integrals equals zero. Either the integrals are all individually equal to zero or else they are finite but cancel. However the latter possibility is ruled out because the variations $\delta\phi_i^*$ are arbitrary. By appropriately picking $\delta\phi_i^*$, we could always spoil cancellation if the various integrals for different i were nonzero. But the equation states that the sum vanishes for every $\delta\phi_i^*$. Therefore, we are forced to conclude that each integral vanishes.

Continuing in the same spirit, we can conclude that the term in brackets in the integrand is zero. For the integral to vanish requires the integrand either to be identically zero or else to have equal positive and negative parts. If the latter were true for some choice of $\delta\phi_i^*$, it would be possible to change $\delta\phi_i^*$ so as to unbalance the cancellation and produce a nonzero integral. Since the integral is zero for all $\delta\phi_i^*$, it must be that the bracketed term vanishes identically. Thus,

$$\left[H_{(1)}^{\text{core}} + \sum_j (2\hat{J}_j(1) - \hat{K}_j(1))\right]\phi_i(1) = \sum_j \epsilon_{ij}\phi_j(1) \tag{A7-39}$$

for all $i = 1$ to n and for a certain set of constants ϵ_{ij}.

The original development of SCF equations was performed by Hartree for simple product wavefunctions. Fock later extended the approach to apply to antisymmetrized wavefunctions. For this reason, the collection of operators in brackets in Eq. (A7-39) is called the Fock operator, symbolized $\hat{F}$, and Eq. (A7-39) becomes

$$\hat{F}(1)\phi_i(1) = \sum_j \epsilon_{ij}\phi_j(1) \tag{A7-40}$$

Equation A7-40 is a differential equation for each MO ϕ_i. But as it stands it is not an eigenvalue equation because, instead of regenerating ϕ_i, we obtain a sum of functions ϕ_j times the various unknown constants ϵ_{ij}. However, there remains a degree of freedom in the problem that can be used to throw Eq. (A7-40) into eigenvalue form. It is pointed out in Appendix 2 that the value of a determinant is unchanged if any row or column, multiplied by a constant, is added to any other row or column. This means that a Slater determinant of "best" MOs is unaffected by such internal rearrangements. In other words, if we were to solve Eq. (A7-40) for a set of "best" MOs, ϕ_i^{b}, we could form various new orthonormal MOs, (e.g., $\phi_i^{\mathrm{b}} + \lambda_{ik}\phi_k^{\mathrm{b}}, \phi_k^{\mathrm{b}} - \lambda_{ki}\phi_i^{\mathrm{b}}$) by mixing them together, and our wavefunction ψ, and all values of observables predicted from ψ, including $\bar{E}$, would be precisely the same.

A transformation that mixes the MOs ϕ without affecting the property of orthonormality is called a unitary transformation (see Chapter 9). Letting U stand for such a transformation, we have that a transformed set of ϕ's, called ϕ', is given by

$$\phi_i' = \sum_j U_{ji}\phi_j, \quad i = 1, \ldots, n \tag{A7-41}$$

In matrix notation, this is

$$\tilde{\Phi}' = \tilde{\Phi}\mathsf{U} \tag{A7-42}$$

where $\tilde{\Phi}'$ and $\tilde{\Phi}$ are row vectors, viz.

$$\tilde{\Phi}' = (\Phi_1' \Phi_2' \cdots \Phi_n'), \tag{A7-43}$$

and U is an $n \times n$ matrix, with

$$\mathsf{U}\mathsf{U}^\dagger = \mathsf{U}^\dagger\mathsf{U} = 1 \tag{A7-44}$$

In terms of these matrices, Eq. (A7-40) is

$$\hat{F}\tilde{\Phi} = \tilde{\Phi}\mathsf{E} \tag{A7-45}$$

where E is an $n \times n$ matrix. If we multiply this from the right by U, we obtain

$$\hat{F}\tilde{\Phi}\mathsf{U} = \tilde{\Phi}\mathsf{E}\mathsf{U} \tag{A7-46}$$

Inserting 1 (in the form $\mathsf{U}\mathsf{U}^\dagger$) between $\tilde{\Phi}$ and E gives

$$\hat{F}\tilde{\Phi}\mathsf{U} = \tilde{\Phi}\mathsf{U}\mathsf{U}^\dagger\mathsf{E}\mathsf{U} \tag{A7-47}$$

or

$$\hat{F}\tilde{\Phi}' = \tilde{\Phi}'\mathsf{U}^\dagger\mathsf{E}\mathsf{U} \tag{A7-48}$$

We can now require that the matrix U be such that $\mathsf{U}^\dagger \mathsf{E} \mathsf{U}$ is a diagonal matrix E'. (This requires that E be a hermitian matrix, which can be shown to be the case.)[2] This requirement defines U, and we have

$$\hat{F}\tilde{\Phi}' = \tilde{\Phi}'\mathsf{E}' \tag{A7-49}$$

which corresponds to

$$\hat{F}\phi_i' = \epsilon_i'\phi_i', \quad i = 1, 2, \ldots, n \tag{A7-50}$$

This equation has the desired eigenvalue form, and is commonly referred to as the *Hartree–Fock* equation. It is discussed at length in Chapter 11.

It is important to bear in mind that our transformation U is for mathematical convenience and has no physical effect. We may imagine that our original basis set spans a certain function space, and that solution of Eq. (A7-50) produces a set of occupied MOs ϕ_i' that span a "best" subspace. Transformations by unitary matrices produce new sets of MOs ϕ'' but these still span the same subspace as ϕ'. However, they are generally not eigenfunctions of $\hat{F}$, and satisfy the less convenient Eq. (A7-40). Nonetheless, there are occasions when it is useful to use some set of MOs other than ϕ', and we can always do this without having to worry about introducing physical changes as long as our converted MO's are related to ϕ' by a unitary transformation.

The Hartree–Fock equation is ordinarily used in quantum chemistry in connection with a basis set of AOs, and it is possible to carry through a derivation of the Hartree–Fock equation for this type of basis. Detailed treatments of this derivation may be found in the paper by Roothaan [2] and in the book by Pople and Beveridge [3].

References

[1] H. Margenau and G. M. Murphy, *The Mathematics of Physics and Chemistry*. Van Nostrand-Reinhold, Princeton, New Jersey, 1956.

[2] C. C. J. Roothaan, *Rev. Mod. Phys.* **23**, 69 (1951).

[3] J. A. Pople and D. L. Beveridge, *Approximate Molecular Orbital Theory*. McGraw-Hill, New York, 1970.

[2] See Roothaan [2].

Appendix 8

The Virial Theorem for Atoms and Diatomic Molecules

A8-1 Atoms

In Chapter 3 it was shown that, for the ground state of the quantum-mechanical harmonic oscillator, the average value of the kinetic energy is equal to the average value of the potential energy. We now consider how the average electronic kinetic and potential energies are related in an atom. We begin by deriving a rather general expression, and then we discuss how it applies to different levels of calculation. As our first step, we examine the effects of *coordinate scaling* on average values. In order to follow this discussion, it is useful to recall that one can manipulate variables and limits in an integral as follows:

$$\int_a^b f(x)\,dx = \int_a^b f(y)\,dy = \int_{\eta x=a}^{\eta x=b} f(\eta x)\,d(\eta x) = \eta \int_{x=a/\eta}^{x=b/\eta} f(\eta x)\,dx \qquad \text{(A8-1)}$$

Let $\psi(\mathbf{r}_1, \mathbf{r}_2, \ldots, \mathbf{r}_n)$ be a normalized function of the space coordinates of n electrons. Let

$$\bar{T} = \langle \psi | \hat{T} | \psi \rangle \qquad \text{(A8-2)}$$

$$\bar{V} = \langle \psi | \hat{V} | \psi \rangle \qquad \text{(A8-3)}$$

where $\hat{T}$ and $\hat{V}$ are, respectively, the kinetic and potential energy operators for some system, and are independent of spin.

We introduce a scale factor η into ψ. This factor affects the lengths of the vectors $\mathbf{r}_i$ but not their directions. That is,

$$\psi_\eta \equiv \psi(\eta\mathbf{r}_1, \eta\mathbf{r}_2, \ldots, \eta\mathbf{r}_n) \qquad \text{(A8-4)}$$

If $\eta > 1$, ψ_n is more contracted in $3n$-dimensional space than ψ. For $\eta < 1$, ψ_n is more diffuse.

We must check to see if our scaled function ψ_n is normalized. We know that

$$1 = \int \psi^*(\mathbf{r}_1, \ldots, \mathbf{r}_n)\psi(\mathbf{r}_1, \ldots, \mathbf{r}_n)\,dv$$

$$= \int \psi^*(\eta\mathbf{r}_1, \ldots, \eta\mathbf{r}_n)\psi(\eta\mathbf{r}_1, \ldots, \eta\mathbf{r}_n)\,d(\eta v) \qquad \text{(A8-5)}$$

because we have simply relabeled all variables $\mathbf{r}$ by $\eta\mathbf{r}$, including the volume element, just as in Eq. A8-1. We now factor η out of the volume element and divide the limits of integration by η, just as in Eq. A8-1. However, the limits are zero and infinity, so they are unaffected. The volume element $d(\eta v)$ is given by

$$d(\eta v) = (\eta r_1)^2 \sin\theta_1 d(\eta r_1)\, d\theta_1 d\phi_1 (\eta r_2)^2 \sin\theta_2 d(\eta r_2)\, d\theta_2 d\phi_2 \cdots \quad \text{(A8-6)}$$

and so η^3 appears for each electron. Thus, we are led to

$$1 = \eta^{3n} \int \psi_\eta^* \psi_\eta \, dv \quad \text{(A8-7)}$$

Therefore, our normalization constant for ψ_η is $\eta^{3n/2}$, and our *normalized*, scaled function is

$$\psi_\eta = \eta^{3n/2} \psi(\eta \mathbf{r}_1, \eta \mathbf{r}_2, \ldots, \eta \mathbf{r}_n) \quad \text{(A8-8)}$$

[Compare this with the specific example encountered in Eq. (7-7).] We now inquire as to the values of $\bar{V}_\eta$ and $\bar{T}_\eta$, where

$$\bar{T}_\eta = \langle \psi_\eta | \hat{T} | \psi_\eta \rangle \quad \text{(A8-9)}$$

$$\bar{V}_\eta = \langle \psi_\eta | \hat{V} | \psi_\eta \rangle \quad \text{(A8-10)}$$

For an n-electron atom,

$$\hat{T} = -\frac{1}{2} \sum_{i=1}^{n} \nabla_i^2 \quad \text{(A8-11)}$$

$$\hat{V} = -\sum_{i=1}^{n} (Z/r_i) + \sum_{i=1}^{n-1} \sum_{j=i+1}^{n} 1/r_{ij} \quad \text{(A8-12)}$$

Therefore

$$\bar{V}_\eta = \eta^{3n} \int \psi^*(\eta \mathbf{r}_1, \ldots) \left[\sum_{i=1}^{n} \left(\frac{-Z}{r_i} \right) + \sum_{i=1}^{n-1} \sum_{j=i+1}^{n} \frac{1}{r_{ij}} \right] \psi(\eta \mathbf{r}_1, \ldots)\, dv \quad \text{(A8-13)}$$

We could make the integral equal to $\bar{V}$ if we could get the scale factor into all the r terms in the operator and also into dv. The volume element dv requires η^{3n}, which is already present in Eq. A8-13 from the normalization constants. To get η into the operator, we need to multiply the operator by η^{-1}. Multiplying Eq. A8-13 by $\eta\eta^{-1}$ gives, then

$$\bar{V}_\eta = \eta \int \psi^*(\eta \mathbf{r}_1, \ldots) \left[\sum_{t=1}^{n} \left(\frac{-Z}{\eta r_i} \right) + \sum_{t=1}^{n-1} \sum_{j=i+1}^{n} \frac{1}{\eta r_{ij}} \right] \psi(\eta \mathbf{r}_1, \ldots)\, d(\eta v) \quad \text{(A8-14)}$$

or

$$\bar{V}_\eta = \eta \bar{V} \quad \text{(A8-15)}$$

The same approach to $\bar{T}_\eta$ gives

$$\bar{T}_\eta = \eta^2 \bar{T} \quad \text{(A8-16)}$$

This arises from the fact that

$$\nabla^2 = \frac{1}{r^2}\frac{\partial}{\partial r}\left(r^2\frac{\partial}{\partial r}\right) + \frac{1}{r^2\sin\theta}\frac{\partial}{\partial\theta}\left(\sin\theta\frac{\partial}{\partial\theta}\right) + \frac{1}{r^2\sin^2\theta}\frac{\partial^2}{\partial\phi^2} \tag{A8-17}$$

and scaling the r terms here requires multiplying by η^{-2}. Hence, the integral is multiplied by $\eta^2\eta^{-2}$ in the final step.

The *general* result is that, for any quantum-mechanical system where

$$\hat{V} = f(r^{-v}) \tag{A8-18}$$

scaling results in

$$\bar{T}_\eta = \eta^2\bar{T}, \qquad \bar{V}_\eta = \eta^v\bar{V} \tag{A8-19}$$

For atoms, $\hat{V}$ contains r as r^{-1}. For all systems, $\hat{T}$ involves ∇^2, which contains r to the net power of -2. We can now write the expression for the total energy of the atom as given by the scaled function

$$\bar{E}_\eta = \bar{T}_\eta + \bar{V}_\eta = \eta^2\bar{T} + \eta\bar{V} \tag{A8-20}$$

Now we can seek the best value of the scale factor. We do this by minimizing $\bar{E}_\eta$ with respect to variations in η:

$$\partial\bar{E}\eta/\partial\eta = 2\eta\bar{T} + \bar{V} = 0 \tag{A8-21}$$

($\bar{T}$ and $\bar{V}$ are independent of η.)

We are now in a position to make some statements about the average values of $\hat{T}$ and $\hat{V}$ for certain wavefunctions. Let us consider first the *exact* values of $\bar{T}$ and $\bar{V}$. We know that, if ψ were an exact eigenfunction, no further energy lowering would result from rescaling. That is, η equals unity in Eq. A8-21. As a result,

$$2\bar{T} + \bar{V} = 0 \tag{A8-22}$$

or

$$\bar{V} = -2\bar{T} \tag{A8-23}$$

or, since $\bar{T} + \bar{V} = \bar{E}$,

$$E = -\bar{T} = \frac{1}{2}\bar{V} \tag{A8-24}$$

Thus, for an atom, we know that the *exact* nonrelativistic energy is equal to minus the *exact* average kinetic energy and is equal to one half the exact potential energy. Knowing that the exact energy of the ground-state neon atom is -128.925 a.u. enables us to say that $\bar{T} = +128.925$ a.u. and $\bar{V} = -257.850$ a.u. without actually knowing ψ. Moreover, the same relation holds for *any* stable state of an atom.

This same argument holds, not only for exact solutions, but for any trial function that has already been energy-optimized with respect to a scale factor, for then a new scaling parameter η gives no improvement, $\eta = 1$, and all is as above. Thus, any *nonlinear* variation scheme consistent with uniform scaling should ultimately lead to the relations

[Eqs. A8-22–A8-24]. Satisfying these relations is frequently referred to as satisfying the virial relation. Completely optimized single-ζ and double-ζ functions satisfy the virial relation.

It follows that Hartree–Fock atomic wavefunctions must satisfy the virial relation. Such solutions are, by definition, the *best* (lowest energy) attainable in a single determinantal form. "Best" includes all conceivable variation, linear or nonlinear, so all improvements achievable by scale factor variation are already present at the Hartree–Fock level, and $\eta = 1$.

In the event that $\bar{E}$ *can* be lowered by scaling, it is possible to evaluate the optimum η from Eq. A8-21, which gives

$$\eta = -\bar{V}/2\bar{T} \tag{A8-25}$$

One of the useful applications of the virial theorem is as an indicator of closeness of approach to the Hartree–Fock solution for an atom. If the calculation involves nonlinear variation (uniformly applied to all r coordinates), then the resulting wavefunction will satisfy the virial relations no matter how deficient it is as an approximation to the true eigenfunction. However, if the calculation involves only linear variation, as for example, when a linear combination of gaussian functions is used to approximate an AO, then there is no guarantee that the virial relation will be satisfied. If the basis set is extensive enough, however, the Hartree–Fock limit will be approached, and $\bar{V}/\bar{T}$ will approach -2. Strictly speaking, a linear variation calculation on an atom that gives $\bar{V}/\bar{T} = -2$ is simply one that cannot be improved by uniform scaling. Therefore, approach to -2 is not a guarantee of approach to the Hartree-Fock limit. It is a necessary but not a sufficient condition.

A8-2 Diatomic Molecules

The treatment here is very similar to that for atoms. We make the Born–Oppenheimer approximation by assuming that ψ depends parametrically on the internuclear separation R:

$$\psi = \psi(\mathbf{r}_1, \mathbf{r}_2, \ldots, \mathbf{r}_n, R) \tag{A8-26}$$

When we scale $\mathbf{r}_i$, we scale R as well:

$$\psi_\eta = \psi(\eta\mathbf{r}_1, \eta\mathbf{r}_2, \ldots, \eta\mathbf{r}_n, \eta R) \tag{A8-27}$$

Henceforth, we let $\eta R \equiv \rho$. Performing the same variable manipulations as in Section A8-1, we find

$$\bar{T}_\eta \equiv \bar{T}(\eta, \rho) = \eta^2 \bar{T}(1, \rho) \tag{A8-28}$$
$$\bar{V}_\eta \equiv \bar{V}(\eta, \rho) = \eta \bar{V}(1, \rho) \tag{A8-29}$$

Here, $\hat{V}$ may or may not include the internuclear repulsion term. This gives, for the total energy,

$$\bar{E}_\eta = \eta^2 \bar{T}(1, \rho) + \eta \bar{V}(1, \rho) \tag{A8-30}$$

Upon taking the derivative with respect to η, we obtain

$$\frac{\partial \bar{E}_n}{\partial \eta} = 2\eta \bar{T}(1,\rho) + \bar{V}(1,\rho) + \eta^2 \frac{\partial \bar{T}(1,\rho)}{\partial \eta} + \eta \frac{\partial \bar{V}(1,\rho)}{\partial \eta} = 0 \qquad \text{(A8-31)}$$

This differs from the atomic case in that $\bar{V}(1,\rho)$ and $\bar{T}(1,\rho)$ depend on η through ρ. But

$$\frac{\partial}{\partial \eta} = \left(\frac{\partial}{\partial \rho}\right)\left(\frac{\partial \rho}{\partial \eta}\right) = \left(\frac{\partial}{\partial \rho}\right) R \qquad \text{(A8-32)}$$

and so Eq. A8-31 becomes

$$\frac{\partial \bar{E}_\eta}{\partial \eta} = 0 = 2\eta \bar{T}(1,\rho) + \eta \bar{V}(1,\rho) + \eta^2 R \frac{\partial \bar{T}(1,\rho)}{\partial \rho} + \eta R \frac{\partial \bar{V}(1,\rho)}{\partial \rho} \qquad \text{(A8-33)}$$

If we assume that ψ is the exact eigenfunction, then $\eta = 1$, and

$$2\bar{T} + \bar{V} + R\left(\frac{\partial \bar{E}}{\partial R}\right) = 0 \qquad \text{(A8-34)}$$

Indeed, this relation holds for any case in which all improvement in the nature of a scale factor variation has been made, such as, for example, the Hartree-Fock limit. Note that, if $\bar{V}$ contains internuclear repulsion, $\bar{E}$ is the total energy. If not, $\bar{E}$ is the electronic energy.

A8-2.1 Problems

A8-1. Use the methods outlined in this appendix to show that $\bar{V} = \bar{T}$ for any stationary state of the quantum mechanical harmonic oscillator.

A8-2. Evaluate $\bar{V}$ and $\bar{T}$ with $\psi = 1/\sqrt{\pi}\exp(-r)$ for the Li^{2+} ion. From these, establish the optimum scale factor η and write down the expression for the normalized optimized ψ_η and the optimized energy E_η. Compare these results with the eigenfunction for Li^{2+}.

Appendix 9

Bra-ket Notation

Bra-ket, or Dirac, notation is frequently used in the literature because of its economical form. Perhaps the best way to learn how this notation is used is by studying its use in a few familiar relations and proofs. Accordingly, we have outlined a few of these uses. The applications and subtleties of this notation go considerably beyond the treatment summarized here.[1]

"Usual" notation		Dirac notation	
$\int \phi_m^* \phi_n d\tau$	$\equiv$	$\underbrace{\langle \phi_m \mid}_{\text{bra}} \underbrace{\phi_n \rangle}_{\text{ket}} \equiv \langle m \vert n \rangle$	(A9-1)
$\int \phi_m^* A \phi_n d\tau$	$\equiv$	$\langle \phi_m \vert A \vert \phi_n \rangle \equiv \langle m \vert A \vert n \rangle \equiv A_{mn}$	(A9-2)
$\left[\int \phi_m^* \phi_n d\tau\right]^* = \int \phi_n^* \phi_m d\tau$		$\langle \phi_m \vert \phi_n \rangle^* = \langle \phi_n \vert \phi_m \rangle$	
	or		(A9-3)
		$\langle m \vert n \rangle^* = \langle n \vert m \rangle$	

For hermitian operator A:

$$\int \phi_m^* A \phi_n d\tau = \int \phi_n (A \phi_m)^* d\tau \qquad \langle m|A|n \rangle = \langle n|A|m \rangle^*$$

$$= \left[\int \phi_n^* A \phi_m d_T\right]^* \qquad \text{(A9-4)}$$

Any function ψ can be written as a sum of a complete set of orthonormal functions ϕ:

$$\psi = \sum_m c_m \phi_m \qquad |\psi\rangle = \sum_m c_m |\phi_m\rangle \equiv \sum_m c_m |m\rangle$$

$$\int \phi_n^* \psi d\tau = \sum_m c_m \int \phi_n^* \phi_m d\tau = c_n \qquad \langle n|\psi\rangle = \sum_m c_m \langle n|m\rangle = c_n$$

$$c_n = \int \phi_n^* \psi d\tau \qquad c_n = \langle n|\psi\rangle$$

$$\psi = \sum_m c_m \phi_m = \sum_m \int \phi_m^* \psi d\tau \phi_m \qquad |\psi\rangle = \sum_m \langle m|\psi\rangle |m\rangle$$

$$= \sum_m |m\rangle \langle m|\psi\rangle$$

[1] Strictly speaking, for instance, $\langle \phi_m | \phi_n \rangle$ and $\langle m|n \rangle$ are not identical in meaning. The former refers to specific functions, ϕ_m and ϕ_n, which represent state vectors in a specific representation. The latter refers to the state vectors in *any* representation and hence is a more general expression. Distinctions such as this will not be necessary at the level of this text.

Example of Use: Proof that eigenvalues of A (hermitian) are real.

$$A|m\rangle = a_m|m\rangle \tag{A9-5}$$

$$\langle m|A|m\rangle = a_m \underbrace{\langle m|m\rangle}_{\neq 0, \neq \infty}, \tag{A9-6}$$

$$\langle m|A|m\rangle^* = a_m^* \langle m|m\rangle \tag{A9-7}$$

Combining Eqs. A9-4, A9-6, and A9-7, we have

$$(a_m - a_m^*) = 0$$

Appendix 10

Values of Some Useful Constants and Conversion Factors

Values of Some Useful Constants[a]

Quantity	Symbol and/or formula[c]	Value: Atomic units	Value: SI units	Value: Other units	Uncertainty (ppm)
a) Fundamental constants					
Planck's constant	h	2π	6.626176×10^{-34} J sec	4.135701×10^{-15} eV sec	5.4
Planck's constant $h/2\pi$	$\hbar$	1	$1.0545887 \times 10^{-34}$ J sec	6.582173×10^{-16} eV sec	5.4
Rest mass of electron	m_e	1	9.109534×10^{-31} kg	9.109534×10^{-28} gm	5.1
Charge of electron	$-e$	-1	$-1.602189 \times 10^{-19}$ C	$-4.803242 \times 10^{-10}$ esu	2.9
Rest mass of proton	m_p	1.83615×10^3	$1.6726485 \times 10^{-27}$ kg	$1.6726485 \times 10^{-24}$ gm	5.1
Rest mass of neutron	m_n	1.83868×10^3	$1.6749543 \times 10^{-27}$ kg	$1.6749543 \times 10^{-24}$ gm	5.1
Speed of light in vacuum	c	137.039	2.99792458×10^{8} m sec^{-1}	$2.99792458 \times 10^{10}$ cm sec^{-1}	0.004
Avogadro's number[b]	N_A	—	6.0220943×10^{23} mol^{-1}	—	1.05
Bohr magneton	β_e	$\frac{1}{2}$	9.274078×10^{-24} JT^{-1}		
Electron g value	g_e	—	2.00232		
b) Derived quantities					
Bohr radius	a_0	1	$5.2917706 \times 10^{-11}$ m	0.52917706 Å	0.82
Vacuum permittivity	$4\pi\epsilon_0$	1	1.11265×10^{-10} J^{-1} C^2 m^{-1}		
Twice the ionization potential of the hydrogen atom with infinite nuclear mass	$E_a = e^2/4\pi\epsilon_0 a_0$	1	4.359814×10^{-18} J	27.21161 eV; 2 rydbergs	6.6
Electric field strength one bohr radius from proton	$e/4\pi\epsilon_0 a_0^2$	1	5.1423×10^{11} V m^{-1}	1.715270×10^{10} esu cm^{-2}	4.4

Polarizability (of a molecule)	$\alpha = e^2 a_0{}^2 / E_a$	1	1.648776×10^{-41} C^2 m^2 J^{-1}	1.481846×10^{-25} esu^2 cm^2 erg^{-1}	1.4
Bohr magneton	$\mu_B = e\hbar/2m_e$	$\frac{1}{2}$	9.274078×10^{-24} J T^{-1}	5.788378×10^{-9} eV G^{-1}	3.9
Nuclear magneton	$\mu_N = e\hbar/2m_p$	2.723087×10^{-4}	5.050824×10^{-27} J T^{-1}	$3.1524515 \times 10^{-12}$ eV G^{-1}	3.9
Electric dipole moment of electron–proton separated by one Bohr radius	ea_0	1	8.478418×10^{-30} C m	2.541765×10^{-20} esu m = 2.541765 debyes	3.8
Time for 1s electron in hydrogen atom to travel one bohr radius	$t = \hbar / E_a$	1	2.41888×10^{-17} sec	—	12
Atomic unit of velocity	a_0/t	1	2.18769×10^{6} m sec^{-1}	2.18767×10^{8} cm sec^{-1}	13
Atomic unit of volume	$a_0{}^3$	1	1.481846×10^{-31} m^3	0.14818 Å^3	2.0
Atomic unit of probability density	$a_0{}^{-3}$	1	6.748340×10^{30} m^{-3}	6.748340 Å^{-3}	3.0

[a]From Cohen and Taylor [1]. C = coulomb, J = joule, V = volt, T = tesla, G = gauss, Å = angstrom.
[b]See Ref. [2].
[c]Formula appropriate for atomic units.

Energy Conversion Factors[a]

	eV	joule[b]	kcal/mole	Hz	m^{-1}	°K	a.u.
eV	1	$1.6021892 \times 10^{-19}$	23.060362	2.4179696×10^{14}	8.065479×10^{5}	1.160450×10^{4}	3.674901×10^{-2}
joule[b]	6.2414601×10^{18}	1	1.4393033×10^{20}	1.5091661×10^{33}	5.034037×10^{24}	7.242902×10^{22}	2.293675×10^{17}
kcal/mole	4.336445×10^{-2}	6.947805×10^{-21}	1	1.0485393×10^{13}	3.497551×10^{4}	5.032223×10^{2}	1.593601×10^{-3}
Hz	$4.1357012 \times 10^{-15}$	$6.6261759 \times 10^{-34}$	$9.5370770 \times 10^{-14}$	1	3.3356412×10^{-9}	4.799274×10^{-11}	1.519829×10^{-16}
m^{-1}	1.239852×10^{-6}	1.986477×10^{-25}	2.859144×10^{-5}	2.9979243×10^{8}	1	1.438786×10^{-2}	4.556333×10^{-8}
°K	8.617347×10^{-5}	1.380662×10^{-23}	1.987191×10^{-3}	2.083648×10^{10}	6.950303×10^{1}	1	3.166790×10^{-6}
a.u.	27.21161	4.359816×10^{-18}	6.275098×10^{2}	6.579686×10^{15}	2.194747×10^{7}	3.157772×10^{5}	1

[a] To convert *from* units in the left hand column *to* units in the top row, multiply by the factor in the row-column position, e.g., 1 kcal/mole = 1.0485393×10^{13} Hz.
[b] 1 joule = 10^{7} erg.

References

[1] E. R. Cohen and B. N. Taylor, *J. Phys. Chem. Ref. Data* 2, No. 4, 663 (1973).

[2] A. L. Robinson, *Science*, **185**, 1037 (1974).

See also P. J. Mohr and B. N. Taylor, *Rev. Mod. Phys.* 72 (2000) for revised values.

Appendix 11

Group Theoretical Charts and Tables

A11-1 Flow Scheme for Group Symbols

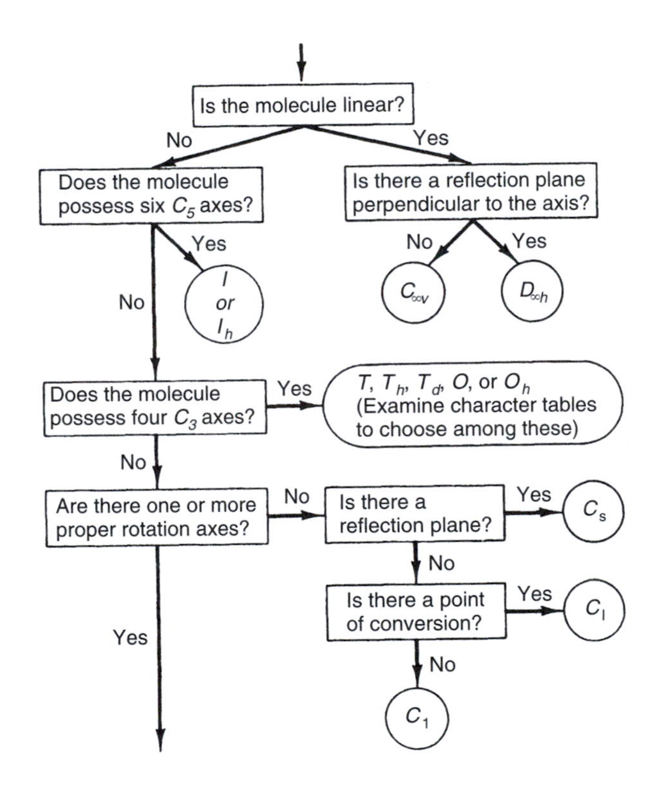

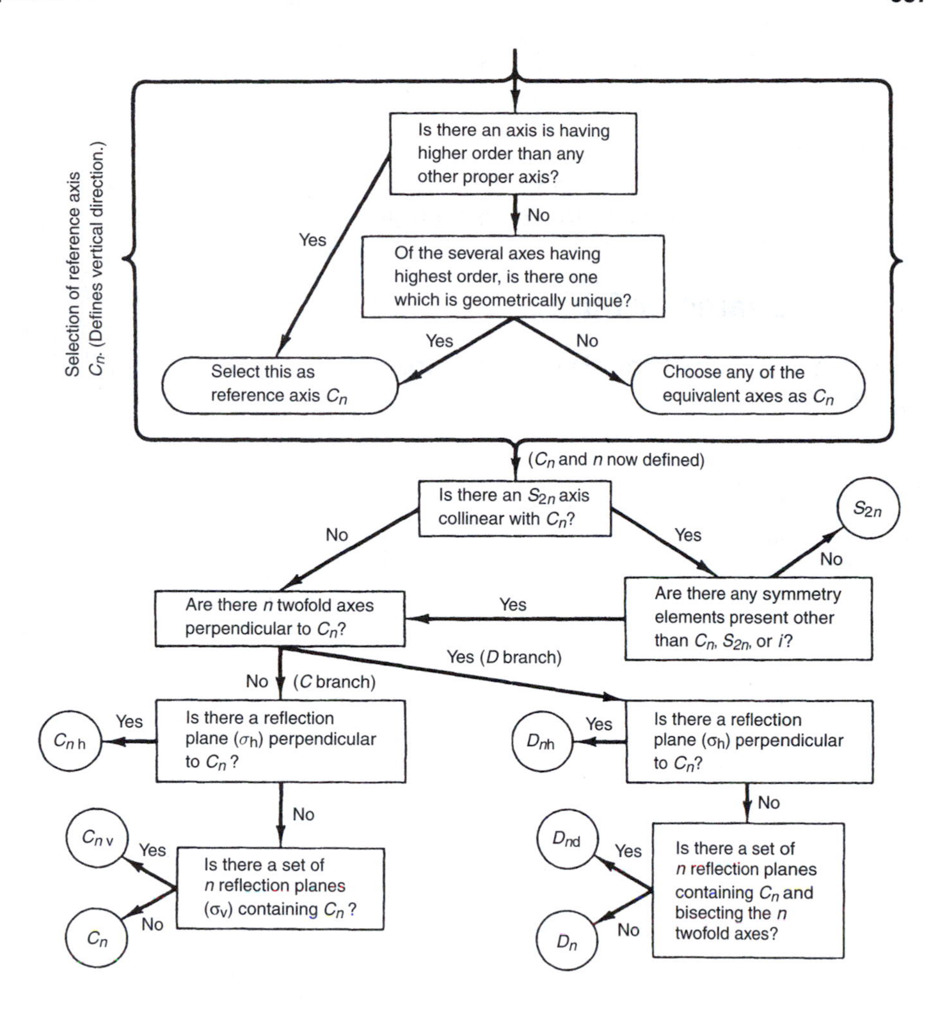

A11-2 Meaning of Labels for Representations

Symbol	Interpretation
Main Symbol	
A	One-dimensional representation symmetric for rotation by $2\pi/n$ about principal axis. (For c_1, c_s, c_i, which have no principal axis, this symbol merely means a one-dimensional representation.)
B	One-dimensional representation but antisymmetric for rotation by $2\pi/n$ about principal axis
E	Two-dimensional representation
T (or F)	Three-dimensional representation
G	Four-dimensional representation
Subscripts	
1	Symmetric for perpendicular C_2 rotations (or else σ_v or σ_d reflections)
2	Antisymmetric for perpendicular C_2 rotations (or else σ_v or σ_d reflections)

g	Symmetric for inversion
u	Antisymmetric for inversion

Superscripts

′	Symmetric for σ_h reflection
″	Antisymmetric for σ_h reflection

A11-3 Character Tables

A11-3.A Special High-Symmetry Groups: $C_{\infty v}$, $D_{\infty h}$, I, I_h, T, T_h, T_d, O, O_h

$C_{\infty v}$	E	$2C_\infty^{\Phi}$	$\cdots$	$\infty\sigma_v$		
$A_1 \equiv \Sigma^+$	1	1	$\cdots$	1	z	x^2+y^2, z^2
$A_2 \equiv \Sigma^-$	1	1	$\cdots$	-1	R_z	
$E_1 \equiv \Pi$	2	$2\cos\Phi$	$\cdots$	0	$(x, y); (R_x, R_y)$	(xz, yz)
$E_2 \equiv \Delta$	2	$2\cos 2\Phi$	$\cdots$	0		(x^2-y^2, xy)
$E_3 \equiv \Phi$	2	$2\cos 3\Phi$	$\cdots$	0		
$\cdots$	$\cdots$	$\cdots$	$\cdots$	$\cdots$		

$D_{\infty h}$	E	$2C_\infty^{\Phi}$	$\cdots$	$\infty\sigma_v$	i	$2S_\infty^{\Phi}$	$\cdots$	∞C_2		
Σ_g^+	1	1	$\cdots$	1	1	1	$\cdots$	1		x^2+y^2, z^2
Σ_g^-	1	1	$\cdots$	-1	1	1	$\cdots$	-1	R_z	
Π_g	2	$2\cos\Phi$	$\cdots$	0	2	$-2\cos\Phi$	$\cdots$	0	(R_x, R_y)	(xz, yz)
Δ_g	2	$2\cos 2\Phi$	$\cdots$	0	2	$2\cos 2\Phi$	$\cdots$	0		(x^2-y^2, xy)
$\cdots$	$\cdots$	$\cdots$	$\cdots$	$\cdots$	$\cdots$	$\cdots$	$\cdots$	$\cdots$		
Σ_u^+	1	1	$\cdots$	1	-1	-1	$\cdots$	-1	z	
Σ_u^-	1	1	$\cdots$	-1	-1	-1	$\cdots$	1		
Π_u	2	$2\cos\Phi$	$\cdots$	0	-2	$2\cos\Phi$	$\cdots$	0	(x, y)	
Δ_u	2	$2\cos 2\Phi$	$\cdots$	0	-2	$-2\cos 2\Phi$	$\cdots$	0		
$\cdots$	$\cdots$	$\cdots$	$\cdots$	$\cdots$	$\cdots$	$\cdots$	$\cdots$	$\cdots$		

I	E	$12C_5$	$12C_5^2$	$20C_3$	$15C_2$		
A	1	1	1	1	1		$x^2+y^2+z^2$
T_1	3	$\frac{1}{2}(1+\sqrt{5})$	$\frac{1}{2}(1-\sqrt{5})$	0	-1	$(x, y, z); (R_x, R_y, R_z)$	
T_2	3	$\frac{1}{2}(1-\sqrt{5})$	$\frac{1}{2}(1+\sqrt{5})$	0	-1		
G	4	-1	-1	1	0		
H	5	0	0	-1	1		$(2z^2-x^2-y^2, x^2-y^2, xy, yz, zx)$

I_h	E	$12C_5$	$12C_5^2$	$20C_3$	$15C_2$	i	$12S_{10}$	$12S_{10}^3$	$20S_6$	15σ		
A_g	1	1	1	1	1	1	1	1	1	1		$x^2+y^2+z^2$
T_{1g}	3	$\frac{1}{2}(1+\sqrt{5})$	$\frac{1}{2}(1-\sqrt{5})$	0	−1	3	$\frac{1}{2}(1-\sqrt{5})$	$\frac{1}{2}(1+\sqrt{5})$	0	−1	(R_x, R_y, R_z)	
T_{2g}	3	$\frac{1}{2}(1-\sqrt{5})$	$\frac{1}{2}(1+\sqrt{5})$	0	−1	3	$\frac{1}{2}(1+\sqrt{5})$	$\frac{1}{2}(1-\sqrt{5})$	0	−1		
G_g	4	−1	−1	1	0	4	−1	−1	1	0		
H_g	5	0	0	−1	1	5	0	0	−1	1		$(2z^2-x^2-y^2, x^2-y^2, xy, yz, zx)$
A_u	1	1	1	1	1	−1	−1	−1	−1	−1		
T_{1u}	3	$\frac{1}{2}(1+\sqrt{5})$	$\frac{1}{2}(1-\sqrt{5})$	0	−1	−3	$-\frac{1}{2}(1-\sqrt{5})$	$-\frac{1}{2}(1+\sqrt{5})$	0	1	(x, y, z)	
T_{2u}	3	$\frac{1}{2}(1-\sqrt{5})$	$\frac{1}{2}(1+\sqrt{5})$	0	−1	−3	$-\frac{1}{2}(1+\sqrt{5})$	$-\frac{1}{2}(1-\sqrt{5})$	0	1		
G_u	4	−1	−1	1	0	−4	1	1	−1	0		
H_u	5	0	0	−1	1	−5	0	0	1	−1		

T	E	$4C_3$	$4C_3^2$	$3C_2$		$\epsilon = \exp(2\pi i/3)$
A	1	1	1	1		$x^2+y^2+z^2$
E	$\begin{Bmatrix}1\\1\end{Bmatrix}$	ϵ ϵ^*	ϵ^* ϵ	1 1		$(2z^2-x^2-y^2, x^2-y^2)$
T	3	0	0	−1	(R_x, R_y, R_z); (x, y, z)	(xy, xz, yz)

T_h	E	$4C_3$	$4C_3^2$	$3C_2$	i	$4S_6$	$4S_6^5$	$3\sigma_h$		$\epsilon = \exp(2\pi i/3)$
A_g	1	1	1	1	1	1	1	1		$x^2+y^2+z^2$
A_u	1	1	1	1	−1	−1	−1	−1		
E_g	1 1	ϵ ϵ^*	ϵ^* ϵ	1 1	1 1	ϵ ϵ^*	ϵ^* ϵ	1 1		$(2z^2-x^2-y^2,$ $x^2-y^2)$
E_u	1 1	ϵ ϵ^*	ϵ^* ϵ	1 1	−1 −1	$-\epsilon$ $-\epsilon^*$	$-\epsilon^*$ $-\epsilon$	−1 −1		
T_g	3	0	0	−1	1	0	0	−1	(R_x, R_y, R_z)	(xz, yz, xy)
T_u	3	0	0	−1	−1	0	0	1	(x, y, z)	

T_d	E	$8C_3$	$3C_2$	$6S_4$	$6\sigma_d$		
A_1	1	1	1	1	1		$x^2+y^2+z^2$
A_2	1	1	1	−1	−1		
E	2	−1	2	0	0		$(2z^2-x^2-y^2, x^2-y^2)$
T_1	3	0	−1	1	−1	(R_x, R_y, R_z)	
T_2	3	0	−1	−1	1	(x, y, z)	(xy, xz, yz)

O	E	$6C_4$	$3C_2(=C_4^2)$	$8C_3$	$6C_2$		
A_1	1	1	1	1	1		$x^2+y^2+z^2$
A_2	1	−1	1	1	−1		
E	2	0	2	−1	0		$(2z^2-x^2-y^2,$ $x^2-y^2)$
T_1	3	1	−1	0	−1	(R_x, R_y, R_z); (x, y, z)	
T_2	3	−1	−1	0	1		(xy, xz, yz)

O_h	E	$8C_3$	$6C_2$	$6C_4$	$3C_2$ $(=C_4^2)$	i	$6S_4$	$8S_6$	$3\sigma_h$	$6\sigma_d$		
A_{1g}	1	1	1	1	1	1	1	1	1	1		$x^2+y^2+z^2$
A_{2g}	1	1	−1	−1	1	1	−1	1	1	−1		
E_g	2	−1	0	0	2	2	0	−1	2	0		$(2z^2-x^2-y^2,$ $x^2-y^2)$
T_{1g}	3	0	−1	1	−1	3	1	0	−1	−1	(R_x, R_y, R_z)	
T_{2g}	3	0	1	−1	−1	3	−1	0	−1	1		(xz, yz, xy)
A_{1u}	1	1	1	1	1	−1	−1	−1	−1	−1		
A_{2u}	1	1	−1	−1	1	−1	1	−1	−1	1		
E_u	2	−1	0	0	2	−2	0	1	−2	0		
T_{1u}	3	0	−1	1	−1	−3	−1	0	1	1	(x, y, z)	
T_{2u}	3	0	1	−1	−1	−3	1	0	1	−1		

A11-3.B Groups with No Axis of Symmetry: C_1, C_i, C_s

C_1	E
A	1

C_s	E	σ_h		
A'	1	1	x, y, R_z	$x^2, y^2,$ z^2, xy
A''	1	−1	z, R_x, R_y	yz, xz

C_i	E	i		
A_g	1	1	R_x, R_y, R_z	$x^2, y^2, z^2,$ xy, xz, yz
A_u	1	−1	x, y, z	

A11-3.C The S_{2n} Groups

S_4	E	S_4	C_2	S_4^3		
A	1	1	1	1	R_z	x^2+y^2, z^2
B	1	−1	1	−1	z	x^2-y^2, xy
E	$\begin{Bmatrix}1\\1\end{Bmatrix}$	$\begin{matrix}i\\-i\end{matrix}$	$\begin{matrix}-1\\-1\end{matrix}$	$\begin{matrix}-i\\i\end{matrix}$	$(x, y); (R_x, R_y)$	(xz, yz)

S_6	E	C_3	C_3^2	i	S_6^5	S_6		$\epsilon=\exp(2\pi i/3)$
A_g	1	1	1	1	1	1	R_z	x^2+y^2, z^2
E_g	$\begin{Bmatrix}1\\1\end{Bmatrix}$	$\begin{matrix}\epsilon\\\epsilon^*\end{matrix}$	$\begin{matrix}\epsilon^*\\\epsilon\end{matrix}$	$\begin{matrix}1\\1\end{matrix}$	$\begin{matrix}\epsilon\\\epsilon^*\end{matrix}$	$\begin{matrix}\epsilon^*\\\epsilon\end{matrix}$	(R_x, R_y)	$(x^2-y^2, xy); (xz, yz)$
A_u	1	1	1	−1	−1	−1	z	
E_u	$\begin{Bmatrix}1\\1\end{Bmatrix}$	$\begin{matrix}\epsilon\\\epsilon^*\end{matrix}$	$\begin{matrix}\epsilon^*\\\epsilon\end{matrix}$	$\begin{matrix}-1\\-1\end{matrix}$	$\begin{matrix}-\epsilon\\-\epsilon^*\end{matrix}$	$\begin{matrix}-\epsilon^*\\-\epsilon\end{matrix}$	(x, y)	

S_8	E	S_8	C_4	$S_8{}^3$	C_2	$S_8{}^5$	$C_4{}^3$	$S_8{}^7$		$\epsilon = \exp(2\pi i/8)$
A	1	1	1	1	1	1	1	1	R_z	x^2+y^2, z^2
B	1	-1	1	-1	1	-1	1	-1	z	
E_1	1	ϵ	i	$-\epsilon^*$	-1	$-\epsilon$	$-i$	ϵ^*	(x, y);	
	1	ϵ^*	$-i$	$-\epsilon$	-1	$-\epsilon^*$	i	ϵ	(R_x, R_y)	
E_2	1	i	-1	$-i$	1	i	-1	$-i$		(x^2-y^2, xy)
	1	$-i$	-1	i	1	$-i$	-1	i		
E_3	1	$-\epsilon^*$	$-i$	ϵ	-1	ϵ^*	i	$-\epsilon$		(xz, yz)
	1	$-\epsilon$	i	ϵ^*	-1	ϵ	$-i$	$-\epsilon^*$		

A11-3.D The C_n Groups

C_2	E	C_2		
A	1	1	z, R_z	x^2, y^2, z^2, xy
B	1	-1	x, y, R_x, R_y	yz, xz

C_3	E	C_3	$C_3{}^2$		$\epsilon = \exp(2\pi i/3)$
A	1	1	1	z, R_z	x^2+y^2, z^2
E	1	ϵ	ϵ^*	$(x, y); (R_x, R_y)$	$(x^2-y^2, xy); (yz, xz)$
	1	ϵ^*	ϵ		

C_4	E	C_1	C_2	$C_4{}^3$		
A	1	1	1	1	z, R_z	x^2+y^2, z^2
B	1	-1	1	-1		x^2-y^2, xy
E	1	i	-1	$-i$	$(x, y); (R_x, R_y)$	(yz, xz)
	1	$-i$	-1	i		

C_5	E	C_5	$C_5{}^2$	$C_5{}^3$	$C_5{}^4$		$\epsilon = \exp(2\pi i/5)$
A	1	1	1	1	1	z, R_z	x^2+y^2, z^2
E_1	1	ϵ	ϵ^2	ϵ^{2*}	ϵ^*	$(x, y); (R_x, R_y)$	(yz, xz)
	1	ϵ^*	ϵ^{2*}	ϵ^2	ϵ		
E_2	1	ϵ^2	ϵ^*	ϵ	ϵ^{2*}		(x^2-y^2, xy)
	1	ϵ^{2*}	ϵ	ϵ^*	ϵ^2		

C_6	E	C_6	C_3	C_2	$C_3{}^2$	$C_6{}^5$		$\epsilon = \exp(2\pi i/6)$
A	1	1	1	1	1	1	z, R_z	x^2+y^2, z^2
B	1	−1	1	−1	1	−1		
E_1	1	ϵ	$-\epsilon^*$	−1	$-\epsilon$	ϵ^*	(x, y)	(xz, yz)
	1	ϵ^*	$-\epsilon$	−1	$-\epsilon^*$	ϵ	(R_x, R_y)	
E_2	1	$-\epsilon^*$	$-\epsilon$	1	$-\epsilon^*$	$-\epsilon$		(x^2-y^2, xy)
	1	$-\epsilon$	$-\epsilon^*$	1	$-\epsilon$	$-\epsilon^*$		

C_7	E	C_7	$C_7{}^2$	$C_7{}^3$	$C_7{}^4$	$C_7{}^5$	$C_7{}^6$		$\epsilon = \exp(2\pi i/7)$
A	1	1	1	1	1	1	1	z, R_z	x^2+y^2, z^2
E_1	1	ϵ	ϵ^2	ϵ^3	ϵ^{3*}	ϵ^{2*}	ϵ^*	(x, y)	(xz, yz)
	1	ϵ^*	ϵ^{2*}	ϵ^{3*}	ϵ^3	ϵ^2	ϵ	(R_x, R_y)	
E_2	1	ϵ^2	ϵ^{3*}	ϵ^*	ϵ	ϵ^3	ϵ^{2*}		(x^2-y^2, xy)
	1	ϵ^{2*}	ϵ^3	ϵ	ϵ^*	ϵ^{3*}	ϵ^2		
E_3	1	ϵ^3	ϵ^*	ϵ^2	ϵ^{2*}	ϵ	ϵ^{3*}		
	1	ϵ^{3*}	ϵ	ϵ^{2*}	ϵ^2	ϵ^*	ϵ^3		

C_8	E	C_8	C_4	C_2	$C_4{}^3$	$C_8{}^3$	$C_8{}^5$	$C_8{}^7$		$\epsilon = \exp(2\pi i/8)$
A	1	1	1	1	1	1	1	1	z, R_z	x^2+y^2, z^2
B	1	−1	1	1	1	−1	−1	1		
E_1	1	ϵ	i	−1	$-i$	$-\epsilon^*$	$-\epsilon$	ϵ^*	$(x, y); (R_x, R_y)$	(xz, yz)
	1	ϵ^*	$-i$	−1	i	$-\epsilon$	$-\epsilon^*$	ϵ		
E_2	1	i	−1	1	−1	$-i$	i	$-i$		(x^2-y^2, xy)
	1	$-i$	−1	1	−1	i	$-i$	i		
E_3	1	$-\epsilon$	i	−1	$-i$	ϵ^*	ϵ	$-\epsilon^*$		
	1	$-\epsilon^*$	$-i$	−1	i	ϵ	ϵ^*	$-\epsilon$		

A11-3.E The C_{nv} Groups

C_{2v}	E	C_2	$\sigma_v(xz)$	$\sigma_v{}'(yz)$		
A_1	1	1	1	1	z	x^2, y^2, z^2
A_2	1	1	−1	−1	R_z	xy
B_1	1	−1	1	−1	x, R_y	xz
B_2	1	−1	−1	1	y, R_x	yz

C_{3v}	E	$3\sigma_v$	$2C_3$		
A_1	1	1	1	z	x^2+y^2, z^2
A_2	1	−1	1	R_z	
E	2	0	−1	$(x, y); (R_x, R_y)$	$(x^2-y^2, xy); (xz, yz)$

C_{4v}	E	$2C_4$	C_2	$2\sigma_v$	$2\sigma_d$		
A_1	1	1	1	1	1	z	x^2+y^2, z^2
A_2	1	1	1	−1	−1	R_z	
B_1	1	−1	1	1	−1		x^2-y^2
B_2	1	−1	1	−1	1		xy
E	2	0	−2	0	0	$(x, y); (R_x, R_y)$	(xz, yz)

C_{5v}	E	$2C_5$	$2C_5{}^2$	$5\sigma_v$		
A_1	1	1	1	1	z	x^2+y^2, z^2
A_2	1	1	1	−1	R_z	
E_1	2	2 cos 72°	2 cos 144°	0	$(x, y); (R_x, R_y)$	(xz, yz)
E_2	2	2 cos 144°	2 cos 72°	0		(x^2-y^2, xy)

C_{6v}	E	$2C_6$	$2C_3$	C_2	$3\sigma_v$	$3\sigma_d$		
A_1	1	1	1	1	1	1	z	x^2+y^2, z^2
A_2	1	1	1	1	−1	−1	R_z	
B_1	1	−1	1	−1	1	−1		
B_2	1	−1	1	−1	−1	1		
E_1	2	1	−1	−2	0	0	$(x, y); (R_x, R_y)$	(xz, yz)
E_2	2	−1	−1	2	0	0		(x^2-y^2, xy)

A11-3.F The C_{nh} Groups

C_{2h}	E	C_2	i	σ_h		
A_g	1	1	1	1	R_z	x^2, y^2, z^2, xy
B_g	1	−1	1	−1	R_x, R_y	xz, yz
A_u	1	1	−1	−1	z	
B_u	1	−1	−1	1	x, y	

C_{3h}	E	C_3	$C_3{}^2$	σ_h	S_3	$S_3{}^5$		$\epsilon = \exp(2\pi i/3)$
A'	1	1	1	1	1	1	R_z	x^2+y^2, z^2
E'	$\begin{Bmatrix} 1 \\ 1 \end{Bmatrix}$	ϵ ϵ^*	ϵ^* ϵ	1 1	ϵ ϵ^*	ϵ^* ϵ	(x, y)	(x^2-y^2, xy)
A''	1	1	1	−1	−1	−1	z	
E''	$\begin{Bmatrix} 1 \\ 1 \end{Bmatrix}$	ϵ ϵ^*	ϵ^* ϵ	−1 −1	$-\epsilon$ $-\epsilon^*$	$-\epsilon^*$ $-\epsilon$	(R_x, R_y)	(xz, yz)

C_{4h}	E	C_4	C_2	$C_4{}^3$	i	$S_4{}^3$	σ_h	S_4		
A_g	1	1	1	1	1	1	1	1	R_z	x^2+y^2, z^2
B_g	1	−1	1	−1	1	−1	1	−1		x^2-y^2, xy
E_g	1	i	−1	$-i$	1	i	−1	$-i$	(R_x, R_y)	(xz, yz)
	1	$-i$	−1	i	1	$-i$	−1	i		
A_u	1	1	1	1	−1	−1	−1	−1	z	
B_u	1	−1	1	−1	−1	1	−1	1		
E_u	1	i	−1	$-i$	−1	$-i$	1	i	(x, y)	
	1	$-i$	−1	i	−1	i	1	$-i$		

C_{5h}	E	C_5	$C_5{}^2$	$C_5{}^3$	$C_5{}^4$	σ_h	S_5	$S_5{}^7$	$S_5{}^3$	$S_5{}^9$		$\epsilon=\exp(2\pi i/5)$
A'	1	1	1	1	1	1	1	1	1	1	R_z	x^2+y^2, z^2
E_1'	1	ϵ	ϵ^2	ϵ^{2*}	ϵ^*	1	ϵ	ϵ^2	ϵ^{2*}	ϵ^*	(x, y)	
	1	ϵ^*	ϵ^{2*}	ϵ^2	ϵ	1	ϵ^*	ϵ^{2*}	ϵ^2	ϵ		
E_2'	1	ϵ^2	ϵ^*	ϵ	ϵ^{2*}	1	ϵ^2	ϵ^*	ϵ	ϵ^{2*}		(x^2-y^2, xy)
	1	ϵ^{2*}	ϵ	ϵ^*	ϵ^2	1	ϵ^{2*}	ϵ	ϵ^*	ϵ^2		
A''	1	1	1	1	1	−1	−1	−1	−1	−1	z	
E_1''	1	ϵ	ϵ^2	ϵ^{2*}	ϵ^*	−1	$-\epsilon$	$-\epsilon^2$	$-\epsilon^{2*}$	$-\epsilon^*$	(R_x, R_y)	(xz, yz)
	1	ϵ^*	ϵ^{2*}	ϵ^2	ϵ	−1	$-\epsilon^*$	$-\epsilon^{2*}$	$-\epsilon^2$	$-\epsilon$		
E_2''	1	ϵ^2	ϵ^*	ϵ	ϵ^{2*}	−1	$-\epsilon^2$	$-\epsilon^*$	$-\epsilon$	$-\epsilon^{2*}$		
	1	ϵ^{2*}	ϵ	ϵ^*	ϵ^2	−1	$-\epsilon^{2*}$	$-\epsilon$	$-\epsilon^*$	$-\epsilon^2$		

C_{6h}	E	C_6	C_3	C_2	$C_3{}^2$	$C_6{}^5$	i	$S_3{}^5$	$S_6{}^5$	σ_h	S_6	S_3		$\epsilon=\exp(2\pi i/6)$
A_g	1	1	1	1	1	1	1	1	1	1	1	1	R_z	x^2+y^2, z^2
B_g	1	−1	1	−1	1	−1	1	−1	1	−1	1	−1		
E_{1g}	1	ϵ	$-\epsilon^*$	−1	$-\epsilon$	ϵ^*	1	ϵ	$-\epsilon^*$	−1	$-\epsilon$	ϵ^*	(R_x, R_y)	(xz, yz)
	1	ϵ^*	$-\epsilon$	−1	$-\epsilon^*$	ϵ	1	ϵ^*	$-\epsilon$	−1	$-\epsilon^*$	ϵ		
E_{2g}	1	$-\epsilon^*$	$-\epsilon$	1	$-\epsilon^*$	$-\epsilon$	1	$-\epsilon^*$	$-\epsilon$	1	$-\epsilon^*$	$-\epsilon$		(x^2-y^2, xy)
	1	$-\epsilon$	$-\epsilon^*$	1	$-\epsilon$	$-\epsilon^*$	1	$-\epsilon$	$-\epsilon^*$	1	$-\epsilon$	$-\epsilon^*$		
A_u	1	1	1	1	1	1	−1	−1	−1	−1	−1	−1	z	
B_u	1	−1	1	−1	1	−1	−1	1	−1	1	−1	1		
E_{1u}	1	ϵ	$-\epsilon^*$	−1	$-\epsilon$	ϵ^*	−1	$-\epsilon$	ϵ^*	1	ϵ	$-\epsilon^*$	(x, y)	
	1	ϵ^*	$-\epsilon$	−1	$-\epsilon^*$	ϵ	−1	$-\epsilon^*$	ϵ	1	ϵ^*	$-\epsilon$		
E_{2u}	1	$-\epsilon^*$	$-\epsilon$	1	$-\epsilon^*$	$-\epsilon$	−1	ϵ^*	ϵ	−1	ϵ^*	ϵ		
	1	$-\epsilon$	$-\epsilon^*$	1	$-\epsilon$	$-\epsilon^*$	−1	ϵ	ϵ^*	−1	ϵ	ϵ^*		

A11-3.G The D_n Groups

D_2	E	$C_2(z)$	$C_2(y)$	$C_2(x)$		
A	1	1	1	1		x^2, y^2, z^2
B_1	1	1	−1	−1	z, R_z	xy
B_2	1	−1	1	−1	y, R_y	xz
B_3	1	−1	−1	1	x, R_x	yz

D_3	E	$2C_3$	$3C_2$		
A_1	1	1	1		x^2+y^2, z^2
A_2	1	1	−1	z, R_z	
E	2	−1	0	$(x, y); (R_x, R_y)$	$(x^2-y^2, xy); (xz, yz)$

D_4	E	$2C_4$	$C_2(=C_4^2)$	$2C_2'$	$2C_2''$		
A_1	1	1	1	1	1		x^2+y^2, z^2
A_2	1	1	1	−1	−1	z, R_z	
B_1	1	−1	1	1	−1		x^2-y^2
B_2	1	−1	1	−1	1		xy
E	2	0	−2	0	0	$(x, y); (R_x, R_y)$	(xz, yz)

D_5	E	$2C_5$	$2C_5^2$	$5C_2$		
A_1	1	1	1	1		x^2+y^2, z^2
A_2	1	1	1	−1	z, R_z	
E_1	2	$2\cos 72°$	$2\cos 144°$	0	$(x, y); (R_x, R_y)$	(xz, yz)
E_2	2	$2\cos 144°$	$2\cos 72°$	0		(x^2-y^2, xy)

D_6	E	$2C_6$	$2C_3$	C_2	$3C_2'$	$3C_2''$		
A_1	1	1	1	1	1	1		x^2+y^2, z^2
A_2	1	1	1	1	−1	−1	z, R_z	
B_1	1	−1	1	−1	1	−1		
B_2	1	−1	1	−1	−1	1		
E_1	2	1	−1	−2	0	0	$(x, y); (R_x, R_y)$	(xz, yz)
E_2	2	−1	−1	2	0	0		(x^2-y^2, xy)

A11-3.H The D_{nd} Groups

D_{2d}	E	$2S_4$	C_2	$2C_2'$	$2\sigma_d$		
A_1	1	1	1	1	1		x^2+y^2, z^2
A_2	1	1	1	−1	−1	R_z	
B_1	1	−1	1	1	−1		x^2-y^2
B_2	1	−1	1	−1	1	z	xy
E	2	0	−2	0	0	(x, y); (R_x, R_y)	(xz, yz)

D_{3d}	E	$2C_3$	$3C_2$	i	$2S_6$	$3\sigma_d$		
A_{1g}	1	1	1	1	1	1		x^2+y^2, z^2
A_{2g}	1	1	−1	1	1	−1	R_z	
E_g	2	−1	0	2	−1	0	(R_x, R_y)	(x^2-y^2, xy), (xz, yz)
A_{1u}	1	1	1	−1	−1	−1		
A_{2u}	1	1	−1	−1	−1	1	z	
E_u	2	−1	0	−2	1	0	(x, y)	

D_{4d}	E	$2S_8$	$2C_4$	$2S_8{}^3$	C_2	$4C_2'$	$4\sigma_d$		
A_1	1	1	1	1	1	1	1		x^2+y^2, z^2
A_2	1	1	1	1	1	−1	−1	R_z	
B_1	1	−1	1	−1	1	1	−1		
B_2	1	−1	1	−1	1	−1	1	z	
E_1	2	$\sqrt{2}$	0	$-\sqrt{2}$	−2	0	0	(x, y)	
E_2	2	0	−2	0	2	0	0		(x^2-y^2, xy)
E_3	2	$-\sqrt{2}$	0	$\sqrt{2}$	−2	0	0	(R_x, R_y)	(xz, yz)

D_{5d}	E	$2C_5$	$2C_5{}^2$	$5C_2$	i	$2S_{10}^3$	$2S_{10}$	$5\sigma_d$		
A_{1g}	1	1	1	1	1	1	1	1		x^2+y^2, z^2
A_{2g}	1	1	1	−1	1	1	1	−1	R_z	
E_{1g}	2	2 cos 72°	2 cos 144°	0	2	2 cos 72°	2 cos 144°	0	(R_x, R_y)	(xz, yz)
E_{2g}	2	2 cos 144°	2 cos 72°	0	2	2 cos 144°	2 cos 72°	0		(x^2-y^2, xy)
A_{1u}	1	1	1	1	−1	−1	−1	−1		
A_{2u}	1	1	1	−1	−1	−1	−1	1	z	
E_{1u}	2	2 cos 72°	2 cos 144°	0	−2	−2 cos 72°	−2 cos 144°	0	(x, y)	
E_{2u}	2	2 cos 144°	2 cos 72°	0	−2	−2 cos 144°	−2 cos 72°	0		

D_{6d}	E	$2S_{12}$	$2C_6$	$2S_4$	$2C_3$	$2S_{12}^5$	C_2	$6C_2'$	$6\sigma_d$		
A_1	1	1	1	1	1	1	1	1	1		x^2+y^2, z^2
A_2	1	1	1	1	1	1	1	−1	−1	R_z	
B_1	1	−1	1	−1	1	−1	1	1	−1		
B_2	1	−1	1	−1	1	−1	1	−1	1	z	
E_1	2	$\sqrt{3}$	1	0	−1	$-\sqrt{3}$	−2	0	0	(x, y)	
E_2	2	1	−1	−2	−1	1	2	0	0		(x^2-y^2, xy)
E_3	2	0	−2	0	2	0	−2	0	0		
E_4	2	−1	−1	2	−1	−1	2	0	0		
E_5	2	$-\sqrt{3}$	1	0	−1	$\sqrt{3}$	−2	0	0	(R_x, R_y)	(xz, yz)

A11-3.I The D_{nh} Groups

D_{2h}	E	$C_2(z)$	$C_2(y)$	$C_2(x)$	i	$\sigma(xy)$	$\sigma(xz)$	$\sigma(yz)$		
A_g	1	1	1	1	1	1	1	1		x^2, y^2, z^2
B_{1g}	1	1	−1	−1	1	1	−1	−1	R_z	xy
B_{2g}	1	−1	1	−1	1	−1	1	−1	R_y	xz
B_{3g}	1	−1	−1	1	1	−1	−1	1	R_x	yz
A_u	1	1	1	1	−1	−1	−1	−1		
B_{1u}	1	1	−1	−1	−1	−1	1	1	z	
B_{2u}	1	−1	1	−1	−1	1	−1	1	y	
B_{3u}	1	−1	−1	1	−1	1	1	−1	x	

D_{3h}	E	$2C_3$	$3C_2$	σ_h	$2S_3$	$3\sigma_v$		
A_1'	1	1	1	1	1	1		x^2+y^2, z^2
A_2'	1	1	−1	1	1	−1	R_z	
E'	2	−1	0	2	−1	0	(x, y)	(x^2-y^2, xy)
A_1''	1	1	1	−1	−1	−1		
A_2''	1	1	−1	−1	−1	1	z	
E''	2	−1	0	−2	1	0	(R_x, R_y)	(xz, yz)

D_{4h}	E	$2C_4$	C_2	$2C_2'$	$2C_2''$	i	$2S_4$	σ_h	$2\sigma_v$	$2\sigma_d$		
A_{1g}	1	1	1	1	1	1	1	1	1	1		x^2+y^2, z^2
A_{2g}	1	1	1	−1	−1	1	1	1	−1	−1	R_z	
B_{1g}	1	−1	1	1	−1	1	−1	1	1	−1		x^2-y^2
B_{2g}	1	−1	1	−1	1	1	−1	1	−1	1		xy
E_g	2	0	−2	0	0	2	0	−2	0	0	(R_x, R_y)	(xz, yz)
A_{1u}	1	1	1	1	1	−1	−1	−1	−1	−1		
A_{2u}	1	1	1	−1	−1	−1	−1	−1	1	1	z	
B_{1u}	1	−1	1	1	−1	−1	1	−1	−1	1		
B_{2u}	1	−1	1	−1	1	−1	1	−1	1	−1		
E_u	2	0	−2	0	0	−2	0	2	0	0	(x, y)	

D_{5h}	E	$2C_5$	$2C_5^2$	$5C_2$	σ_h	$2S_5$	$2S_5^3$	$5\sigma_v$		
A_1'	1	1	1	1	1	1	1	1		x^2+y^2, z^2
A_2'	1	1	1	−1	1	1	1	−1	R_z	
E_1'	2	2 cos 72°	2 cos 144°	0	2	2 cos 72°	2 cos 144°	0	(x, y)	
E_2'	2	2 cos 144°	2 cos 72°	0	2	2 cos 144°	2 cos 72°	0		(x^2-y^2, xy)
A_1''	1	1	1	1	−1	−1	−1	−1		
A_2''	1	1	1	−1	−1	−1	−1	1	z	
E_1''	2	2 cos 72°	2 cos 144°	0	−2	−2 cos 72°	−2 cos 144°	0	(R_x, R_y)	(xz, yz)
E_2''	2	2 cos 144°	2 cos 72°	0	−2	−2 cos 144°	−2 cos 72°	0		

D_{6h}	E	$2C_6$	$2C_3$	C_2	$3C_2'$	$3C_2''$	i	$2S_3$	$2S_6$	σ_h	$3\sigma_d$	$3\sigma_v$		
A_{1g}	1	1	1	1	1	1	1	1	1	1	1	1		x^2+y^2, z^2
A_{2g}	1	1	1	1	−1	−1	1	1	1	1	−1	−1	R_z	
B_{1g}	1	−1	1	−1	1	−1	1	−1	1	−1	1	−1		
B_{2g}	1	−1	1	−1	−1	1	1	−1	1	−1	−1	1		
E_{1g}	2	1	−1	−2	0	0	2	1	−1	−2	0	0	(R_x, R_y)	(xz, yz)
E_{2g}	2	−1	−1	2	0	0	2	−1	−1	2	0	0		(x^2-y^2, xy)
A_{1u}	1	1	1	1	1	1	−1	−1	−1	−1	−1	−1		
A_{2u}	1	1	1	1	−1	−1	−1	−1	−1	−1	1	1	z	
B_{1u}	1	−1	1	−1	1	−1	−1	1	−1	1	−1	1		
B_{2u}	1	−1	1	−1	−1	1	−1	1	−1	1	1	−1		
E_{1u}	2	1	−1	−2	0	0	−2	−1	1	2	0	0	(x, y)	
E_{2u}	2	−1	−1	2	0	0	−2	1	1	−2	0	0		

D_{8h}	E	$2C_8$	$2C_8{}^3$	$2C_4$	C_2	$4C_2{}'$	$4C_2{}''$	i	$2S_8$	$2S_8^3$	$2S_4$	σ_h	$4\sigma_d$	$4\sigma_v$		
A_{1g}	1	1	1	1	1	1	1	1	1	1	1	1	1	1		x^2+y^2, z^2
A_{2g}	1	1	1	1	1	−1	−1	1	1	1	1	1	−1	−1	R_z	
B_{1g}	1	−1	−1	1	1	1	−1	1	−1	−1	1	1	1	−1		
B_{2g}	1	−1	−1	1	1	−1	1	1	−1	−1	1	1	−1	1		
E_{1g}	2	$\sqrt{2}$	$-\sqrt{2}$	0	−2	0	0	2	$\sqrt{2}$	$-\sqrt{2}$	0	−2	0	0	(R_x, R_y)	(xz, yz)
E_{2g}	2	0	0	−2	2	0	0	2	0	0	−2	2	0	0		(x^2-y^2, xy)
E_{3g}	2	$-\sqrt{2}$	$\sqrt{2}$	0	−2	0	0	2	$-\sqrt{2}$	$\sqrt{2}$	0	−2	0	0		
A_{1u}	1	1	1	1	1	1	1	−1	−1	−1	−1	−1	−1	−1		
A_{2u}	1	1	1	1	1	−1	−1	−1	−1	−1	−1	−1	1	1	z	
B_{1u}	1	−1	−1	1	1	1	−1	−1	1	1	−1	−1	−1	1		
B_{2u}	1	−1	−1	1	1	−1	1	−1	1	1	−1	−1	1	−1		
E_{1u}	2	$\sqrt{2}$	$-\sqrt{2}$	0	−2	0	0	−2	$-\sqrt{2}$	$\sqrt{2}$	0	2	0	0	(x, y)	
E_{2u}	2	0	0	−2	2	0	0	−2	0	0	2	−2	0	0		
E_{3u}	2	$-\sqrt{2}$	$\sqrt{2}$	0	−2	0	0	−2	$\sqrt{2}$	$-\sqrt{2}$	0	2	0	0		

Appendix 12

Hints for Solving Selected Problems

Chapter 1

1-1. Use Eq. (1-25).

1-7.

$$P.E.(t) = -\int_0^{\psi(x,t)} m\left[\partial^2\Psi(x,t)/\partial t^2\right]d\Psi(x,t) = \frac{1}{2}m\omega^2\Psi^2(x,t).$$

Next integrate $P.E.(t)$ over one complete cycle $(0 - t')$.

Chapter 2

2-9. $\sin x \sin y = \frac{1}{2}[\cos(x-y) - \cos(x+y)]$.

2-10. What kind of function has $\lambda_{II} = \infty$? When could such a function join smoothly onto a sine function in region I?

Chapter 3

3-1. Imagine an auto runs from A to B at 30 mph and from B to C at 60 mph. Sketch the distribution function for the auto. Then reason how you arrived at this function and apply similar reasoning to the harmonic oscillator.

3-6. a) Seek points where $H_2(y)$ equals zero.

b) Seek places where $d\psi^2/dy = 0$; then evaluate ψ^2 at these points and compare.

3-15.

$$\int_1^{\infty} \exp(-y^2)dy \sim 0.10\left[\exp(-1.05^2) + \exp(-1.15^2) + \exp(-1.25^2) + \cdots\right]$$

to convergence.

3-19. Consider how the solutions for the harmonic oscillator would meet the conditions imposed by this new potential.

Chapter 4

4-3. The classical turning point occurs when the total energy equals the potential energy.

4-11. $E = V(r)$ when r equals classical turning point.

4-12. Ignore all but the θ and ϕ dependences in Eq. (4-30). Do not forget to *square* these dependences, and do not forget to include the $3^{1/2}$ term of $3d_z^2$.

4-17. $x = r \sin\theta \cos\phi$.

4-18. a) Do not forget to include θ dependence of dv.

4-26. Refer to Problem 2-11.

4-37. μ should be in units of kg/molecule, and you have masses in a.m.u. These can be taken directly as g/mole and then converted.

Chapter 5

5-4. Square ψ and integrate, using the fact that 1s, 2s are orthonormal.

Chapter 6

6-9. b) $\Psi^*\Psi$ must be same at $t = 0$ and $t = 1/\nu$.

6-11. Use the Schmidt orthogonalization method.

6-13. a) Means: is ψ an eigenfunction of the momentum operator?

6-18. Wavefunction is not normalized.

6-24. Use that $\exp(ikx) = \cos(kx) + i\sin(kx)$.

Chapter 7

7-4. a) Note that, for symmetric ψ_π, $\int \psi\phi dx = 2\int_0^{L/2} \psi_n \phi\, dx$. For antisymmetric ψ_n, the integral can be evaluated by inspection. A useful integral is $\int x \sin x\, dx = \sin x - x\cos x$.

c) Use the fact that $\overline{E} = \Sigma_n c_n^2 E_n$. The series can be estimated with a small calculator (tedious) or else by summing the first few terms and integrating over a function that envelopes the higher terms.

7-11. Use the fact that $\phi = \Sigma_i c_i \psi_i$ and $\overline{E} = \Sigma_i c_i^* c_i E_i$.

7-13. b) Take limit as $F \to 0$ rather than simply evaluating at $F = 0.1$. Note that $(1 + x)^m = 1 + mx + [m(m-1)/2!]x^2 + \cdots$.

7-16. Do not forget that overlap between ϕ_a and ϕ_b must enter normality condition: $c_a^2 + c_b^2 + 2c_a c_b S = 1$

7-17. $\int \phi^2(-1/r_a)dv = -1 + (\zeta + 1)\exp(-2\zeta)$ using the method of Appendix 3.

7-19. Note that ψ_+ and ψ_- are degenerate at $R = \infty$.

7-20. In the limit of $R \to 0$, H_{AA} is indeterminate. Use l'Hospital's rule (i.e., take d/dR on the numerator and denominator and evaluate at $R = 0$). Alternatively, you can expand $\exp(-2R)$ in powers of R and evaluate at $R = 0$.

7-27. Note that (f) and (g) have both AOs on center a. When $\hat{H}$ is present in the integral, you are restricted to considering symmetry operations that do not affect $\hat{H}$.

7-34. Use the Schmidt orthogonalization procedure (Chapter 6) to construct 2s′.

Chapter 8

8-15. b) Simply note where HOMO is bonding, antibonding, nonbonding, and recognize that some of the effect of this MO will be lost upon ionization.

Chapter 12

12-2. Recall that perturbation should be greatest for states with ψ^2 largest in region of perturbation.

12-7. To evaluate the first-order correction to the energy, you can recognize that $\langle 1s| - 1/r|1s\rangle$ is identical to the potential energy of the H atom. The virial theorem tells you the value of this quantity at once.

12-16. Use the fact that $E_\pi = \alpha \Sigma_i^{\text{MOs}} n_i + 2\beta \Sigma_i^{\text{MOs}} \Sigma_{k<l}^{\text{neighbors}} p_{kl}$.

Chapter 13

13-9. Notice which MOs are degenerate when assigning symmetries. For the final part of the question, notice that the molecular x axis corresponds to the group theoretic z axis.

Chapter 14

14-13. Do not forget that the cyclobutadiene molecule differs from two acetylenes in both π and σ systems. There are a total of eight orbitals to be sketched for each side of this reaction.

Chapter 15

15-2. The curve in Fig. 15-2b is proportional to the reciprocal of the slope of the curve in Fig. 15-2a.

15-3. It is helpful to use that $\exp(in\pi) = 1$ for even n, -1 for odd n, and that $\exp[i(a+b)] = \exp(ia)\exp(ib)$.

15-4. Remember that E is degenerate due to states at negative j.

Appendix 13

Answers to Problems[1]

Chapter 1

1-1. [See Hint.] $(A+D+C)\cos(kx)+(B+iC-iD)\sin(kx)$.

1-2. $\psi(x) =$ same as Eq. (1-32) except cos instead of sin when n is odd.

1-3. $\alpha^2=\beta^2=(2\pi/\lambda)^2$.

1-4. Work functions: $\mathrm{Cs}=1.9\,\mathrm{eV}, \mathrm{Zn}=3.7\,\mathrm{eV}. h=4.13\times 10^{-13}$ eVs.

1-5. a) 0.055 nm. b) 3.31×10^{-25} nm.

1-6. Argument of cosine must change by 2π. Cycle time $=2\pi/\omega$.

1-7. [See Hint.] Integrating $\mathrm{PE}(t)=\frac{1}{2}m\omega^2\Psi^2(x,t)$ over a cycle and dividing by t' to give average PE per unit time gives $\mathrm{PE}=m\omega^2\psi^2(x)/4$.[$m$ is really ρdx.] An identical result comes from integrating $\mathrm{KE}(t)=\frac{1}{2}m\omega^2\psi^2(x)\sin^2(\omega t)$ over the same cycle.

1-8. a) No. Becomes infinite at $x=\pm\infty$. b) Same as a). c) Yes. d) No. Becomes infinite at $x=-\infty$. e) Yes.

1-9. $\psi=\sin x$ or $\cos x$ are examples.

1-10. We need establish only one of the extreme profiles for the string. Then, as $\cos(\omega t)$ oscillates between $+1$ and -1, the string oscillates between the two extremes. In other words, $\sin(x)$ and $-\sin(x)$ are the same solution at different times. (They differ by a phase factor.)

1-11. Only c), d) and f). [The latter is most easily seen after recognizing that the function equals $\exp(4ix)$.] For c), the eigenvalue is zero.

Multiple Choice (MC): d c e e b

Chapter 2

2-1. $\mathrm{J^2s^2kg^{-1}m^{-2}=(kg\,m^2s^{-2})^2\,sec^2\,kg^{-1}m^{-2}=kg\,m^2s^{-2}=J}$.

2-2. $A=\left[\frac{L}{n\pi}\int_0^L \sin^2(\frac{n\pi x}{L})d(\frac{n\pi x}{L})\right]^{-1/2}=\left[\frac{L}{n\pi}\int_0^{n\pi}\sin^2 y dy\right]^{-1/2}$
$=\left[\frac{L}{n\pi}\cdot\frac{n\pi}{2}\right]^{-1/2}=\sqrt{\frac{2}{L}}$

[1]Hints for some of these are given in Appendix 12.

2-3. Recognize that, since $\sin^2+\cos^2=1$, $\int_0^L(\sin^2+\cos^2)dx$ is the area of a rectangle of height 1 and width L, i.e., has area L. Half of this area goes with $\sin^2$. For cases where $n>1$, there are more wiggles, but the value of the integral over $\sin^2$ still equals $L/2$.

2-4. $0.6090, 0.1955, \frac{1}{3}, \frac{1}{3}$.

2-5. a) Use height times width of this narrow rectangle in ψ^2. Height $=\left[(\sqrt{2/L})\sin(0.5\pi L/L)\right]^2=(2/L)\sin^2(\pi/2)=2/L$. Width $=0.01L$. Area $=0.02$. This is 2% of the probability density in 1% of the box width. That is twice the classical probability, which is uniform in the box.
b) $(2/L)\sin^2(\pi/3)$ times $0.01L=0.015$. 1.5 times classical.

2-6. a) 1/5. (There are five equal hills in ψ_5^2, and $0\le x\le L/5$ contains one of them.
b) Smaller. ψ_1^2 (Fig. 2-5) is small in this region (less than $1/L$ in value) and integrates to less than $1/5$.

2-7. a) S. b) −A. c) SS. d) AA. e) −AS. f) AASASSA. g) −AASASAA. Rule: Product antisymmetric when *odd* number of antisymmetric functions is present.

2-8. All are zero by symmetry except d), e), h). i) is integral of $2\sin^2 x\cos x$ which is SA. j) is integral of $\sin^2 x\cos x$ which is SS from $-\pi$ to π, but SA in each half of the range (i.e., like i). If each half must give zero, so must the whole.

2-9. [See Hint.] $(2/L)\int_0^L\sin(n\pi x)\sin(m\pi x)dx=$ (using Hint) $(1/\pi)\int_0^\pi\{\cos[(n-m)y]-\cos[(n+m)y]\}dy=0$, for n and m integers and $n\neq m$.

2-10. [See Hint.] $\lambda_{II}=\infty$, and so ψ_{II} is a constant. A constant has a zero derivative, and ψ_I must arrive at $x=L$ with zero derivative if successful junction is to be made. This requires that an odd number of quarter-waves fit between 0 and L (so wave arrives with either a peak or a valley at $x=L$). This requires that $[(2n+1)/4]\lambda_I=L$; $\lambda_I=4L/(2n+1)=h/\sqrt{2mU}$; $U=(2n+1)^2h^2/32mL^2$ is the relation between U and L that is required for a state to exist at $E=U$. (Strictly speaking, one can only approach $\lambda=\infty$ as a limit, and this problem is physically meaningless. However, it makes a good exercise.)

2-11. Given that $H\psi_1=E_1\psi_1$, $H\psi_2=E_2\psi_2$, $E_1=E_2=E$ and $\phi=c_1\psi_1+c_2\psi_2$. Then $H\phi=c_1H\psi_1+c_2H\psi_2=c_1E_1\psi_1+c_2E_2\psi_2=E(c_1\psi_1+c_2\psi_2)=E\phi$.
Q.E.D.

2-12. a) ψ should oscillate on right with same λ as on left. b) ψ should be symmetric or antisymmetric (and λ should be same in each side). c) ψ should be smooth (i.e., have no cusp) at finite barrier. d) ψ should be a decaying exponential at right. e) Same as d). ψ should not become infinite.

2-13. a) ψ_4 is two sine waves.

b) E_4 is the same as E_2 in a half-box of width $L: E_4=2^2h^2/8mL^2=h^2/8mL^2$. Or, one can also calculate it as the $n=4$ solution with width $=2L$.

2-14. For exponential eigenfunctions, $\psi^*\psi = \left[\exp(\mathrm{i}j\phi)\right]^*[\exp(\mathrm{i}j\phi)\right] = \exp(-ij\phi)$ $\exp(ij\phi) = \exp(0) = 1$. This gives a "rectangle" of height 1 and width 2π, hence area 2π and normalizing constant $1/\sqrt{2\pi}$. For sine or cosine functions, we get a squared function that oscillates from 0 to 1, gives us half the area, namely π, and yields normalizing constant $1/\sqrt{\pi}$.

2-15. $\exp(i\sqrt{2}\phi)$ does not join onto itself when $\phi \to \phi + 2\pi$.

2-16. a) $H = -(h^2/8\pi^2 I)d^2/d\phi^2$. b) $H\psi = (9h^2/8\pi^2 I)\psi$. c) $[(h/2\pi i)d/d\phi]\psi \neq$ constant times ψ. *Not* a constant of motion.

2-17. The barrier forces solutions to vanish at $\phi = 0$. Of our four choices $\sin(k\phi)$, $\cos(k\phi)$, $\exp(\pm ik\phi)$ only the $\sin(k\phi)$ set has this property. Therefore a) $E = 0$ is lost. b) All degenerate levels become nondegenerate. c) No, only sine solutions exist. d) No, sine solutions are not eigenfunctions of the angular momentum operator.

2-18. a) $E_{nx,ny} = (h^2/8m)(n_x^2/L_x^2 + n_y^2/L_y^2)] = (n_x^2 + 4n_y^2)h^2/8mL_x^2$.
b) Zero point energy $= E_{1,1} = 5h^2/8mL_x^2$
c)

n_x	1	2	3	1	4	2	3	5	4
n_y	1	1	1	2	1	2	2	1	2
$E/(h^2/8mL_x^2)$	5	8	13	17	20	20	25	29	32
degeneracy	1	1	1	1	2		1	1	1

d)

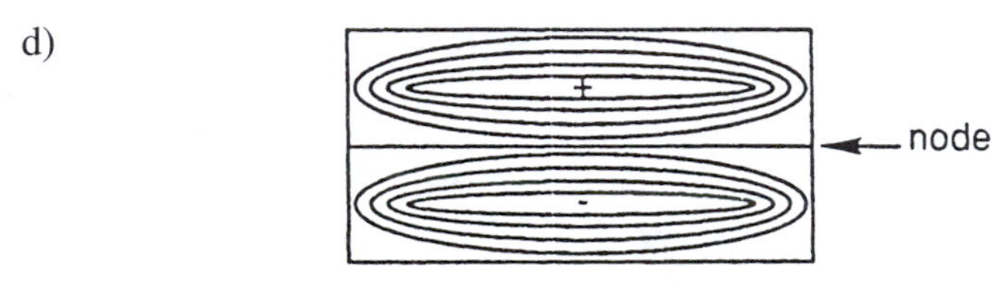

e) i) All eigenvalues increase by $10J$. (ii) No effect on eigenfunctions.

2-19. Both $3h^2/4mL_x^2$. Accidental degeneracy results when nodes that are not equivalent by symmetry nevertheless give the same energy.

2-20. a) Yes. The two eigenfunctions being mixed are degenerate ($5^2 + 1^2 + 1^2 = 3^2 + 3^2 + 3^2$), so the linear combination remains an eigenfunction. (See Problem 2-11.) b) Instead of integrating, we can take the value of $\psi^2\Delta V$ because ΔV is small enough to make ψ^2 essentially constant in it. $\psi^2 = (\sqrt{2/L})^6$ $\sin^2(\pi/2)\sin^2(\pi/2)\sin^2(\pi/2) = 8/L^3$; $\Delta V = 0.001L^3$; $\psi^2\Delta V = 0.008$. The classical value (uniform distribution) is 0.001, so the quantum-mechanical probability for finding the particle at the center is 8 times greater than classical. (Not coincidentally, this is the cube of the answer for the one-dimensional analog in Problem 2-5.)

2-21. $\Delta E = [(6+n)^2 - (5+n)^2]h^2/8mL^2$

$$\lambda = (8mcl^2/h)(2n+10)^2/(2n+11) = 637(2n+10)^2/(2n+11)\text{Å}$$

n:	0	1	2	3
λ:	5791	7056	8323	9592

2-22. Momentum is a constant of motion if $[(h/2\pi i)d/dx]\psi =$ constant times ψ. This applies for b) and d), and the momentum equals $3h/2\pi$ and $-3h/2\pi$, respectively. Kinetic energy is equal to $k^2h^2/8\pi^2m$, or $9h^2/8\pi^2m$ for all four cases.

2-23. The z component of angular momentum is a constant of motion if $[(h/2\pi i)d/d\phi]\psi =$ constant times ψ. This applies for b) and d), and the angular momentum equals $-3h/2\pi$ and $3h/2\pi$, respectively. Kinetic energy equals $9h^2/8\pi^2 I$ in all four cases.

2-24. ψ and $d\psi/dx$ are real at $x = a$. $(\psi = \psi^*)_{x=a}$; $A\exp(ika) + B\exp(-ika) = A^*\exp(-ika) + B^*\exp(ika)$; so $A\exp(ika) - A^*\exp(-ika) = B^*\exp(ika) - B\exp(-ika)$. $[(d\psi/dx) = (d\psi/dx)^*]_{x=a}$; $A\exp(ika) + A^*\exp(-ika) = B^*\exp(ika) + B\exp(-ika)$; adding these two equations gives $A\exp(ika) = B^*\exp(ika)$, so $A = B^*$. Subtracting the two equations gives $A^*\exp(-ika) = B\exp(-ika)$, so $A^* = B$. (Or we could take the complex conjugate of the previous equation.) Then, $|A|^2/|B|^2 = A^*A/B^*B = A^*A/A^*A = 1$, so $|A| = |B|$.

2-25. Matching values at $x = 0$ gives $A + B = C$. Matching slopes gives $kA - kB = k'C$. Using the first of these to eliminate C from the second gives the first part of (2-73). Using the first to eliminate B from the second gives the second part of (2-73).

2-26. Now A is zero, $B = C + D$, $-kB = k'C - k'D$, $C/D = (k-k')/(k+k')$, $B/D = 2k'/(k+k')$.

2-27. a) $|C|^2/|A|^2$ gives the relative spatial *densities* of transmitted to impinging particles, but we need relative *fluxes*, which depend on both particle densities and velocities. The relative velocities are the same as the relative momenta (since masses are the same), and these are in turn the same as the relative k values. So k'/k gives the relative velocities of impinging and transmitting particles. For reflecting particles, the analogous ratio is k/k, so it does not appear.

b) $[k'|C|^2/k|A|^2] + [|B|^2/|A|^2] = 4k'k/(k+k')^2 + (k-k')^2/(k+k')^2 = (k+k')^2/(k+k')^2 = 1$.

2-28. 100% transmission occurs when an integral number of de Broglie half-waves fit in $\Delta x = d$; $n\lambda/2 = d$; also, $\lambda = h/p = h/\sqrt{2T} = h/\sqrt{2m(E-U)}$. Equating these expressions for λ: $4d^2/n^2 = h^2/[2m(E-U)]$; $(E/U) - 1 = n^2h^2/(8d^2mU)$. Substituting for U on the right: $(E/U) - 1 = n^2\pi^2/16$. When $n = 1$, $E/U = 1.617$. When $n = 2$, $E/U = 3.467$. (Compare with Fig. 2-18b.)

2-29. ΔE_1 0.805 cm^{-1}, ΔE_2 36 cm^{-1}.

MC: c d e a e b d b d a c d d b a

Chapter 3

3-1. [See Hint.] $P(x)$ is proportional to $1/v(x)$, which is $[dx(t)/dt]^{-1}$, which is $\left[-\sqrt{k/m}L\sin(\sqrt{k/m}t)\right]^{-1}$. This is proportional to

$$\left[\sin^2(\sqrt{k/m}t)\right]^{-1/2} = \left[1-\cos^2(\sqrt{k/m}t)\right]^{-1/2} = \left[1-x^2/L^2\right]^{-1/2}$$

The *normalized* probability distribution function is $(\pi\sqrt{L^2-x^2})^{-1}$.

3-2. a) $x(t) = 0.100\cos(\sqrt{2}t)$, t in seconds, x in meters. b) -0.0453 m. c) 1.00×10^{-2} J. d) 2.00×10^{-3} J. (e, f) 5.00×10^{-3} J. g) $0.126\,\mathrm{ms}^{-1}$. h) ± 0.100 m. i) $0.225\,\mathrm{s}^{-1}$

3-3. $x = \pm\left[(n+1/2)h/\pi)/\sqrt{km}\right]^{1/2}$.

3-4. a) $\left[(\sqrt{\beta/\pi})1/(2^n n!)\right]^{1/2}$ is the normalizing factor. It keeps the total probability density equal to one. $H_n(y)$ is a Hermite polynomial. It provides the nodes in the wavefunction. $\mathrm{Exp}(-y^2/2)$ forces ψ to decay at large values of x. It gives the correct asymptotic behavior.

b)
$$\psi_0(y) = \sqrt[4]{\beta/\pi}\exp(-y^2/2)$$
$$\psi_1(y) = \sqrt[4]{\beta/\pi}(1/\sqrt{2})2y\exp(-y^2/2)$$
$$\psi_2(y) = \sqrt[4]{\beta/\pi}(1/\sqrt{8})(4y^2-2)\exp(-y^2/2)$$

3-5.
$$\begin{aligned} H\psi_0 &= \left\{-\left[h^2/(8\pi^2 m)\right]d^2/dx^2 + kx^2/2\right\}(\beta/\pi)^{1/4}\exp(-\beta x^2/2) \\ &= \left[h^2\beta/(8\pi^2 m) - \beta^2 h^2 x^2/(8\pi^2 m) + kx^2/2\right]\psi_0 \\ &= \left[(h/4\pi)\sqrt{k/m} - kx^2/2 + kx^2/2\right]\psi_0 \\ &= (h/2)(1/2\pi)\sqrt{k/m})\psi_0 = (hv/2)\psi_0. \end{aligned}$$

3-6. [See Hint.] a) $y = \pm\sqrt{1/2}$. b) $y = \pm\sqrt{5/2}$.

3-7. The first function is asymmetric; the second becomes infinite in both limits of y.

3-8. a) Approaches zero. Decaying exponential overwhelms polynomial. b) Antisymmetric. Polynomial is antisymmetric and exponential is symmetric. c) Value $= 0$, slope $= 120$.

3-9. a) $c_0 = 7$. b) $c_1 = 0$.

3-10. a) *Not*, because missing y^3 term should cause the polynomial to terminate, not permit y^5 term.
b) *Not*, because of mixed symmetry. (Powers 5, 3, 1, 0.)
c) OK.

3-11.

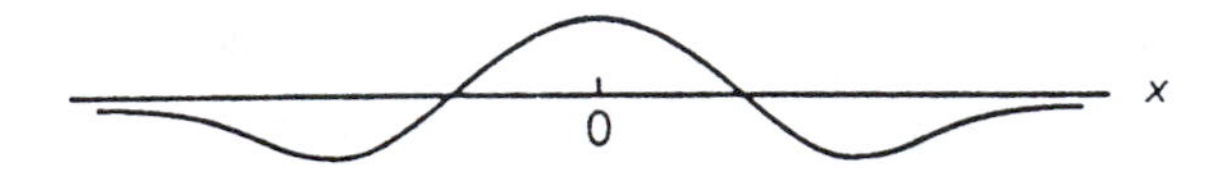

3-12. a) Zero (integrand antisymmetric). b) $\sqrt{\pi}/2$ (symmetric).

3-13. Zero (integrand antisymmetric).

3-14. a) $x_{tp} = \pm\left[3h/(2\pi\sqrt{mk})\right]^{1/2} = \pm\sqrt{3/\beta}$. b) ψ^2 is maximum at $y = \pm 1$, or $x = \pm 1/\sqrt{\beta} = \pm h/(2\pi\sqrt{mk})$. c) 1/2. d) 0.0144 [from $(2/e)\sqrt{\beta/\pi}$ times $0.02\sqrt{3h}/(2\pi\sqrt{km})^{1/2}$].

3-15. [See Hint.] $2\int_a^\infty \sqrt{\beta/\pi}\exp(-\beta x^2)dx = (2/\sqrt{\pi})\int_1^\infty \exp(-y^2)dy \sim 0.156$, where $a = 1/\sqrt{\beta}$.

3-16, 17. $H_2(y) = 4y^2 - 2$.

3-18. Zero point energy $= \frac{3}{2}h\nu$; 10; 3.

3-19. [See Hint.] The barrier requires $\psi = 0$ at $x \le 0$ but does not affect $\hat{H}$ at $x > 0$. Therefore, antisymmetric solutions of harmonic oscillator are still good. Results are $\psi_{x>0} = \sqrt{2}\psi_{\text{harm,osc}}, n = 1, 3, 5, \ldots$, and $E = (n + \frac{1}{2})h\nu, n = 1, 3, 5, \ldots$ $\psi_{x\le 0} = 0$.

3-20. a) 40. b) -1.

3-21. $\overline{V}_5 = E_5/2 = (11/4)h\nu = \overline{T}_5$.

3-22. For each unit of energy going into vibration, half goes into kinetic energy, which registers as a rise in temperature, and half is "hidden" as potential energy. None is hidden as potential energy in rotation or translation.

3-23. Using nominal nuclear masses gives k values of 958, 512, 408, 311 N m^{-1}, respectively. Decreasing bond stiffness implies decreasing bond strengths. (Observed D_0 values are, in kJ mol^{-1}, 564, 428, 363, 295, respectively.)

MC: e e d, c c

Chapter 4

4-1. $E_2 - E_1 = 2.46737 \times 10^{15}$ Hz using m_e. Using μ gives 0.999455 times this value, a difference of 545 ppm.

4-2. The radial distribution function vanishes at $r = 0$ because $4\pi r^2$ is zero there and again at $r = \infty$ because the decaying exponential of ψ^2 overwhelms the finite power term $4\pi r^2$.

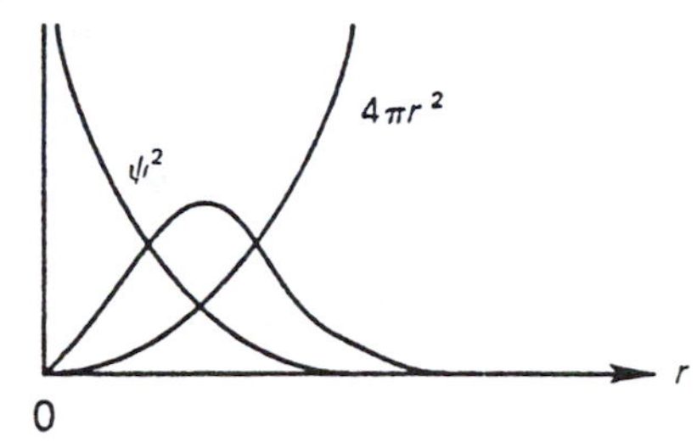

4-3. [See Hint.] $E_{1s} = -\frac{1}{2}$ a.u. $V(r) = -1/r$ a.u. Equal when $r = 2$ a.u. $\int_2^\infty \psi^2 dv = 0.238$, and so 23.8%.

4-4. a) $1/Z$ a.u. b) $3/2Z$ a.u. c) 0 (by inspection of ψ^2). d) $-Z^2$ a.u. Differs from $-Z$ times the reciprocal of b) because average of $1/r \neq 1/$(average of r).

d) is lower because $1/r$ blows up at small r, contributing large (negative) amounts of potential energy.

4-5. $(1/\sqrt{2})\int_0^\infty \exp(-r)(2-r)\exp(-r/2)r^2dr = \ldots 0$. (Do not forget r^2 from dv.)

4-6. $\pi^{-1/2}$ is the normalizing factor.

4-7. $\bar{x} = \int_{-\infty}^{+\infty} \psi^2 x\, dx$. ψ^2 is symmetric, x is antisymmetric, and $\bar{x} = 0$.

4-8. a) $\int_0^L \psi_n (x - L/2)^2 \psi_n\, dx$. b) Expect it to increase as n increases, approaching a limiting value. This because ψ^2 favors the box center for lower states and approaches classical (uniform) distribution as n increases. c) $L^2(1/12 - 1/2n^2\pi^2)$, $n = 1 : 0.03267L^2$, $n = 2 : 0.070668L^2$. Approaching $L^2/12$, which is the classical value achieved if ψ^2 in a) is replaced by $1/L$.

4-9. For reflection in the xy plane, $2p_z$ is antisymmetric, and $3d_{xy}$ is symmetric.

4-10. a) $4/Z$ a.u. b) $5/Z$ a.u. c) $2/Z$ a.u. d) $-Z^2/4$ a.u.

4-11. [See Hint.] $r_{tp} = 2n^2/Z$.

4-12. [See Hint.] The sum of squares of angular dependencies equals $\frac{4}{3}$. Use trigonometric identities to remove angle dependence.

4-13. a) $-1/2$ a.u. b) -2 a.u., $-1/2$ a.u. c) 25 d) 2, 1 e) -2 a.u.

4-14. a) $l = m = 0$. (No angular dependence: must be s.)
b) 2 (quadratic in r).
c) $-1/18$ a.u. (2 nodes means it is 3s, so $n = 3$.)
d) 18 a.u.

4-15. a) Yes, at $r = 6$ a.u. b) $3p_y$($r\sin\theta\sin\phi = y$, so it is p_y, and there is one radial node).

4-16. a) Looks OK. b) No. Blows up at large r. c) No. Lacks exponential decay function in r.

4-17. [See Hint.] $\int 1s^2 r\sin\theta\cos\phi\, dv = 0$ (because $\cos\phi$ is antisymmetric in each subrange 0–π, π–2π). The average value of x *should* be zero because the electron is equally likely to be found at equal $\pm x$ positions due to the spherical symmetry of ψ^2.

4-18. [See Hint.] a) $\theta_{mp} = 35°15'$, $144°45'$. b) Angular nodes come where $(3\cos^2\theta - 1) = 0$. $\theta = 54.74°, 125.26°$.

4-19. $xy = (r\sin\theta\cos\phi)(r\sin\theta\sin\phi) = r^2\sin^2\theta\cos\phi\sin\phi = (r^2\sin^2\theta\sin 2\phi)/2$.

4-20. Equation (4-45) predicts 2/3 and 0 for these integrals. Actual integration over x^2 and over $5x^4/2 - 3x^2/2$ gives 2/3 and 0. Integral over all space for $3p_0 3d_{-1}$ involves integral from 0 to π of $P_1^0(\cos\theta)P_2^{-2}(\cos\theta)\sin\theta\, d\theta$, which is same as integral from 1 to -1 of $P_1^0(x)P_2^{-1}(x)\,dx$, so the integral vanishes.

4-21. $\left[15/(2\sqrt{10\pi}\right](1 - \cos^2\theta)\cos\theta\exp(-2i\phi)$, f_{-2}.

4-22. $\hat{L}_z Y_{l,m}(\theta,\phi) = (\hbar/i)(d/d\phi)\Theta_{l,m}(\theta)\exp(im\phi) = m\hbar\Theta_{l.m}(\theta)\exp(im\phi) = m\hbar\, Y_{l,m}(\theta,\phi)$.

4-23.
$$\hat{L}_x 2p_0 = \hat{L}_x R(r)\cos\theta = -iR(r)\sin\theta\sin\phi \neq \text{constant} \times 2p_0$$
$$\hat{L}_y 2p_0 = iR(r)\sin\theta\cos\phi \neq \text{constant} \times 2p_0$$
$$\hat{L}_x 1s = 0 = \hat{L}_y 1s = \hat{L}_z 1s = \hat{L}^2 1s \cdots = 0 \times 1s$$

Since the vector has zero length, its x, y, z components must also have zero length. The question of vector orientation becomes meaningless.

4-24. $[-1/(\sin\theta)(d/d\theta)\sin\theta d/d\theta]R(r)\cos\theta = 2R(r)\cos\theta$, $l(l+1) = 2$, $l = 1$.

4-25.

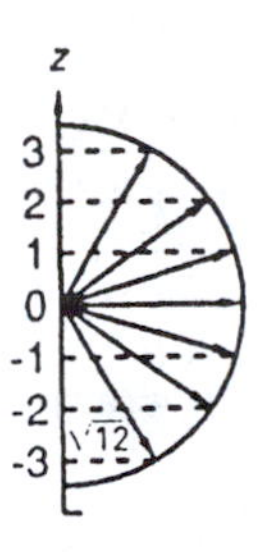

Same for 6f

4-26. [See Hint.] Will be eigenfunctions in cases b) and e). Only these correspond to mixing degenerate eigenfunctions.

4-27. a) $\sqrt{6}\hbar$ or $\sqrt{6}$ a.u. b) d_{-2}.

4-28. a) 0. b) $-3\hbar$ or -3 a.u.

4-29. a) 0. b) $2\hbar^2$ or 2 a.u. c) 0 (same as if all x's were z's). d) 0 (by symmetry). e) 0.

4-30. a) $6\psi_{3,2,1}$. b) $2\psi_{2p_x}$. c) $(-1/18)\psi_{3p_x}$. d) $-\psi_{2p_{-1}}$.

4-31. a) Yes, $-1/18$. e) Yes, 0 (same as $\hat{L}_z\psi_{3p_z}$). f) Yes, 2. All others, No.

4-32. 1.44×10^{-3} a.u.

4-33. 2.80×10^{10} Hz, 11.4 ppm.

4-34. a) $-1/32$. b) 16. c) 0, $\sqrt{2}$, $\sqrt{6}$, $\sqrt{12}$. d) 7. e) 4.

4-35. $\mu r^2 = ? = m_1 r_1^2 + m_2 r_2^2$. Chug, chug, Bingo.

4-36. Using Eq. (4-56), we write Eq. (4-68) as $(\hat{L}^2/\hbar^2)\psi = (-2IE/\hbar^2)\psi$; $(\hat{L}^2/2I) = E\psi$; since there is no potential term, E is all kinetic, so this completes the demonstration.

4-37. [See Hint.] First find μ in a.m.u., then divide by 6.0221×10^{26} to convert to kg/molecule. $\mu = 1.6529 \times 10^{-27}$ kg. $r = \sqrt{I/\mu} = 1.4144 \times 10^{-10}$ m.

4-38. $10.976\,\text{cm}^{-1}$, $21.953\,\text{cm}^{-1}$, $32.929\,\text{cm}^{-1}$.

4-39. 1.129×10^{-10} m($\mu = 1.1385 \times 10^{-26}$ kg, $I = 1.4504 \times 10^{-46}$ kg m^2).

4-40. $J = 3$, so $m_J = 3, 2, 1, 0, -1, -2, -3$. The states split into seven equally spaced levels.

4-41. $\sqrt{3}/2$, or 0.866 a.u.

MC: **e b b e d e b c c**

Chapter 5

5-1. $-\frac{1}{2}(\nabla_1^2+\nabla_2^2+\nabla_3^2)-3/r_1-3/r_2-3/r_3+1/r_{12}+1/r_{23}+1/r_{13}$.

5-2. $\bar{r}_{1s}=3/2Z=$ (for He^+) $\frac{3}{4}$ a.u., $\bar{r}_{2s}=6/Z=$ (for He^+) 3 a.u.

5-3. $E=2.343\times10^5$ eV compared with an IE of 13.6 eV. It shows that λ small enough to locate the electron with useful precision involves photons with energy sufficient to excite the electron completely out of the system.

5-4. [See Hint.] $\int\psi^2dv=\frac{1}{2}\int\int[1s(1)^2 2s(2)^2+2(1s(1)2s(1)2s(2)1s(2))+2s(1)^2 1s(2)^2]dv(1)dv(2)=\frac{1}{2}(1\cdot1+2\cdot0\cdot0+1\cdot1)=1$.

5-5. $\psi_a(2\leftrightarrow1)=(1/\sqrt{2})[1s(2)2s(1)-2s(2)1s(1)]=-\psi_a$.

5-6. Upon substitution and expansion, complete cancellation occurs.

5-7. $(1/\sqrt{6})[1s2p1s(\alpha\beta\beta-\beta\beta\alpha)+1s1s2p(\beta\alpha\beta-\alpha\beta\beta)+2p1s1s(\beta\beta\alpha-\beta\alpha\beta)]$.

5-8. $\int 1s^*2s dv\int 1s^*1s dv\int 2s^*1s dv\int\alpha^*\alpha d\omega\int\beta^*\alpha d\omega\int\alpha^*\beta d\omega=0\cdot1\cdot0\cdot1\cdot0\cdot0=0$.

5-9. For $r_1=1$, $r_2=2$, $r_3=0$ get

$$\psi(1,2,0)=(1/\sqrt{6})[1\bar{s}(r=1)2p_z(r=2)1s(r=0)+2p_z(r=1)1s(r=2)1\bar{s}(r=0)$$
$$-2p_z(r=1)1\bar{s}(r=2)1s(r=0)-1s(r=1)2p_z(r=2)1\bar{s}(r=0)]$$

The other cases are the same except for factor of -1. Thus, ψ^2 is identical for all three cases, and no physical distinction exists.

5-10. Replace each α in the lowest row with β.

5-11. a) $\alpha\alpha,\beta\beta,\gamma\gamma,\alpha\beta+\beta\alpha,\alpha\gamma+\gamma\alpha,\beta\gamma+\gamma\beta$
b) $\alpha\beta-\beta\alpha,\alpha\gamma-\gamma\alpha,\beta\gamma-\gamma\beta$

5-12. a) Yes. Antisymmetric for exchange of any two electrons.
b) $-1/2(\nabla_1^2+\nabla_2^2+\nabla_3^2)-3/r_1-3/r_2-3/r_3+1/r_{1,2}+1/r_{2,3}+1/r_{1,3}$.
c) No. It is a product of one-electron orbitals, hence an independent-electron solution, but H is not separable.
d) $(-9/2)(1+1/4+1/9)=-6.125$ a.u. e) $-3/2$ a.u.

5-13. Equation (5-41) with $U_1=1s$, $U_2=1\bar{s}$, $U_3=2s$, $U_4=2\bar{s}$.

5-14. F, $1s^2 2s^2 2p^5$, $Z=9$, $\xi_{1s}=8.7$, $\xi_{2s}=\xi_{2p}=2.6$.

5-15. Let $\hat{A}\phi=a\phi$ and $\int\phi^*\phi d\tau=1$. $(\hat{A})_{av}=\int\phi^*\hat{A}\phi d\tau=\int\phi^*a\phi\, d\tau=a\int\phi^*\phi\, d\tau=a$.

5-16. Li^{2+} is a hydrogenlike ion, and hence should have all states of same n degenerate. Li differs in that potential seen by electron is not of form $-Z/r$, due to screening of nucleus by other electrons. Hence, degeneracy is lost. The 2s AO of Li is lower than the 2p due to the fact that the 2s electron spends a larger fraction of time near nucleus where it experiences full nuclear charge.

5-17. $\sqrt{\frac{1}{2}\left(\frac{1}{2}+1\right)}=\sqrt{\frac{3}{4}}$ a.u.

5-18. n electrons of α spin give one state. Each time one α is changed to a β we get a different state. There are $n\alpha$ spins available to change.

5-19. Symmetric combination gives $e^2 + 8e$ either way. Antisymmetric gives $e^2 - 8e$ one way and $-(e^2 - 8e)$ the other.

5-20. a) $^2s_{1/2}$(2 states). b) $^2p_{3/2}, ^2p_{3/2}, ^2s_{1/2}$(8 states).

5-21. a) Not satisfactory since electron 1 is identified with the 1s AO, etc. $[1s(1)3d_2(2) - 3d_2(1)1s(2)]\alpha(1)\alpha(2)$.
b) $^3D_3$.

5-22. 120.

5-23. a) $^2P_{3/2}, ^2p_{1/2}$. b) 2. c) 4, 2.

5-24. a) $\alpha(1)\beta(2) - \beta(1)\alpha(2)$. b) -22.5 a.u. c) -22.5 a.u. $+ J_{1s,2p_1} + K_{1s,2p_1}$.
d) Eigenvalue for S^2 is 0.

5-25. All in a.u.: a) 6. b) 2. c) 12. d) $2, 1, 0, -1, -2$. e) $1, 0, -1$. f) $3, 2, 1, 0, -1, -2, -3$.

5-26. a) In both cases, maximum net z components are: spin $= 1$, orbital$= 1$.
b) $^3P_0$ below $^3P_1$ below $^3P_2$ for p^2 case, reverse for p^4.
c) Each pairing reduces multiplicity to $m + 1$, where m is number of unpaired electrons, which equals the number of holes. Each pair's contribution to M_L due to their orbital's m_l value is either totally canceled by a pair at $-m_l$ or else half-canceled by an unpaired electron at $-m_l$. Therefore, the uncanceled portions reflect the presence of unpaired electrons, hence holes. Therefore, the amount of uncanceled m_l is equal to minus that which we could assign to the holes.

5-27. a) $^2S_{1/2}$. b) $^4S_{3/2}$. c) $^1S_0$. d) $^3F_2$.

5-28. a) 12. b) $^3P_2, ^3P_1, ^3P_0, ^1P_1$.

5-29. a) $^2P_{1/2}$ below $^2P_{3/2}$. b) $^4S_{3/2}$ below $^2D_{3/2}, ^2D_{5/2}$ below $^2P_{1/2}, ^2P_{3/2}$. (Rules do not allow us to sort by J in this case since shell is exactly half filled.)

5-30. Striving for maximum multiplicity means avoiding pairing electrons.

5-31. a) 15. b) 35.

5-32. No. Impossible for doublet and singlet combinations to arise from same number of electrons. Each pairing reduces number of unpaired electrons by two, so allowed multiplicities are all even or all odd.

5-33. a) 20. b) 12. c) 6. d) 60. e) 100.

5-34. a) 28. b) $^4F_{9/2}, ^4F_{7/2}, ^4F_{5/2}, ^4F_{3/2}$.

5-35. $^3D_3$. Also $^3D_2, ^3D_1$.

5-36. $2g^2 - g$, 15, 45.

5-37. Splittings equal 9.274×10^{-24} J times g, with g equal to 4/3, 7/6, 1/2, 1, respectively.

MC: d e e d

Chapter 6

6-1. $\int_{-\infty}^{\infty}\psi^*(d^2/dx^2)\phi dv \stackrel{?}{=} \int_{-\infty}^{\infty}\phi(d^2/dx^2)\psi^* dv$. Use $\int_{-\infty}^{\infty} v du = uv|_{-\infty}^{\infty} - \int_{-\infty}^{\infty} u dv$. On the left, let $v=\psi^*$, $u=d\phi/dx$, $du=d^2\phi/dx^2$, $dv=d\psi^*/dx$. On the right, let $v=\phi$, $u=d\psi^*/dx$, etc. The uv term vanishes since ψ^* and ϕ each vanish at limits. The remaining integrals are identical.

6-2. Each equals $-4\sqrt{8}/27$.

6-3. $\int\psi^*\psi dv=1$, $\int\chi_i^*\chi_j dv=\delta_{i,j}$, $\psi=\Sigma_i c_i\chi_i$. Then

$$\int\psi^*\psi dv = 1 = \int \Sigma_i c_i^*\chi_i^*\Sigma_j c_j\chi_j dv = \Sigma_i\Sigma_j c_i^* c_j\delta_{ij} = \Sigma_i c_i^* c_i.$$

Q.E.D.

6-4. $\psi=\Sigma_i c_i\mu_i$, $\int\mu_i^*\mu_j dv=\delta_{i.j}$, want c_k:

$$\int\mu_k^*\psi dv = \int\mu_k^*\Sigma_i c_i\mu_i dv = \Sigma_i c_i\int\mu_k^*\mu_i dv = \Sigma_i c_i\delta_{k,i} = c_k$$

6-5. a) $$(1/\pi)\int_0^{2\pi}\cos 2\phi(\hbar/i)(d/d\phi)\cos 2\phi d\phi = -(2/\hbar\pi i)\int_0^{2\pi}\cos 2\phi\sin 2\phi d\phi$$

$$\propto\int_0^{2\pi}\text{sym}\cdot\text{antisym} = 0$$

b) $$\psi = (1/\sqrt{2})\left[(1/\sqrt{2\pi})\exp(2i\phi)\right] + (1/\sqrt{2})\left[(1/\sqrt{2\pi})\exp(-2i\phi)\right]$$

Terms in [] are normalized eigenfunctions of L_z with eigenvalues of $+2\hbar$ and $-2\hbar$. So $(L_z)_{av} = (1/\sqrt{2})^2(2\hbar) + (1/\sqrt{2})^2(-2\hbar) = 0$.

6-6. $[x, p_x] = [x(\hbar/i)(d/dx) - (\hbar/i)(d/dx)x]f(x) = (\hbar/i)\left(xf' - f - xf'\right) = -(\hbar/i)f$

$$\Delta x\cdot\Delta p_x \geq \frac{1}{2}\left|\int\psi^*(-\hbar/i)\psi d\tau\right| = |-\hbar/2i| = \hbar/2$$

6-7. ϕ must be identical to the eigenfunction ψ_0.

6-8. No. The existence of *some* real eigenvalues does not guarantee that the operator satisfies the definition of hermiticity:

$$(d/dr)\exp(-ar) = -a\exp(-ar)$$

but

$$\int_0^{\infty}\exp(-ar)(d/dr)\exp(-br)r^2 dr \neq \int_0^{\infty}\exp(-br)(d/dr)\exp(-ar)r^2 dr$$

if $a \neq b$

6-9. a) Each side equals $-(1/2\sqrt{2})\psi_{1s}\exp(it/2) - (1/8\sqrt{2})\psi_{2p_0}\exp(it/8)$.

b) [See Hint.] $(\Psi^*\Psi)_{t=0} = (1/2)(1s^2 + 21s2p_0 + 2p_0^2)$; $(\Psi^*\Psi)_{t=1/\nu} = (1/2)(1s^2 + 2\cos(3/8\nu)1s2p_0 + 2p_0^2)$. Equal when $3/8\nu = 2\pi$; $\nu = 3/16\pi$. $\Delta E = (1/2)(1 - 1/4) = 3/8$ a.u.; $\Delta E = h\nu = 2\pi h\nu/2\pi = 2\pi\nu\hbar = 2\pi\nu$ in a.u. $\nu = \Delta E/2\pi = (3/8)/2\pi = 3/16\pi$.

6-10. $(1-S^2)^{-1/2}$.

6-11. [See Hint.] $S=\sqrt{3}/2$, and so $\phi=(2/\sqrt{3\pi})(r-\frac{3}{2})\exp(-r)$.

6-12. a) $8\sqrt{8}/27=0.838$. b) 0 (by symmetry).

6-13. [See Hint.] a) $\hat{p}_x\psi_0=-(\beta\hbar x/i)\psi_0\neq$ constant times ψ_0. Momentum is hence *not* a constant of motion.
b) $\int\psi_0^*\hat{p}_x\psi_0 dx=-(\beta\hbar/i)\int\psi_0^* x\psi_0 dx=\int(\text{sym})(\text{anti})(\text{sym})=0$ (Must be zero since otherwise motion in one direction would involve greater momentum than motion in the other.)

6-14. First pair: No. $x^2(d^2/dx^2)+x^3(d^3/dx^3)\neq 2x(d/dx)+4x^2(d^2/dx^2)+x^3(d^3/dx^3)$. Second pair: Yes. Both arrangements give $2x^2(d^2/dx^2)+x^3(d^3/dx^3)$.

6-15. a) 6 a.u. (since $l=2$). b) 0 (since real form of ψ involves equal mix of $\pm m_l$).

6-16. a) Because these operators commute, yes, there must be a set of simultaneous eigenfunctions.
b) Mixing degenerate-energy cases gives functions that remain eigenfunctions for the energy operator but not for the momentum operator. (These are the sine and cosine versions.)
c) Exactly knowable because knowledge of momentum gives knowledge of an eigenfunction which in turn gives knowledge of energy.

6-17. $E=(1/3)(-1/2-1/8-1/18)=-0.2268$ a.u.

6-18. [See Hint]. Normalized $\psi=0.26726[1s+2(2p_1)+3(3d_2)]$. $\langle\hat{L}_z\rangle=0.26726^2[0+4(1)+9(2)]=1.571$ a.u.

6-19. Proofs are in Sections a) 6-8, b) 6-9, c) 6-11.

6-20. $|\Psi|^2=(1/2)\left\{|\psi_1|^2+|\psi_2|^2+2|\psi_1\psi_2|\cos[(E_2-E_1)t/\hbar]\right\}$.

6-21. Cycle time $=mL^2/h=mL^2/2\pi$ a.u.

6-22. $c_n=(2\sqrt{2}/L)\int_0^{L/2}\sin(2\pi x/L)\sin(n\pi x/L)dx$.

a) $\sqrt{2}\psi_2$ and Ψ are same function in range $0\leq x\leq L/2$, guaranteeing $c_2=\sqrt{2}$. No other c can be larger or the resulting function will give a total probability density greater than 1.
b) ψ_4 and Ψ have opposite symmetry for $0\leq x\leq L/2$.
c) As ψ_i becomes more oscillatory, the positive and negative portions of its product with Ψ will cancel more effectively.

6-23. a) $c_1=0.838$, $c_2=0.2048$.
b) Whereas only s-type AOs appear in Eq. (6-41), explicit account of changing potential would yield a nonspherically symmetric potential so that p-type AOs would enter too.

6-24. [See Hint.] c_k is small if there is effective cancellation between positive and negative portions of the product of $\exp(-ax^2)$ and $\cos(kx)$. Larger k makes $\cos(kx)$ more oscillatory and makes cancellation more effective. The broader $\exp(-ax^2)$ is, the more effective is this cancellation for a given value of k.

6-25. a) Real V means real $\hat{H}$. $\hat{H}\Psi(x,y,z,t) = -(\hbar/i)(\partial/\partial t)\Psi(x,y,z,t)$. Complex conjugate of equation gives $\hat{H}\Psi^*(x,y,z,t) = (\hbar/i)(\partial/\partial t)\Psi^*(x,y,z,t)$. Transform $t \to -t$ throughout (does not affect equality), then recognize $\partial/\partial(-t) = -\partial/\partial t$, so $\hat{H}\Psi^*(x,y,z,-t) = -(\hbar/i)(\partial/\partial t)\Psi^*(x,y,z,-t)$. Q.E.D.

b) $\Psi(x,y,z,t)$ becomes $\psi(x,y,z)\exp(-iEt/\hbar)$. $\Psi^*(x,y,z,-t)$ becomes $\psi^*(x,y,z)\exp[iE(-t)/\hbar] = \psi^*(x,y,z)\exp(-iEt/\hbar)$. Carrying these through Eqs. (6-3,4,5) shows that $\hat{H}\psi^* = E\psi^*$.

c) If ψ and ψ^* are independent, we have two solutions with the same energy, i.e., degeneracy. If E is *not* degenerate, ψ and ψ^* are not different, so $\psi = \psi^*$ and is real.

d) $2p_{-1}$ becomes $-2p_{+1}$. No change for $2p_0$.

e) No. While all complex eigenfunctions must be degenerate, real eigenfunctions can be degenerate too. Cases we have seen are (1) accidental degeneracies such as $\psi_{3,3,3}$ and $\psi_{5,1,1}$ in the 3-dimensional cubic box and (2) real eigenfunctions constructed as linear combinations of complex ones, as $2p_x$ and $2p_y$ from $2p_{+1}$ and $2p_{-1}$.

MC: c a d

Chapter 7

7-1. a) Yes. $c_1 = \sqrt{2/L}\int_0^L f(x)\sin(\pi x/L)dx$, $c_2 = \sqrt{2/L}\int_0^L f(x)\sin(2\pi x/L)\,dx$. $c_1 = 0$ by symmetry. c_2 is positive.

b) No. Both a) and b) are continuous, smooth, single-valued functions, but b) does not go to zero at $x = 0, L$ as do all the box eigenfunctions.

7-2. a)

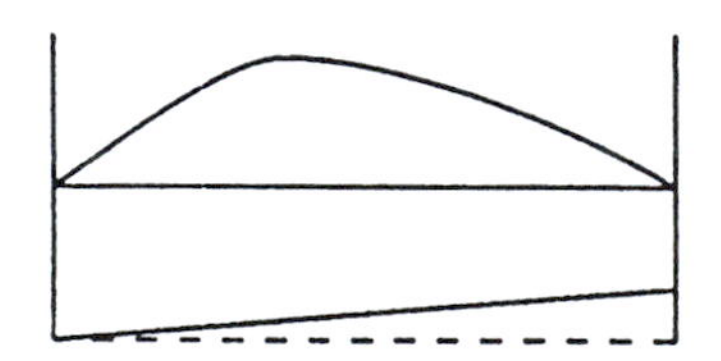

b) $1.3h^2/8mL^2$.

7-3. $c_1 = \sqrt{0.4} = 0.632$, $c_3 = 0$ by symmetry.

7-4. a) [See Hint.] For $n = \text{odd}$,

$$c_n = 2\int_0^{L/2} \phi\psi\,dx = \pm\frac{4\sqrt{6}}{n^2\pi^2}\begin{cases} +\text{for } n = 1, 5, 9, 13, \ldots \\ -\text{for } n = 3, 7, 11, 15, \ldots \end{cases}$$

For $n = \text{even}$, $c_n = 0$ (by symmetry) (i.e., ϕ is symmetric and so contains only symmetric ψ_n).

b)
$$\phi_{\text{approx}}(x = L/2) = \sum_{n=1}^{m} c_n\psi_n(x = L/2)$$

$$= (4\sqrt{6}/\pi^2)\sqrt{2/L}\sum_{n=1(\text{odd})}^{m}(1/n^2)$$

m:	1	3	5	7	9	...	135
$\sqrt{L}\phi_{\text{approx}}(x=L/2)$:	1.40395	1.55994	1.61609	1.64475	1.66208		1.72689

$\sqrt{L}\phi(x=L/2)=\sqrt{3}=1.73205$.

c) [See Hint.]

$$\begin{aligned}\bar{E} &= \sum_{\text{odd } n}(4\sqrt{6}/n^2\pi^2)^2(n^2h^2/8mL^2)=(12h^2/\pi^4 mL^2)\sum_{\text{odd } n}(1/n^2)\\ &\le (12h^2/\pi^4 mL^2)\left[\sum_{\text{odd } n=1}^{m}(1/n^2)+\frac{1}{2}\int_{m+1}^{\infty}(1/x^2)dx\right]\\ &= [(h^2/8mL^2)(96/\pi^4)(1.23386)]_{m=9}\\ &= [1.21602(h^2/8mL^2)]_{m=9};\ [1.21432(h^2/8mL^2)]_{m=135}\end{aligned}$$

7-5. Normalizing factor $=(2\alpha/\pi)^{3/4}$; $\bar{E}=(3\alpha/2)-2\sqrt{2\alpha}/\sqrt{\pi}$; $\alpha=8/9\pi$; $\bar{E}(\text{min})=-4/3\pi=-0.4244$ a.u.; $\bar{r}=1.5$ a.u.; $r_{\text{mp}}=\sqrt{9\pi}/4=1.329$ a.u. (For ψ_{exact}, $E=-0.5$ a.u., $\bar{r}=1.5$ a.u., $r_{\text{mp}}=1.0$ a.u.)

7-6. a) and b) See text and Eqs. (7-16)–(7-20). c) $\alpha=\frac{5}{3}$, $\bar{E}=-0.370$ a.u.

7-7. a) $1/\sqrt{8}$. b) -0.292 a.u.

7-8. a) Zero, because the function is symmetric for xy reflection whereas $2p_z$ is antisymmetric. b) -0.4215 a.u., (by assuming c_{3s} is $\sqrt{0.05}$ and all higher terms vanish).

7-9. a) $\zeta=5/3$, $\bar{E}=-0.2777$ a.u.
b) $c_1=0.897$.
c) $\chi=2.26(\phi-0.897\psi_{1s})$.
d) No. χ cannot have $\bar{E}$ lower than the $n=2$ value, which is the lowest-energy case orthogonal to $n=1$ and which has $E=-0.125$ a.u.

7-10. -0.75 a.u.

7-11. [See Hint.] The energy $\bar{E}=-0.375$ a.u. $=(0.9775)^2(-\frac{1}{2}$ a.u.)+higher-energy contributions. But this leading term equals -0.478 a.u., and so the net value of the higher energy terms must be positive. Therefore, at least one of them must correspond to a state with positive energy—a continuum state.

7-12. Slater's rules give $\zeta=1.7$, whereas the variation method gives $\zeta=27/16=1.6875$.

7-13. a) $S_{11}=1$, $S_{12}=0$, $S_{22}=1$, $H_{11}=-\frac{1}{2}$, $H_{12}=-F$, $H_{22}=0$. $\bar{E}=-\frac{1}{4}-\frac{1}{4}\sqrt{1+16F^2}$; for $F=0.1$, $\bar{E}=-0.51926$ a.u. This trial form is superior because $z\cdot\psi_{1s}$ is more contracted than ψ_{2p_z}, closer in size to ψ_{1s}, hence interferes constructively and destructively with ψ_{1s} more effectively.
b) [See Hint.] $\lim(F\to 0)$ of $\sqrt{1+16F^2}=1+8F^2$, in lim, $\bar{E}=-\frac{1}{2}-2F^2$, $-\frac{1}{2}\alpha F^2=-2F^2$; $\alpha=4$. $E(e^2/a_0)\leftrightarrow\alpha F^2$ (α units) $\cdot(e/a_0^2)^2$, α units $=a_0^3$. (See Appendix 10.)

7-14. Since 2s is isoenergetic with 2p states, these should mix freely in response to field. Hence, 2s is more polarizable.

7-15. $\bar{E}(\min) = -12/5 = -2.4$ a.u. $\psi = \sqrt{2/5}(\phi_a + \phi_b)$.

7-16. [See Hint.] $\bar{E}(\text{lowest}) = -2.030$ a.u. $\psi = 1.045\phi_a - 0.179\phi_b$.

7-17. [See Hint.] $\bar{E}_{\text{elec}} = (\zeta^2/2) - 2 + 2(\zeta + 1)\exp(-2\zeta)$; $\zeta_{\text{best}} = 0.9118$, $\bar{E}_{\text{elec}} = -0.9668$ a.u.; $\bar{E}_{\text{elec}} + 1/R = \bar{E}_{\text{tot}} = -0.4668$ a.u. Since this energy exceeds that of $H + H^+$ (-0.5 a.u. at $R = \infty$), this function does not demonstrate the existence of a bound state.

7-18. $1\sigma_u$ becomes a 2p AO of He^+, so $E = -0.5$ a.u.

7-19. [See Hint.] Since they are degenerate, ψ_+ and ψ_- may be mixed. The sum gives $1s_a$, describing the case in which the electron is at a. The difference gives $1s_b$.

7-20. [See Hint.] Both equal $-\frac{3}{2}$ a.u. This is higher than the lowest He^+ eigenvalue because these are hydrogen atom 1s functions instead of He^+ functions.

7-21. k must be greater than S.

7-22. a) $\hat{H} = -\frac{1}{2}\nabla^2 - (1/r_H) - 2/r_{He}$.
b) Separated atoms: lowest energy for $H^+ + He^+(1s) = -2$ a.u. For the united atom; $Li^{2+}(1s) = -4.5$ a.u.

7-23. a) π_u bonding. b) σ_u antibonding. c) σ_g bonding. d) δ_g bonding. e) π_u bonding.

7-24.

$1s\sigma_g$	$2p_z\sigma_u$	$3p_y\pi_u$	$3d_{xy}\delta_g$
$2s\sigma_g$	$2p_x\pi_u$	$3d_{z^2}\sigma_g$	$3d_{xz}\pi_g$

7-25. a) Antibonding. b) Bonding. c) Bonding.

7-26. $2p_x$, $2p_y$, $3p_x$, $3p_y$, $3d_{xy}$, $3d_{yz}$.

7-27. [See Hint.] The integrals that vanish by symmetry are b), c), e), and f); g) does not vanish. The AOs are orthogonal due to different symmetry for reflection in the xy plane at a. But $\hat{H}$ is not invariant to this reflection. In f), the relevant reflection is through the yz plane; $\hat{H}$ is invariant to this one.

7-28. Sketches show that several planes qualify, but symmetries are opposite no matter which is chosen. For instance, if the xz plane is selected, $\delta_{x^2-y^2}$ is symmetric, δ_{xy} is antisymmetric, π_{xz} is symmetric, π_{yz} antisymmetric.

7-29. 0. (It is a σ MO.)

7-30. a) σ_u antibonding. b) π_u bonding. c) π_g antibonding. d) δ_g bonding.

7-31. a) 4. b) 3 (a triplet). c) (1) increase, (2) decrease. d) 2 (a doublet).

7-32. a) $1\sigma_g^2 1\sigma_u^2 2\sigma_g^2 2\sigma_u^2 3\sigma_g^2 1\pi_u^4 1\pi_g^1$. b) 5. c) O_2^+ has larger D_0. d) $^2\Pi_g$. e) All σ MOs.

7-33. For He_2, the second MO ($\sigma_u 1s$) correlates with third united atom AO ($2p_\sigma$). For LiH, the second MO (σ) correlates with second united atom AO (2s). Thus, this MO is less antibonding in heteronuclear case.

7-34. [See Hint.] $2s' = 1.0295\ 2s - 0.2447\ 1s$; $2\sigma_g = 0.0136\ 1s_A - 0.6523\ 2s'_A - 0.0854\ 2p_{\sigma,A}$ and similarly for B.

7-35. $E_{elec} = E_{sepatoms} - V_{nn} - D_e$; $D_e = -E_{elec} - V_{nn} + E_{sepatoms} = (1.1026 - 0.500 - 0.500)\text{a.u.} = 0.1026$ a.u.

MC: a a d b d

Chapter 8

8-1. For ψ_{prod}, $E = E_1 + E_2 + E_3$. For ψ_{det}, $E = \frac{1}{6}(E_1 + E_2 + E_3)$six times. Energies of products in ψ are $E_1 + E_2 + E_3$ and $E_1 + E_2 + E_4$. These can be factored out to give $\hat{H}\psi = E\psi$ only if $E_3 = E_4$.

8-2. a)

$$\longrightarrow \begin{vmatrix} x & 1 & 0 & 0 & 0 & 0 \\ 1 & x & 1 & 0 & 1 & 0 \\ 0 & 1 & x & 1 & 0 & 0 \\ 0 & 0 & 1 & x & 1 & 0 \\ 0 & 1 & 0 & 1 & x & 1 \\ 0 & 0 & 0 & 0 & 1 & x \end{vmatrix}$$

b)

$$\longrightarrow \begin{vmatrix} x & 1 & 1 & 0 & 0 & 0 \\ 1 & x & 1 & 0 & 0 & 0 \\ 1 & 1 & x & 1 & 0 & 0 \\ 0 & 0 & 1 & x & 1 & 1 \\ 0 & 0 & 0 & 1 & x & 1 \\ 0 & 0 & 0 & 1 & 1 & x \end{vmatrix}$$

c) $\begin{vmatrix} x & 1 \\ 1 & x \end{vmatrix}$ (only 2 unsaturated carbons, so the same as ethylene).

d) Same as c). Same as *two* ethylenes since the two π systems are noninteracting due to spatial separation.

e) Same as c). Same as *two* ethylenes since the two π systems are orthogonal and noninteracting.

8-3. $\begin{vmatrix} x & 1 & 1 & 1 \\ 1 & x & 0 & 0 \\ 1 & 0 & x & 0 \\ 1 & 0 & 0 & x \end{vmatrix} = 0$, $\quad x^4 - 3x^2 = 0$ See Appendix 6 for results.

8-4. $\frac{1}{3}$, since only χ_3 is common to both MOs.

8-5. $S = 1/\sqrt{2}$, $1/\sqrt{1 - S^2} = \sqrt{2}$, ϕ_2'(normalized) $= \sqrt{2}(\phi_2 - \phi_1/\sqrt{2})$. This results in all $|c| = 1/2$, with negative values on same side. That is, the node for ϕ_2' is vertical, while that for ϕ_1 is horizontal.

8-6. a) $\alpha - 2\beta$ (using octagon in circle).

b)

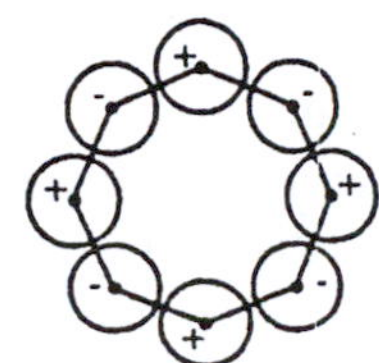

(since totally antibonding).

8-7. a) $E = \alpha + 2\beta(\sqrt{2}/3\sqrt{3}) + 0 + 0 + 0) = \alpha + 0.544\beta$.

b) $p_{12} = 0.272$. All others zero.

c) $E = \alpha + \beta$. (It is an ethylene pi bond, distributed over three locations, or three-thirds of a double bond. Or one can use $a = 1/\sqrt{6}$ and equations for p and E.)

8-8. Highest two levels have same magnitudes as lowest two, with coefficients multiplied by -1 on alternating atoms (here taken to be second and fourth).

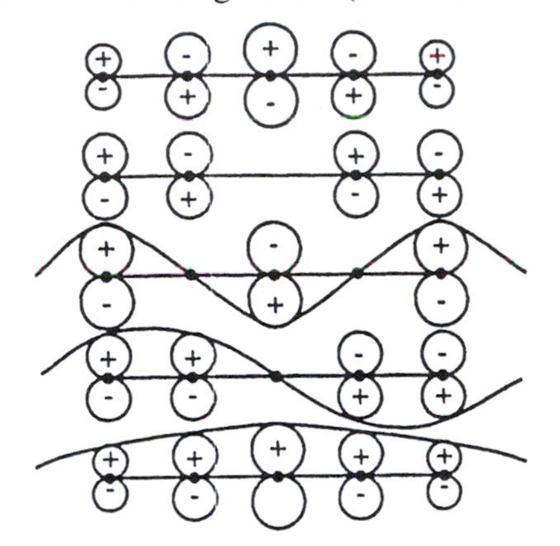

8-9. a) No. This is an odd alternant, so expect paired energy levels with an MO at $E = \alpha$. The unpaired electron is therefore not in a degenerate level. b) Yes. Pentagon in circle forces degeneracies. Second level has three electrons. c) Yes. Square in circle shows that second level is degenerate. It contains one electron. d) Yes. The degenerate level of benzene now has an odd number of electrons.

8-12. Bond orders: CH_2–CH, 0.8944 → 0.6708; CH–CH, 0.4472 → 0.5854. Bond lengths in A (Eq. 8-61): CH_2–CH, 1.354 → 1.392; CH–CH, 1.436 → 1.408; $\Delta CH_2\text{–}CH = +0.038$A, $\Delta CH\text{–}CH = -0.028$ Å.

8-13. For benzene, $c_{\mu i}$ should be taken as $1/\sqrt{6}$ since all carbons are equivalent.

8-14. Only fluoranthene deviates markedly because it is nonalternant. Hence, its LUMO and HOMO energies are not symmetrically disposed about $E = \alpha$.

8-15. a) Oxidation potential ~ 0.97 V, reduction potential ~ 1.41 V. b) [See Hint]. To shorten: 4–10, 9–10, 8–9; to lengthen: 3–10, 1–9, 4–5, 7–8; otherwise no change.

8-16. $E = 18\alpha + 21.877\beta$. Error $= 0.0015\beta$ per π electron.

8-17. a) Naphthalene; E_π (from Table 8-2) $= 10\alpha + 13.128\beta$, from HMO $= 10\alpha + 13.6832\beta$. The difference $= 0.055\beta$ per π electron, aromatic. Perylene; E_π (Table 8-2) $= 20\alpha + 27.2796\beta$, E_π(HMO) $= 20\alpha + 28.2453\beta$; $\Delta E_\pi = 0.048\beta$ per π electron, aromatic.

b) RE for perylene is slightly less than double that for naphthalene. The central ring does not appear to be contributing.

c) These two bonds are single in all formal (nonpolar) structures.

d) The calculated length = 1.433 Å. The HMO length is too short. The actual length is more consistent with these being "truly" single bonds..

8-18. The fourth molecule in Fig. 8-24 should strive for six electrons in each ring. This would make the left side (i.e., the seven-membered ring) net positive. The other two molecules become net negative on the left. Electron densities corroborate this. For the fourth molecule, charge densities exceed unity in the five-membered ring and are less elsewhere. The fifth molecule has only one π electron density less than unity, and this is for the methylene carbon.

8-19. a) 9. b) 10. c) 4. d) 6. e) 10.

8-20. Since the formal structure always shows C–O single bonds, $C_1{=}C_2$, and $C_3{=}C_4$ double bonds, and C_2–C_3 as single, we can use the single-, double-bond distinctions of Table 8-3. These give

$$\begin{vmatrix} x & 1.1 & 0 & 0 & 0.8 & 0 & 0 & 0 \\ 1.1 & x & 0.9 & 0 & 0 & 0.3 & 0 & 0 \\ 0 & 0.9 & x-0.1 & 1.1 & 0 & 0 & 0.8 & 0 \\ 0 & 0 & 1.1 & x & 0.8 & 0 & 0 & 0 \\ 0.8 & 0 & 0 & 0.8 & x+2.0 & 0 & 0 & 0 \\ 0 & 0.3 & 0 & 0 & 0 & x+1.5 & 0 & 0 \\ 0 & 0 & 0.8 & 0 & 0 & 0 & x-0.1 & 3.0 \\ 0 & 0 & 0 & 0 & 0 & 0 & 3.0 & x-0.5 \end{vmatrix}$$

Otherwise, the positions with 1.1 and 0.9 become 1.0.

8-21. a) Left. b) Left. c) Right.

8-22. $q_r = 1$ at all centers, and so it does not predict some centers best for nucleophilic, and hence worst for electrophilic substitution, so it doesn't apply to this question. Since the HOMO and LUMO have identical absolute coefficients (by the pairing theorem), the same site is most favored for both nucleophilic and electrophilic substitution. L_r must be identical for nucleophilic, radical, or electrophilic substitution because an interrupted even alternant produces an odd alternant, for which cationic, neutral, and anionic π energies (β part) are the same. Thus, HOMO and LUMO indices are consistent with coincidence of active sites for nucleophilic and electrophilic substitutions, and L_r is consistent with the coincidence of these with active sites for radical addition.

8-23. No. Both types should prefer the most polarizable site, since that is the site to which charge is most easily attracted or from which it is most easily repelled.

8-24. a) $F_1 = 0.0684$, $F_2 = 0.4618$, $F_4 = 0.9737$.

b)

Index	Values		Preferred site
q_r	$q_2 = 0.818$	$q_4 = 1.488$	4
HOMO	$c_2^2 = 0.1356$	$c_4^2 = 0.6646$	4
L_r^+	$L_2^+ = 2.134\beta$	$L_4^+ := 0.962\beta$	4
π_{rr}	$\pi_{22} = -0.4340$	$\pi_{44} = -0.4019$	2

c) Only protons on C_2 and C_3 will produce ESR splitting in simplest theory, since the singly occupied MO of the radical anion is zero elsewhere. d) Net bonding, because energy is below $E = \alpha$ and this happens only when bonding interactions dominate.

8-25. The correlation is fairly good except for styrene, which is way off. But styrene is the only member of the set where addition is not occurring at a ring position. Because the geometric constraints are so different, the relation between free valence and transition-state energy is presumably rather different for styrene.

8-26. The five-membered ring because, as it strives for six electrons in order to satisfy the $4n+2$ rule, it becomes negatively charged. (The seven-membered ring also strives for six electrons, becoming positive.)

8-27. Fulvene a) should experience greater change. It is not an alternant hydrocarbon, whereas benzyl b) is. The MO into which the electron goes in benzyl is necessarily nonbonding, with zero coefficients at positions 1, 3, and 5. (See Appendix 5 for discussion, Appendix 6 for coefficient values.)

8-28. a) 11. b) 2 and 4. c) 3, 6, and 8 (1, 9, and 10 have no attached hydrogen).

8-29. a) is easier to ionize because the HOMO is zero at the nitrogen position (in the all-carbon analog). This means ionization does not remove electronic charge from the more electronegative atom in case a) but does in case b).

MC: c b

Chapter 9

9-1. a) 212 b) $\begin{pmatrix} 6a & 6b & 6c \\ 7a & 7b & 7c \end{pmatrix}$ c) $\begin{pmatrix} 25 & 13 \\ i-7 & 18 \end{pmatrix}$ d) $\begin{pmatrix} 1 & 0 \\ 0 & 1 \end{pmatrix}$ e) $\begin{pmatrix} 3i+16 & 2i+28 \\ 31 & 51 \\ -12 & -21 \end{pmatrix}$

f) product not defined g) 1

9-2. H is defined to be hermitian if $H_{ij} = H^*_{ji}$. $H_{ji} = \int \chi_j^* \hat{H} \chi_i d\tau$, and so $H^*_{ji} = \int \chi_j \hat{H}^* \chi_i^* d\tau$. But if $\hat{H}$ is hermitian, this must equal $\int \chi_i^* \hat{H} \chi_j d\tau \equiv H_{ij}$. Therefore, $H^*_{ji} = H_{ij}$ and H is hermitian.

9-4. $AC = \widetilde{CA}$. But $\widetilde{CA} = \tilde{A}\tilde{C}$, so $AC = \tilde{A}\tilde{C}$. This must be true in this example because A and C are symmetric. That is, $A = \tilde{A}$, $C = \tilde{C}$.

9-5. a) $|B - \lambda_i 1| = |T^{-1}AT - \lambda_i 1| = |T^{-1}AT - \lambda_i T^{-1}1T| = |T^{-1}(A - \lambda_i 1)T| = |T^{-1}||A - \lambda_i 1||T| = |TT^{-1}(A - \lambda_i 1)| = |A - \lambda_i 1|$

b) For diagonal B, value of $|B - \lambda_1 1|$ is product of diagonal elements. For this to vanish, at least one such element must vanish. This will occur whenever λ_i equals a diagonal element of B. Therefore, the latent roots are the diagonal values.

9-6. If a latent root is zero, then the product of latent roots is zero. But this product is the value of the determinant of the matrix. If the determinant of the matrix is zero, there is no inverse.

9-7. a)

$$\mathrm{tr}(\mathsf{ABC}) = \sum_{i=1}^{n}(\mathsf{ABC})_{ii} = \sum_i\sum_j\sum_k a_{ij}b_{jk}c_{ki}$$
$$= \sum_i\sum_j\sum_k c_{ki}a_{ij}b_{jk} \text{ which is } \sum_k(\mathsf{CAB})_{kk} = \mathrm{tr}(\mathsf{CAB})$$
$$= \sum_i\sum_j\sum_k b_{jk}c_{ki}a_{ij} \text{ which is } \sum_j(\mathsf{BCA})_{jj} = \mathrm{tr}(\mathsf{BCA})$$
$$= \sum_i\sum_j\sum_k (c_{ki}b_{jk}a_{ij}) \text{ which is } \textit{not} \sum_i(\mathsf{CBA})_{ii}, \text{ hence } \neq \mathrm{tr}(\mathsf{CBA}).$$

b) $\mathrm{tr}(\mathsf{T}^{-1}\mathsf{AT}) = \mathrm{tr}(\mathsf{TT}^{-1}\mathsf{A}) = \mathrm{tr}(\mathsf{A})$.

9-8.

$$(\text{norm } \tilde{\mathsf{T}}\mathsf{AT})^2 = \sum_{i,j}(\widetilde{\tilde{\mathsf{T}}\mathsf{AT}})_{ij}(\tilde{\mathsf{T}}\mathsf{AT})_{ji} = \sum_{i,j}(\tilde{\mathsf{T}}\tilde{\mathsf{A}}\mathsf{T})_{ij}(\tilde{\mathsf{T}}\mathsf{AT})_{ji}$$
$$= \sum_{i,j,k,l}(\tilde{\mathsf{T}})_{ik}(\tilde{\mathsf{A}})_{kl}(\mathsf{T})_{lj}(\tilde{\mathsf{T}})_{jl}(\mathsf{A})_{lk}(\mathsf{T})_{ki}$$
$$= \sum_{k,l}\left[(\tilde{\mathsf{A}})_{kl}(\mathsf{A})_{lk}\underbrace{\sum_i(\tilde{\mathsf{T}})_{ik}(\mathsf{T})_{ki}}_{1}\underbrace{\sum_j(\mathsf{T})_{lj}(\tilde{\mathsf{T}})_{jl}}_{1}\right]$$
$$= \sum_{k,l}(\tilde{\mathsf{A}})_{kl}(\mathsf{A})_{lk} = (\text{norm } \mathsf{A})^2$$

9-9. a) $\mathrm{tr} = 0$, $\det = 2$, $\text{norm} = \sqrt{6}$; therefore, $a+b+c=0$, $abc=2$, $a^2+b^2+c^2=6$; solutions: $2, -1, -1$.

b) Solutions $1, 1, -1$. c) $0, 1+\sqrt{3}, 1-\sqrt{3}$.

9-10. Both vectors transform to $\begin{pmatrix} 3\cos\theta \\ -3\sin\theta \end{pmatrix}$. Hence, reversal is not possible and transformation is singular. This is verified by fact that the determinant vanishes.

9-11. The matrix is already diagonalized. This means the eigenvector matrix is the 3×3 unit matrix.

9-12. Let $\mathsf{T}^{-1}\mathsf{AT} = \mathsf{D}_A$ (diagonal) and $\mathsf{T}^{-1}\mathsf{BT} = \mathsf{D}_B$ (diagonal). Then $\mathsf{D}_A\mathsf{D}_B = \mathsf{D}_B\mathsf{D}_A$ (diagonal matrices commute); $\mathsf{T}^{-1}\mathsf{ATT}^{-1}\mathsf{BT} = \mathsf{T}^{-1}\mathsf{BTT}^{-1}\mathsf{AT}$; $\mathsf{T}^{-1}\mathsf{ABT} = \mathsf{T}^{-1}\mathsf{BAT}$; $\mathsf{TT}^{-1}\mathsf{ABTT}^{-1} = \mathsf{TT}^{-1}\mathsf{BATT}^{-1}$; $\mathsf{AB} = \mathsf{BA}$.

9-13. In the second case, C is not unitary, since $\mathsf{C}^\dagger\mathsf{SC} = 1$. The ordinary procedures for diagonalizing H have *built in* the requirement that $\mathsf{C}^\dagger\mathsf{C} = 1$. The problem would be to find a matrix C that simultaneously diagonalizes H and satisfies $\mathsf{C}^\dagger\mathsf{SC} = 1$.

$$\int \alpha^\dagger\beta\, d\omega \to (1\ 0)\begin{pmatrix}0\\1\end{pmatrix} = 0; \quad \int \alpha^\dagger\alpha\, d\omega \to (1\ 0)\begin{pmatrix}1\\0\end{pmatrix} = 1$$
$$\hat{S}_z\alpha = \frac{1}{2}\begin{pmatrix}1 & 0\\0 & -1\end{pmatrix}\begin{pmatrix}1\\0\end{pmatrix} = \frac{1}{2}\begin{pmatrix}1\\0\end{pmatrix} = \frac{1}{2}\alpha$$

Chapter 10

10-1.

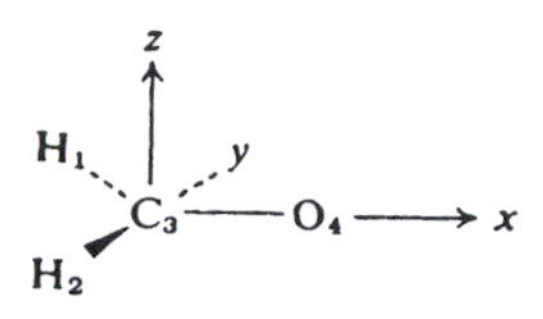

10-2.

AO no.	Atom	AO type	AO no.	Atom	AO type
1	H_1	1s	6	C_3	$2p_y$
2	H_2	1s	7	O_4	2s
3	C_3	2s	8	O_4	$2p_x$
4	C_3	$2p_x$	9	O_4	$2p_x$
5	C_3	$2p_x$	10	O_4	$2p_y$

10-3. $E = -0.756$ a.u.; MO 9

$\phi_9 = -0.27\ 1s_1 - 0.27\ 1s_2 - 0.49\ 2s_C + 0.22\ 2p_{xC} + 0.31\ 2s_0 + 0.33\ 2p_{x0}$

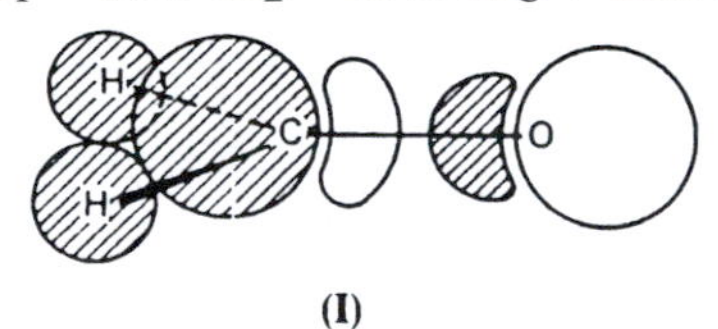

(I)

A σ MO, mainly C–H_2 bonding and lone pair (nonbonding) on oxygen. Shows some C–O antibonding character [see **I**).

$E = -0.611$ a.u.; MO 8

$\phi_8 = -0.21\ 1s_1 + 0.21\ 1s_2 - 0.32\ 2p_{yC} - 0.76\ 2p_{yO}$

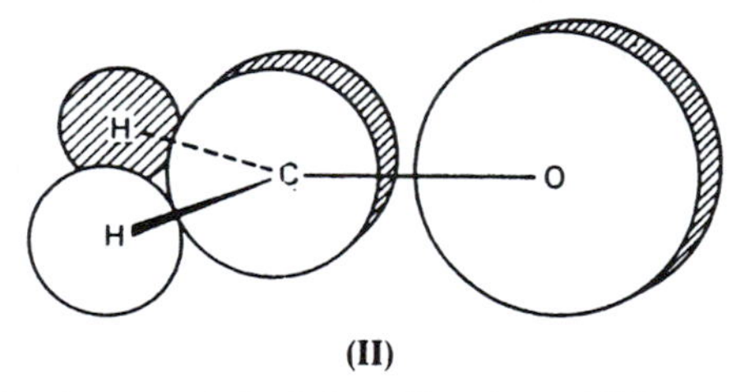

(II)

A σ MO, CH_2 and C–O bonding [see (**II**)].

$E = -0.597$ a.u.; MO no. 7

$\phi_7 = 0.24\ 2p_{zC} + 0.92\ 2p_{zO}$

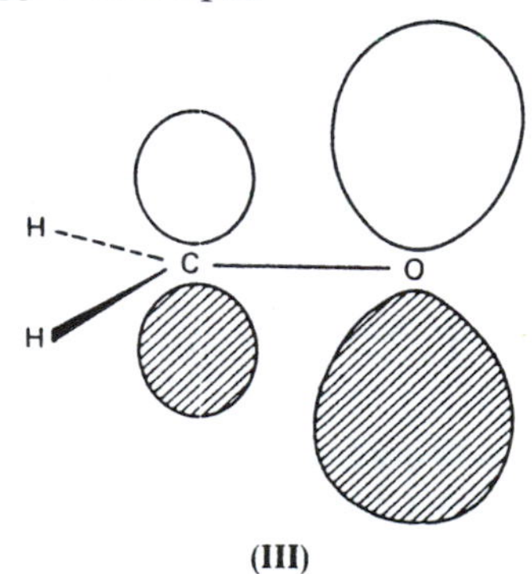

(III)

A π MO, mostly on oxygen, but somewhat delocalized to give some C–O bonding [see (**III**)].

10-4. The π MOs are 4 and 7. All others are σ.

10-5. C and O $2p_\pi$ AOs are 4 and 8. The 4, 8 overlap population is seen from the data to be 0.1936. Since MO 7 is C–O bonding, loss of an electron should cause the C–O bond to lengthen.

10-6.
$$E_7 = (0.2456)^2(-10.67\,\text{eV}) + (0.9181)^2(-15.85\,\text{eV})$$
$$+2(0.2456)(0.9181)(1.75)(0.2146)(-10.67\,\text{eV} - 15.85\,\text{eV})/2$$
$$= -16.25\,\text{eV} = -0.5972\ \text{a.u.}$$

10-7. The sum of the elements in the upper triangle = 12. (Use of *all* elements would count overlap populations twice.)

10-8. If column 7 is the gross populations of MO no. 7, then it should turn out that $0.2175 = 2\left[c_{47}^2 + (0.5)(2)c_{47}c_{87}S_{48}\right]$: $2\left[(0.2456)^2 + (0.2456)(0.9181)(0.2146)\right] = 0.2175$.

10-9. These must be AOs because MO charges must be 0, 1, or 2. AO 4, for example, gets its charge from MO 7. We have just seen (previous problem) that this is 0.2175. (AO 4 also appears in MO 4, and the "charge matrix" gives a value of 1.7825 for this. But this does not appear in the gross population because MO no. 4 is unoccupied in the ground state configuration.)

10-10. The net charges are the AO charges plus the nuclear charges (after cancellation of some nuclear charge by inner-shell electrons). These results indicate high polarity, with oxygen being the negative end of the dipole. The predicted polarity is unrealistically high because EHMO neglects interelectronic repulsion which would tend to counteract such extreme charge imbalance.

10-11. Number MOs = number AOs = 1 on each H and 4 (valence) on each C = 22.

Chapter 11

11-1.

	Koopmans	ΔSCF	Experiment
$2s \longrightarrow 2p$	1.0800	1.0830	0.989
$1s \longrightarrow 2p$	31.9220	31.1921	31.19

11-2.

	Koopmans–SCF (eV) (electron relaxation)	SCF – observed (eV) (electron correlation)
$^2B_1$	2.71	−1.54
$^2A_1$	2.52	−1.40
$^2B_2$	1.86	−0.90

11-3. For electron affinities, these errors should reinforce, rather than cancel, because adding an electron should *increase* electron correlation.

11-4. $ad - cb + \lambda(af - be) = ad + \lambda af - bc - \lambda be.$

11-5. Neither ψ_1 nor ψ_2 is already the best function in our function space. Hence, we cannot argue that mixing will bring no improvement.

11-6. This must be true to enable a_1 to be factored from the expanded form of $\hat{A}\psi$.

11-7. For a given choice of basis functions, there are two integrals:

$$\langle\chi_a(1)\chi_b(2)|\chi_c(1)\chi_d(2)\rangle \quad \text{and} \quad \langle\chi_a(1)\chi_b(2)|\chi_d(1)\chi_c(2)\rangle$$

There are five ways to choose a function for each position. Thus, the number of integrals is $2\times 5^4 = 1,250$.

11-8. a) and c) would be prevented from contributing.

11-9.
$$\hat{H} = -\frac{1}{2}\sum_{i=1}^{10}\nabla_i^2 - \sum_{i=1}^{10}\left(\frac{1}{r_{i,\mathrm{H}_1}} + \frac{1}{r_{i,\mathrm{H}_2}} + \frac{8}{r_{i,\mathrm{O}}}\right) + \sum_{i=1}^{9}\sum_{j=i+1}^{10}\frac{1}{r_{ij}}$$

11-10. a) $(1/\sqrt{2})|\sigma_g(1)\bar{\sigma}_g(2)|$. b) $E_{\mathrm{elec}} = -1.804$ a.u. c) $E_{\mathrm{tot}} = -1.090$ a.u. d) $D_e = 0.090$ a.u. e) IE (Koopmans) $= 0.619$ a.u. f) KE $+ V_{\mathrm{ne}} = -1.185$ a.u.

Chapter 12

12-1. $E_0 + W_0^{(1)} = \langle\psi|H_0|\psi\rangle + \langle\psi|H'|\psi\rangle = \langle\psi|H|\psi\rangle \geq W_0$.

12-2. [See Hint.] $E_1^{(1)} > E_3^{(1)} > E_2^{(1)}$.

12-3. a) δ. b) δ. c) $-\delta/2$. d) $\delta/2$. e) $-\delta/2$. f) 0.

12-4. a) $W_2^{(1)} = 3\delta/4$.
(b-1) Expect $c_{21}^{(1)}$ to cause ψ_1 to shift to right. Since ψ_2 is positive on left of box, negative on right, $c_{21}^{(1)}$ should be negative. Since ψ_2 is above ψ_1, it should cause energy to be depressed, and $c_{21}^{(1)}$ leads to a negative contribution to $W_1^{(2)}$.
(b-2) Because $c_{ij}^{(1)} = -c_{ji}^{(1)}$, $c_{12}^{(1)}$ must be positive, giving a $\phi_2^{(1)}$ that causes charge to shift left and a contribution to $W_2^{(2)}$ that is positive.

12-5.
$$c_{41}^{(1)} = \frac{\langle\psi_1|H'|\psi_4\rangle}{E_1 - E_4} = \frac{(2/L)\int_0^L \sin(\pi x/L)(Ux/L)\sin(4\pi x/L)dx}{-15\pi^2/2L^2} = \frac{64UL^2}{15^3\pi^4}$$

$$c_{21}^{(1)} = \frac{32UL^2}{27\pi^4}, \quad \frac{c_{41}^{(1)}}{c_{21}^{(1)}} = \frac{2}{125} = 1.6\%$$

12-6. The effect is zero, to first order, because ψ^2 is symmetric for every state and the perturbation is antisymmetric.

12-7. [See Hint.] $E = E_0 + E_1 = -\frac{1}{2}$ a.u. $+ \langle 1s|-1/r|1s\rangle = -\frac{3}{2}$ a.u. Correction seeks to make ψ less diffuse, yet still spherical. $c_{2s,1s}^{(1)}$ should be positive (augments hydrogen 1s at small r and cancels at large r) and c_{2p_0}, $1s^{(1)}$ should be zero (wrong symmetry).

12-8. $W^{(0)} + W^{(1)} = -2$ a.u. $+ 2$ a.u. $= 0$. Reasonable, since must be above -0.5 a.u. (See Problem 12-1.)

12-9. a) Zero, because the perturbation is antisymmetric for x, y reflection, while the 1s function is symmetric. b) $c^{(1)}_{2s,1s}$ vanishes. The relevant integral suffers the same symmetry disagreement as in part a). Or, in another view, the 2s AO will affect only the diffuseness of ψ, whereas the perturbation affects polarity. $c^{(1)}_{2p_z,1s}$ should be positive. The perturbation lowers the potential for positive z, the wavefunction should skew that way, so $2p_z$ should enter with its positive lobe reinforcing 1s at positive z. c) It is negative, reflecting the lower energy of the electron distribution as a result of first-order polarization.

12-10. $E_\pi = 4\alpha + 4.9624\beta + (0.1)(q_4)$; $q_4 = 1.4881$; $E_\pi = 4\alpha + 5.1112\beta$.

12-11. The effect is least at C_6 since q_6 is smallest. First-order result: $E = 6\alpha + 7.777\beta$. Computed result: $E = 6\alpha + 7.8546\beta$.

12-12. $\Pi_{kk} = 4\sum_{j=1}^{m}\sum_{i=m+1}^{n} c_{kj}^2 c_{ki}^2/(E_j - E_i)$. This must be negative. ∂q_k must be negative if $\partial\alpha_k$ is positive. This means that making atom k less attractive causes electron density to decrease there. This makes sense.

12-13. Since the total charge must be conserved, the sum of all atom-atom polarizabilities, including $\pi_{1,1}$ must be zero. Since the sum over the polarizabilities in Example 12-5 is 0.44277, $\pi_{1,1} = -0.44277$.

12-14. a) $W^{(1)} = h\beta$ (because all $q = 1$).
b) The lowest-energy one, which is the MO where the central atoms find the largest $|c|$.

12-15. a) $E_\pi = 9\alpha + 12.1118\beta$.
b) Atom 1 seems likely to have the greater self-atom polarizability because, in the cation, the product of squares of HOMO, LUMO coefficients vanishes at atom 2.

12-16. [See Hint.] For butadiene: $\Delta E_\pi = 4\beta[(0.3718)^2 - (0.6015)^2] = -0.894\beta$. For hexatriene: $\Delta E_\pi = 4\beta[(0.2319)^2 - (0.4179)^2 + (0.5211)^2] = 0.603\beta$. The energy of hexatriene is lowered, that of butadiene is raised, and so hexatriene benefits.
For cyclobutadiene: $E_0 + E^{(1)} = 4\alpha + 3.5778\beta$, E (Hückel) $= 4\alpha + 4.000\beta$.
For benzene: $E_0 + E^{(1)} = 6\alpha + 7.591\beta$, E (Hückel) $= 6\alpha + 8.000\beta$.

12-17. a)

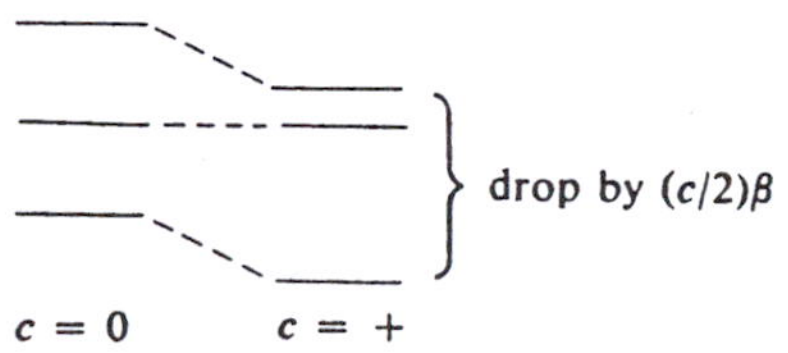

drop by $(c/2)\beta$

$c = 0$ $\quad$ $c = +$

b)

$$\phi_1^{(1)} = \frac{\langle\psi_2|H'|\psi_1\rangle}{E_1 - E_2}\psi_2 + \frac{\langle\psi_3|H'|\psi_1\rangle}{E_1 - E_3}\psi_3$$

Since H' depends only on density at C_2, it comes out of the integral. [See **(IV)**.]

$$\phi_1^{(1)} = 0 + (-\tfrac{1}{2}c\beta/2\sqrt{2}\beta)\psi_3 = -0.1768c\psi_3, \quad \phi_2^{(1)} = 0$$
$$\phi_3^{(1)} = +0.1768c\psi_1$$

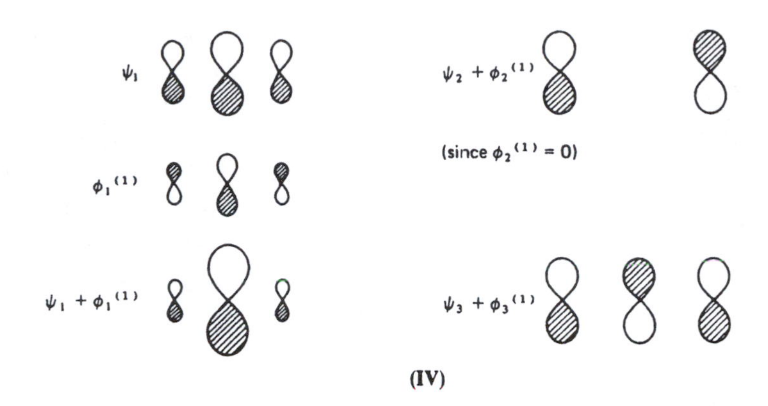

(IV)

Thus, $\psi_1 + \phi_1^{(1)}$ has more density at C_2 than did ψ_1, $\psi_2 + \phi_2^{(1)}$ is identical to ψ_2, and $\psi_3 + \phi_3^{(1)}$ has lost density at C_2.

12-18. $\pi_{1,2} = -0.1768\beta^{-1}$; $\pi_{1,3} = -0.265\beta^{-1}$. Both atoms 2 and 3 lose charge, but atom 3 loses more than atom 2.

12-19. a) Nondegenerate MO, $\phi_1 = (1/\sqrt{3})(\chi_1 + \chi_2 + \chi_3)$ is already correct. The correct zeroth-order degenerate MOs must give zero interaction element with H'. For this H', this means that the overlap between these MOs must be zero at C_2. Thus, one MO must have a node at C_2 $\left[\phi_2 = (1/\sqrt{2})(\chi_1 - \chi_3)\right]$ and the other must be orthogonal to it $\left[\phi_3 = (1/\sqrt{6})(2\chi_2 - \chi_1 - \chi_3)\right]$.

If one uses data from Appendix 6, one obtains, for degenerate MOs, $\psi_1 = -0.8165\chi_1 + 0.4082\chi_2 + 0.4082\chi_3$, and $\psi_2 = 0.7071\chi_2 - 0.7071\chi_3$. These give $H'_{11} = 0.1666c\beta$, $H'_{22} = 0.5c\beta$, $H'_{12} = 0.2886c\beta$. Since $H'_{12} \neq 0$, one knows ψ_1 and ψ_2 are not correct zeroth-order wavefunctions. Solving the determinantal equation gives $E_1 = 0$, $E_2 = 0.6666c\beta$. These are the first-order corrections to the energies of the two upper levels. Solving for coefficients gives, for $E = 0$, $c_1 = -0.866$, $c_2 = 0.500$, and so the proper zeroth-order wavefunction having a zero first-order correction is $\phi_1^{(0)} = -0.866\psi_1 + 0.500\psi_2 = 0.7071\chi_1 - 0.7071\chi_3$. For $E = 0.6666c\beta$, $c_1 = 0.500$, $c_2 = 0.866$, and so $\phi_2^{(0)} = -0.4082\chi_1 + 0.8165\chi_2 - 0.4082\chi_3$. These are the same functions arrived at intuitively above. b) For the above zeroth-order MOs, the densities at C_2 are respectively $\frac{1}{3}, 0, \frac{2}{3}$. Therefore, the energy of the lowest level drops by $c\beta/3$, that for one of the originally degenerate levels drops by $2c\beta/3$, and the other level is unaffected (to first order).

12-20. $\alpha + 2.1\beta, \alpha + 0.818\beta, \alpha + 0.618\beta, \alpha - 1.418\beta, \alpha - 1.618\beta$.

12-21. Answers to part a) and b) follow answers for parts c) and d).
c) $E = \alpha \pm 2\beta(0.5)^2(2) = \alpha \pm \beta$. These agree exactly with benzene orbital energies, which is reasonable since the unperturbed fragment nonbonding orbitals combine to give exactly the benzene MOs.
d) $\alpha + \sqrt{2}\beta + 2\beta(0.353)^2(2) = \alpha + 1.913\beta$ compared to $\alpha + 2\beta$ for benzene.

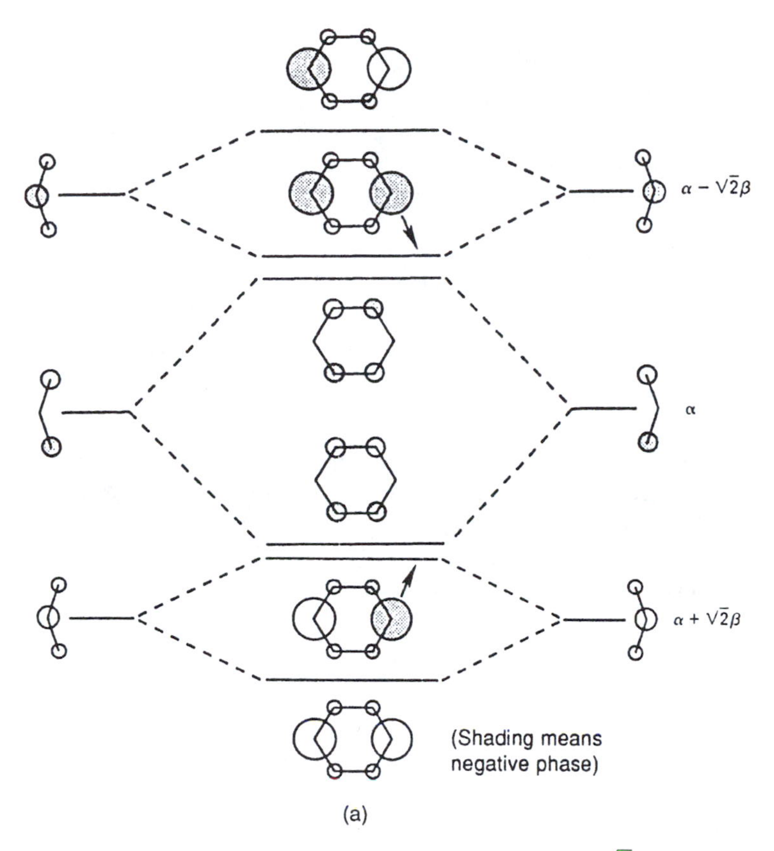

(a)

The allyl coefficients are renormalized by dividing by $\sqrt{2}$.

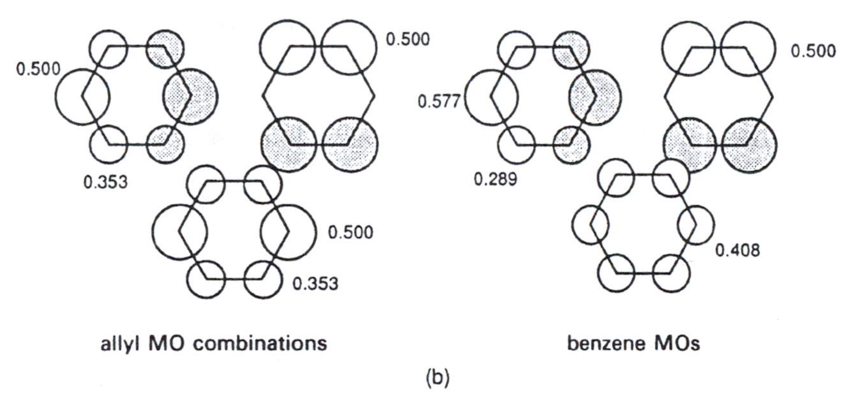

(b)

12-22. a) Only $\pi_3 \leftarrow \pi_2$ is dipole allowed.

b) It is polarized along the outer C–C bonds. If we pretend the molecule is linear, this is the line connecting the carbons. If we adopt a more realistic structure, it depends on whether we choose a *cis* or *trans* structure:

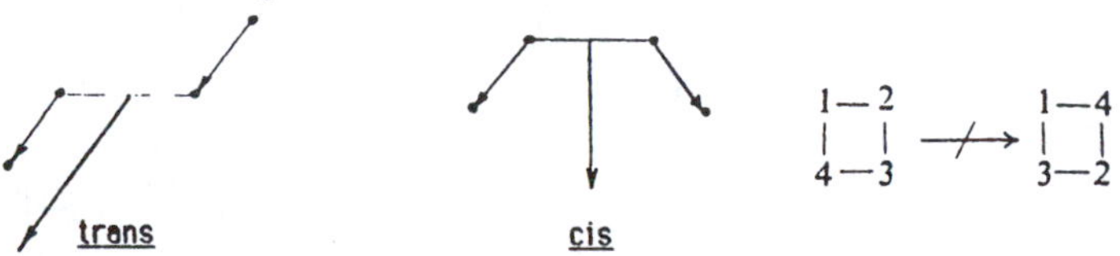

12-23. $\langle\phi_{\pm}^{(0)}|-z|\phi_{\pm}^{(0)}\rangle = \pm\langle 2s|r\cos\theta|2p_Z\rangle = \pm 3$ a.u. For $\phi = \cos(\alpha)2s + \sin(\alpha)2p_Z$, maximum dipole occurs when $\cos\alpha = 1/\sqrt{2}$, $\sin\alpha = \pm 1/\sqrt{2}$, that is, when $\phi = \phi_{\pm}^{(0)}$. This is reasonable since the mixing of degenerate states like these requires no energy "expense" in terms of the unperturbed hamiltonian. As soon as the slightest external field appears, the mixing will occur to the above extent to produce maximum dipoles.

12-24. Since, for MO 4, $c_3 = c_7$, this MO is symmetric for reflection through the yz plane. The operator x is antisymmetric for this reflection. Therefore, if integral $\langle\phi_4|x|\phi_?\rangle$ is to be nonzero, $\phi_?$ must be antisymmetric for this reflection. Of the empty MOs shown, only ϕ_5 satisfies this requirement. There is no other symmetry operation present that will independently cause the integral to vanish, and $\phi_4 \to \phi_5$ is likely to be the observed transition.

12-25. a) The HOMO is antisymmetric for reflection through the yz plane. (Assume that the coordinate origin is in the center of the 5–10 bond.) The x operator is also antisymmetric. Therefore, the allowed transition should be to an MO that is symmetric for this reflection. There are three such MOs, at energies of $+1.000$, $+1.303$, $+2.303$ in units of $-\beta$. But the middle of these disagrees in symmetry with the HOMO for reflection through the xz plane. Therefore, only the other two transitions are allowed.

b) Here, the integrals containing y will vanish by symmetry except for transitions to $+0.618$ and $+1.303$, but the latter state disagrees in symmetry with the HOMO for the yz plane reflection. Hence, only $-0.618 \to +0.618$ is y allowed.

c) No π–π transition is z allowed.

d) Transitions to 1.303 and 1.618 are not allowed for any polarization.

12-26. The field polarizes the atom, which means that the 2s state acquires 2p character. But the 2p $\to$ ls transition is allowed, and so the atom now relaxes to the 1s state.

12-27. $4 \leftarrow 7$ allowed, polarized along C–O axis. $3 \leftarrow 7$ forbidden. $2 \leftarrow 7$ and $1 \leftarrow 7$ allowed, polarized perpendicular to molecular plane.

12-28. a) $\alpha, \alpha \to \alpha, \alpha + 0.5\beta$.

b) $\alpha, \alpha \to \alpha + 0.25\beta, \alpha + 0.25\beta$. If we view this as a sequence of substitutions, then case a) is clearly a case of cooperative perturbations since the proper zeroth-order orbital that is strongly lowered by the first substitution is lowered again by the second. For case b), the second substitution affects the *other* one of the proper zeroth-order orbital energies. To this order, a) should be more stable since the lowering of energy due to the two electrons in this level is β, whereas the most that can come from b) is $\beta/2$. [Analysis of character tables (Chapter 13) enables us to see that the degeneracy for case b) must become split at higher levels of perturbation because a C_{2v} molecule has only one-dimensional representations, hence cannot have degenerate orbitals.]

12-29. Lowest energy rises by 0.0667β. One of the next pair rises by 0.1β and the other drops by 0.0333β.

12-30. x,y components $= 0$. z component $= \frac{0.293}{Z}$ a.u. $= \frac{0.745}{Z}$D $= \frac{2.485\times 10^{-30}}{Z}$Cm.

MC: c a c b

Chapter 13

13-1. No. "Come about 180°" is needed for closure.

13-2. It means that for each operation there is a right inverse as well as a left inverse. If $AB = E$, then A is the left inverse of B *and* B is the right inverse of A.

13-3. a) C_4. b) 4. c) Probably only three, since one tends to think of left face and right face as being in same class. However, in the C_4 group they are not, because there is no operation in the group to interchange them (such as reflection through a plane containing the z axis).

13-4. There are many possible arrangements counted in 4! that are not physically achievable through operations in the group. For example,

$$\begin{matrix} 1 - 2 \\ | \quad\ | \\ 4 - 3 \end{matrix} \nrightarrow \begin{matrix} 1 - 4 \\ | \quad\ | \\ 3 - 2 \end{matrix}$$

For C_{3v}, all possible arrangements are accessible.

13-5. a) C_{2v}. b) D_{2h}. c) D_{4h}. d) T_d.

13-6. a) $\mathsf{U}^\dagger\mathsf{U} = \mathsf{U}\mathsf{U}^\dagger = \mathbf{1}$ (upon explicit multiplication).

b) Upon explicit multiplication, $\mathsf{U}^\dagger\mathsf{A}\mathsf{U} = \begin{pmatrix} 1 & 0 \\ 0 & -1 \end{pmatrix}$.

13-7. a) D_{6h} (yes). b) C_{2v} (no). c) D_{2h} (no). d) D_{3d} (yes). e) C_{3v} (yes).

13-8. a) D_{3h}. b) (1) a_2''; (2)e''; (3)a_2'; (4)a_1'.

13-9. [See Hint.] D_{2d},

MO number:	1	2	3	4	5 6	7 8	9 10	11	12 13	14	15	16
MO symmetry:	b_2	a_1	a_1	b_2	e	e	e	b_2	e	a_1	b_2	a_1

The highest occupied MO is 9,10; the lowest empty MO is 8,7, and so the transition is $e \rightarrow e$.

$$\langle\phi_9|x \text{ or } y|\phi_8\rangle = \int e \otimes e \otimes e = \int e \oplus e \oplus e \oplus e = 0$$

$$\langle\phi_9|z|\phi_8\rangle = \int e \otimes b_2 \otimes e = \int a_1 \oplus a_2 \oplus b_1 \oplus b_2 \neq 0$$

Transition is allowed for (group theory) z polarized light. This means x polarized for the coordinate system shown.

13-10. a) and b) can be checked against C_{2v} character table.
c) Hydrogens generate characters 2 0 2 0, which is $a_1 \oplus b_1$.
d) The unnormalized symmetry combinations are: a_1, $1s_A + 1s_B$; b_1, $1s_A - 1s_B$.

13-11. a) and b) can be checked against the C_{4v} character table.
c) Hydrogens generate characters 4 0 0 2 0 or 4 0 0 0 2, depending on how σ_v and σ_d are selected. Assuming that σ_v contains corner ammonia molecules gives the former set. This resolves to $a_1 \oplus b_1 \oplus e$. (The other choice gives b_2 instead of b_1, but reversal of choice of σ_v and σ_d has the effect of interchanging the symbols b_1 and b_2, and there is no real difference involved in this choice.) For a situation where the nitrogens are numbered as shown in the figure, the unnormalized symmetry orbitals are: a_1, $2s_1 + 2s_2 + 2s_3 + 2s_4$; b_1, $2s_1 - 2s_2 + 2s_3 - 2s_4$; e, $2s_1 - 2s_3$ and $2s_2 - 2s_4$; or $2s_1 + 2s_2 - 2s_3 - 2s_4$ and $2s_1 - 2s_2 - 2s_3 + 2s_4$. (Other combinations are also possible, but these two sets are the most convenient.)

13-12. a) Four operations means order 4. Four classes means four representations. Hence, each representation must be one dimensional, therefore having character +1 or −1 for every operation. One of these must be +1 everywhere (A_1). The others must all have +1 in the first column and −1 in two of the other three columns. The result is as given in Appendix 13 for C_{2v}.
b) Twelve operations gives order 12. Six classes means six representations. This must mean two 2×2 and four 1×1. There is but one unique set of orthonormal character vectors that fit this framework. Check against the D_{3h} character table.

13-13. Two two-dimensional representations and two one-dimensional representations.

13-14. Check against C_{2v}.

13-15. Order = 12. There are six classes.

13-16.

$$\frac{1}{2}\begin{pmatrix} 1 & 1 & -1 & 1 \\ 1 & -1 & 1 & 1 \\ 1 & -1 & -1 & -1 \\ 1 & 1 & 1 & -1 \end{pmatrix}\begin{pmatrix} \mathrm{s} \\ \mathrm{p}_x \\ \mathrm{p}_y \\ \mathrm{p}_z \end{pmatrix} = \mathrm{sp}^3 \text{ set}$$

$$\begin{pmatrix} 1/\sqrt{3} & 0 & \sqrt{2}/\sqrt{3} & 0 \\ 1/\sqrt{3} & 1/\sqrt{2} & -1/\sqrt{6} & 0 \\ 1/\sqrt{3} & -1/\sqrt{2} & -1/\sqrt{6} & 0 \\ 0 & 0 & 0 & 1 \end{pmatrix}\begin{pmatrix} \mathrm{s} \\ \mathrm{p}_x \\ \mathrm{p}_y \\ \mathrm{p}_z \end{pmatrix} = \mathrm{sp}^2 \text{ set}$$

For each matrix T, $\mathsf{T}^\dagger\mathsf{T} = \mathsf{1}$.

13-17. Yes. Since various hybridized sets are equivalent, the final result of the calculation is independent of one's choice.

13-18. a) $A_2 \oplus 2B_1 \oplus E$. Yes.
b) $A_g \oplus 2B_{3u}$. No.
c) $A_{1g} \oplus A_{2g} \oplus E_g \oplus A_{1u} \oplus A_{2u} \oplus E_u$. No.

13-19. a) $e \otimes b_2 \otimes e = a_1 \oplus a_2 \oplus b_1 \oplus b_2$. Need not vanish.
b) $e \otimes e \otimes e = e \oplus e \oplus e \oplus e$. Must vanish.
c) $a_1 \otimes b_2 \otimes b_2 = a_1$. Need not vanish.

d) $a_1 \otimes b_2 \otimes a_1 = b_2$. Must vanish.
e) $a_1 \otimes e \otimes b_2 = e$. Must vanish.

13-20. $e_g \to a_{2u}, e_u$ are x, y allowed. $e_g \to e_u$ is z allowed.

13-21. The symmetry of the molecule is C_{2v}. All π MOs are antisymmetric for C_2 so must be bases for b_1 or b_2 representations. The function y (for the molecule) is z (in the table), and so it is a basis for the a_1 representation. For the product $\phi_4 y \phi_n$ to contain a_1 symmetry, it is necessary, then, that ϕ_n have the same symmetry as ϕ_4. ϕ_6 and ϕ_7 do have the same symmetry; therefore, $\phi_4 \to \phi_6, \phi_7$ are y allowed (where y is coincident with the symmetry axis).

13-22. a) A nonspecial point above the top face has seven equivalent positions above that face (into which it can be sent by various symmetry operations, so there are eight equivalent positions above that face. Because the two sides of the square are identical, there are another eight equivalent positions below. Therefore, we expect a group of order 16. (See D_{4h}.) b) Like the square, each face of the cube has eight equivalent points. The inside of a face is not equivalent to the outside, so each face is limited to eight equivalent points, not 16. Six faces times eight points per face yields a predicted group order of 48. (O_h)

MC: d c a e d a d a b

Chapter 14

14-1. $\langle 1s_A | \hat{H}_{hyd} - 1/r_B | 1s_A \rangle = -\frac{1}{2} - (1/R) + [(1/R) + 1]\exp(-2R)$ where R is distance between nuclei. For $R = 2$ a.u., $E_{elec} = -0.9725$ a.u., $E_{tot} = -0.4725$ a.u. For $R = 1$ a.u., $E_{elec} = -1.2293$ a.u., $E_{tot} = -0.2293$ a.u. For $R = 3$ a.u., $E_{elec} = -0.8300$ a.u., $E_{tot} = -0.4967$ a.u.

14-2. Expected QMOT behavior is reversed for very low H_{AA}. The "antibonding" MO is lower than the "bonding" for $H_{AA} = -20$. This happens because the loss of energy involved in removing charge from the atoms is not compensated by that gained by putting the charge into the bond region.

14-3. The molecular electron affinities tend to be greater than for the atoms when the extra molecular electron goes into a bonding MO, smaller if into an antibonding MO. QMOT leads us to expect the MO energy to be lowered (raised) from separated atom levels if the interaction is bonding (antibonding). Koopmans' theorem then leads to predictions in qualitative agreement with these data.

14-4. The observed ground-state angles are more consistent with the triplet-state configuration. The first excited state should then be a singlet and have a smaller angle. (Accurate calculations give 103°.)

14-5. It should become more bent. (Also, the O–H bonds should lengthen, although this is not the kind of geometric change that the figure explicitly treats.)

14-6. The computed results will largely agree with QMOT ideas. However, it is possible that the $1\sigma_g - 1a_1$ level will rise where Walsh's rules predict it will fall. [Whether this occurs depends upon details of EHMO parameter choices.] Upon analysis, this turns out to result from a situation similar to that examined in Problem 14-2. That is, the increase in H–H overlap population comes at the expense of population elsewhere. If the energy associated with this "population elsewhere" is low (i.e., if the original MO energy is low enough) the energy cost is greater than the gain due to increased H–H overlap population. This inversion of behavior for low levels is ignored in the Walsh approach. However, it does not occur for a given MO until it is fairly deeply buried under higher filled MOs. Also, since the higher MOs dominate the behavior of the molecule anyway, the inversion does not affect our prediction.

14-7. Since the exact answer depends on details of your EHMO program, allow us to take this opportunity to toast your good health.

14-8. If $S = S_0 \cos\theta, dS/d\theta = -S_0 \sin\theta$. ΔS for $\theta = 0$ to $\theta = 30°$ equals $-(1 - 0.866)S_0 = -0.134S_0$; for $90 \rightarrow 60°$, $\Delta S = -S_0(0 - 0.5) = 0.5S_0$.

14-9. For a diagram and discussion, see Gimarc [1]. Those with six valence electrons (the first three in the list) are planar. Those with eight electrons (the last four) are pyramidal.

14-10. Sketches and discussions of AB_2 molecules may be found in the literature [2,3]. Molecules with 16 or fewer valence electrons ($BeCl_2, C_3, CO_2, N_3^-$) should be linear. Those with more than 16 should be bent (NO_2, O_3, F_2O).

14-11. Bending should increase end-to-end antibonding. These MOs favor the linear form [see (**V**)].

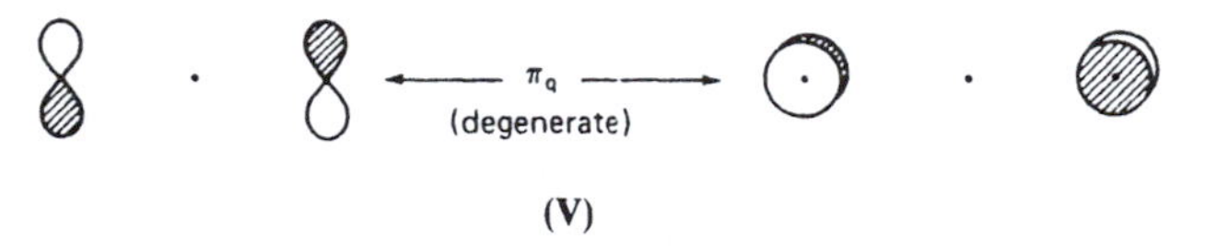

(**V**)

Bending should increase end-to-end bonding. These MOs favor the bent form [see (**VI**)].

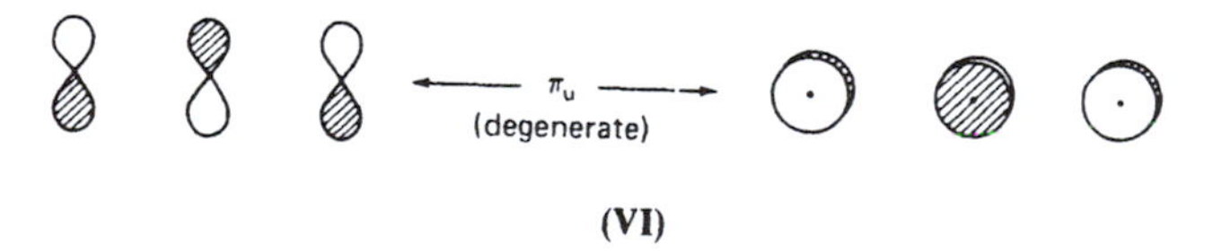

(**VI**)

Therefore, $\pi_g \rightarrow \pi_u$ should make molecule more bent.

14-12. For allyl, draw the three π MOs. For the cyclic molecule, draw a C–C bond (lowest in energy), a single p-π AO on the negative carbon (intermediate energy) and the C–C antibond (high energy). Conrotary motion preserves a C_2 axis, disrotatory preserves a reflection plane. For the C_2 axis, the allyl MO symmetries are (in order of increasing energy) A, S, A. For the cyclopropenyl anion, they are S, A, A. For the reflection plane, the same MOs have symmetries S, A, S; and S, S, A. The resulting predictions for the four-electron anion are that thermal closure goes conrotatory, photochemical goes disrotatory.

14-13. [See Hint.] The diagram is given in (**VII**). σ and π refer to symmetry (S or A) for reflection through the molecular plane. The second symmetry symbol refers to reflections through a plane between the two acetylenes. The third symbol refers to a reflection orthogonal to the first two (i.e., bisecting both acetylenes at their bond midpoints). The reaction does not appear favorable for a thermal mode. Photochemically it is less unfavorable, but still corresponds to a rather highly excited product. A square planar intermediate is unlikely for either mode.

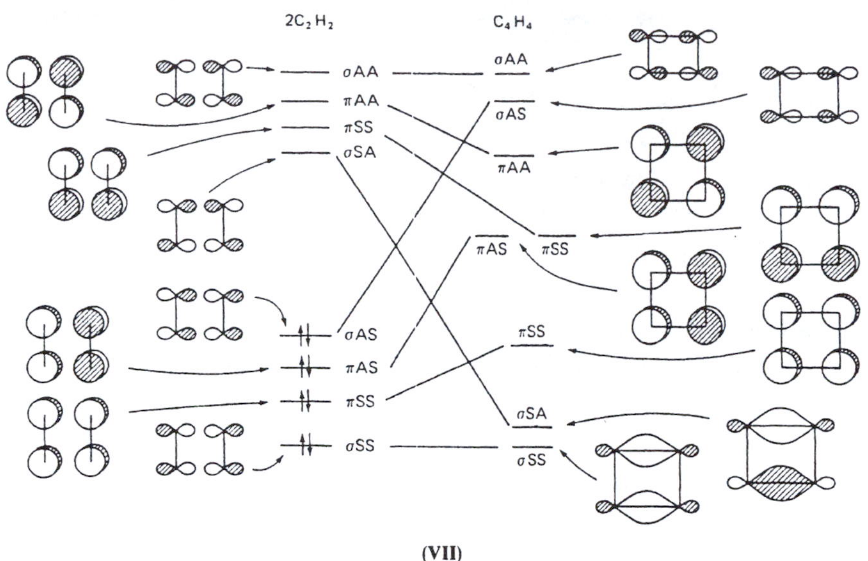

(VII)

14-14. No states can differ in symmetry in this case, and no crossings should occur.

14-15. See Woodward and Hoffmann [4, pp. 23, 24].

14-16. See Woodward and Hoffmann [4, pp. 70, 117].

Chapter 15

15-1. $(h/2\pi)k = p_x = \sqrt{2mE} = \sqrt{2mT} = h(1/\lambda)$.

15-2. [See Hint.] $dk/dE = (dE/dk)^{-1} = 4\pi^2 m/kh^2 \propto 1/k \propto 1/\sqrt{E}$ so plot E vs $1/\sqrt{E}$.

15-3. [See Hint.]

a) For example, $\phi_2 = (1/\sqrt{6})\left[p_{\pi 1} + \exp(2\pi i/3)p_{\pi 2} - \exp(\pi i/3)p_{\pi 3} + p_{\pi 4} + \exp(2\pi i/3)p_{\pi 5} - \exp(\pi i/3)p_{\pi 6}\right]. E_2 = \alpha - \beta$.

b) $j = 0$ gives $\exp(0) = 1$ at every atom, $j = 6$ gives $\exp(2(n-1)\pi i) = 1$ at every atom.

15-4. [See Hint.]

a) DOS, "normalized" by dividing by N, and in units of $|\beta|^{-1}$, is $1/(\pi \sin(h\pi))$, where $h = j/N$ and $0 \le h \le 1/2$. $E = \alpha - 2|\beta|\cos(h\pi)$. Therefore, DOS $= 1/\{\pi \sin(\arccos[(E-\alpha)/2|\beta|])\}$.

b) $2N$ states spread uniformly over an energy range of 4β gives a "normalized" DOS of $0.5|\beta|^{-1}$.

15-5. For example, $j = -1 \rightarrow (1/\sqrt{6})\left[p_{\pi 1} + \exp(-\pi i/3)p_{\pi 2} + \exp(-2\pi i/3)p_{\pi 3} - p_{\pi 4} - \exp(-\pi i/3)p_{\pi 5} - \exp(-2\pi i/3)p_{\pi 6}\right]$. Sum of $j = \pm 1$ functions divided by $\sqrt{2}$ gives one real function, and difference divided by $i\sqrt{2}$ gives the other.

15-6. Requires that $\exp(-2\pi iqnN/n)$ equal 1. This requires that the argument equal πi times an even number, which is the case if qN is an integer. Since q and N are integers, so is qN.

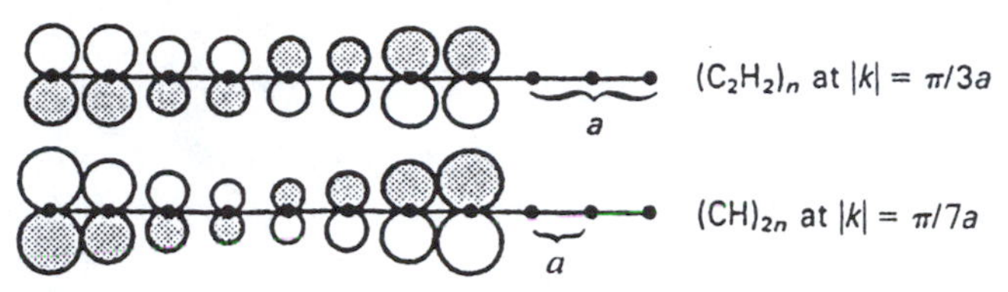

15-7. $(CH)_{2n}$ has lower energy. Variation would not affect $(CH)_{2n}$, but would mix π and π^* versions of $(C_2H_2)_n$ to produce COs like those for $(CH)_{2n}$. (Note that, for convenience, we have arranged phases so that the node is identically placed in these diagrams.)

15-8. The HOMO and LUMO at $|k| = \pi/3a$ respond oppositely to distortion that lengthens every third bond and/or shortens the others.

15-9. Symmetry for reflection through a plane perpendicular to the polymer axis is not useful for this analysis. Symmetry for reflection through a plane containing the atoms divides the COs into σ and π types. Only σ–π crossings are allowed.

15-10. The σ bands involve 2s and 2p AOs. One band pair includes 2s AOs with 2p AOs that are oriented parallel to the screw axis. The other band includes no 2s AOs and only 2p AOs perpendicular to the screw axis. The former functions form a basis that is symmetric for the symmetry operation, whereas the latter form a basis that is antisymmetric:

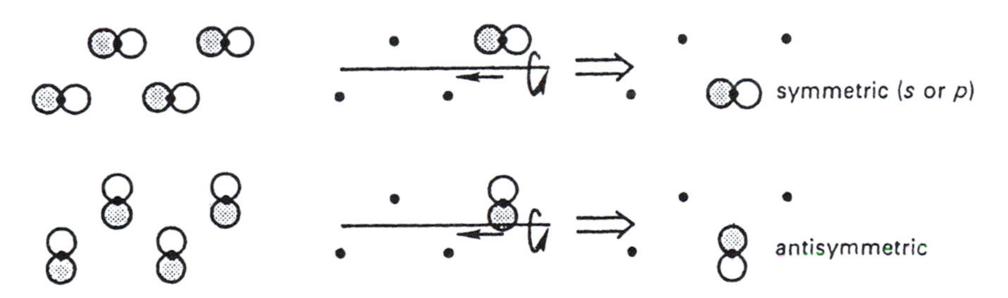

15-11. At high pressures, intermolecular distance becomes the same as intramolecular distance. This yields a zero "gap" at the Fermi level (same as polyacetylene with uniform bond lengths).

15-12.

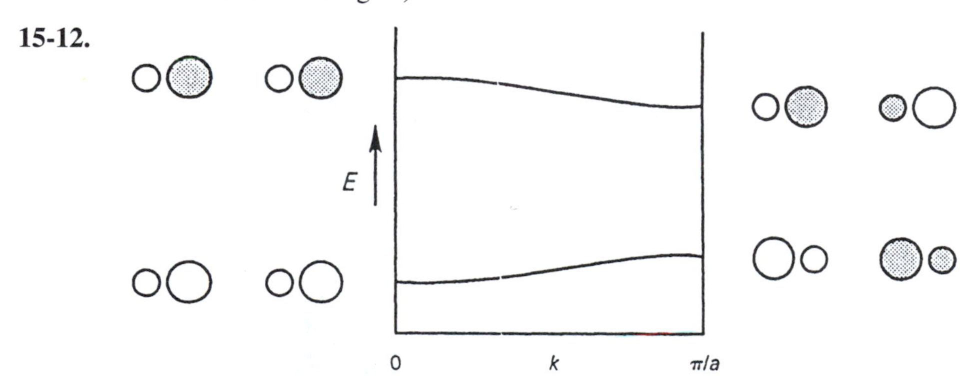

15-13. a) $\epsilon_{\text{HOMO}} = -8.19\,\text{eV}$. b) $\epsilon_{\text{LUMO}} = -4.09\,\text{eV}$. c) $E_{\text{gap}} = 4.10\,\text{eV}$. d) Gap should open (both up and down) so ϵ_{HOMO} becomes lower and IE is higher. ϵ_{LUMO} is higher and EA is less negative. Computed values: IE $= 8.39\,\text{eV}$, EA $= -2.26\,\text{eV}$, $E_{\text{gap}} = 6.13\,\text{eV}$.

15-14. In Chapter 8 we show correlations between ϵ_{HOMO} and IE and between ϵ_{LUMO} and EA, with *different* values for β in each case. The difference in interelectronic repulsion and exchange for neutral and anionic species is handled implicitly in this manner by simple Hückel theory.

15-15. Only π_4 has the proper symmetry to combine with π_1 to make $c_1 = c_4 \neq c_2 = c_3 = c_5 = c_6$. Mixing causes the bands to split apart. π_1 is pushed down by π_4, and π_4 is pushed up by π_1. The coefficients decrease on carbons 1 and 4 in π_4, consistent with a rise in energy due to less bonding.

15-16. No. Atoms 2 and 6 have zero coefficients in the appropriate COs (see $\pi_a^{(0)}$ of Fig. 15-26), but atom 5 has a nonzero coefficient.

15-17. The band diagram is identical to the one in Fig. 15-23 except that the $\alpha + \beta$ level at $k = 0$ has an intended correlation with $\alpha - \beta$ at $k = \pi/a$, and *vice versa*. However, these have the same symmetry (anti for σ_2), hence avoid crossing. The resulting diagram is identical to that in Fig. 15-23 except that band π_2 has an intermediate hill and π_5 an intermediate valley—height and depth unknown until a variational calculation is performed.

15-18. Edge energies without substitution at $k = 0 : \alpha + 2.5\beta, \alpha, \alpha - \beta, \alpha - 1.5\beta$. At $k = \pi/a : \alpha + 1.5\beta, \alpha + \beta, \alpha, \alpha - 2.5\beta$. (Proper zeroth-order MOs for the $E = \alpha$ level of cyclobutadiene are the versions having nodes through atoms rather than through bonds.) The lines at $\alpha, \alpha - \beta$ for $k = 0$ cross. This is allowed by symmetry disagreement for reflection through the plane containing the polymer axis and perpendicular to the molecular plane. Edge energies are lowered by substitution by the following amounts as we move up the diagram on the $k = 0$ side: $\beta/8, \beta/4, 0, \beta/8$. On the $k = \pi/a$ side: $\beta/8, 0, \beta/4, \beta/8$. Symmetry is maintained so the crossing is still allowed. For only one substitution, the energy lowerings are half as great. Symmetry is lost, and the crossing becomes avoided.

15-19. Like Fig. 15-28a with the middle horizontal row shaded. Higher in energy than Γ, but lower than X or M.

15-20.

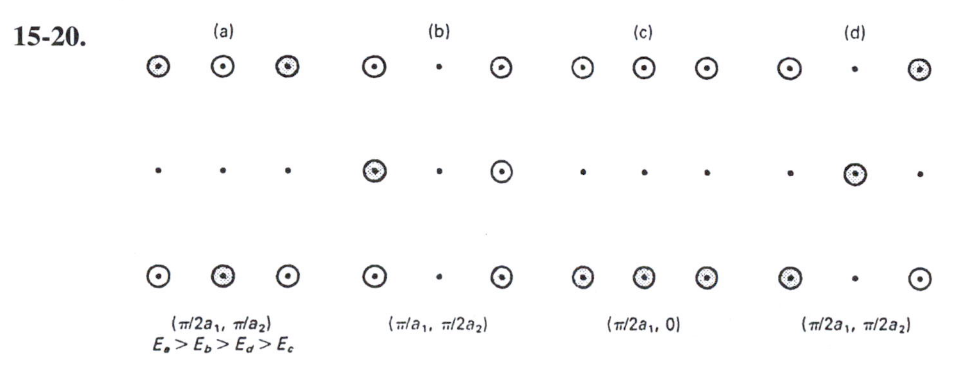

15-21.

The energy level for X is lower than those for Γ and X'. For X, all interactions are bonding. For Γ and X', nearest-neighbor interactions are antibonding and longer-range interactions of both bonding and anti bonding types occur. Judging the relative energies of Γ and X' is complicated since we need to evaluate relative magnitudes of interactions between p AOs on the same long edge of the rectangle and also on opposite corners. Since p AO interactions depend on distance *and* angle, this is not trivial, as it is for s-type AOs.

15-22. All are nondegenerate.

15-23. The RFBZ is triangular. Referring to Fig. 15-27c, the RFBZ shown there becomes square, but symmetry now makes points on opposite sides of the $\Gamma - M$ line equivalent (e.g., X and X' become equivalent), so only the Γ-X-M-Γ line is needed.

15-24.

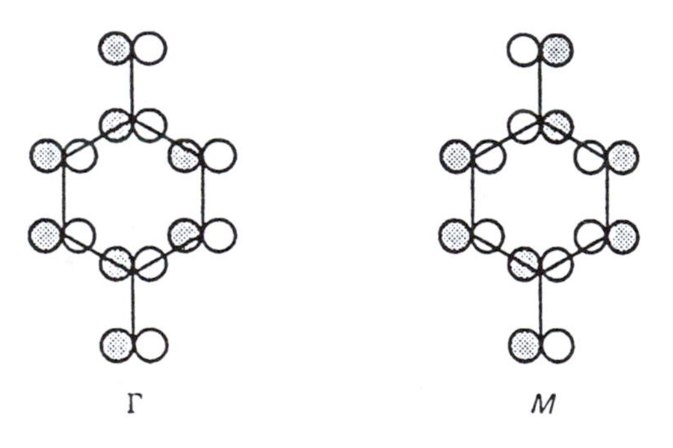

Both are nonbonding and would have the same energy.

15-25.

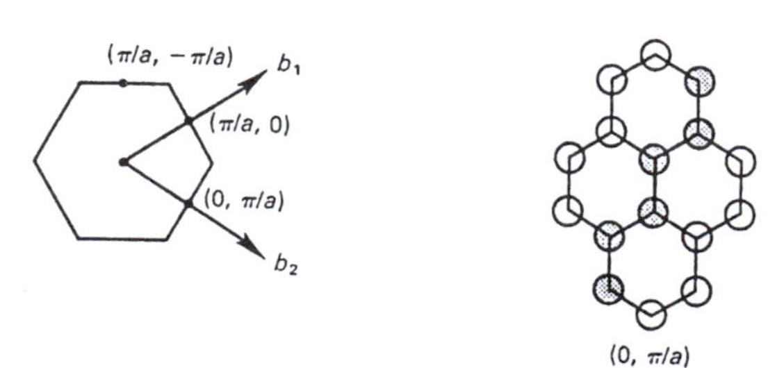

COs at $(0, \pi/a)$ and $(\pi/a, 0)$ are the same except that the nodal planes are rotated by 60°.

15-26. The crystal is symmetrical for reflection through a hexagonal sheet. Plane waves experiencing the same potential and having the same $|k|$ are degenerate.

15-27. At the top of the RFBZ, unit cell functions are stacked with sign reversal. If the upper and lower phases of the unit cell functions agree, this means that we must have a bonding interaction *in* the cell, but we will also get antibonding between cells, hence no splitting. On the other hand, if the upper and lower phases are opposite, we must have antibonding in the cell but we now get bonding between them, again for no splitting. The same is true for sigma bonds.

15-28.

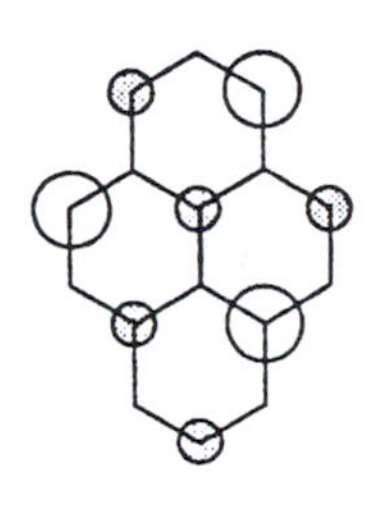

$\pi(M) + \pi^*(M)$

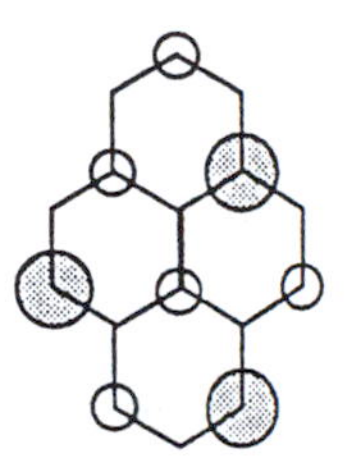

$\pi(M) - \pi^*(M)$

Appendix 2

A2-2. a) -2. b) 1. c) 0. d) $x = \pm\sqrt{2}$.

A2-4. Determinant of coefficients vanishes, and so nontrivial roots *do* exist.

A2-5. $c = 2,\ -1 \pm \sqrt{7}$ are the roots.

A2-6. a) If two rows or columns differ by factor c, then multiplication of the smaller row or column gives $|M'| = c|M|$ (by rule 1), and M' has two identical rows or columns. Interchanging these gives M'', and $|M'| = -|M''|$. But $M'' = M'$ because the interchanged rows or columns are identical. Therefore, $|M'| = -|M'|$, and $|M'| = 0$. Therefore, $|M| = c|M'| = 0$.

Appendix 4

A4-1. From Eq. (A4-47), $L_z L_+ = L_+(L_z + 1)$. Then $L_z L_+ Y_{l,m} = L_+(L_z + 1)Y_{l,m} = L_+(k_m + 1)Y_{l,m} = (k_m + 1)L_+ Y_{l,m}$.

A4-2. $L_+ L_- = (L_x + iL_y)(L_x - iL_y) = L_x^2 + L_y^2 + iL_y L_x - iL_x L_y$. Equation (A3-31) tells us that $iL_y L_x - iL_x L_y = -i^2 L_z = L_z$, so $L_+ L_- = L_x^2 + L_y^2 + L_z = L_x^2 + L_y^2 + L_z^2 - L_z^2 + L_z = L^2 - L_z^2 + L_z$. Therefore, $L^2 = L_+ L_- + L_z^2 - L_z$.

A4-3. Equation (A4-55) indicates that the result of L_+ on $Y_{l,m}$ remains an eigenfunction of L^2 with the same eigenvalue, i.e., that $L^2 L_+ Y_{l,m} = L^2 C_+ Y_{l,m+1} = k_l C_+ Y_{l,m+1}$. But also, $L_+ L^2 Y_{l,m} = L_+ k_l Y_{l,m} = C_+ k_l Y_{l,m+1}$, so Eq. (A4-55) indicates that L_+ and L^2 commute. This is easily verified since L_+ is a linear combination of L_x and L_y, both of which commute with L^2. We can establish Eq. (A4-55) by evaluating the result of the *reverse* order of operations and using the fact that L_+ and L^2 commute: $L_+ L^2 Y_{l,m} = L_+ k_l Y_{l,m} = k_l L_+ Y_{l,m} = k_l C_+ Y_{l,m+1}$. But $L_+ L^2 Y_{l,m} = L^2 L_+ Y_{l,m}$, so $L^2 L_+ Y_{l,m} = k_l C_+ Y_{l,m+1}$. Q.E.D.

A4-4. $L_x = (1/2)(L_+ + L_-)$, $(1/2)\langle Y_{l,m}|L_+ + L_-|Y_{l,m}\rangle = (1/2)[C_+\langle Y_{l,m}|Y_{l,m+1}\rangle + C_-\langle Y_{l,m}|Y_{l,m-1}\rangle = 0 + 0$.

A4-5. a) $(10)\begin{pmatrix}0\\1\end{pmatrix} = 0$.

b) $S_+\beta = (S_x + iS_y)\beta = \begin{pmatrix}0+0 & \frac{1}{2} - \frac{i^2}{2}\\ \frac{1}{2} + \frac{i^2}{2} & 0+0\end{pmatrix}\begin{pmatrix}0\\1\end{pmatrix} = \begin{pmatrix}1\\0\end{pmatrix} = \alpha$

c) $S_+\alpha = \begin{pmatrix}0 & 1\\0 & 0\end{pmatrix}\begin{pmatrix}1\\0\end{pmatrix} = \begin{pmatrix}0\\0\end{pmatrix}$

d) $[S_x, S_y] = \frac{1}{4}\left\{\begin{pmatrix}0 & 1\\1 & 0\end{pmatrix}\begin{pmatrix}0 & -i\\i & 0\end{pmatrix} - \begin{pmatrix}0 & -i\\i & 0\end{pmatrix}\begin{pmatrix}0 & 1\\1 & 0\end{pmatrix}\right\} = \frac{1}{4}\begin{pmatrix}2i & 0\\0 & -2i\end{pmatrix} = iS_z$.

A4-6. $S^2 = S_x^2 + S_y^2 + S_z^2 = \frac{1}{4}\left\{\begin{pmatrix}0 & 1\\1 & 0\end{pmatrix}\begin{pmatrix}0 & 1\\1 & 0\end{pmatrix} + \begin{pmatrix}0 & -i\\i & 0\end{pmatrix}\begin{pmatrix}0 & -i\\i & 0\end{pmatrix} + \begin{pmatrix}1 & 0\\0 & -1\end{pmatrix}\begin{pmatrix}1 & 0\\0 & -1\end{pmatrix}\right\}$

$= \frac{3}{4}\begin{pmatrix}1 & 0\\0 & 1\end{pmatrix}$; $S^2\alpha = \frac{3}{4}\begin{pmatrix}1 & 0\\0 & 1\end{pmatrix}\begin{pmatrix}1\\0\end{pmatrix} = \frac{3}{4}\begin{pmatrix}1\\0\end{pmatrix} = \frac{3}{4}\alpha$.

Appendix 8

A8-1. $\hat{T} = -\frac{1}{2}d^2/dx^2$, $\hat{V} = \frac{1}{2}kx^2$. Equation (A8-19) gives $\bar{E}\eta = \eta^2\overline{T} + \eta^{-2}\overline{V}$. $\partial\bar{E}/\partial\eta = 0 = 2\eta\overline{T} - 2\eta^{-3}\overline{V}$. For an exact solution, $\eta = 1$, and $\overline{T} = \overline{V}$.

A8-2. $\overline{V} = -3$ a.u., $\overline{T} = \frac{1}{2}$ a.u., $\eta = -\overline{V}/2\overline{T} = 3$, $\psi_\eta = \sqrt{\eta^3/\pi}\exp(-\eta r) = \sqrt{27/\pi}\exp(-3r)$, $E_\eta = \eta^2\overline{T} + \eta\overline{V} = \frac{9}{2} - 9 = -4.5$ a.u.

References

[1] B. M. Gimarc, *Accounts Chem. Res*. **7**, 384 (1974).

[2] R. S. Mulliken, *Rev. Mod. Phys*. **4**, 1 (1932).

[3] A. D. Walsh, *J. Chem. Soc*. 2260 (1953).

[4] R. B. Woodward and R. Hoffmann, *The Conservation of Orbital Symmetry*. Academic Press, New York, 1970.

Index